Organic Structures from Spectra

Third Edition

Organic Structures from Spectra

Third Edition

L. D. Field
University of Sydney, Australia

S. Sternhell
University of Sydney, Australia

J. R. Kalman
University of Technology, Sydney, Australia

JOHN WILEY & SONS, LTD

National 01243 779777
International (+44) 1243 779777

e-mail (for orders and customer service enquiries): cs-books@wiley.co.uk

Visit our Home Page on http://www.wileyeurope.com
or
http://www.wiley.com

Other Wiley Editorial Offices

John Wiley & Sons, Inc., 605 Third Avenue,
New York, NY 10158-0012, USA

Wiley-VCH Verlag GmbH, Pappelallee 3,
D-69469 Weinheim, Germany

John Wiley & Sons (Australia) Ltd, 33 Park Road, Milton,
Queensland 4064, Australia

John Wiley & Sons (Asia) Pte Ltd, 2 Clementi Loop #02-01,
Jin Xing Distripark, Singapore 0512

John Wiley & Sons (Canada) Ltd, 22 Worcester Road,
Rexdale, Ontario M9W 1L1, Canada

British Library Cataloguing in Publication Data

A catalogue record for this book is available from the British Library

ISBN 0 470 84361 6 (Hardback) 0 470 84362 4 (Paperback)

Printed and bound in Great Britain by Antony Rowe Ltd., Chippenham, Wilts
This book is printed on acid-free paper responsibly manufactured from sustainable forestry, in which at least two trees are planted for each one used for paper production.

CONTENTS

PREFACE

The derivation of structural information from spectroscopic data is now an integral part of Organic Chemistry courses at all Universities. At the undergraduate level, the principal aim of such courses is to teach students to solve simple structural problems efficiently by using combinations of the major techniques (UV, IR, NMR and MS), and over more than 20 years we have evolved a course at the University of Sydney, which achieves this aim quickly and painlessly. The text is tailored specifically to the needs and philosophy of this course. As we believe our approach to be successful, we hope that it may be of use in other institutions.

The course is taught at the beginning of the third year, at which stage students have completed an elementary course of Organic Chemistry in first year and a mechanistically-oriented intermediate course in second year. Students have also been exposed in their Physical Chemistry courses to elementary spectroscopic theory, but are, in general, unable to relate it to the material presented in this course.

The course consists of 9 lectures outlining the theory, instrumentation and the structure-spectra correlations of the major spectroscopic techniques and the text of this book corresponds to the material presented in the 9 lectures. The treatment is both elementary and condensed and, not surprisingly, the students have great difficulties in solving even the simplest problems at this stage. The lectures are followed by a series of 8, 2-hour problem solving seminars with 5 to 6 problems being presented per seminar. At the conclusion of the course the great majority of the class is quite proficient and has achieved a satisfactory level of understanding of all methods used. Clearly, the real teaching is done during the problem seminars, which are organised in a manner modelled on that used at the E.T.H. Zurich.

The class (60 - 100 students, attendance is compulsory) is seated in a large lecture theatre in alternate rows and the problems for the day are identified. The students are permitted to work either individually or in groups and may use any written or printed aids they desire. Students solve the problems on their individual copies of this book thereby transforming it into a set of worked examples and we find that most students voluntarily complete many more problems than are set. Staff (generally 4 or 5) wander around giving help and tuition as needed, the empty alternate rows of seats making it possible to speak to each student individually. When an important general point needs to be made, the staff member in charge gives a very brief exposition at the board. There is a $1^1/_2$ hour examination consisting essentially of 4 problems and the results are in general very satisfactory. Moreover, the students themselves find this a rewarding course since the practical skills acquired are obvious to them. There is also a real sense of achievement and understanding since the challenge in solving the graded problems builds confidence even though the more difficult examples are quite demanding.

Our philosophy can be summarised as follows:

(a) Theoretical exposition must be kept to a minimum, consistent with gaining of an understanding of the parts of the technique actually used in solving the problems. Our experience indicates that both mathematical detail and description of advanced techniques merely confuse the average student.

(b) The learning of data must be kept to a minimum. We believe that it is more important to learn to use a restricted range of data well rather than to achieve a nodding acquaintance with more extensive sets of data.

(c) Emphasis is placed on the concept of identifying "structural elements" and the logic needed to produce a structure out of the structural elements.

We have concluded that the best way to learn how to obtain "structures from spectra" is to practise on simple problems. This book was produced principally to assemble a collection of problems that we consider satisfactory for that purpose.

Problems 1 - 250 are of the standard "structures from spectra" type and are arranged roughly in order of increasing difficulty. A number of problems are groups of isomers which differ only in the connectivity of the structural elements and these problems are ideally set together (*e.g.* problems 2 and 3, 20 and 21; 23 and 24; 32 and 33; 34 and 35; 36 and 37; 44, 45 and 46; 47, 48 and 49; 52 and 53; 61 and 62; 82 and 83; 84, 85 and 86; 87 and 88; 91, 92 and 93; 98 and 99; 109, 110, 111, 113 and 114; 126 and 127; 148 and 149; 150 and 151; 155 and 156; 199, 200 and 201; 245 and 246). A number of problems exemplify complexities arising from the presence of chiral centres (*e.g.* problems 29, 145, 166, 197, 228, 231, 232, 233, 235, 239, 242, 243 and 244), or of restricted rotation about peptide or amide bonds (*e.g.* problems 129 and 130), while other problems deal with structures of compounds of biological, environmental or industrial significance (*e.g.* problems 18, 19, 72, 98, 100, 101, 112, 123, 124, 129, 131, 145, 157, 170, 188, 219, 227, 229, 231, 232, 233, 234, 235, 242, 243 and 244).

Problems 251 - 256 are again structures from spectra but with the data presented in a textual form such as might be encountered when reading the experimental section of a paper or a report. Problems 257 - 277 deal with more detailed analysis of NMR spectra - this tends to be a stumbling block for many students.

We have also included two worked solutions (to problems 73 and 98) as Appendix 1 as an illustration of a logical approach to solving problems. However, with the exception that we insist that students perform all routine measurements first, we do not recommend a mechanical attitude to problem solving - intuition has an important place in solving structures from spectra as it has elsewhere in chemistry.

Bona fide instructors may obtain a list of solutions by writing to the authors or
EMAIL: L.Field@chem.usyd.edu.au or FAX: (61-2)-9351-6650

We wish to thank Dr Ian Luck, Mrs Jacqueline Morgan, Dr Hsiulin Li, Dr Jacques Nemorin, Dr Xiaomin Song, Mr Michael Smyth, Mr Simon Mann and Ms Michelle Morris in the School of Chemistry, University of Sydney, all of whom helped to assemble the samples and spectra in this book. Thanks are also due to the graduate students who supplied us with many of the compounds used in the problems.

L D Field

S Sternhell

J R Kalman January 2002

LIST OF TABLES

LIST OF FIGURES

1

INTRODUCTION

1.1 GENERAL PRINCIPLES OF ABSORPTION SPECTROSCOPY

The basic principles of absorption spectroscopy are summarised below. These are most obviously applicable to UV and IR spectroscopy and are simply extended to cover NMR spectroscopy. Mass Spectrometry is somewhat different and is not a type of absorption spectroscopy.

Spectroscopy is the study of the quantised interaction of energy (typically electromagnetic energy) with matter. In Organic Chemistry, we typically deal with molecular spectroscopy *i.e.* the spectroscopy of atoms that are bound together in molecules.

A schematic absorption spectrum is given in Figure 1.1. The absorption spectrum is a plot of absorption of energy (radiation) against its wavelength (λ) or frequency (ν).

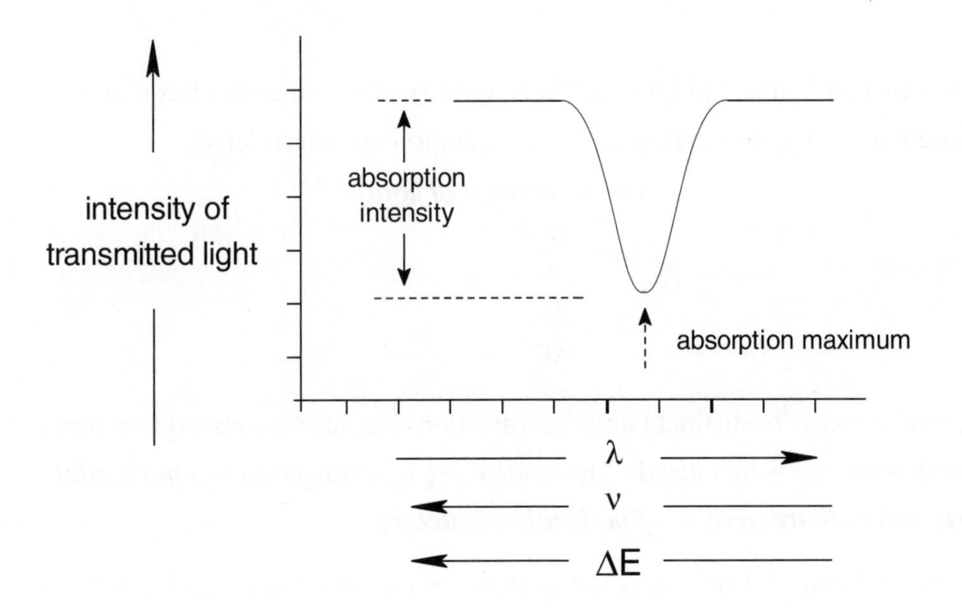

Figure 1.1 Schematic Absorption Spectrum

Any absorption band can be characterised primarily by two parameters:

(a) the wavelength at which maximum absorption occurs

(b) the intensity of absorption at this wave length compared to base-line (or background) absorption

A spectroscopic transition is the energy required to take a molecule from one state to a state of a higher energy. For any spectroscopic transition between energy states (*e.g.* E_1 and E_2 in Figure 1.2), the change in energy (ΔE) is given by:

$$\Delta E = h\nu$$

where h is the Planck's constant and ν is the frequency of the electromagnetic energy absorbed. Therefore $\nu \propto \Delta E$.

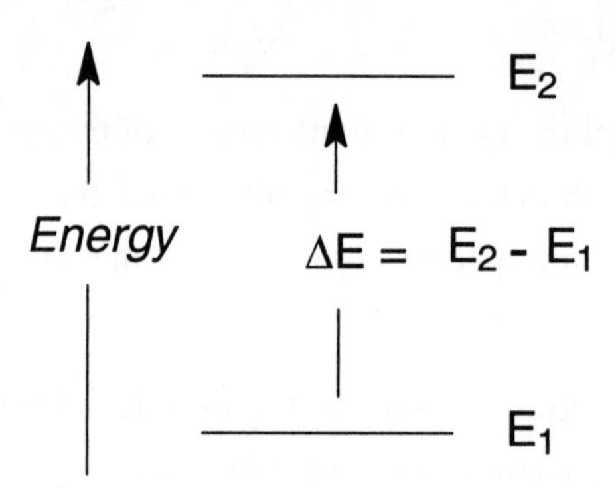

Figure 1.2 Definition of a Spectroscopic Transition

It follows that the x-axis in Figure 1.1 is an **energy** scale, since the frequency, wavelength and energy of electromagnetic radiation are interrelated:

$$\nu\lambda = c \text{ (speed of light)}$$

$$\lambda = \frac{c}{\nu}$$

$$\lambda \propto \frac{1}{\Delta E}$$

A spectrum consists of distinct bands or transitions because the absorption (or emission) of energy is quantised. The energy gap of a transition is a ***molecular property*** and is ***characteristic of molecular structure***.

The y-axis in Figure 1.1 measures the intensity of the absorption band which is proportional to the number of molecules observed (the Beer-Lambert Law) and to the probability of the transition between the energy levels. The absorption intensity is also a molecular property and both the frequency and the intensity of a transition can provide structural information.

1.2 CHROMOPHORES

In general, any spectral feature, *i.e.* a band or group of bands, is due not to the whole molecule, but to an identifiable part of the molecule, which we loosely call a *chromophore.*

A chromophore may correspond to a functional group (*e.g.* a hydroxyl group or the double bond in a carbonyl group). However, it may equally well correspond to a single atom within a molecule or to a group of atoms (*e.g.* a methyl group) which is not normally associated with chemical functionality.

The detection of a chromophore permits us to deduce the presence of a *structural fragment* or a *structural element* in the molecule. The fact that it is the chromophores and not the molecules as a whole that give rise to spectral features is fortunate, otherwise spectroscopy would only permit us to identify known compounds by direct comparison of their spectra with authentic samples. This "fingerprint" technique is often useful for establishing the identity of known compounds, but the direct determination of molecular structure building up from the molecular fragments is far more powerful.

1.3 CONNECTIVITY

Even if it were possible to identify enough structural elements in a molecule to account for the molecular formula, it may not be possible to deduce the structural formula from a knowledge of the structural elements alone. For example, it could demonstrated that a substance of molecular formula C_3H_5OCl contains the structural elements:

$-CH_3$

$-Cl$

$>C{=}O$

$-CH_2-$

and this leaves two possible structures:

$$\underset{\mathbf{1}}{CH_3-\overset{}{\underset{\|}{C}}(=O)-CH_2-Cl} \quad \text{and} \quad \underset{\mathbf{2}}{CH_3-CH_2-C(=O)-Cl}$$

Not only the presence of various structural elements but also their juxtaposition must be determined to establish the structure of a molecule. Fortunately, spectroscopy often gives valuable information concerning the *connectivity* of structural elements and in the above example it would be very easy to determine whether there is a ketonic carbonyl group (as in **1**) or an acid chloride (as in **2**). In addition, it is possible to determine independently whether the methyl ($-CH_3$) and methylene ($-CH_2-$) groups are separated (as in **1**) or adjacent (as in **2**).

1.4 SENSITIVITY

Sensitivity is generally taken to signify the limits of detectability of a chromophore. Some methods (*e.g.* ^{1}H NMR) detect all chromophores accessible to them with equal sensitivity while in other techniques (*e.g.* UV) the range of sensitivity towards different chromophores spans many orders of magnitude. In terms of overall sensitivity, *i.e.* the amount of sample required, it is generally observed that:

$$\text{MS} > \text{UV} > \text{IR} > {}^1\text{H NMR} > {}^{13}\text{C NMR}$$

but considerations of relative sensitivity toward different chromophores may be more important.

1.5 PRACTICAL CONSIDERATIONS

The 5 major spectroscopic methods (MS, UV, IR, ^{1}H NMR and ^{13}C NMR) have become established as the principal tools for the determination of the structures of organic compounds, because between them they detect a wide variety of structural elements.

The instrumentation and skills involved in the use of all five major spectroscopic methods are now widely spread, but the ease of obtaining and interpreting the data from each method under real laboratory conditions varies.

In very general terms:

(a) While the ***cost*** of each type of instrumentation differs greatly (NMR instruments cost between $50,000 and several million dollars), as an overall guide, MS and NMR instruments are much more costly than UV and IR spectrometers. With increasing cost goes increasing difficulty in maintenance, thus compounding the total outlay.

(b) In terms of ***ease of usage*** for routine operation, most UV and IR instruments are comparatively straightforward. NMR Spectrometers are also common as "hands-on" instruments in most chemistry laboratories but the users require some training, computer skills and expertise. Similarly some Mass Spectrometers are now designed to be used by researchers as "hands-on" routine instruments. However, the more advanced NMR Spectrometers and most Mass Spectrometers are sophisticated instruments that are technically difficult to operate and are usually operated by specialists.

(c) The ***scope*** of each method can be defined as the amount of useful information it provides. This is a function not only of the total amount of information obtainable but also how difficult the data is to interpret. The scope of each method varies from problem to problem and each method has its aficionados and experienced experts, but the overall utility undoubtedly decreases in the order:

$$\text{NMR} > \text{MS} > \text{IR} > \text{UV}$$

with the combination of ^{1}H and ^{13}C NMR providing the most useful information.

(d) The theoretical background needed for each method varies with the nature of the experiment, but the minimum overall amount of theory needed decreases in the order:

$$\text{NMR} \gg \text{MS} > \text{UV} \approx \text{IR}$$

2

ULTRAVIOLET (UV) SPECTROSCOPY

2.1 BASIC INSTRUMENTATION

Basic instrumentation for both UV and IR spectroscopy consists of an energy *source*, a *sample cell,* a *dispersing device* (prism or grating) and a *detector,* arranged as schematically shown in Figure 2.1.

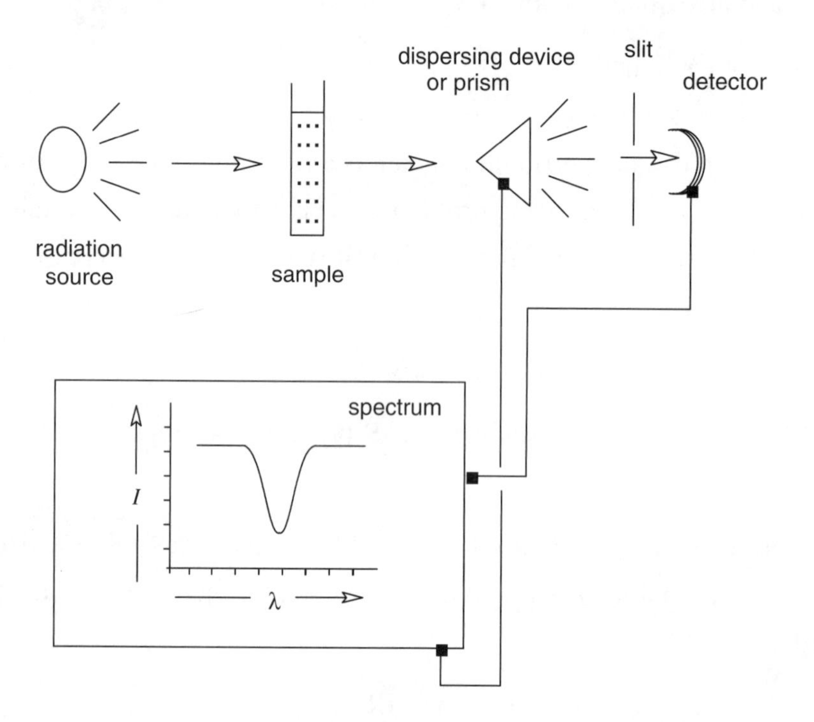

Figure 2.1 Schematic Representation of an IR or UV Spectrometer

The drive of the dispersing device is synchronised with the x-axis of the recorder so that the latter indicates the wavelength of radiation reaching the detector via the slit. The signal from the detector is transmitted to the y-axis of the recorder indicating how much radiation is absorbed by the sample at any particular wavelength.

In practice *double-beam* instruments are used where the absorption of a *reference cell*, containing only solvent, is subtracted from the absorption of the sample cell. Such instruments also cancel out absorption of the atmosphere in the optical path and the solvent.

The energy source, the materials from which the dispersing device and the detector are constructed must be appropriate for the range of wavelength scanned and as transparent as possible to the radiation. For UV measurements, the cells and optical components are typically made of quartz and ethanol, hexane, water or dioxan are usually chosen as solvents.

2.2 THE NATURE OF ULTRAVIOLET SPECTROSCOPY

The term "UV spectroscopy" generally refers to *electronic transitions* occurring in the region of the electromagnetic spectrum (λ in the range 200-380 nm) accessible to standard UV spectrometers.

Electronic transitions are also responsible for absorption in the visible region (380-800 nm) which is easily accessible instrumentally but of less importance in the solution of structural problems, because most organic compounds are colourless. An extensive region at wavelengths shorter than ~ 200 nm ("vacuum ultraviolet") also corresponds to electronic transitions, but this region is not readily accessible with standard instruments.

UV spectra used for determination of structures are invariably obtained in solution.

2.3 QUANTITATIVE ASPECTS OF ULTRAVIOLET SPECTROSCOPY

The y-axis of a UV spectrum may be calibrated in terms of the intensity of transmitted light (*i.e.* percentage of transmission or absorption), as is shown in Figure 2.2, or it may be calibrated on a logarithmic scale *i.e.* in terms of *absorptivity* (A) defined in Figure 2.2.

Absorptivity is proportional to concentration and path length (the Beer-Lambert Law). For purposes of characterisation of a chromophore, intensity of absorption is expressed in terms of *molar absorptivity* (ε) given by:

$$\varepsilon = \frac{M\,A}{C\,l}$$

where M is the molecular weight, C the concentration (in grams per litre) and l is the path length through the sample in centimetres.

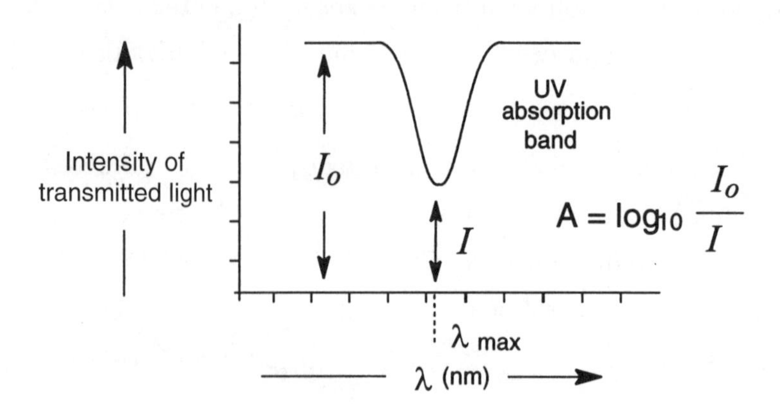

Figure 2.2 Definition of Absorptivity (A)

UV absorption bands (Figure 2.2) are characterised by the wavelength of the absorption maximum (λ_{max}) and ε. The values of ε associated with commonly encountered chromophores vary between 10 and 10^5. For convenience, extinction coefficients are usually tabulated as $\log_{10}(\varepsilon)$ as this gives numerical values which are easier to manage. The presence of small amounts of strongly absorbing impurities may lead to errors in the interpretation of UV data.

2.4 CLASSIFICATION OF UV ABSORPTION BANDS

UV absorption bands have fine structure due to the presence of vibrational sub-levels, but this is rarely observed in solution due to collisional broadening. As the transitions are associated with changes of electron orbitals, they are often described in terms of the orbitals involved, *e.g.*

$$\sigma \rightarrow \sigma^*$$
$$\pi \rightarrow \pi^*$$
$$n \rightarrow \pi^*$$
$$n \rightarrow \sigma^*$$

where n denotes a non-bonding orbital, the asterisk denotes an antibonding orbital and σ and π have the usual significance.

Another method of classification uses the symbols:

B (for benzenoid)
E (for ethylenic)
R (for radical-like)
K (for conjugated - from the German "konjugierte")

A molecule may give rise to more than one band in its UV spectrum, either because it contains more than one chromophore or because more than one transition of a single chromophore is observed. However, UV spectra typically contain far fewer features (bands) than IR, MS or NMR spectra and therefore have a lower information content. The ultraviolet spectrum of acetophenone in ethanol contains 3 easily observed bands:

O
||
C
CH_3

acetophenone

λ_{max} (nm)	ε	$\log_{10}(\varepsilon)$	Assignment	
244	12,600	4.1	$\pi \rightarrow \pi^*$	K
280	1,600	3.2	$\pi \rightarrow \pi^*$	B
317	60	1.8	$n \rightarrow \pi^*$	R

2.5 SPECIAL TERMS IN UV SPECTROSCOPY

Auxochromes (auxiliary chromophores) are groups which have little UV absorption by themselves, but which often have significant effects on the absorption (both λ_{max} and ε) of a chromophore to which they are attached. Generally, auxochromes are atoms with one or more lone pairs *e.g.* -OH, -OR, $-NR_2$, -halogen.

If a structural change, such as the attachment of an auxochrome, leads to the absorption maximum being shifted to a longer wavelength, the phenomenon is termed a *bathochromic shift.* A shift towards shorter wavelength is called a *hypsochromic shift.*

2.6 IMPORTANT CHROMOPHORES

Most of the reliable and useful data is due to relatively strongly absorbing chromophores ($\varepsilon > 200$) which are mainly indicative of conjugated or aromatic systems. Examples listed below encompass most of the commonly encountered effects.

(1) Dienes and Polyenes

Extension of conjugation in a carbon chain is always associated with a pronounced shift towards longer wavelength, and usually towards greater intensity (Table 2.1).

Table 2.1 The Effect of Extended Conjugation on UV Absorption

alkene	λ max (nm)	ε	$\log_{10}(\varepsilon)$
$CH_2=CH_2$	165	10,000	4.0
$CH_3\text{-}CH_2\text{-}CH=CH\text{-}CH_2\text{-}CH_3$ (*trans*)	184	10,000	4.0
$CH_2=CH\text{-}CH=CH_2$	217	20,000	4.3
$CH_3\text{-}CH=CH\text{-}CH=CH_2$ (*trans*)	224	23,000	4.4
$CH_2=CH\text{-}CH=CH\text{-}CH=CH_2$ (*trans*)	263	53,000	4.7
$CH_3\text{-}(CH=CH)_5\text{-}CH_3$ (*trans*)	341	126,000	5.1

When there are more than 8 conjugated double bonds, the absorption maximum of polyenes is such that they absorb light strongly in the visible region of the spectrum.

Empirical rules (Woodward's Rules) of good predictive value are available for the prediction of the positions of the absorption maxima in conjugated alkenes and conjugated carbonyl compounds.

Configuration and the presence of substituents also influence UV absorption by the diene chromophore. For example:

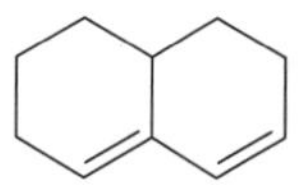

λ_{max} = 214 nm
ε = 16,000
$\log_{10}(\varepsilon)$ = 4.2

λ_{max} = 253 nm
ε = 8,000
$\log_{10}(\varepsilon)$ = 3.9

(2) Carbonyl compounds

All carbonyl derivatives exhibit weak ($\varepsilon < 100$) absorption between 250 and 350 nm, and this is only of marginal use in determining structure. However, conjugated carbonyl derivatives always exhibit strong absorption (Table 2.2).

Table 2.2 UV Absorption Bands in Common Carbonyl Compounds

compound	structure	λ_{max} (nm)	ε	$\log_{10}(\varepsilon)$
acetaldehyde	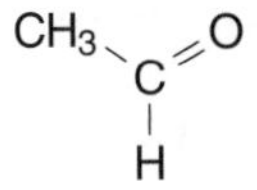	293	12	1.1
		(hexane solution)		
acetone	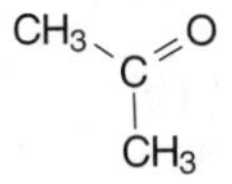	279	15	1.2
		(hexane solution)		
butenal		207	12,000	4.1
		328	20	1.3
		(ethanol solution)		
(E)-pent-3-en-2-one	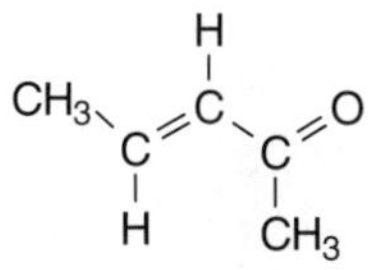	221	12,000	4.1
		312	40	1.6
		(ethanol solution)		
4-methylpent-3-en-4-one	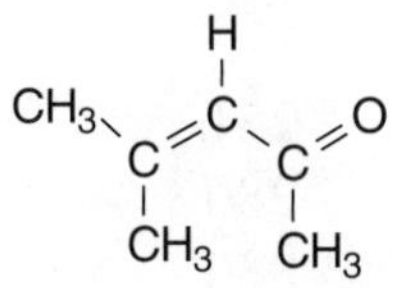	238	12,000	4.1
		316	60	1.8
		(ethanol solution)		
cyclohex-2-en-1-one	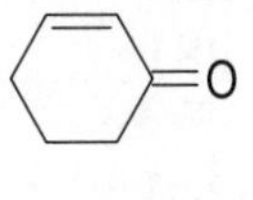	225	7,950	3.9
benzoquinone	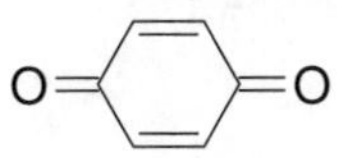	247	12,600	4.1
		292	1,000	3.0
		363	250	2.4

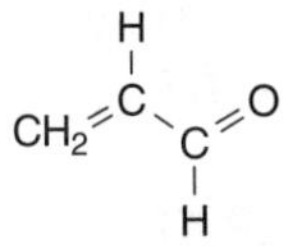

(3) Benzene derivatives

Benzene derivatives exhibit medium to strong absorption in the UV region. Bands usually have characteristic fine structure and the intensity of the absorption is strongly influenced by substituents. Examples listed in Table 2.3 include weak auxochromes (-CH_3, -Cl, -OMe), groups which increase conjugation (-CH=CH_2, -C(=O)-R, -NO_2) and auxochromes whose absorption is pH dependent (-NH_2 and -OH).

Table 2.3 UV Absorption Bands in Common Benzene Derivatives

compound	structure	λ_{max} (nm)	ε	$\log_{10}(\varepsilon)$
Benzene	C_6H_6	184 204 256	60,000 7,900 200	4.8 3.9 2.3
Toluene	C_6H_5-CH_3	208 261	8,000 300	3.9 2.5
Chlorobenzene	C_6H_5-Cl	216 265	8,000 240	3.9 2.4
Anisole	C_6H_5-OCH_3	220 272	8,000 1,500	3.9 3.2
Styrene	C_6H_5-CH=CH_2	244 282	12,000 450	4.1 2.7
Acetophenone	C_6H_5-C(=O)-CH_3	244 280	12,600 1,600	4.1 3.2
Nitrobenzene	C_6H_5-NO_2	251 280 330	9,000 1,000 130	4.0 3.0 2.1
Aniline	C_6H_5-NH_2	230 281	8,000 1,500	3.9 3.2
Anilinium ion	C_6H_5-NH_3^+	203 254	8,000 160	3.9 2.2
Phenol	C_6H_5-OH	211 270	6,300 1,500	3.8 3.2
Phenoxide ion	C_6H_5-O^-	235 287	9,500 2,500	4.0 3.4

Aniline and phenoxide ion have strong UV absorptions due to the overlap of the lone pair on the nitrogen (or oxygen) with the π-system of the benzene ring. This may be expressed in the usual Valence Bond terms:

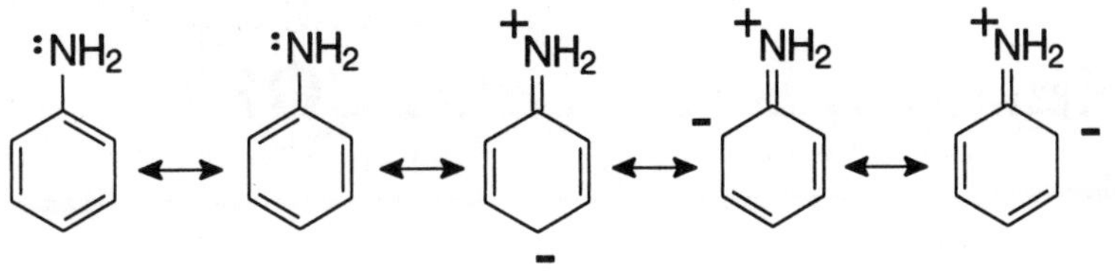

The striking changes in the ultraviolet spectra accompanying protonation of aniline and phenoxide ion are due to loss (or substantial reduction) of the overlap between the lone pairs and the benzene ring.

2.7 THE EFFECT OF SOLVENTS

Solvent polarity may affect the absorption characteristics, in particular λ_{max}, since the polarity of a molecule usually changes when an electron is moved from one orbital to another. Solvent effects of up to 20 nm may be observed with carbonyl compounds. Thus the $n \rightarrow \pi^*$ absorption of acetone occurs at 279 nm in *n*-hexane, 270 nm in ethanol, and at 265 nm in water.

3

INFRARED (IR) SPECTROSCOPY

3.1 ABSORPTION RANGE AND THE NATURE OF IR ABSORPTION.

Infrared absorption spectra are calibrated in wavelengths expressed in micrometers:

$$1\mu m = 10^{-6}\ m$$

or in frequency-related *wave numbers* (cm^{-1}) which are reciprocals of wavelengths:

$$\text{wave number } \bar{\nu}\ (cm^{-1}) = \frac{1 \times 10^4}{\text{wavelength (in } \mu m)}$$

The range accessible for standard instrumentation is usually:

$$\bar{\nu} = 4000 \text{ to } 666\ cm^{-1}$$

$$\text{or } \lambda = 2.5 \text{ to } 15\ \mu m$$

Infrared absorption intensities are rarely described quantitatively, except for the general classifications of s (strong), m (medium) or w (weak).

The transitions responsible for IR bands are due to molecular *vibrations, i.e.* to periodic motions involving stretching or bending of bonds. Polar bonds are associated with strong IR absorption ***while symmetrical bonds may not absorb at all***.

Clearly the vibrational frequency, *i.e.* the position of the IR bands in the spectrum, depends on the nature of the bond. Shorter and stronger bonds have their stretching vibrations at the higher energy end (shorter wavelength) of the IR spectrum than the longer and weaker bonds. Similarly, bonds to lighter atoms (*e.g.* hydrogen), vibrate at higher energy than bonds to heavier atoms.

IR bands often have rotational sub-structure, but this is normally resolved only in spectra taken in the gas phase.

3.2 EXPERIMENTAL ASPECTS OF INFRARED SPECTROSCOPY

The basic layout of an IR spectrometer is the same as for an UV spectrometer (Figure 2.1), except that all components must now match the different energy range of electromagnetic radiation. Very few substances are transparent over the whole of the IR range: sodium and potassium chloride and sodium and potassium bromide are most common. The cells used for obtaining IR spectra in solution typically have NaCl windows and liquids can be examined as films on NaCl plates. Solution spectra are generally obtained in chloroform or carbon tetrachloride but this leads to loss of information at longer wavelengths where there is considerable absorption of energy by the solvent. Alternatively, organic solids may be examined as mulls (fine suspensions) in heavy oils. The oils absorb infrared radiation but only in well-defined regions of the IR spectrum. Solids may also be examined as dispersions in compressed KBr or KCl discs.

To a first approximation, the absorption frequencies due to the important IR chromophores are the same in solid and liquid states.

3.3 GENERAL FEATURES OF INFRARED SPECTRA

Molecules generally consist of large assemblages of bonds and each bond may have several IR-active *vibrational modes.* IR spectra are therefore complex and have many features.

The characteristic frequencies of IR vibrations are influenced strongly by small changes in molecular structure, thus making it difficult to deduce the presence of structural fragments from IR data alone.

IR spectra are very useful for identifying compounds by direct comparison with spectra from authentic samples (*"fingerprinting"*), but IR spectra are limited in deducing structures from spectroscopic data. Fortunately, some groups of atoms form chromophores that are readily recognised from IR spectra.

An IR chromophore will tend to be useful for the determination of structure if it meets some of the following criteria:

(a) The chromophore should not absorb in the *most crowded region* of the spectrum (600-1400 cm^{-1}) where strong overlapping stretching absorptions from C-X single bonds (X = O, N, S, P and halogens) make assignment difficult.

(b) The chromophores should be *strongly absorbing* to avoid confusion with weak harmonics. However, in otherwise empty regions *e.g.* 1800-2500 cm^{-1}, even weak absorptions can be assigned with confidence.

(c) The absorption frequency must be structure dependent in an *interpretable* manner. This is particularly true of the very important bands due to the C=O stretching vibrations, which generally occur between 1630 and 1850 cm^{-1}.

3.4 IMPORTANT IR CHROMOPHORES

(1)	***-O-H Stretch***	Not hydrogen-bonded ("free")	3600 cm^{-1}
		Hydrogen-bonded	3100 - 3200 cm^{-1}

This difference between hydrogen bonded and free OH frequencies is clearly related to the weakening of the O-H bond as a consequence of hydrogen bonding.

(2) ***Carbonyl groups*** always give rise to **strong** absorption between 1630 and 1850 cm^{-1} due to C=O stretching vibrations. Moreover, carbonyl groups in different functional groups are associated with well-defined regions of IR absorption (Table 3.1).

Even though the ranges for individual types often overlap, it may be possible to make a definite decision from information derived from other regions of the IR spectrum. Thus esters also exhibit strong C-O stretching absorption between 1200 and 1300 cm^{-1} while carboxylic acids exhibit O-H stretching absorption generally near 3000 cm^{-1}.

The characteristic shift toward lower frequency associated with the introduction of α, β–unsaturation can be rationalised by considering the Valence Bond description of an enone:

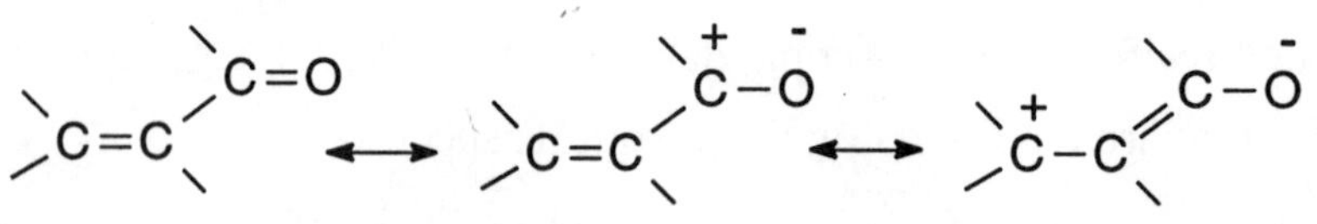

C=C–C=O ⟷ C=C–C⁺–O⁻ ⟷ C⁺–C=C–O⁻

A **B** **C**

The additional structure **C**, which cannot be drawn for an unconjugated carbonyl derivative, implies that the carbonyl band in an enone has more single bond character and is therefore weaker. The involvement of a carbonyl group in hydrogen bonding reduces the frequency of the carbonyl stretching vibration by about 10 cm^{-1}. This can be rationalised in a manner analogous to that proposed above for free and H-bonded O-H vibrations.

Table 3.1 Carbonyl IR Absorption Frequencies in Common Functional Groups

Carbonyl group	Structure	$\bar{\nu}$ (cm^{-1})
Ketones	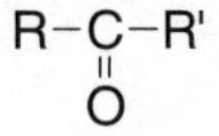 R–C(=O)–R'	1700 - 1725
Aldehydes	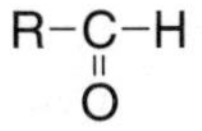 R–C(=O)–H	1720 - 1740
Aryl aldehydes or ketones, α, β-unsaturated aldehydes or ketones	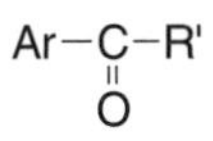 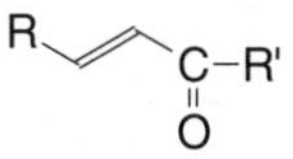 Ar–C(=O)–R' ; R–CH=CH–C(=O)–R' ; R' = alkyl, aryl, or H	1660 - 1715
Cyclopentanones	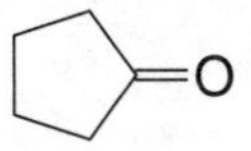 cyclopentanone (ring C=O)	1740 - 1750
Cyclobutanones	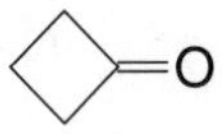 cyclobutanone (ring C=O)	1760 - 1780
Carboxylic acids	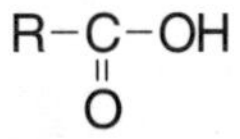 R–C(=O)–OH	1700 - 1725
Esters	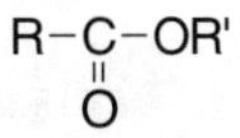 R–C(=O)–OR'	1735 - 1750
Phenolic Esters	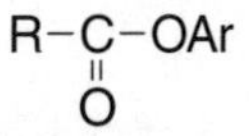 R–C(=O)–OAr	1770 - 1800
Aryl or α, β–unsaturated Esters	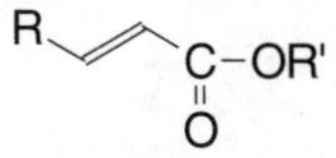 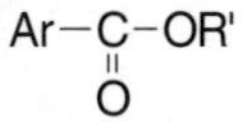 R–CH=CH–C(=O)–OR' ; Ar–C(=O)–OR'	1715 - 1730
δ-Lactones	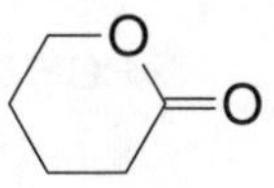 six-membered ring lactone (O, C=O)	1735 - 1750
γ-Lactones	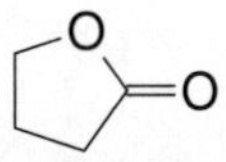 five-membered ring lactone (O, C=O)	1760 - 1780
Amides	 R–C(=O)–NR'R"	1630 - 1690
Acid chlorides	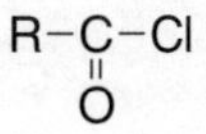 R–C(=O)–Cl	1785 - 1815
Acid anhydrides (two bands)	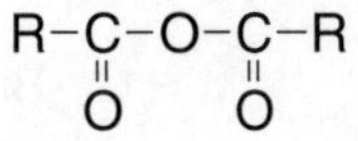 R–C(=O)–O–C(=O)–R	1740 - 1850
Carboxylates	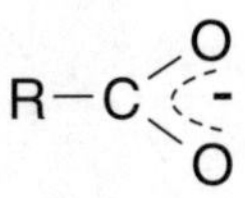 R–C(O)(O)$^-$	1550 – 1610 1300 - 1450

(3) ***Other polar functional groups.*** Many functional groups have characteristic IR absorptions and can be identified with the aid of IR data. This is particularly useful for groups that do not contain magnetic nuclei and are thus not readily identified by NMR spectroscopy.

Some of the most common functional groups that are readily identified by IR spectroscopy are listed in Table 3.2.

Table 3.2 Characteristic IR Absorption Frequencies for Common Functional Groups

Functional group	Structure	$\bar{\nu}$ (cm^{-1})	intensity
Amine	>N–H	3300 - 3500	
Imine	>C=N–	1480 - 1690	
Nitro	$-N^{+}(=O)O^{-}$	1500 - 1650 1250 - 1400	strong medium
Sulfoxide	>S=O	1010 - 1070	strong
Sulfone	O=S=O	1300 - 1350 1100 - 1150	strong strong
Sulfonamides and Sulfonate esters	$-SO_2-N<$ $-SO_2-O-$	1140 - 1180 1300 - 1370	strong strong
Alcohols	–C–OH	1000 - 1260	strong
Ethers	–C–OR	1085 - 1150	strong
Alkyl fluorides	–C–F	1000 - 1400	strong
Alkyl chlorides	–C–Cl	580 - 780	strong
Alkyl bromides	–C–Br	560 - 800	strong
Alkyl iodides	–C–I	500 - 600	strong

While the essentially symmetrical carbon-carbon double bonds have only weak infrared absorption, the more polar carbon-carbon double bonds in enol ethers and enones may absorb strongly between 1600 and 1700 cm^{-1}.

(4) ***Chromophores Absorbing in the Region Between 1900 and 2600 cm^{-1}.*** The absorptions listed in Table 3.3 often yield useful information because, though usually of only weak or medium intensity, they occur in regions largely devoid of absorption by commonly occurring chromophores.

Table 3.3 IR Absorption Frequencies in the Region 1900 – 2600 cm^{-1}

Functional group	Structure	$\bar{\nu}$ (cm^{-1})
Alkyne	—C≡C—	2100 - 2300
Nitrile	—C≡N	~ 2250
Cyanate	—N=C=O	~ 2270
Thiocyanate	—N=C=S	~ 2150 (broad)
Allene	>C=C=C<	~ 1950

4

MASS SPECTROMETRY

It is possible to determine the masses of individual ions in the gas phase. Strictly speaking, it is only possible to measure their mass/charge ratio (m/e), but as multi charged ions are very much less abundant than those with a single electronic charge ($e = 1$), m/e is for all practical purposes equal to the mass of the ion, m. The principal experimental problems in mass spectrometry are firstly to volatilise the substrate (which implies high vacuum) and secondly to ionise the neutral molecules to charged species.

4.1 IONISATION PROCESSES

The most common method of ionisation involves *electron impact* (EI) and there are two general courses of events following a collision of a molecule M with an electron e. By far the most probable event involves electron ejection which yields an odd-electron positively charged *cation radical* $[M]^{+\cdot}$ of the same mass as the initial molecule M.

$$M + e \rightarrow \quad [M]^{+\cdot} \quad + \ 2e$$

The cation radical produced is known as the *molecular ion* and its mass gives a direct measure of the molecular weight of a substance. An alternative, far less probable process, also takes place and it involves the capture of an electron to give a negative *anion radical*, $[M]^{-\cdot}$.

$$M + e \rightarrow \quad [M]^{-\cdot}$$

Electron impact mass spectrometers are generally set up to detect only positive ions, but negative-ion mass spectrometry is also possible.

The energy of the electron responsible for the ionisation process can be varied. It must be sufficient to knock out an electron and this threshold, typically about 10-12 eV, is known as the *appearance potential.* In practice much higher energies (~70 eV) are used and this large excess energy (1 eV = 95 kJ mol^{-1}) causes further *fragmentation* of the molecular ion.

The two important types of fragmentation are:

$$[M]^{+\cdot} \rightarrow A^{+} \text{ (even electron cation)} + B^{\cdot} \text{ (radical)}$$

or

$$[M]^{+\cdot} \rightarrow C^{+\cdot} \text{ (cation radical)} + D \text{ (neutral molecule)}$$

As only species bearing a positive charge will be detected, the mass spectrum will show signals due not only to $[M]^{+\cdot}$ but also due to A^{+}, $C^{+\cdot}$ and to fragment ions resulting from subsequent fragmentation of A^{+} and $C^{+\cdot}$.

As any species may fragment in a variety of ways, the typical mass spectrum consists of many signals. The mass spectrum consists of a plot of masses of ions against their relative abundance.

The most important alternative ionisation method to electron impact, is *chemical ionisation* (CI). In CI mass spectrometry, an intermediate substance (generally methane) is introduced at a higher concentration than that of the substance being investigated. The carrier gas is ionised by electron impact and the substrate is then ionised by collisions with these ions. CI is a milder ionisation method than EI and leads to less fragmentation of the molecular ion. All of the subsequent discussion of mass spectrometry is limited to positive-ion electron-impact mass spectrometry.

4.2 INSTRUMENTATION

In the most common type of mass spectrometer (Figure 4.1), the positively charged ions of mass, m, and charge, e (generally $e = 1$) are subjected to an accelerating voltage V and passed through a magnetic field H which causes them to be deflected into a curved path of radius r. The quantities are connected by the relationship:

$$\frac{m}{e} = \frac{H^2 r^2}{2V}$$

The values of H and V are known, r is determined experimentally and e is assumed to be unity thus permitting us to determine the mass m. In practice the magnetic field is scanned so that streams of ions of different mass pass sequentially to the detecting system (ion collector). The whole system (Figure 4.1) is under high vacuum (better than 10^{-6} Torr) to permit the volatilisation of the sample and so that the passage of ions is not impeded. The introduction of the sample into the ion chamber at high vacuum requires a complex sample inlet system.

The magnetic scan is synchronised with the x-axis of a recorder and calibrated to appear as *mass number* (strictly *m/e*). The amplified current from the ion collector gives the relative abundance of ions on the y-axis. The signals are usually pre-processed by a computer that assigns a relative abundance of 100% to the strongest peak (*base peak*).

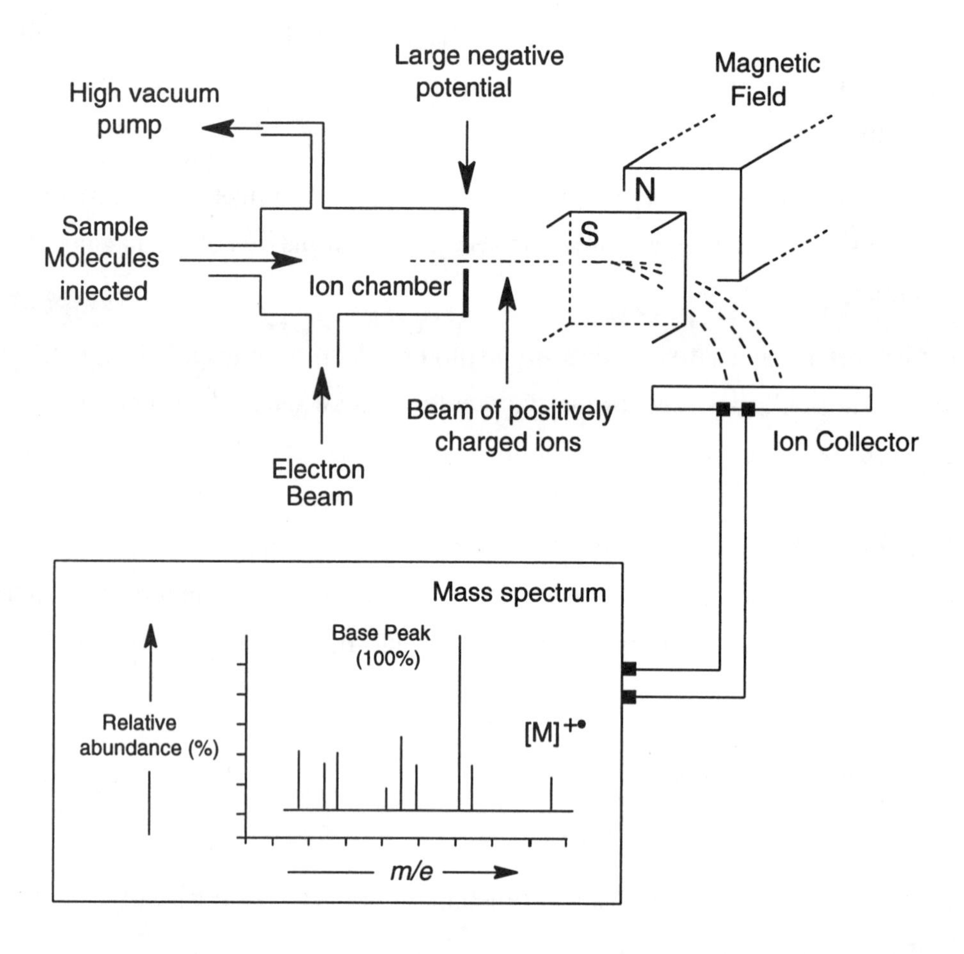

Figure 4.1 Schematic Diagram of an Electron-Impact Mass Spectrometer

4.3 MASS SPECTRAL DATA

As well as giving the molecular weight of a substance, the molecular ion of a compound may provide additional information. The "nitrogen rule" states that a molecule with an even molecular weight must contain no nitrogen atoms or an even number of nitrogen atoms. This means that a molecule with an odd molecular weight must contain an odd number of nitrogen atoms.

(1) ***High resolution mass spectra.*** The **mass of an ion** can be easily determined to the nearest unit value. Thus the position of $[M]^{+\cdot}$ gives a direct measure of molecular weight. By use of *double focussing*, the mass of an ion may be determined to an accuracy of approximately ± 0.0001 of a mass unit. It is not usually possible to assign a molecular formula to a compound on the basis of the integer *m/e* value of its parent ion. For example, a parent ion at *m/e* 72 could be due to a compound whose molecular formula is C_4H_8O or one with a molecular formula $C_3H_4O_2$ or one with a molecular formula $C_3H_8N_2$. However, if the mass spectrum is recorded with an accuracy of approximately ± 0.0001 of a mass unit *(a high resolution mass spectrum)* then the mass of the parent ion, or any fragment, can be recorded to much better than integer precision. Since the masses of the atoms of each element are known to high accuracy, molecules that may have the same mass when measured only to the nearest integer mass unit, can be distinguished when the mass is measured with high precision. The accurate masses of ^{12}C, ^{16}O, ^{14}N and ^{1}H are 12.0000 (by definition) 15.9949, 14.0031, and 1.0078 (Table 4.1) so ions with the formulas $C_4H_8O^{+\cdot}$, $C_3H_4O_2^{+\cdot}$ or $C_3H_8N_2^{+\cdot}$ would have accurate masses 72.0573, 72.0210, and 72.0686 so these could easily be distinguished by high resolution mass spectroscopy. Additionally, if the mass of any fragment in the mass spectrum can be accurately determined, then there is usually only one combination of elements which can give rise to that signal since there are only a limited number of elements and their masses are accurately known. By examining a mass spectrum at sufficiently high resolution, one can obtain the exact composition of *each ion* in a mass spectrum, unambiguously. Most importantly, determining the accurate mass of $[M]^{+\cdot}$ *gives the molecular formula of the compound.*

For some elements where there is more than one isotope of high natural abundance *e.g.* bromine has two abundant isotopes - ^{79}Br 49 % and ^{81}Br 51 %; chlorine also has two abundant isotopes- ^{37}Cl 25 % and ^{35}Cl 75% (Table 4.1). The presence of elements that contain significant proportions ($\geq$ 1%) of minor isotopes is often obvious simply by inspection of ions near the molecular ion. The relative intensities of the $[M]^{+\cdot}$, $[M+1]^{+\cdot}$ and $[M+2]^{+\cdot}$ ions show a characteristic pattern depending on the elements that make up the ion. For any molecular ion (or fragment) which contains a bromine atom, the mass spectrum will contain two peaks separated by two *m/e* units, one for the ions which contain ^{79}Br and one for the ions which contain ^{81}Br. For bromine-containing ions, the relative intensities of the two ions will be approximately the same since the natural abundances of ^{79}Br and ^{81}Br are approximately the same. Similarly, for any molecule (or fragment of an molecule) which contains a chlorine atom, the mass spectrum will contain two fragments separated by two *m/e* units, one for the ions which contain ^{37}Cl and one for the ions which contain ^{35}Cl.

For chlorine-containing ions, the relative intensities of the two ions will be approximately the 1:3 since this reflects the natural abundances of ^{37}Cl and ^{35}Cl.

Table 4.1 Accurate Masses of Selected Isotopes

Isotope	Natural Abundance (%)	Mass
^{1}H	99.98	1.00783
^{12}C	98.9	12.0000
^{13}C	1.1	13.00336
^{14}N	99.6	14.0031
^{16}O	99.8	15.9949
^{19}F	100.0	18.99840
^{31}P	100.0	30.97376
^{32}S	95.0	31.9721
^{33}S	0.75	32.9715
^{34}S	4.2	33.9679
^{35}Cl	75.8	34.9689
^{37}Cl	24.2	36.9659
^{79}Br	50.7	78.9183
^{81}Br	49.3	80.9163

(2) Molecular Fragmentation. The **fragmentation pattern** is a molecular fingerprint. The mass spectrum (see Figure 4.1) consists, in addition to the molecular ion peak, of a number of peaks at lower mass number and these result from fragmentation of the molecular ion. The principles determining the mode of fragmentation are reasonably well understood, and it is possible to derive structural information from the fragmentation pattern in several ways.

(a) the appearance of prominent peaks at certain mass numbers can be correlated empirically with certain structural elements (Table 4.2), *e.g.* a prominent peak at m/e = 43 is a strong indication of the presence of a CH_3-CO- fragment in the molecule.

Information can also be obtained from *differences* between the masses of two peaks. Thus a prominent fragment ion that occurs 15 mass numbers below the molecular ion, suggests strongly the loss of a CH_3- group and therefore that a methyl group was present in the substance examined.

(b) The knowledge of the principles governing the **mode of fragmentation** of ions makes it possible to confirm the structure assigned to a compound and, quite often, to determine the juxtaposition of structural fragments and to distinguish between isomeric substances. For example, the mass spectrum of benzyl methyl ketone, Ph-CH_2-CO-CH_3 contains a strong peak at m/e = 91 due to the stable ion Ph-CH_2^+, but this ion is absent in the mass spectrum of the isomeric propiophenone Ph-CO-CH_2CH_3 where the structural elements Ph- and -CH_2- are separated. Instead, a prominent peak occurs at m/e = 105 due to the presence of the structural element Ph-CO-.

Catalogues (and now electronic databases) of the mass spectral fragmentation patterns of known molecules are now available and can be rapidly searched by computer. The pattern and intensity of fragments in the mass spectrum is characteristic of an individual compound so comparison of the experimental mass spectrum of a compound with those in a library can be used to positively identify it, if its spectrum has been recorded previously. It is now common to couple an instrument for separating a mixture of organic compounds *e.g.* a gas chromatograph, directly to the input of a mass spectrometer. In this way as each individual compound is separated from the mixture, its mass spectrum can be recorded and compared with the library of known compounds and identified immediately if it is a known compound.

(c) **Meta-stable peaks** in a mass spectrum arise if the fragmentation process

$$a^+ \rightarrow b^+ + c \text{ (neutral)}$$

takes place within the ion-accelerating region of the mass spectrometer (Figure 4.1). Ion peaks corresponding to the masses of a^+ and to b^+ (m_a and m_b) may be accompanied by a broader peak at mass m^*, such that:

$$m^* = \frac{m_b^2}{m_a}$$

This often permits positive identification of a particular fragmentation path.

Table 4.2 Common Fragments and their Masses

Fragment	Mass	Fragment	Mass	Fragment	Mass
CH_3-	15	CH_3CH_2-	29	$H-C(=O)-$	29
NO	30	$-CH_2OH$	31	$CH_2=CH-CH_2$	41
$CH_3-C(=O)-$	43	$HO-C(=O)-$	45	$-NO_2$	46
C_4H_7	55	C_4H_9	57	$CH_3CH_2-C(=O)-$	57
$CH_2=C(OH)_2$	60	C_5H_5	65	C_6H_5 (phenyl)	77
$C_6H_5-CH_2-$ (C_7H_7)	91	$NC_5H_4-CH_2-$ (C_6H_6N)	92	$C_6H_5-C(=O)-$ (C_7H_5O)	105
$CH_3-C_6H_4-C(=O)-$ (C_8H_7O)	119	I—	127		

4.4 REPRESENTATION OF FRAGMENTATION PROCESSES

As fragmentation reactions in a mass spectrometer involve the breaking of bonds, they can be represented by the standard "arrow notation" used in organic chemistry. For some purposes a radical cation (*e.g.* a generalised ion of the molecular ion) can be represented without attempting to localise the missing electron:

$$[M]^{+\cdot} \quad \text{or} \quad [H_3C\text{-}CH_2\text{-}O\text{-}R]^{+\cdot}$$

However, to show a fragmentation process it is generally necessary to indicate "from where the electron is missing" even though no information about this exists. In the case of the molecular ion corresponding to an alkyl ethyl ether, it can be reasonably inferred that the missing electron resided on the oxygen. The application of standard arrow notation permits us to represent a commonly observed process, *viz.* the loss of a methyl fragment from the $[H_3C\text{-}CH_2\text{-}O\text{-}R]^{+\cdot}$ molecular ion:

$$CH_3\text{-}CH_2\text{-}\ddot{O}^{\cdot+}\text{-}R \rightarrow \dot{C}H_3 + H_2C{=}\overset{+}{O}\text{-}R \longleftrightarrow H_2\overset{+}{C}\text{-}O\text{-}R$$

4.5 FACTORS GOVERNING FRAGMENTATION PROCESSES

Three factors dominate the fragmentation processes:

(a) **Weak bonds** tend to be broken most easily

(b) **Stable fragments** (not only ions, but also the accompanying radicals and molecules) tend to be formed most readily

(c) Some fragmentation processes depend on the ability of molecules to assume cyclic transition states

Favourable fragmentation processes naturally occur more often and ions thus formed give rise to strong peaks in the mass spectrum.

4.6 EXAMPLES OF COMMON TYPES OF FRAGMENTATION

There are a number of common types of cleavage which are characteristic of various classes of organic compounds. These result in the loss of well-defined fragments which are characteristic of certain functional groups or structural elements.

(1) ***Cleavage at Branch Points.*** Cleavage of aliphatic carbon skeletons at branch points is favoured as it leads to the formation of more substituted carbocations:

$$CH_3-C(CH_3)_2-CH_2-CH_2-CH_3 \longrightarrow CH_3-\overset{\bullet +}{C}(CH_3)_2-CH_2-CH_2-CH_3 \longrightarrow \dot{C}H_3 + {}^{+}C(CH_3)_2-CH_2-CH_2-CH_3$$

neutral fragment

stable tertiary carbocation

(2) ***β - Cleavage.*** Chain cleavage tends to occur β to heteroatoms, double bonds and aromatic rings because relatively stable, delocalised carbocations result in each case.

(a)

$$R-\ddot{X}-C-C- \xrightarrow{-e^-} R-\overset{+\bullet}{X}-C-C- \longrightarrow R-X-\overset{+}{C} \longleftrightarrow R-\overset{+}{X}=C \quad + \quad \dot{C}-$$

X = O, N, S, halogen

resonance stablised carbocation

neutral fragment

(b)

$$C=C-C-C- \xrightarrow{-e^-} \overset{+}{C}-\dot{C}-C-C- \longrightarrow C=C-\overset{+}{C}- \longleftrightarrow \overset{+}{C}-C=C- \quad + \quad \dot{C}-$$

resonance stablised carbocation

neutral fragment

(c)

C—C— $- e^-$ C+ •C—

neutral fragment

C C C +

resonance stablised carbocation

(3) ***Cleavage α to carbonyl groups.*** Cleavage tends to occur α to carbonyl groups to give stable acylium cations. R may be an alkyl, -OH or -OR group.

O C C R $- e^-$ +O• C C R → +O≡C–R •C—

neutral fragment

resonance stablised carbocation O=C+–R

(4) ***Cleavage α to heteroatoms.*** Cleavage of chains may also occur α to heteroatoms, *e.g.* in the case of ethers:

R–Ö–C— $- e^-$ R–Ö+•–C— → R–Ö: +C—

free radical carbocation

(5) ***retro Diels-Alder reaction.*** Cyclohexene derivatives may undergo a retro Diels-Alder reaction:

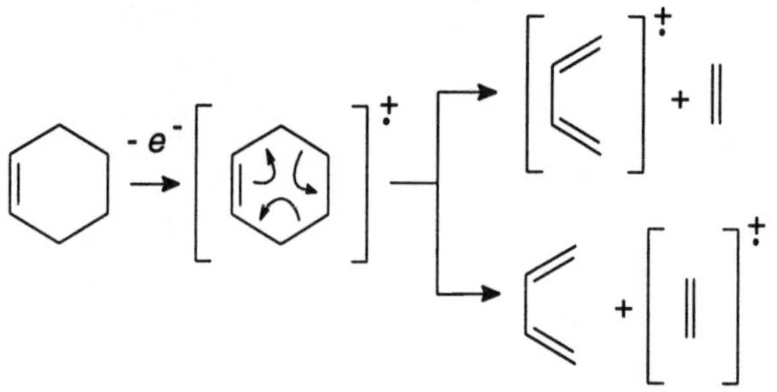

(6) ***The McLafferty rearrangement.*** Compounds where the molecular ion can assume the appropriate 6-membered cyclic transition state usually undergo a cyclic fragmentation, known as the **McLafferty rearrangement.** This rearrangement involves a transfer of a γ hydrogen atom to an oxygen and is often observed with ketones, acids and esters:

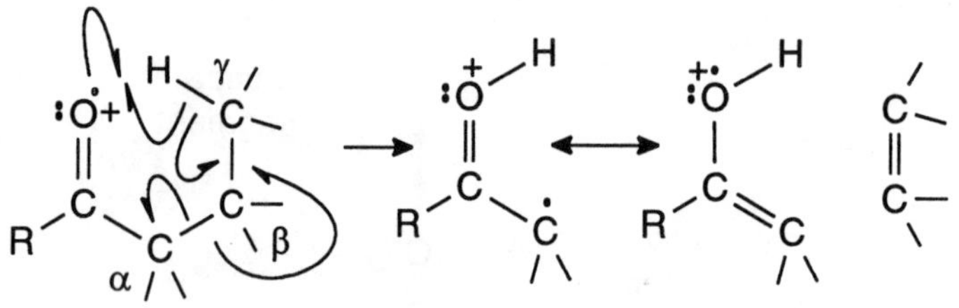

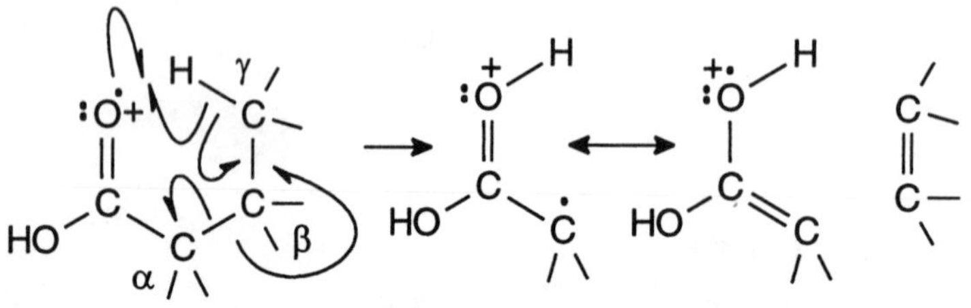

With primary carboxylic acids R-CH_2-COOH, this fragmentation leads to a characteristic peak at *m/e* = 60

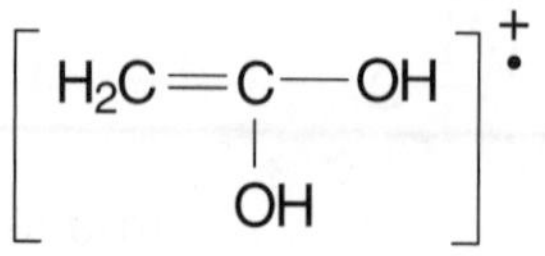

$$\left[H_2C{=}C(OH){-}OH \right]^{+\bullet}$$

With carboxylic esters, two types of McLafferty rearrangements may be observed and ions resulting from either fragmentation pathway are observed in the mass spectrum:

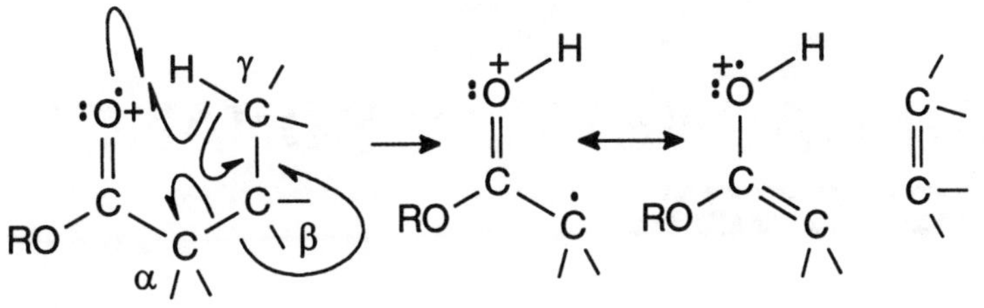

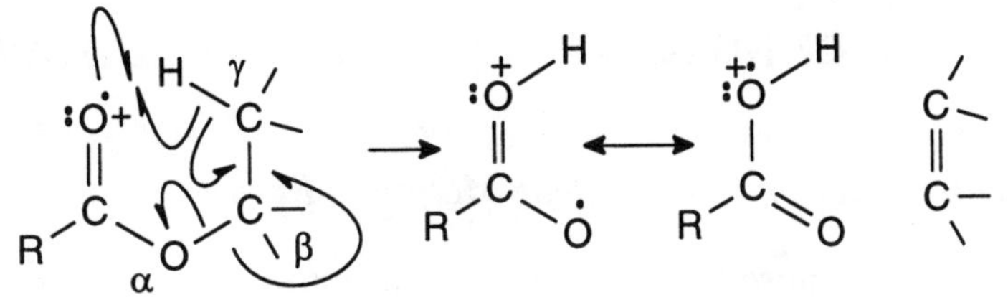

5

NUCLEAR MAGNETIC RESONANCE (NMR) SPECTROSCOPY

5.1 THE PHYSICS OF NUCLEAR SPINS AND NMR INSTRUMENTS

(1) The Larmor Equation and Nuclear Magnetic Resonance

All nuclei have charge because they contain protons and some of them also behave as if they spin. A spinning charge is equivalent to a conductor carrying a current and therefore will be associated with a magnetic field **H** (Figure 5.1). Such nuclear magnetic dipoles are characterised by nuclear magnetic **spin quantum numbers** which are designated by the letter **I** and can take up values equal to 0, $^1/_2$, 1,$^3/_2$ ... *etc.*

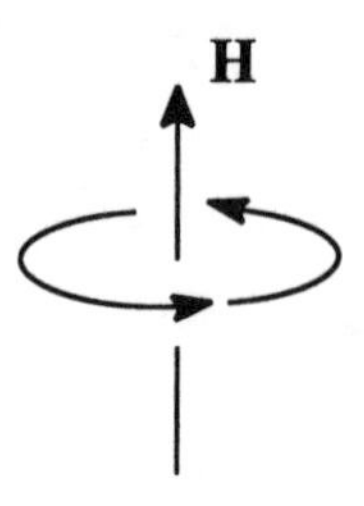

Figure 5.1 A Spinning Charge Generates a Magnetic Field and Behaves Like a Small Magnet

It is useful to consider three types of nuclei:

***Type 1*:** Nuclei with **I** = 0. These nuclei do not interact with the applied magnetic field and **are not NMR chromophores**. Nuclei with **I** = 0 have an even number of protons and even number of neutrons and have no net spin. This means that nuclear spin is a property characteristic of certain **isotopes** rather than of certain elements. The most prominent examples of nuclei with **I** = 0 are ^{12}C and ^{16}O, the dominant isotopes of carbon and oxygen. Both oxygen and carbon of also have isotopes that can be observed by NMR spectroscopy.

Type 2: Nuclei with **I** = $^1/_2$. These nuclei have a non-zero magnetic moment and are NMR visible and have no nuclear electric quadrupole (Q). The two most important nuclei for NMR spectroscopy belong to this category: ^{1}H (ordinary hydrogen) and ^{13}C (a non-radioactive isotope of carbon occurring to the extent of 1.06% at natural abundance). Also, two other commonly observed nuclei ^{19}F and ^{31}P have **I** = $^1/_2$. Together, NMR data for ^{1}H and ^{13}C account for well over 90% of all NMR observations in the literature and the discussion and examples in this book all refer to these two nuclei. However, the spectra of all nuclei with **I** = $^1/_2$ can be understood easily on the basis of common theory.

Type 3: Nuclei with **I** > $^1/_2$. These nuclei have both a magnetic moment and an electric quadrupole. This group includes some common isotopes (*e.g.* ^{2}H and ^{14}N) but they are more difficult to observe and spectra are generally very broad. This group of nuclei will not be discussed further.

The most important consequence of nuclear spin is that in a uniform magnetic field, a nucleus of spin **I** may assume 2**I** + 1 orientations. For nuclei with **I** = $^1/_2$, there are just 2 permissible orientations (since 2 x $^1/_2$ + 1 = 2). These two orientations will be of unequal energy (by analogy with the parallel and antiparallel orientations of a bar magnet in a magnetic field) and it is possible to induce a spectroscopic transition (spin-flip) by the absorption of a quantum of electromagnetic energy (ΔE) of the appropriate frequency (ν):

$$\nu = \frac{\Delta E}{h} \qquad (5.1)$$

In the case of NMR, the energy required to induce the nuclear spin flip also depends on the strength of the applied field, $\boldsymbol{H_o}$. It is found that

$$\nu = K\boldsymbol{H_o} \qquad (5.2)$$

where K is a constant characteristic of the nucleus observed. Equation 5.2 is known as the **Larmor equation** and is the fundamental relationship in NMR spectroscopy. Unlike other forms of spectroscopy, in NMR the frequency of the absorbed electromagnetic radiation is not an absolute value for any particular transition, but has a different value depending on the strength of the applied magnetic field. For every value of $\boldsymbol{H_o}$, there is a matching value of ν corresponding to the condition of *resonance* according to Equation 5.2, and this is the origin of the term *Resonance* in Nuclear Magnetic Resonance Spectroscopy. Thus for ^{1}H and ^{13}C, *resonance* *frequencies* corresponding to magnitudes of applied magnetic field (H_o) commonly found in commercial instruments are given in Table 5.1.

Table 5.1 Resonance Frequencies of ^{1}H and ^{13}C Nuclei in Magnetic Fields of Different Strengths

ν ^{1}H (MHz)	ν ^{13}C (MHz)	H_o (Tesla)
60	15.087	1.4093
90	22.629	2.1139
100	25.144	2.3488
200	50.288	4.6975
400	100.577	9.3950
500	125.720	11.744
600	150.864	14.0923
750	188.580	17.616
800	201.154	18.790
900	226.296	21.128

In common jargon, NMR spectrometers are commonly known by the frequency they use to observe ^{1}H *e.g.* as "60 MHz", "200 MHz" or "400 MHz" instruments, even if the spectrometer is set to observe a nucleus other than ^{1}H.

All the frequencies listed in Table 5.1 correspond to the radio frequency region of the electromagnetic spectrum and inserting these values into Equation 5.1 gives the size of the energy gap between the states in an NMR experiment. A resonance frequency of 100 MHz corresponds to an energy gap of approximately 4 x 10^{-5} kJmol^{-1}. This is an extremely small value on the chemical energy scale and this means that NMR spectroscopy is, for all practical purposes, a ground-state phenomenon.

Any absorption signal observed in a spectroscopic experiment must originate from excess of the population in the lower energy state, the so called *Boltzmann excess*, which is equal to N_β-N_α, where N_β and N_α are the populations in the lower (β) and upper (α) energy states.

For molar quantities, the general Boltzmann relation (Equation 5.3) shows that:

$$\frac{N_\beta}{N_\alpha} = e^{\frac{\Delta E}{RT}} \tag{5.3}$$

Clearly, as the energy gap (ΔE) approaches zero, the right hand side of Equation 5.3 approaches 1 and the Boltzmann excess becomes very small. For the NMR experiment, the population excess in the lower energy state is typically of the order of 1 in 10^5 which renders NMR spectroscopy an **inherently insensitive** spectroscopic technique. Equations 5.1 and 5.2 show that the energy gap (and therefore ultimately the Boltzmann excess and sensitivity), increases with increasing applied magnetic field. This is one of the reasons why it is desirable to use high magnetic fields in NMR spectrometers.

(2) Nuclear Relaxation

Even at the highest fields, the NMR experiment would not be practicable if mechanisms did not exist to restore the Boltzmann equilibrium that is perturbed as the result of the absorption of electromagnetic radiation in making an NMR measurement. These mechanisms are known by the general term of **relaxation** and are not confined to NMR spectroscopy. Because of the small magnitude of the Boltzmann excess in the NMR experiment, relaxation is more critical and more important in NMR than in other forms of spectroscopy.

If relaxation is too efficient (*i.e.* it takes a short time for the nuclear spins to relax after being excited in an NMR experiment) the lines observed in the NMR spectrum are very broad. If relaxation is too slow (*i.e.* it takes a long time for the nuclear spins to relax after being excited in an NMR experiment) the spins in the sample quickly *saturate* and only a very weak signal can be observed.

The most important relaxation processes in NMR involve interactions with other nuclear spins that are in the state of random thermal motion. This is called *spin-lattice relaxation* and results in a simple exponential recovery process after the spins are disturbed in an NMR experiment. The exponential recovery is characterised by a time constant T_1 that can be measured for different types of nuclei. For organic liquids and samples in solution, T_1 is typically of the order of several seconds. In the presence of paramagnetic impurities or in very viscous solvents, relaxation of the spins can be very efficient and NMR spectra obtained become broad.

Nuclei in solid samples typically relax very efficiently and give rise to very broad spectra. NMR spectra of solid samples can only be acquired using specialised spectroscopic equipment and solid state NMR spectroscopy will not be discussed further.

(3) The Acquisition of an NMR spectrum

As the NMR phenomenon is not observable in the absence of an applied magnetic field, a magnet is an essential component of any NMR spectrometer. Magnets for NMR may be permanent magnets (as in many low field routine instruments), electromagnets, or in most modern instruments they are based on superconducting solenoids, cooled by liquid helium. All magnets used for NMR spectroscopy share the following characteristics:

(a) The magnetic field must be **strong**. This is partly due to the fact that the sensitivity of the NMR experiment increases as the strength of the magnet increases, but more importantly it ensures adequate **dispersion** of signals and, in the case of ^{1}H NMR, also very important **simplification** of the spectrum.

(b) The magnetic field must be extremely **homogeneous** so that all portions of the sample experience exactly the same magnetic field. Any inhomogeneity of the magnetic field will result in broadening and distortion of spectral bands. For determining of the structure of organic compounds, the highest attainable degree of magnetic field homogeneity is desirable, because useful information may be lost if the width of the NMR spectral lines exceeds about 0.2 Hz. Clearly, 0.2 Hz in, say, 100 MHz implies a homogeneity of about 2 parts in 10^9, and this is a very stringent requirement over the whole volume of an NMR sample.

(c) The magnetic field must be very **stable,** so that it does not drift during the acquisition of the spectrum, which may take from several seconds to several hours.

5.2 CONTINUOUS WAVE (CW) NMR SPECTROSCOPY

Inspection of the Larmor equation (Equation 5.2) shows that for any nucleus the condition of resonance may be achieved by keeping the field constant and changing (or sweeping) the frequency or, alternatively, by keeping the frequency constant and sweeping the field. A schematic diagram of a frequency sweep CW NMR spectrometer is given in Figure 5.2.

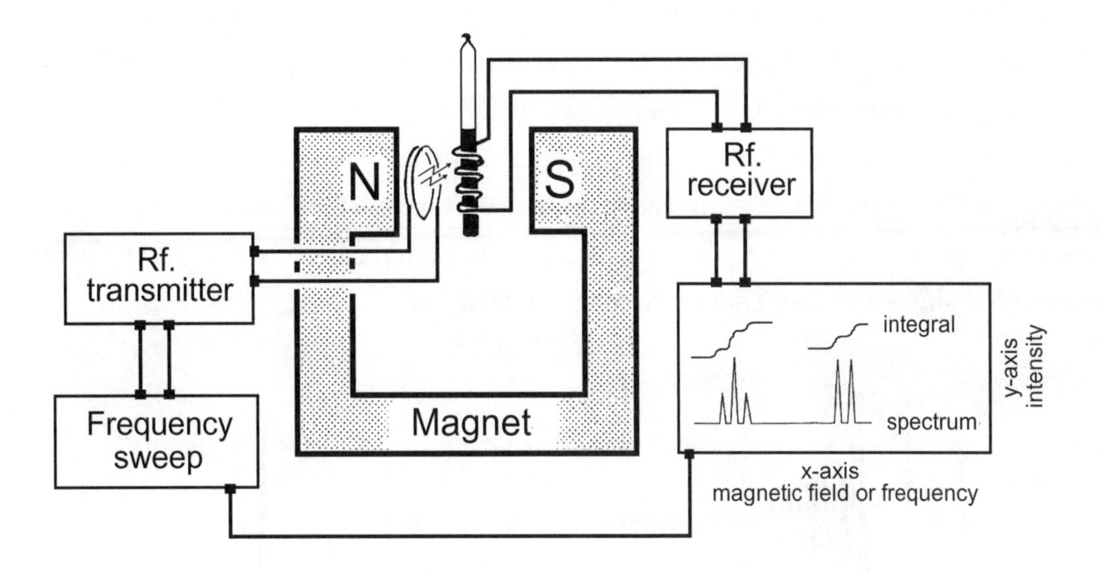

Figure 5.2 Schematic Representation of a CW NMR Spectrometer

An NMR spectrum is effectively a graph of the intensity of absorption of Rf radiation (y-axis) against the frequency of the Rf radiation (x-axis). Since frequency and magnetic field strength are linked by the Larmor equation, the x-axis could also be calibrated in units of magnetic field strength. In a CW NMR spectrometer, the x-axis of the output device (usually a pen plotter) is coupled to the frequency sweep so that the response of the sample is displayed as the frequency of the Rf transmitter varies. The integral of the spectrum represents the area of the absorption peak. In most NMR spectrometers, the integral is represented as a line plotted over the spectrum. Whenever a peak is encountered, the vertical displacement of the integral line is proportional to the area of the peak.

5.3 FOURIER-TRANSFORM (FT) NMR SPECTROSCOPY

As an alternative to the CW method, an intense short pulse of electromagnetic energy can be used to excite the nuclei in an NMR sample. The first property of pulsed NMR spectroscopy is that all of the nuclei are excited simultaneously whereas the CW NMR experiment requires a significant period of time (usually several minutes) to sweep or scan through a range of frequencies. Following the radiofrequency pulse, the magnetism in the sample is sampled as a function of time and, for a single resonance, the detected signal decays exponentially. The detected signal is called a *free induction decay* or FID (Figure 5.3a) and this type of spectrum (known as a *time-domain* spectrum) is converted into the more usual *frequency-domain* spectrum (Figure 5.3b) by performing a mathematical operation known as *Fourier transformation (FT).* Because the signal needs mathematical processing, pulsed NMR spectrometers require a computer and as well as performing the Fourier transformation, the computer also provides a convenient means of storing NMR data and performing secondary data processing and analysis.

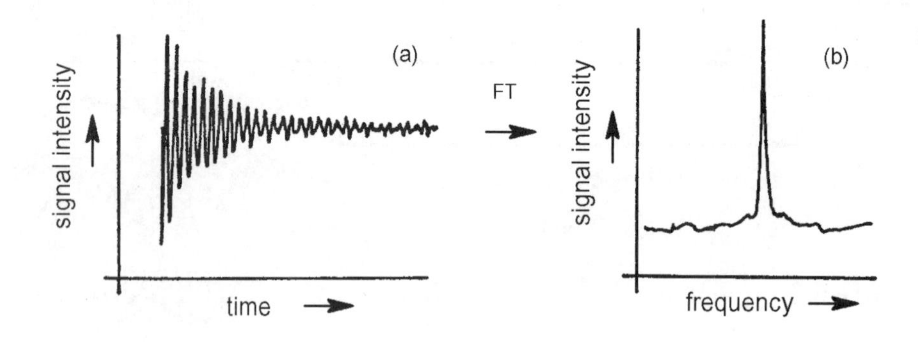

Figure 5.3 Time Domain and Frequency Domain NMR Spectra

Most NMR spectra consist of a number of signals and their time-domain spectra appear as a superposition of a number of traces of the type shown in Figure 5.3. Such spectra are quite uninterpretable by inspection, but Fourier transformation converts them into ordinary frequency-domain spectra. The time-scale of the FID experiment is of the order of seconds during which the magnetisation may be sampled many thousands of time. Data sampling is accomplished by a dedicated computer that is also used to perform the Fourier transformation.

The principal advantage of FT NMR spectroscopy is a great *increase in sensitivity per unit time* of the experiment. A CW scan generally takes of the order of one hundred times as long as the collection of the equivalent FID. During the time it would have taken to acquire one CW spectrum, the mini computer can accumulate many FID scans and add them up in its memory. The sensitivity (signal-to-noise ratio) of the NMR spectrum is proportional to the square root of the number of scans which are added together, so the quality of NMR spectra is vastly improved as more scans are added. It is the increase in sensitivity brought about by the introduction of FT NMR spectroscopy that has permitted the routine observation of ^{13}C NMR spectra.

In addition, the FID can be manipulated mathematically to enhance sensitivity (*e.g.* for routine ^{13}C NMR) at the expense of resolution, or to enhance resolution (often important for ^{1}H NMR) at the expense of sensitivity. Furthermore, it is possible to devise **sequences** of Rf pulses that result, after suitable mathematical manipulation, in NMR spectroscopic data that are of great value. Such methods (*e.g.* two-dimensional NMR, mathematical enhancement and massage of data) are, in the most part, beyond the scope of this book however some aspects are discussed in Chapter 7.

5.4 CHEMICAL SHIFT IN ^{1}H NMR SPECTROSCOPY

It is clear that NMR spectroscopy could be used to detect certain nuclei (*e.g.* ^{1}H, ^{13}C, ^{19}F, ^{31}P) and, also to estimate them quantitatively. The real usefulness of NMR spectroscopy in chemistry is based on secondary phenomena, the *chemical shift* and *spin-spin coupling* and, to a lesser extent, on effects related to the *time-scale* of the NMR experiment. Both the chemical shift and spin-spin coupling reflect the **chemical environment** of the nuclear spins whose spin-flips are observed in the NMR experiment and these can be considered as chemical effects in NMR spectroscopy.

A ^{1}H NMR spectrum is a graph of resonance frequency (chemical shift) vs. the intensity of Rf absorption by the sample. The spectrum is usually calibrated in dimensionless units called "Parts Per Million" (abbreviated to ppm) although the horizontal scale is a frequency scale, the units are "normalised" so that the scale has the same numbers **irrespective of the strength of the magnetic field** in which the measurement was made. The scale in ppm is termed a δ scale and NMR spectra are usually referenced to the resonance of some standard substance (a reference compound) whose frequency is chosen as 0.0 ppm. The frequency difference between the resonance of a nucleus and the resonance of the reference compound is termed the **chemical shift**.

Tetramethylsilane, $(CH_3)_4Si$, (abbreviated commonly as TMS) is the usual reference compound chosen for both ^{1}H and ^{13}C NMR and it is normally added directly to the solution of the substance to be examined. TMS has the following advantages as a reference compound:

(a) it is a relatively inert low boiling (b.p. 26.5°C) liquid which can be easily removed after use;

(b) it gives a sharp single signal in both ^{1}H and ^{13}C because the compound has only one type of hydrogen and one type of carbon;

(c) the chemical environment of both carbon and hydrogen in TMS is unusual due to the presence of silicon and hence the TMS signal occurs outside the normal range observed for organic compounds so the reference signal is unlikely to overlap a signal from the substance examined;

(d) the chemical shift of TMS is not substantially affected by complexation or solvent effects because the molecule doesn't contain any polar groups.

Chemical shifts can be measured in Hz but are more usually expressed in ppm.

$$\text{chemical shift in ppm} = \frac{10^6 \text{ x chemical shift from TMS in Hz}}{\text{spectrometer frequency in Hz}}$$

Note that for a spectrometer operating at 200 MHz, 1 ppm corresponds to 200 Hz *i.e.* for a spectrometer operating at x MHz, 1.00 ppm corresponds to x Hz.

For the majority of organic compounds, the chemical shift range for ^{1}H covers approximately the range 0-10 ppm (from TMS) and for ^{13}C covers approximately the range 0-220 ppm (from TMS). By convention, the δ scale runs (with increasing values) from right-to-left; for ^{1}H.

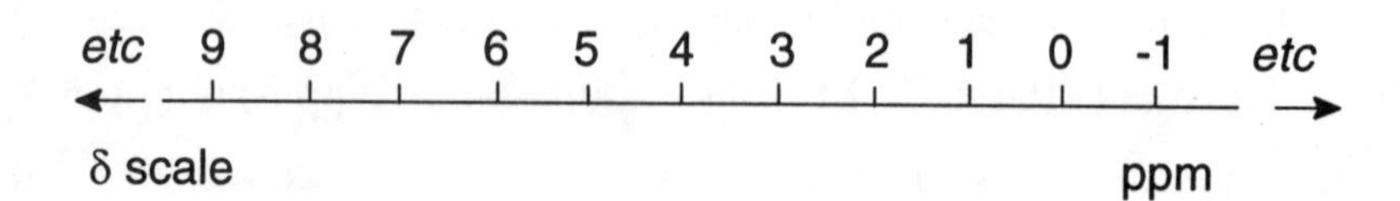

Each ^{1}H nucleus is **shielded or screened** by the electrons that surround it. Consequently each nucleus feels the influence of the main magnetic field to a different extent, depending on the efficiency with which it is screened. Each ^{1}H nucleus with a different chemical environment has a slightly different shielding and hence a different chemical shift in the ^{1}H NMR spectrum. Conversely, the number of different signals in the ^{1}H NMR spectrum reflects the number of chemically distinct environments for ^{1}H in the molecule. Unless two ^{1}H environments are precisely identical (by symmetry) *their chemical shifts must be different.* When two nuclei have identical molecular environments and hence the same chemical shift, they are termed *chemically equivalent* or *isochronous* nuclei. Non-equivalent nuclei that fortuitously have chemical shifts that are so close that their signals are indistinguishable are termed *accidentally equivalent* nuclei.

The chemical shift of a nucleus reflects the molecular structure and it can therefore be used to obtain structural information. Further, as hydrogen and carbon (and therefore ^1H and ^{13}C nuclei) are universal constituents of organic compounds the amount of structural information available from ^1H and ^{13}C NMR spectroscopy greatly exceeds in value the information available from other forms of molecular spectroscopy. **Every hydrogen and carbon atom in an organic molecule is "a chromophore"** for NMR spectroscopy.

For ^1H NMR, the intensity of the signal (which may be measured by electronically measuring the area under individual resonance signals) is directly proportional to the number of nuclei undergoing a spin-flip and **proton NMR spectroscopy is a quantitative method.**

Any effect which alters the density or spatial distribution of electrons around a ^1H nucleus will alter the degree of shielding and hence it's chemical shift. ^1H chemical shifts are sensitive to both the hybridisation of the atom to which the ^1H nucleus is attached (*sp*^{2}, *sp*^{3} *etc.*) and to electronic effects (the presence of neighbouring electronegative/electropositive groups). The chemical shift of a nucleus may also be affected by the presence in its vicinity of a *magnetically anisotropic* group (*e.g.* an aromatic ring or carbonyl group).

Nuclei tend to be deshielded by groups with withdraw electron density. Deshielded nuclei resonate at higher δ values (away from TMS). Conversely shielded nuclei resonate at lower δ values (towards TMS).

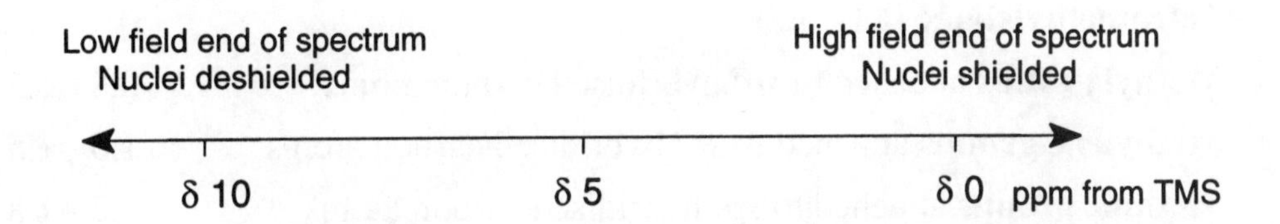

Electron withdrawing substituents (-OH, -OCOR, -OR, $-NO_2$, halogen) attached to an aliphatic carbon chain cause a **downfield shift** of 2-4 ppm when present at C_α and have less than half of this effect when present at C_β.

When *sp*^{2} hybridised carbon atoms (carbonyl groups, olefinic fragments, aromatic rings) are present in an aliphatic carbon chain they cause a downfield shift of 1-2 ppm when present at C_α. They have less than half of this effect when present at C_β. Tables 5.2 and 5.3 give characteristic shifts for ^1H nuclei in some representative organic compounds.

Table 5.2 Typical ^1H Chemical Shift Values in Selected Organic Compounds

Compound	δ ^{1}H (ppm from TMS)
CH_4	0.23
CH_3Cl	3.05
CH_2Cl_2	5.33
$CHCl_3$	7.24
CH_3CH_3	0.86
$CH_2=CH_2$	5.25
benzene	7.26
CH_3CHO	2.20 (CH_3), 9.80 (-CHO)
$CH_3CH_2CH_2Cl$	1.06 (CH_3), 1.81(-CH_2-), 3.47(-CH_2-Cl)

Table 5.3 Typical ^1H Chemical Shift Ranges in Organic Compounds

Group	δ ^{1}H (ppm from TMS)
Tetramethylsilane $(CH_3)_4Si$	0
Methyl groups attached to sp^3 hybridised carbon atoms	0.8 - 1.2
Methylene groups attached to sp^3 hybridised carbon atoms	1.0 - 1.5
Methine groups attached to sp^3 hybridised carbon atoms	1.2 - 1.8
Acetylenic protons	2 – 3.5
Olefinic protons	5 - 8
Aromatic and heterocyclic protons	6 - 9
Aldehydic protons	9 - 10

Note that **labile** protons (-OH, -NH_2, -SH *etc.*) have no characteristic chemical shift ranges. However, such resonances can always be positively identified by *in situ* exchange with D_2O, which causes them to disappear from the spectrum.

Table 5.4 gives characteristic chemical shifts for protons in common alkyl derivatives. Table 5.5 gives characteristic chemical shifts for the olefinic protons in common substituted alkenes. To a first approximation, the shifts induced by substituents attached an alkene are additive. So, for example, an olefinic proton which is *trans* to a –CN group and has a geminal alkyl group will have a chemical shift of approximately 6.25 ppm [5.25 + 0.55(*trans*–CN) + 0.45(*gem*-alkyl)].

Table 5.4 ^{1}H Chemical Shifts (δ) for Protons in Common Alkyl Derivatives

	CH_3—X	CH_3CH_2—X		$(CH_3)_2CH$—X	
X	—CH_3	—CH_3	—CH_2—	—CH_3	>CH—
—H	0.23	0.86	0.86	0.91	1.33
—CH=CH_2	1.71	1.00	2.00	1.00	1.73
—Ph	2.35	1.21	2.63	1.25	2.89
—Cl	3.06	1.33	3.47	1.55	4.14
—Br	2.69	1.66	3.37	1.73	4.21
—I	2.16	1.88	3.16	1.89	4.24
—OH	3.39	1.18	3.59	1.16	3.94
—OCH_3	3.24	1.15	3.37	1.08	3.55
—OCO—CH_3	3.67	1.21	4.05	1.22	4.94
—CO—CH_3	2.09	1.05	2.47	1.08	2.54
—CO—Ph	2.55	1.18	2.92	1.22	3.58
—CO—OCH_3	2.01	1.12	2.28	1.15	2.48
—NH_2	2.47	1.10	2.74	1.03	3.07
—NH—$COCH_3$	2.71	1.12	3.21	1.13	4.01
—C≡N	1.98	1.31	2.35	1.35	2.67
—NO_2	4.29	1.58	4.37	1.53	4.44

Table 5.5 Approximate [1]H Chemical Shifts (δ) for Olefinic Protons C=C-H

$$\delta_{C=C\text{-}H} = 5.25 + \sigma_{gem} + \sigma_{cis} + \sigma_{trans}$$

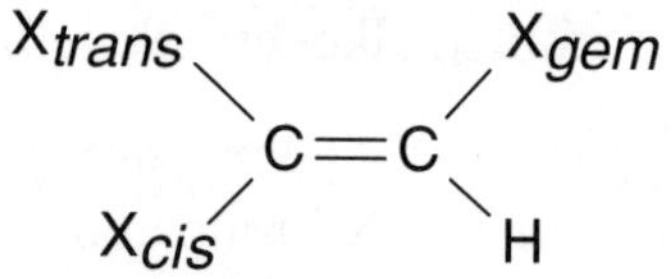

X	σ_{gem}	σ_{cis}	σ_{trans}
—H	0.0	0.0	0.0
—alkyl	0.45	-0.22	-0.28
—aryl	1.38	0.36	-0.07
—CH=CH$_2$	1.00	-0.09	-0.23
—CO−R	1.10	1.12	0.87
—CO−OH	0.80	0.98	0.32
—CO−OR	0.78	1.01	0.46
—C≡N	0.27	0.75	0.55
—Cl	1.08	0.18	0.13
—Br	1.07	0.45	0.55
—OR	1.22	-1.07	-1.21
—NR$_2$	0.80	-1.26	-1.21

Table 5.6 gives characteristic ^{1}H chemical shifts for the aromatic protons in benzene derivatives. To a first approximation, the shifts induced by substituents are additive. So, for example, an aromatic proton which has a $-NO_2$ group in the *para* position and a –Br group in the *ortho* position will appear at approximately 7.82 ppm [(7.26 + 0.38(p-NO_2) + 0.18(o-Br)].

Tables 5.7 gives characteristic chemical shifts for ^{1}H nuclei in some polynuclear aromatic compounds and heteroaromatic compounds.

Table 5.6 **^{1}H Chemical Shifts (δ) for Aromatic Protons in Benzene Derivatives Ph-X in ppm Relative to Benzene at δ 7.26 ppm (positive sign denotes a downfield shift)**

X	*ortho*	*meta*	*para*
—H	0.0	0.0	0.0
—CH_3	-0.20	-0.12	-0.22
—CH=CH_2	0.06	-0.03	-0.10
—CO–OR	0.71	0.11	0.21
—CO–R	0.62	0.14	0.21
—OCO–R	-0.25	0.03	-0.13
—OCH_3	-0.48	-0.09	-0.44
—OH	-0.56	-0.12	-0.45
—Cl	0.03	-0.02	-0.09
—Br	0.18	-0.08	-0.04
—C≡N	0.36	0.18	0.28
—NO_2	0.95	0.26	0.38
—NR_2	-0.66	-0.18	-0.67
—NH_2	-0.75	-0.25	-0.65

Table 5.7 **^{1}H Chemical Shifts (δ) in some Polynuclear Aromatic Compounds and Heteroaromatic Compounds**

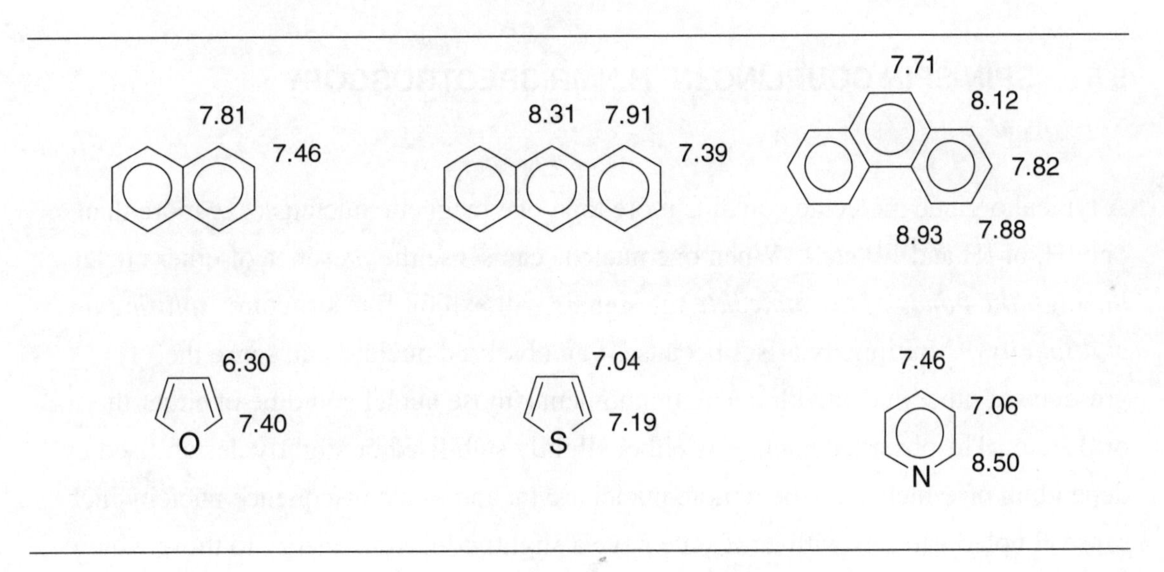

Solvents for NMR Spectroscopy. NMR spectra are almost invariably obtained in solution. The solvents of choice:

(a) should have adequate dissolving power.

(b) should not associate strongly with solute molecules as this is likely to produce appreciable effects on chemical shifts. This requirement must sometimes be sacrificed to achieve adequate solubility.

(c) should be essentially free of interfering signals. Thus for ^{1}H NMR, the best solvents are proton-free.

(d) should preferably contain deuterium, ^{2}H, which provides a convenient "locking" signal for stabilisation of the magnetic field.

The most commonly used solvent is **deuterochloroform**, $CDCl_3$, which is an excellent organic solvent and is only weakly associated with most organic substrates. $CDCl_3$ contains no protons and has a deuterium atom.

Almost all deuterated solvents are not 100% deuterated and they contain a residual protonated impurity. With the sensitivity of modern NMR instruments, the signal from residual protons in the deuterated solvent is usually visible in the ^{1}H NMR spectrum. For many spectra, the signal from residual protons can be used as a reference signal (instead of adding TMS) since the chemical shifts of most common solvents are known accurately. Solvents that are miscible with water (and are difficult to "dry" completely) *e.g.* CD_3COCD_3, CD_3SOCD_3, D_2O, also commonly contain a small amount of residual water. The residual water typically appears as a broad resonance in the region 3 – 5 ppm in the ^{1}H NMR spectrum.

5.5 SPIN-SPIN COUPLING IN ^{1}H NMR SPECTROSCOPY

A typical organic molecule contains more than one magnetic nucleus (*e.g.* more than one ^{1}H, or ^{1}H and ^{31}P *etc.*). When one nucleus can sense the presence of other nuclei ***through the bonds of the molecule*** the signals will exhibit fine structure (*splitting or multiplicity*). Multiplicity arises because if an observed nucleus can sense the presence of other nuclei with magnetic moments, those nuclei could be in either the α or β state. The observed nucleus is either slightly stabilised or slightly destabilised by depending on which state the remote nuclei are in, and as a consequence nuclei which sense coupled partners with an α state have a slightly different energy to those which sense coupled partners with a β state.

The additional fine structure caused by spin-spin coupling is not only the principal cause of difficulty in interpreting ^{1}H NMR spectra, but also provides valuable structural information when correctly interpreted. The **coupling constant** (related to the size of the splittings in the multiplet) is given the symbol J and is measured in Hz. By convention, a superscript before the symbol 'J' represents the number of intervening bonds between the coupled nuclei. Labels identifying the coupled nuclei are usually indicated as subscripts after the symbol 'J' *e.g.* $^{2}J_{ab} = 2.7$ Hz would indicate a coupling of 2.7 Hz between nuclei a and b which are separated by two intervening bonds.

Because J depends only on the number, type and spatial arrangement of the bonds separating the two nuclei, it is a property of the molecule and is **independent of the applied magnetic field**. The magnitude of J, or even the mere presence of detectable interaction, constitutes valuable structural information.

Two important observations that relate to ^{1}H - ^{1}H spin-spin coupling:

(a) No **inter-molecular** spin-spin coupling is observed. Spin-spin coupling is transmitted through the bonds of a molecule and doesn't occur between nuclei in different molecules.

(b) The effect of coupling falls off as the number of bonds between the coupled nuclei increases. ^{1}H - ^{1}H coupling is generally unobservable across more than 3 intervening bonds. Unexpectedly large couplings across many bonds may occur if there is a particularly favourable bonding pathway *e.g.* extended π-conjugation or a particularly favourable rigid σ-bonding skeleton (Table 5.8).

Table 5.8 Typical ^{1}H – ^{1}H Coupling Constants

Group	J (Hz)
$CH_3CH_2CH_2CH_3$	$^{2}J_{HH} \approx -16$
$CH_3CH_2CH_2CH_3$	$^{3}J_{HH} = 7.2$
$CH_3CH_2CH_2CH_3$	$^{4}J_{HH} = 0.3$
$H_2C{=}C{=}C{=}CH_2$	$^{5}J_{HH} = 7$
$H_2C{=}CH{-}CH{=}CH_2$	$^{5}J_{HH} = 1.3$
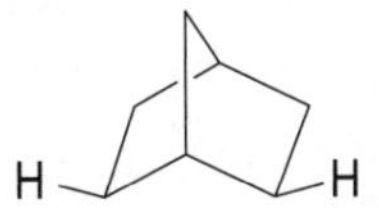 H H	$^{4}J_{HH} = 1.5$

Signal Multiplicity - the n+1 rule. Spin-spin coupling gives rise to multiplet splittings in ^{1}H NMR spectra. The NMR signal of a nucleus coupled to ***n*** equivalent hydrogens will be split into a multiplet with (***n***+1) lines. For simple multiplets, the spacing between the lines (in Hz) is the coupling constant. The relative intensity of the lines in multiplet will be given by the binomial coefficients of order '***n***' (Table 5.9).

Table 5.9 Relative Line Intensities for Simple Multiplets

n	multiplicity *n+1*	relative line intensities	multiplet name
0	1	1	singlet
1	2	1 : 1	doublet
2	3	1 : 2 : 1	triplet
3	4	1 : 3 : 3 : 1	quartet
4	5	1 : 4 : 6 : 4 : 1	quintet
5	6	1 : 5 : 10 : 10 : 5 : 1	sextet
6	7	1 : 6 : 15 : 20 : 15 : 6 : 1	septet
7	8	1 : 7 : 21 : 35 : 35 : 21 : 7 : 1	octet
8	9	1 : 8 : 28 : 56 : 70 : 56 : 28 : 8 : 1	nonet

These simple multiplet patterns give rise to characteristic "fingerprints" for common fragments of organic structures.

A methyl group, $-CH_3$, (isolated from coupling to other protons in the molecule) will always occur as a singlet. A CH_3-CH_2- group, (isolated from coupling to other protons in the molecule) will appear as a quartet ($-CH_2-$) and a triplet (CH_3-). Table 5.10 shows the schematic appearance of the NMR spectra of various common molecular fragments encountered in organic molecules.

Table 5.10 Characteristic Multiplet Patterns for Common Organic Fragments

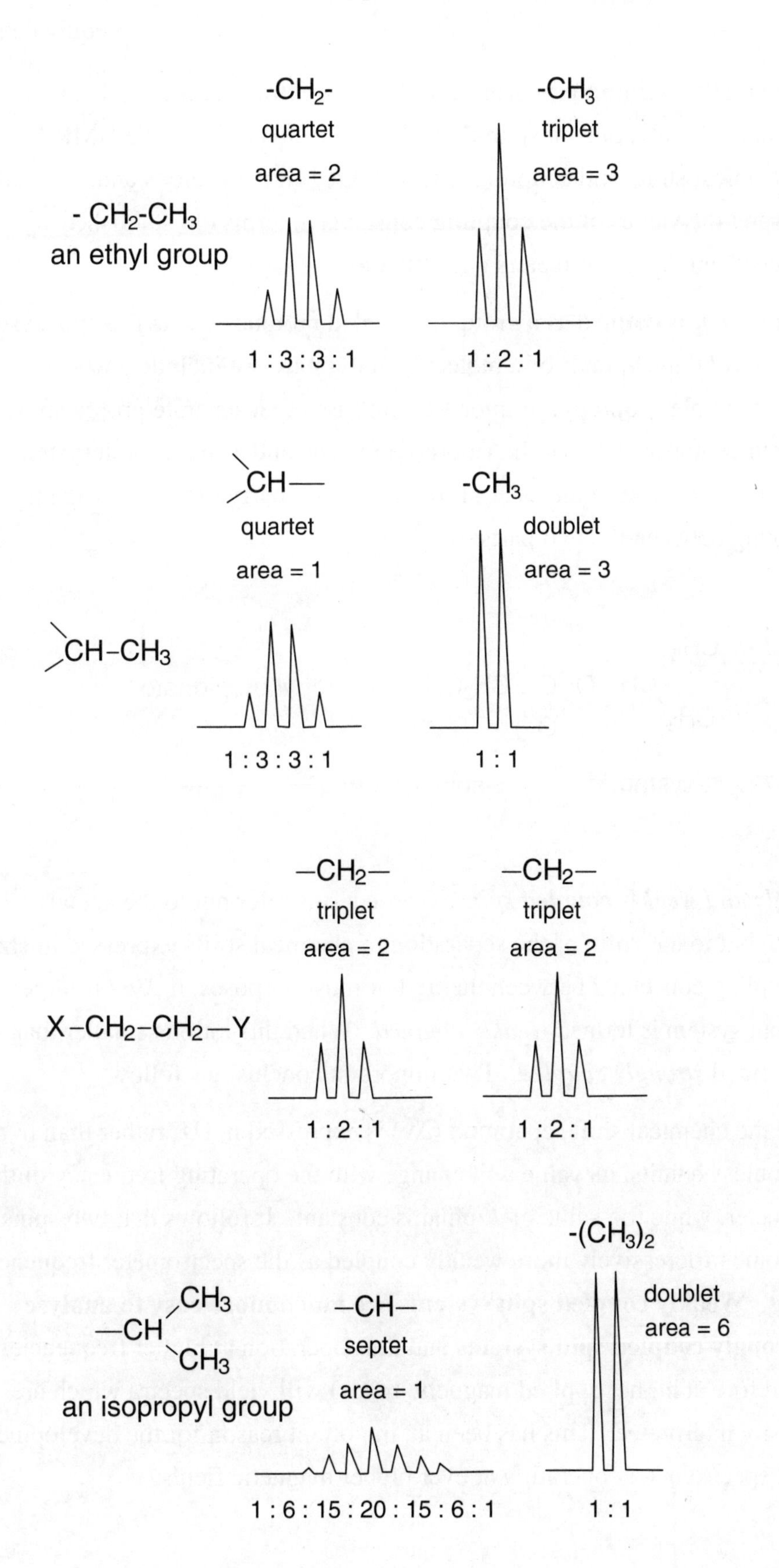

5.6 ANALYSIS OF ^{1}H NMR SPECTRA

To obtain structurally useful information from NMR spectra, one must solve two separate problems. Firstly, one must **analyse** the spectrum to obtain the NMR parameters (chemical shifts and coupling constants) for all the protons and, secondly, one must interpret the values of the coupling constants in terms of established relationships between these parameters and structure.

(1) ***A spin system*** is defined as a group of coupled protons. Clearly, a spin system cannot extend beyond the bounds of a molecule, but it may not include a whole molecule. For example, isopropyl propionate comprises **two** separate proton spin systems, a seven-proton system for the isopropyl residue and a five-proton system for the propionate residue because the ester group provides a barrier (5 bonds) against effective coupling between the two parts.

$(CH_3)_2CH$–O–C(=O)–CH_2CH_3 Isopropyl propionate

7-spin system 5-spin system

(2) ***Strongly and weakly coupled spins.*** These terms refer not to the actual magnitude of J, but to the **ratio** of the separation of chemical shifts expressed in Hz ($\Delta\nu$) to the coupling constant J between them. For most purposes, if $\Delta\nu/J$ is larger than ~3, the spin system is termed *weakly coupled.* When this ratio is smaller than ~3, the spins are termed *strongly coupled.* Two important conclusions follow:

(a) Because the chemical shift separation ($\Delta\nu$) is expressed in Hz, rather than in the dimensionless δ units, its value will change with the operating frequency of the spectrometer, while the value of J remains constant. It follows that two spins will become progressively more weakly coupled as the spectrometer frequency increases. **Weakly coupled spin systems are much more easy to analyse than strongly coupled spin systems** and thus operation at higher frequencies (and therefore at higher applied magnetic fields) will yield spectra which are more easily interpreted. This has been an important reason for the development of NMR spectrometers operating at ever higher magnetic fields.

(b) Within a spin system, some pairs of nuclei or groups of nuclei may be strongly coupled and others weakly coupled. Thus a spin-system may be *partially strongly coupled.*

(3) Magnetic equivalence. A group of protons is *magnetically equivalent* when they not only have the same chemical shift (chemical equivalence) but also have identical spin-spin coupling to each individual nucleus **outside** the group.

(4) Conventions used in naming spin systems. Consecutive letters of the alphabet (*e.g.* A, B, C D, ...) are used to describe groups of protons which are strongly coupled. Subscripts are used to give the number of protons that are magnetically equivalent. Primes are used to denote protons that are chemically equivalent but not magnetically equivalent. A break in the alphabet indicates weakly coupled groups. For example:

ABC denotes a strongly coupled 3-spin system

AMX denotes a weakly coupled 3-spin system

ABX denotes a *partially* strongly coupled 3-spin system

A_3BMXY denotes a spin system in which the three magnetically equivalent A nuclei are strongly coupled to the B nucleus, but weakly coupled to the M, X and Y nuclei. The nucleus X is strongly coupled to the nucleus Y but weakly coupled to all the other nuclei. The nucleus M is weakly coupled to all the other 6 nuclei.

AA'XX' is a 4-spin system described by two chemical shift parameters (for the nuclei A and X) but where $J_{AX} \neq J_{AX'}$. A and A' (as well as X and X') are pairs of nuclei which are chemically equivalent but magnetically non-equivalent.

The process of deriving the NMR parameters (δ and J) from a set of multiplets in a spin system is known as *the analysis of the NMR spectrum.* In principle, **any** spectrum arising from a spin system, however complicated, can be analysed but some will require calculations performed by a computer.

Fortunately, in a very large number of cases multiplets can be correctly analysed by inspection and direct measurements. Spectra of that type are known as *first order spectra* and **they arise from weakly coupled spin systems.** At high applied magnetic fields, a large proportion of ^{1}H NMR spectra are nearly pure first-order and there is a tendency for simple molecules, *e.g.* those exemplified in the problems in this text, to exhibit first-order spectra even at moderate fields.

5.7 RULES FOR SPECTRAL ANALYSIS

Rule 1 A group of $\boldsymbol{n}$ magnetically equivalent protons will split a resonance of an interacting group of protons into $\boldsymbol{n}$+1 lines. For example, the resonance due to the A protons in an A_nX_m system will be split into $\boldsymbol{m}$+1 lines, while the resonance due to the X protons will be split into $\boldsymbol{n}$+1 lines. More generally, splitting by n nuclei of spin quantum number I, results in 2$\boldsymbol{n}$**I**+1 lines. This simply reduces to $\boldsymbol{n}$+1 for protons where **I** = ½.

Rule 2 The spacing (measured in Hz) of the lines in the multiplet will be equal to the coupling constant. In the above example all spacings in both parts of the spectrum will be equal to J_{AX}.

Rule 3 The true chemical shift of each group of interacting protons lies in the centre of the (always symmetrical) multiplet.

Rule 4 The relative intensities of the lines within each multiplet will be in the ratio of the binomial coefficient (Table 5.9). In the case of higher multiplets the outside components of multiplets are relatively weak and may be lost in the instrumental noise, *e.g.* a septet may appear as a quintet if the outer lines are not visible. The intensity relationship is the first to be significantly distorted in non-ideal cases, but this does not lead to serious errors in spectral analysis.

Rule 5 When a group of magnetically equivalent protons interacts with more than one group of protons, its resonance will take the form of a *multiplet of multiplets*. For example, the resonance due to the A protons in a system $A_nM_pX_m$ will have the multiplicity of $(p+1)(m+1)$. The multiplet patterns are chained *e.g.* a proton coupled to 2 different protons will be split to a doublet by coupling to the first proton then each of the component of the doublet will be split further by coupling to the second proton resulting in a symmetrical multiplet with 4 lines (a doublet of doublets).

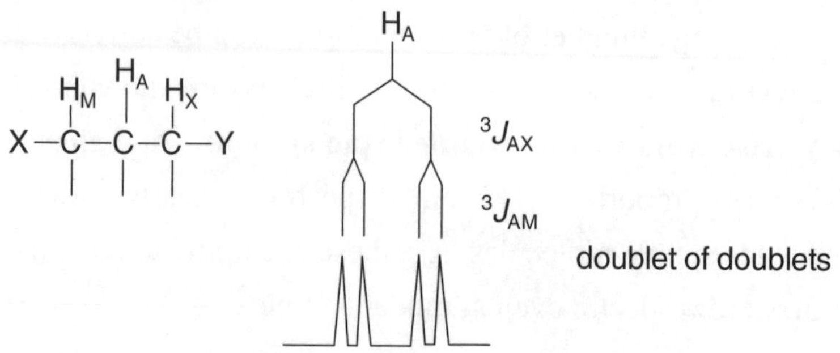

H_M H_A H_X

X—C—C—C—Y

H_A

$^3J_{AX}$

$^3J_{AM}$

doublet of doublets

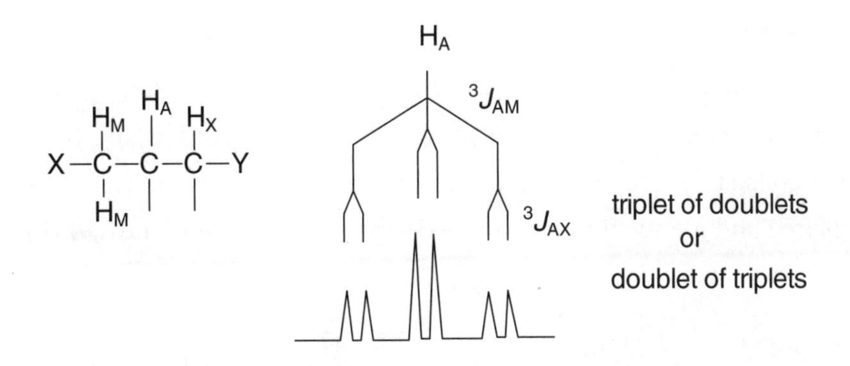

The appropriate coupling constants will control splitting and relative intensities will obey rule 4.

Rule 6 Protons that are magnetically equivalent do not split each other. Any system A_n will give rise to a singlet.

Rule 7 Spin systems that contain groups of chemically equivalent protons that are not magnetically equivalent cannot be analysed by first-order methods.

Rule 8 Not only strongly coupled, but also *partially* strongly coupled spectra **cannot** be analysed by first-order methods.

(1) Splitting Diagrams

The knowledge of the rules listed above, permits the development of a simple procedure for the analysis of any spectrum which is suspected of being first order. The first step consists of drawing a *splitting diagram*, from which the line spacings can be measured and identical (hence related) splittings can be identified (Figure 5.4).

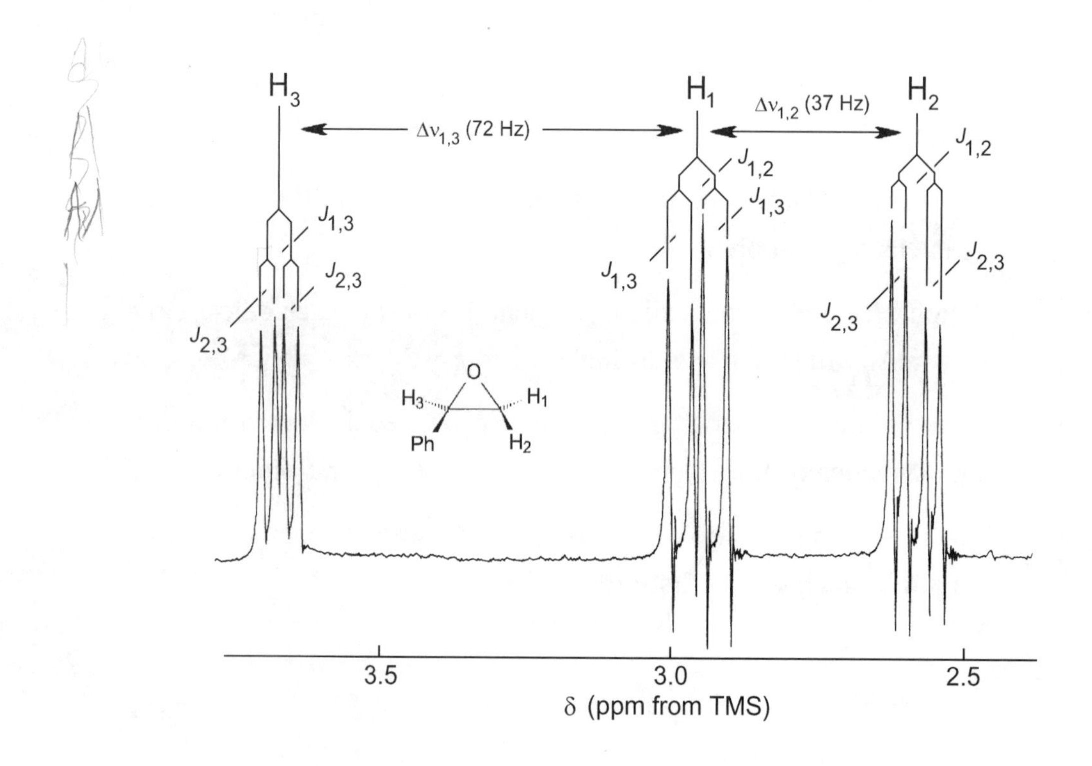

Figure 5.4 A Portion of the ^{1}H NMR Spectrum of Styrene Epoxide (100 MHz as a 5% solution in CCl_4)

The section of the spectrum of styrene oxide (Figure 5.4) clearly contains the signals from 3 separate protons (identified as H_1, H_2 and H_3) with H_1 at δ 2.95, H_2 at δ 2.58 and H_3 at δ 3.67 ppm. Each signal appears as a doublet of doublets and the chemical shift of each proton is simply obtained by locating the centre of the multiplet. The pair of nuclei giving rise to each splitting is clearly indicated by the splitting diagram above each multiplet with $^2J_{H1\text{-}H2} = 5.9$ Hz, $^3J_{H1\text{-}H3} = 4.0$ Hz and $^3J_{H2\text{-}H3} = 2.5$ Hz.

The validity of a first order analysis can be verified by calculating the ratio $\Delta\nu/J$ for each pair of nuclei and establishing that it is greater than 3.

From Figure 5.4

$$\frac{\Delta\nu_{12}}{J_{12}} = \frac{37}{5.9} = 6.3 \qquad \frac{\Delta\nu_{13}}{J_{13}} = \frac{72}{4.0} = 18.0 \qquad \frac{\Delta\nu_{23}}{J_{23}} = \frac{109}{2.5} = 43.6$$

Each ratio is greater than 3 so a first order analysis is justified and the 100 MHz spectrum of the aliphatic protons of styrene oxide is indeed a first order spectrum and could be labelled as an AMX spin system.

The 60 MHz ^{1}H spectrum of a 4 spin AMX_2 system is given in Figure 5.5. This system contains 3 separate proton signals (in the intensity ratios 1:1:2, identified as H_A, H_M and H_X). The multiplicity of H_A is a triplet of doublets, the multiplicity of H_M is a triplet of doublets and the multiplicity of H_X is a doublet of doublets. Again, the nuclei giving rise to each splitting are clearly indicated by the splitting diagram above each multiplet and the chemical shifts of each multiplet are simply obtained by measuring the centres of each multiplet.

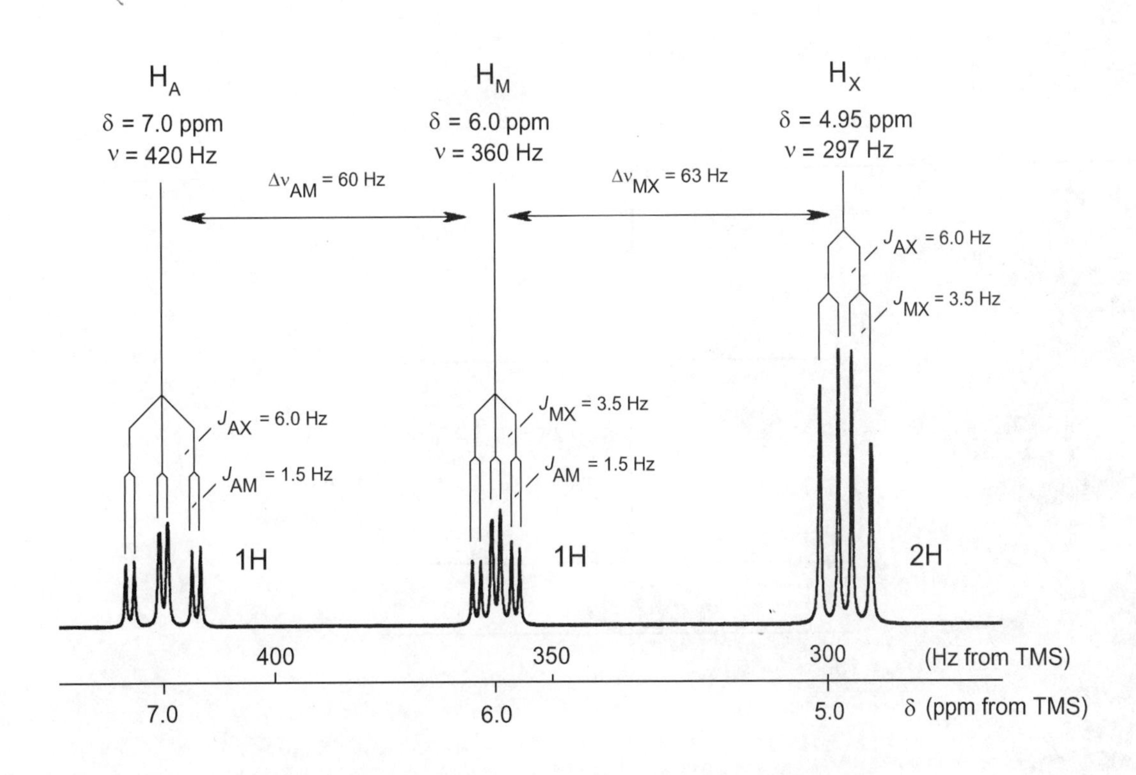

Figure 5.5 The 60 MHz ^{1}H NMR Spectrum of a 4-Spin AMX_2 Spin System

A spin system comprising just two protons (*i.e.* an AX or an AB system) is always exceptionally easy to analyse because, independent of the value of the ratio of $\Delta\nu/J$, the spectrum always consists of just four lines with each pair of lines separated by the coupling constant J. The only distortion from the first-order pattern consists of the gradual reduction of intensities of the outer lines in favour of the inner lines, a characteristic "sloping" or "tenting" towards the coupling partner. A series of simulated spectra of two-spin systems are shown in Figure 5.6.

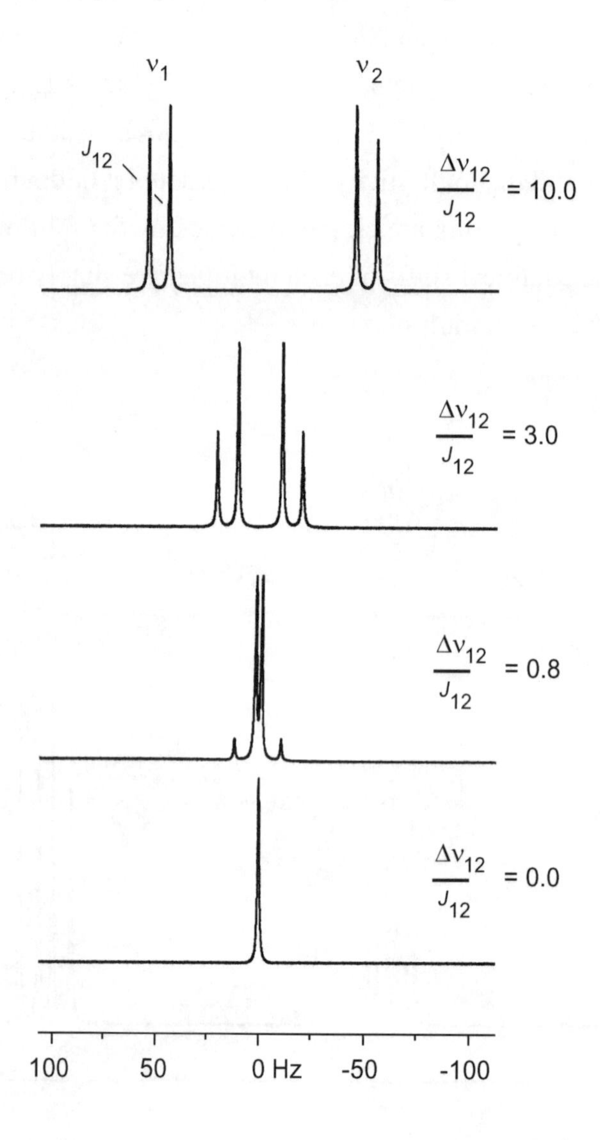

Figure 5.6 Simulated ^{1}H NMR Spectra of a 2-Spin System as the Ratio $\Delta\nu/J$, is Varied from 10.0 to 0.0

(2) Spin Decoupling

In the signal of a proton that is a multiplet due to spin-spin coupling, it is possible to remove the splitting effects by irradiating the sample with an additional Rf source at the exact resonance frequency of the proton giving rise to the splitting. The additional radiofrequency causes rapid flipping of the irradiated nuclei and as a consequence nuclei coupled to them cannot sense them as being in either an α or β state for long enough to cause splitting. The irradiated nuclei are said to be **decoupled** from other nuclei in the spin system. Decoupling simplifies the appearance of complex multiplets by removing some of the splittings. In addition, decoupling is a powerful tool for assigning spectra because the skilled spectroscopist can use a series of decoupling experiments to sequentially identify which nuclei are coupled.

In a 4-spin AM_2X spin system, the signal for proton H_A would appear as a doublet of triplets (with the triplet splitting due to coupling to the 2 M protons and the doublet splitting due to coupling to the X proton). Irradiation at the frequency of H_X reduces the multiplicity of the A signal to a triplet (with the remaining splitting due to J_{AM}) and irradiation at the frequency of H_M reduces the multiplicity of the A signal to a doublet (with the remaining splitting due to J_{AX}) (Figure 5.7).

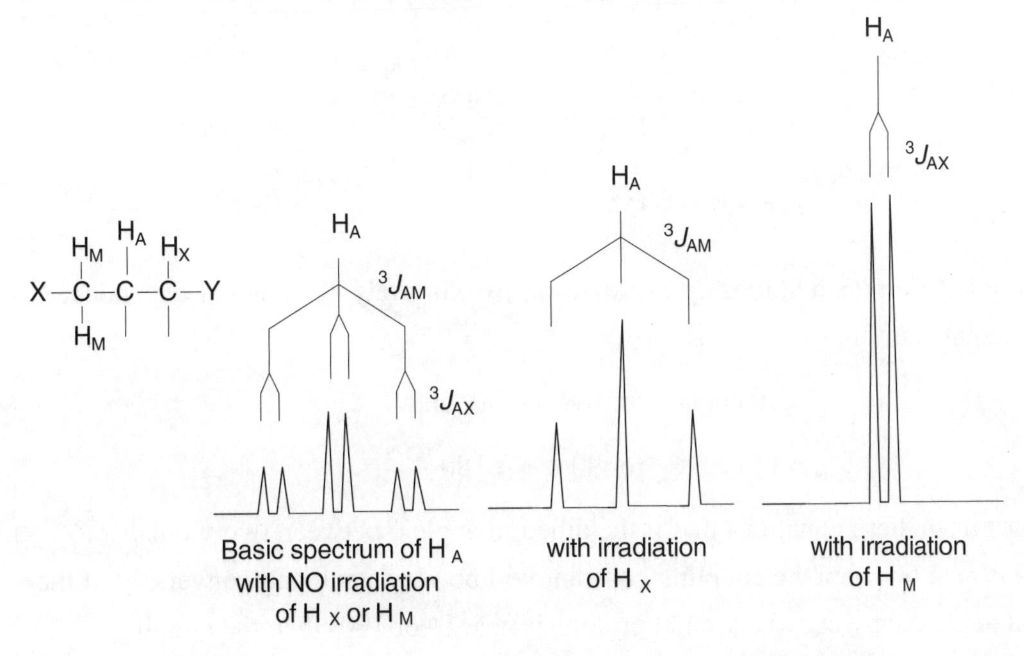

Figure 5.7 Selective Decoupling in a Simple 4-Spin System

(3) Correlation of $^1H - {}^1H$ Coupling Constants with Structure

Interproton spin-spin coupling constants are of obvious value in obtaining structural data about a molecule, in particular information about the connectivity of structural elements and the relative disposition of various protons.

Non-aromatic Spin Systems.

In saturated systems, the magnitude of the *geminal* coupling constant $^2J_{H\text{-}C\text{-}H}$ (two protons attached to the same carbon atom) is typically between 10 and 16 Hz but values between 0 and 22 Hz have been recorded in some unusual structures.

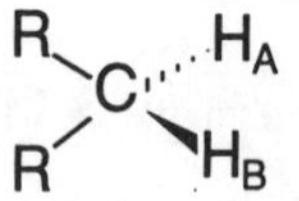

$^2J_{AB}$ = 10 -16 Hz.

The *vicinal* coupling (protons on adjacent carbon atoms) $^3J_{H\text{-}C\text{-}C\text{-}H}$ can have values 0 - 16 Hz depending mainly on the dihedral angle ϕ.

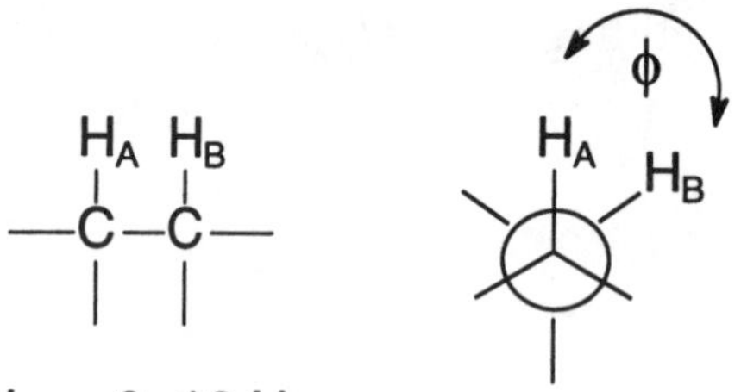

$^3J_{AB}$ = 0 -16 Hz.

The so-called Karplus relationship expresses **approximately**, the angular dependence of the vicinal coupling constant as:

$$^3J_{H\text{-}C\text{-}C\text{-}H} = 10\cos^2\phi \quad \text{for } 0 < \phi < 90^\circ \text{ and}$$

$$^3J_{H\text{-}C\text{-}C\text{-}H} = 15\cos^2\phi \quad \text{for } 90 < \phi < 180^\circ$$

It follows from these equations that if the dihedral angle ϕ between two vicinal protons is near 90° then the coupling constant will be very small and conversely, if the dihedral angle ϕ between two vicinal protons is near 0° or 180° then the coupling constant will be relatively large. The Karplus relationship is of great value in determining the stereochemistry of organic molecules but must be treated with caution because vicinal coupling constants also depend markedly on the nature of substituents. In systems that assume an average conformation, such as a flexible hydrocarbon chain, $^3J_{H\text{-}H}$ generally lies between 6 and 8 Hz.

The coupling constants in unsaturated (olefinic) systems depend on the nature of the substituents attached to the C=C but for the vast majority of substituents, the ranges for ${}^3J_{H\text{-}C=C\text{-}H(cis)}$ and ${}^3J_{H\text{-}C=C\text{-}H(trans)}$ do not overlap. This means that the stereochemistry of the double bond can be determined by measuring the coupling constant between vinylic protons. Where the C=C bond is in a ring, the ${}^3J_{H\text{-}C=C\text{-}H}$ coupling reflects the ring size.

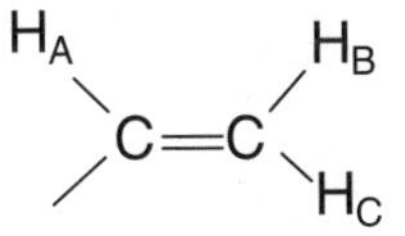

${}^3J_{AB(cis)}$ = 6 - 11 Hz

${}^3J_{AC(trans)}$ = 12 - 19 Hz

${}^2J_{BC(gem)}$ = 0 - 3 Hz

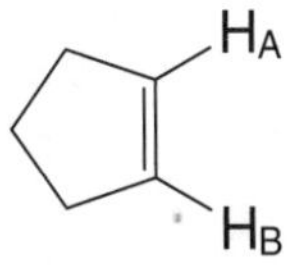

${}^3J_{AB(cis)}$ = 5 - 7 Hz

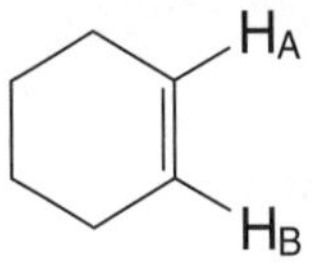

${}^3J_{AB(cis)}$ = 9 - 11 Hz

The magnitude of the long-range allylic coupling, (${}^4J_{AB}$) is controlled by the dihedral angle between the C-H_A bond and the plane of the double bond in a relationship reminiscent of the Karplus relation.

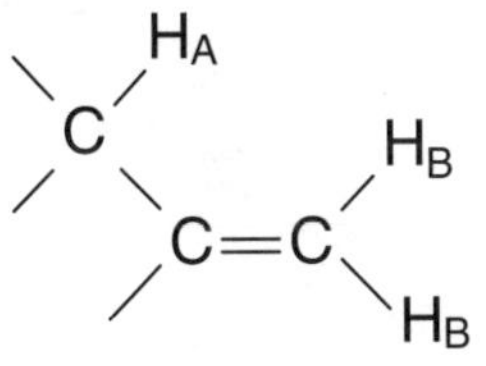

${}^4J_{AB}$ = 0 - 3 Hz

Aromatic Spin Systems

In aromatic systems, the coupling constant between protons attached to an aromatic ring is characteristic of the relative position of the coupled protons *i.e.* whether they are *ortho*, *meta* or *para*.

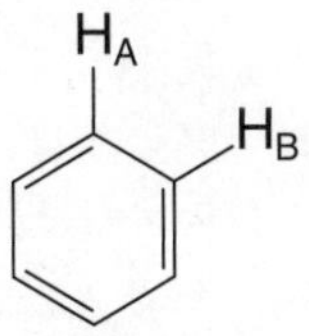

${}^3J_{AB(ortho)}$ = 6 - 10 Hz ${}^4J_{AB(meta)}$ = 1 - 3 Hz ${}^5J_{AB(para)}$ = 0 - 1.5 Hz

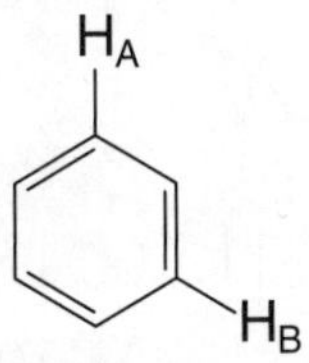

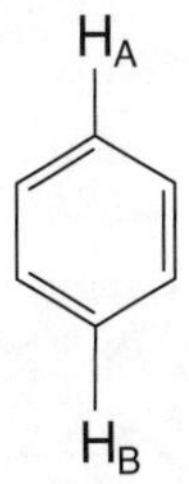

Similarly in condensed polynuclear aromatic compounds and heterocyclic compounds, the magnitude of the coupling constants between protons in the aromatic rings reflects the relative position of the coupled protons.

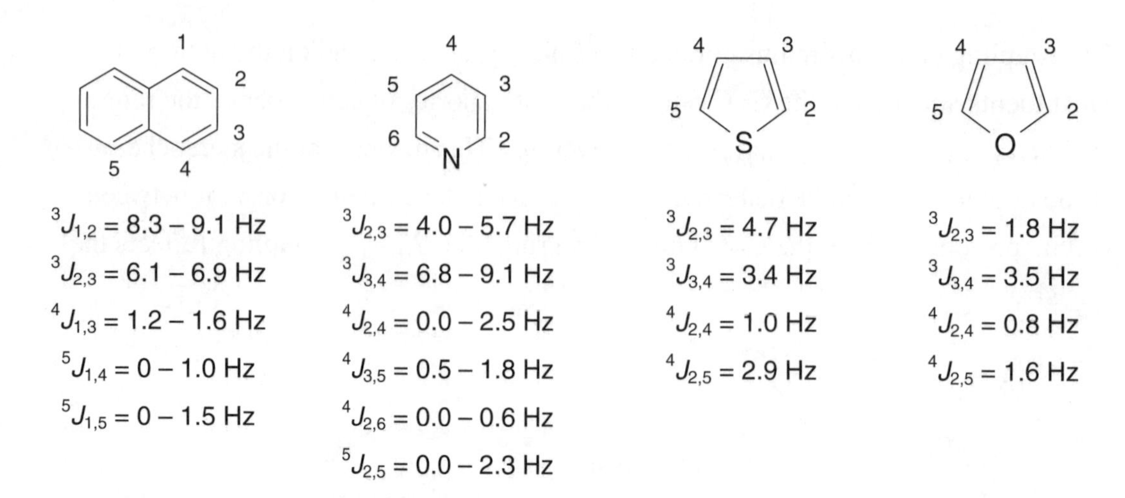

The splitting patterns of the protons in the aromatic region of the ^{1}H spectrum are frequently used to establish the substitution pattern of an aromatic ring. For example, a trisubstituted aromatic ring has 3 remaining protons. There are 3 possible arrangements for the 3 protons - they can have relative positions 1,2,3-; 1,2,4-; or 1,3,5- and each has a **characteristic splitting pattern**.

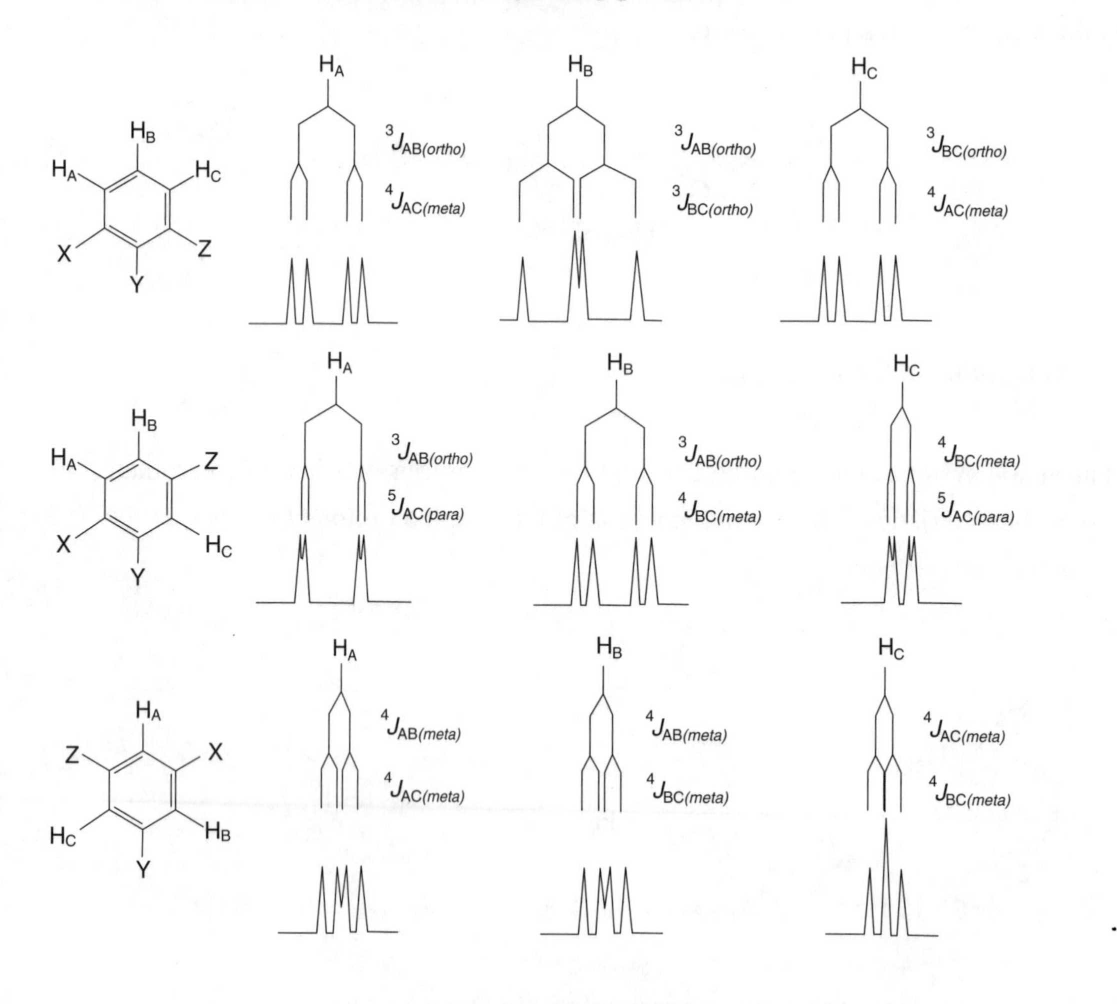

***para*-Disubstituted benzenes**

para-Disubstituted benzenes have characteristically "simple" and symmetrical ^{1}H NMR spectra in the aromatic region. Superficially, the spectra of p-disubstituted benzenes always appear as two strong doublets with the line positions symmetrically disposed about a central frequency. The spectra are in fact far more complex (many lines make up the pattern for the NMR spectrum when it is analysed in detail) but the symmetry of the pattern of lines makes 1,4-disubstituted benzenes very easy to recognise from their ^{1}H NMR spectra. The ^{1}H NMR spectrum of *p*-nitrophenylacetylene is given in Figure 5.8. The expanded section shows the 4 strong prominent signals in the aromatic region, characteristic of 1,4-substitution on a benzene ring.

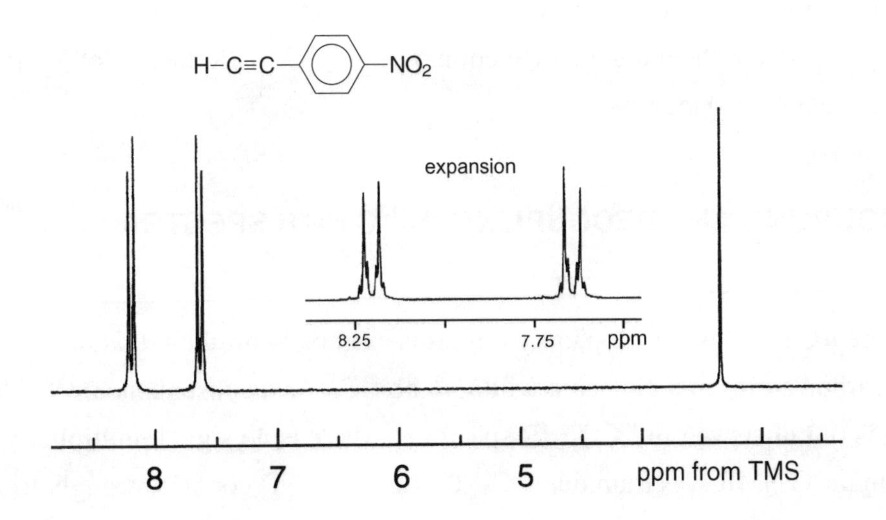

Figure 5.8 ^{1}H NMR Spectrum of *p*-Nitrophenylacetylene (200 MHz as a 10% solution in $CDCl_3$)

6

^{13}C NMR SPECTROSCOPY

The most abundant isotope of carbon (^{12}C) cannot be observed by NMR. ^{13}C is a rare nucleus (1.1% natural abundance) and its low concentration coupled with the fact that ^{13}C has a relatively low resonance frequency, leads to its relative insensitivity as an NMR-active nucleus (about 1/6000 as sensitive as ^{1}H). However, with the increasing availability of routine pulsed FT NMR spectrometers, it is now common to acquire many spectra and add them together (Section 5.3), so ^{13}C NMR spectra of good quality can be obtained readily.

6.1 COUPLING AND DECOUPLING IN ^{13}C NMR SPECTRA

Because the ^{13}C nucleus is isotopically rare, it is extremely unlikely that any two adjacent carbon atoms in a molecule will *both* be ^{13}C. As a consequence, **^{13}C-^{13}C coupling is not observed** in ^{13}C NMR spectra *i.e.* there is no signal multiplicity or splitting in a ^{13}C NMR spectrum due ^{13}C-^{13}C coupling. ^{13}C couples strongly to any protons that may be attached ($^{1}J_{CH}$ is typically about 125 Hz for saturated carbon atoms in organic molecules). It is the usual practice to irradiate the ^{1}H nuclei during ^{13}C acquisition so that all ^{1}H are fully decoupled from the ^{13}C nuclei (usually termed **broad band decoupling** or **noise decoupling**). **^{13}C NMR spectra usually appear as a series of singlets** (when ^{1}H is fully decoupled) and *each distinct ^{13}C environment in the molecule gives rise to a separate signal.*

If ^{1}H is **not decoupled** from the ^{13}C nuclei during acquisition, the signals in the ^{13}C spectrum appear as multiplets where the major splittings are due to the $^{1}J_{C\text{-}H}$ couplings (about 125 Hz for sp^3 hybridised carbon atoms, about 160 Hz for sp^2 hybridised carbon atoms, about 250 Hz for sp hybridised carbon atoms). CH_3- signals appear as quartets, -CH_2- signals appear as triplets, -CH- groups appear as doublets and quaternary C (no attached H) appear as singlets. The **multiplicity information**, taken together with chemical shift data, is useful in identifying and assigning the ^{13}C resonances.

In ^{13}C spectra acquired without proton decoupling, there is usually much more *"long range"* coupling information visible in the fine structure of each multiplet. The fine structure arises from coupling between the carbon and protons that are not directly bonded to it (*e.g.* from $^{2}J_{C\text{-}C\text{-}H}$, $^{3}J_{C\text{-}C\text{-}C\text{-}H}$). The magnitude of long range C-H coupling is typically < 10 Hz and this is much less than $^{1}J_{C\text{-}H}$. Sometimes a more detailed analysis of the long-range C-H couplings can be used to provide additional information about the structure of the molecule.

In most ^{13}C spectra, ^{13}C nuclei which have directly attached protons receive a significant (but not easily predictable) signal enhancement when the protons are decoupled and as a consequence, peak intensity does not, in general, reflect the number of ^{13}C nuclei giving rise to the signal. It is not usually possible to integrate ^{13}C spectra directly unless specific precautions have been taken.

SFORD : Off- resonance decoupling. Another method for obtaining ^{13}C NMR spectra (still retaining the multiplicity information) involves the application of a strong decoupling signal at a single frequency *just outside* the range of proton resonances. This has the effect of incompletely or partially decoupling protons from the ^{13}C nuclei. The technique is usually referred to as *off resonance decoupling* or SFORD (**S**ingle **F**requency **O**ff **R**esonance **D**ecoupling). When an SFORD spectrum is acquired, the effect on the ^{13}C spectrum is to *reduce* the values of splittings due to all carbon-proton coupling (Figure 6.1d). The multiplicity due to the larger one-bond C-H couplings remains, making it possible to distinguish by inspection whether a carbon atom is a part of a methyl group (quartet), methylene group (triplet), a methine (CH) group (doublet) or a quaternary carbon (a singlet) just as in the fully proton-coupled spectrum. It should be noted that because the protons are partially decoupled, the magnitude of the splittings observed in the signals in an SFORD ^{13}C NMR spectrum is not the C-H coupling constant (the splittings are always less than the real coupling constants). SFORD spectra are useful only to establish signal multiplicity.

6.2 DETERMINING ^{13}C SIGNAL MULTIPLICITY USING DEPT

With most modern NMR instrumentation, the DEPT experiment (**D**istortionless **E**nhancement by **P**olarisation **T**ransfer) is the most commonly used method to determine the multiplicity of ^{13}C signals. The DEPT experiment is a pulsed NMR experiment which requires a series of programmed Rf pulses to both the ^{1}H and ^{13}C nuclei in a sample. The resulting ^{13}C DEPT spectrum contains only signals arising from protonated carbons (non protonated carbons do not give signals in the ^{13}C DEPT spectrum). The signals arising from carbons in CH_3 and CH groups (*i.e.* those with an odd number of attached protons) appear oppositely phased from those in CH_2 groups (*i.e.* those with an even number of attached protons) so signals from CH_3 and CH groups point upwards while signals from CH_2 groups point downwards (Figure 6.1b).

In more advanced applications, the ^{13}C DEPT experiment can be used to separate the signals arising from carbons in CH_3, CH_2 and CH groups. This is termed spectral editing and can be used to produce separate ^{13}C sub-spectra of just the CH_3 carbons, just the CH_2 carbons or just the CH carbons.

Figure 6.1 shows various ^{13}C spectra of methyl cyclopropyl ketone. The ^{13}C spectrum acquired with full proton decoupling (Figure 6.1a) shows 4 singlet peaks, one for each of the 4 different carbon environments in the molecule. The DEPT spectrum (Figure 6.1b) shows only the 3 resonances for the protonated carbons. The carbon atoms that have an odd number of attached hydrogens (CH and CH_3 groups) point upwards and those with an even number of attached hydrogen atoms (the signals CH_2 groups) point downwards. Note that the carbonyl carbon does not appear in the DEPT spectrum since it has no attached protons.

In the carbon spectrum with no proton decoupling (Figure 6.1c), all of the resonances of protonated carbons appear as multiplets and the multiplet structure is due to coupling to the attached protons. The CH_3 (methyl) group appears as a quartet, the CH_2 (methylene) groups appear as a triplet and the CH (methine) group appears as a doublet while the carbonyl carbon (with no attached protons) appears as a singlet. In Figure 6.1c, all of the $^{1}J_{C\text{-}H}$ coupling constants could be measured directly from the spectrum. The SFORD spectrum (Figure 6.1d) shows the expected multiplicity for all of the resonances but the multiplets are narrower due to partial decoupling of the protons and the splittings are less than the true values of $^{1}J_{C\text{-}H}$.

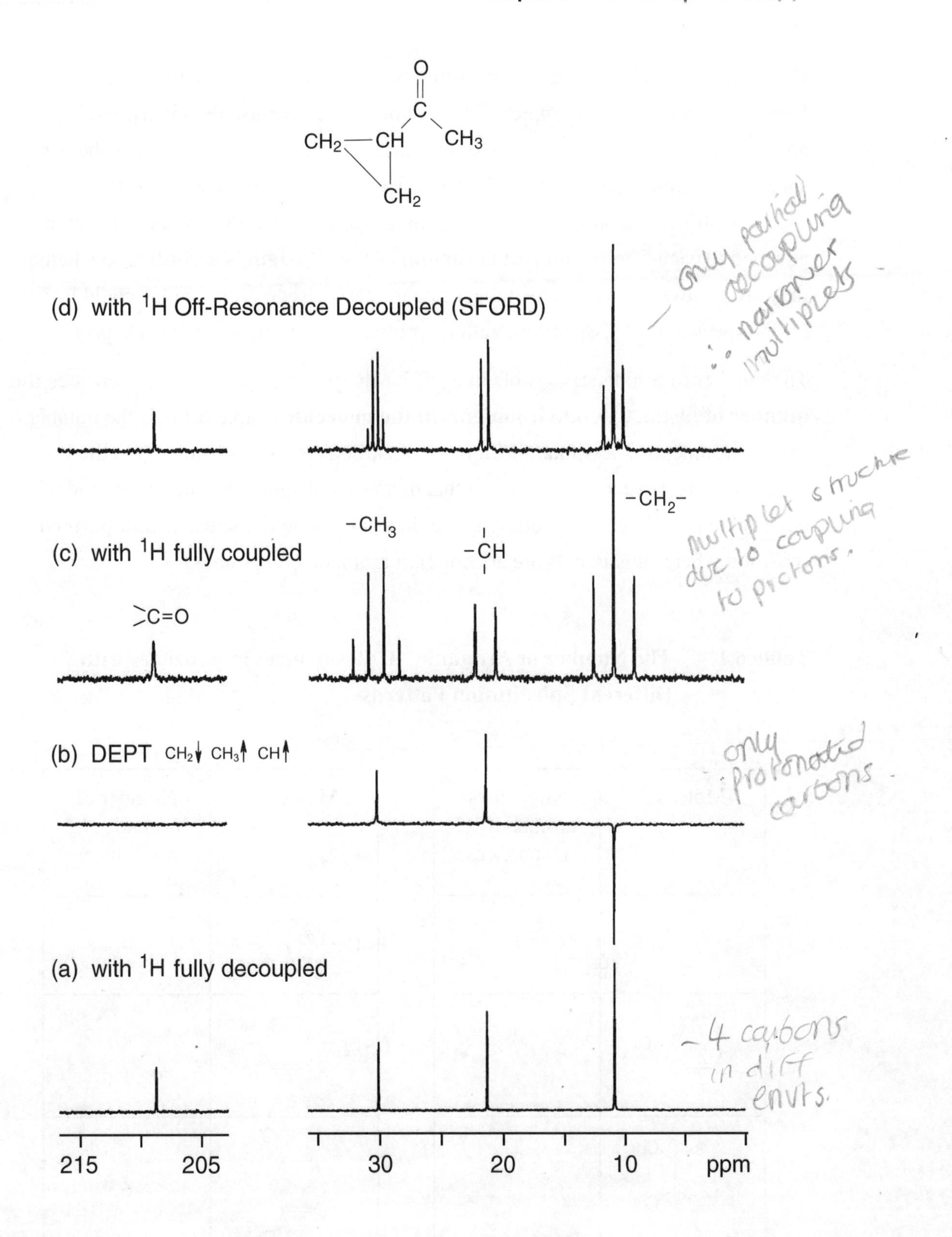

Figure 6.1 **^{13}C NMR Spectra of Methyl Cyclopropyl Ketone ($CDCl_3$ Solvent, 100 MHz). *(a)* Spectrum with Full Broad Band Decoupling of ^1H ; *(b)* DEPT Spectrum *(c)* Spectrum with no Decoupling of ^1H; *(d)* SFORD Spectrum**

For purposes of assigning a ^{13}C spectrum, two ^{13}C spectra are usually obtained. Firstly, a spectrum with complete ^{1}H decoupling to maximise the intensity of signals and provide sharp singlets signals to minimise any signal overlap. This is the best spectrum to **count the number of resonances** and accurately determine their chemical shifts. Secondly, a spectrum which is sensitive to the number of protons attached to each C to permit partial **sorting of the ^{13}C signals** according to whether they are methyl, methylene, methine or quaternary carbon atoms. This could be a DEPT spectrum, a ^{13}C spectrum with no proton decoupling or a SFORD spectrum.

The number of resonances visible in a ^{13}C NMR spectrum immediately indicates **the number of distinct ^{13}C environments in the molecule** (Table 6.1). If the number of ^{13}C environments is less than the number of carbons in the molecule, then the molecule must have some symmetry that dictates that some ^{13}C nuclei are in identical environments. This is particularly useful in establishing the **substitution pattern** (position where substituents are attached) in aromatic compounds.

Table 6.1 The Number of Aromatic ^{13}C Resonances in Benzenes with Different Substitution Patterns

Molecule	Number of aromatic ^{13}C resonances	Molecule	Number of aromatic ^{13}C resonances
	1	Cl—⟨ring⟩—Cl	2
⟨ring⟩—Cl	4	Br—⟨ring⟩—Cl	4
Cl, Cl	3	Cl, Br	6
Cl, Cl	4	Cl, Br	6

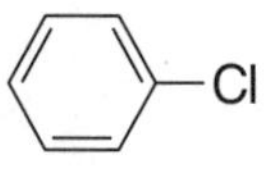

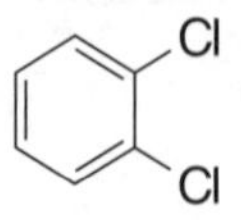

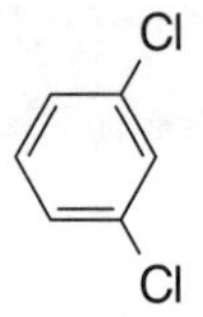

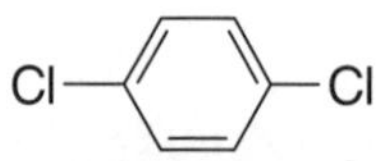

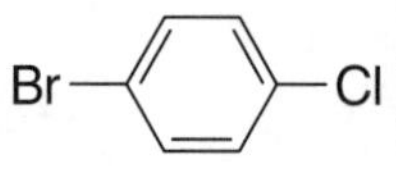

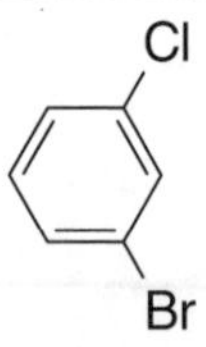

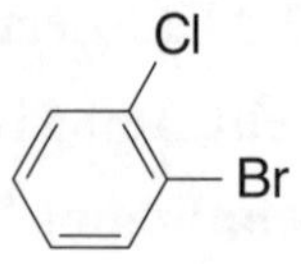

6.3 SHIELDING AND CHARACTERISTIC CHEMICAL SHIFTS IN ^{13}C NMR SPECTRA

The general trends of ^{13}C chemical shifts somewhat parallel those in ^{1}H NMR spectra. However, ^{13}C nuclei have access to a greater variety of hybridisation states (bonding geometries and electron distributions) than ^{1}H nuclei and both hybridisation and changes in electron density have a significantly larger effect on ^{13}C nuclei than ^{1}H nuclei. As a consequence, the ^{13}C chemical shift scale spans some 250 ppm, *cf.* the 10 ppm range commonly encountered for ^{1}H chemical shifts (Tables 6.2 and 6.3).

Table 6.2 Typical ^{13}C Chemical Shift Values in Selected Organic Compounds

Compound	δ ^{13}C (PPM from TMS)
CH_4	-2.1
CH_3CH_3	7.3
CH_3OH	50.2
CH_3Cl	25.6
CH_2Cl_2	52.9
$CHCl_3$	77.3
$CH_3CH_2CH_2Cl$	11.5 (CH_3)
	26.5 ($-CH_2-$)
	46.7 ($-CH_2-Cl$)
$CH_2=CH_2$	123.3
CH_3CHO	31.2 ($-CH_3$)
	200.5 (-CHO)
CH_3COOH	20.6 ($-CH_3$), 178.1 (-COOH)
CH_3COCH_3	30.7 ($-CH_3$), 206.7 (-CO-)
benzene	128.0
pyridine (3, 2, 4, N)	150.2 (C-2)
	123.9 (C-3)
	135.9 (C4)

Table 6.3 Typical ^{13}C Chemical Shift Ranges in Organic Compounds

Group	^{13}C shift (ppm)
TMS	0.0
$-CH_3$ (with only -H or -R at C_α and C_β)	0 - 30
$-CH_2$ (with only -H or -R at C_α and C_β)	20 - 45
-CH (with only -H or -R at C_α and C_β)	30 - 60
C quaternary (with only -H or -R at C_α and C_β)	30 - 50
$O-CH_3$	50 - 60
$N-CH_3$	15 - 45
C≡C	70 - 95
C=C	105 - 145
C (aromatic)	110 - 155
C (heteroaromatic)	105 - 165
-C≡N	115 - 125
C=O (acids, acyl halides, esters, amides)	155 - 185
C=O (aldehydes, ketones)	185 - 225

In ^{1}H NMR spectroscopy, the signal due to $CHCl_3$ (an isotopic impurity in $CDCl_3$) is invariably present at δ 7.27 and provides a secondary internal standard against which chemical shifts can be measured. In ^{13}C NMR spectroscopy the ^{13}C signal due to the carbon in $CDCl_3$ appears as a triplet centred at δ 77.3 with peaks intensities in the ratio 1:1:1 (due to spin-spin coupling between ^{13}C and ^{2}H). This resonance serves as a convenient reference for the chemical shifts of ^{13}C NMR spectra recorded in this solvent.

Table 6.4 gives characteristic chemical shifts for ^{13}C for some sp^3-hybridised carbon atoms in common functional groups. Table 6.5 gives characteristic chemical shifts for ^{13}C for some sp^2-hybridised carbon atoms in substituted alkenes.

Table 6.4 ^{13}C Chemical Shifts (δ) for *sp*^{3} Carbons in Alkyl Derivatives

	CH_3—X	CH_3CH_2—X		$(CH_3)_2CH$—X	
X	—CH_3	—CH_3	—CH_2—	—CH_3	>CH—
—H	-2.3	7.3	7.3	15.4	15.9
—CH=CH_2	18.7	13.4	27.4	22.1	32.3
—Ph	21.4	15.8	29.1	24.0	34.3
—Cl	25.6	18.9	39.9	27.3	53.7
—OH	50.2	18.2	57.8	25.3	64.0
—OCH_3	60.9	14.7	67.7	21.4	72.6
—OCO−CH_3	51.5	14.4	60.4	21.9	67.5
—CO−CH_3	30.7	7.0	35.2	18.2	41.6
—CO−OCH_3	20.6	9.2	27.2	19.1	34.1
—NH_2	28.3	19.0	36.9	26.5	43.0
—NH−$COCH_3$	26.1	14.6	34.1	22.3	40.5
—C≡N	1.7	10.6	10.8	19.9	19.8
—NO_2	61.2	12.3	70.8	20.8	78.8

Table 6.5 ^{13}C Chemical Shifts (δ) for *sp*^{2} Carbons in Vinyl Derivatives CH_2=CH-X

X	CH_2=	=CH—
—H	123.3	123.3
—CH_3	115.9	136.2
—Ph	112.3	135.8
—CH=CH_2	116.3	136.9
—CO−CH_3	128.0	137.1
—CO−OCH_3	130.3	129.6
—Cl	117.2	126.1
—C≡N	137.5	108.2
—$N(CH_3)_2$	91.3	151.3

Table 6.6 gives characteristic ^{13}C chemical shifts for the aromatic carbons in benzene derivatives. To a first approximation, the shifts induced by substituents are additive. So, for example, an aromatic carbon which has a $-NO_2$ group in the *para* position and a –Br group in the *ortho* position will appear at approximately 137.9 ppm [(128.5 + 6.1(*p*-NO_2) + 3.3(*o*-Br)]. Tables 6.7 gives characteristic shifts for ^{13}C nuclei in some polynuclear aromatic compounds and heteroaromatic compounds.

Table 6.6 Approximate ^{13}C Chemical Shifts (δ) for Aromatic Carbons in Benzene Derivatives Ph-X in ppm relative to Benzene at δ 128.5 ppm (a positive sign denotes a downfield shift)

X	*ipso*	*Ortho*	*meta*	*para*
—H	0.0	0.0	0.0	0.0
$—NO_2$	19.9	-4.9	0.9	6.1
$—CO-OCH_3$	2.0	1.2	-0.1	4.3
$—CO-CH_3$	8.9	0.1	-0.1	4.4
$—C\equiv N$	-16.0	3.5	0.7	4.3
—Br	-5.4	3.3	2.2	-1.0
$—CH=CH_2$	8.9	-2.3	-0.1	-0.8
—Cl	5.3	0.4	1.4	-1.9
$—CH_3$	9.2	0.7	-0.1	-3.0
$—OCO-CH_3$	22.4	-7.1	0.4	-3.2
$—OCH_3$	33.5	-14.4	1.0	-7.7
$—NH_2$	18.2	-13.4	0.8	-10.0

Table 6.7 Characteristic ^{13}C Chemical Shifts (δ) in some Polynuclear Aromatic Compounds and Heteroaromatic Compounds

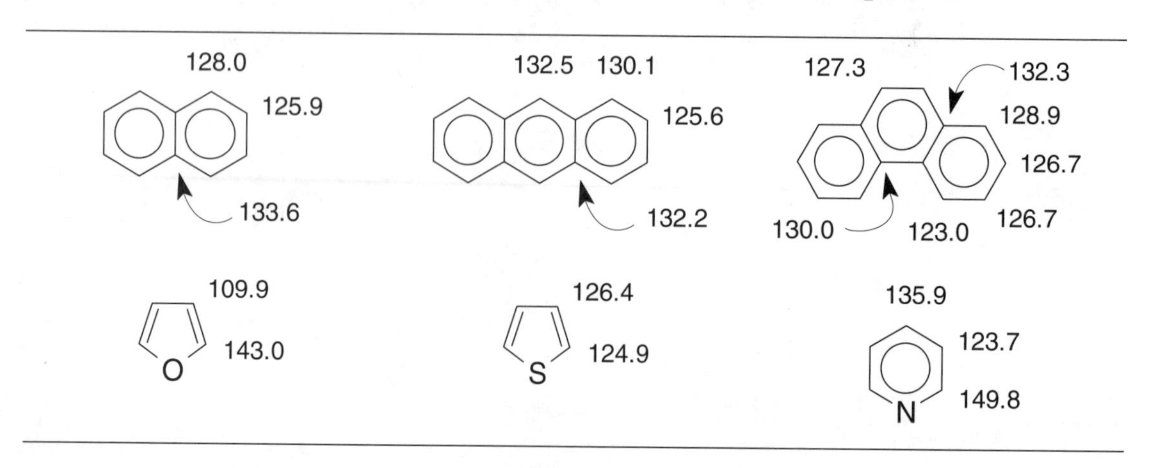

7

MISCELLANEOUS TOPICS

This section deals briefly with a number of topics in NMR spectroscopy, which while not directly pertinent to the solution of the problems collected in this book, give some idea of the power of NMR spectroscopy in solving complex structural problems.

7.1 DYNAMIC NMR SPECTROSCOPY: THE NMR TIME-SCALE

Two magnetic nuclei situated in different molecular environments must give rise to separate signals in the NMR spectrum, say $\Delta\nu$ Hz apart (Figure 7.1a). However, if some process interchanges the environments of the two nuclei at a rate (k) much faster than $\Delta\nu$ times per second, the two nuclei will be observed as a single signal at an intermediate frequency (Figures 7.1d and 7.1e). When the rates (k) of the exchange process are comparable to $\Delta\nu$, *exchange broadened* spectra (Figure 7.1b) are observed. From the exchange broadened spectra, the rate constants for the exchange process (and hence the activation parameters $\Delta G^{\neq}$, $\Delta H^{\neq}$, $\Delta S^{\neq}$) can be derived. Where signals coalesce (Figure 7.1c) from being two separate signals to a single averaged signal, the rate constant for the exchange can be approximated as $k = \pi.\Delta\nu / \sqrt{2}$.

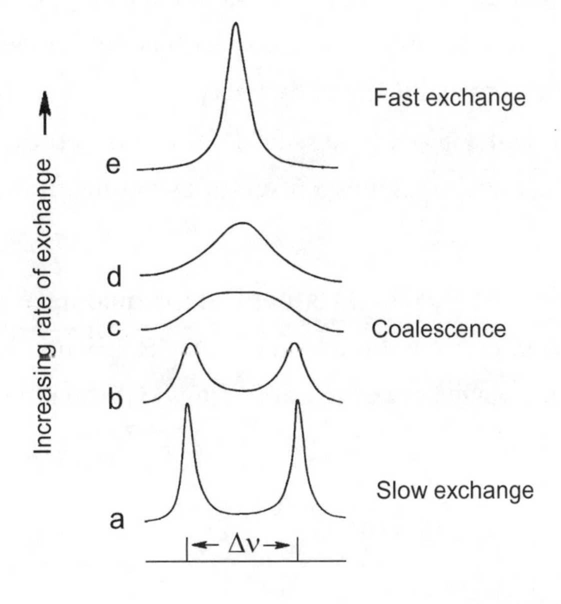

Figure7.1 Schematic NMR Spectra of Two Exchanging Nuclei

In practice, a compound where an exchange process operates can give rise to a series of spectra of the type shown in Figure 7.1, if the NMR spectra are recorded at different temperatures. Changing the sample temperature alters the rate constant for the exchange (increasing the temperature increases the rate of an exchange process) and the spectra will have a different appearance depending on whether the rate constant, k (expressed in sec^{-1}) is large or small compared to the chemical shift differences between exchanging nuclei ($\Delta\nu$ expressed in Hz). *Molecules where there are exchange processes taking place may also give rise to different NMR spectra in different NMR spectrometers* because $\Delta\nu$ depends on the strength of the magnetic field. An exchange process must be faster if the sample is in a spectrometer with a strong magnetic field to give rise to exchange broadening than in a spectrometer with a weak magnetic field. An NMR spectrum which shows exchange broadening will tend to give a slow exchange spectrum if the spectrum is re-run in a spectrometer with a stronger magnetic field.

The averaging effects of exchange apply to any dynamic process that takes place in a molecule (or between molecules). However, many processes occur at rates that are too fast or too slow to give rise to visible broadening of NMR spectra. The **NMR time-scale** happens to coincide with the rates of a number of common chemical processes that give rise to variation of the appearance of NMR spectra with temperature and these include:

(1) ***Conformational exchange processes.*** Conformational processes can give rise to exchange broadening in NMR spectra when a molecule exchanges between two or more conformations. Fortunately most conformational processes are so fast on the NMR time-scale that normally only averaged spectra are observed. In particular, in molecules which are not unusually sterically bulky, the rotation about C-C single bonds is normally fast on the NMR time scale so, for example, the 3 hydrogen atoms of a methyl group appear as a singlet as a result of averaging of the various rotational conformers.

In molecules where there are very bulky groups, steric hindrance can slow the rotation about single bonds and give rise to broadening in NMR spectra. In molecules containing rings, the exchange between various ring conformations (*e.g.* chair-boat-chair) can exchange nuclei.

For example, cyclohexane gives a single averaged resonance in the ^{1}H NMR at room temperature, but separate signals are seen for the axial and equatorial hydrogens when spectra are acquired at very low temperature.

H_1, H_2 $\overset{k}{\rightleftarrows}$ H_1, H_2

(2) Intermolecular interchange of labile (slightly acidic) protons. Functional groups such as -OH, -COOH, $-NH_2$ and -SH have labile protons which are exchangeable in solution. These exchangeable protons in these functional groups frequently give rise to broadened resonances in the ^{1}H NMR spectrum. The -OH protons of a mixture of two different alcohols may give rise to either an averaged signal or to separate signals depending on the rate of exchange and this depends on many factors including temperature, the polarity of the solvent, the concentrations of the solutes and the presence of acidic or basic catalysts.

$$R{-}OH_1 + R'{-}OH_2 \rightleftarrows R{-}OH_2 + R'{-}OH_1$$

Labile protons in -OH, -COOH, $-NH_2$ and -SH groups can be exchanged for deuterium by adding a drop of deuterated water (D_2O) to an NMR sample. Since deuterium is invisible in the ^{1}H NMR spectrum, labile protons disappear from the ^{1}H NMR spectrum when D_2O is added. The N-H protons of primary and secondary amides are slow to exchange and require heating or base catalysis.

$$R{-}O{-}H + D_2O \rightleftarrows R{-}O{-}D + H{-}O{-}D$$

(3) Rotation about partial double bonds.

Exchange broadening is frequently observed in amides due to restricted rotation about the N-C bond of the amide group.

R, Ha, C—N, O, Hb $\overset{k}{\rightleftarrows}$ R, Hb, C—N, O, Ha

The restricted rotation about amide bonds often occurs at a rate that gives rise to observable broadening in NMR spectra.

The restricted rotation in amide bonds results from the double bond character of the C-N bond.

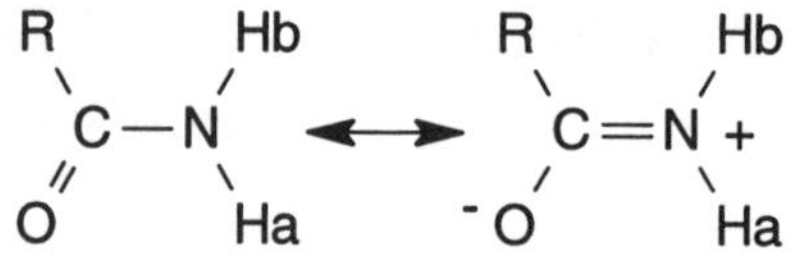

```
R        Hb          R        Hb
 \      /             \      /
  C—N       ←→         C=N +
 //     \             /      \
O        Ha        -O         Ha
```

7.2 THE EFFECT OF CHIRALITY

In an achiral solvent, enantiomers will give identical NMR spectra. However in a chiral solvent or in the presence of a chiral additive to the NMR solvent, enantiomers will have different spectra and this is frequently used to establish the enantiomeric purity of compounds. The resonances of one enantiomer can be integrated against the resonances of the other to quantify the enantiomeric purity of a compound.

In molecules that contain a stereogenic centre, the NMR spectra can sometimes be more complex than would otherwise be expected. Groups such as $-CH_2-$ groups (or any $-CX_2-$ group such as $–C(Me)_2-$ or $-CR_2-$) require particular attention in molecules which contain a stereogenic centre. The carbon atom of a $–CX_2-$ group is termed a **prochiral carbon** if there is a stereogenic centre (a chiral centre) elsewhere in the molecule. A prochiral carbon atom is a carbon in a molecule that would be chiral if one of its substituents was replaced by a different substituent. From an NMR perspective, the important fact is that the presence of stereogenic centre makes the substituents on a prochiral carbon atom **chemically non-equivalent**. So whereas the protons of a $-CH_2-$ group in an acyclic aliphatic compound would normally be expected to be equivalent and resonate at the same frequency in the ^1H NMR spectrum, if there is a stereogenic centre in the molecule, each of the protons of the $-CH_2-$ group will appear at different chemical shifts. Also, since they are non-equivalent, the protons will couple to each other typically with a large coupling of about 15 Hz.

The effect of chirality is particularly important in the spectra of natural products *e.g.* in the spectra of amino acids, proteins or peptides. Many molecules derived from natural sources contain a stereogenic centre and they are typically obtained as a single pure enantiomer. In these molecules you can expect that the resonances for all of the methylene groups (*i.e.* $-CH_2-$ groups) in the molecule will not be simple multiplets but be complicated by the fact that the two protons of the methylene groups will be non-equivalent. Figure 7.2 shows the aliphatic protons in the 1H NMR spectrum of the amino acid cystein ($HSCH_2CHNH_3^+COO^-$). Cystein has a stereogenic centre and the signals of the methylene group appear as separate signals at δ 3.18 and δ 2.92 ppm. Each of the methylene protons is split into a doublet of doublets due to coupling firstly to the other methylene proton and secondly to the proton on the α-carbon (H_c).

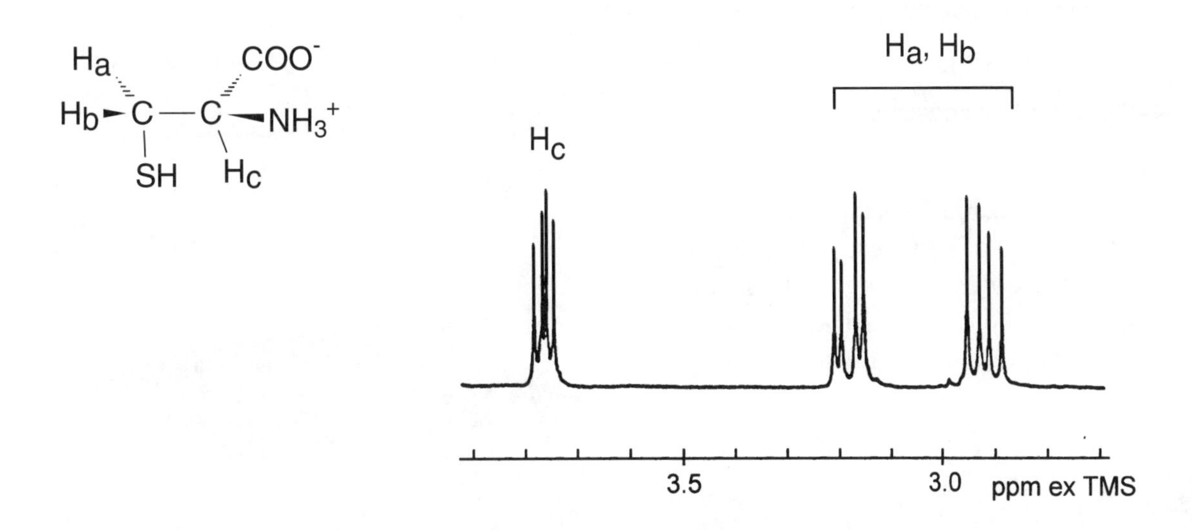

Figure 7.2 **^{1}H NMR Spectrum of the Aliphatic Region of Cystein Indicating Non-equivalence of the Methylene Protons due to the Influence of the Stereogenic Centre**

7.3 THE NUCLEAR OVERHAUSER EFFECT (NOE)

Irradiation of one nucleus while observing the resonance of another may result in a change in the **amplitude** of the observed resonance *i.e.* an enhancement of the signal intensity. This is known as the *nuclear Overhauser effect* (NOE). The NOE is a "through space" effect and its magnitude is inversely proportional to the sixth power of the distance between the interacting nuclei. Because of the distance dependence of the NOE, it is an important method for establishing which groups are close together in space and because the NOE can be measured quite accurately it is a very powerful means for determining the three dimensional structure (and stereochemistry) of organic compounds.

The intensity of ^{13}C resonances may be increased by up to 200% when ^{1}H nuclei which are directly bonded to the carbon atom are irradiated. This effect is very important in increasing the intensity of ^{13}C spectra when they are proton-decoupled. The efficiency of the proton/carbon NOE varies from carbon to carbon and this is a factor that contributes to the generally non-quantitative nature of ^{13}C NMR. While the intensity of protonated carbon atoms can be increased significantly by NOE, non-protonated carbons (quaternary carbon atoms) receive little NOE and are usually the weakest signals in a ^{13}C NMR spectrum.

7.4 TWO-DIMENSIONAL NMR SPECTROSCOPY

Since the advent of pulsed NMR spectroscopy a number of advanced two-dimensional techniques have been devised. These methods yield valuable information for the solution of complex structural problems. The principles behind such techniques are beyond the scope of this book and will not be discussed in detail.

Two-dimensional spectra have the appearance of surfaces, generally with two axes corresponding to chemical shift and the third (vertical) axis corresponding to signal intensity.

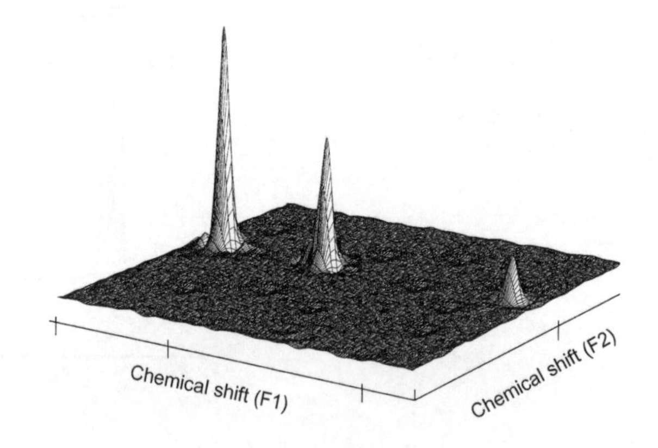

It is usually more useful to plot two-dimensional spectra viewed directly from above (a **contour plot** of the surface) in order to make measurements and assignments.

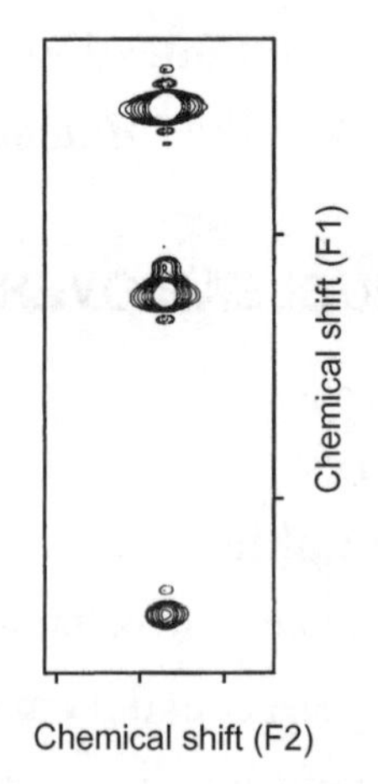

The most important two-dimensional NMR experiments for solving structural problems are COSY (COrrelation SpectroscopY), NOESY (Nuclear Overhauser Enhancement SpectroscopY) and the HSC (Heteronuclear Shift Correlation). Most modern high-field NMR spectrometers have the capability to acquire COSY, NOESY or HSC spectra.

The COSY spectrum shows which pairs of nuclei in a molecule are coupled. The COSY spectrum is a symmetrical spectrum that has the ^{1}H NMR spectrum of the substance as both of the chemical shift axes (F1 and F2). A schematic representation of COSY spectrum is given below.

The COSY spectrum has a diagonal set of peaks (open circles) as well as peaks that are off the diagonal (closed circles). The off diagonal peaks are the important signals since these occur at positions where there is coupling between a proton on the F1 axis and one on the F2 axis. In the schematic spectrum on the right, the off-diagonal signals show that there is spin-spin coupling between H_c and H_d and also between H_b and H_c but the proton labelled H_a has no coupling partners.

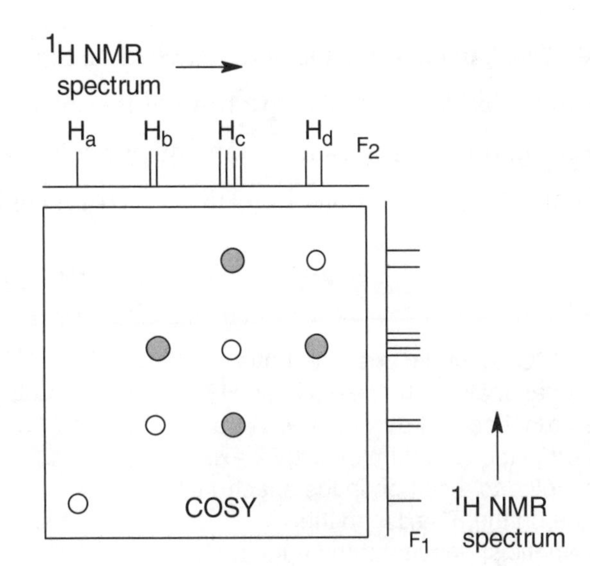

In a single COSY spectrum, all of the spin-spin coupling pathways in a molecule can be identified.

The NOESY spectrum relies on the Nuclear Overhauser Effect and shows that pairs of nuclei in a molecule are close together in space. The NOESY spectrum is very similar in appearance to a COSY spectrum. It is a symmetrical spectrum that has the ^{1}H NMR spectrum of the substance as both of the chemical shift axes (F1 and F2). A schematic representation of NOESY spectrum is given below.

The NOESY spectrum has a diagonal set of peaks (open circles) as well as peaks which are off the diagonal (closed circles). The off diagonal peaks occur at positions where a proton on the F1 axis is close in space to a one on the F2 axis. In the schematic spectrum on the right, the off-diagonal signals show that H_a must be located near H_d and H_b must be located near H_c.

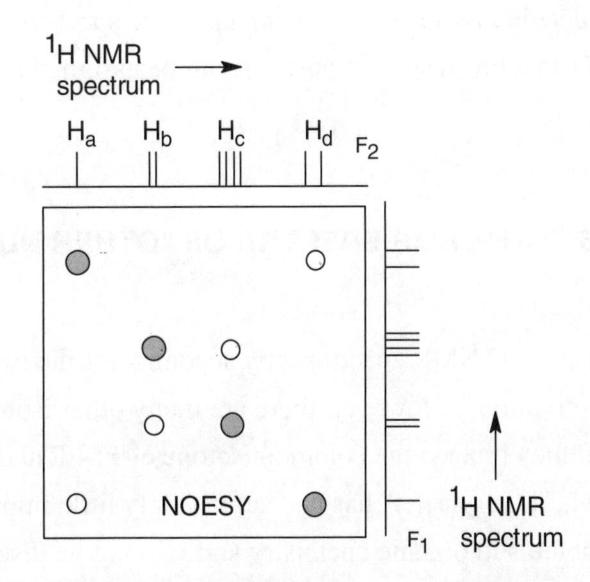

In a NOESY spectrum, all of the nuclei that are close in space can be identified. From the analysis of a NOESY spectrum, the three dimensional structure of a molecule or parts of a molecule can be determined.

The HSC spectrum is the heteronuclear analogue of the COSY spectrum. The HSC spectrum has the ^{1}H NMR spectrum of the substance on one axis (F2) and the ^{13}C spectrum (or the spectrum of some other nucleus) on the second axis (F1). A schematic representation of an HSC spectrum is given below.

The HSC spectrum does not have diagonal peaks. The peaks in an HSC spectrum occur at positions where a proton in the spectrum on the F2 axis is coupled to a carbon in the spectrum on the on the F1 axis. In the schematic spectrum on the right, both H_a and H_b are coupled to C_a, H_d is coupled to C_b, and H_c is coupled to C_c.

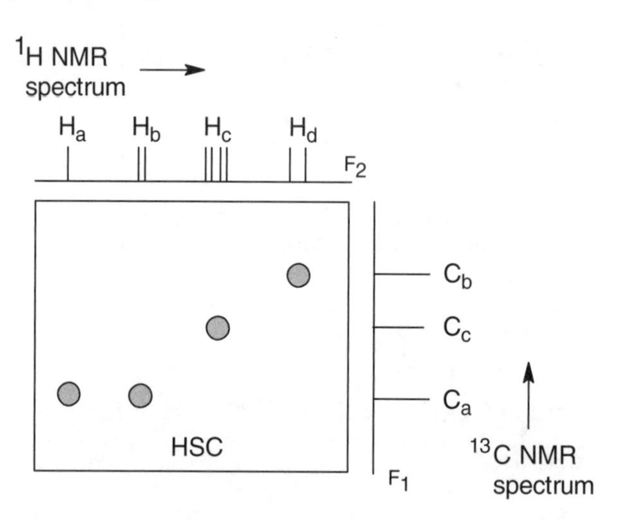

In an HSC spectrum, the correlation between the protons in the ^{1}H NMR spectrum and the carbon nuclei in the ^{13}C spectrum can be obtained. It is usually possible to assign all of the resonances in the ^{1}H NMR spectrum *i.e.* establish which proton in a molecule gives rise to each signal in the spectrum, using spin-spin coupling information. The ^{13}C spectrum can be assigned by correlation to the proton resonances.

7.5 THE NMR SPECTRA OF "OTHER NUCLEI"

^{1}H and ^{13}C NMR spectroscopy accounts for the overwhelming proportion of all NMR observations. However, there are many other isotopes which are NMR observable and they include the common isotopes ^{19}F, ^{31}P and ^{2}H. The NMR spectroscopy of these "other nuclei" has had surprisingly little impact on the solution of structural problems in organic chemistry and will not be discussed here. It is however important to be alert for the presence of other magnetic nuclei in the molecule, because they often cause additional multiplicity in ^{1}H and ^{13}C NMR spectra due to spin-spin coupling.

7.6 SOLVENT INDUCED SHIFTS

Generally solvents chosen for NMR spectroscopy do not associate with the solute. However, solvents which are capable of both association and inducing differential chemical shifts in the solute are sometimes deliberately used to remove accidental chemical equivalence. The most useful solvents for the purpose of inducing *solvent-shifts* are aromatic solvents, in particular hexadeuterobenzene (C_6D_6), and the effect is called *aromatic solvent induced shift* (ASIS). The numerical values of ASIS are usually of the order of 0.1 - 0.5 ppm and they vary with the molecule studied depending mainly on the geometry of the complexation.

8

DETERMINING THE STRUCTURE OF ORGANIC COMPOUNDS FROM SPECTRA

The main purpose of this book is to present a collection of suitable problems to teach and train researchers in the general important methods of spectroscopy.

Problems 1 - 250 are all of the basic "structures from spectra" type, are generally relatively simple and are arranged roughly in order of increasing complexity.

No solutions to the problems are given. It is important to assign NMR spectra as completely as possible and rationalise *all numbered peaks* in the mass spectrum and account for all significant features of the UV and IR spectra.

The next group of problems (251-256) present data in text form rather than graphically. The formal style that is found in the presentation of spectral data in these problems is typical of that found in the experimental of a publication or thesis. This is a completely different type of data presentation and one that students will encounter frequently. Problem 257 deals with molecular symmetry and is a useful exercise to establish how symmetry in a molecule can be established from the number of resonances in ^{1}H and ^{13}C NMR spectra.

The last group of problems (258-277) are of a different type and deal with interpretation of simple ^{1}H NMR spin-spin multiplets. To the best of our knowledge, problems of this type are not available in other collections and they are included here because we have found that the interpretation of multiplicity in ^{1}H NMR spectra is the greatest single cause of confusion in the minds of students.

The spectra presented in problems were obtained under conditions stated on the individual problem sheets. Mass spectra were obtained on an AEI MS-9 spectrometer or a Hewlett Packard MS-Engine mass spectrometer. 60 MHz ^{1}H NMR spectra and 15 MHz ^{13}C NMR spectra were obtained on a Jeol FX60Q spectrometer, 100 MHz ^{1}H NMR spectra were obtained on a Varian XL-100 spectrometer, 200 MHz ^{1}H NMR spectra and 50 MHz ^{13}C NMR spectra were obtained on a Bruker AC-200 spectrometer, 400 MHz ^{1}H NMR spectra and 100 MHz ^{13}C NMR spectra were obtained on Bruker AMX-400 or DRX-400 spectrometers, 600 MHz ^{1}H NMR spectra were obtained on a Bruker AMX-600 or DRX-600 spectrometer and 20 MHz ^{13}C NMR spectra were obtained on a Varian CFT-20 spectrometer.

Ultraviolet spectra were recorded on a Perkin-Elmer 402 UV spectrophotometer or Hitachi 150-20 UV spectrophotometer and Infrared spectra on a Perkin-Elmer 710B or a Perkin-Elmer 1600 series FTIR spectrometer.

The following collections are useful sources of spectroscopic data on organic compounds and some of the data for literature compounds have been derived from these collections:

(a) http://www.aist.go.jp/RIODB/SDBS/ website maintained by the National Institute of Advanced Industrial Science and Technology, Tsukuba, Ibaraki, Japan;

(b) http://webbook.nist.gov/chemistry/ website which is the NIST Chemistry WebBook NIST Standard Reference Database Number 69 (July 2001) Eds P J Linstrom and W G Mallard maintained by the National Institute of Science and Technology, USA.

(c) E Pretch, P Bühlmann and C Affolter, "Structure Determination of Organic Compounds, Tables of Spectral Data", 3rd edition, Springer, Berlin 2000.

While there is no doubt in our minds that the only way to acquire expertise in obtaining "organic structures from spectra" is to practise, some students have found the following **general approach to solving structural problems by a combination of spectroscopic methods** helpful :

(1) Perform all **routine operations :**

(a) Determine the molecular weight from the Mass Spectrum.

(b) Determine relative numbers of protons in different environments from the ^{1}H NMR spectrum.

(c) Determine the number of carbons in different environments and the number of quaternary carbons, methine carbons, methylene carbons and methyl carbons from the ^{13}C NMR spectrum.

(d) Examine the problem for any additional data concerning composition and determine the molecular formula if possible.

(e) Determine the molar absorptivity of UV, if applicable.

(2) Examine each spectrum (IR, mass spectrum, UV, ^{13}C NMR, ^{1}H NMR) in turn for obvious **structural elements:**

(a) Examine the IR spectrum for the presence or absence of carbonyl groups, hydroxyl groups, NH groups, C≡C or C≡N, *etc.*

(b) Examine the mass spectrum for typical fragments *e.g.* $PhCH_2$-, CH_3CO-, CH_3-, *etc.*

(c) Examine the UV spectrum for evidence of conjugation, aromatic rings *etc.*

(d) Examine the ^{1}H NMR spectrum for CH_3- groups, CH_3CH_2- groups, aromatic protons, -CH_nX, exchangeable protons *etc.*

(3) Write down all structural elements you have determined. Note that some are monofunctional (*i.e.* must be end-groups, such as -CH_3, -C≡N, -NO_2) whereas some are bifunctional (*e.g.* -CO-, -CH_2-, -COO-), or trifunctional (*e.g.* CH, N).

Add up the atomic weights of the structural elements you have identified and compare this sum with the molecular weight of the unknown. The difference (if any) may give a clue to the nature of the undetermined structural elements (*e.g.* an ether oxygen). At this stage, elements of symmetry may become apparent.

(4) Try to assemble the structural elements. Note that **there may be more than one way of fitting them together.** Spin-spin coupling data or information about conjugation may enable you to make a definite choice between possibilities.

(5) Return to each spectrum (IR, UV, mass spectrum, ^{13}C NMR, ^{1}H NMR) in turn and *rationalise all major features* (especially all major fragments in the mass spectrum and all features of the NMR spectra) in terms of your proposed structure. Ensure that no spectral features are inconsistent with your proposed structure.

Note on the use of data tables. Tabulated data typically give characteristic absorptions or chemical shifts for representative compounds and these may not correlate *exactly* with those from an unknown compound. The data contained in data tables should always be used indicatively (not mechanically).

9

PROBLEMS

Problem 1

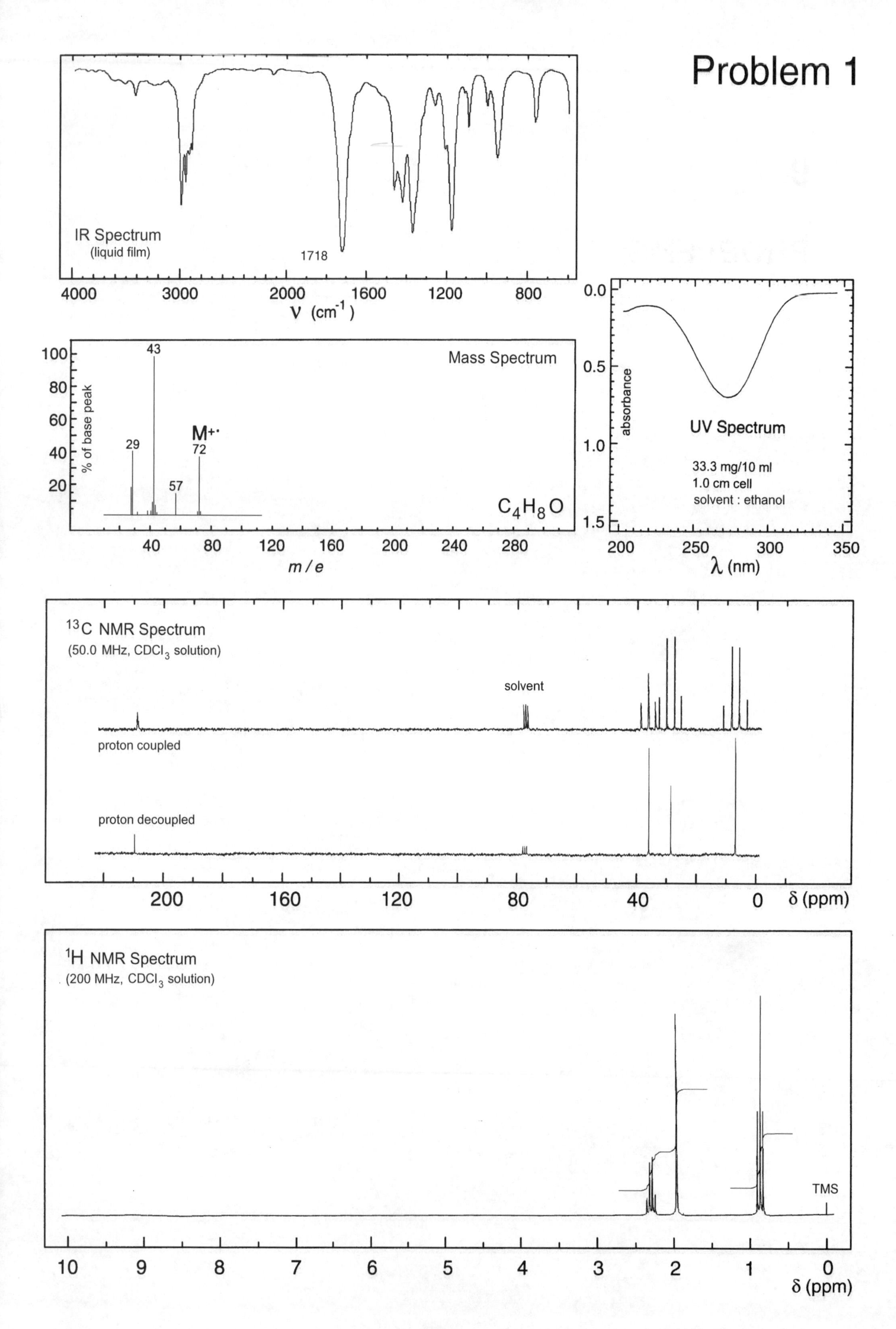

IR Spectrum
(liquid film)
1718
4000
3000
2000
1600
1200
800
ν (cm-1)
Mass Spectrum
100
80
60
40
20
% of base peak
43
29
57
M+·
72
C4H8O
40
80
120
160
200
240
280
m / e
0.0
0.5
1.0
1.5
absorbance
UV Spectrum
33.3 mg/10 ml
1.0 cm cell
solvent : ethanol
200
250
300
350
λ (nm)
13C NMR Spectrum
(50.0 MHz, CDCl3 solution)
solvent
proton coupled
proton decoupled
200
160
120
80
40
0
δ (ppm)
1H NMR Spectrum
(200 MHz, CDCl3 solution)
TMS
10
9
8
7
6
5
4
3
2
1
0
δ (ppm)

Problem 2

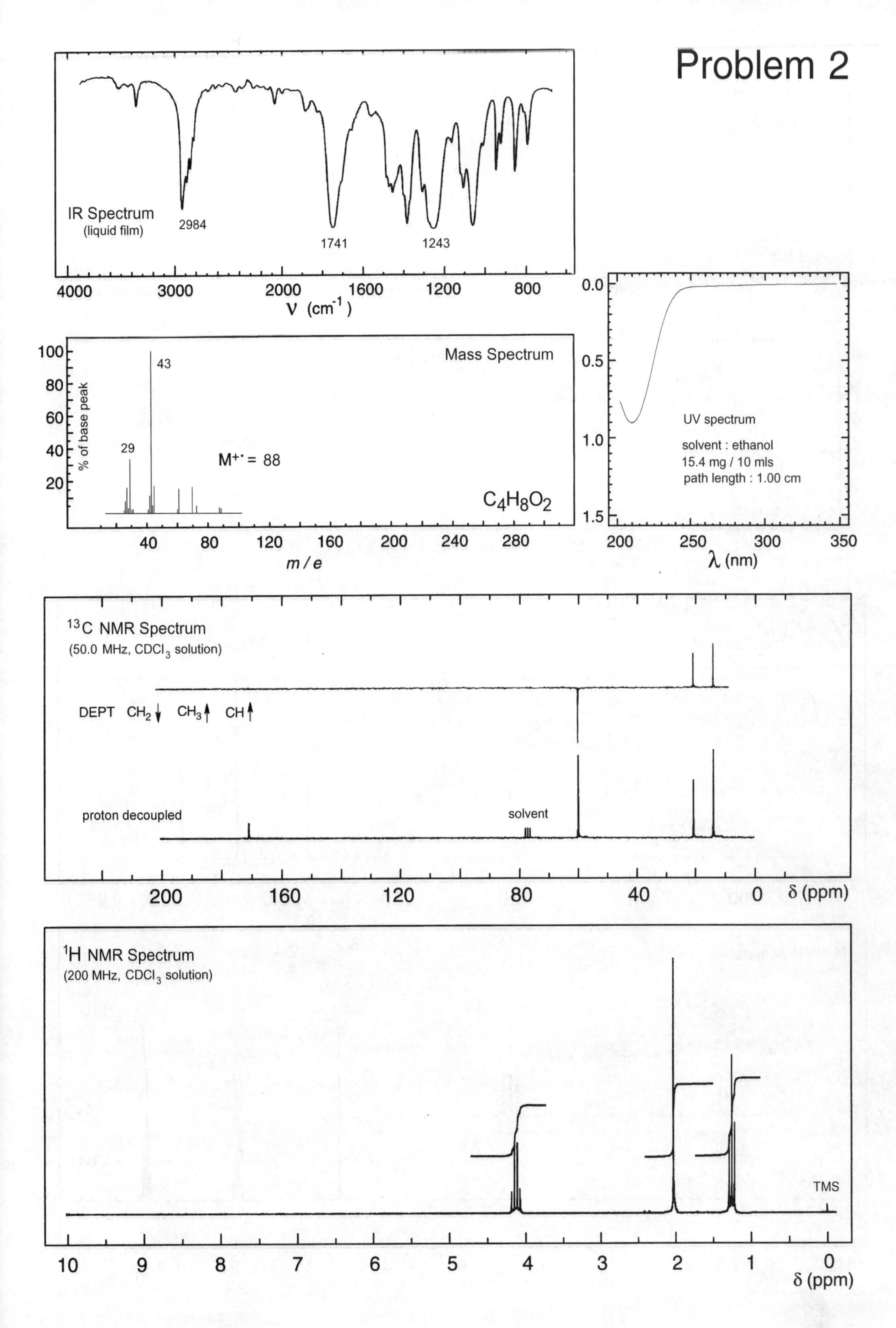

IR Spectrum
(liquid film)
2984
1741
1243
4000
3000
2000
1600
1200
800
ν (cm⁻¹)
Mass Spectrum
100
80
60
40
20
% of base peak
43
29
M⁺˙ = 88
C4H8O2
40
80
120
160
200
240
280
m / e
0.0
0.5
1.0
1.5
UV spectrum
solvent : ethanol
15.4 mg / 10 mls
path length : 1.00 cm
200
250
300
350
λ (nm)
13C NMR Spectrum
(50.0 MHz, CDCl3 solution)
DEPT CH2 ↓ CH3 ↑ CH ↑
proton decoupled
solvent
200
160
120
80
40
0
δ (ppm)
1H NMR Spectrum
(200 MHz, CDCl3 solution)
TMS
10
9
8
7
6
5
4
3
2
1
0
δ (ppm)

Problem 3

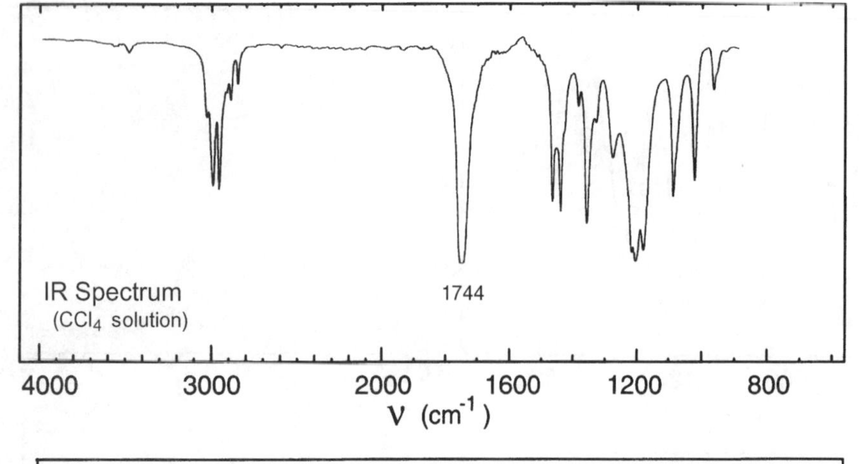

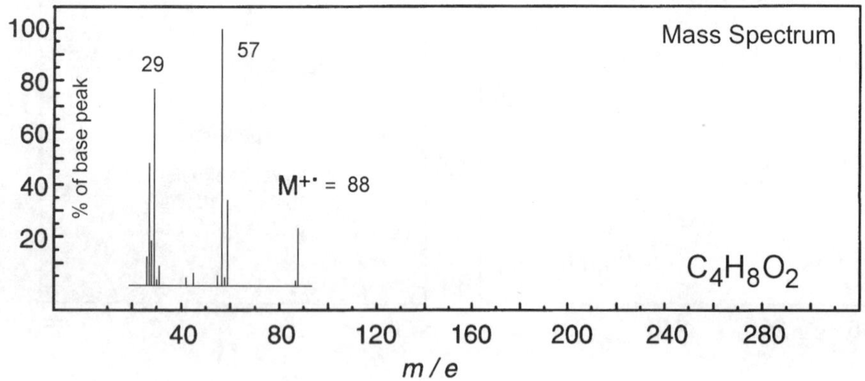

No significant UV absorption above 220 nm

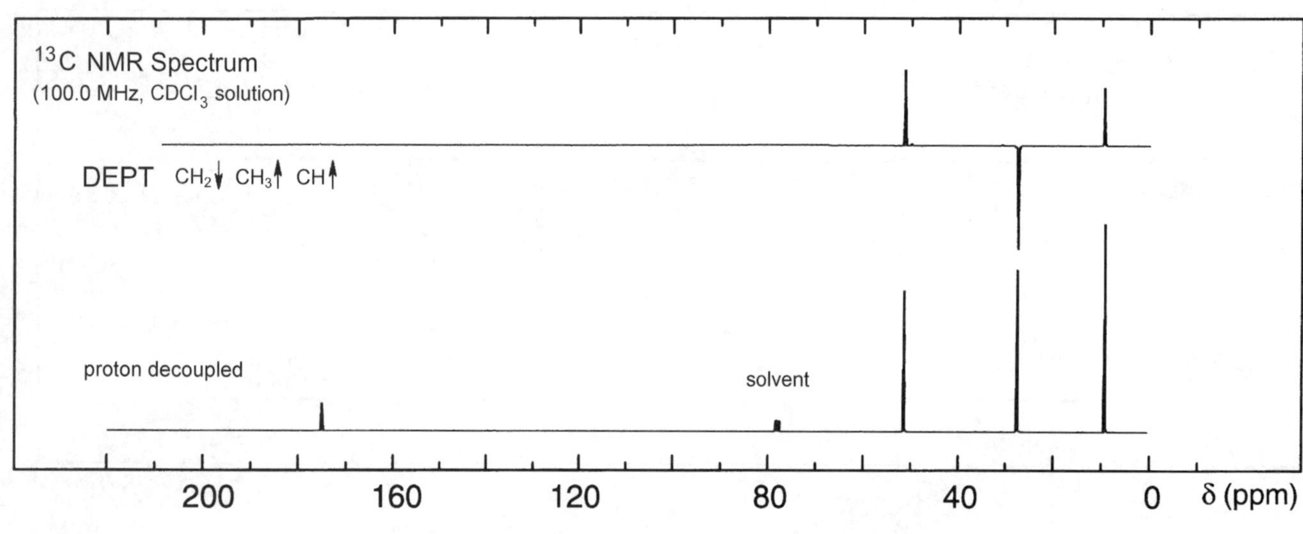

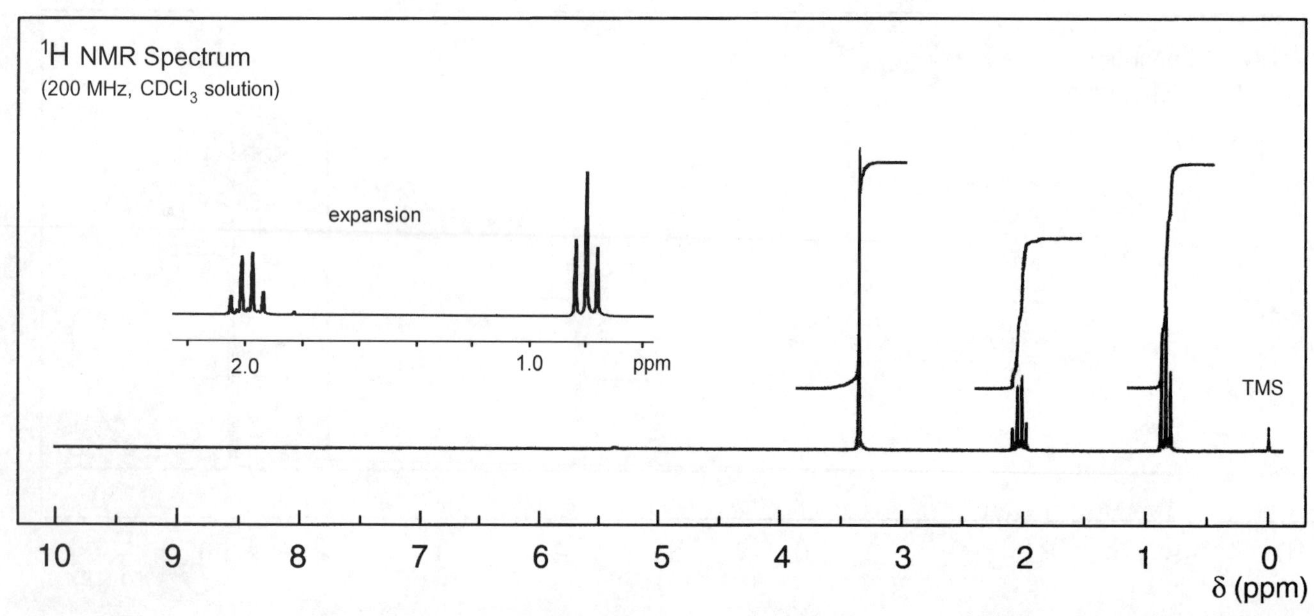

Problem 4

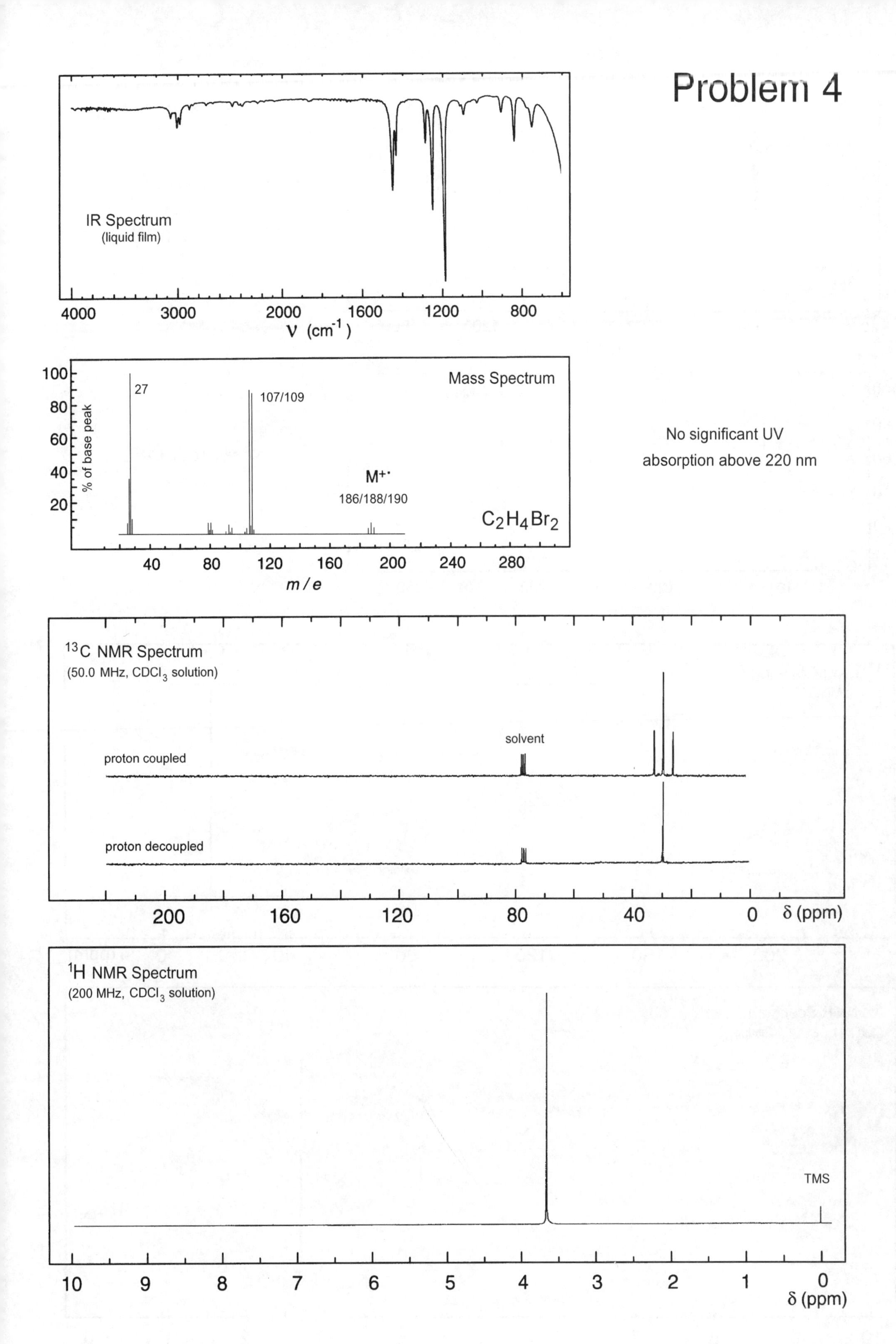

IR Spectrum
(liquid film)
4000
3000
2000
1600
1200
800
ν (cm⁻¹)
Mass Spectrum
% of base peak
100
80
60
40
20
27
107/109
M+·
186/188/190
C2H4Br2
40
80
120
160
200
240
280
m / e
No significant UV
absorption above 220 nm
13C NMR Spectrum
(50.0 MHz, CDCl3 solution)
proton coupled
proton decoupled
solvent
200
160
120
80
40
0
δ (ppm)
1H NMR Spectrum
(200 MHz, CDCl3 solution)
TMS
10
9
8
7
6
5
4
3
2
1
0
δ (ppm)

Problem 5

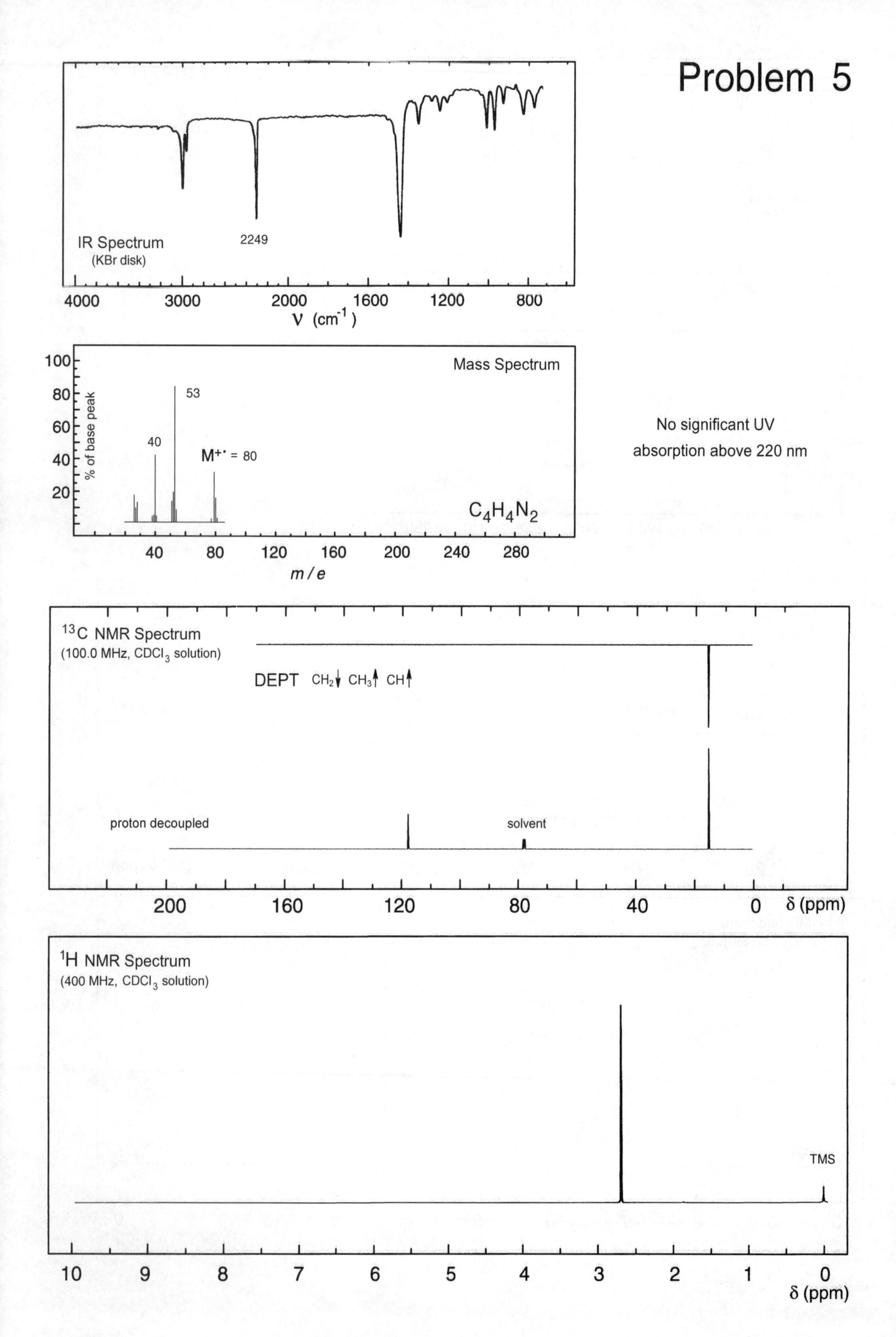

IR Spectrum
(KBr disk)
2249
4000
3000
2000
1600
1200
800
ν (cm⁻¹)
Mass Spectrum
100
80
60
40
20
% of base peak
53
40
M+· = 80
C4H4N2
40
80
120
160
200
240
280
m / e
No significant UV
absorption above 220 nm
13C NMR Spectrum
(100.0 MHz, CDCl3 solution)
DEPT CH2↓ CH3↑ CH↑
proton decoupled
solvent
200
160
120
80
40
0
δ (ppm)
1H NMR Spectrum
(400 MHz, CDCl3 solution)
TMS
10
9
8
7
6
5
4
3
2
1
0
δ (ppm)

Problem 6

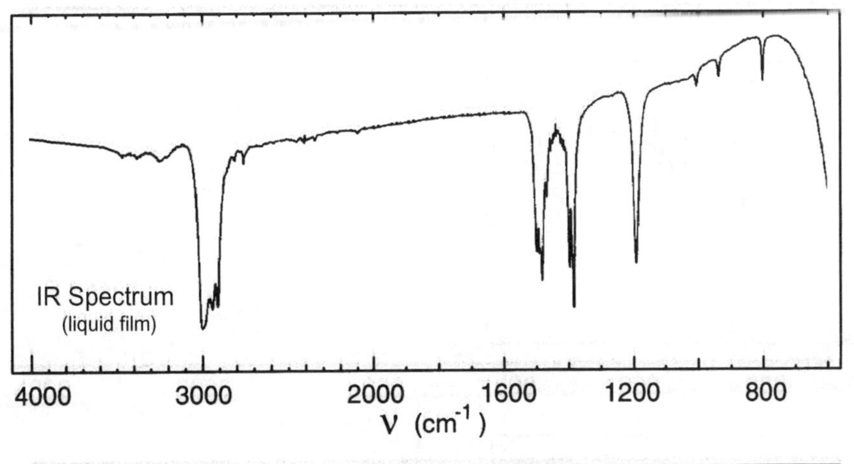

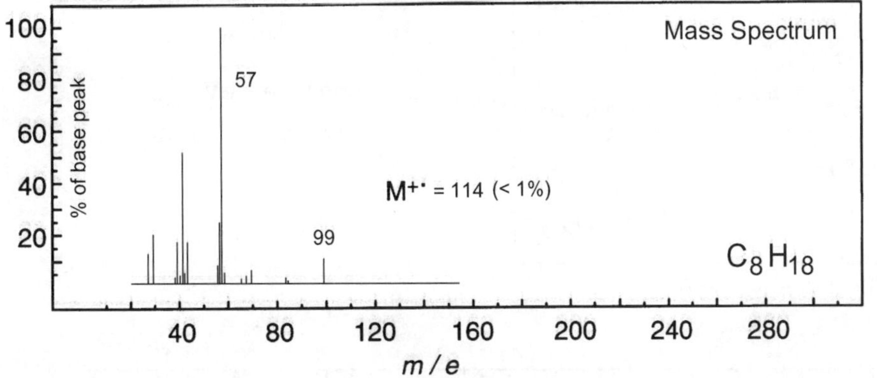

No significant UV absorption above 220 nm

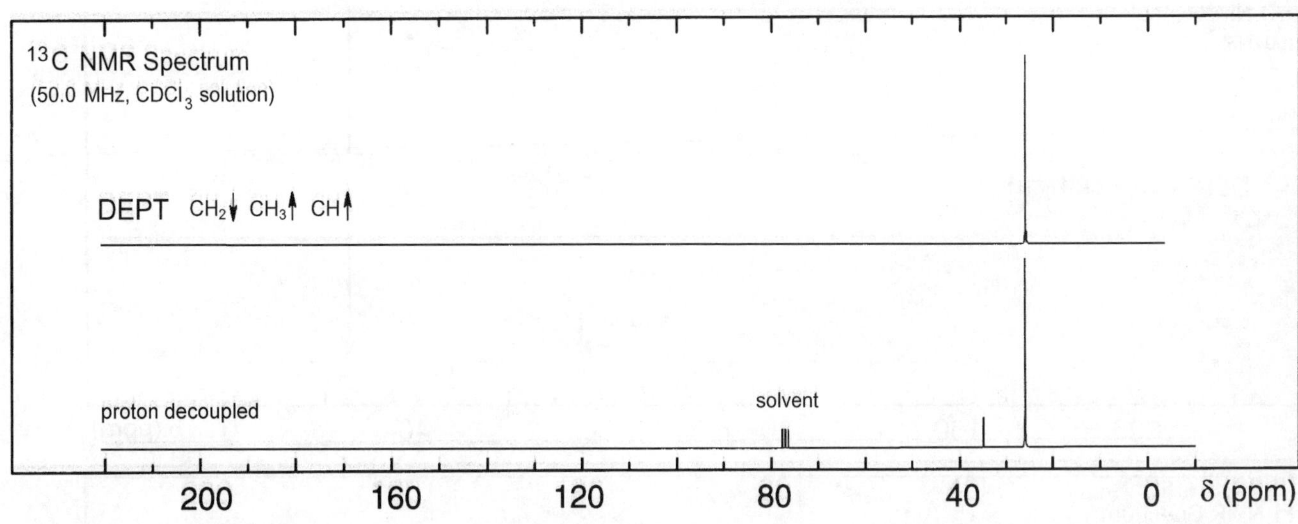

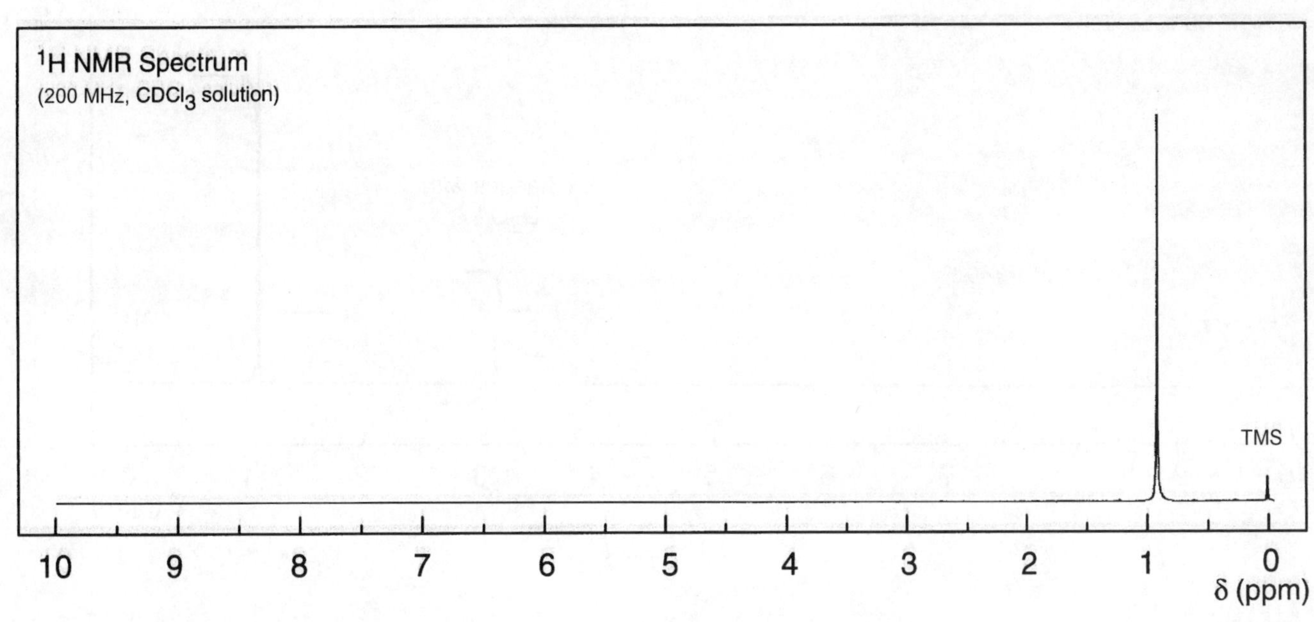

Problem 7

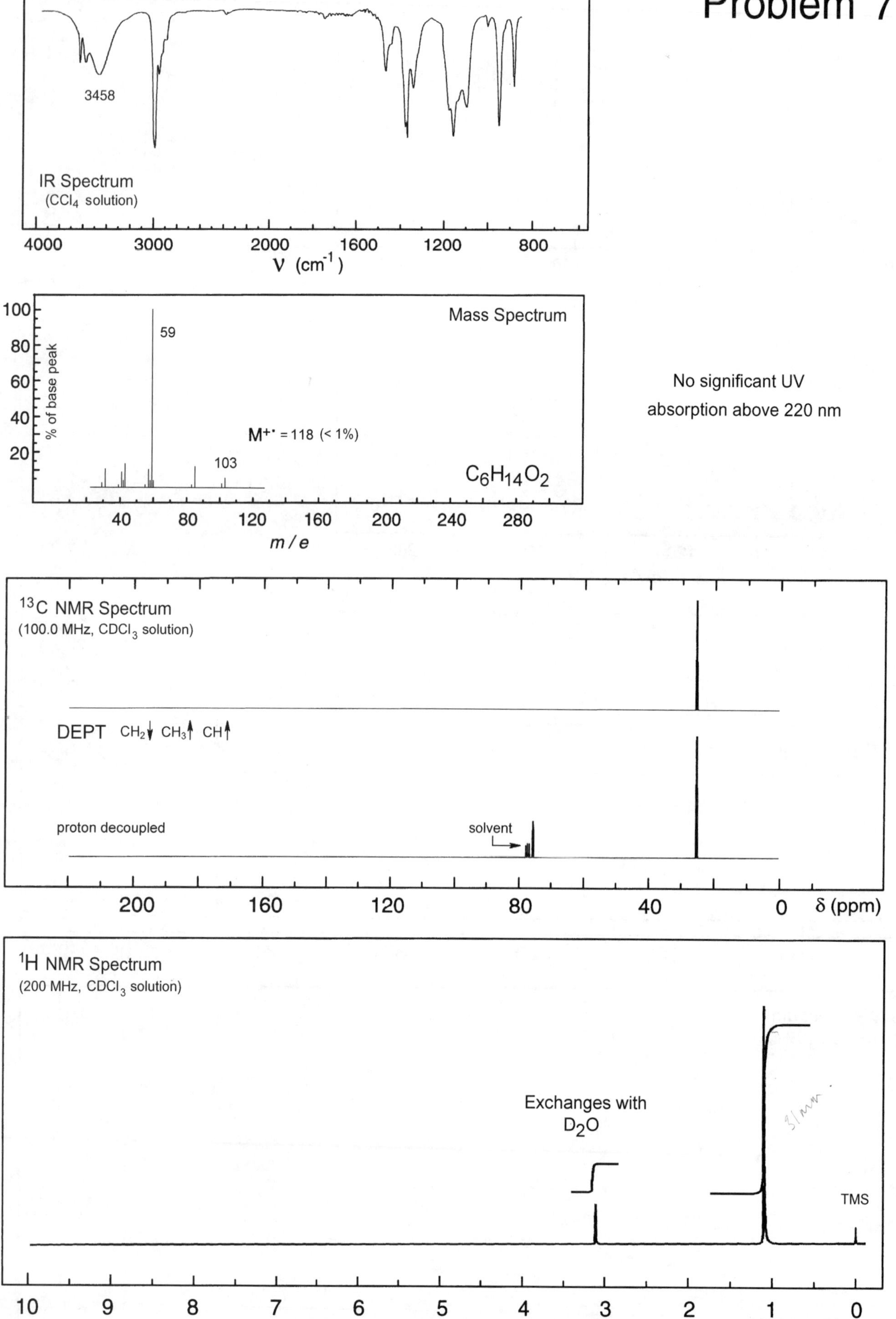
IR Spectrum
(CCl$_4$ solution)
3458
4000 3000 2000 1600 1200 800
ν (cm^{-1})
Mass Spectrum
% of base peak
100 80 60 40 20
59
103
M$^{+\cdot}$ = 118 (< 1%)
$C_6H_{14}O_2$
40 80 120 160 200 240 280
m / e
No significant UV
absorption above 220 nm
^{13}C NMR Spectrum
(100.0 MHz, CDCl$_3$ solution)
DEPT CH$_2$↓ CH$_3$↑ CH↑
proton decoupled
solvent
200 160 120 80 40 0 δ (ppm)
^{1}H NMR Spectrum
(200 MHz, CDCl$_3$ solution)
Exchanges with
D$_2$O
TMS
10 9 8 7 6 5 4 3 2 1 0
δ (ppm)

Problem 8

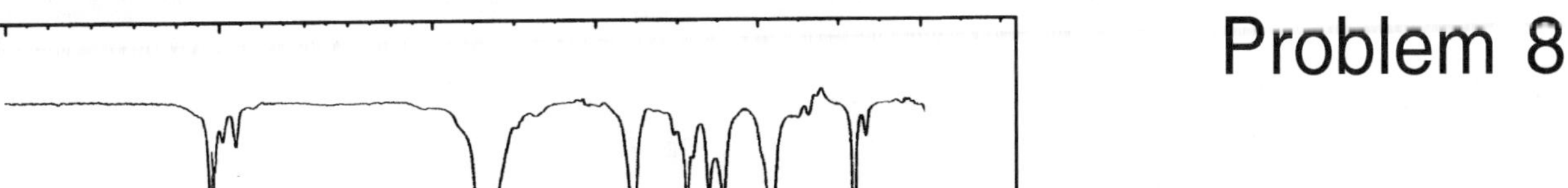

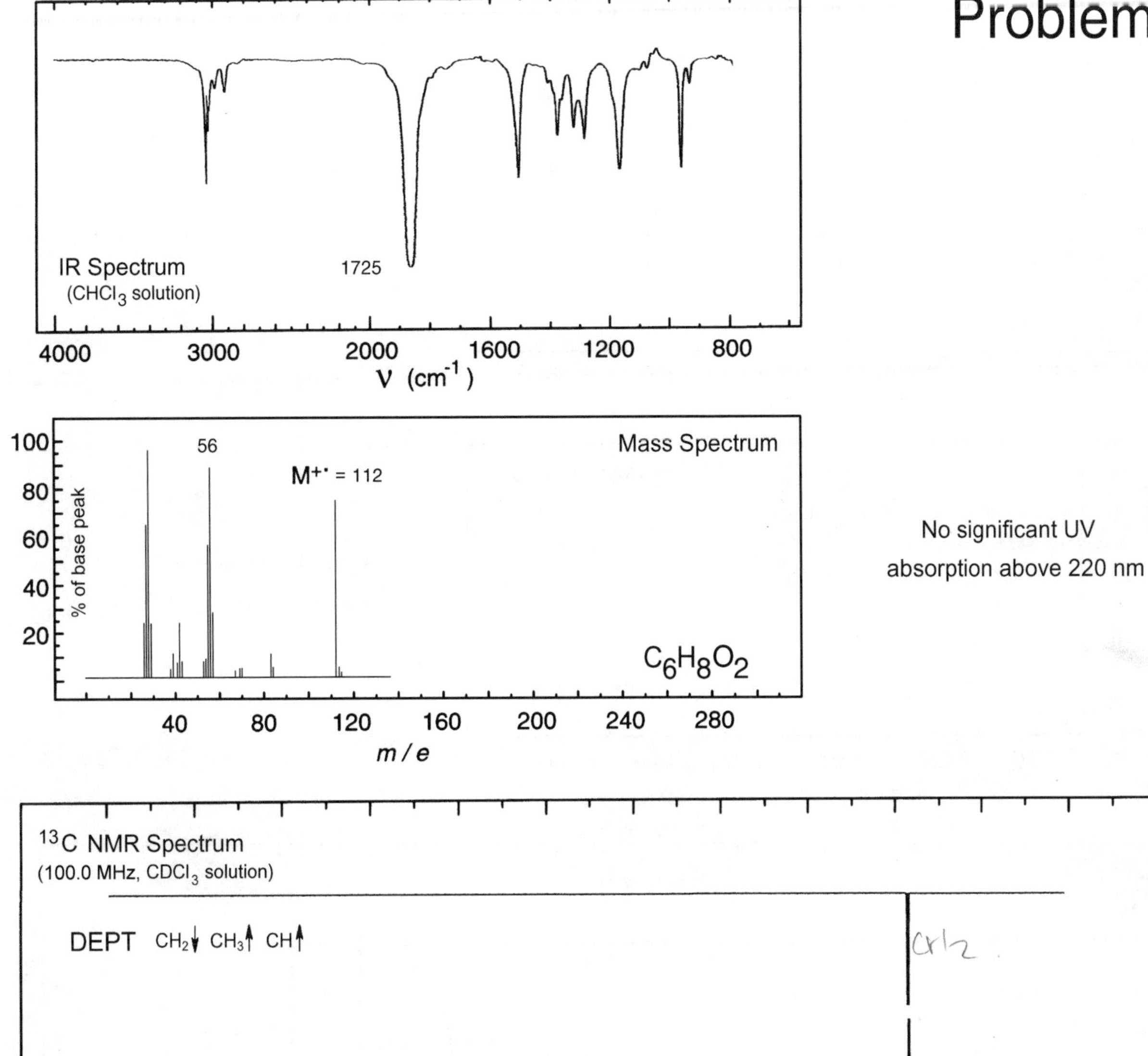

No significant UV absorption above 220 nm

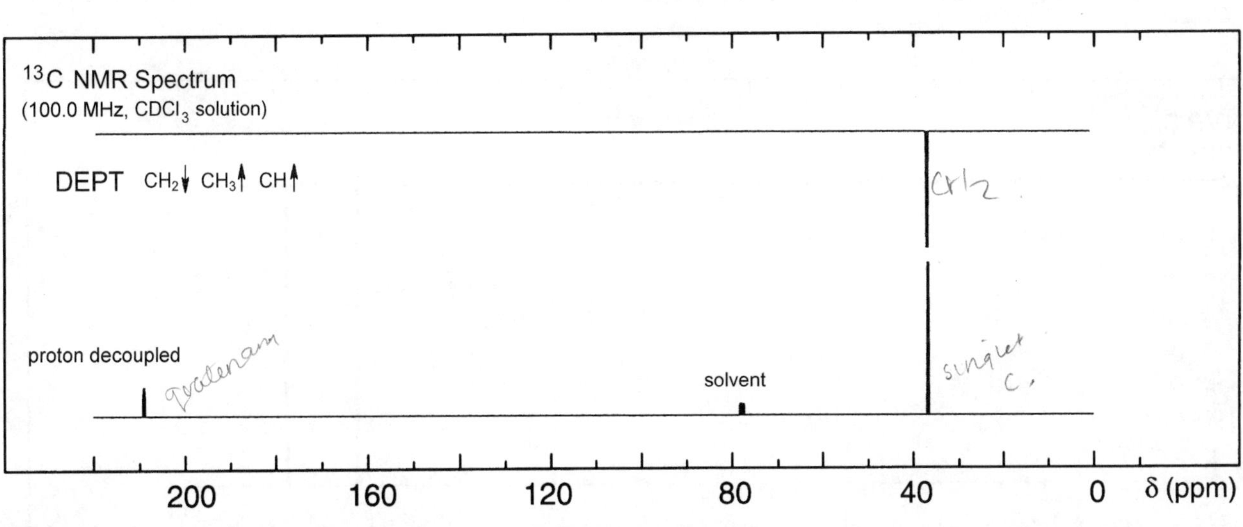

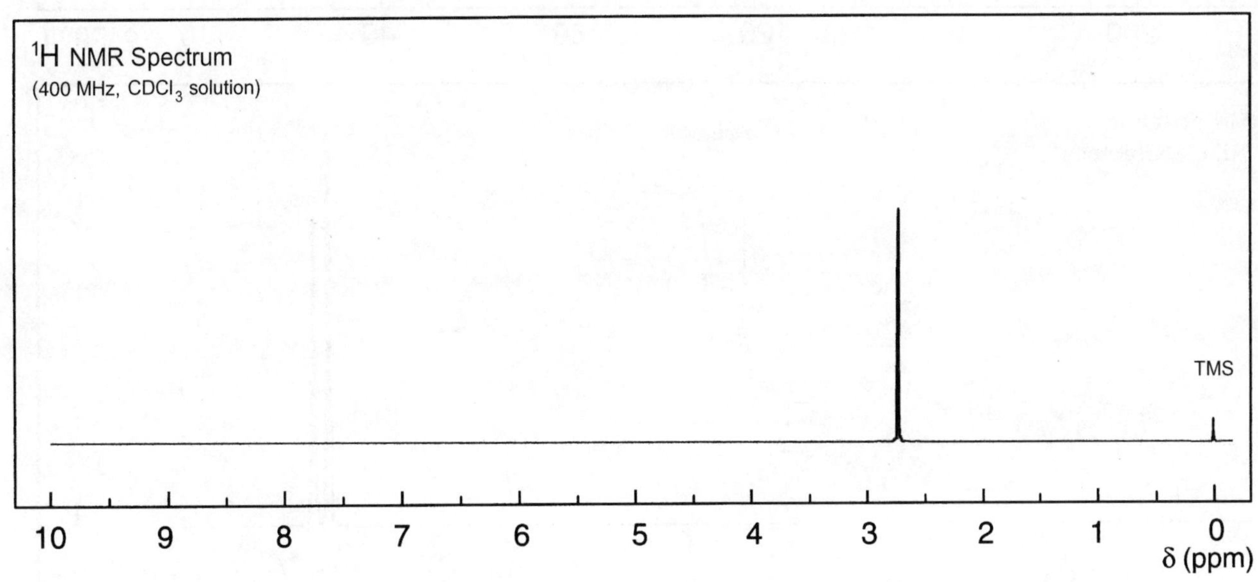

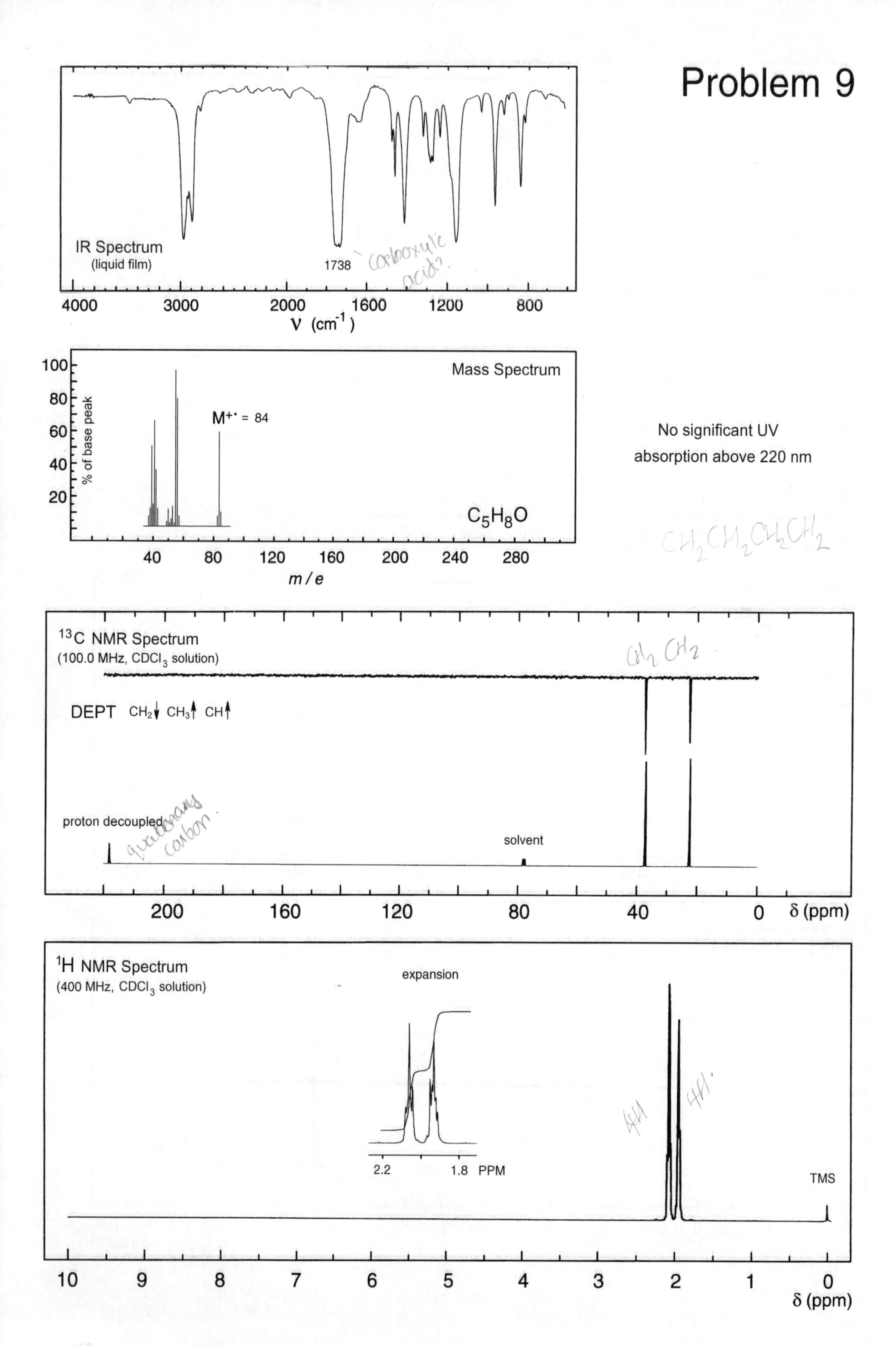
Problem 9
IR Spectrum
(liquid film)
1738
4000
3000
2000
1600
1200
800
ν (cm⁻¹)
Mass Spectrum
% of base peak
100
80
60
40
20
M+· = 84
C5H8O
40
80
120
160
200
240
280
m / e
No significant UV
absorption above 220 nm
13C NMR Spectrum
(100.0 MHz, CDCl3 solution)
DEPT CH2↓ CH3↑ CH↑
proton decoupled
solvent
200
160
120
80
40
0
δ (ppm)
1H NMR Spectrum
(400 MHz, CDCl3 solution)
expansion
2.2
1.8 PPM
TMS
10
9
8
7
6
5
4
3
2
1
0
δ (ppm)

Problem 10

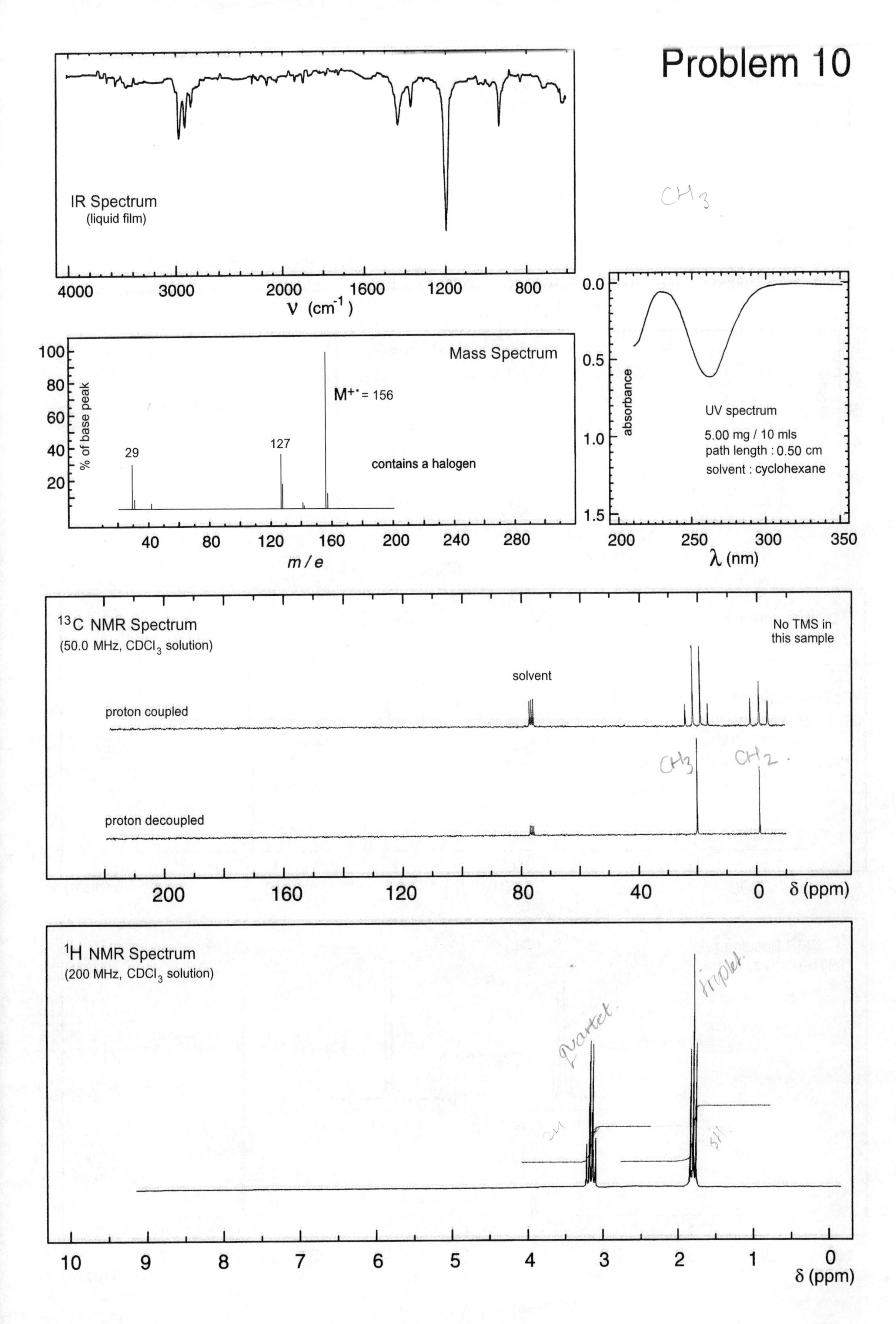
IR Spectrum
(liquid film)
4000
3000
2000
1600
1200
800
ν (cm⁻¹)
Mass Spectrum
M⁺˙ = 156
% of base peak
29
127
contains a halogen
m / e
UV spectrum
5.00 mg / 10 mls
path length : 0.50 cm
solvent : cyclohexane
absorbance
λ (nm)
13C NMR Spectrum
(50.0 MHz, CDCl3 solution)
No TMS in
this sample
solvent
proton coupled
proton decoupled
δ (ppm)
1H NMR Spectrum
(200 MHz, CDCl3 solution)
δ (ppm)

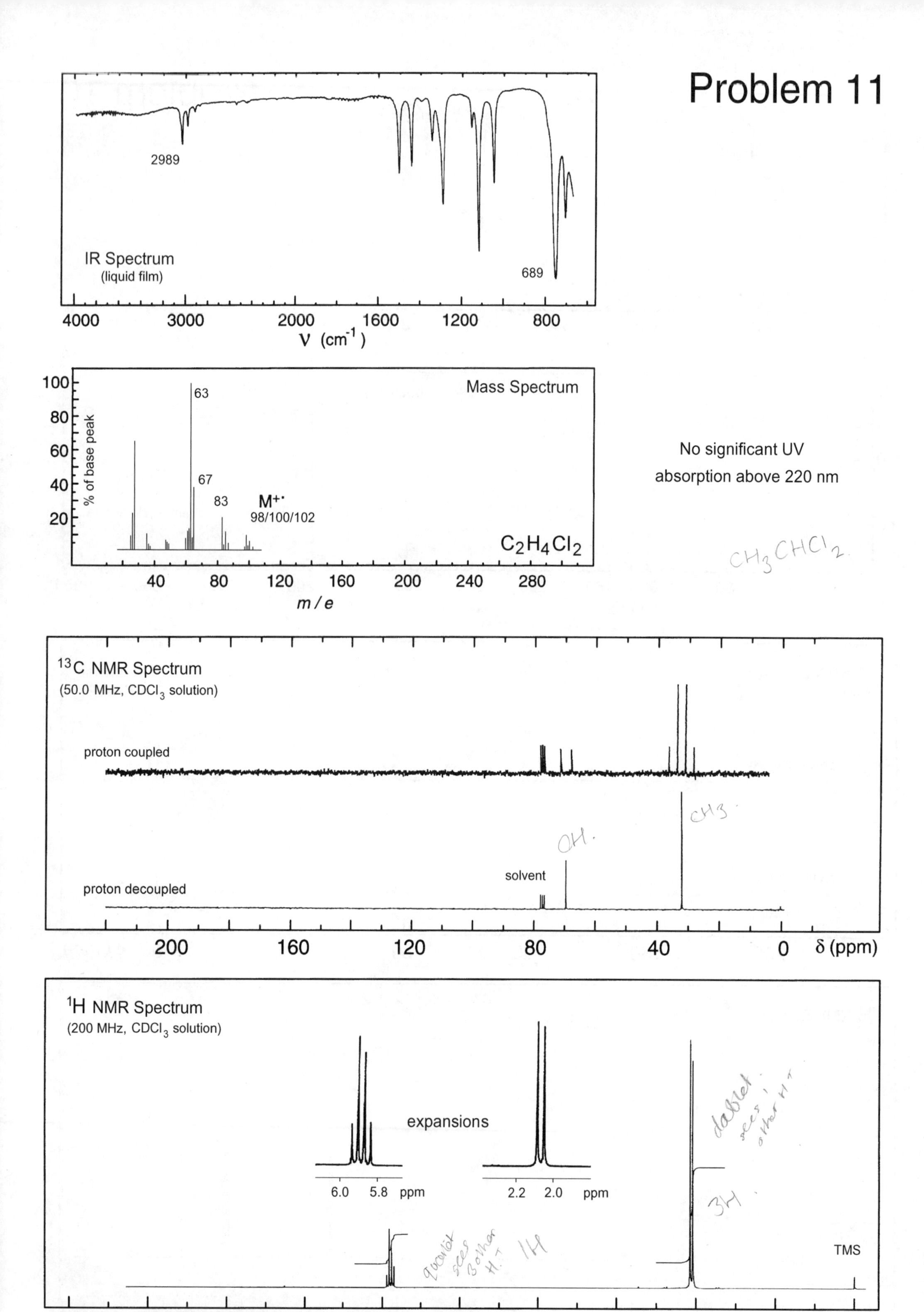
Problem 11
2989
689
IR Spectrum
(liquid film)
4000
3000
2000
1600
1200
800
ν (cm⁻¹)
Mass Spectrum
63
67
83
M+·
98/100/102
C2H4Cl2
% of base peak
m / e
No significant UV
absorption above 220 nm
CH3CHCl2
13C NMR Spectrum
(50.0 MHz, CDCl3 solution)
proton coupled
proton decoupled
solvent
CH3
δ (ppm)
1H NMR Spectrum
(200 MHz, CDCl3 solution)
expansions
ppm
doublet
3H
1H
TMS
δ (ppm)

Problem 12

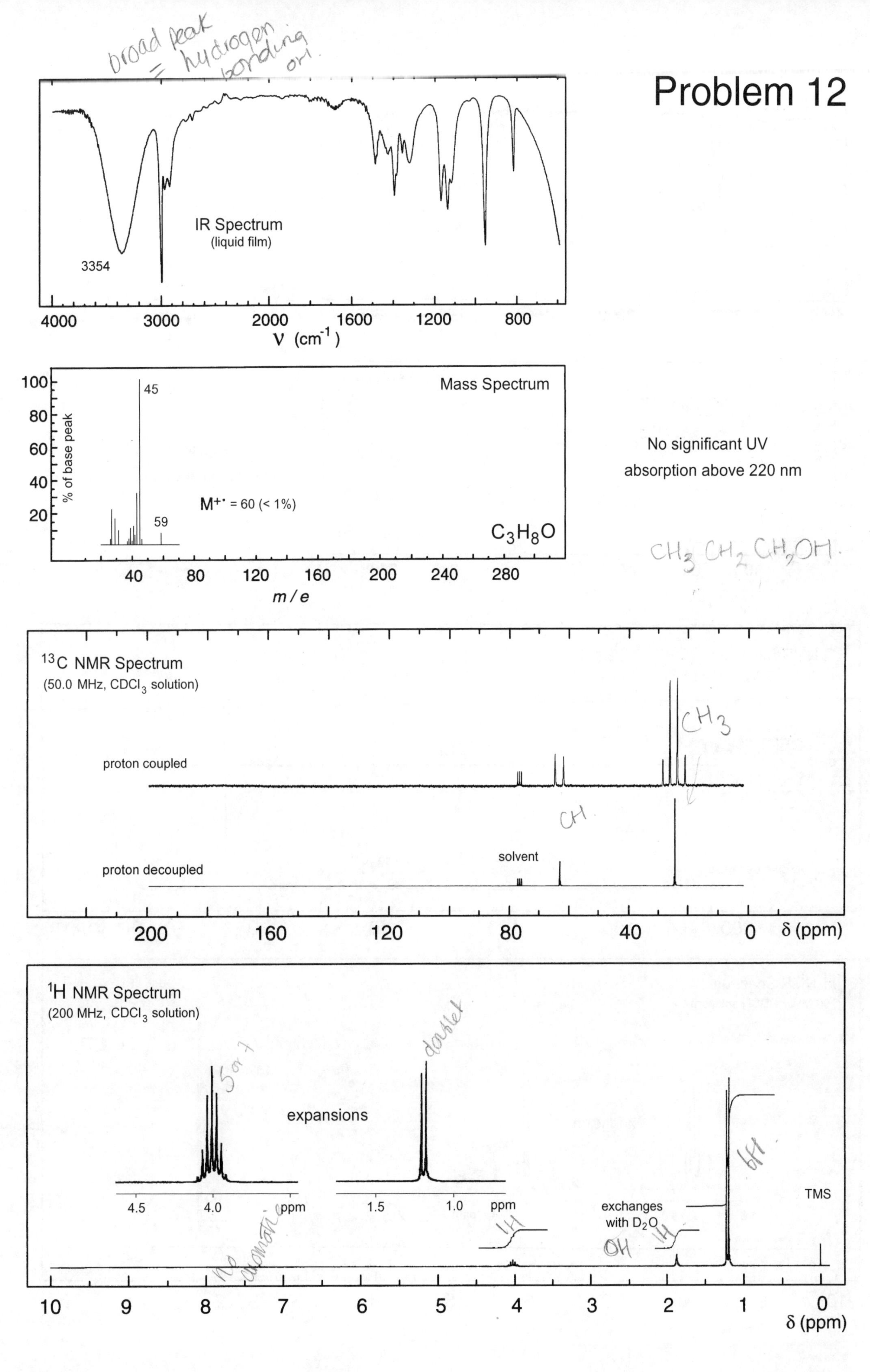

No significant UV absorption above 220 nm

Problem 13

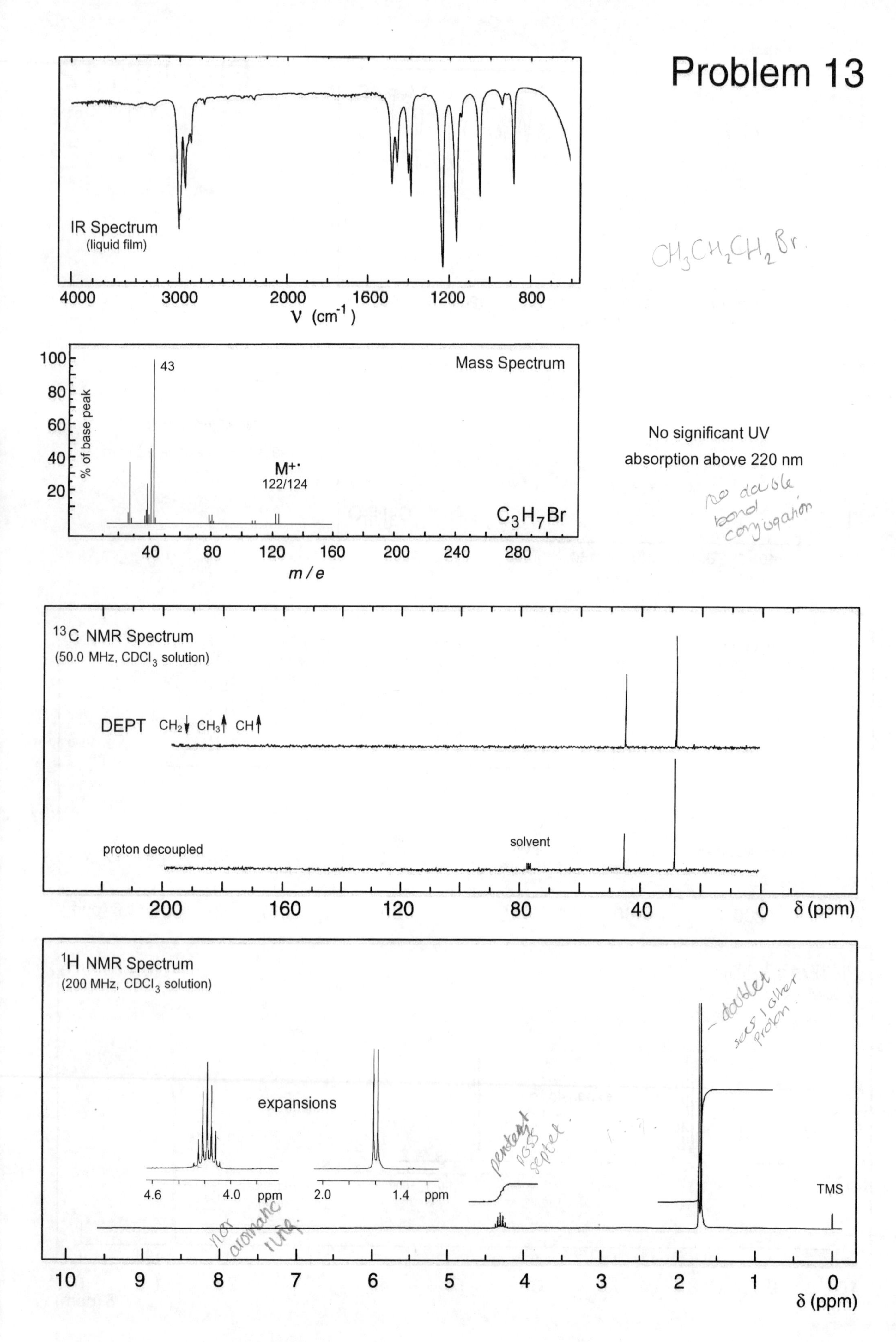

IR Spectrum
(liquid film)
4000
3000
2000
1600
1200
800
ν (cm⁻¹)
CH3CH2CH2Br.
100
80
60
40
20
% of base peak
43
Mass Spectrum
M+·
122/124
C3H7Br
40
80
120
160
200
240
280
m / e
No significant UV
absorption above 220 nm
no double
bond
conjugation
13C NMR Spectrum
(50.0 MHz, CDCl3 solution)
DEPT CH2↓ CH3↑ CH↑
proton decoupled
solvent
200
160
120
80
40
0
δ (ppm)
1H NMR Spectrum
(200 MHz, CDCl3 solution)
expansions
4.6
4.0
ppm
2.0
1.4
ppm
pentet
poss
septet.
- doublet
sees 1 other
proton.
TMS
not
aromatic
ring.
10
9
8
7
6
5
4
3
2
1
0
δ (ppm)

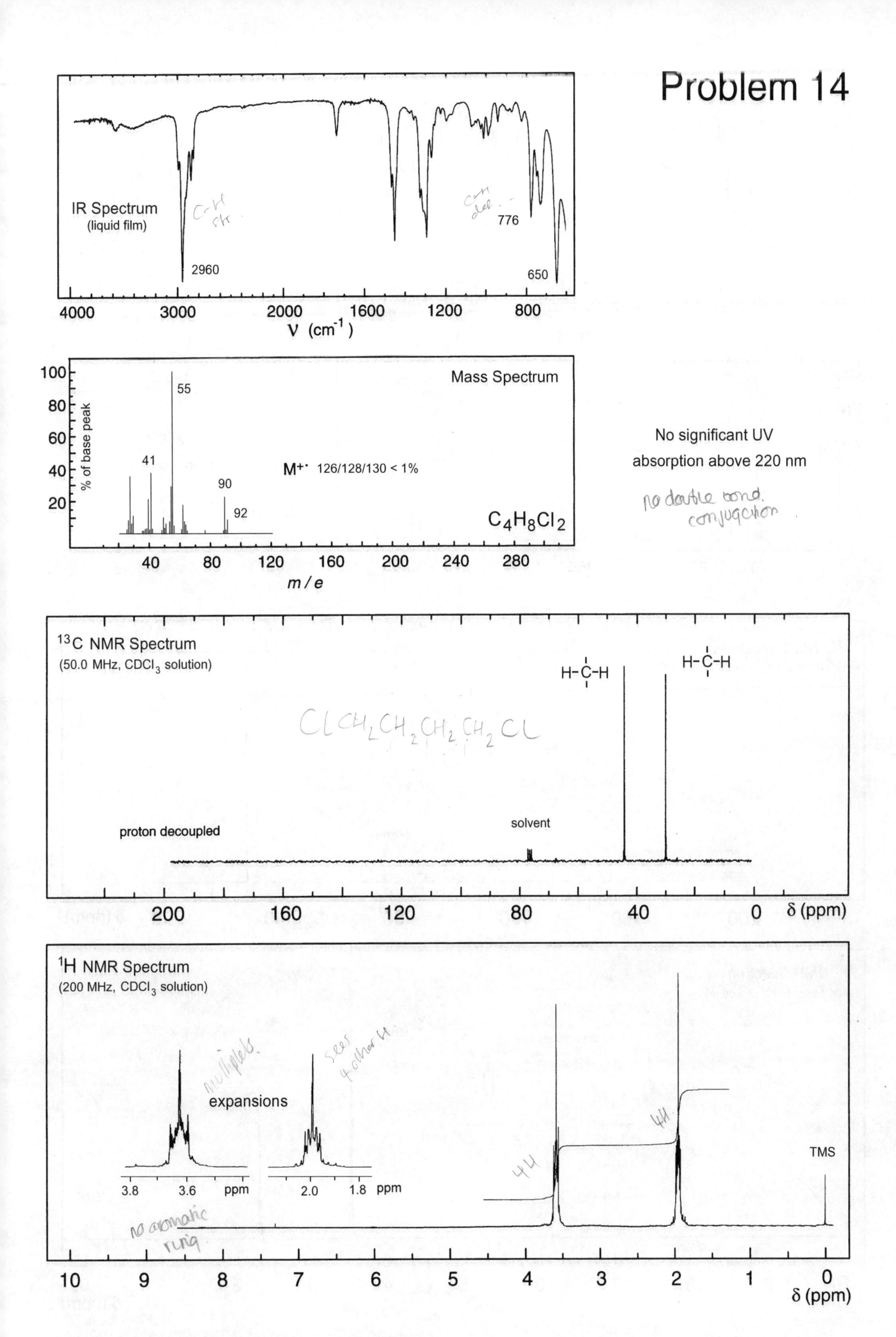
Problem 14
IR Spectrum
(liquid film)
2960
776
650
4000
3000
2000
1600
1200
800
ν (cm⁻¹)
Mass Spectrum
% of base peak
100
80
60
40
20
55
41
90
92
M+· 126/128/130 < 1%
C4H8Cl2
40
80
120
160
200
240
280
m / e
No significant UV
absorption above 220 nm
13C NMR Spectrum
(50.0 MHz, CDCl3 solution)
H-C-H
H-C-H
proton decoupled
solvent
200
160
120
80
40
0
δ (ppm)
1H NMR Spectrum
(200 MHz, CDCl3 solution)
expansions
3.8
3.6
ppm
2.0
1.8
ppm
TMS
10
9
8
7
6
5
4
3
2
1
0
δ (ppm)

Problem 15

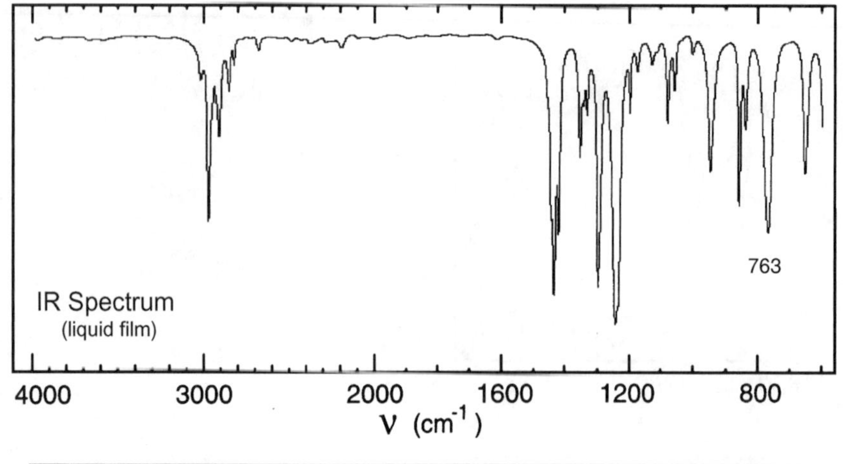

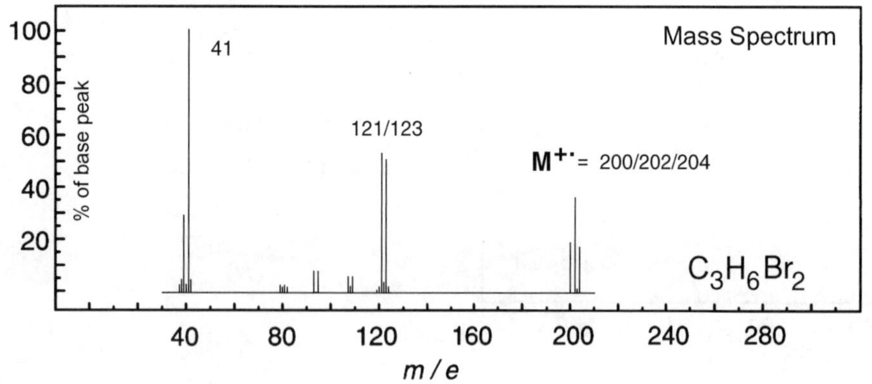

No significant UV absorption above 220 nm

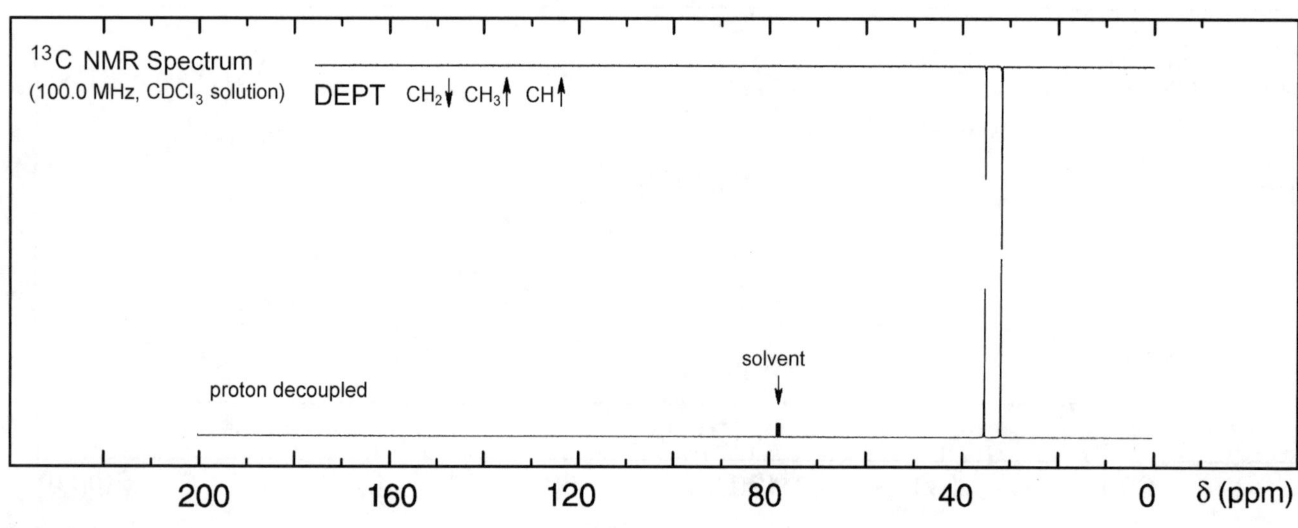

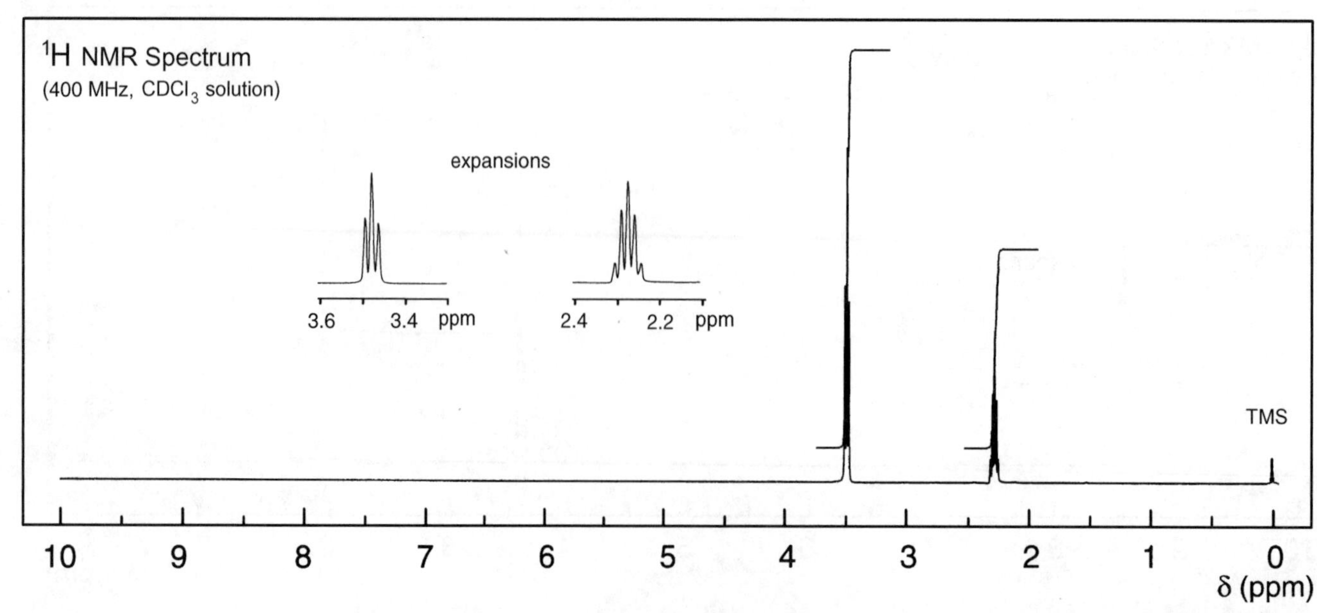

Problem 16

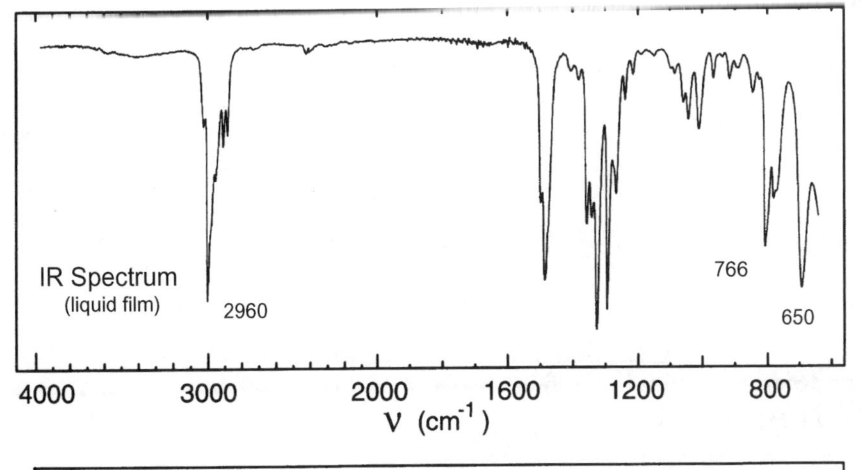

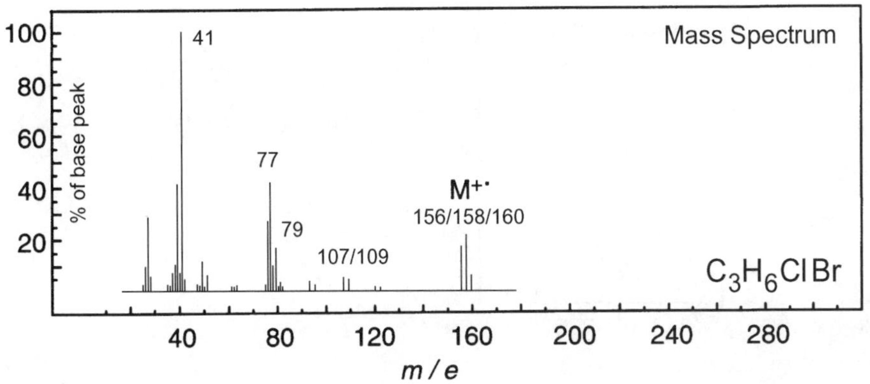

No significant UV absorption above 220 nm

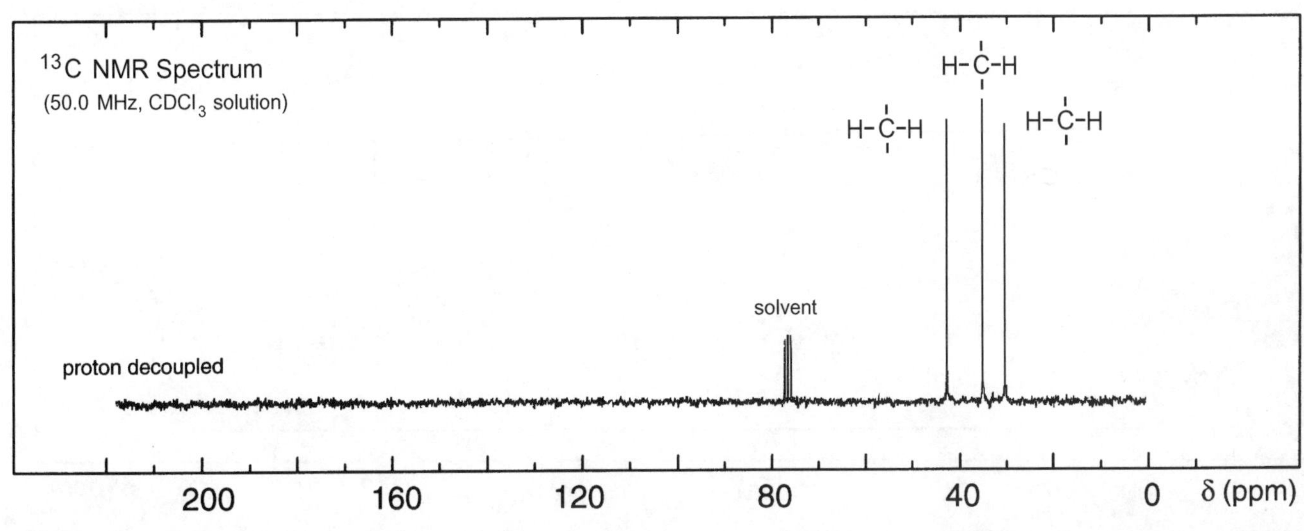

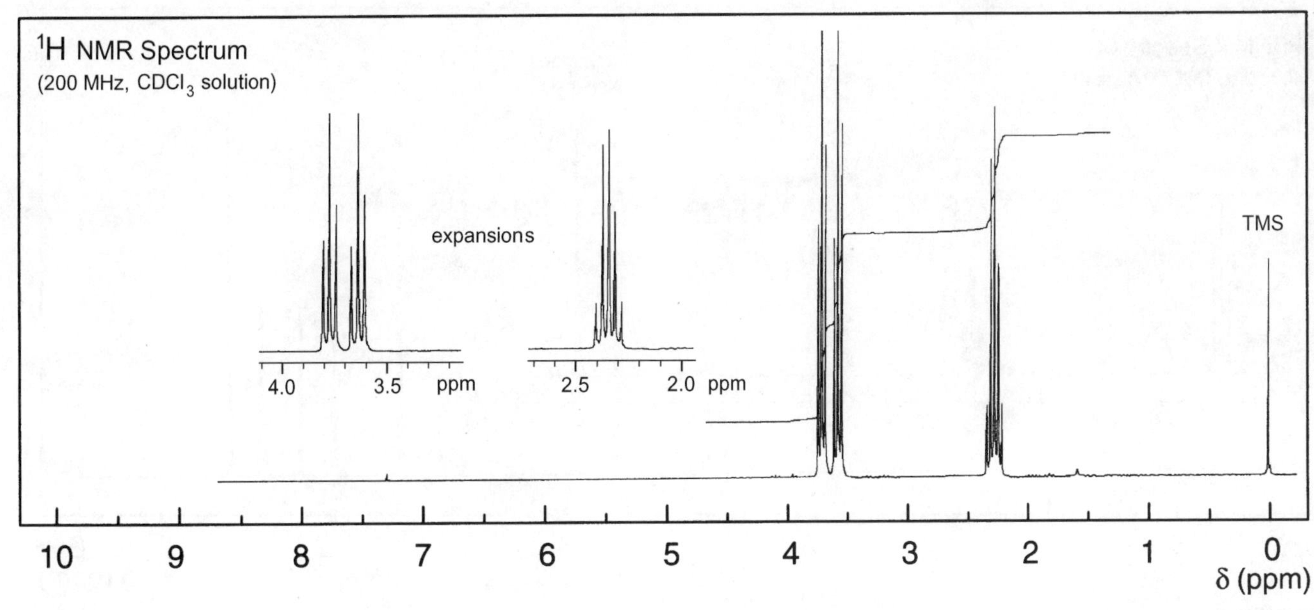

Problem 17

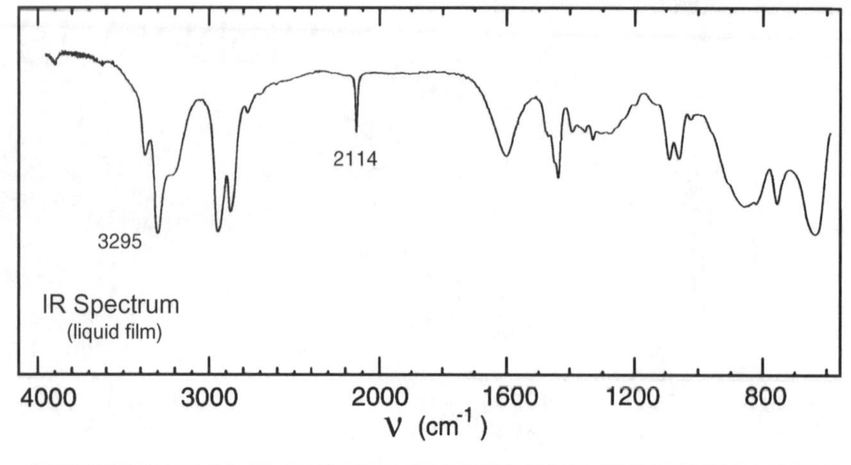

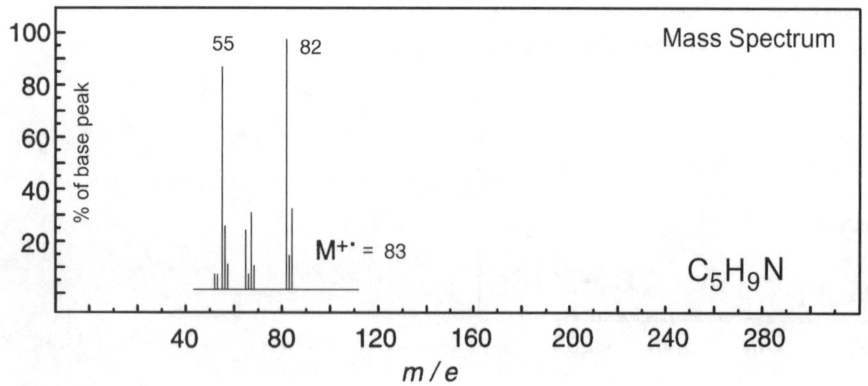

No significant UV absorption above 220 nm

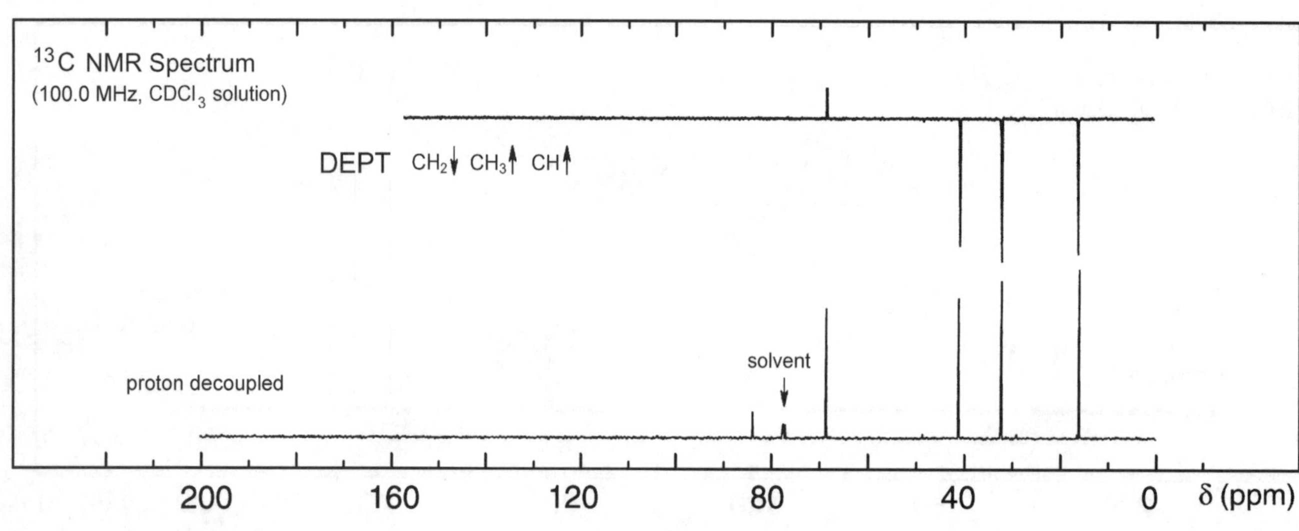

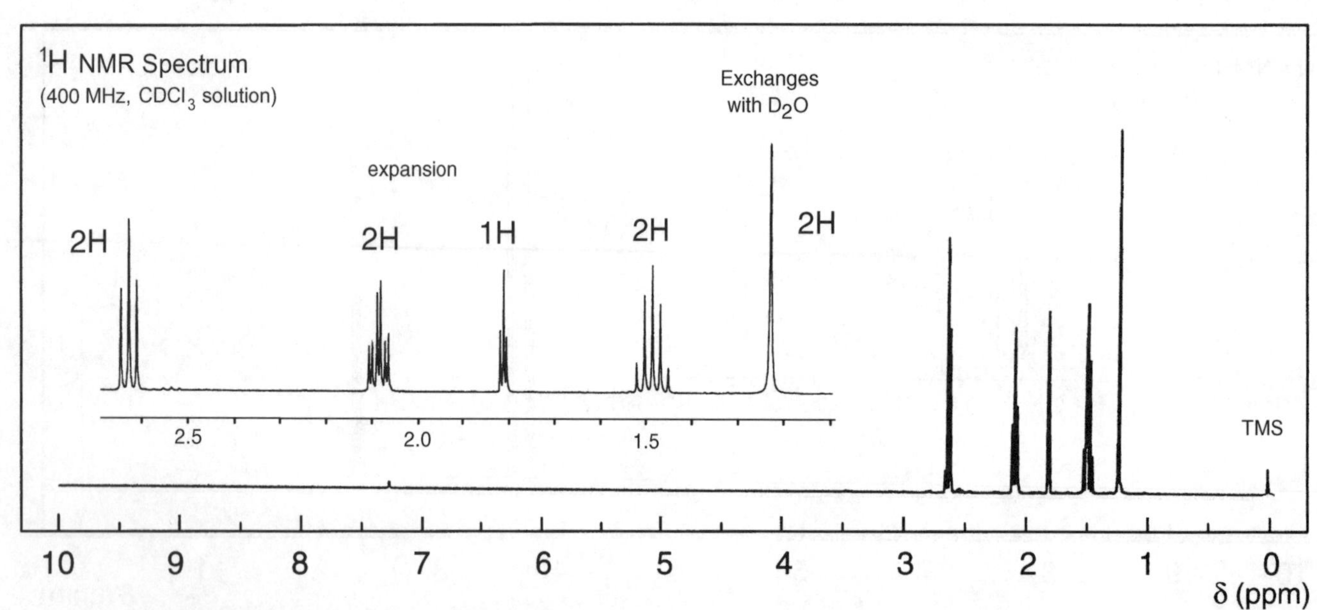

Problem 18

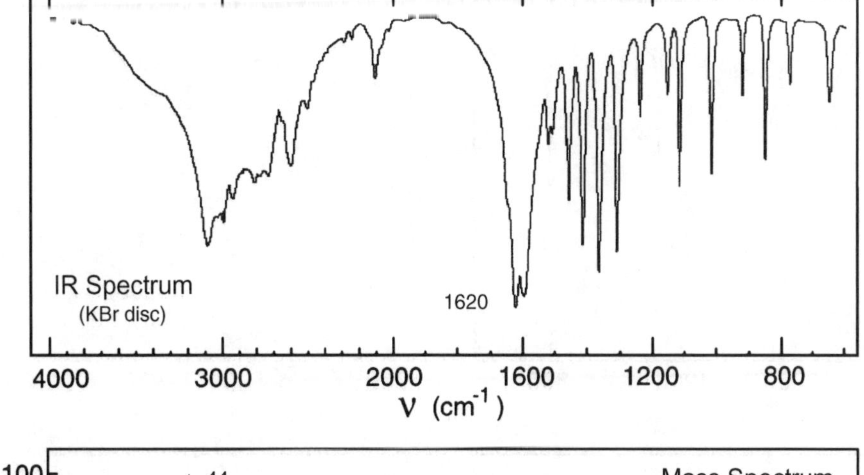

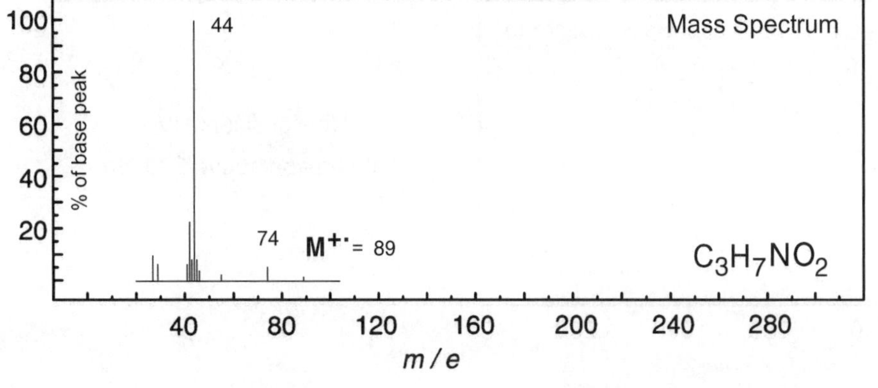

No significant UV absorption above 220 nm

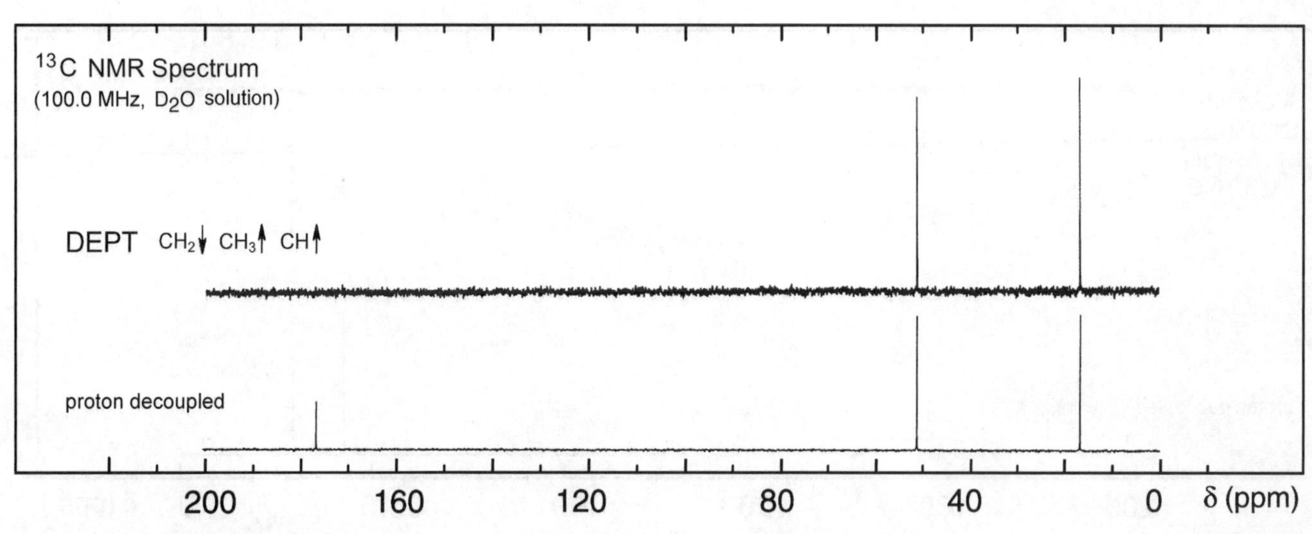

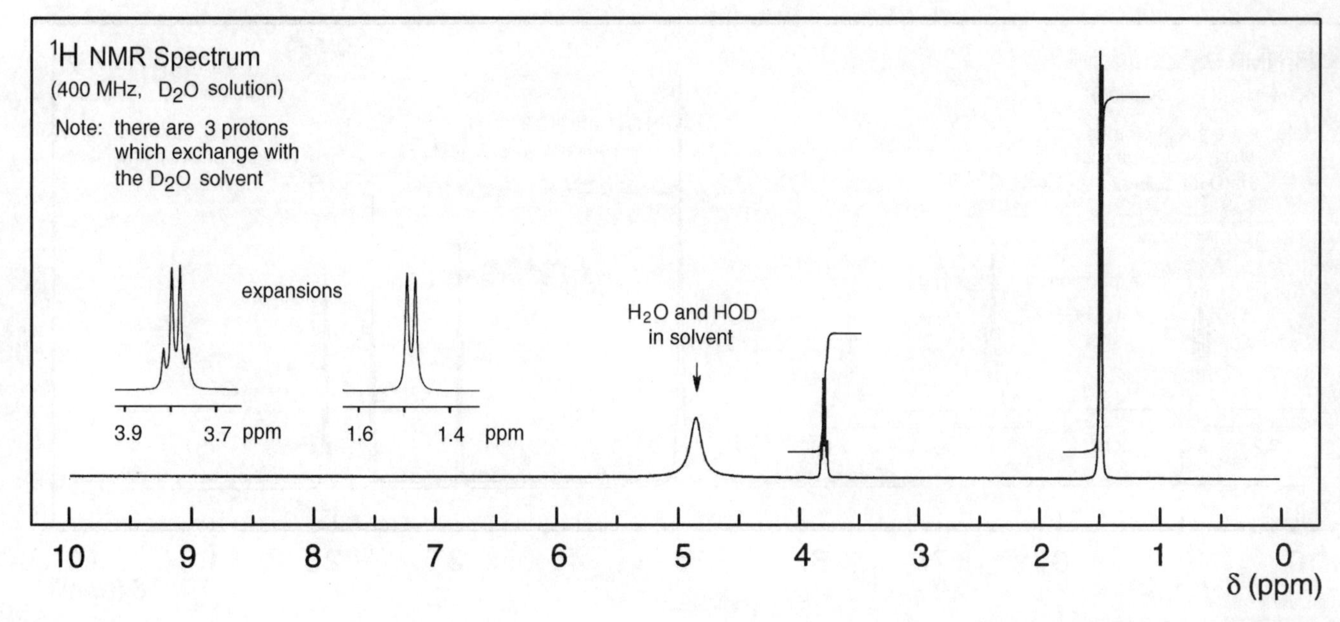

Problem 19

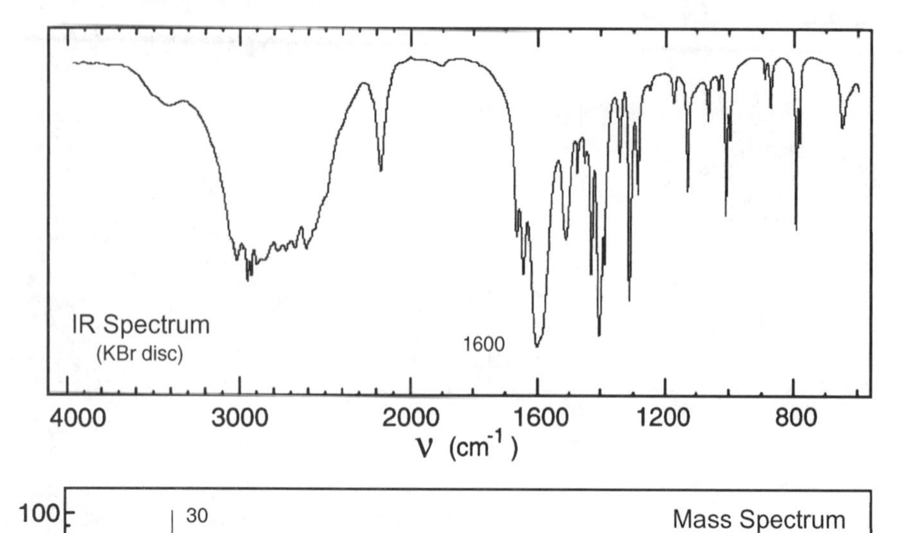

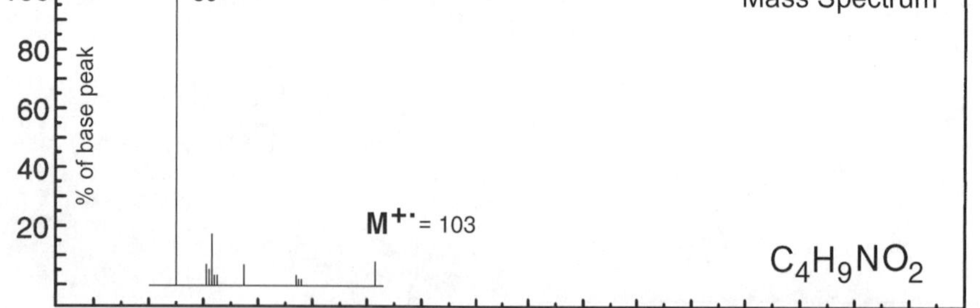

No significant UV absorption above 220 nm

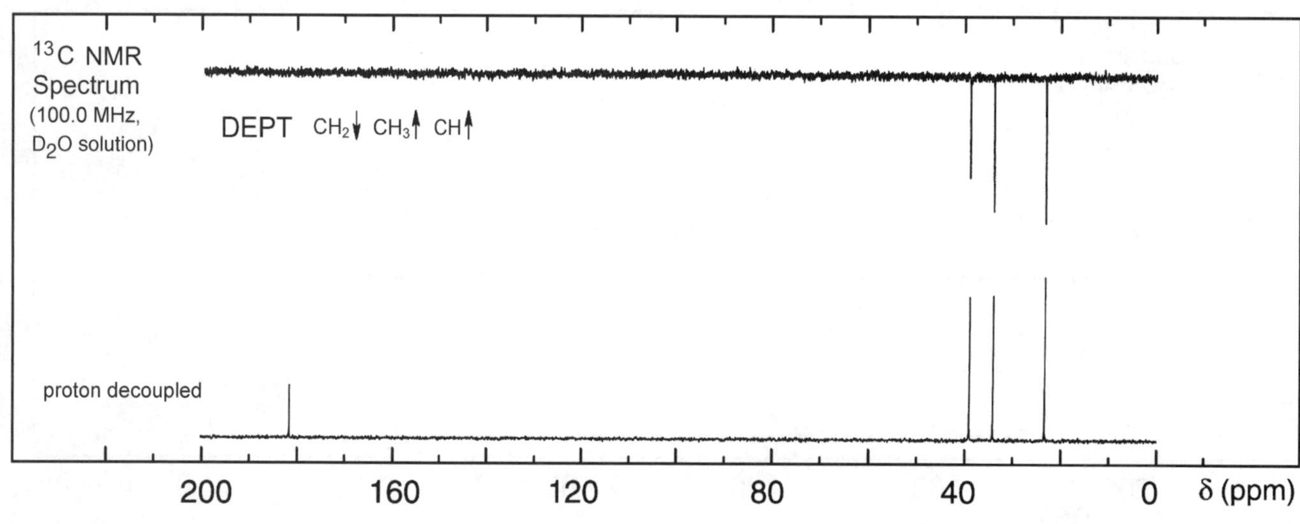

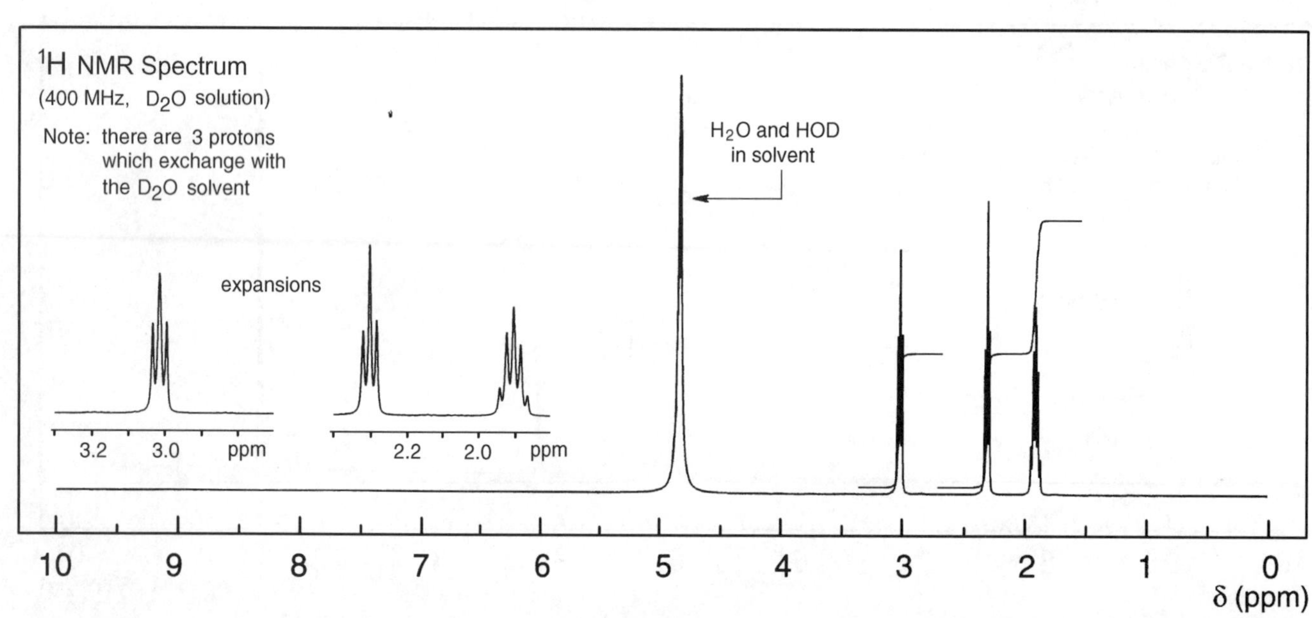

Problem 20

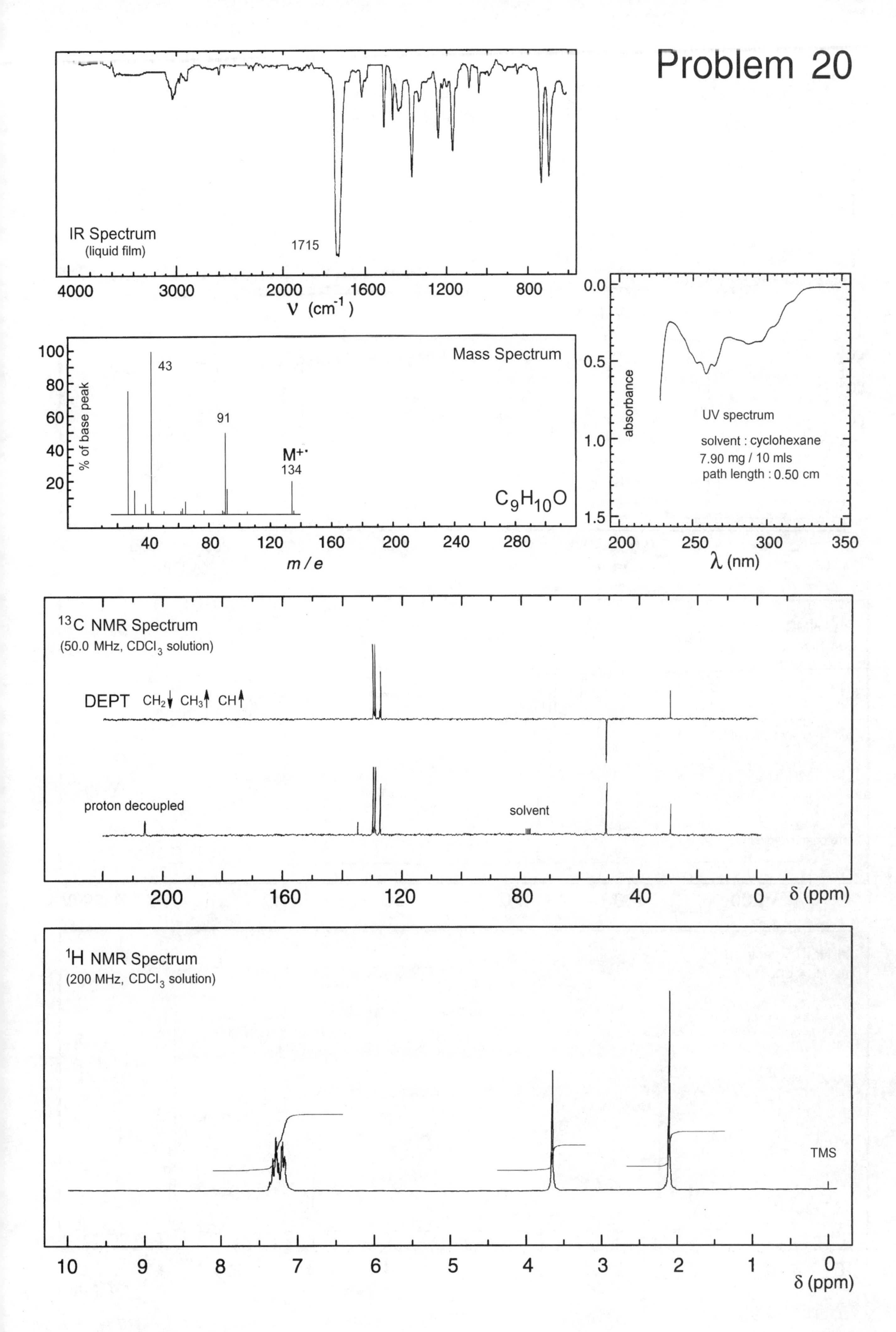

IR Spectrum
(liquid film)
1715
4000
3000
2000
1600
1200
800
ν (cm⁻¹)
Mass Spectrum
% of base peak
100
80
60
40
20
43
91
M+·
134
C9H10O
40
80
120
160
200
240
280
m / e
0.0
0.5
1.0
1.5
absorbance
UV spectrum
solvent : cyclohexane
7.90 mg / 10 mls
path length : 0.50 cm
200
250
300
350
λ (nm)
13C NMR Spectrum
(50.0 MHz, CDCl3 solution)
DEPT CH2↓ CH3↑ CH↑
proton decoupled
solvent
200
160
120
80
40
0
δ (ppm)
1H NMR Spectrum
(200 MHz, CDCl3 solution)
TMS
10
9
8
7
6
5
4
3
2
1
0
δ (ppm)

Problem 21

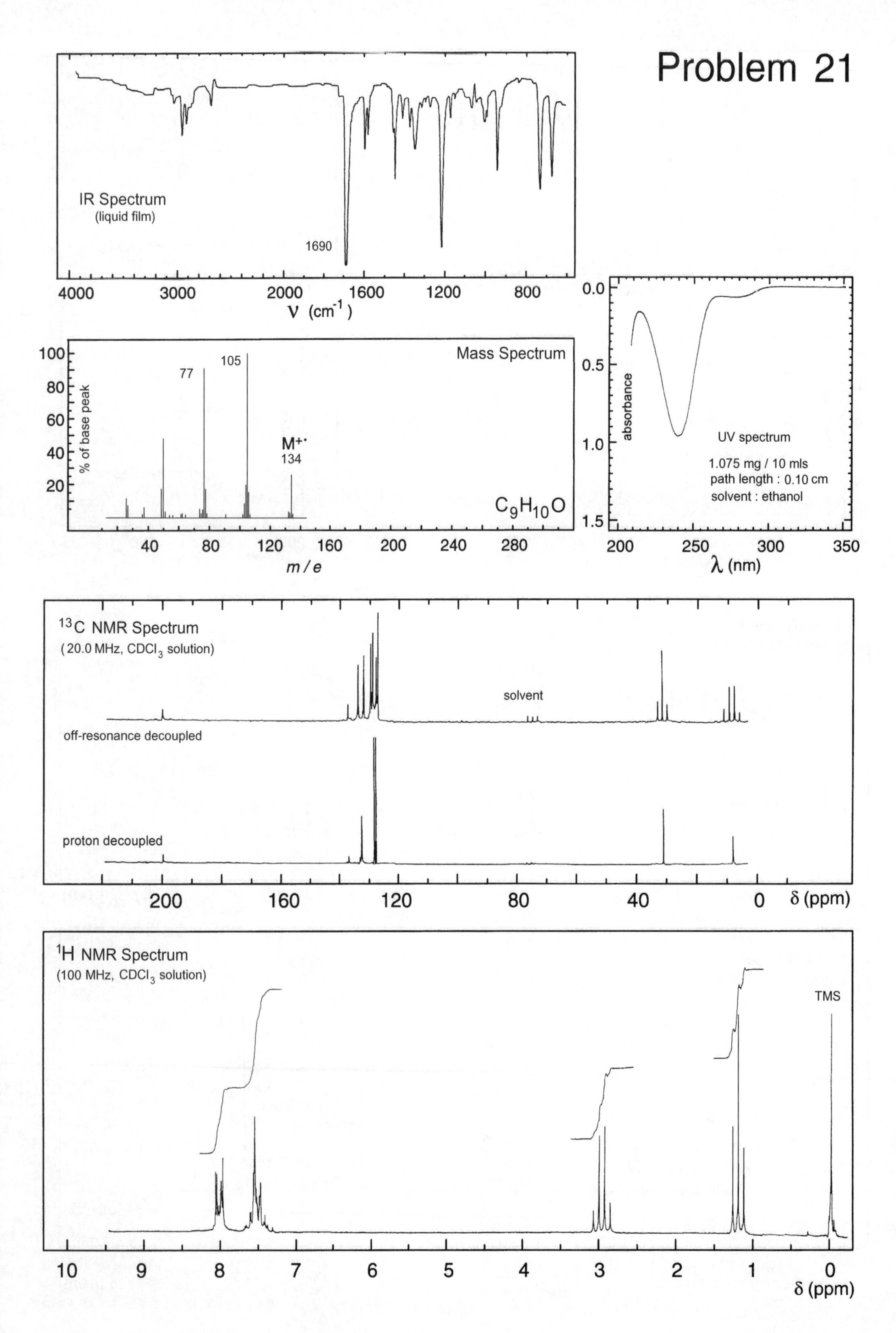

IR Spectrum
(liquid film)
1690
4000
3000
2000
1600
1200
800
ν (cm⁻¹)
Mass Spectrum
% of base peak
100
80
60
40
20
77
105
M+·
134
$C_9H_{10}O$
40
80
120
160
200
240
280
m / e
0.0
0.5
1.0
1.5
absorbance
UV spectrum
1.075 mg / 10 mls
path length : 0.10 cm
solvent : ethanol
200
250
300
350
λ (nm)
¹³C NMR Spectrum
(20.0 MHz, CDCl₃ solution)
solvent
off-resonance decoupled
proton decoupled
200
160
120
80
40
0
δ (ppm)
¹H NMR Spectrum
(100 MHz, CDCl₃ solution)
TMS
10
9
8
7
6
5
4
3
2
1
0
δ (ppm)

Problem 22

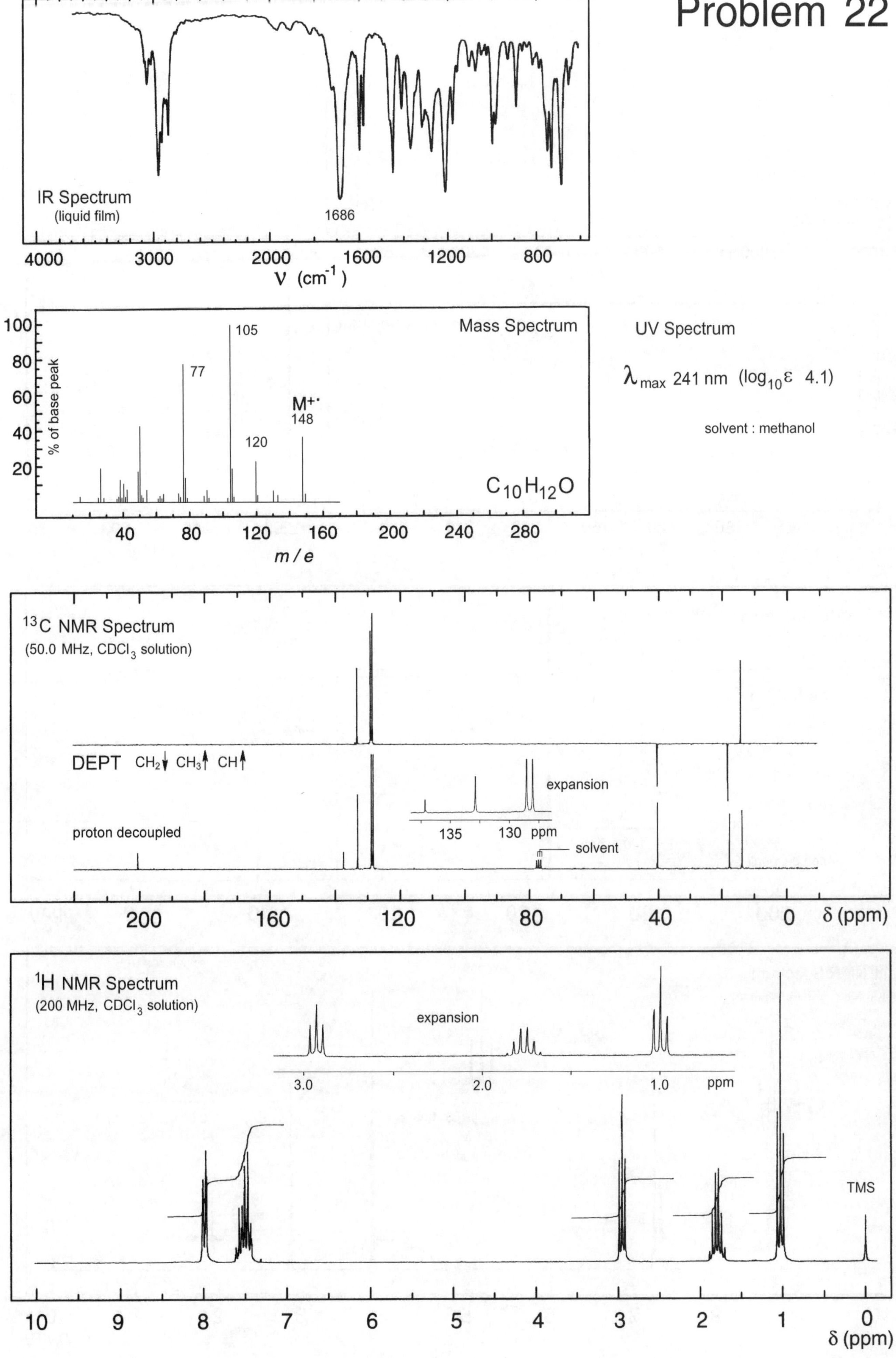
IR Spectrum
(liquid film)
1686
4000
3000
2000
1600
1200
800
ν (cm⁻¹)
Mass Spectrum
100
80
60
40
20
% of base peak
105
77
M+·
148
120
$C_{10}H_{12}O$
40
80
120
160
200
240
280
m / e
UV Spectrum
λ_{max} 241 nm ($\log_{10}\varepsilon$ 4.1)
solvent : methanol
^{13}C NMR Spectrum
(50.0 MHz, $CDCl_3$ solution)
DEPT CH_2↓ CH_3↑ CH↑
expansion
135
130 ppm
proton decoupled
solvent
200
160
120
80
40
0
δ (ppm)
^1H NMR Spectrum
(200 MHz, $CDCl_3$ solution)
expansion
3.0
2.0
1.0
ppm
TMS
10
9
8
7
6
5
4
3
2
1
0
δ (ppm)

Problem 23

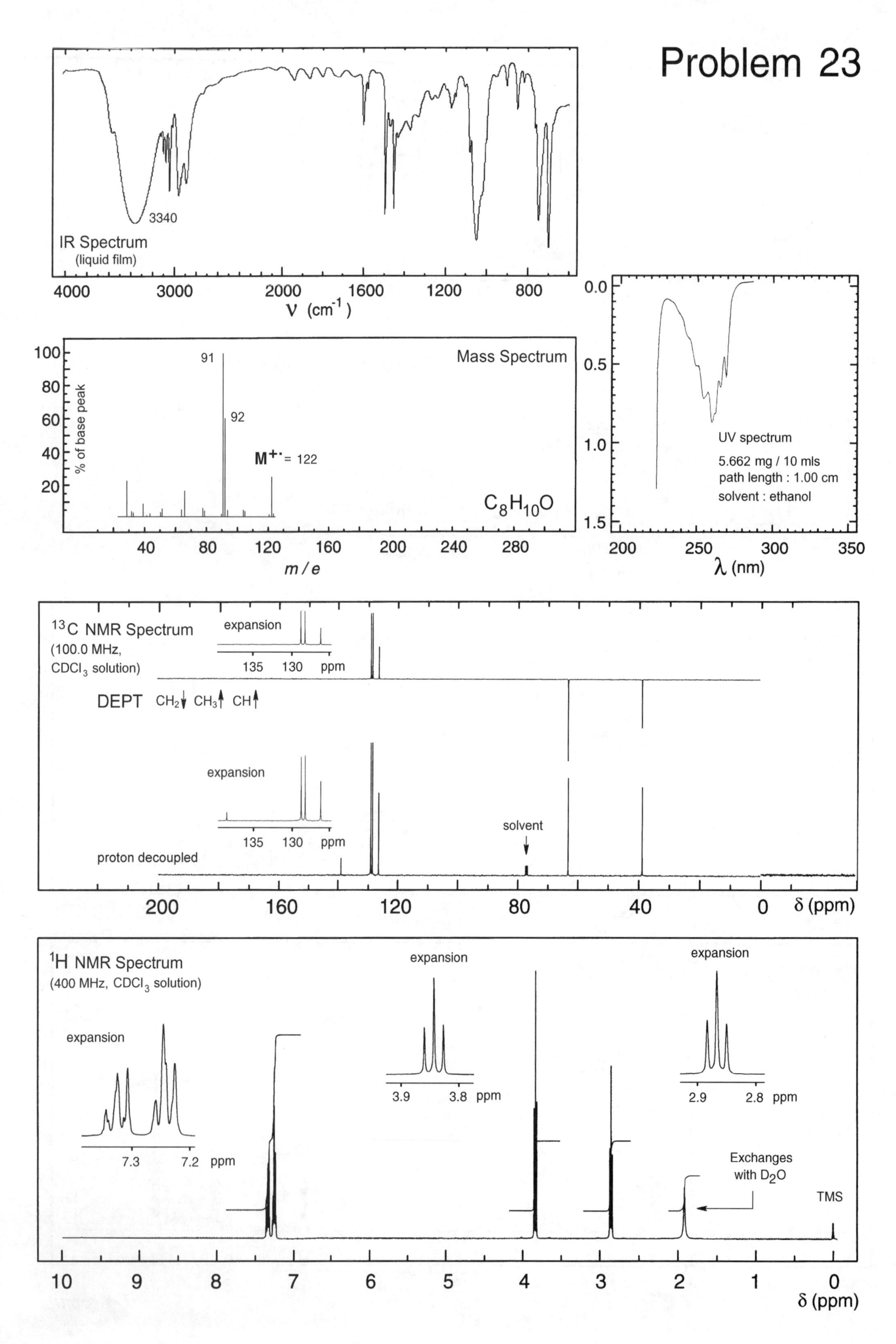

3340
IR Spectrum
(liquid film)
4000
3000
2000
1600
1200
800
ν (cm⁻¹)
Mass Spectrum
% of base peak
91
92
M+· = 122
C8H10O
m / e
UV spectrum
5.662 mg / 10 mls
path length : 1.00 cm
solvent : ethanol
λ (nm)
13C NMR Spectrum
(100.0 MHz,
CDCl3 solution)
expansion
DEPT CH2 CH3 CH
solvent
proton decoupled
δ (ppm)
1H NMR Spectrum
(400 MHz, CDCl3 solution)
expansion
ppm
Exchanges
with D2O
TMS
δ (ppm)

Problem 24

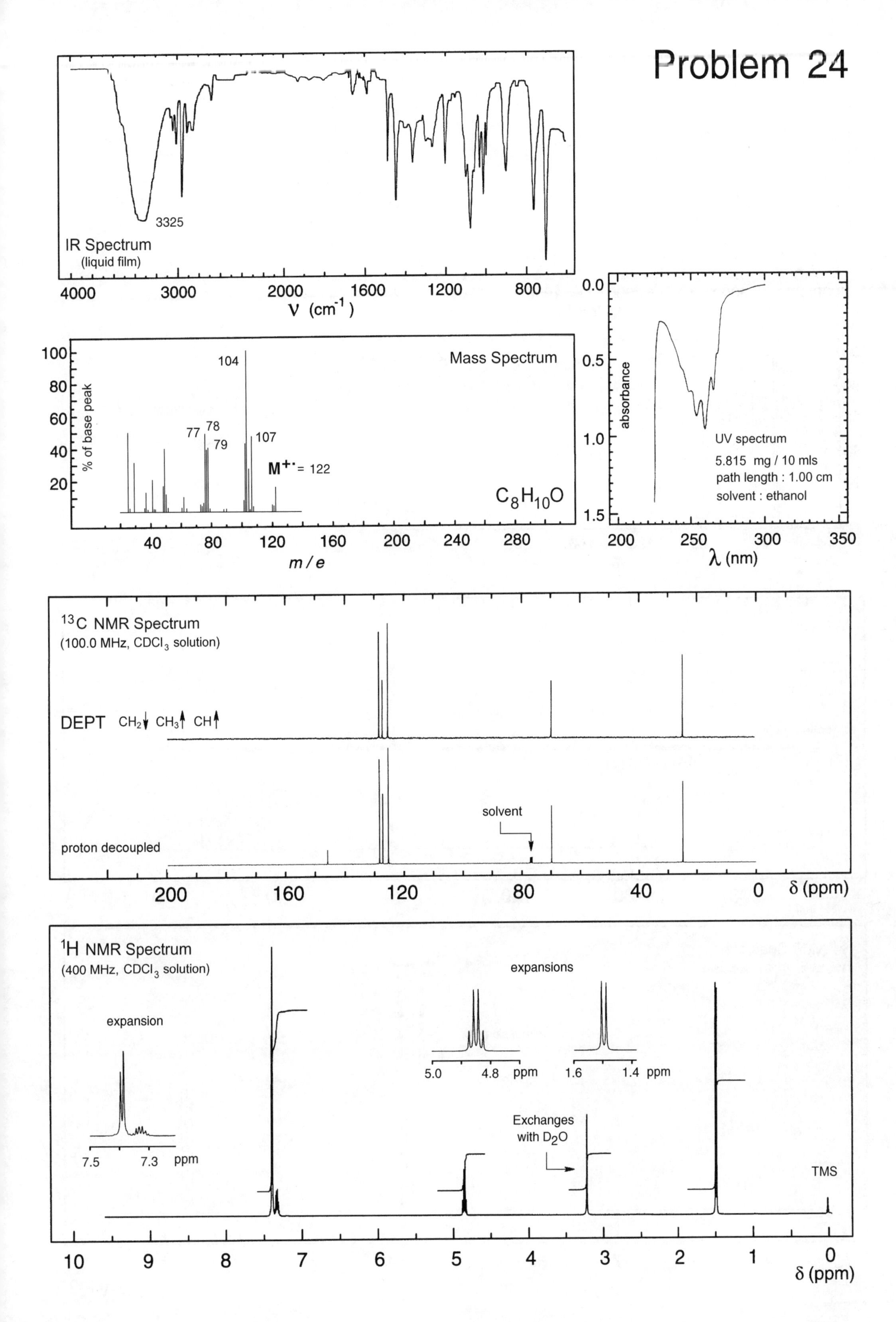

3325
IR Spectrum
(liquid film)
4000
3000
2000
1600
1200
800
ν (cm-1)
Mass Spectrum
104
77
78
79
107
M+· = 122
$C_8H_{10}O$
% of base peak
m / e
UV spectrum
5.815 mg / 10 mls
path length : 1.00 cm
solvent : ethanol
absorbance
λ (nm)
13C NMR Spectrum
(100.0 MHz, CDCl3 solution)
DEPT CH2↓ CH3↑ CH↑
proton decoupled
solvent
δ (ppm)
1H NMR Spectrum
(400 MHz, CDCl3 solution)
expansion
expansions
Exchanges with D2O
TMS
ppm
δ (ppm)

Problem 25

IR spectrum
(liquid film)

1725

4000 3000 2000 1600 1200 800

ν (cm^{-1})

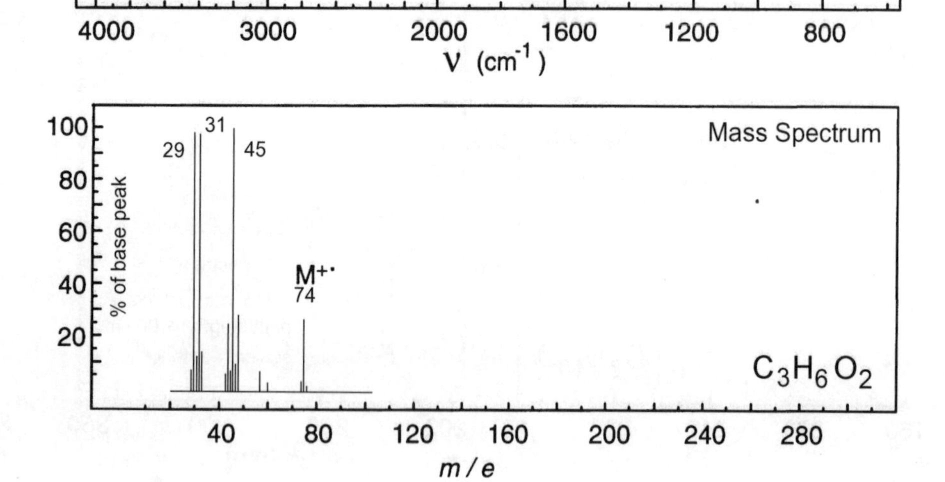

No significant UV absorption above 210 nm

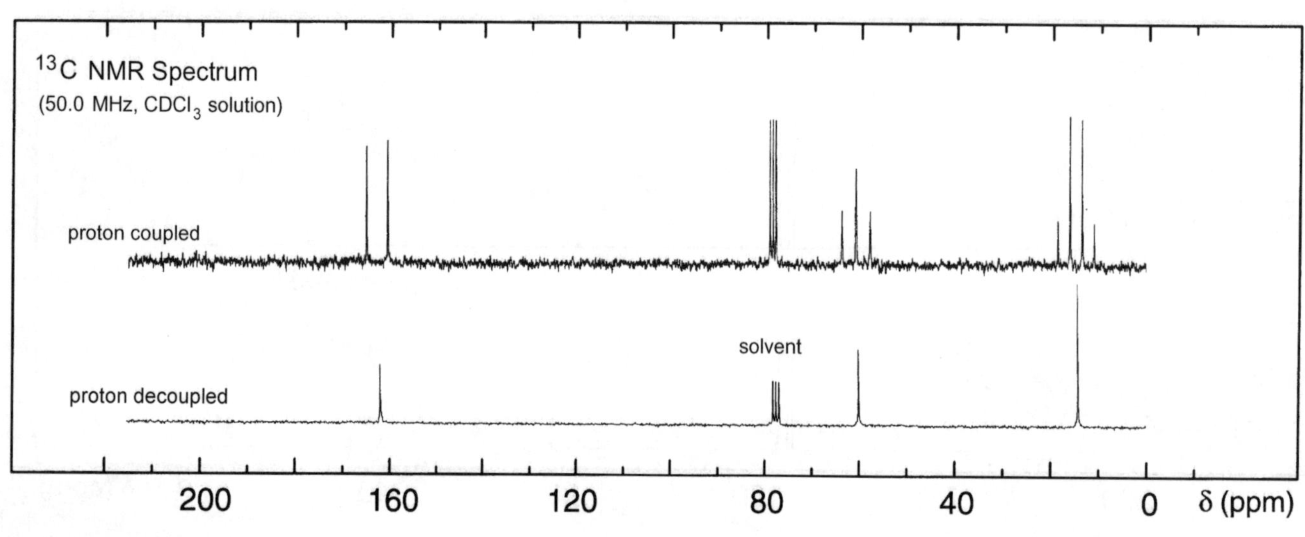

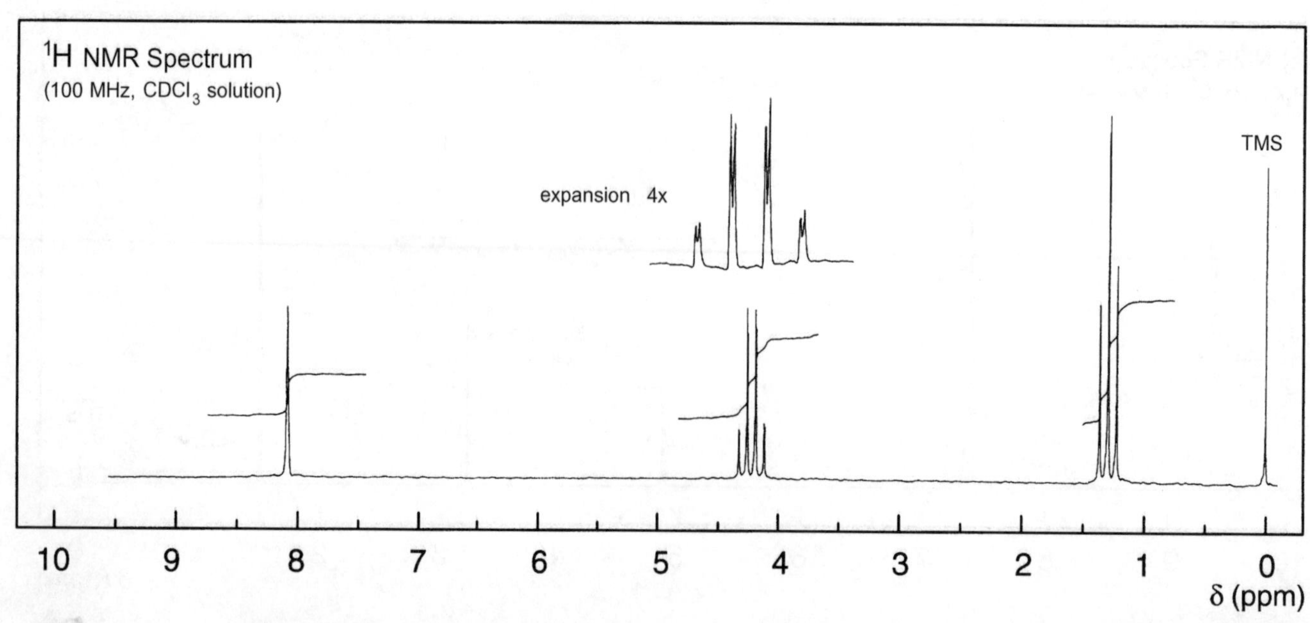

Problem 26

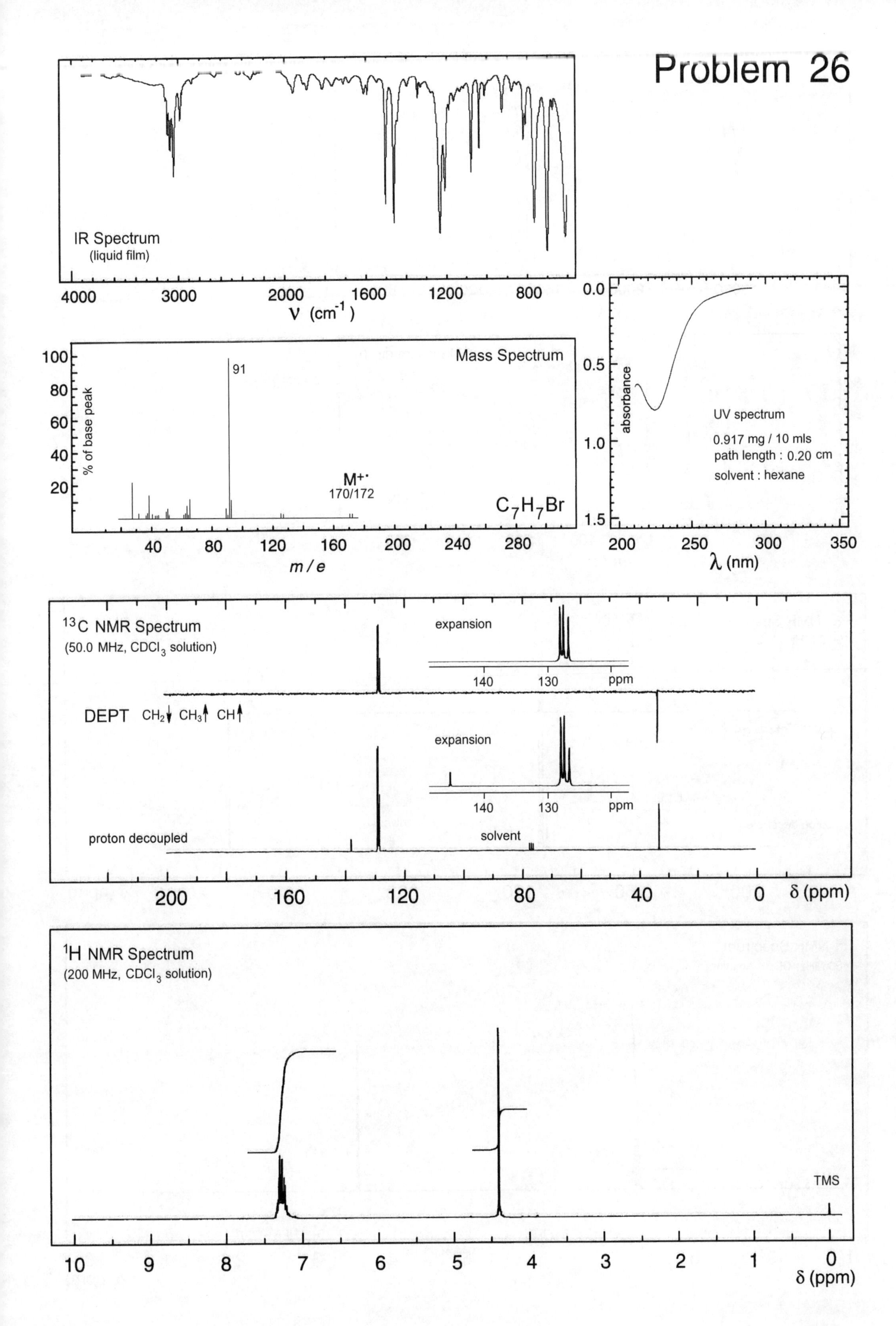

IR Spectrum
(liquid film)
4000
3000
2000
1600
1200
800
ν (cm-1)
Mass Spectrum
100
80
60
40
20
% of base peak
91
M+·
170/172
C7H7Br
40
80
120
160
200
240
280
m / e
0.0
0.5
1.0
1.5
absorbance
UV spectrum
0.917 mg / 10 mls
path length : 0.20 cm
solvent : hexane
200
250
300
350
λ (nm)
13C NMR Spectrum
(50.0 MHz, CDCl3 solution)
expansion
140
130
ppm
DEPT CH2↓ CH3↑ CH↑
proton decoupled
solvent
200
160
120
80
40
0
δ (ppm)
1H NMR Spectrum
(200 MHz, CDCl3 solution)
TMS
10
9
8
7
6
5
4
3
2
1
0
δ (ppm)

Problem 27

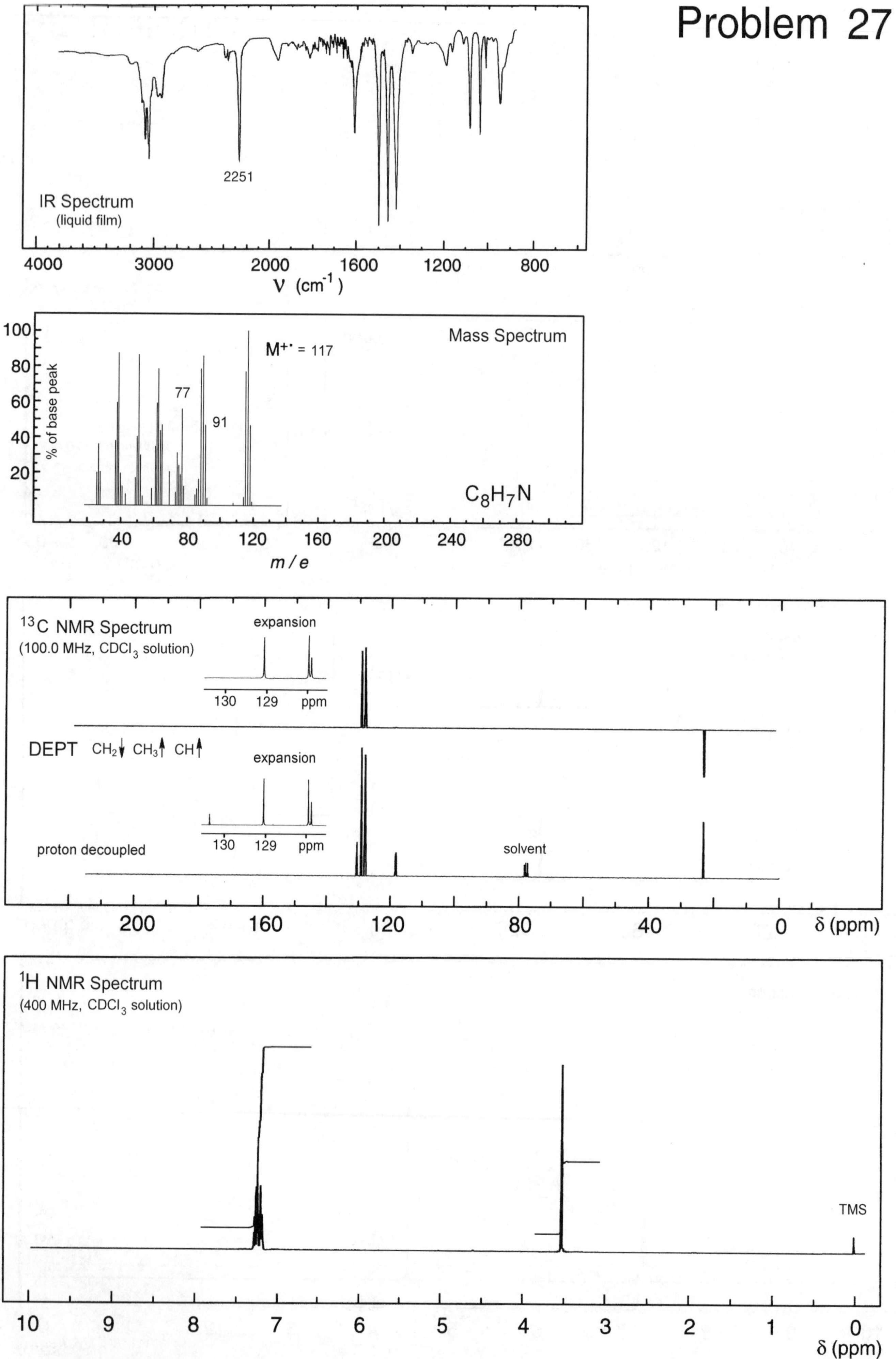

2251
IR Spectrum
(liquid film)
4000
3000
2000
1600
1200
800
ν (cm⁻¹)
Mass Spectrum
M⁺˙ = 117
77
91
% of base peak
C8H7N
m / e
13C NMR Spectrum
(100.0 MHz, CDCl3 solution)
expansion
130
129
ppm
DEPT CH2↓ CH3↑ CH↑
proton decoupled
solvent
δ (ppm)
1H NMR Spectrum
(400 MHz, CDCl3 solution)
TMS
δ (ppm)

Problem 28

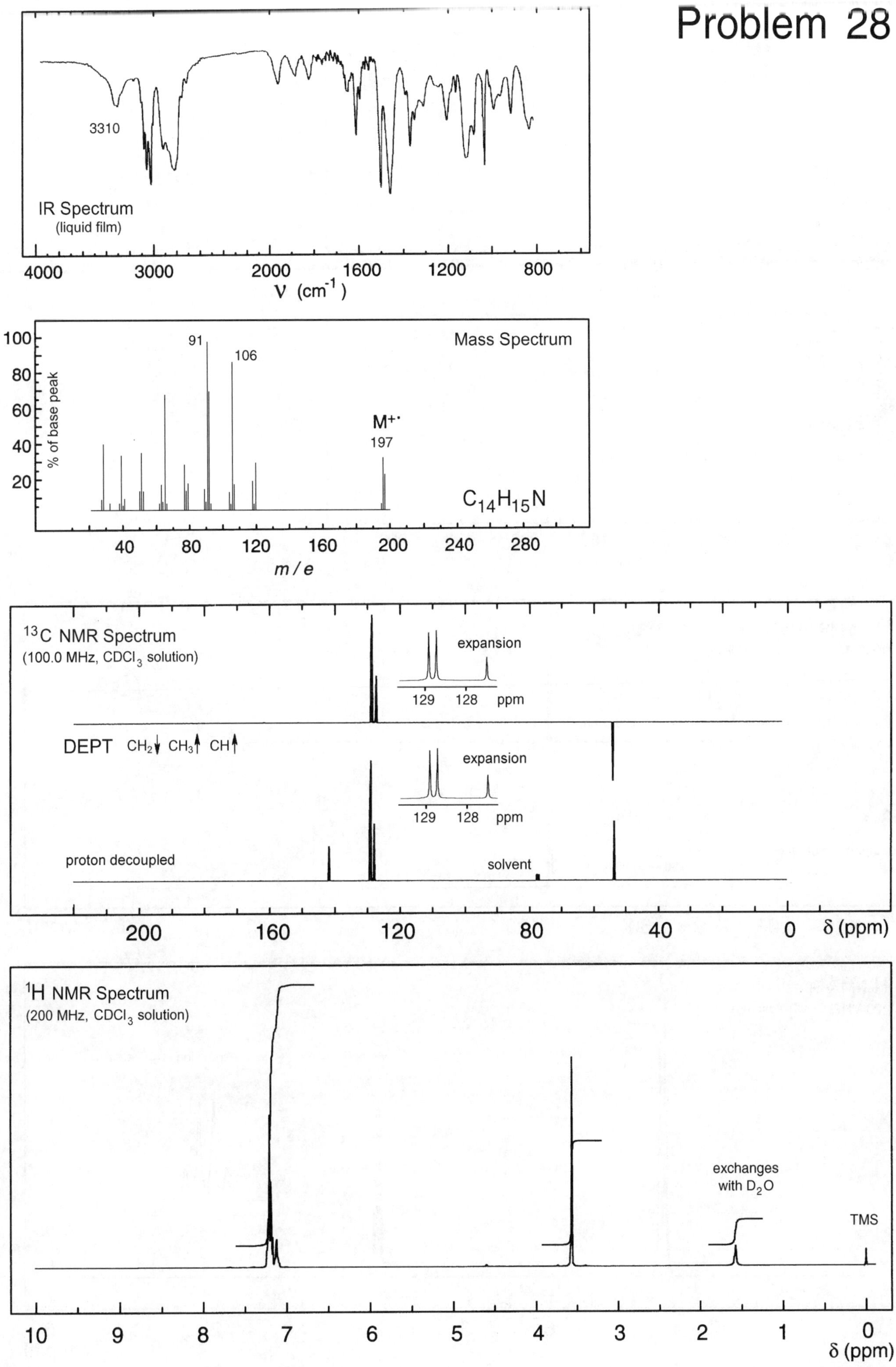

3310
IR Spectrum
(liquid film)
4000
3000
2000
1600
1200
800
ν (cm⁻¹)
Mass Spectrum
100
80
60
40
20
% of base peak
91
106
M+·
197
C14H15N
40
80
120
160
200
240
280
m / e
13C NMR Spectrum
(100.0 MHz, CDCl3 solution)
expansion
129
128
ppm
DEPT CH2↓ CH3↑ CH↑
expansion
129
128
ppm
proton decoupled
solvent
200
160
120
80
40
0
δ (ppm)
1H NMR Spectrum
(200 MHz, CDCl3 solution)
exchanges
with D2O
TMS
10
9
8
7
6
5
4
3
2
1
0
δ (ppm)

Problem 29

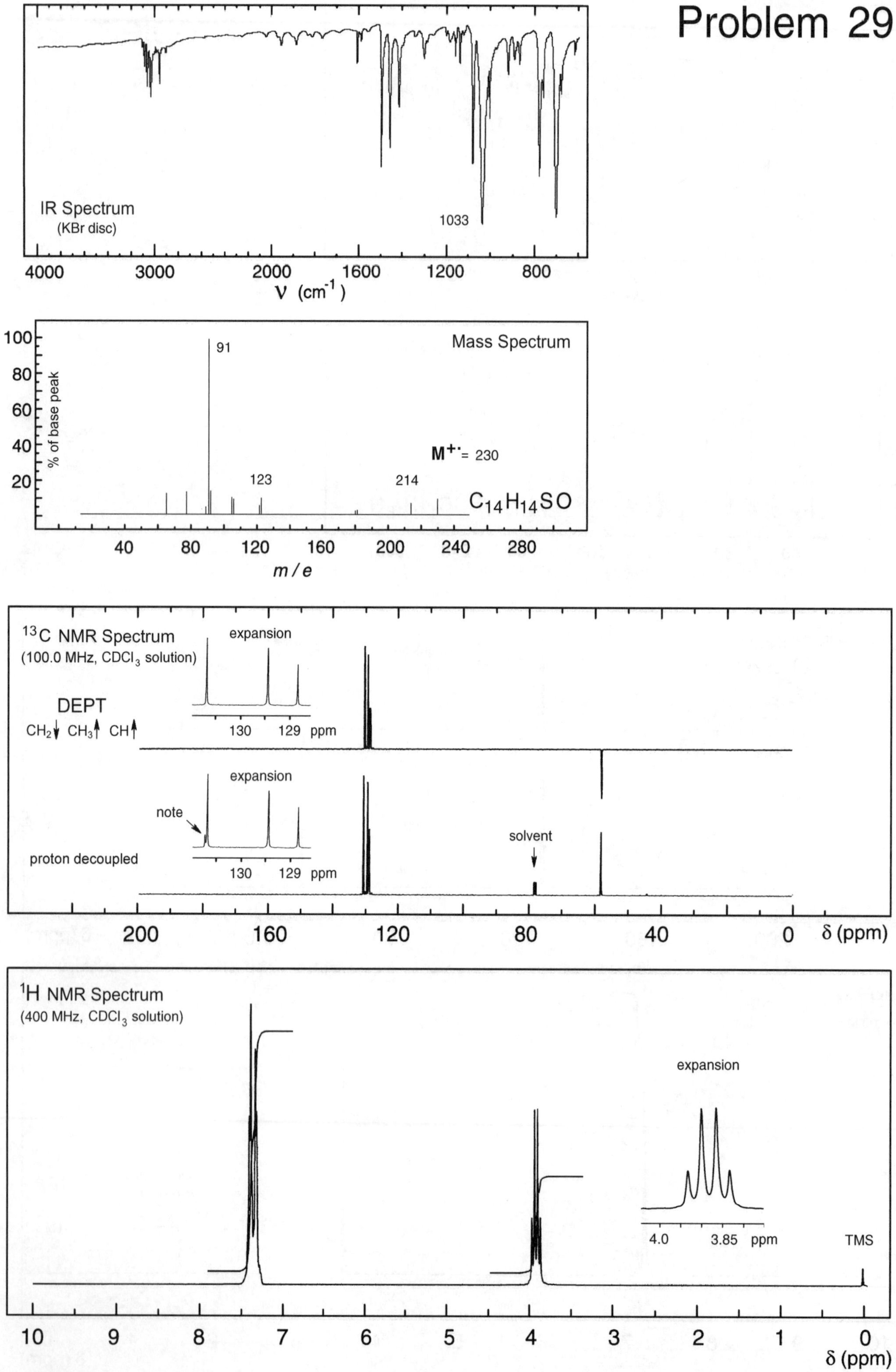

IR Spectrum
(KBr disc)
1033
4000
3000
2000
1600
1200
800
ν (cm-1)
Mass Spectrum
% of base peak
91
123
214
M+· = 230
C14H14SO
m / e
13C NMR Spectrum
(100.0 MHz, CDCl3 solution)
expansion
DEPT
CH2↓ CH3↑ CH↑
130
129
ppm
note
solvent
proton decoupled
δ (ppm)
1H NMR Spectrum
(400 MHz, CDCl3 solution)
4.0
3.85
TMS

Problem 30

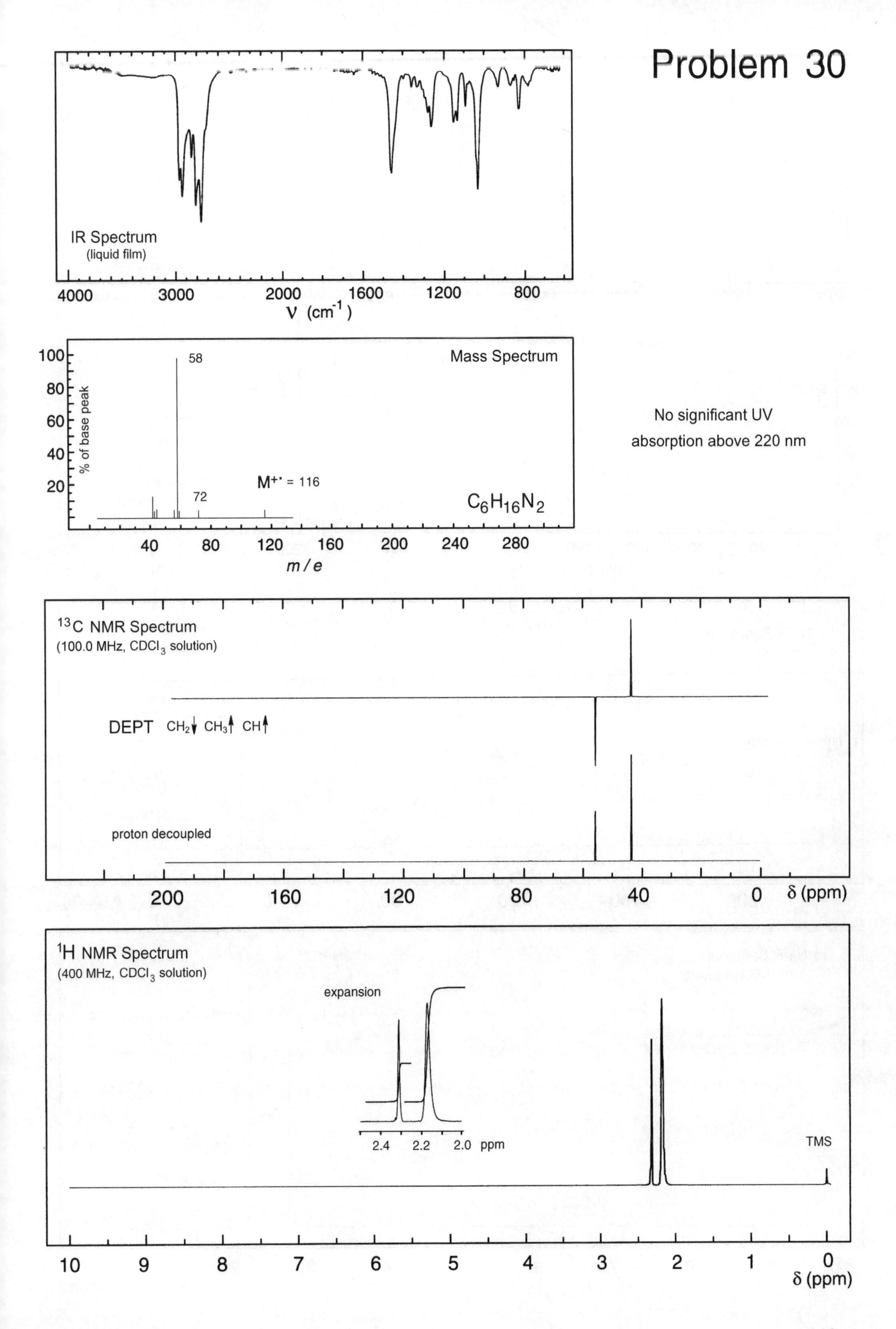

No significant UV absorption above 220 nm

Problem 31

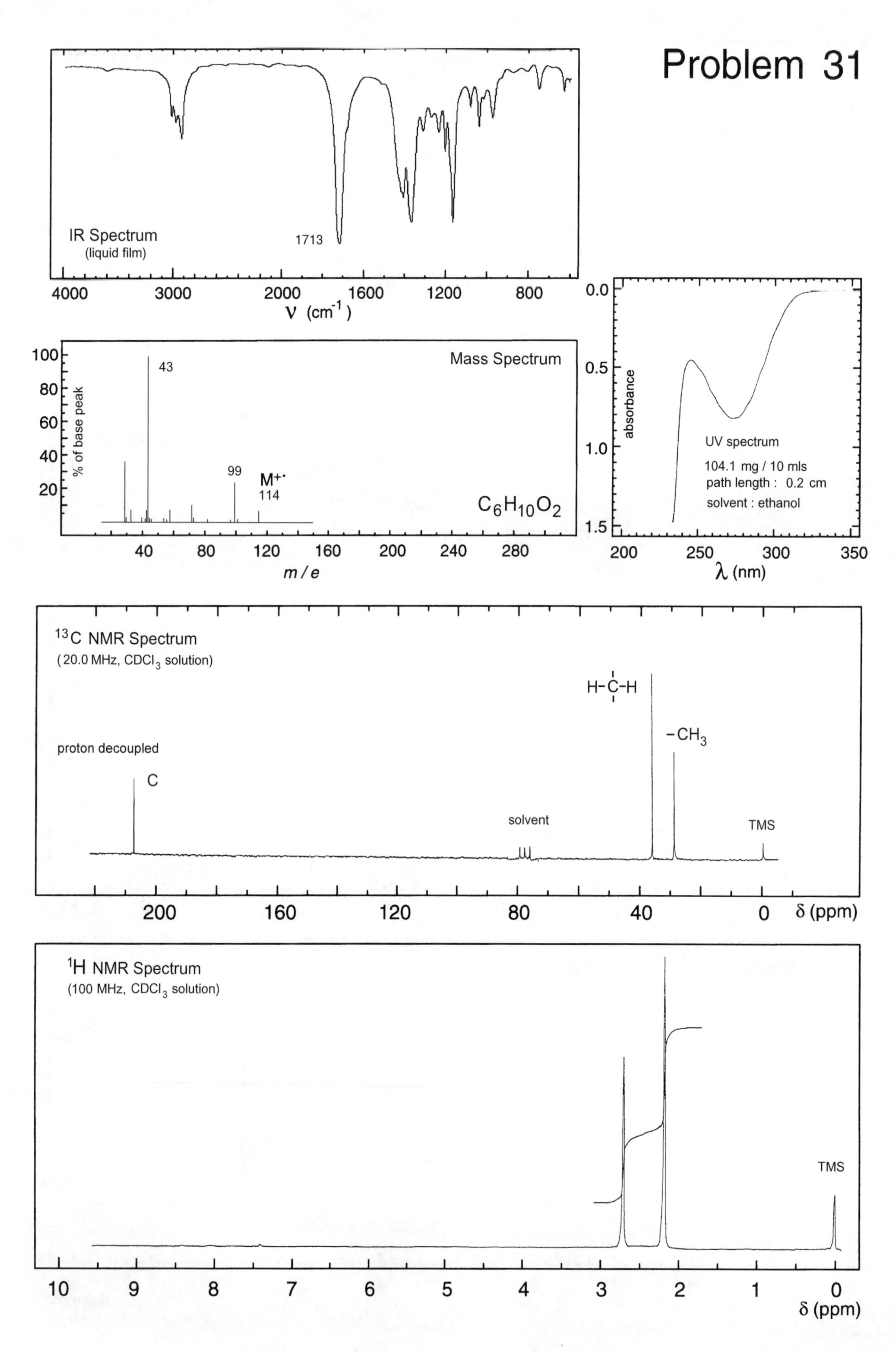

IR Spectrum
(liquid film)
1713
4000
3000
2000
1600
1200
800
ν (cm-1)
Mass Spectrum
% of base peak
100
80
60
40
20
43
99
M+·
114
C6H10O2
40
80
120
160
200
240
280
m / e
0.0
0.5
1.0
1.5
absorbance
UV spectrum
104.1 mg / 10 mls
path length : 0.2 cm
solvent : ethanol
200
250
300
350
λ (nm)
13C NMR Spectrum
(20.0 MHz, CDCl3 solution)
proton decoupled
C
solvent
H-C-H
-CH3
TMS
200
160
120
80
40
0
δ (ppm)
1H NMR Spectrum
(100 MHz, CDCl3 solution)
TMS
10
9
8
7
6
5
4
3
2
1
0
δ (ppm)

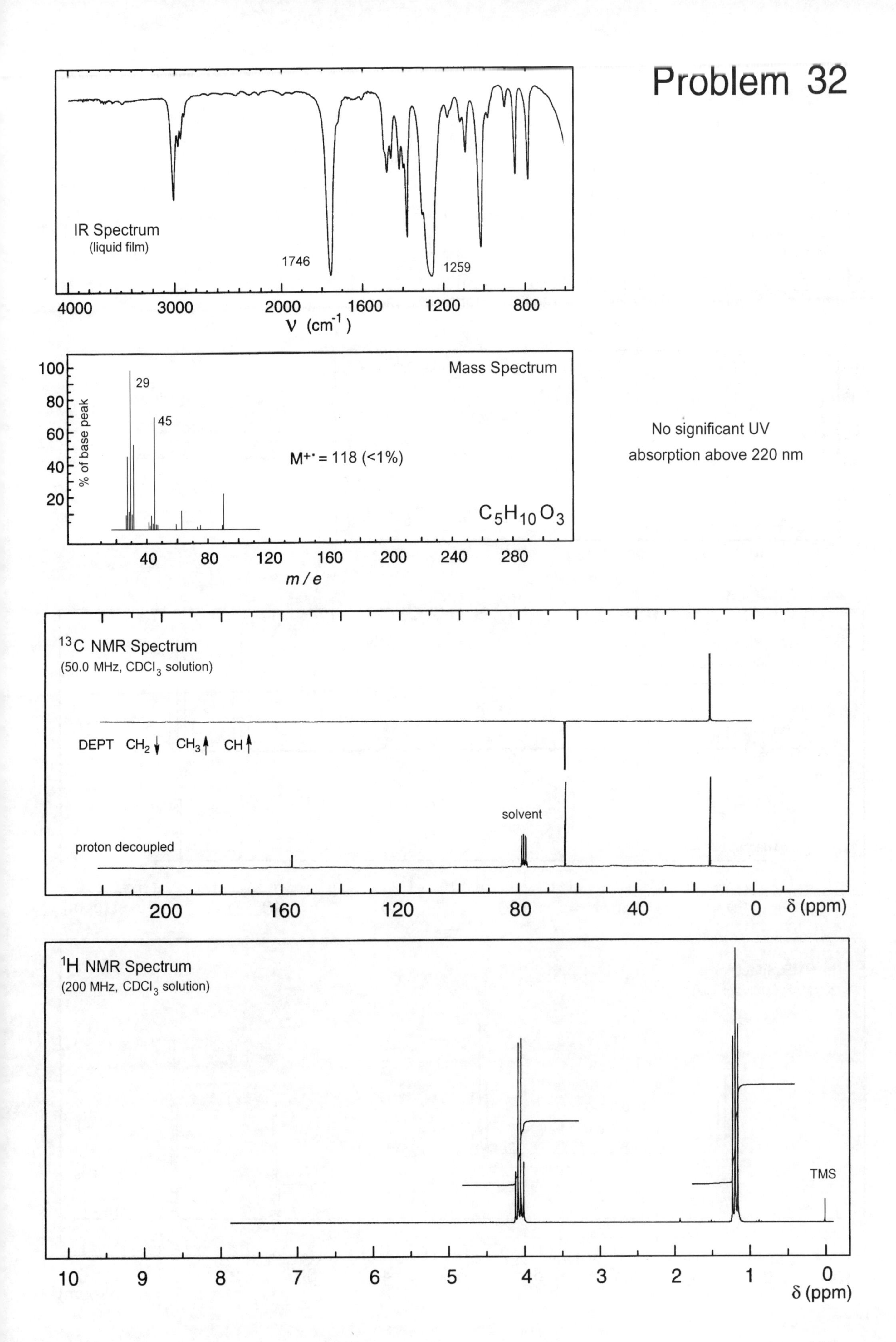
Problem 32
IR Spectrum
(liquid film)
1746
1259
4000
3000
2000
1600
1200
800
ν (cm-1)
Mass Spectrum
100
80
60
40
20
% of base peak
29
45
M+• = 118 (<1%)
C5H10O3
40
80
120
160
200
240
280
m / e
No significant UV
absorption above 220 nm
13C NMR Spectrum
(50.0 MHz, CDCl3 solution)
DEPT CH2 ↓ CH3 ↑ CH ↑
solvent
proton decoupled
200
160
120
80
40
0
δ (ppm)
1H NMR Spectrum
(200 MHz, CDCl3 solution)
TMS
10
9
8
7
6
5
4
3
2
1
0
δ (ppm)

Problem 33

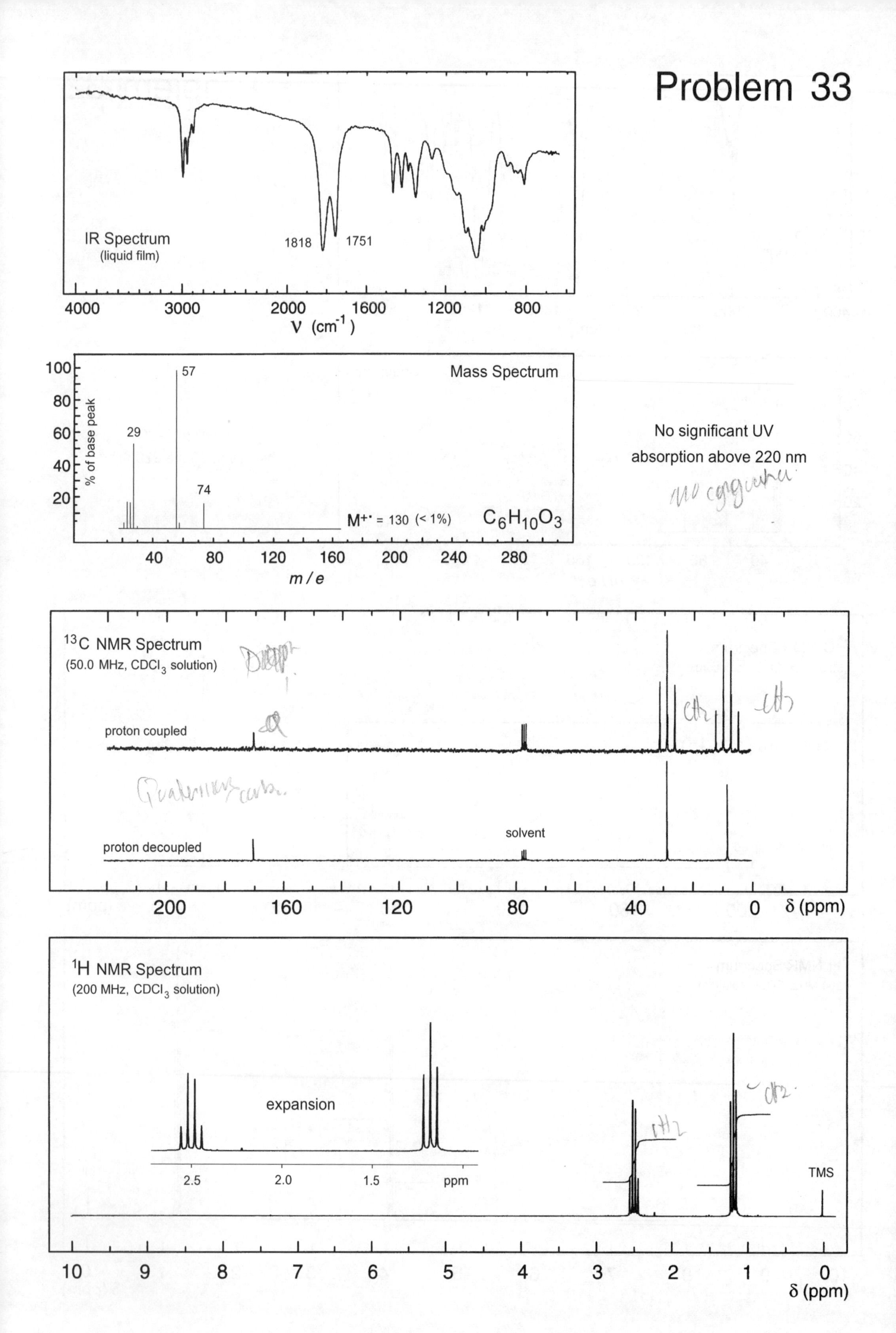

IR Spectrum
(liquid film)
1818
1751
4000
3000
2000
1600
1200
800
ν (cm^{-1})
Mass Spectrum
% of base peak
100
80
60
40
20
57
29
74
M+· = 130 (< 1%)
$C_6H_{10}O_3$
40
80
120
160
200
240
280
m / e
No significant UV absorption above 220 nm
^{13}C NMR Spectrum
(50.0 MHz, $CDCl_3$ solution)
proton coupled
proton decoupled
solvent
200
160
120
80
40
0
δ (ppm)
^{1}H NMR Spectrum
(200 MHz, $CDCl_3$ solution)
expansion
2.5
2.0
1.5
ppm
TMS
10
9
8
7
6
5
4
3
2
1
0
δ (ppm)

Problem 34

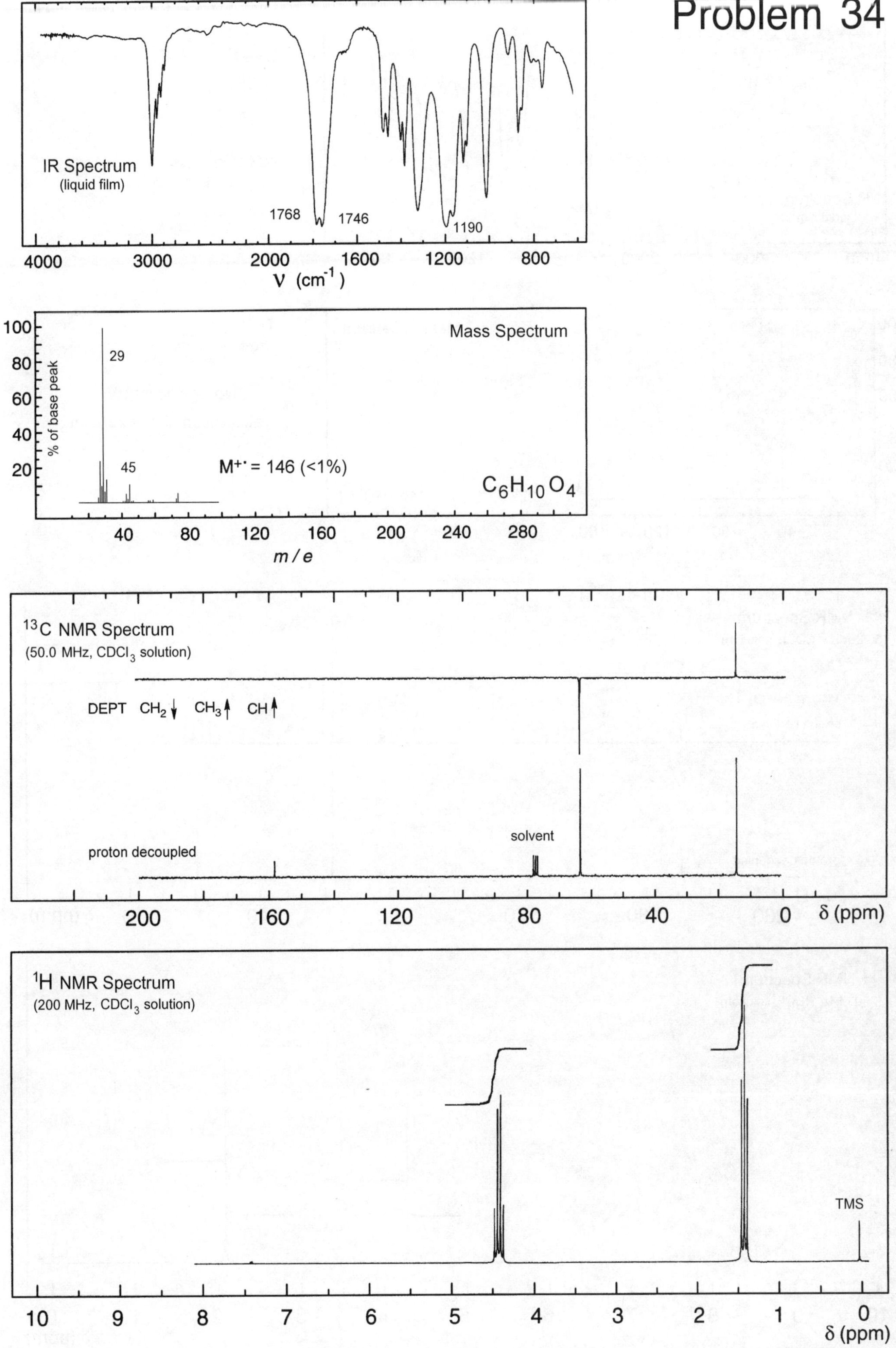

IR Spectrum
(liquid film)
1768
1746
1190
4000
3000
2000
1600
1200
800
ν (cm⁻¹)
Mass Spectrum
% of base peak
100
80
60
40
20
29
45
M⁺˙ = 146 (<1%)
C6H10O4
40
80
120
160
200
240
280
m / e
13C NMR Spectrum
(50.0 MHz, CDCl3 solution)
DEPT CH2 ↓ CH3 ↑ CH ↑
proton decoupled
solvent
200
160
120
80
40
0
δ (ppm)
1H NMR Spectrum
(200 MHz, CDCl3 solution)
TMS
10
9
8
7
6
5
4
3
2
1
0
δ (ppm)

Problem 35

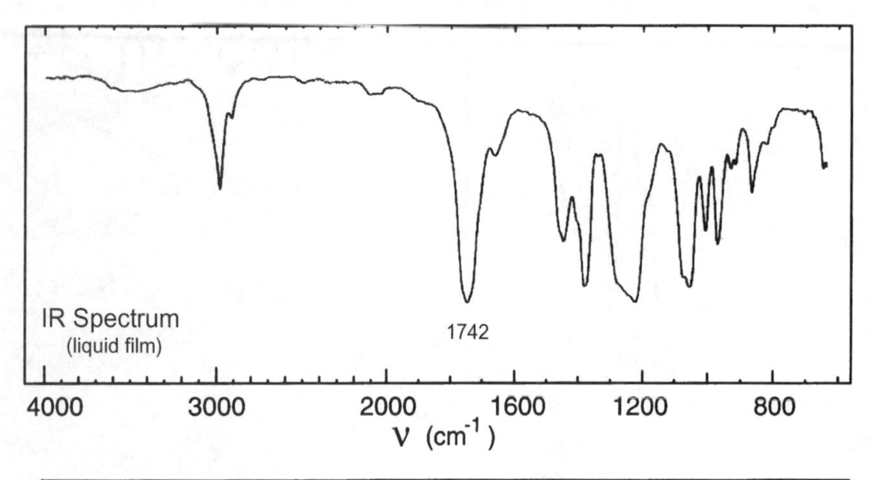

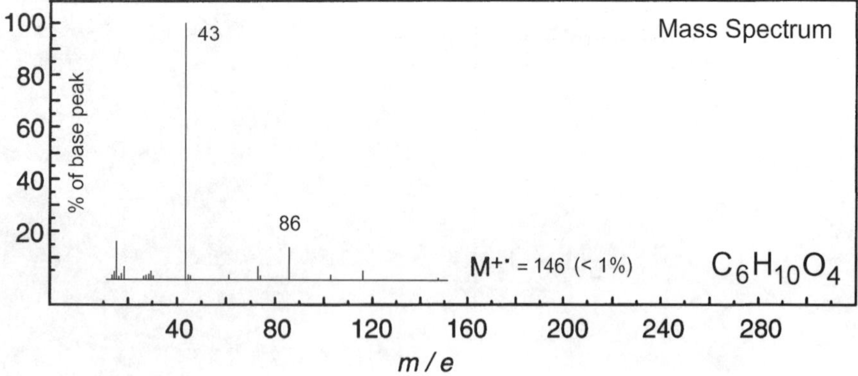

No significant UV absorption above 220 nm

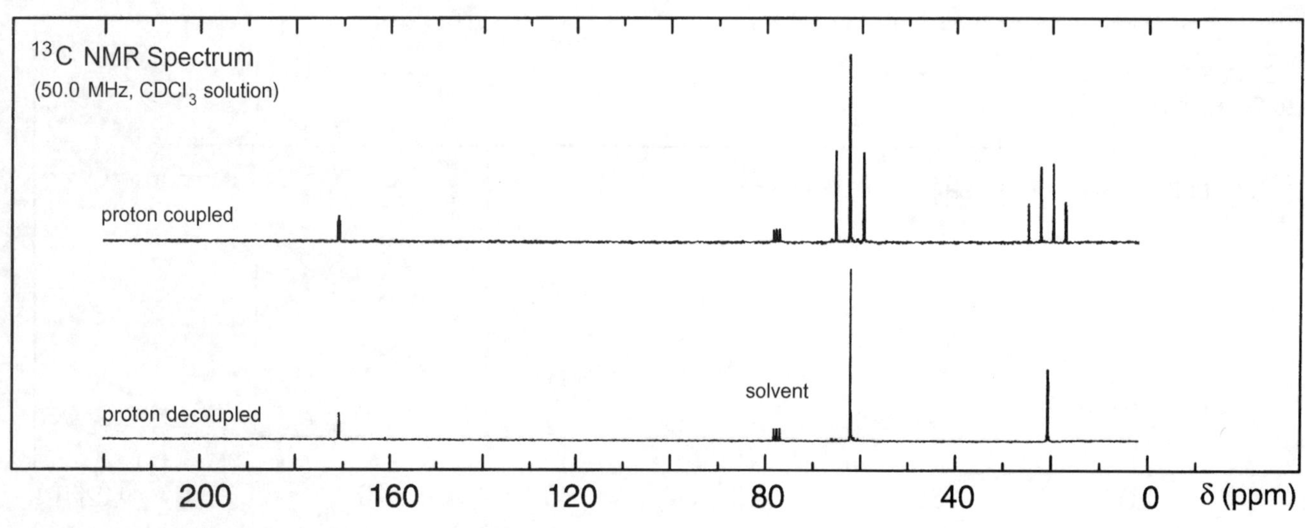

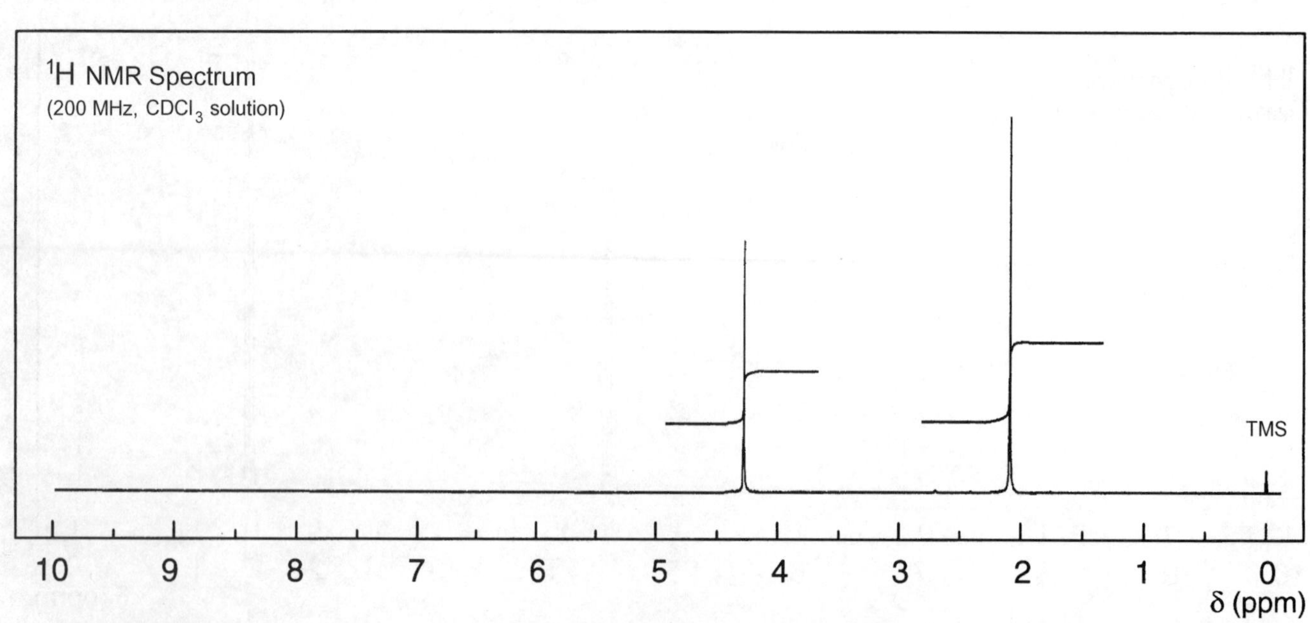

Problem 36

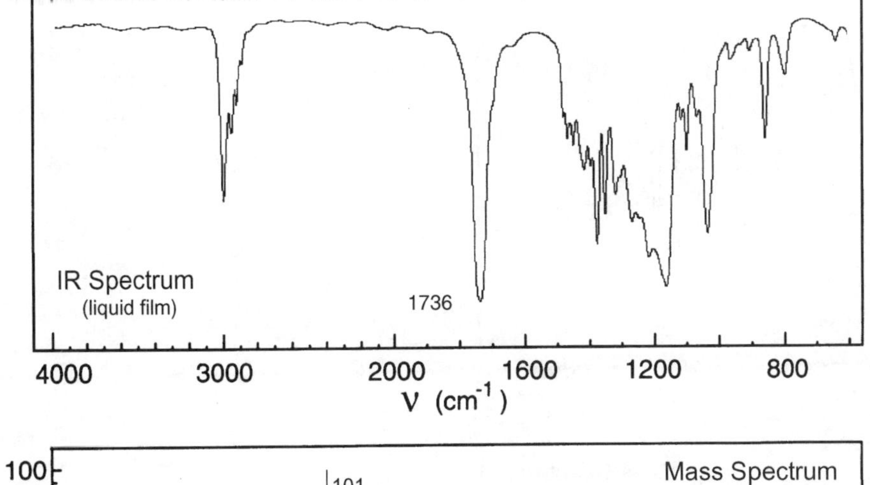

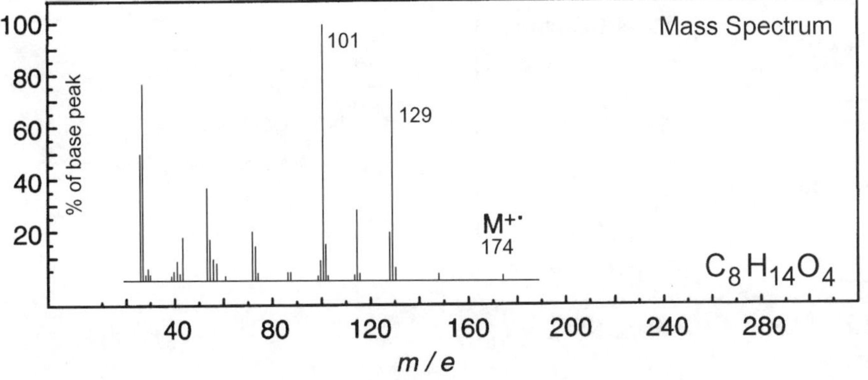

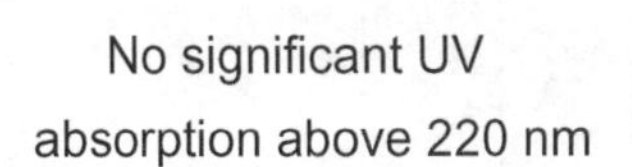
No significant UV absorption above 220 nm

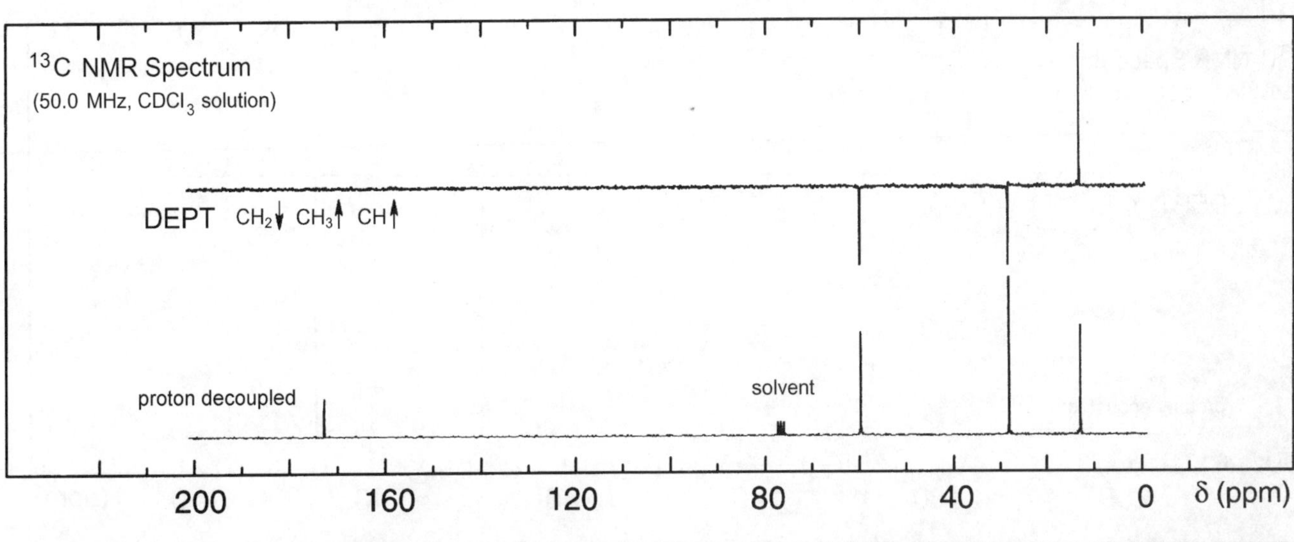

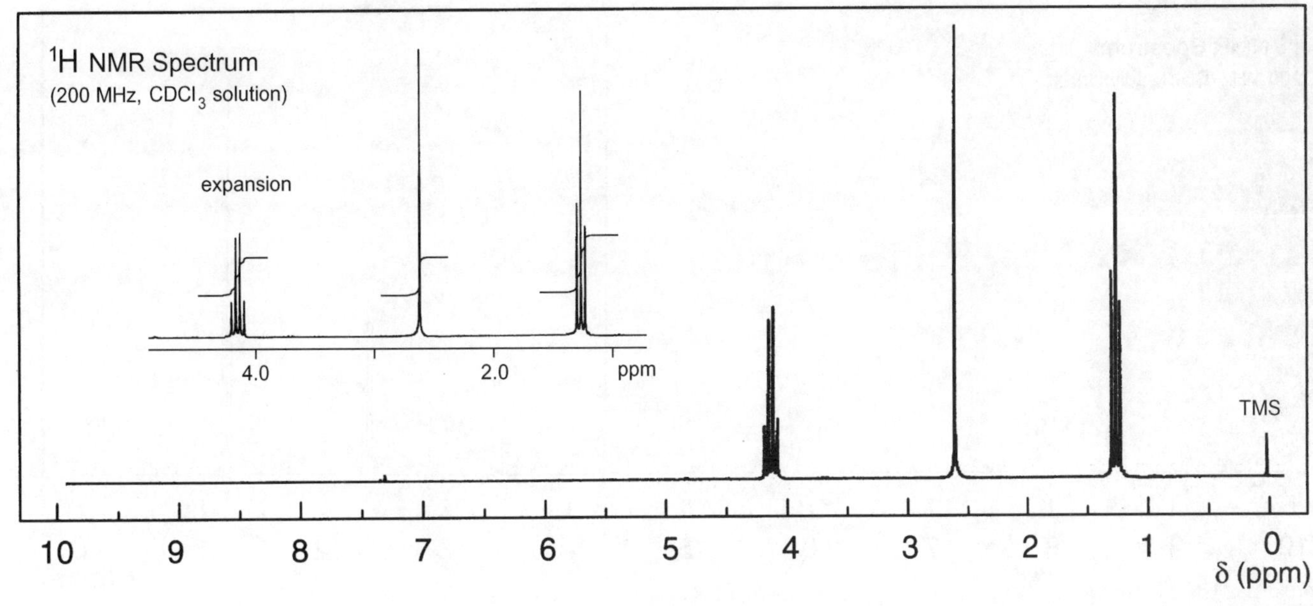

Problem 37

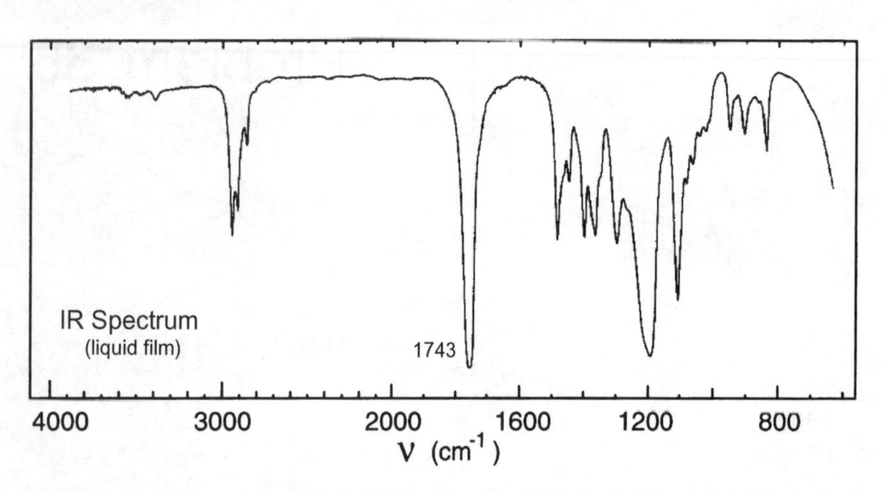

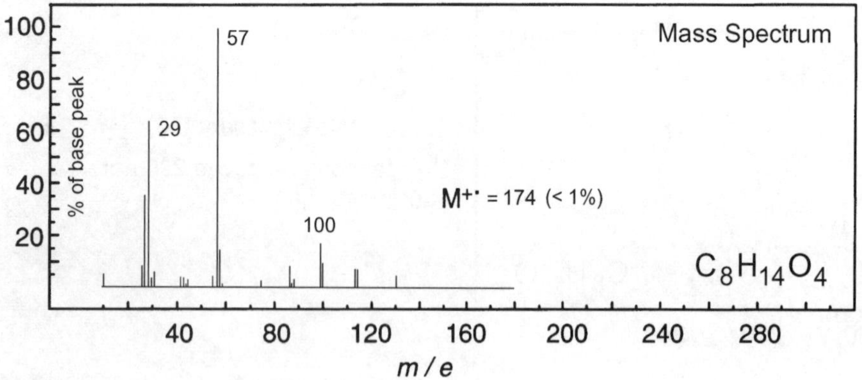

No significant UV absorption above 220 nm

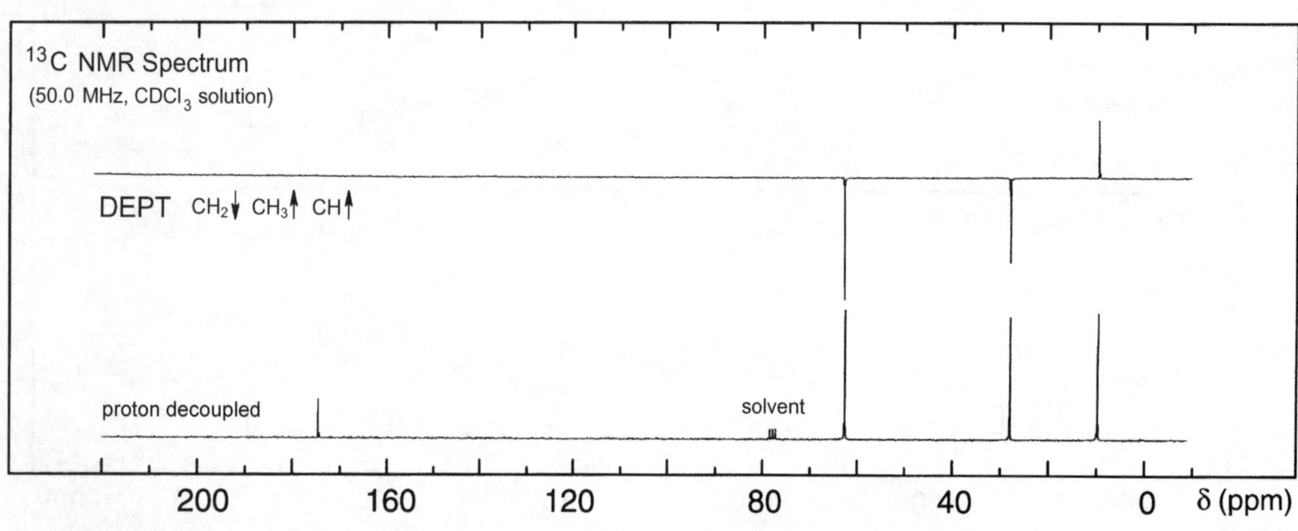

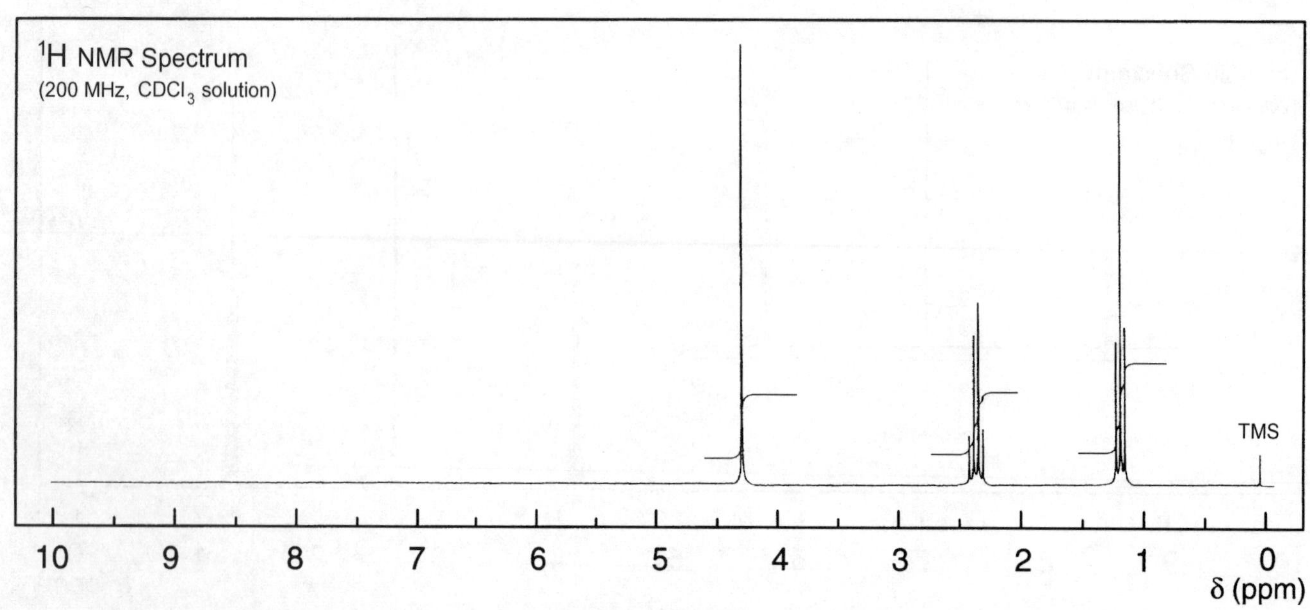

Problem 38

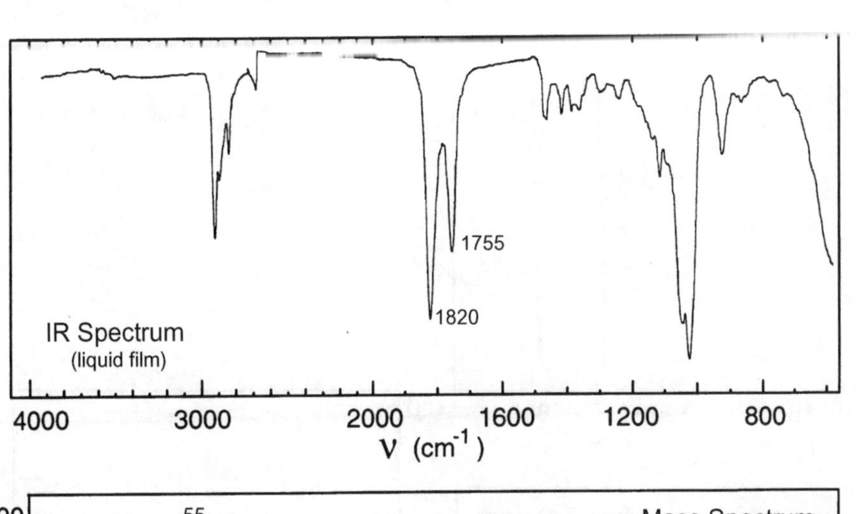

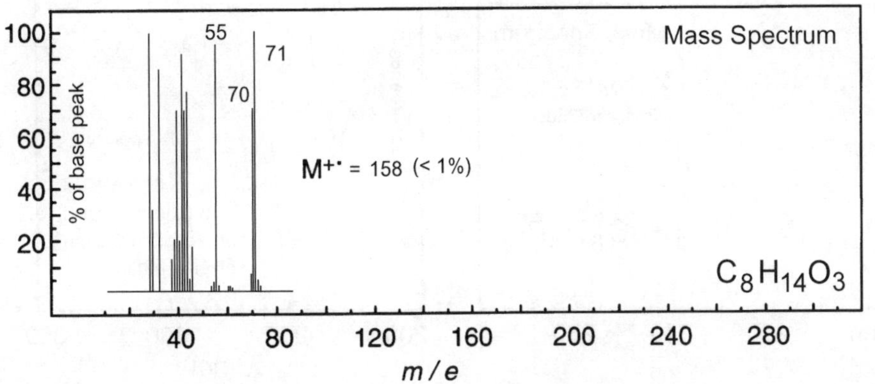

No significant UV absorption above 220 nm

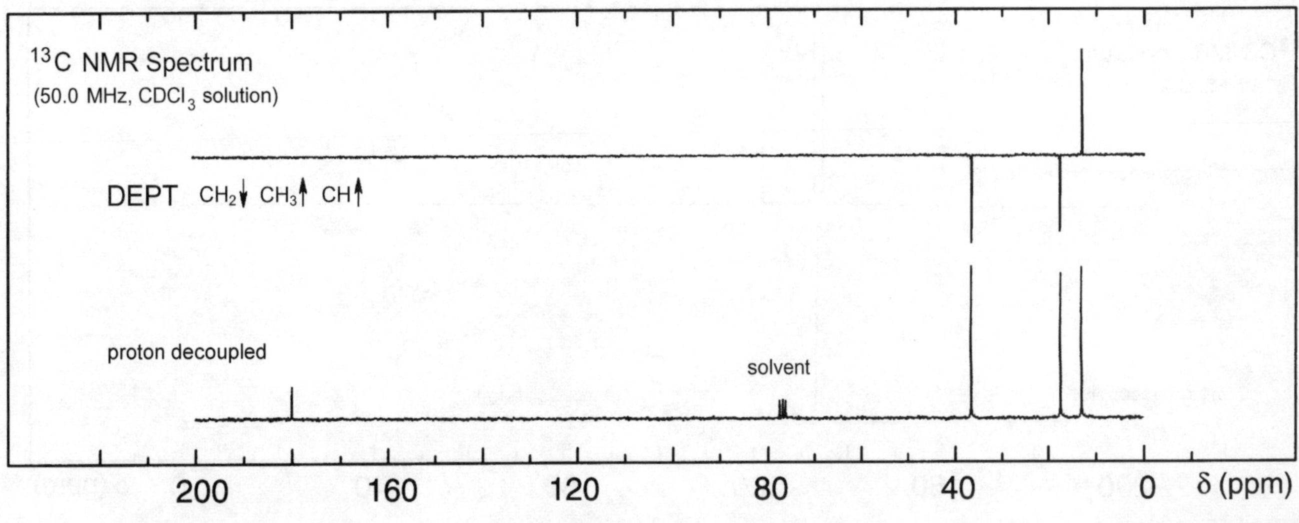

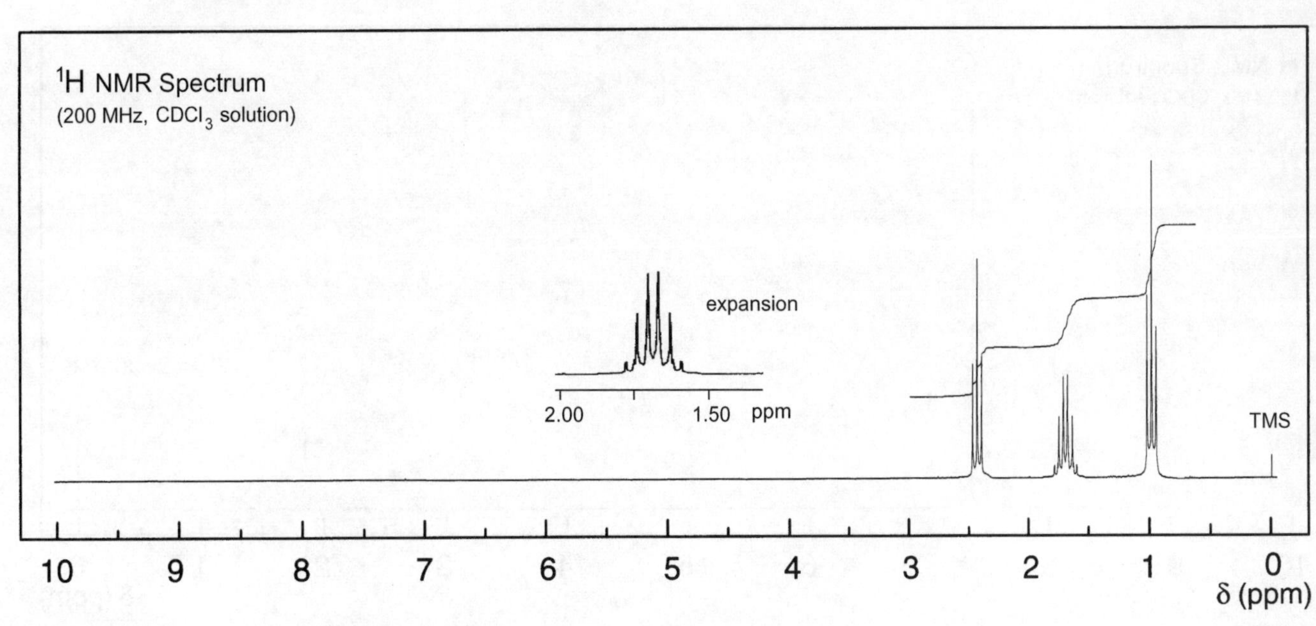

Problem 39

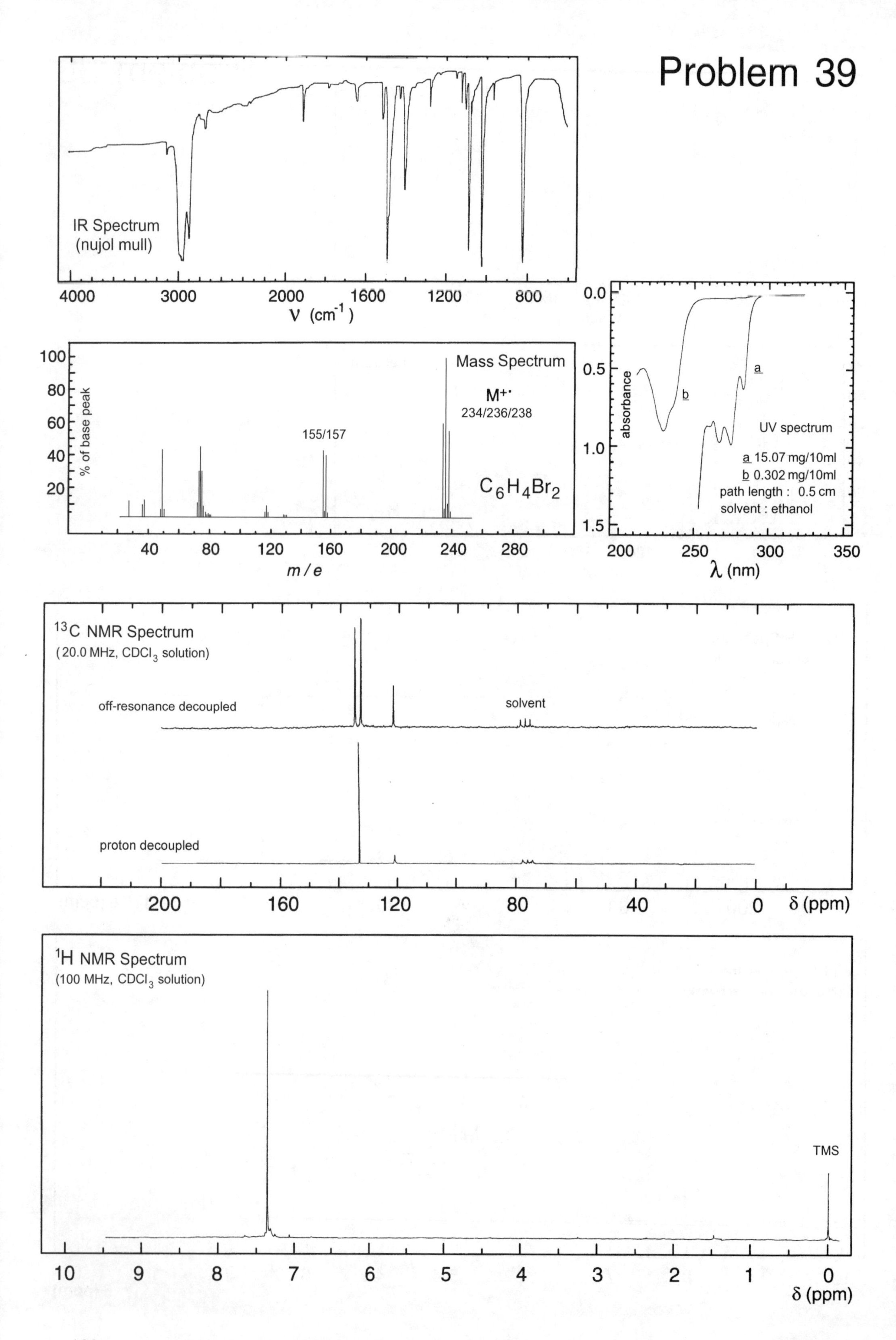

IR Spectrum
(nujol mull)
4000
3000
2000
1600
1200
800
ν (cm^{-1})
Mass Spectrum
% of base peak
100
80
60
40
20
M+·
234/236/238
155/157
$C_6H_4Br_2$
40
80
120
160
200
240
280
m / e
0.0
0.5
1.0
1.5
absorbance
a
b
UV spectrum
a 15.07 mg/10ml
b 0.302 mg/10ml
path length : 0.5 cm
solvent : ethanol
200
250
300
350
λ (nm)
^{13}C NMR Spectrum
(20.0 MHz, $CDCl_3$ solution)
off-resonance decoupled
solvent
proton decoupled
200
160
120
80
40
0
δ (ppm)
^{1}H NMR Spectrum
(100 MHz, $CDCl_3$ solution)
TMS
10
9
8
7
6
5
4
3
2
1
0
δ (ppm)

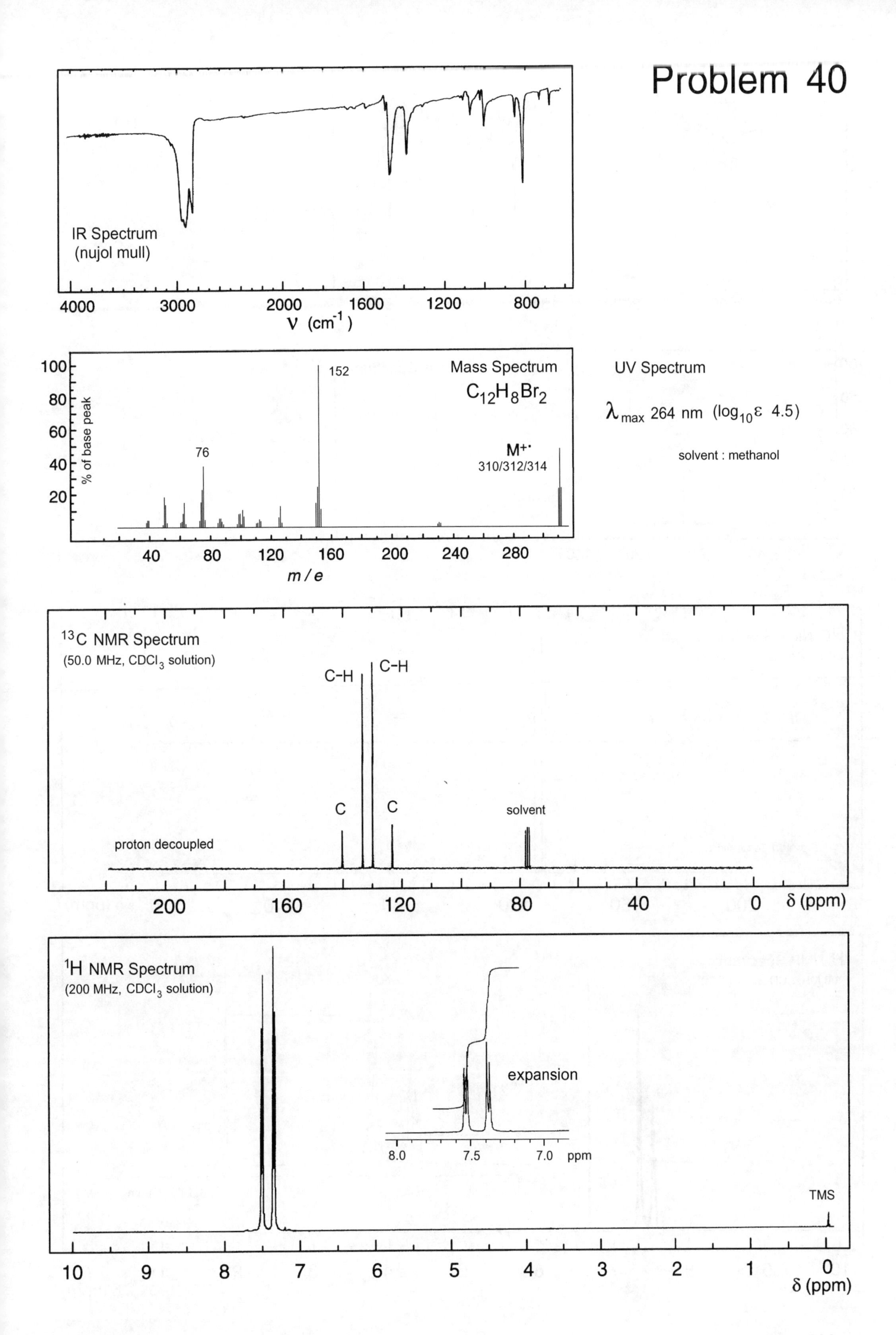
Problem 40
IR Spectrum
(nujol mull)
4000
3000
2000
1600
1200
800
ν (cm⁻¹)
Mass Spectrum
C12H8Br2
100
80
60
40
20
% of base peak
152
76
M+·
310/312/314
40
80
120
160
200
240
280
m / e
UV Spectrum
λmax 264 nm (log10ε 4.5)
solvent : methanol
13C NMR Spectrum
(50.0 MHz, CDCl3 solution)
C-H
C-H
C
C
solvent
proton decoupled
200
160
120
80
40
0
δ (ppm)
1H NMR Spectrum
(200 MHz, CDCl3 solution)
expansion
8.0
7.5
7.0
ppm
TMS
10
9
8
7
6
5
4
3
2
1
0
δ (ppm)

Problem 41

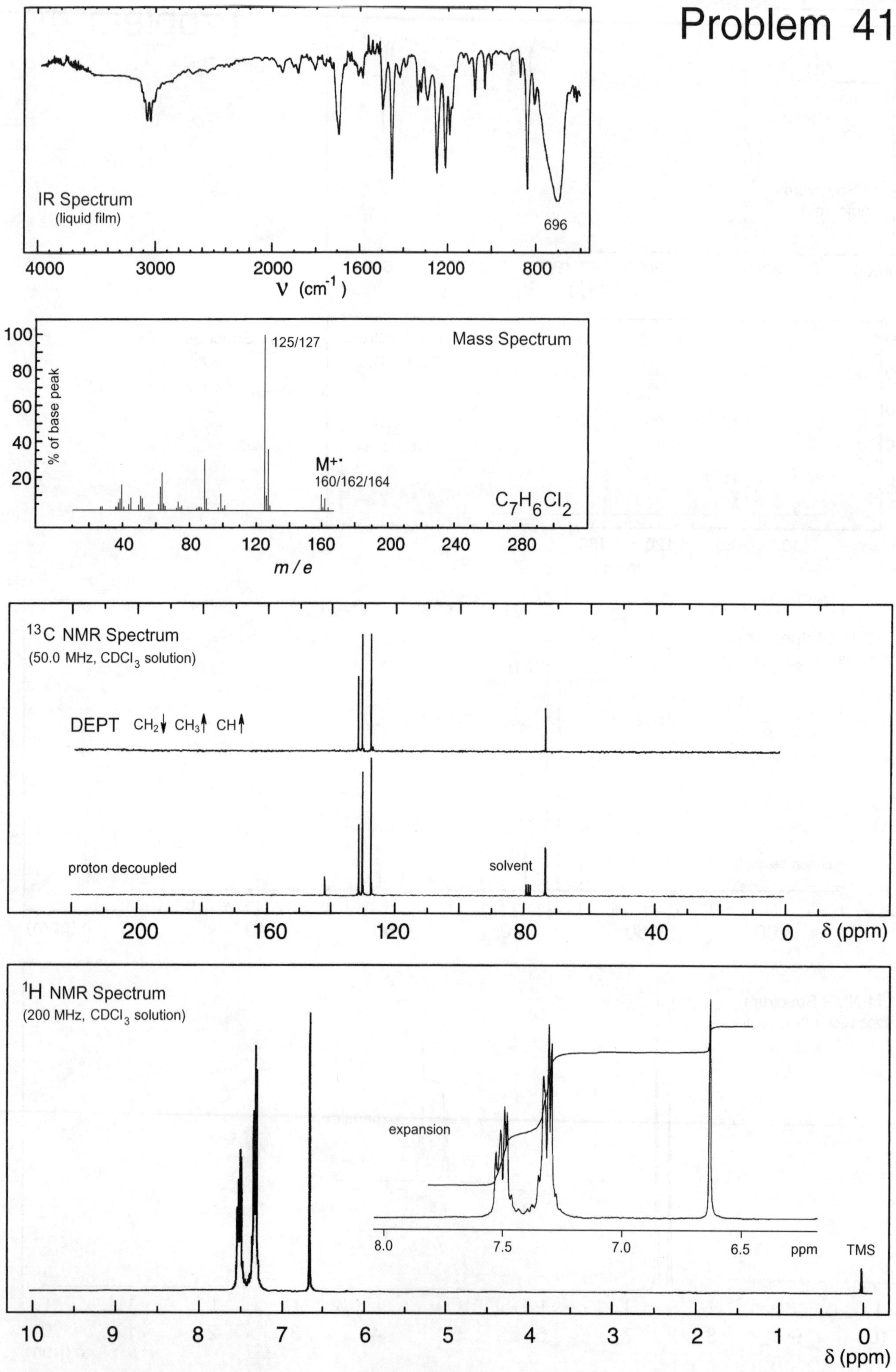

IR Spectrum
(liquid film)
696
4000
3000
2000
1600
1200
800
ν (cm⁻¹)
Mass Spectrum
125/127
M+·
160/162/164
C7H6Cl2
% of base peak
m / e
13C NMR Spectrum
(50.0 MHz, CDCl3 solution)
DEPT CH2↓ CH3↑ CH↑
proton decoupled
solvent
δ (ppm)
1H NMR Spectrum
(200 MHz, CDCl3 solution)
expansion
ppm
TMS
δ (ppm)

Problem 42

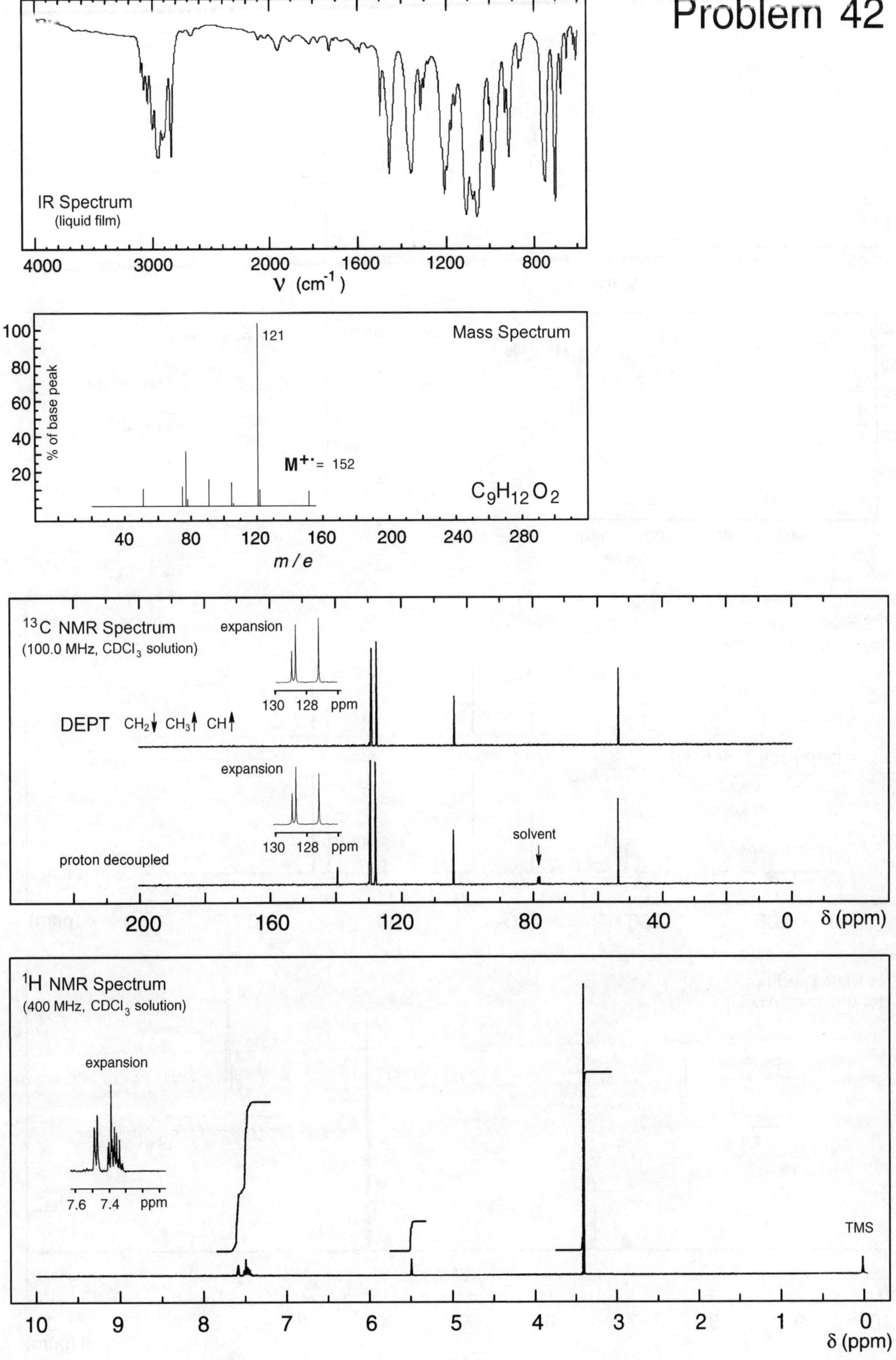

IR Spectrum
(liquid film)
4000
3000
2000
1600
1200
800
ν (cm^{-1})
100
80
60
40
20
% of base peak
121
Mass Spectrum
M$^{+\cdot}$ = 152
$C_9H_{12}O_2$
40
80
120
160
200
240
280
m / e
^{13}C NMR Spectrum
(100.0 MHz, $CDCl_3$ solution)
expansion
130 128 ppm
DEPT CH_2↓ CH_3↑ CH↑
expansion
130 128 ppm
solvent
proton decoupled
200
160
120
80
40
0
δ (ppm)
^{1}H NMR Spectrum
(400 MHz, $CDCl_3$ solution)
expansion
7.6 7.4 ppm
TMS
10
9
8
7
6
5
4
3
2
1
0
δ (ppm)

Problem 43

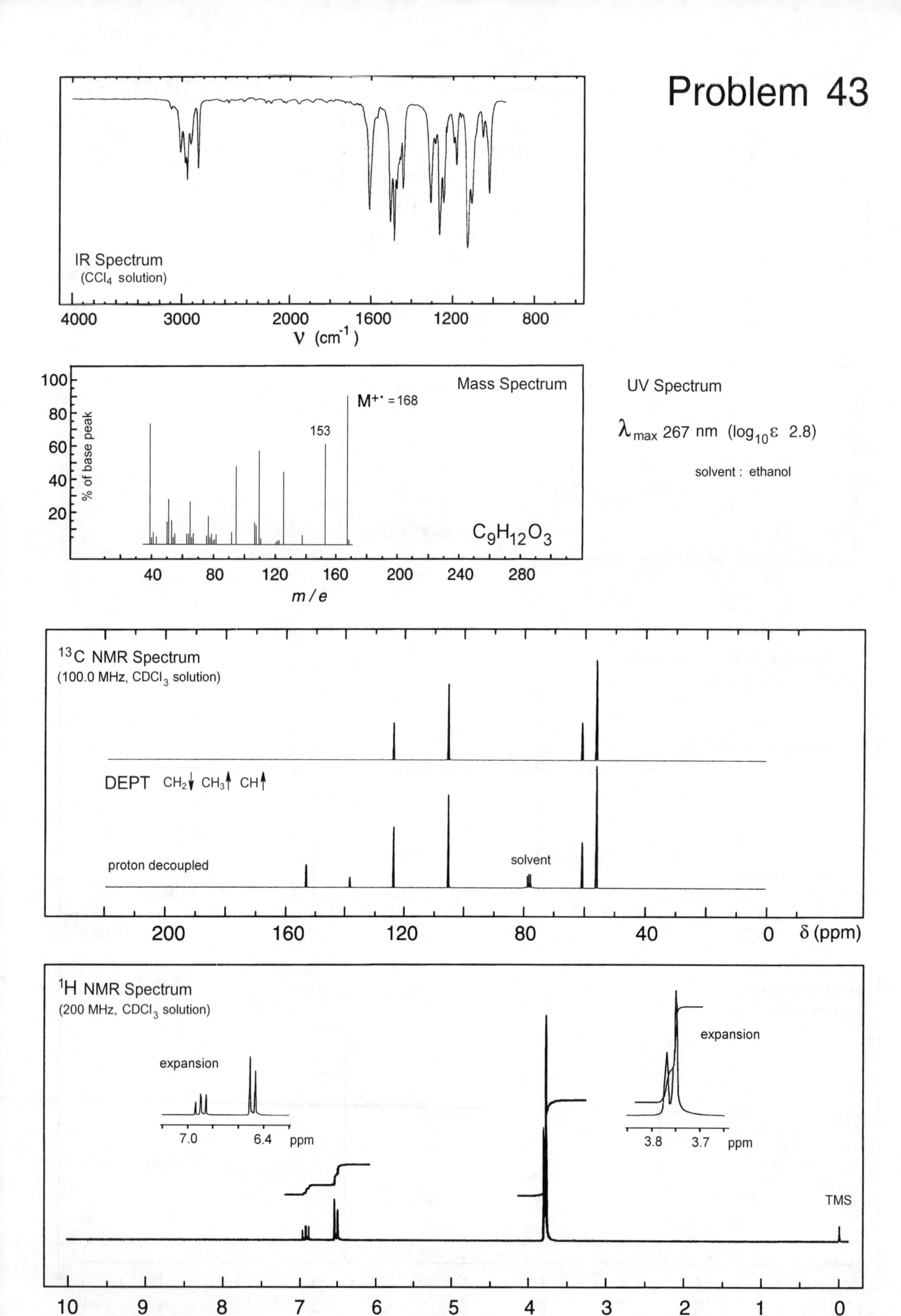

IR Spectrum
(CCl4 solution)
4000
3000
2000
1600
1200
800
ν (cm-1)
Mass Spectrum
100
80
60
40
20
% of base peak
M+· = 168
153
C9H12O3
40
80
120
160
200
240
280
m / e
UV Spectrum
λmax 267 nm (log10ε 2.8)
solvent : ethanol
13C NMR Spectrum
(100.0 MHz, CDCl3 solution)
DEPT CH2↓ CH3↑ CH↑
proton decoupled
solvent
200
160
120
80
40
0
δ (ppm)
1H NMR Spectrum
(200 MHz, CDCl3 solution)
expansion
7.0
6.4
ppm
expansion
3.8
3.7
ppm
TMS
10
9
8
7
6
5
4
3
2
1
0
δ (ppm)

Problem 44

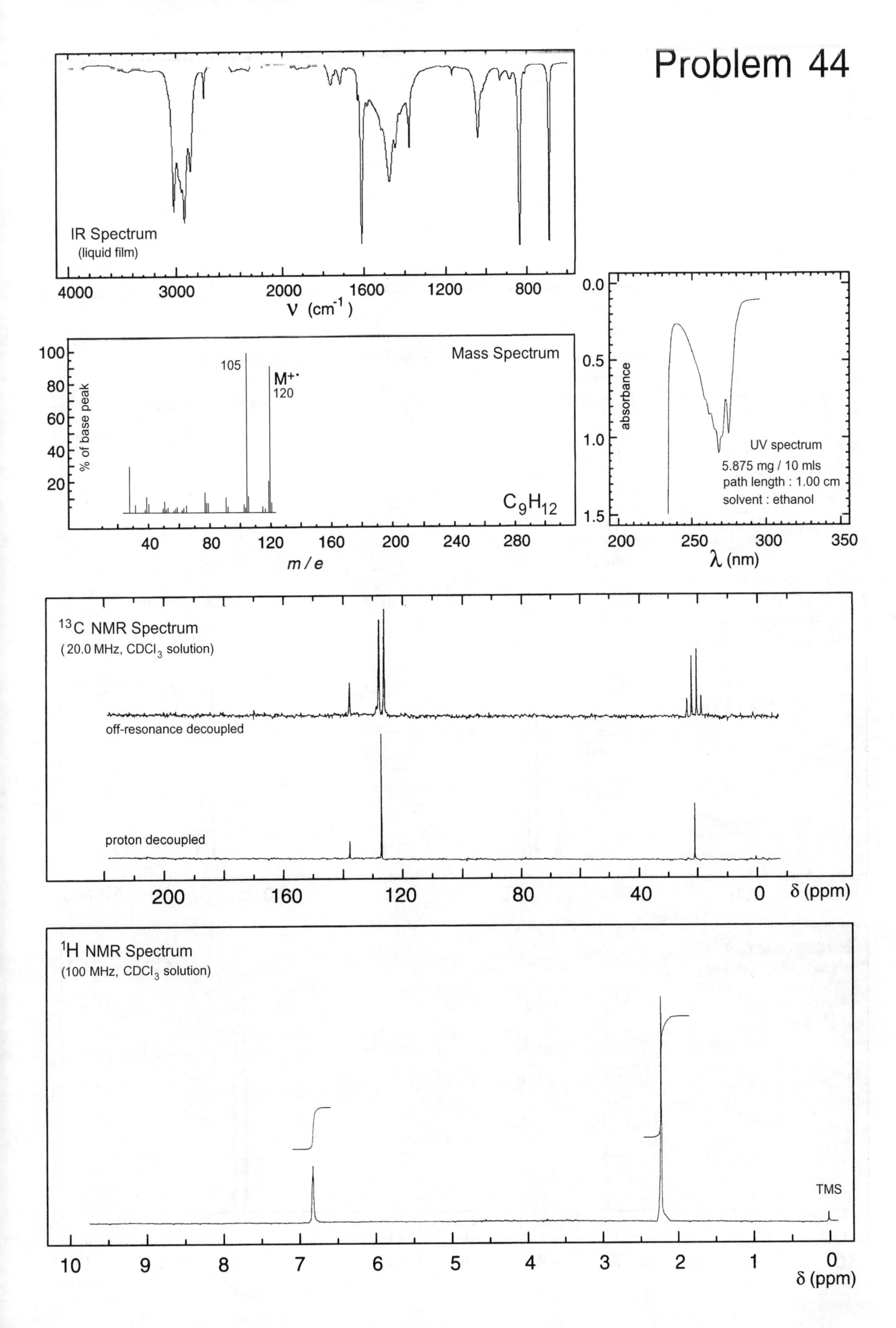

IR Spectrum
(liquid film)
4000
3000
2000
1600
1200
800
ν (cm⁻¹)
Mass Spectrum
100
80
60
40
20
% of base peak
105
M+·
120
C9H12
40
80
120
160
200
240
280
m / e
0.0
0.5
1.0
1.5
absorbance
UV spectrum
5.875 mg / 10 mls
path length : 1.00 cm
solvent : ethanol
200
250
300
350
λ (nm)
13C NMR Spectrum
(20.0 MHz, CDCl3 solution)
off-resonance decoupled
proton decoupled
200
160
120
80
40
0
δ (ppm)
1H NMR Spectrum
(100 MHz, CDCl3 solution)
TMS
10
9
8
7
6
5
4
3
2
1
0
δ (ppm)

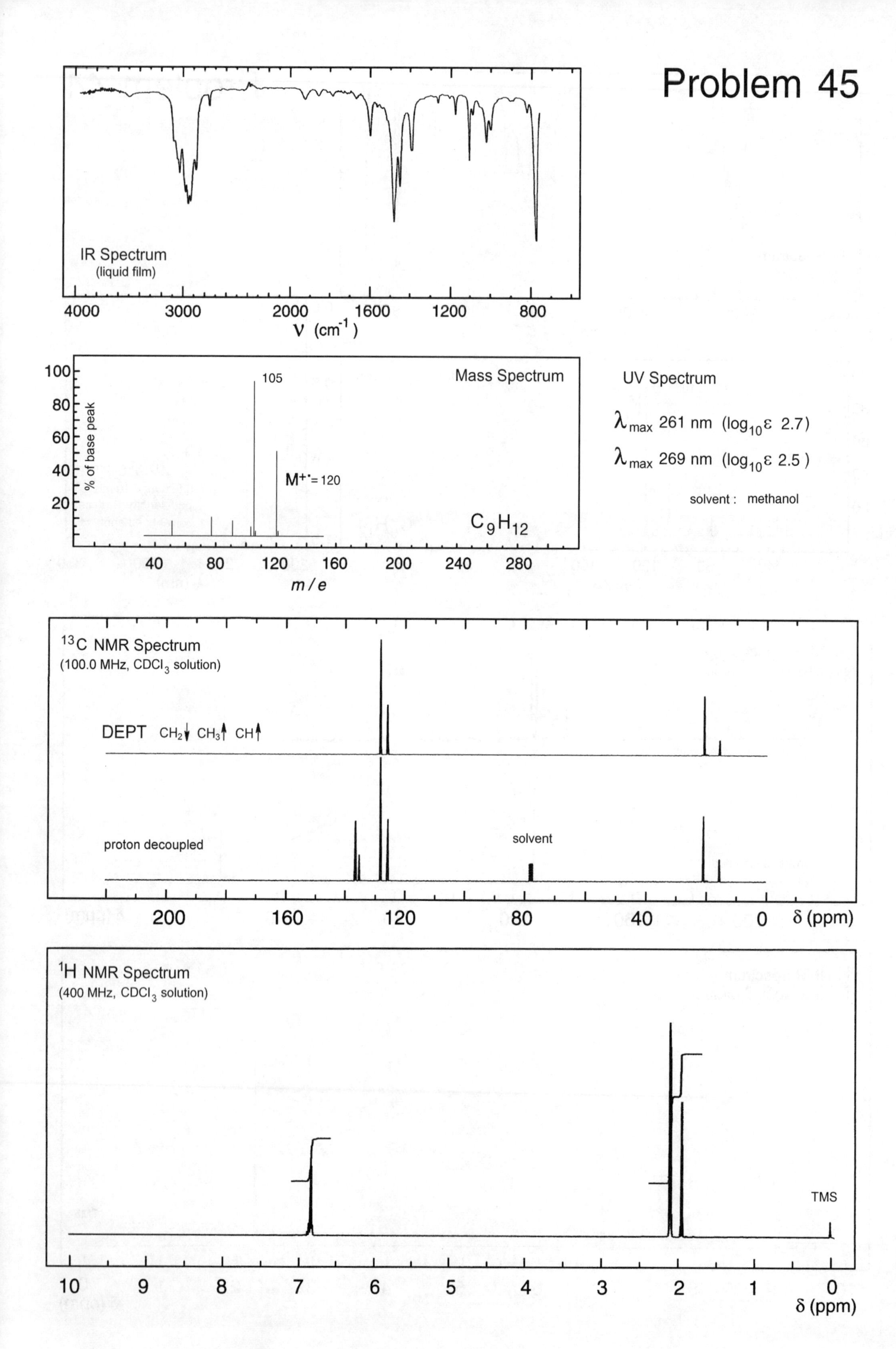
Problem 45
IR Spectrum
(liquid film)
4000
3000
2000
1600
1200
800
ν (cm⁻¹)
100
80
60
40
20
% of base peak
105
Mass Spectrum
M+· = 120
C9H12
40
80
120
160
200
240
280
m / e
UV Spectrum
λmax 261 nm (log10ε 2.7)
λmax 269 nm (log10ε 2.5)
solvent : methanol
13C NMR Spectrum
(100.0 MHz, CDCl3 solution)
DEPT CH2↓ CH3↑ CH↑
proton decoupled
solvent
200
160
120
80
40
0
δ (ppm)
1H NMR Spectrum
(400 MHz, CDCl3 solution)
TMS
10
9
8
7
6
5
4
3
2
1
0
δ (ppm)

Problem 46

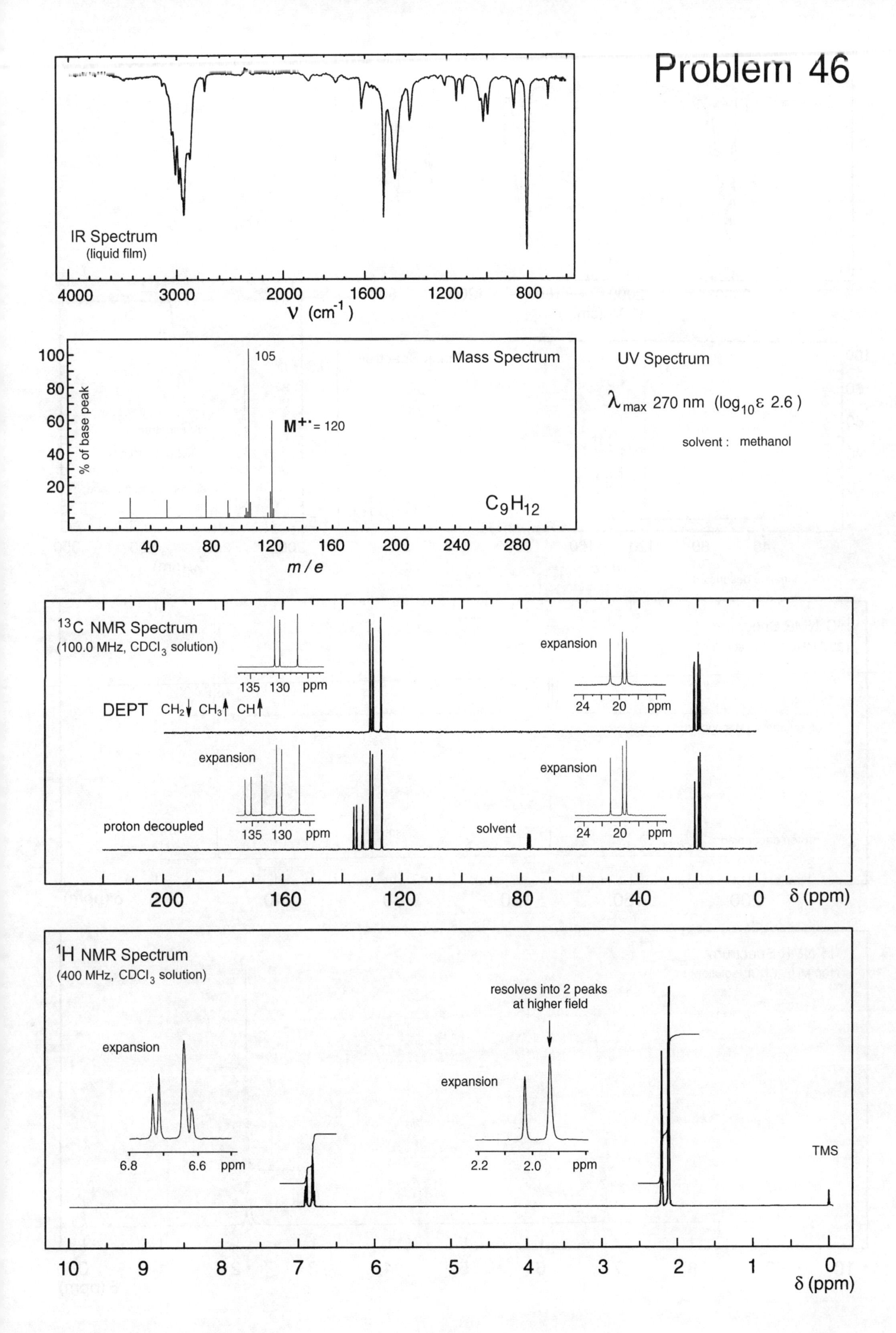

IR Spectrum
(liquid film)
4000
3000
2000
1600
1200
800
ν (cm-1)
Mass Spectrum
% of base peak
100
80
60
40
20
105
M+· = 120
C9H12
40
80
120
160
200
240
280
m / e
UV Spectrum
λmax 270 nm (log10ε 2.6)
solvent : methanol
13C NMR Spectrum
(100.0 MHz, CDCl3 solution)
DEPT CH2↓ CH3↑ CH↑
135 130 ppm
expansion
24 20 ppm
proton decoupled
solvent
200
160
120
80
40
0
δ (ppm)
1H NMR Spectrum
(400 MHz, CDCl3 solution)
resolves into 2 peaks
at higher field
expansion
6.8
6.6
ppm
2.2
2.0
TMS
10
9
8
7
6
5
4
3
2
1
0

Problem 47

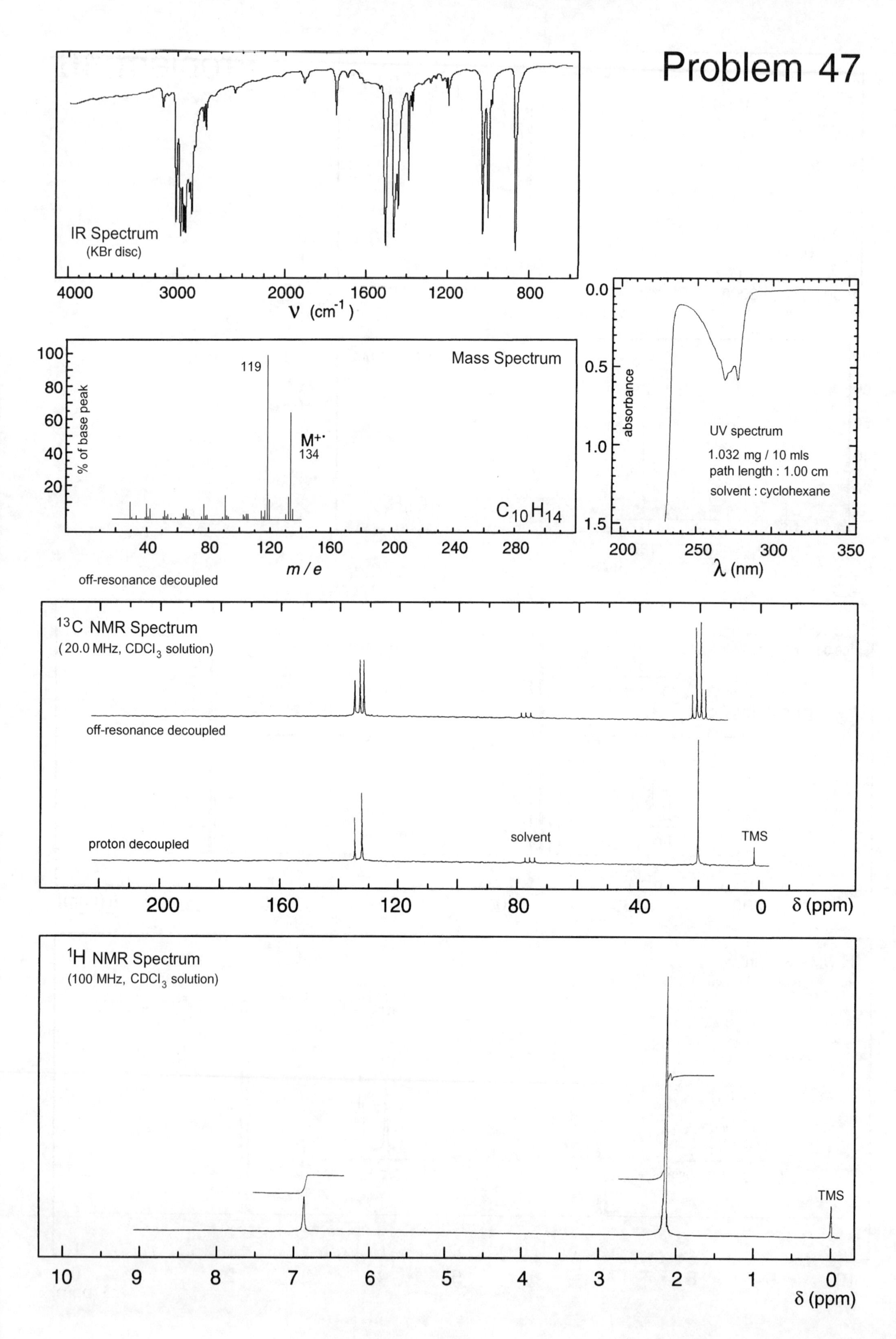

IR Spectrum
(KBr disc)
4000
3000
2000
1600
1200
800
ν (cm^{-1})
Mass Spectrum
100
80
60
40
20
% of base peak
119
M+·
134
$C_{10}H_{14}$
40
80
120
160
200
240
280
m / e
0.0
0.5
1.0
1.5
absorbance
UV spectrum
1.032 mg / 10 mls
path length : 1.00 cm
solvent : cyclohexane
200
250
300
350
λ (nm)
off-resonance decoupled
^{13}C NMR Spectrum
(20.0 MHz, $CDCl_3$ solution)
off-resonance decoupled
proton decoupled
solvent
TMS
200
160
120
80
40
0
δ (ppm)
^{1}H NMR Spectrum
(100 MHz, $CDCl_3$ solution)
TMS
10
9
8
7
6
5
4
3
2
1
0
δ (ppm)

Problem 48

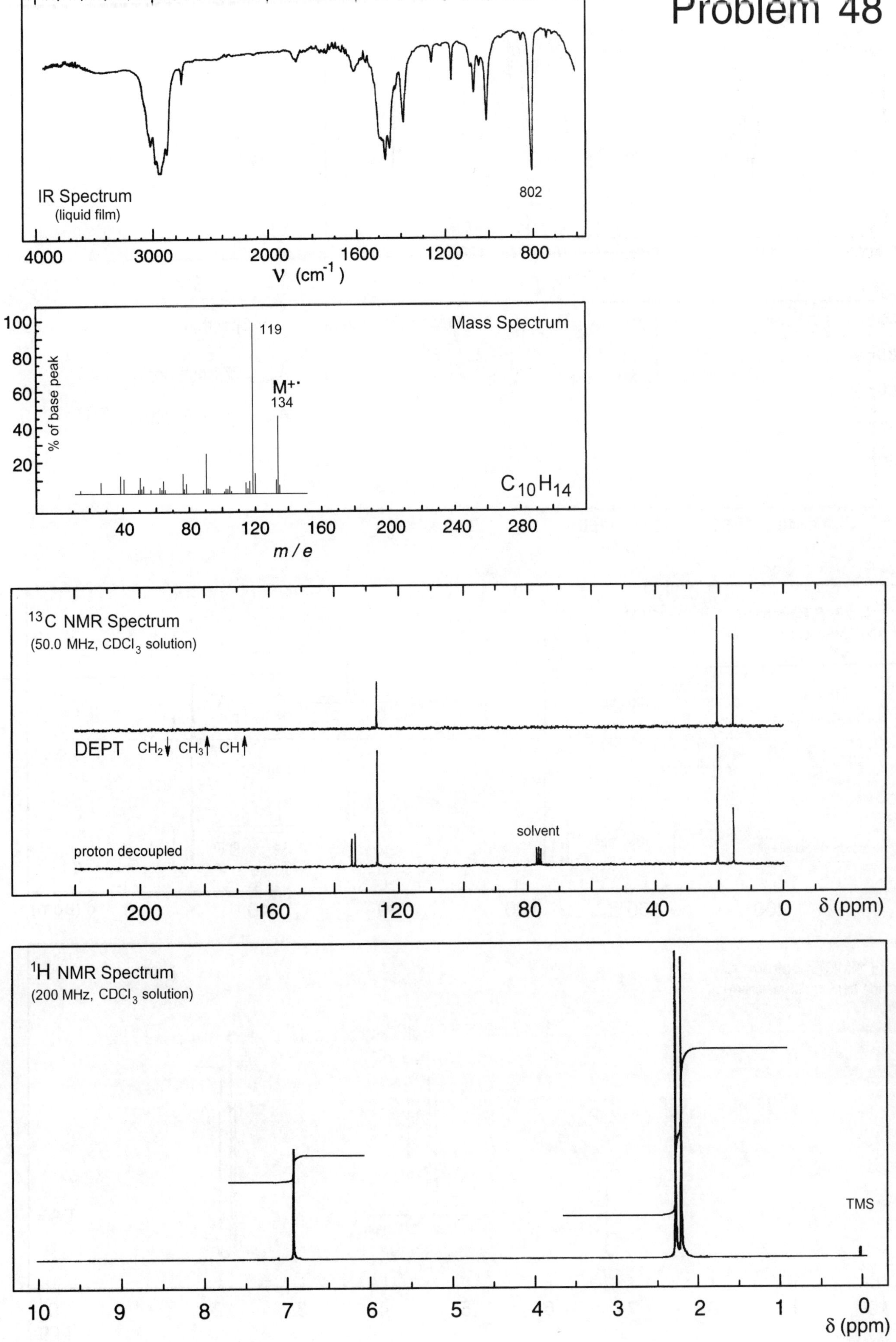

IR Spectrum
(liquid film)
802
4000
3000
2000
1600
1200
800
ν (cm⁻¹)
Mass Spectrum
% of base peak
119
M+·
134
C10H14
m / e
13C NMR Spectrum
(50.0 MHz, CDCl3 solution)
DEPT CH2 CH3 CH
solvent
proton decoupled
δ (ppm)
1H NMR Spectrum
(200 MHz, CDCl3 solution)
TMS
δ (ppm)

Problem 49

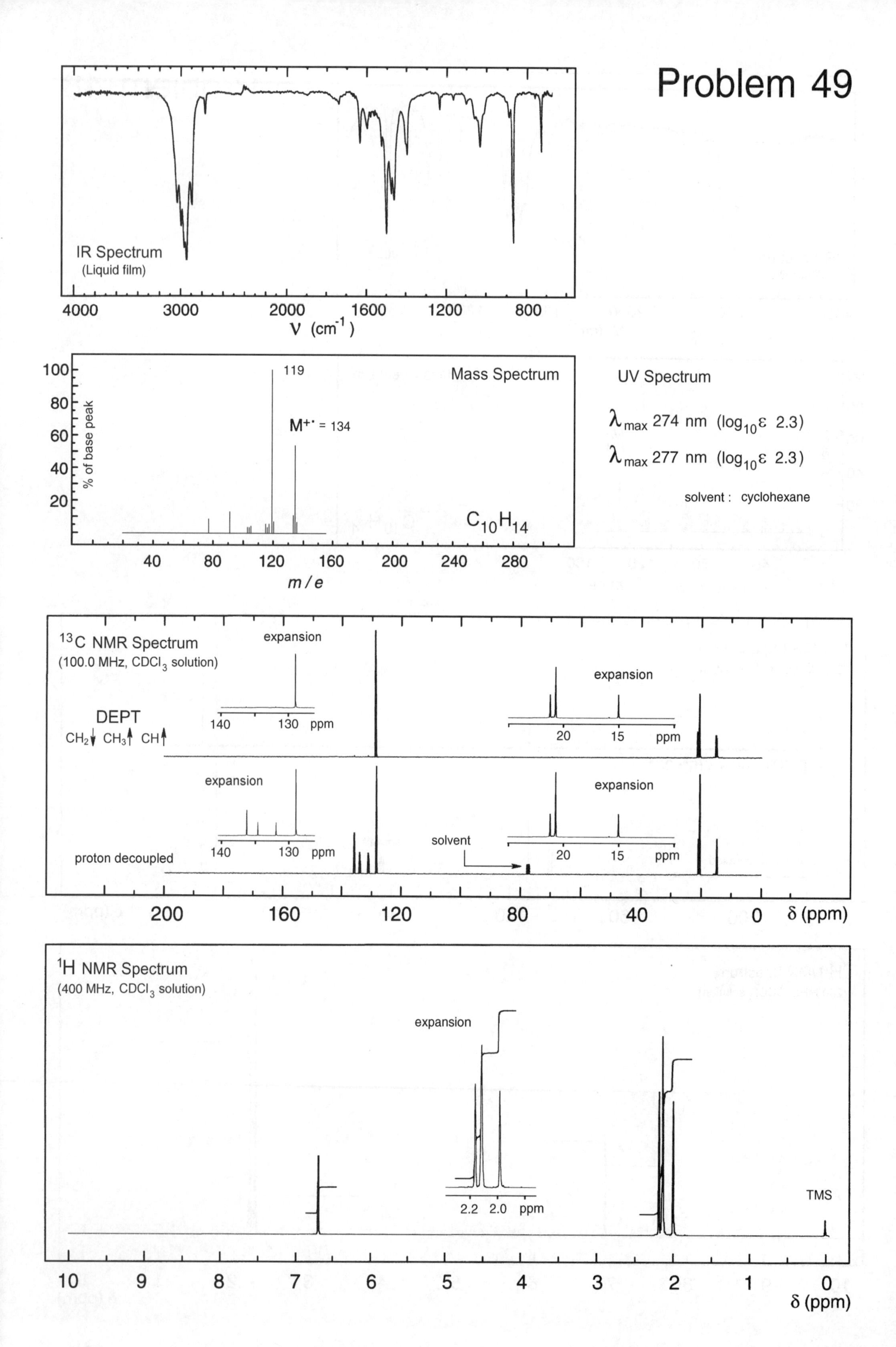

IR Spectrum
(Liquid film)
4000 3000 2000 1600 1200 800
ν (cm⁻¹)
Mass Spectrum
% of base peak
119
M+· = 134
C10H14
m / e
UV Spectrum
λmax 274 nm (log10ε 2.3)
λmax 277 nm (log10ε 2.3)
solvent : cyclohexane
13C NMR Spectrum
(100.0 MHz, CDCl3 solution)
DEPT
CH2↓ CH3↑ CH↑
expansion
proton decoupled
solvent
ppm
δ (ppm)
1H NMR Spectrum
(400 MHz, CDCl3 solution)
expansion
2.2 2.0 ppm
TMS
δ (ppm)

Problem 50

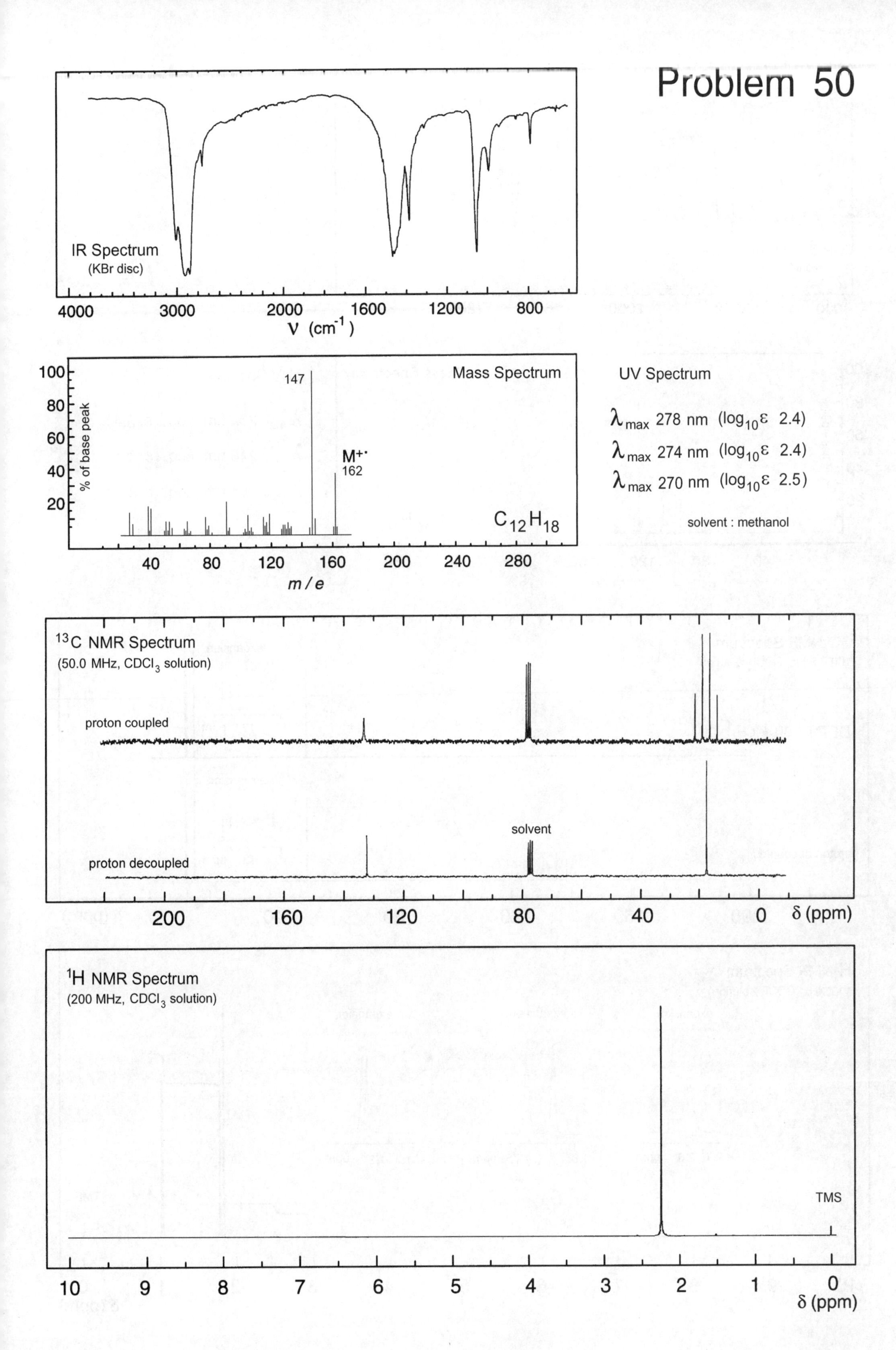

IR Spectrum
(KBr disc)
4000
3000
2000
1600
1200
800
ν (cm⁻¹)
Mass Spectrum
100
80
60
40
20
% of base peak
147
M+·
162
C12H18
40
80
120
160
200
240
280
m / e
UV Spectrum
λmax 278 nm (log10ε 2.4)
λmax 274 nm (log10ε 2.4)
λmax 270 nm (log10ε 2.5)
solvent : methanol
13C NMR Spectrum
(50.0 MHz, CDCl3 solution)
proton coupled
solvent
proton decoupled
200
160
120
80
40
0
δ (ppm)
1H NMR Spectrum
(200 MHz, CDCl3 solution)
TMS
10
9
8
7
6
5
4
3
2
1
0
δ (ppm)

Problem 51

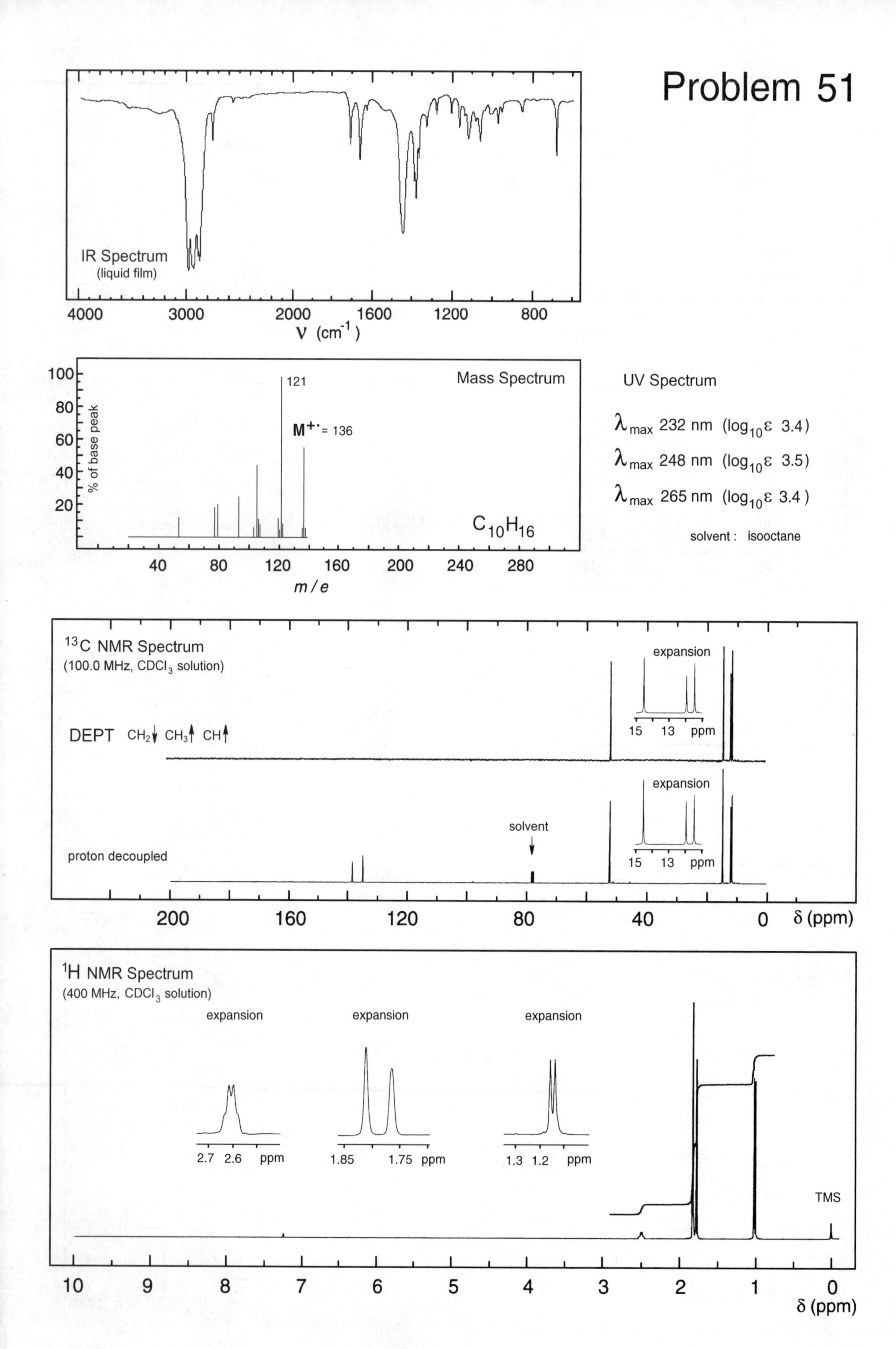
IR Spectrum
(liquid film)
4000 3000 2000 1600 1200 800
ν (cm-1)
Mass Spectrum
100 80 60 40 20
% of base peak
121
M+· = 136
C10H16
40 80 120 160 200 240 280
m / e
UV Spectrum
λmax 232 nm (log10ε 3.4)
λmax 248 nm (log10ε 3.5)
λmax 265 nm (log10ε 3.4)
solvent : isooctane
13C NMR Spectrum
(100.0 MHz, CDCl3 solution)
DEPT CH2↓ CH3↑ CH↑
expansion
15 13 ppm
proton decoupled
solvent
200 160 120 80 40 0 δ (ppm)
1H NMR Spectrum
(400 MHz, CDCl3 solution)
expansion
2.7 2.6 ppm
1.85 1.75 ppm
1.3 1.2 ppm
TMS
10 9 8 7 6 5 4 3 2 1 0
δ (ppm)

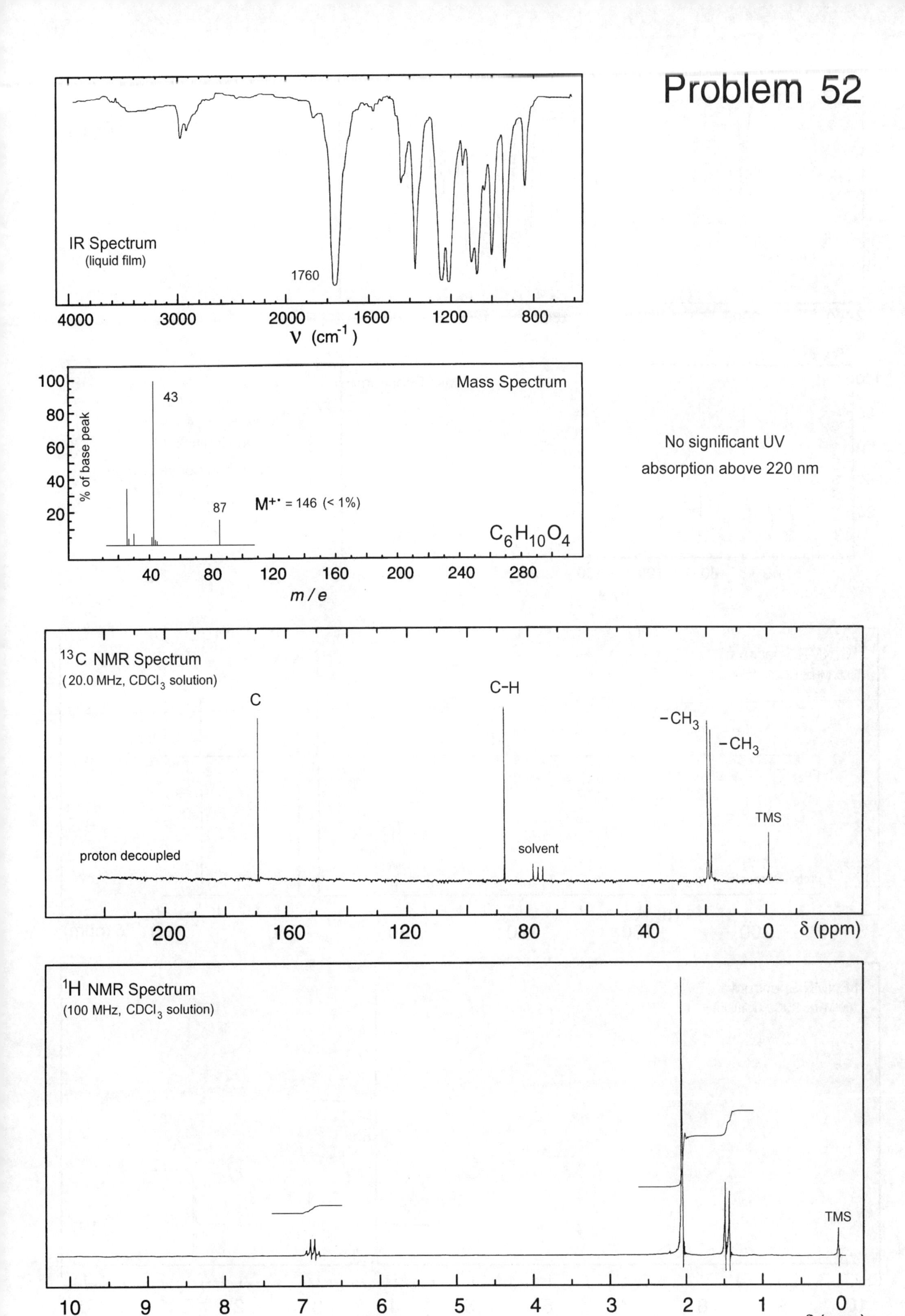

Problem 52
IR Spectrum
(liquid film)
1760
4000
3000
2000
1600
1200
800
ν (cm⁻¹)
Mass Spectrum
100
80
60
40
20
% of base peak
43
87
M⁺˙ = 146 (< 1%)
C6H10O4
40
80
120
160
200
240
280
m / e
No significant UV
absorption above 220 nm
13C NMR Spectrum
(20.0 MHz, CDCl3 solution)
C
C-H
-CH3
-CH3
TMS
solvent
proton decoupled
200
160
120
80
40
0
δ (ppm)
1H NMR Spectrum
(100 MHz, CDCl3 solution)
TMS
10
9
8
7
6
5
4
3
2
1
0
δ (ppm)

Problem 53

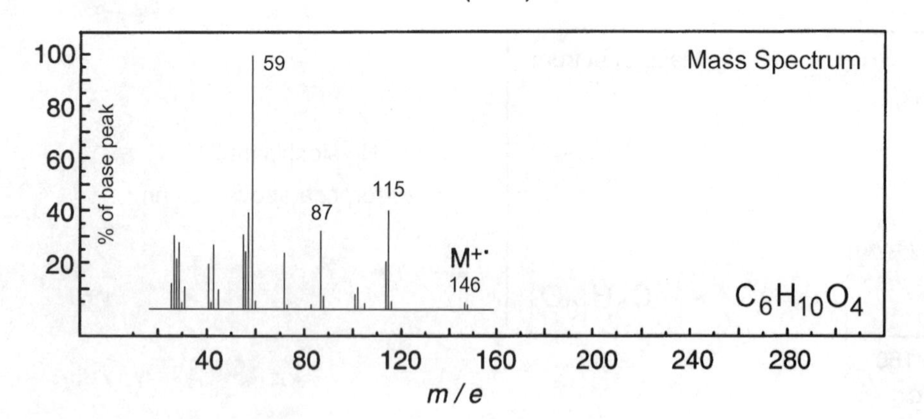

No significant UV absorption above 220 nm

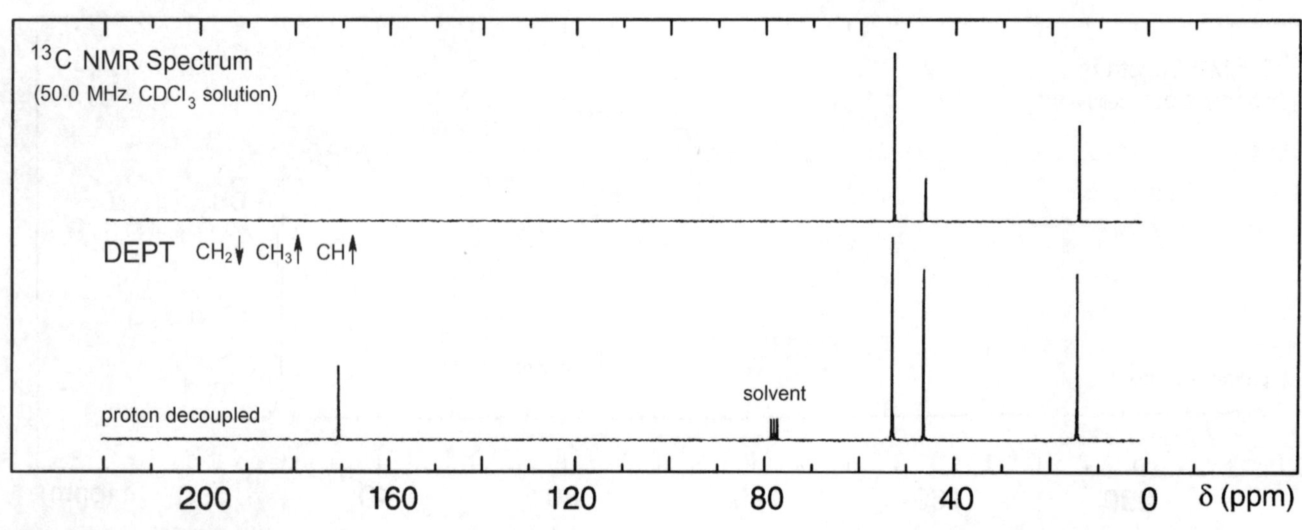

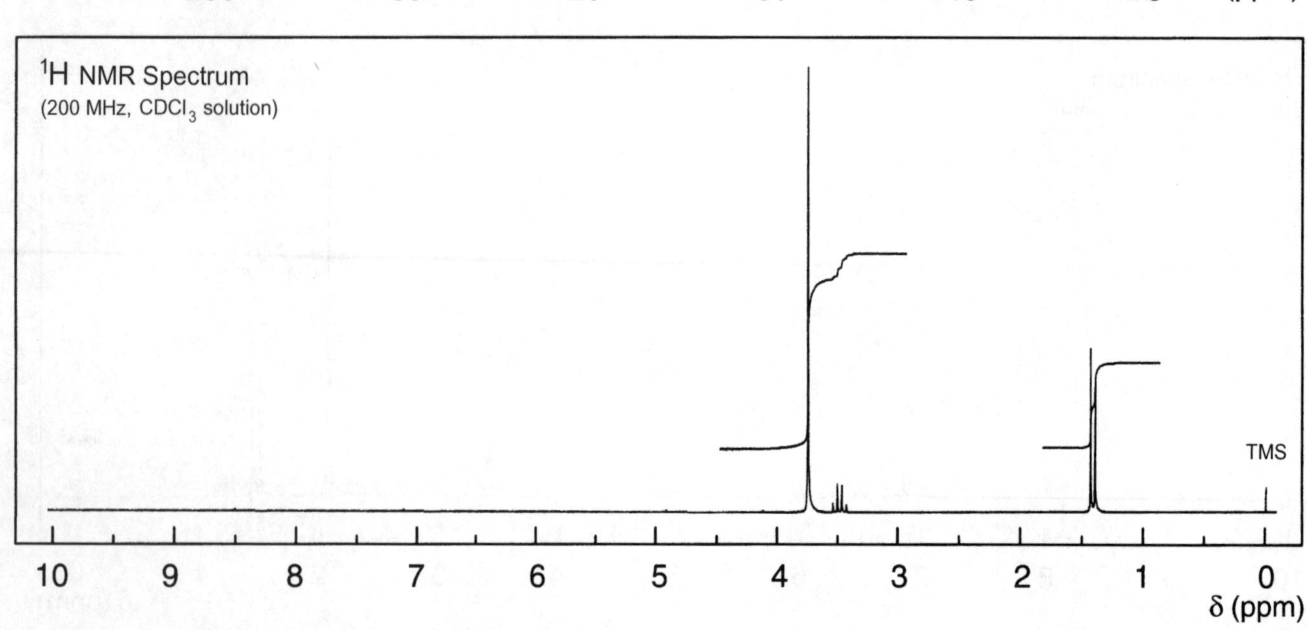

Problem 54

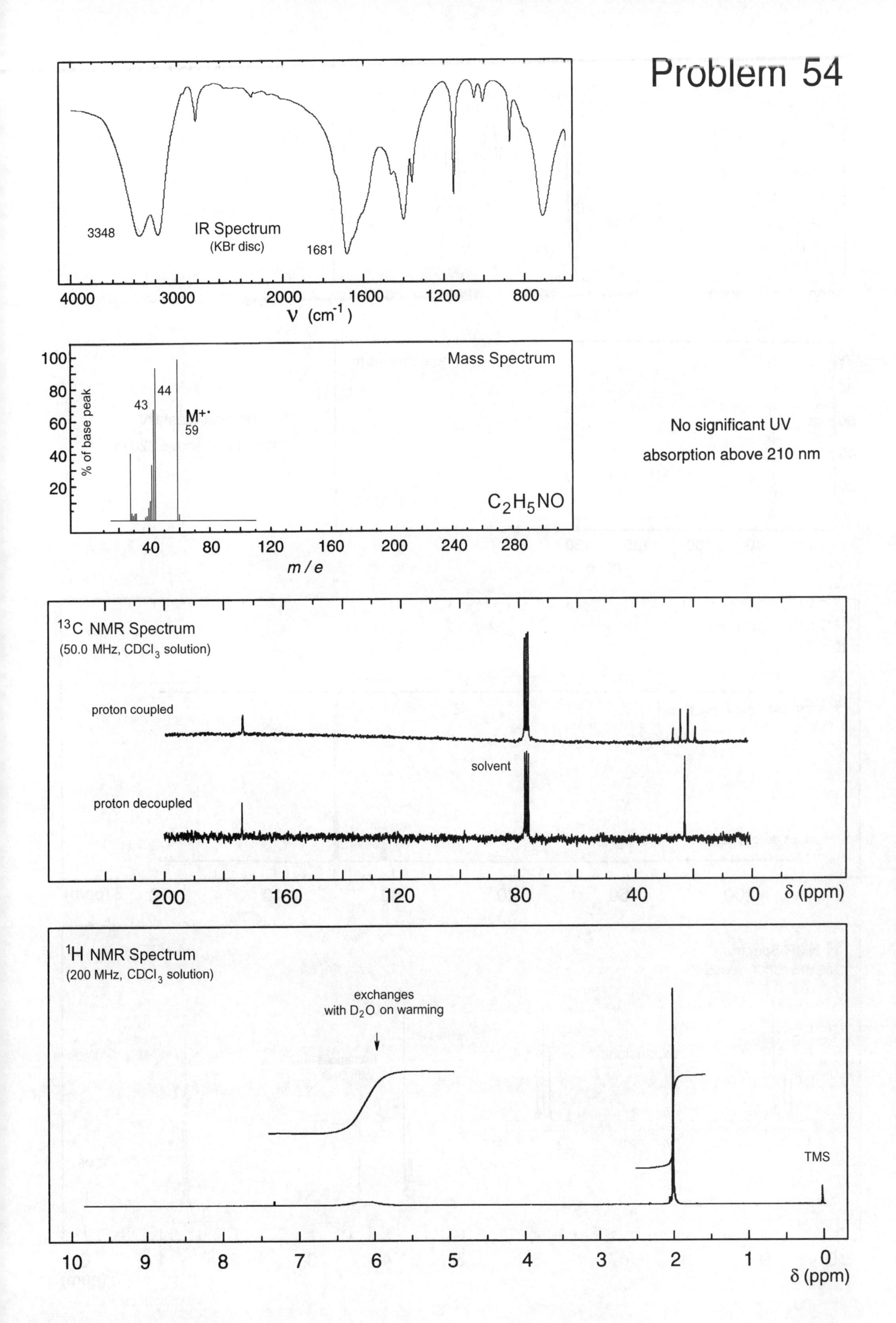

No significant UV absorption above 210 nm

Problem 55

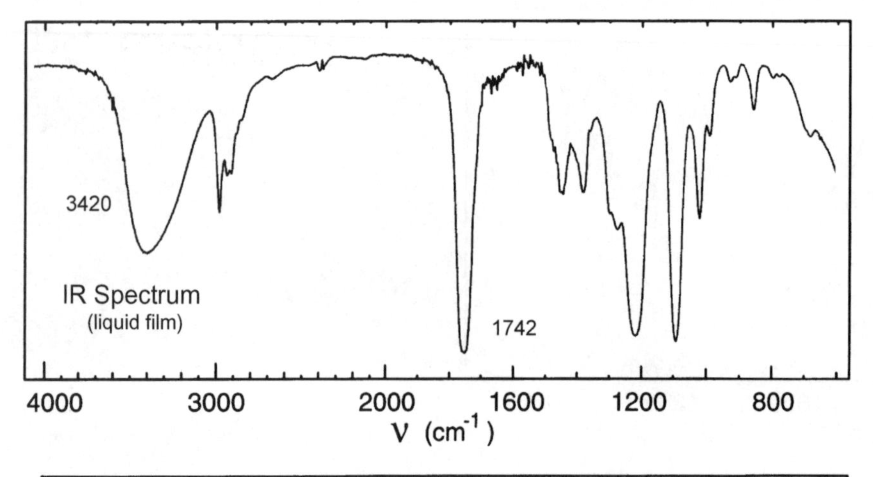

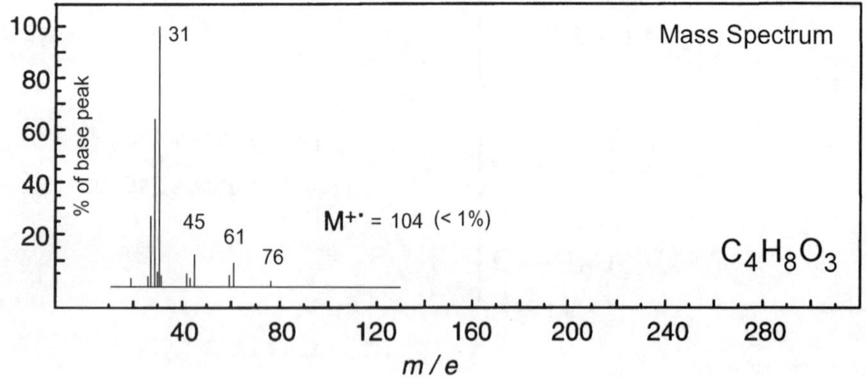

No significant UV absorption above 220 nm

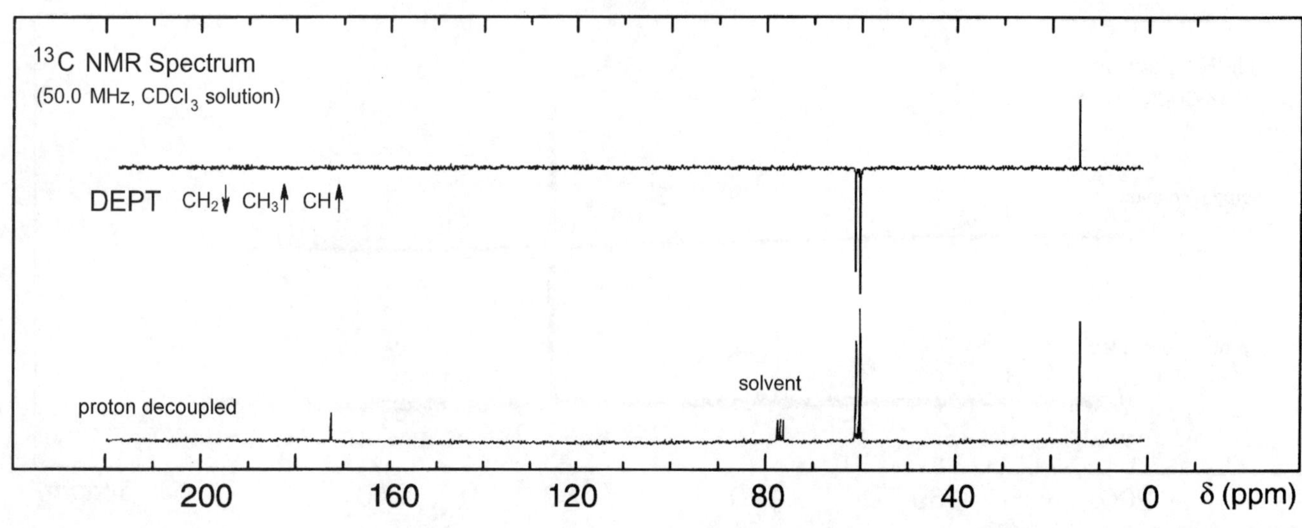

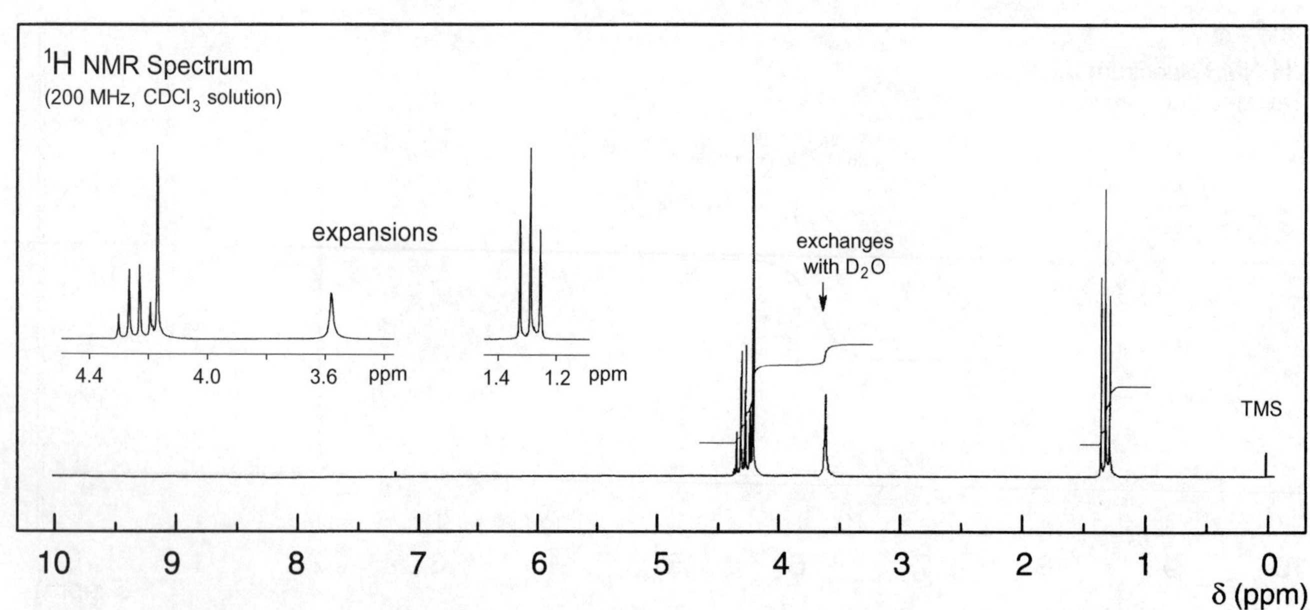

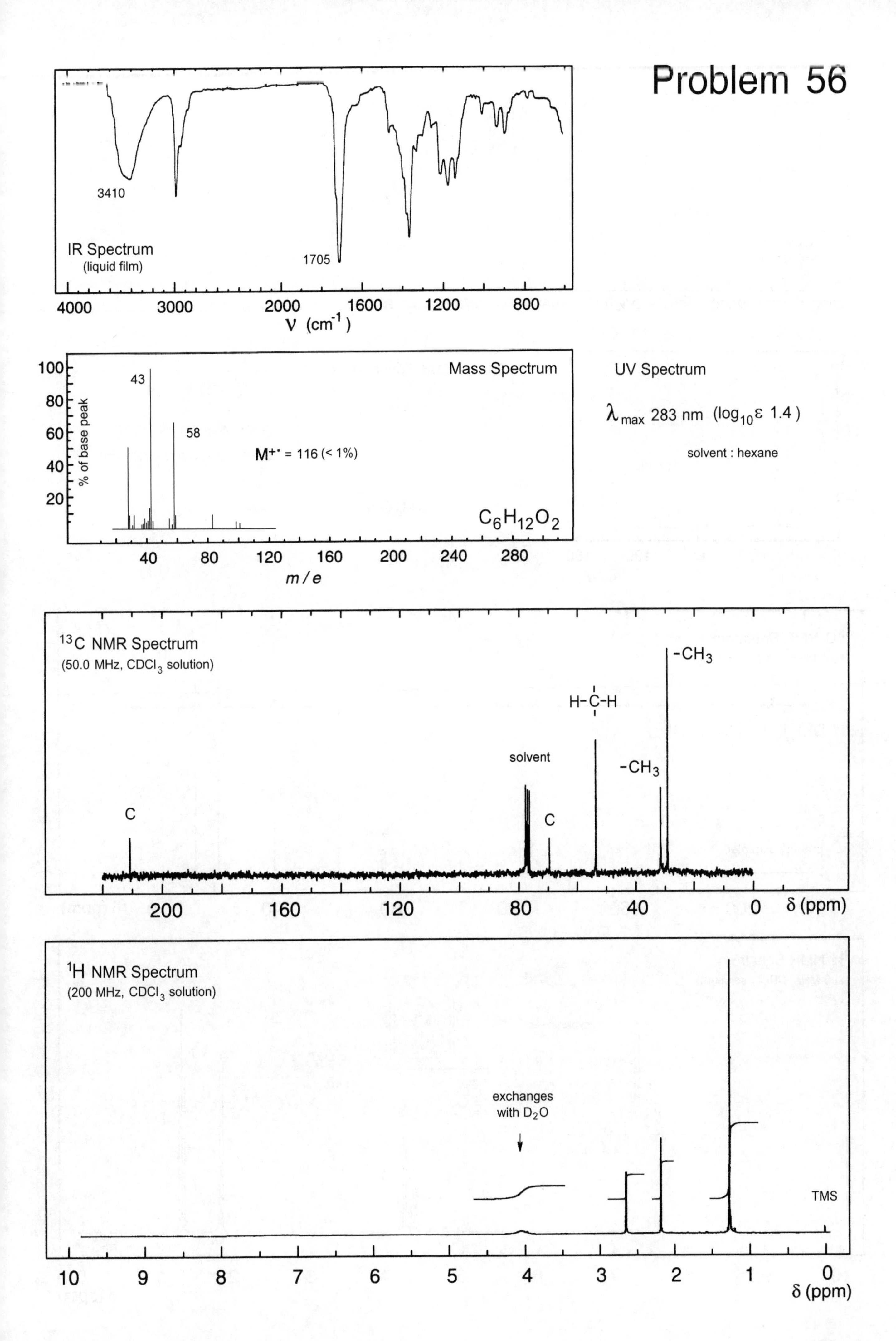
Problem 56
3410
IR Spectrum
(liquid film)
1705
4000
3000
2000
1600
1200
800
ν (cm⁻¹)
Mass Spectrum
100
80
60
40
20
% of base peak
43
58
M⁺˙ = 116 (< 1%)
C6H12O2
40
80
120
160
200
240
280
m / e
UV Spectrum
λmax 283 nm (log10ε 1.4)
solvent : hexane
13C NMR Spectrum
(50.0 MHz, CDCl3 solution)
-CH3
H-C-H
solvent
-CH3
C
C
200
160
120
80
40
0
δ (ppm)
1H NMR Spectrum
(200 MHz, CDCl3 solution)
exchanges
with D2O
TMS
10
9
8
7
6
5
4
3
2
1
0
δ (ppm)

Problem 57

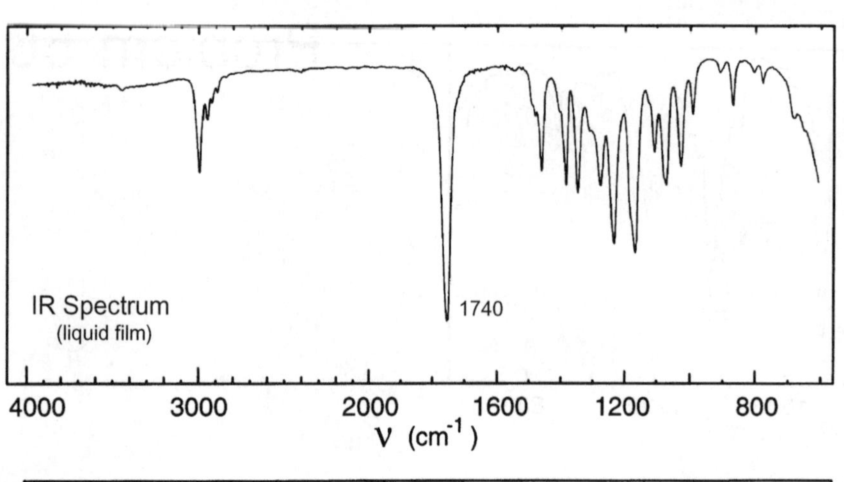

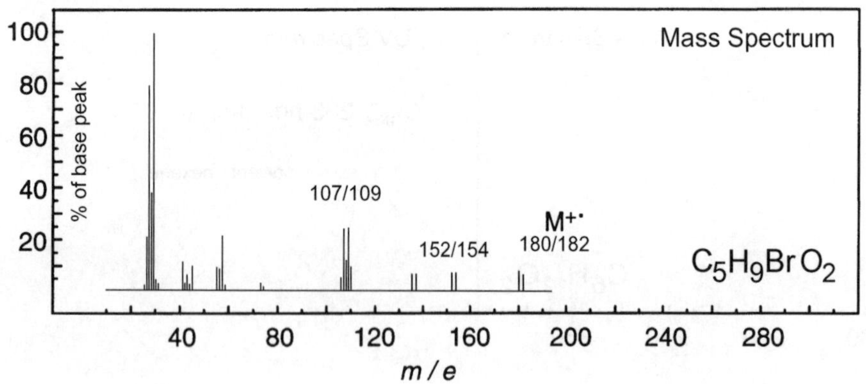

No significant UV absorption above 220 nm

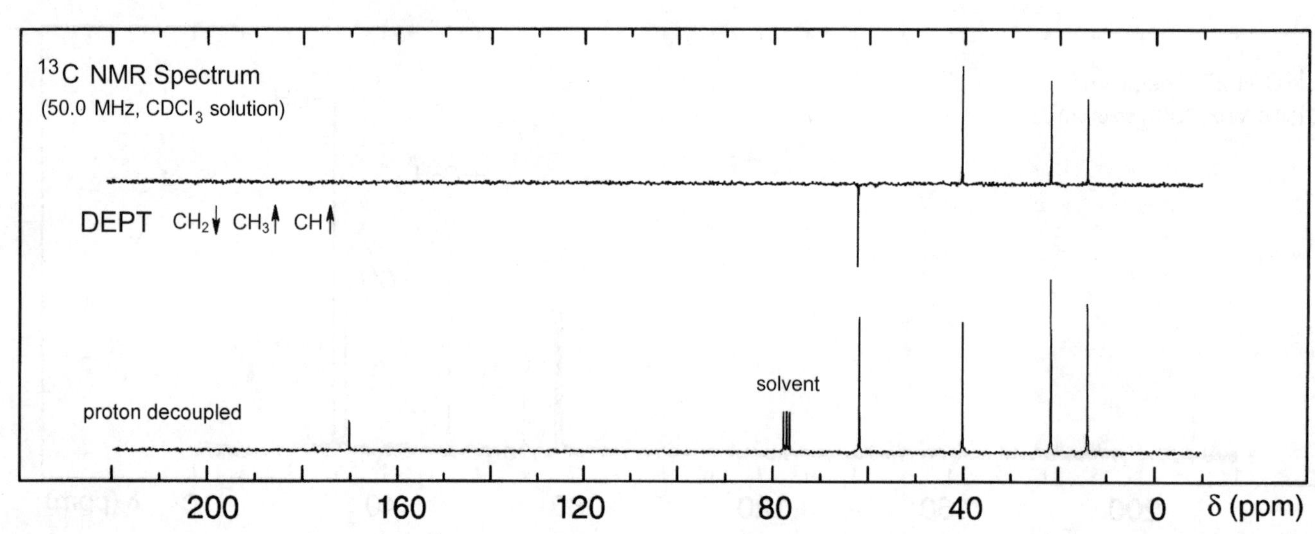

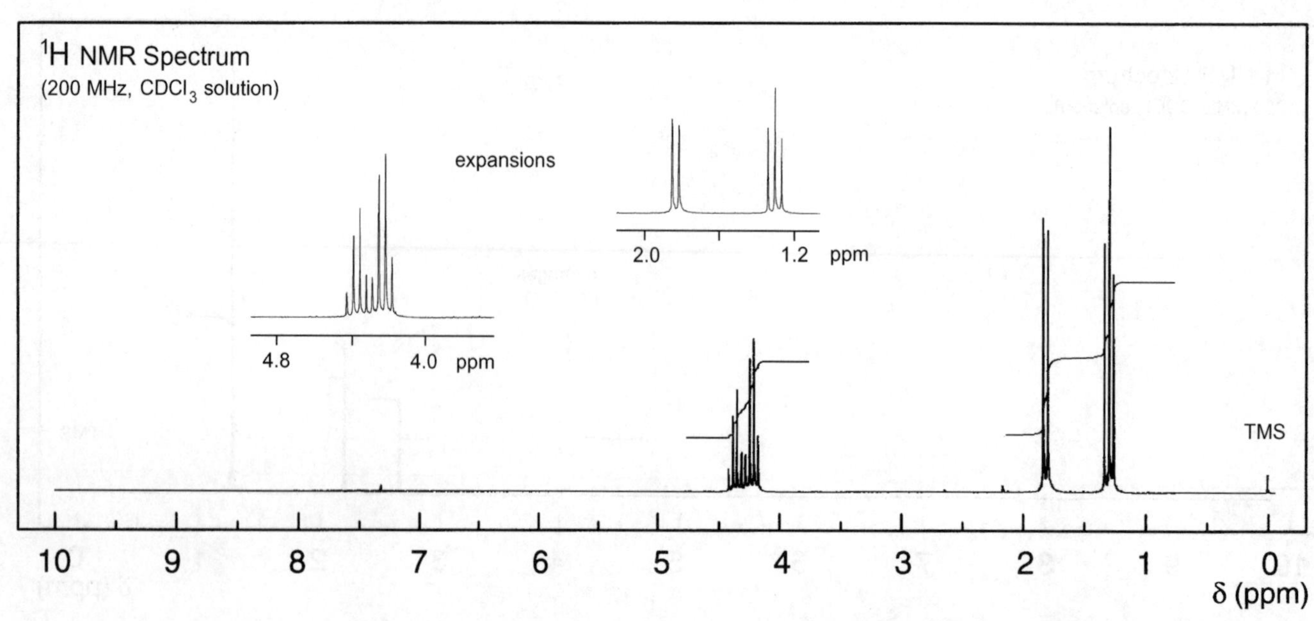

Problem 58

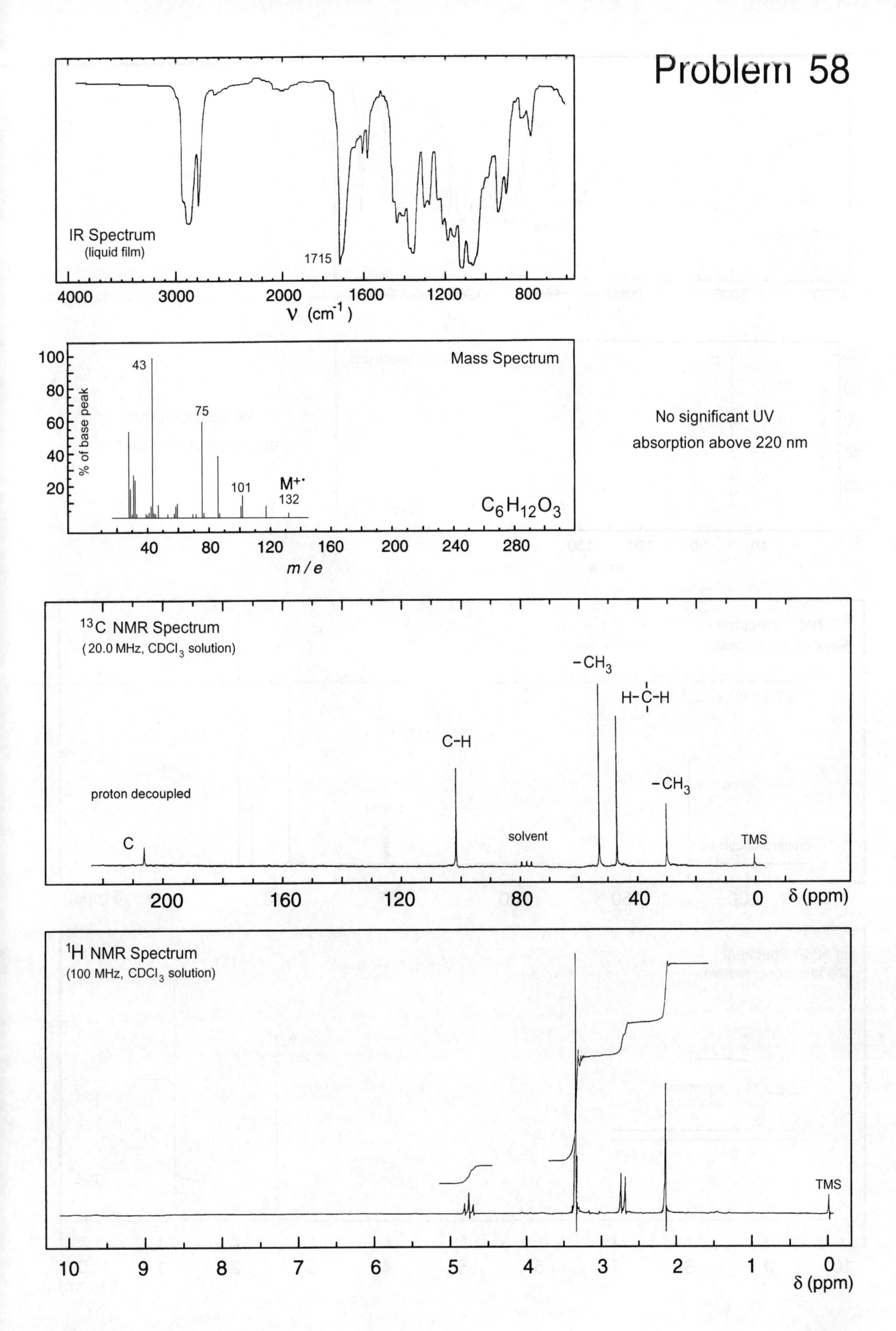

IR Spectrum
(liquid film)
1715
4000
3000
2000
1600
1200
800
ν (cm⁻¹)
Mass Spectrum
% of base peak
100
80
60
40
20
43
75
101
M+·
132
C6H12O3
40
80
120
160
200
240
280
m / e
No significant UV
absorption above 220 nm
13C NMR Spectrum
(20.0 MHz, CDCl3 solution)
proton decoupled
C
C-H
solvent
-CH3
H-C-H
-CH3
TMS
200
160
120
80
40
0
δ (ppm)
1H NMR Spectrum
(100 MHz, CDCl3 solution)
TMS
10
9
8
7
6
5
4
3
2
1
0
δ (ppm)

Problem 59

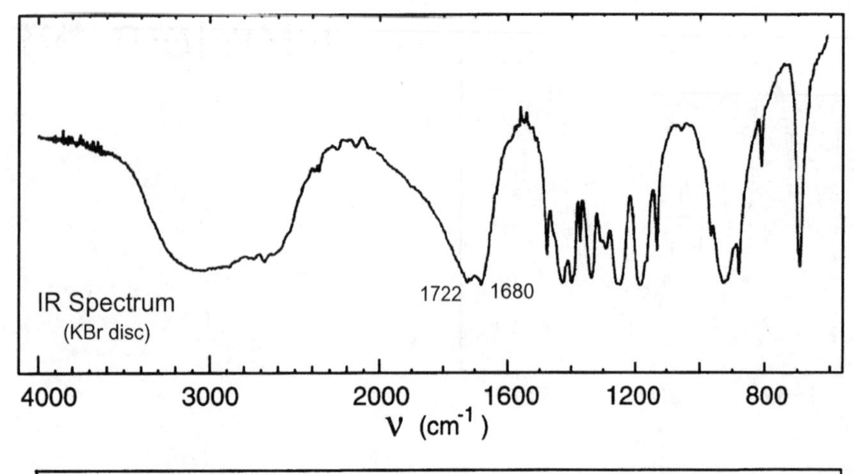

No significant UV absorption above 220 nm

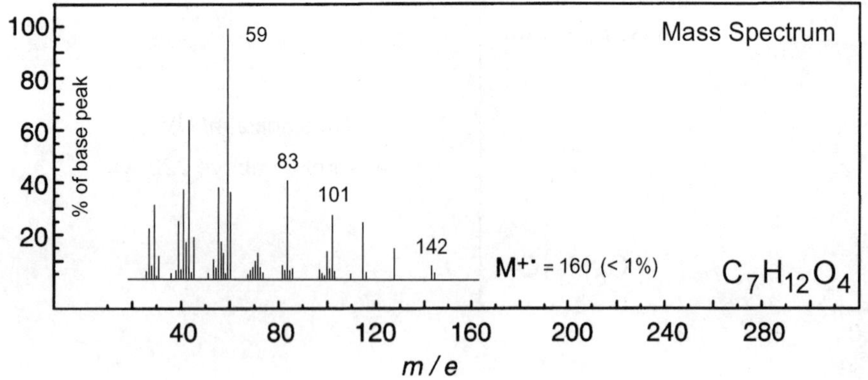

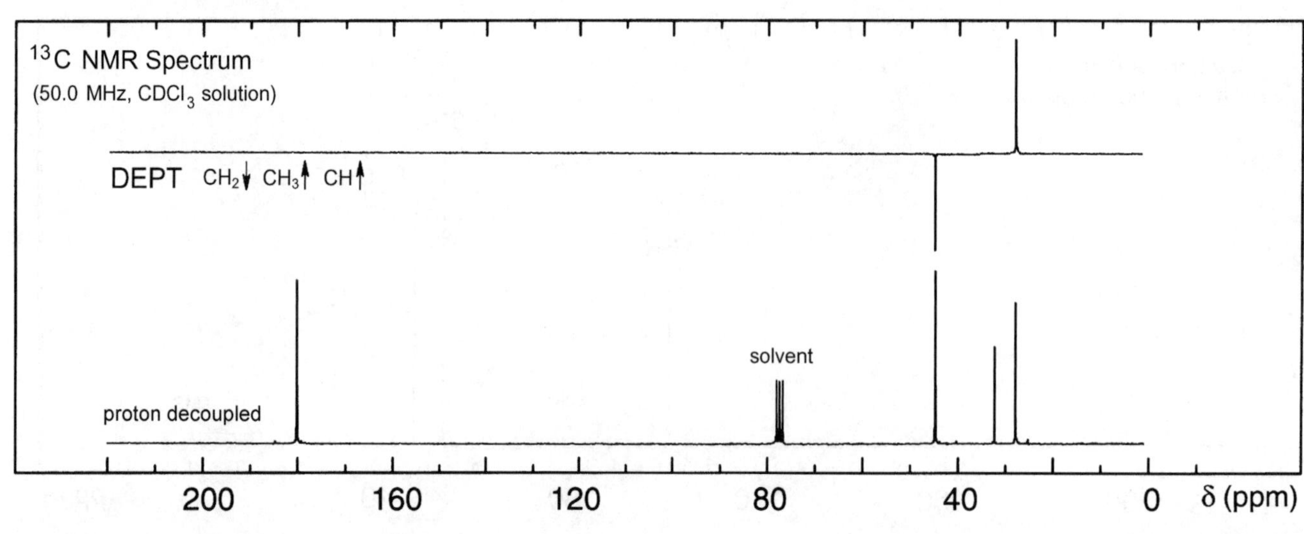

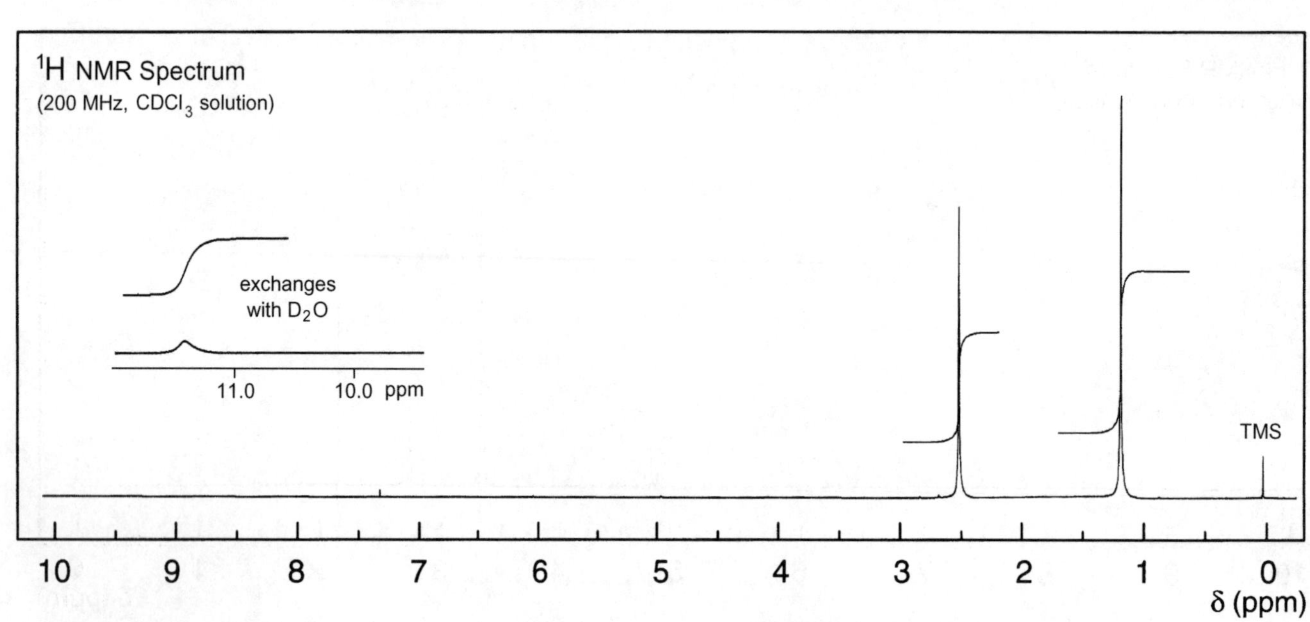

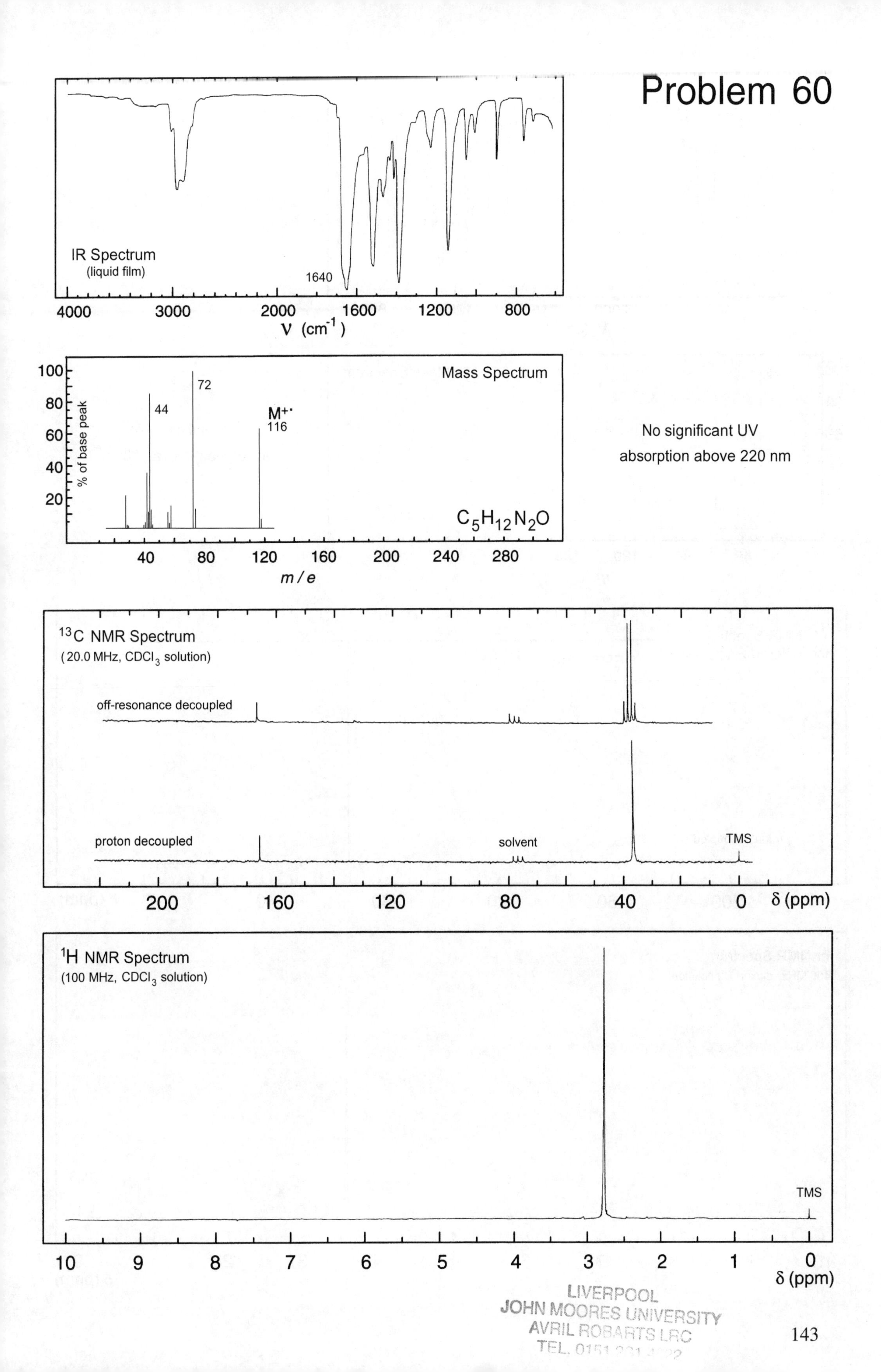
Problem 60
IR Spectrum
(liquid film)
1640
4000
3000
2000
1600
1200
800
ν (cm-1)
Mass Spectrum
% of base peak
100
80
60
40
20
72
44
M+·
116
C5H12N2O
40
80
120
160
200
240
280
m / e
No significant UV
absorption above 220 nm
13C NMR Spectrum
(20.0 MHz, CDCl3 solution)
off-resonance decoupled
proton decoupled
solvent
TMS
200
160
120
80
40
0
δ (ppm)
1H NMR Spectrum
(100 MHz, CDCl3 solution)
TMS
10
9
8
7
6
5
4
3
2
1
0
δ (ppm)

Problem 61

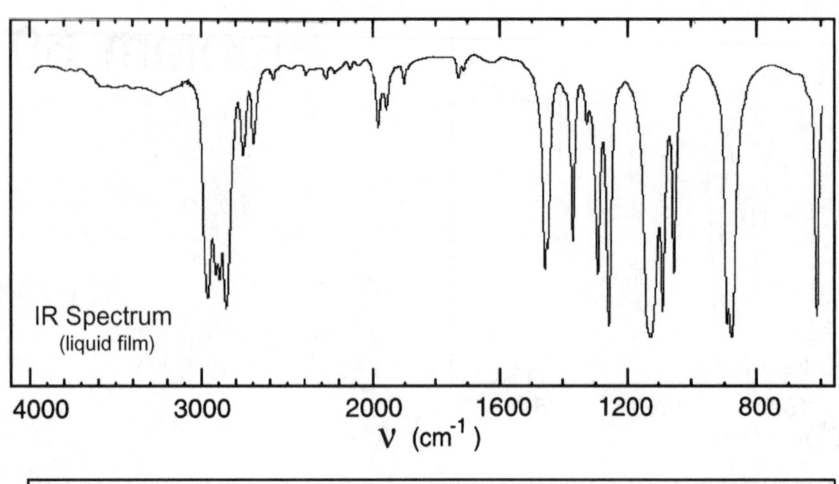

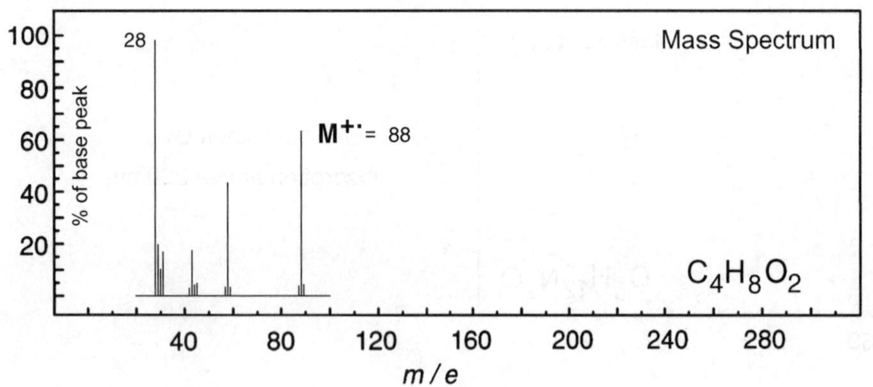

No significant UV absorption above 220 nm

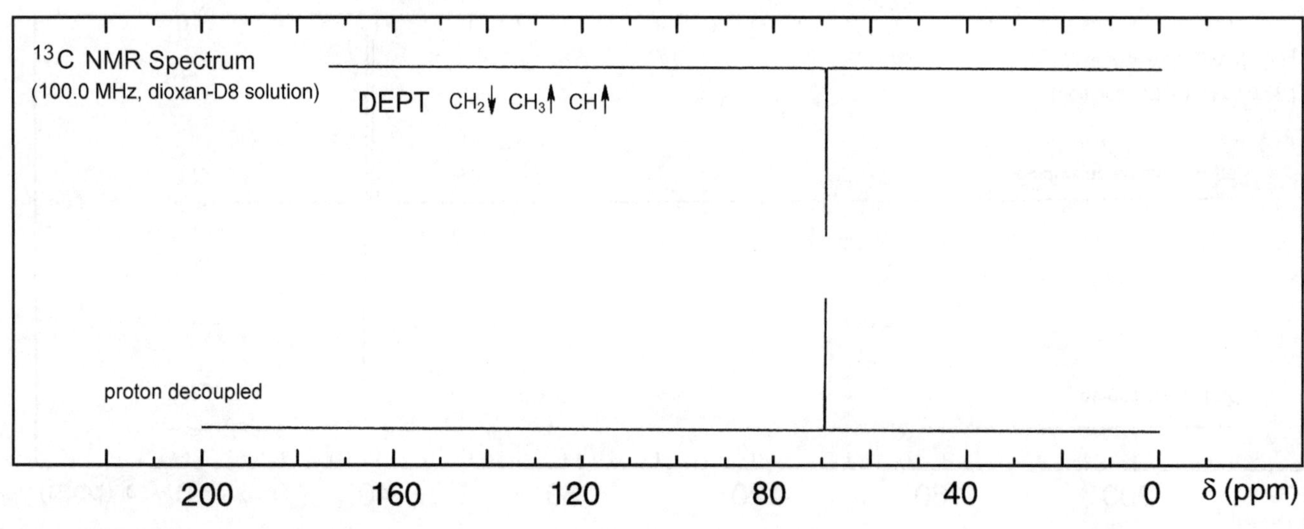

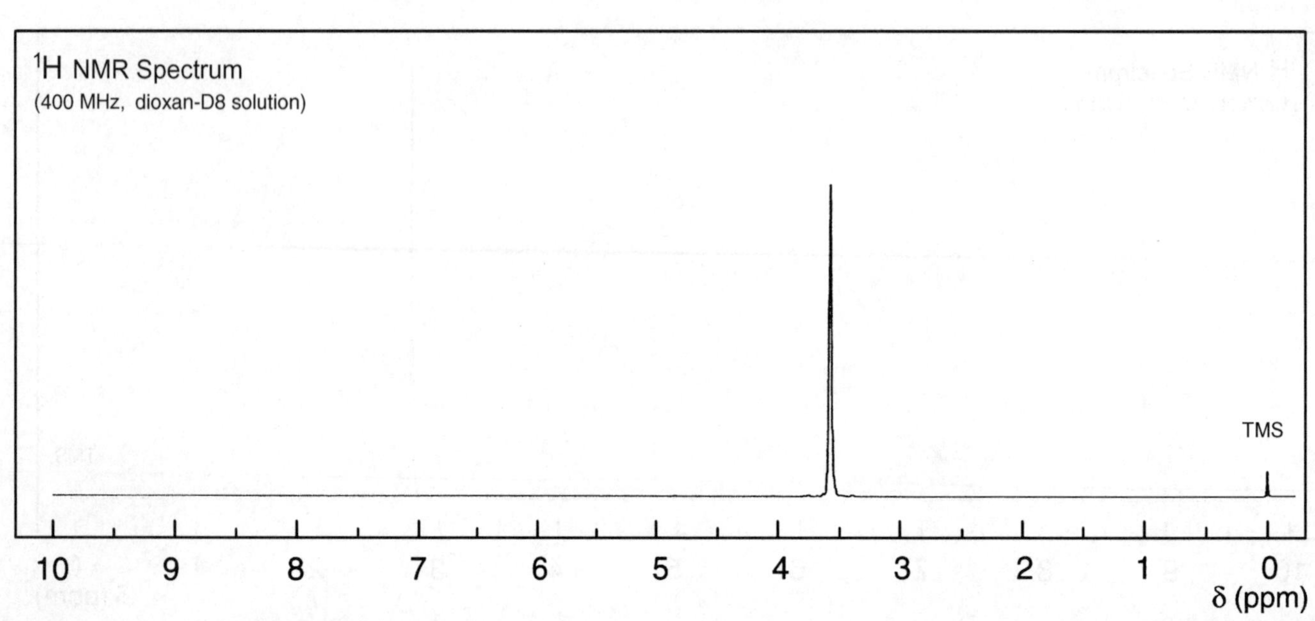

Problem 62

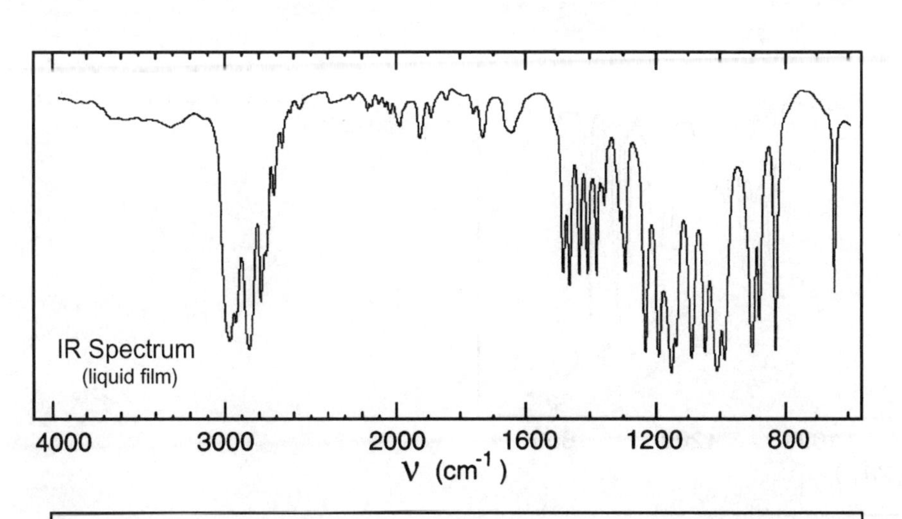

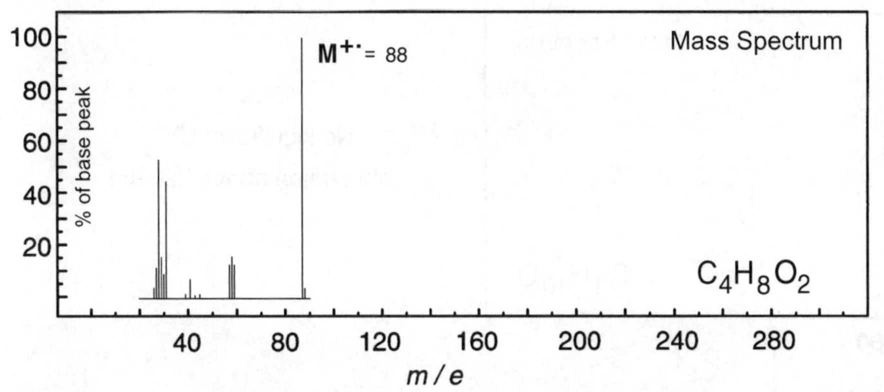

No significant UV absorption above 220 nm

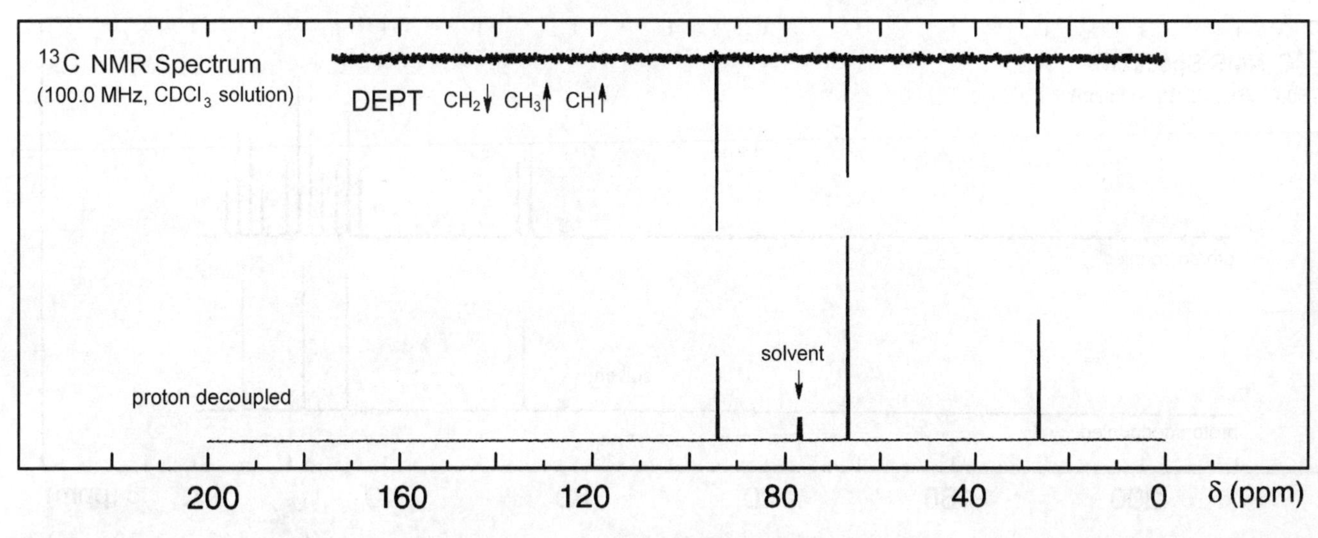

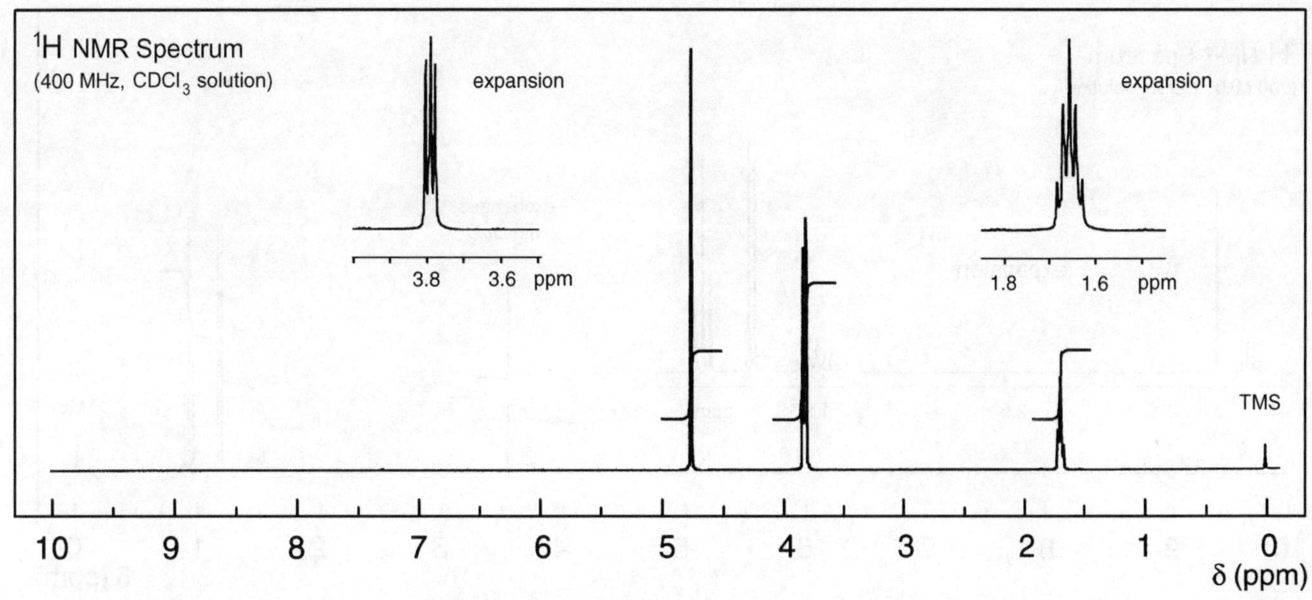

Problem 63

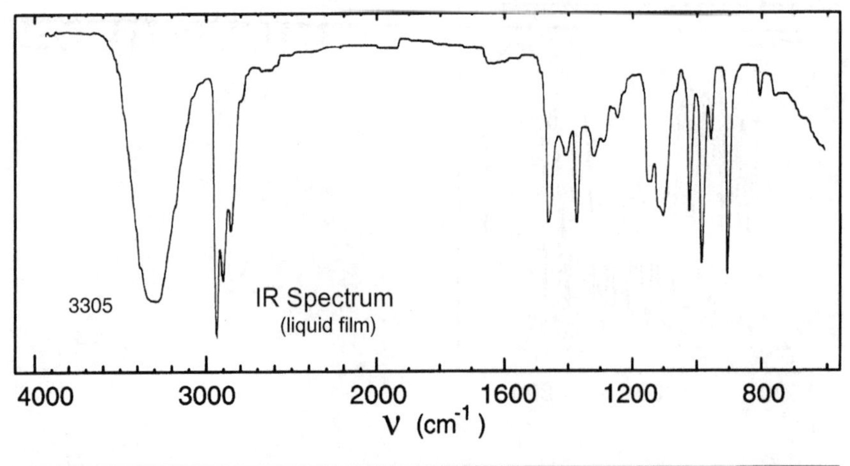

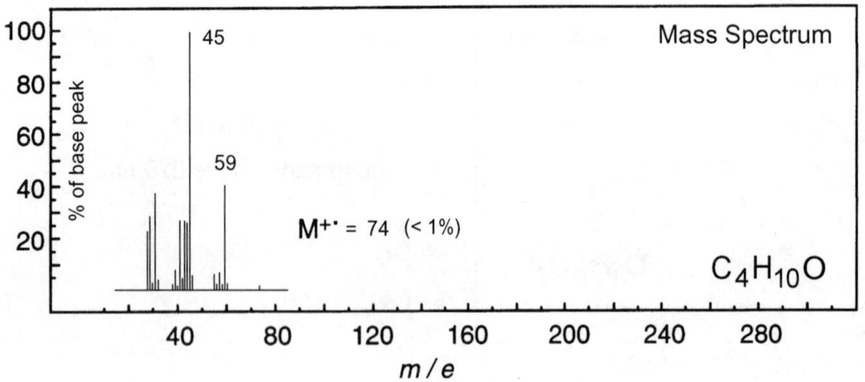

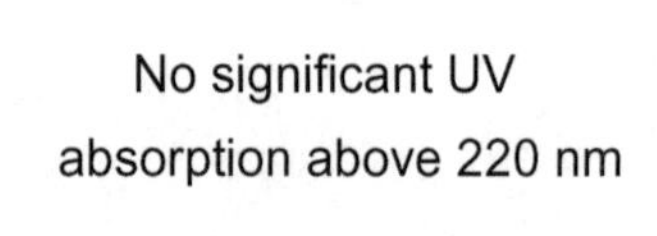
No significant UV absorption above 220 nm

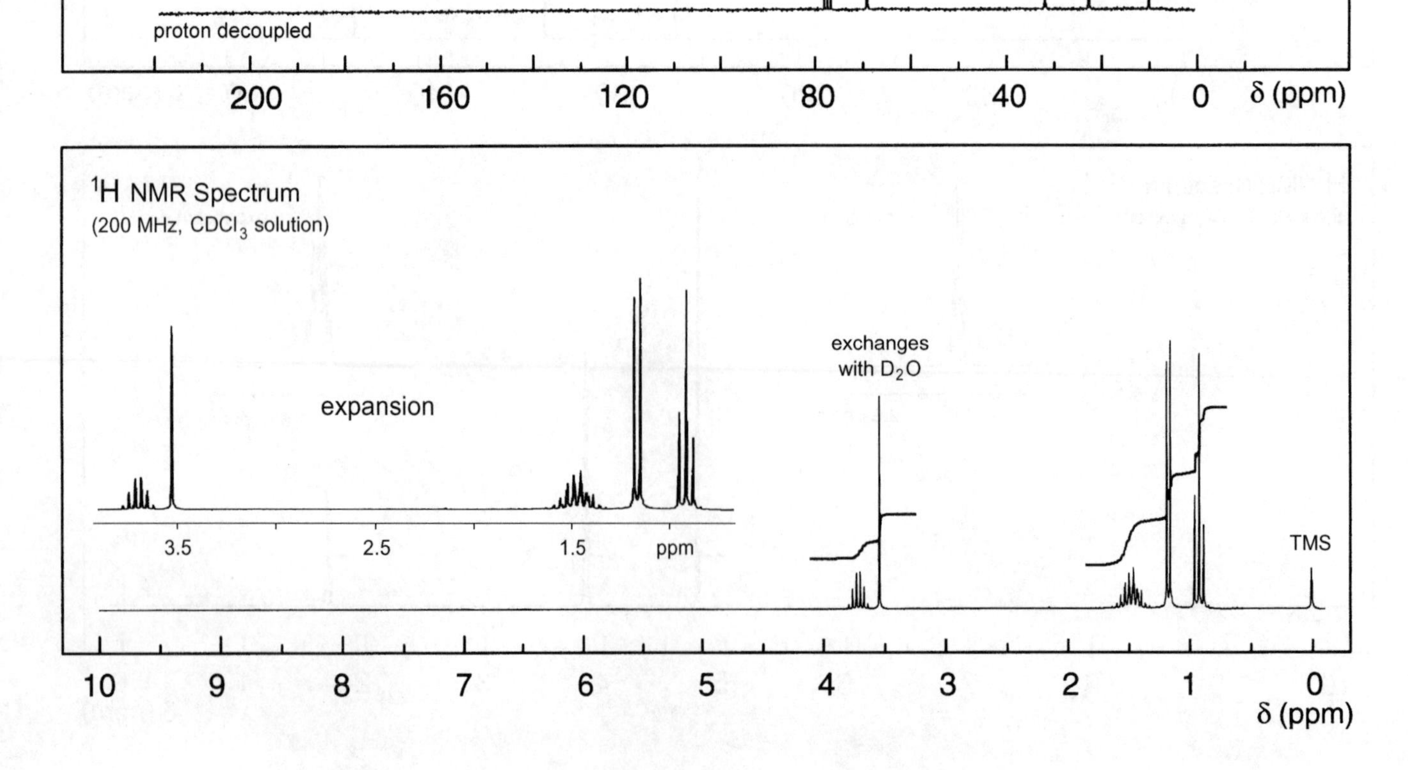

Problem 64

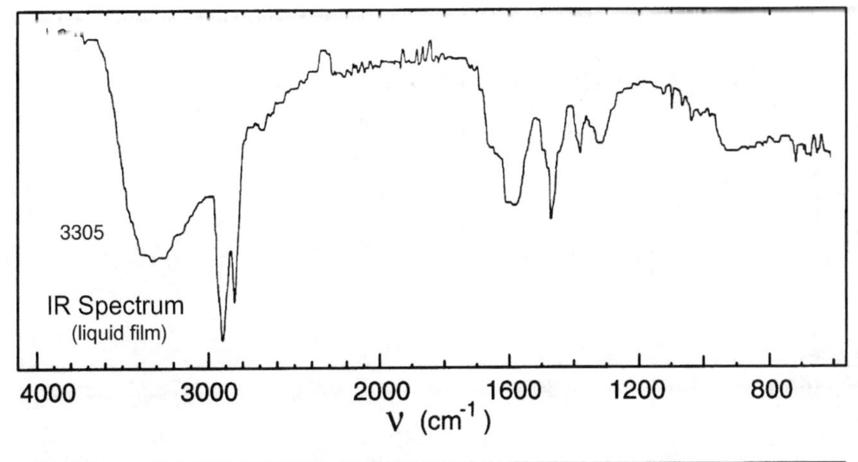

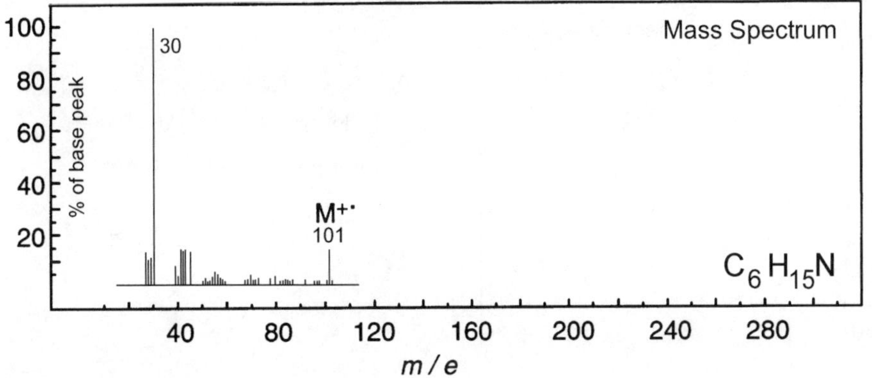

No significant UV absorption above 220 nm

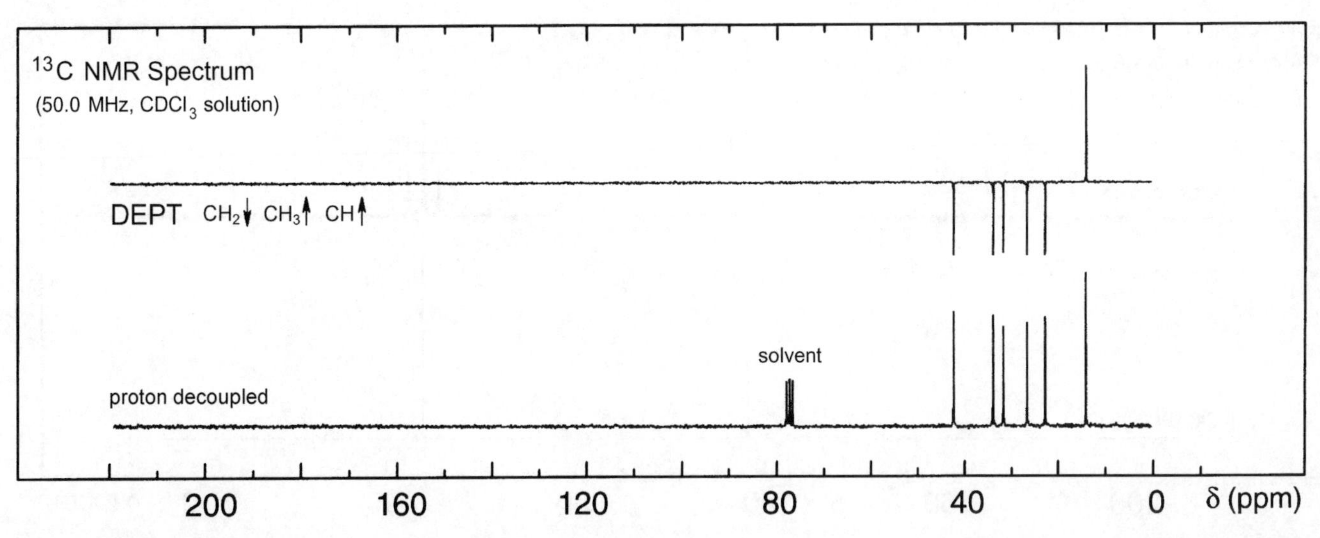

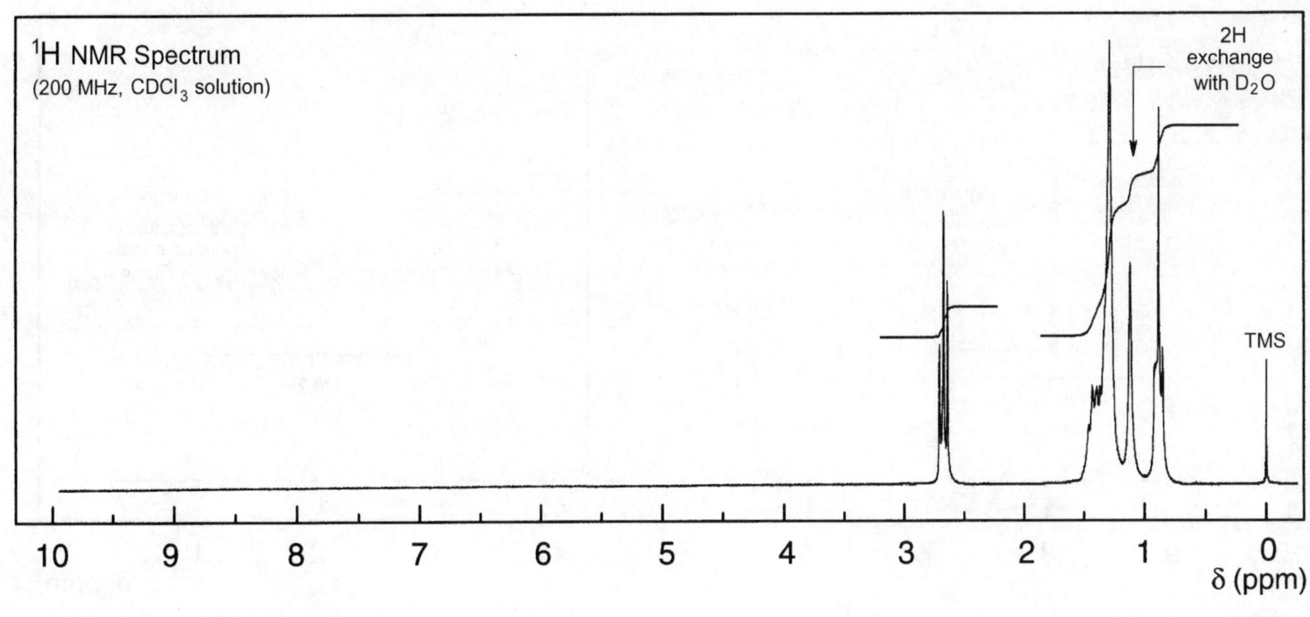

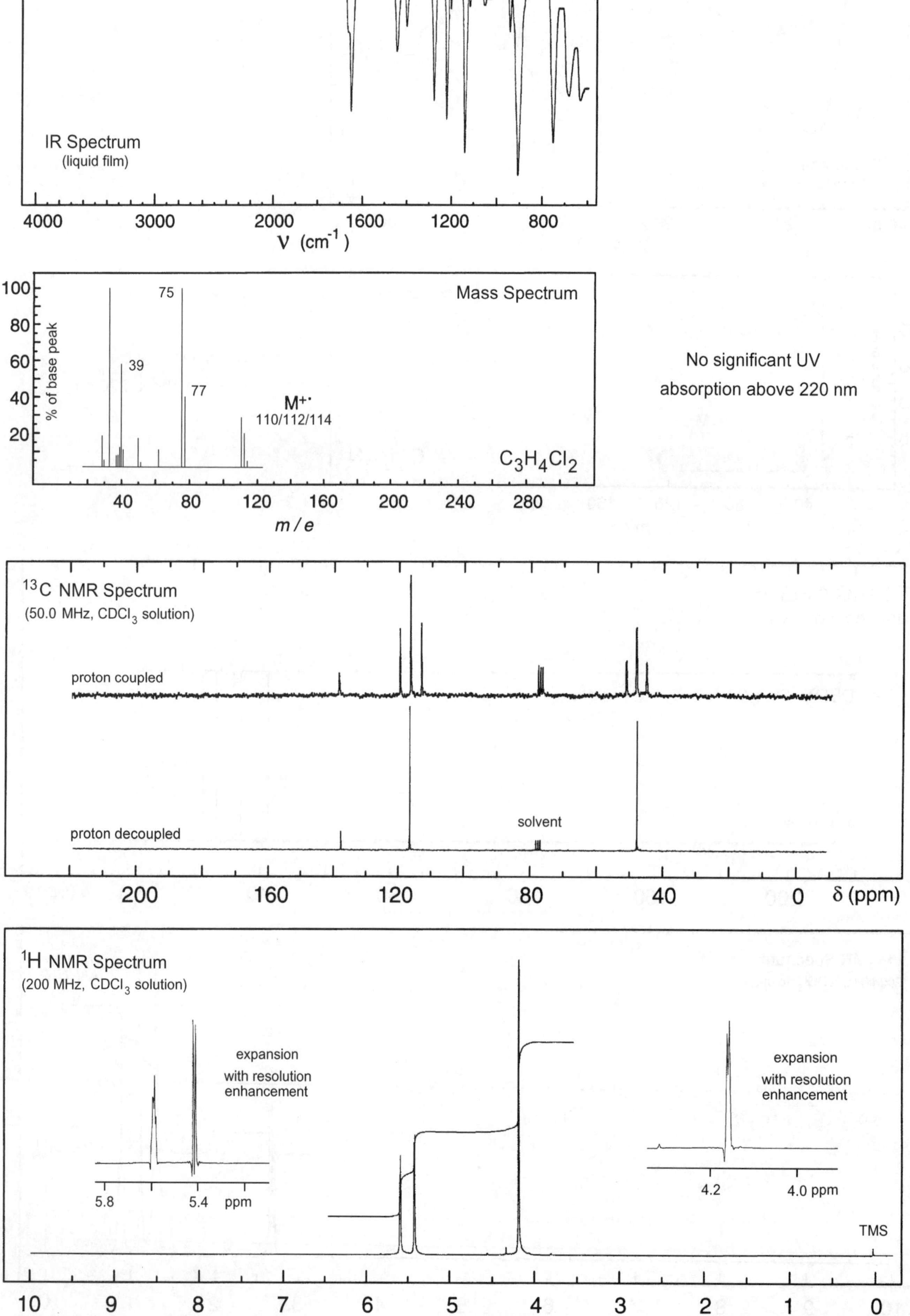
Problem 65
IR Spectrum
(liquid film)
4000
3000
2000
1600
1200
800
ν (cm⁻¹)
Mass Spectrum
100
80
60
40
20
% of base peak
75
39
77
M+·
110/112/114
$C_3H_4Cl_2$
40
80
120
160
200
240
280
m / e
No significant UV
absorption above 220 nm
13C NMR Spectrum
(50.0 MHz, $CDCl_3$ solution)
proton coupled
proton decoupled
solvent
200
160
120
80
40
0
δ (ppm)
1H NMR Spectrum
(200 MHz, $CDCl_3$ solution)
expansion
with resolution
enhancement
5.8
5.4
ppm
expansion
with resolution
enhancement
4.2
4.0 ppm
TMS
10
9
8
7
6
5
4
3
2
1
0
δ (ppm)

Problem 66

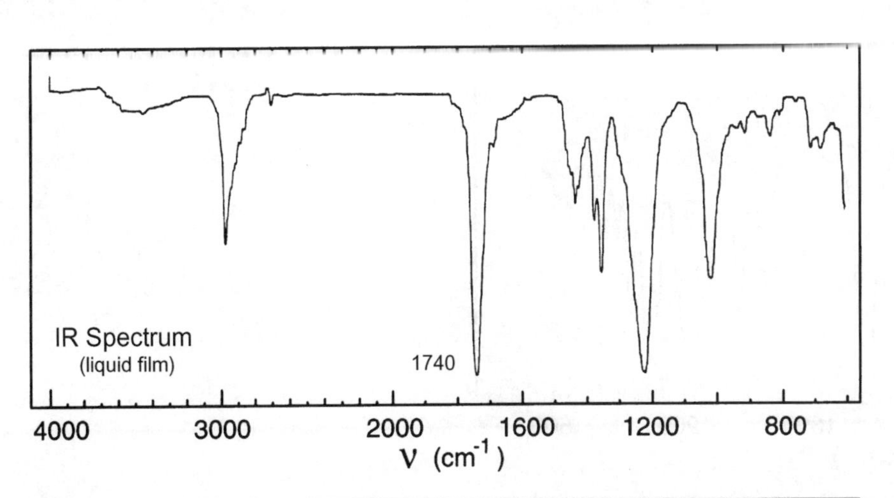

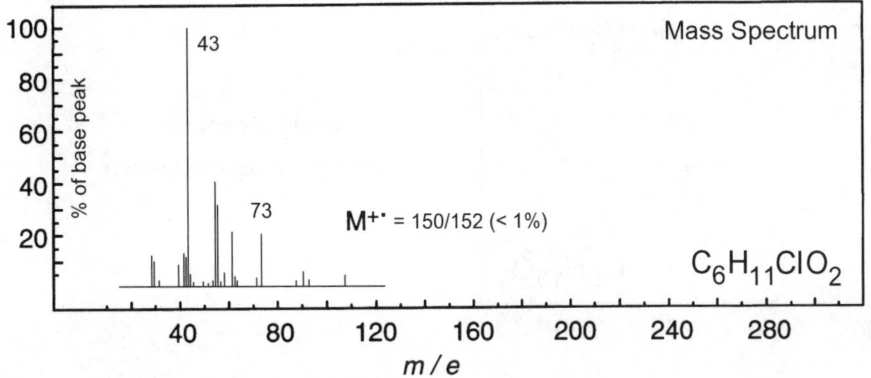

No significant UV absorption above 220 nm

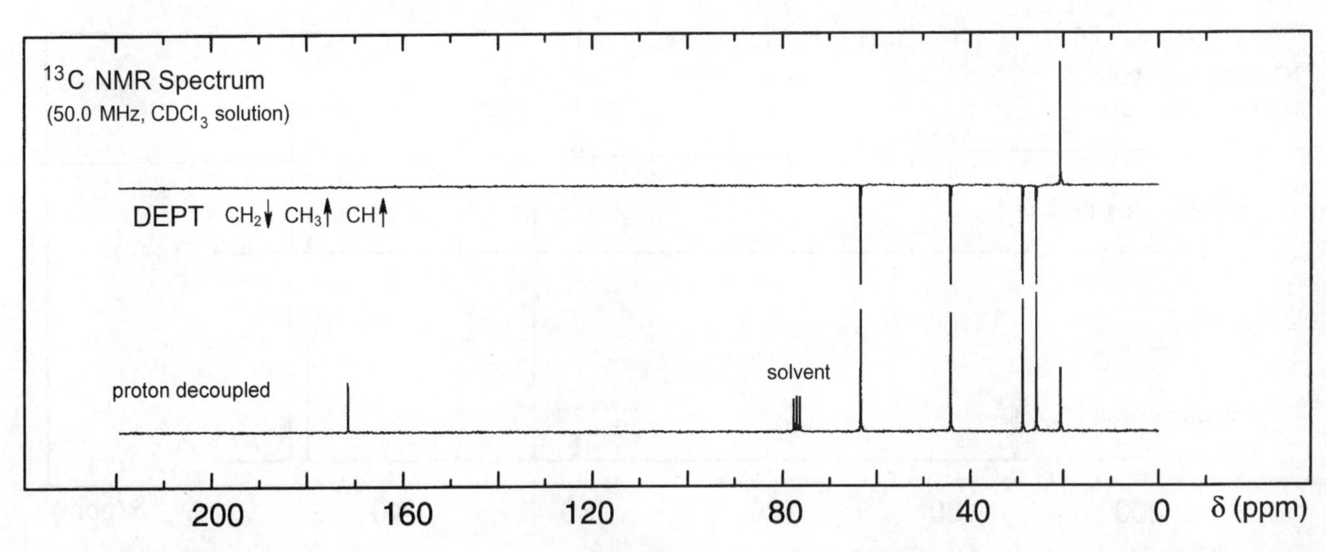

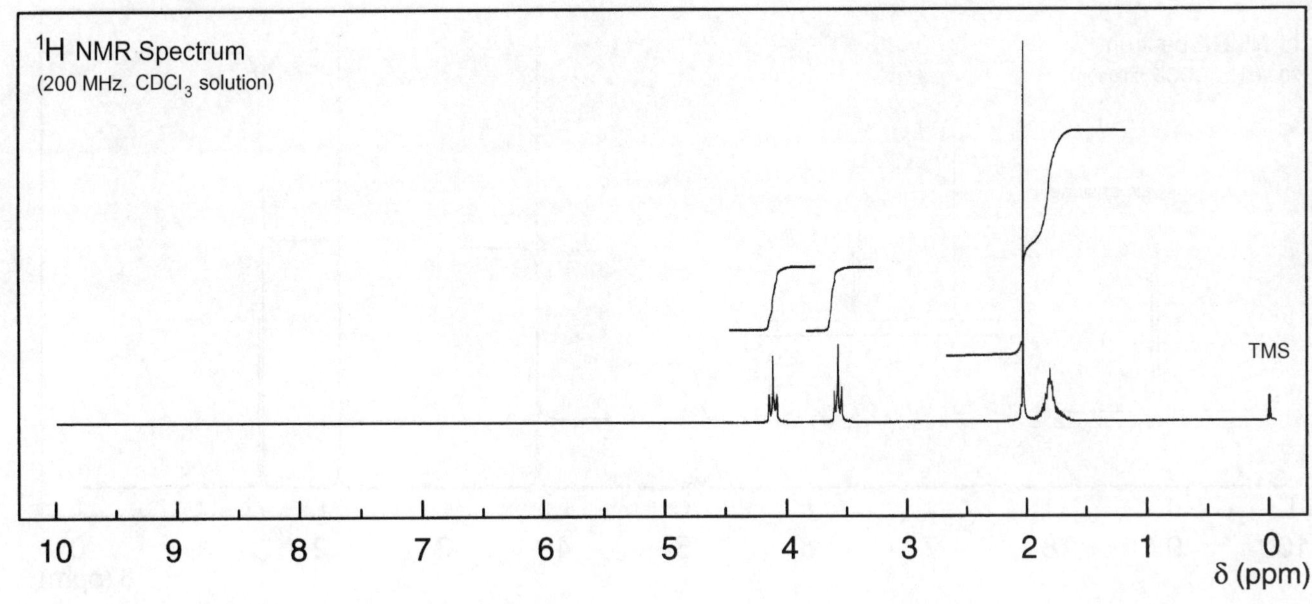

Problem 67

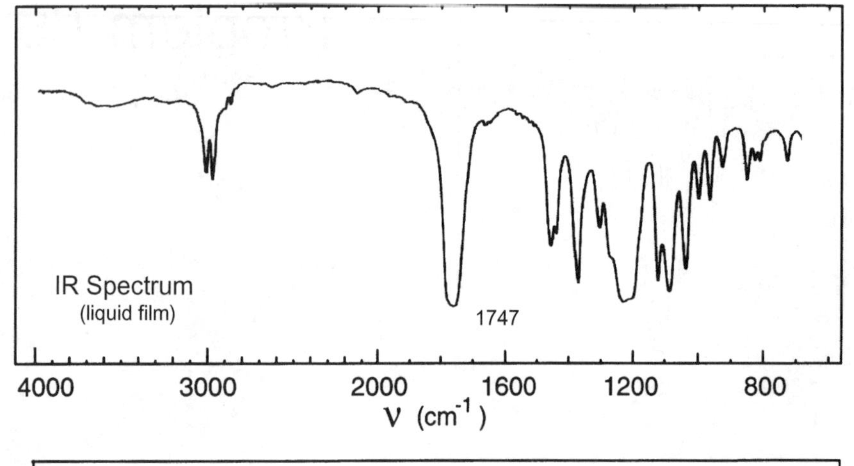

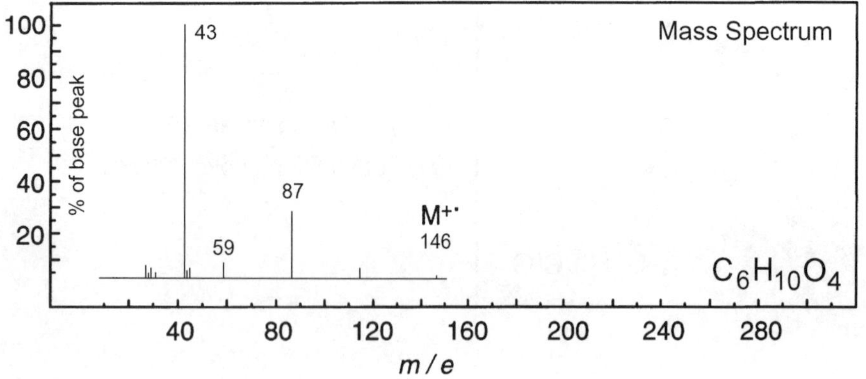

No significant UV absorption above 220 nm

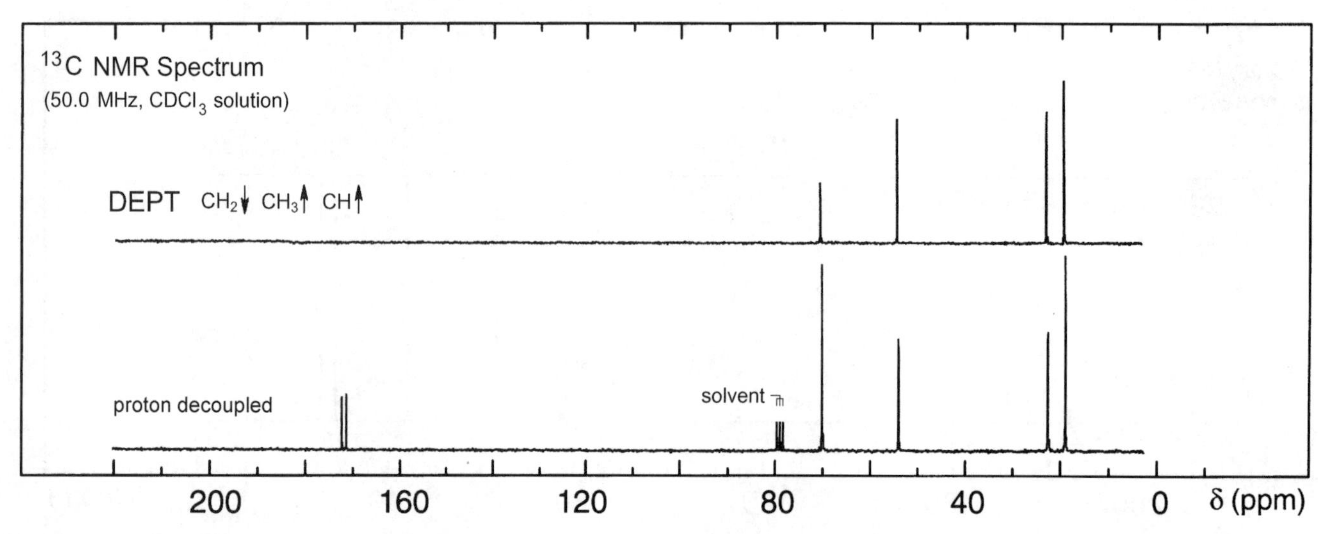

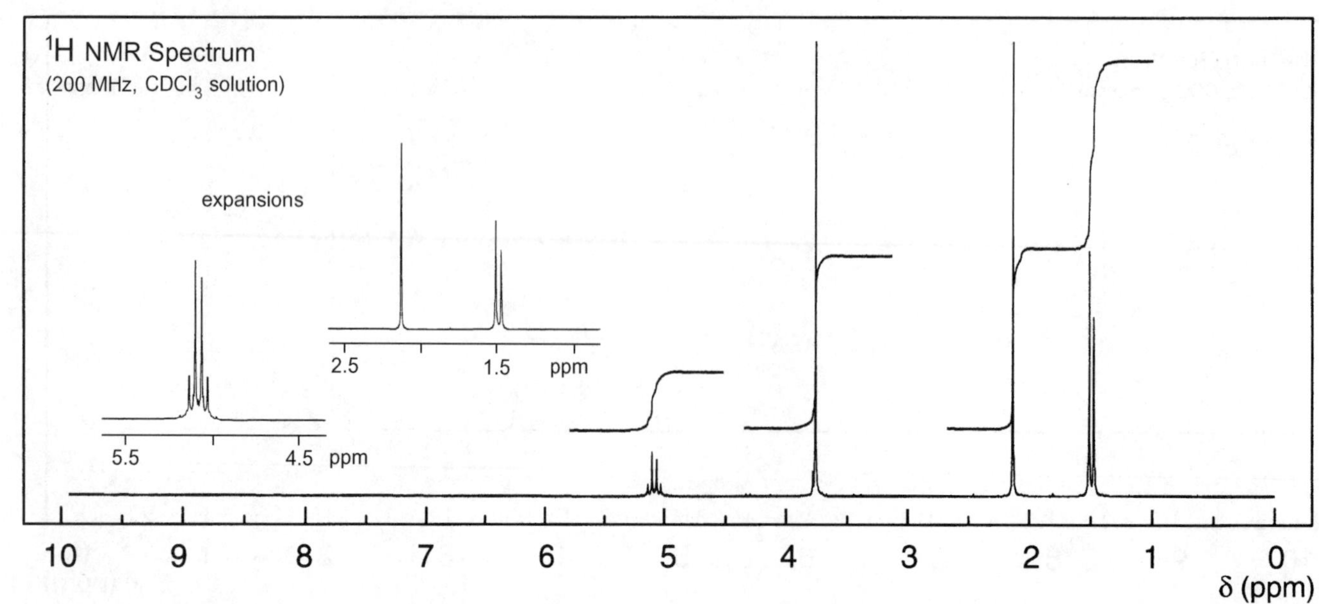

Problem 68

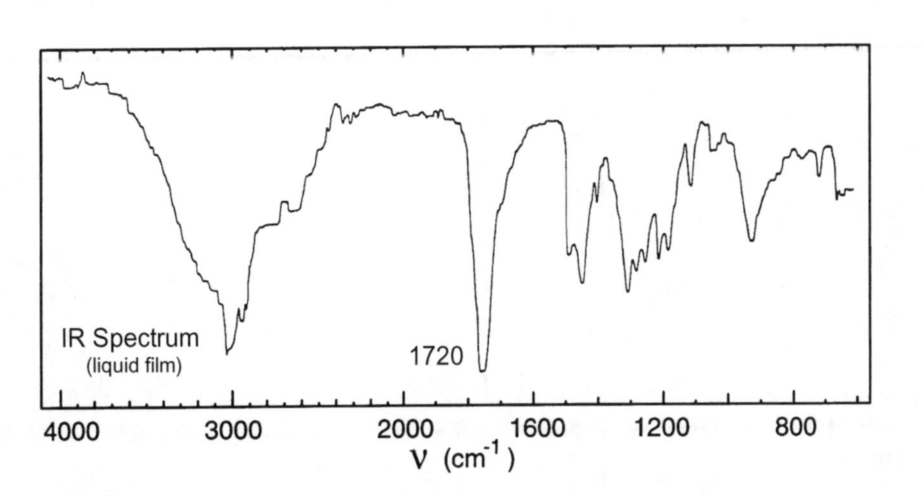

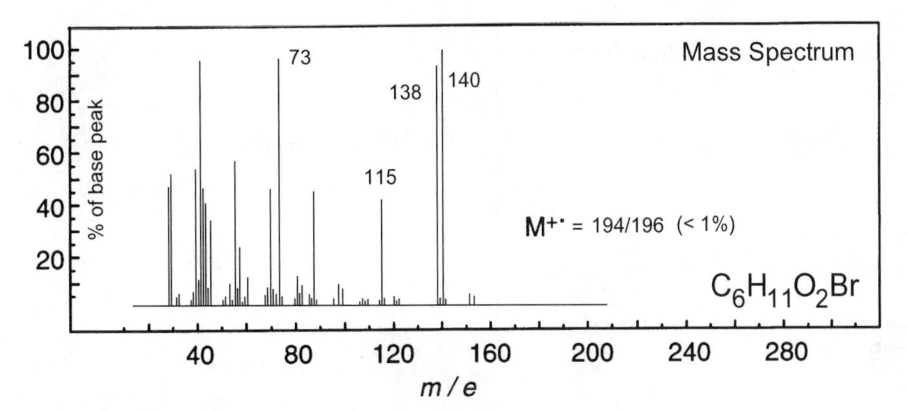

No significant UV absorption above 220 nm

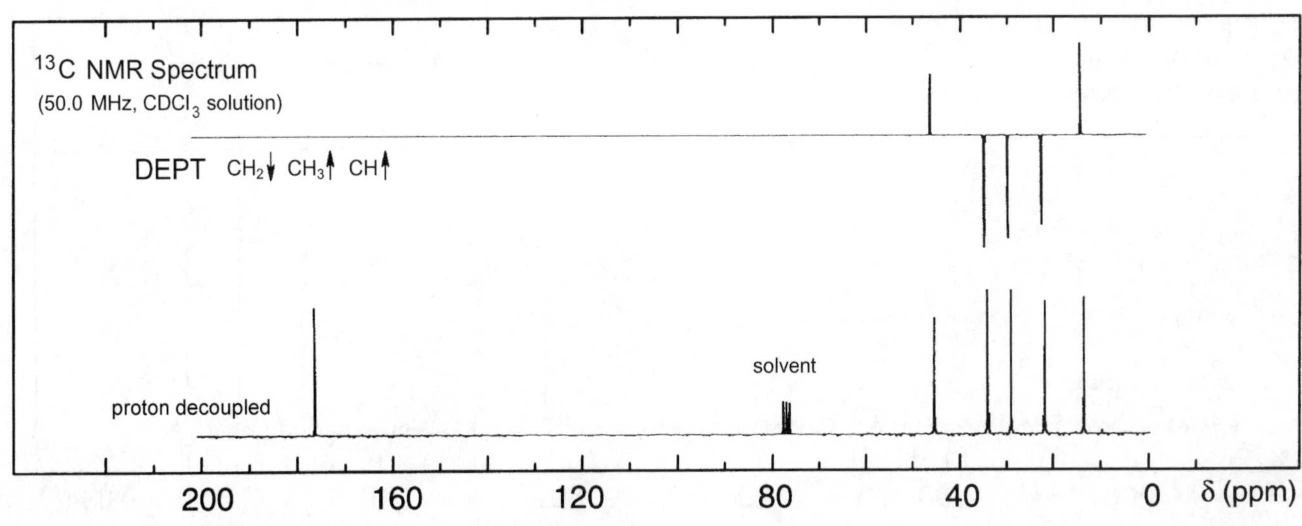

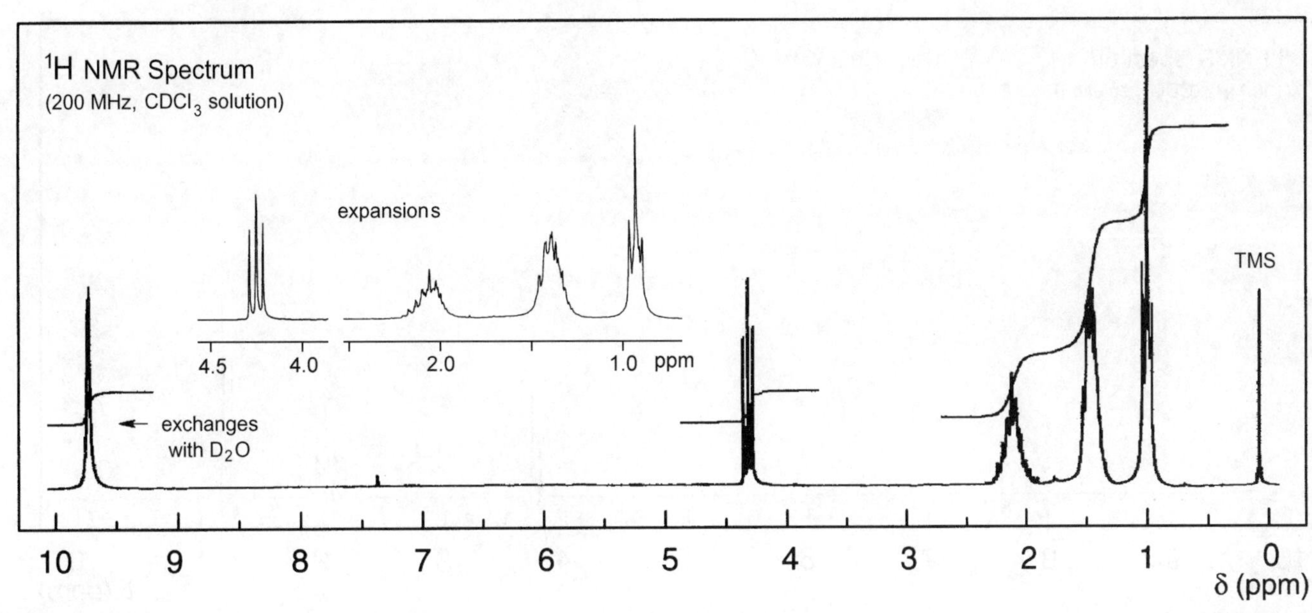

Problem 69

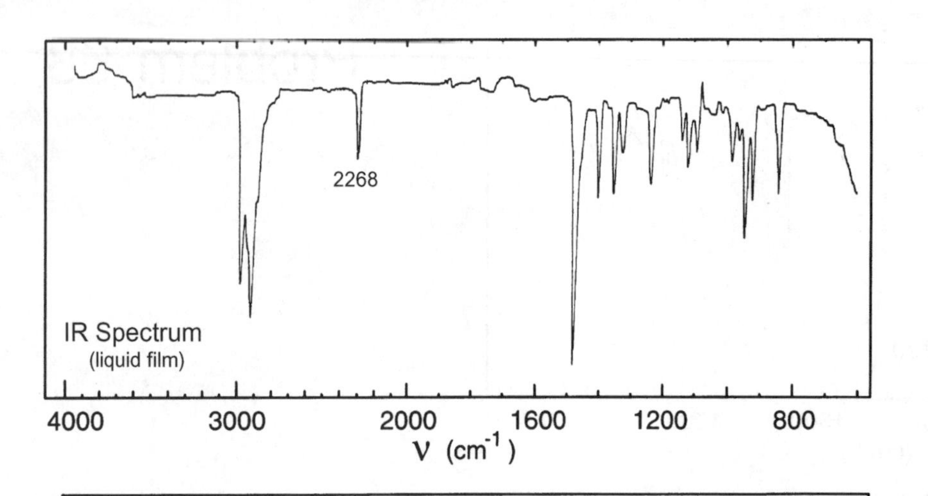

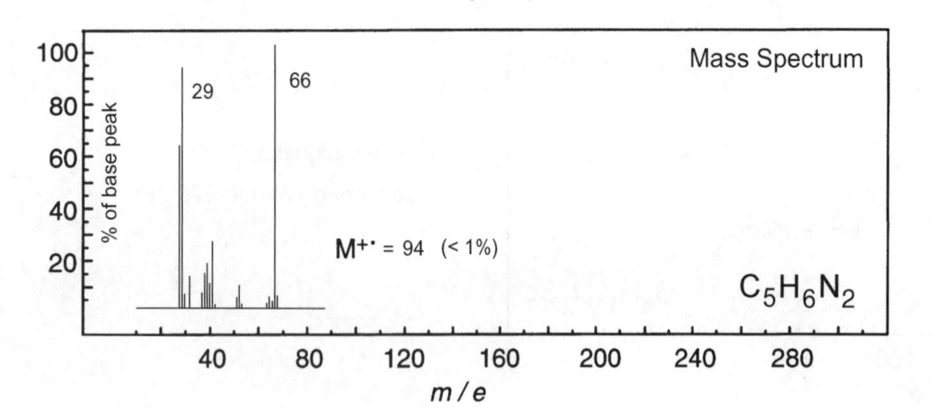

No significant UV absorption above 220 nm

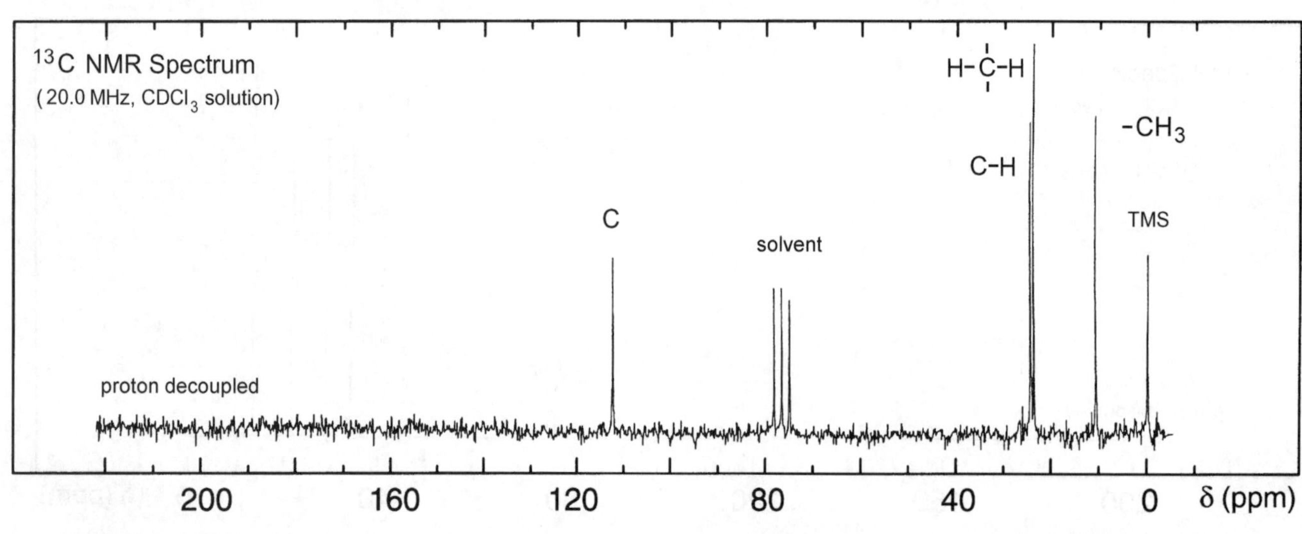

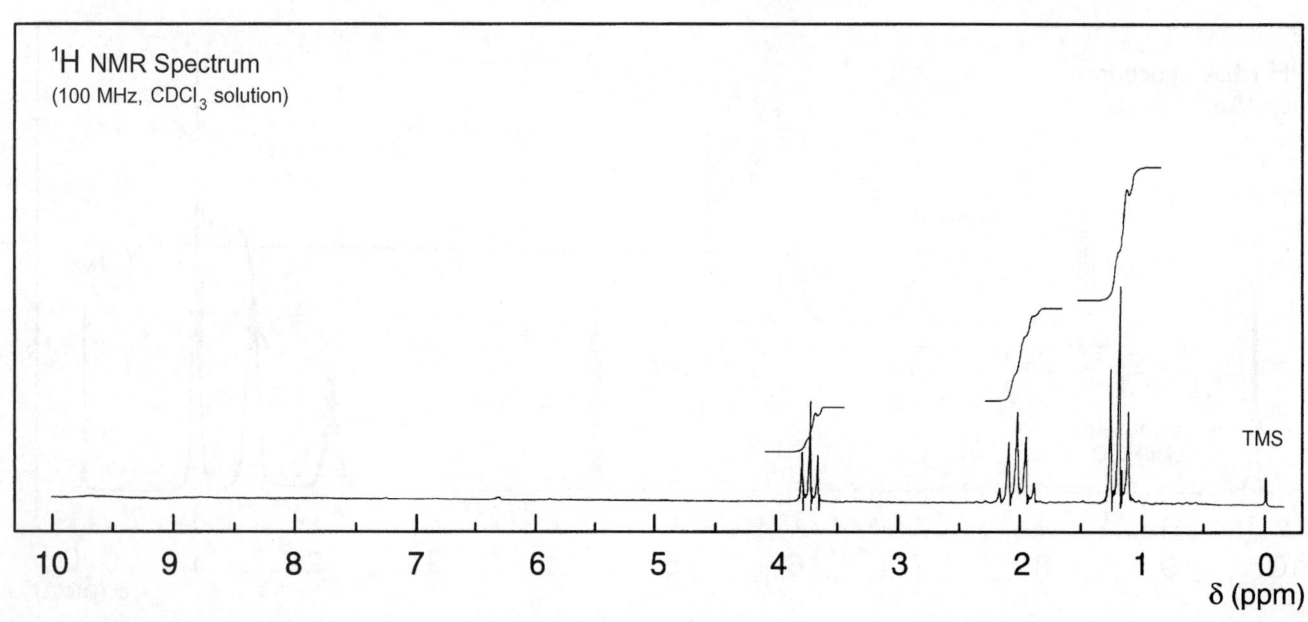

Problem 70

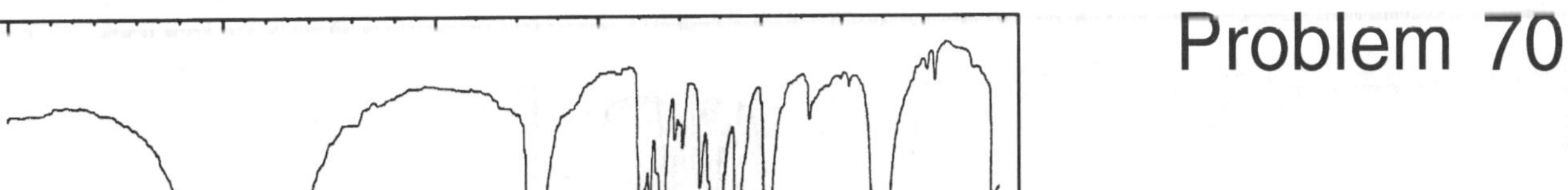

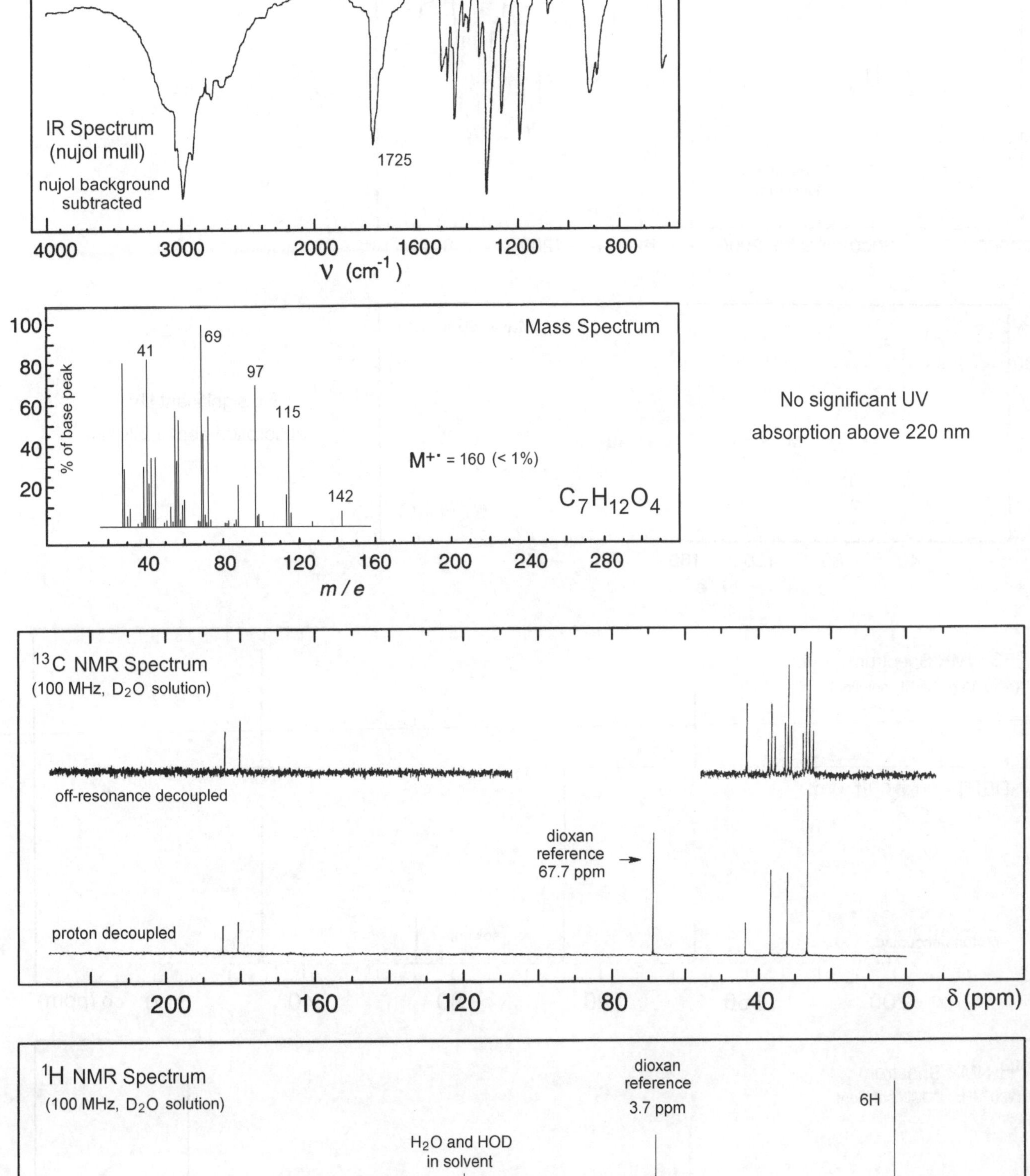

No significant UV absorption above 220 nm

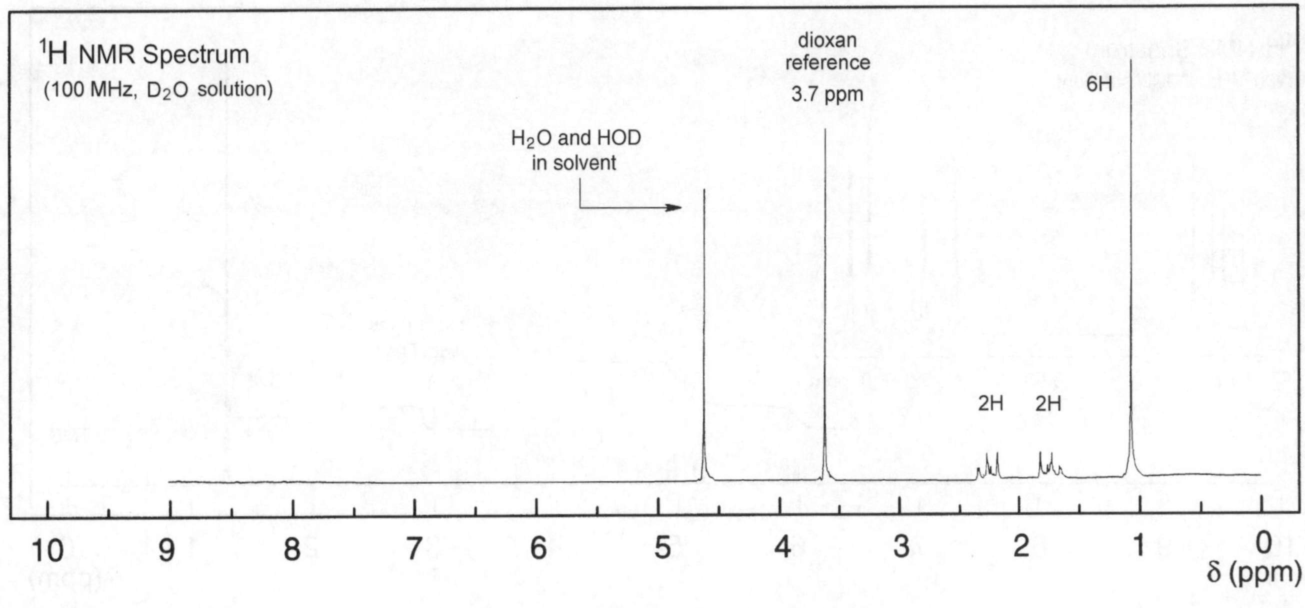

Problem 71

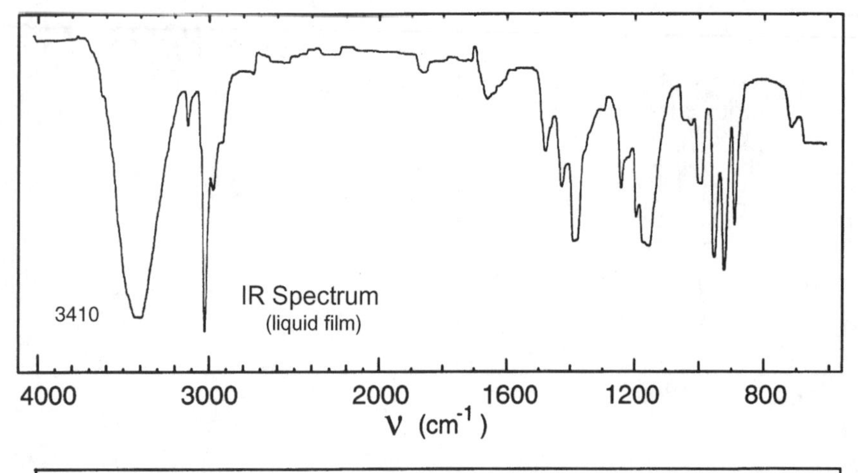

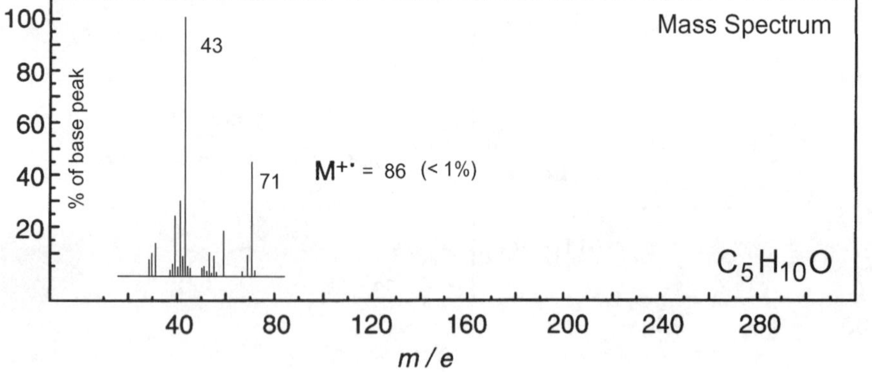

No significant UV absorption above 220 nm

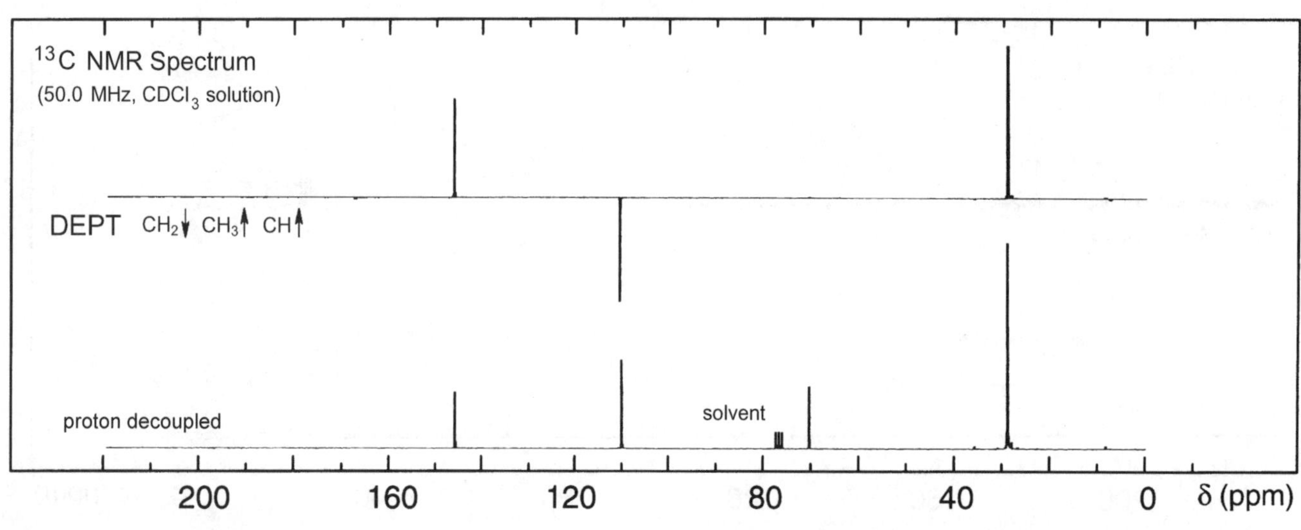

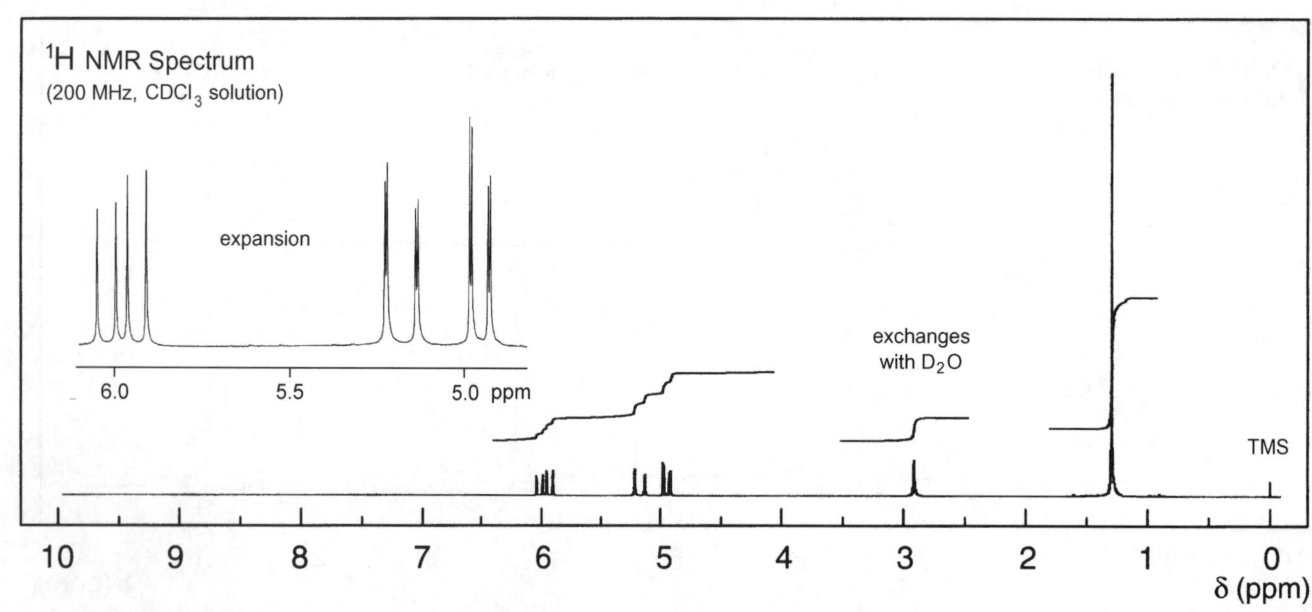

Problem 72

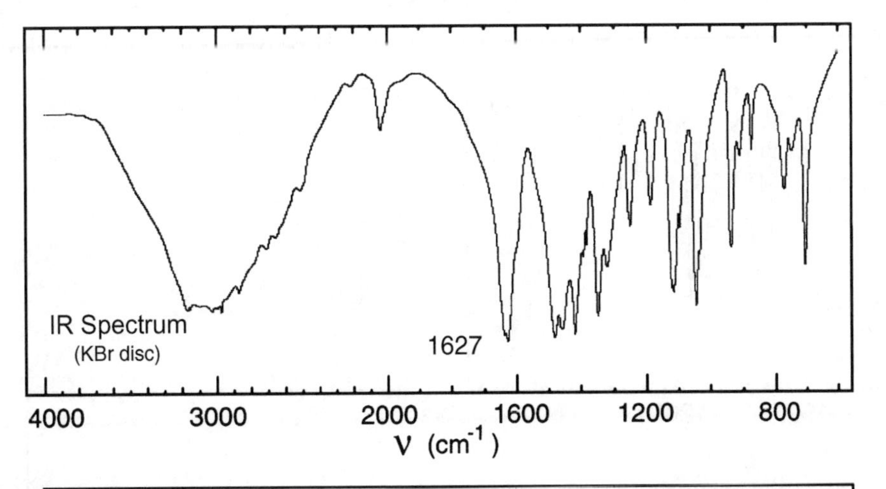

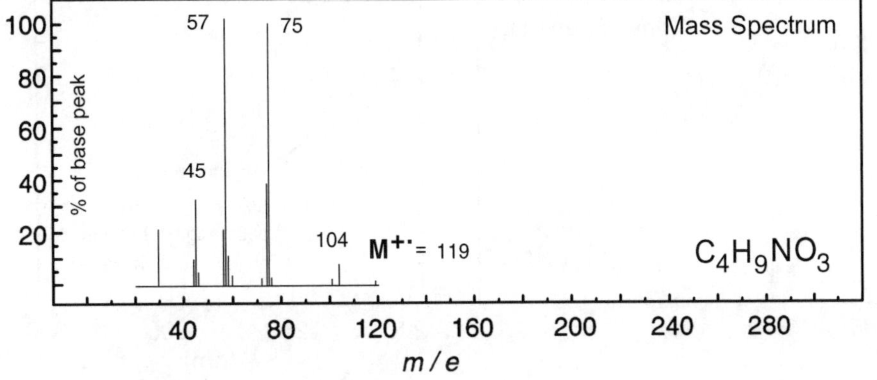

No significant UV absorption above 220 nm

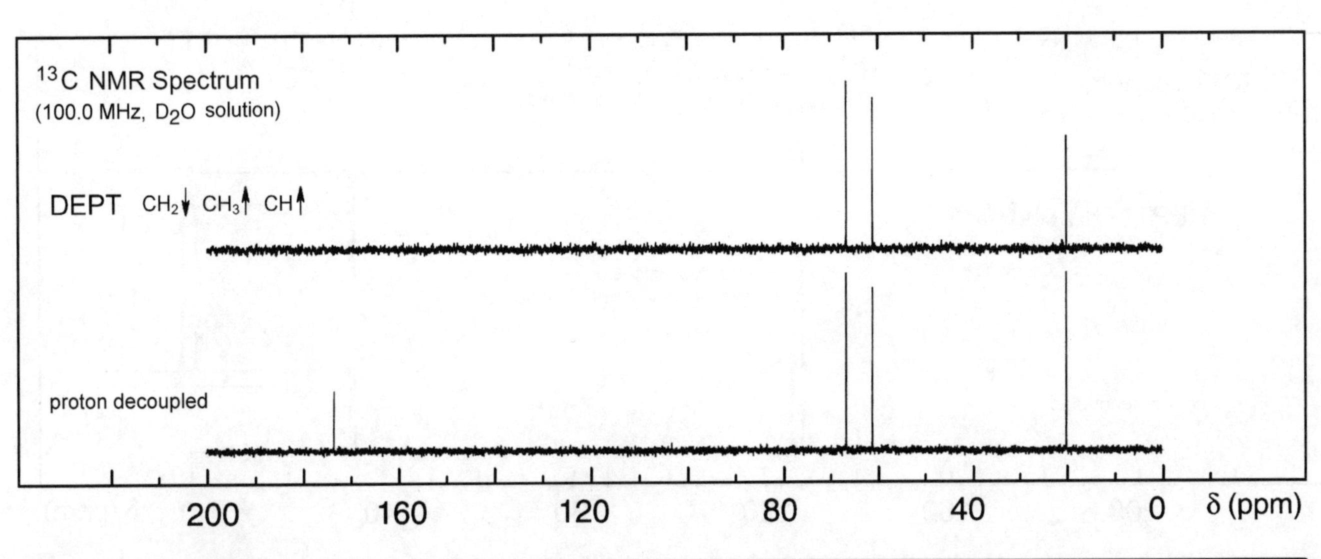

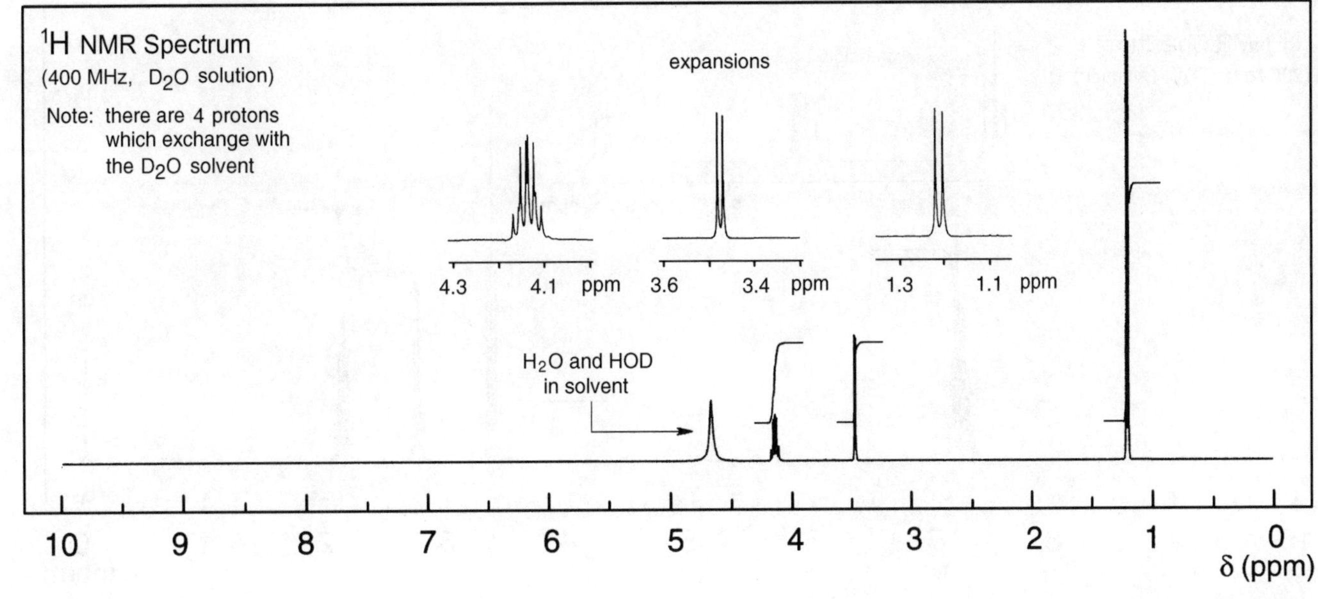

Problem 73

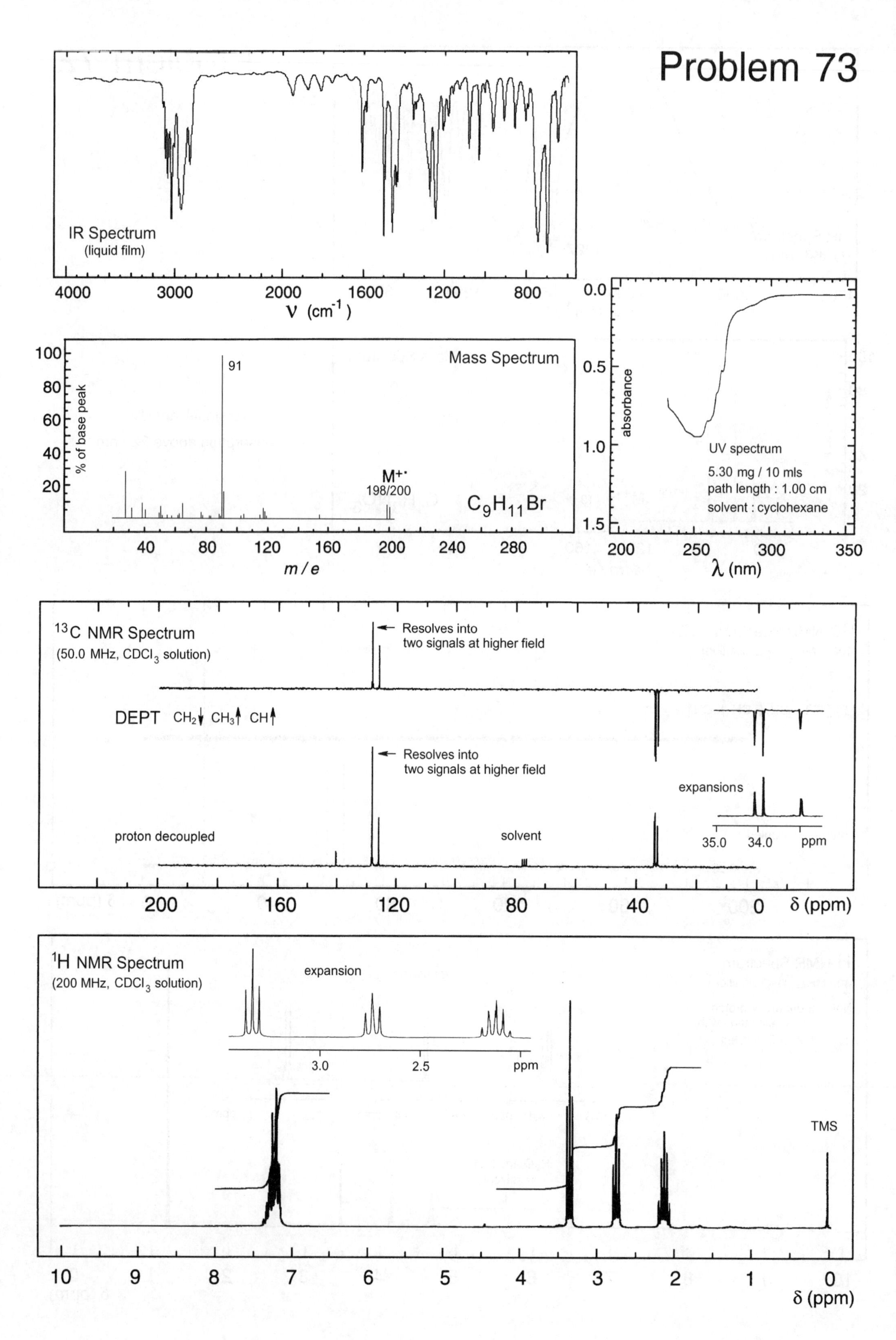

IR Spectrum
(liquid film)
4000
3000
2000
1600
1200
800
ν (cm⁻¹)
Mass Spectrum
% of base peak
100
80
60
40
20
91
M+·
198/200
$C_9H_{11}Br$
40
80
120
160
200
240
280
m / e
0.0
0.5
1.0
1.5
absorbance
UV spectrum
5.30 mg / 10 mls
path length : 1.00 cm
solvent : cyclohexane
200
250
300
350
λ (nm)
13C NMR Spectrum
(50.0 MHz, CDCl3 solution)
Resolves into
two signals at higher field
DEPT CH2↓ CH3↑ CH↑
Resolves into
two signals at higher field
expansions
proton decoupled
solvent
35.0
34.0
ppm
200
160
120
80
40
0
δ (ppm)
1H NMR Spectrum
(200 MHz, CDCl3 solution)
expansion
3.0
2.5
ppm
TMS
10
9
8
7
6
5
4
3
2
1
0
δ (ppm)

Problem 74

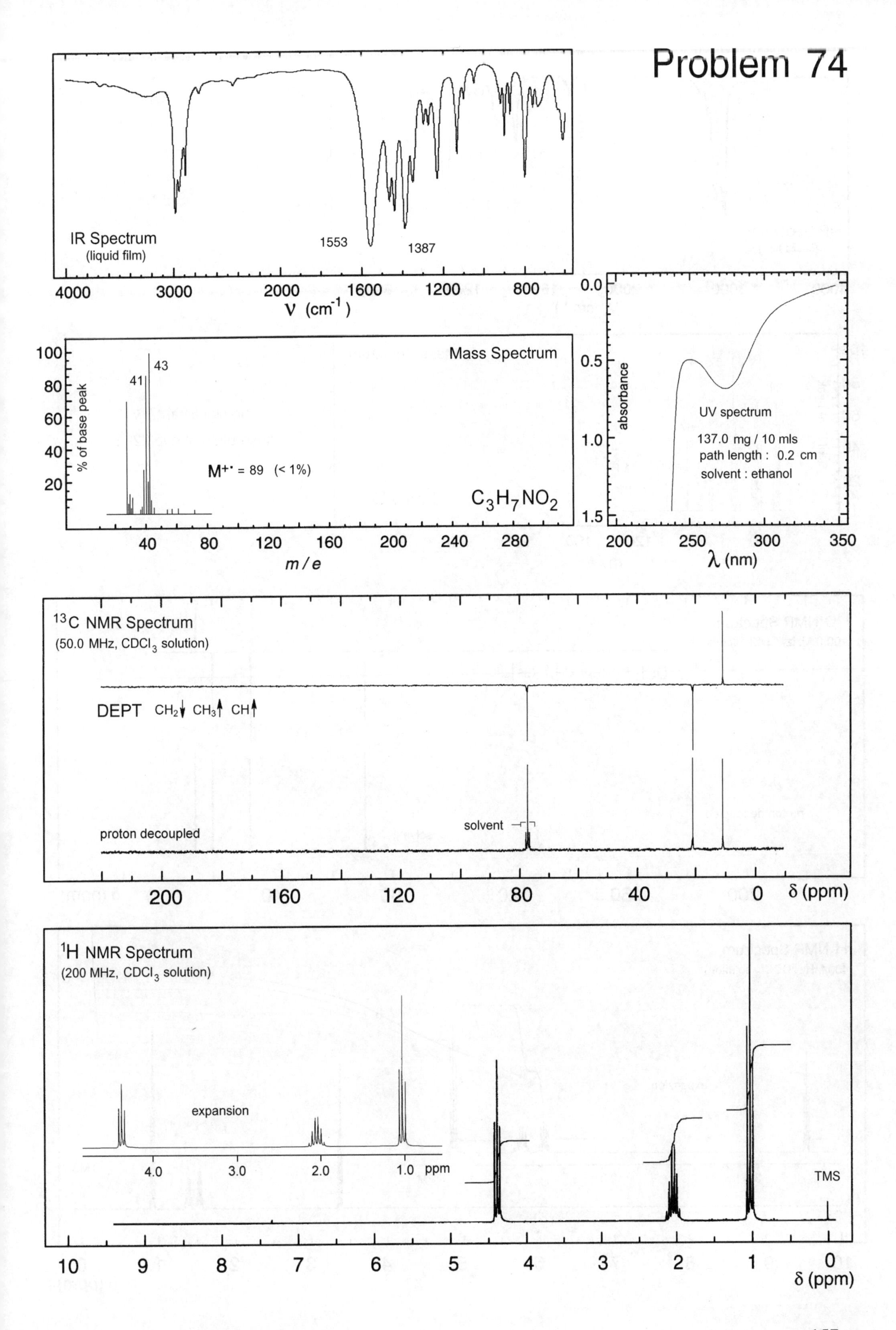

IR Spectrum
(liquid film)
1553
1387
4000
3000
2000
1600
1200
800
ν (cm-1)
Mass Spectrum
% of base peak
100
80
60
40
20
41
43
M+• = 89 (< 1%)
C3H7NO2
40
80
120
160
200
240
280
m / e
absorbance
0.0
0.5
1.0
1.5
UV spectrum
137.0 mg / 10 mls
path length : 0.2 cm
solvent : ethanol
200
250
300
350
λ (nm)
13C NMR Spectrum
(50.0 MHz, CDCl3 solution)
DEPT CH2↓ CH3↑ CH↑
proton decoupled
solvent
200
160
120
80
40
0
δ (ppm)
1H NMR Spectrum
(200 MHz, CDCl3 solution)
expansion
4.0
3.0
2.0
1.0 ppm
TMS
10
9
8
7
6
5
4
3
2
1
0
δ (ppm)

Problem 75

IR Spectrum
(liquid film)

1115

4000 3000 2000 1600 1200 800

ν (cm^{-1})

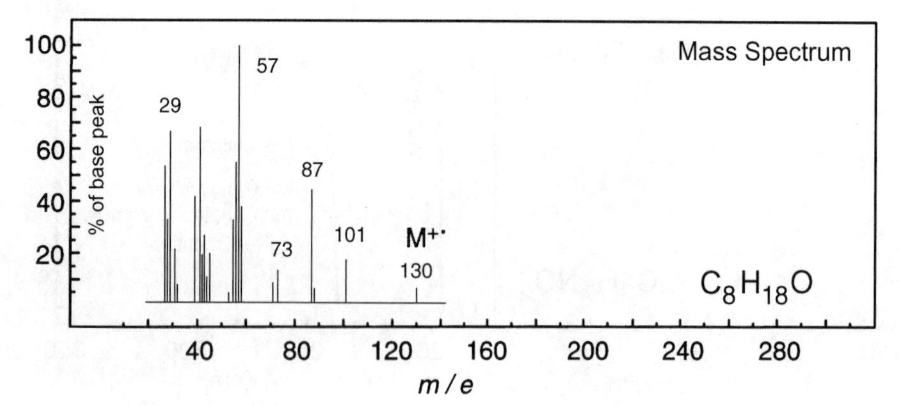

No significant UV absorption above 220 nm

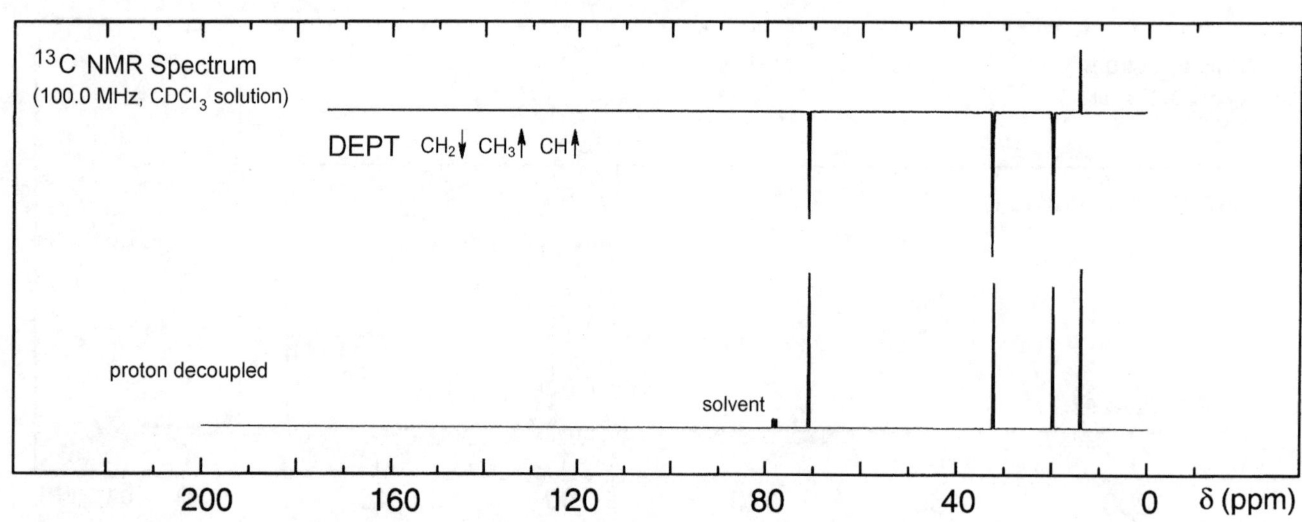

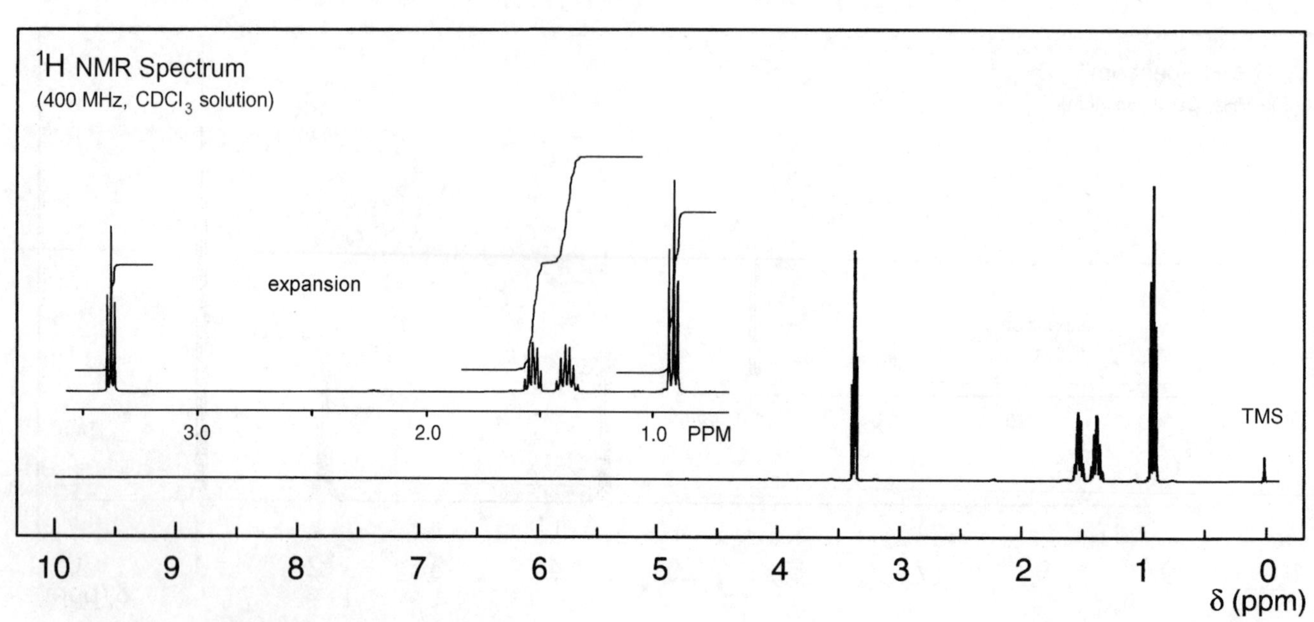

Problem 76

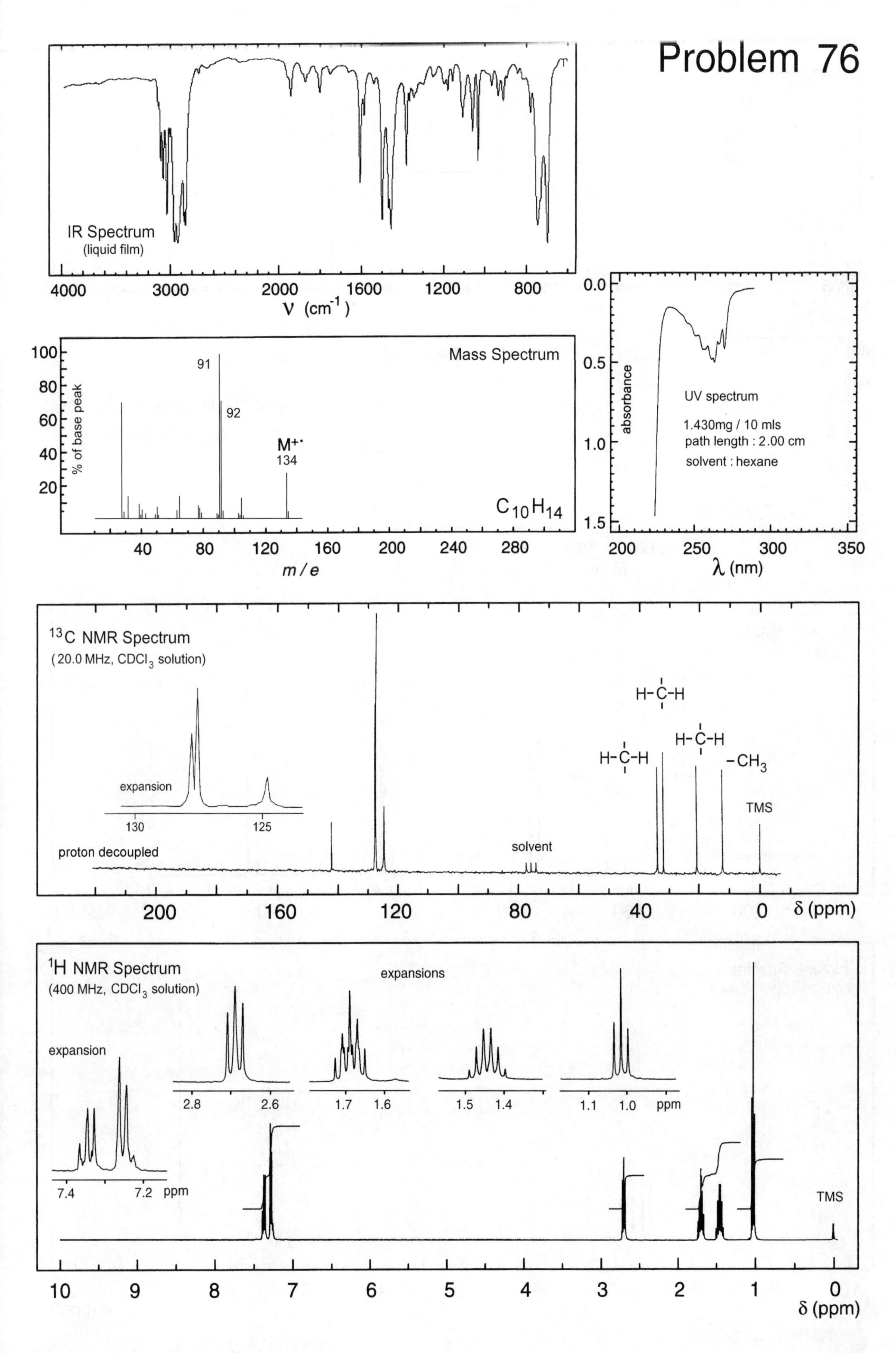

IR Spectrum
(liquid film)
4000
3000
2000
1600
1200
800
ν (cm-1)
Mass Spectrum
91
92
M+·
134
C10H14
% of base peak
m / e
UV spectrum
1.430mg / 10 mls
path length : 2.00 cm
solvent : hexane
absorbance
λ (nm)
13C NMR Spectrum
(20.0 MHz, CDCl3 solution)
expansion
proton decoupled
solvent
H-C-H
-CH3
TMS
δ (ppm)
1H NMR Spectrum
(400 MHz, CDCl3 solution)
expansions
expansion
ppm
TMS
δ (ppm)

Problem 77

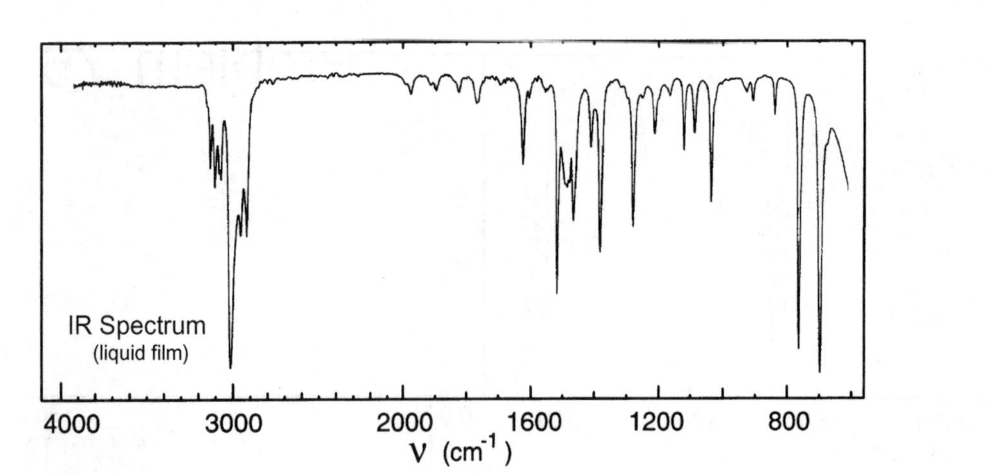

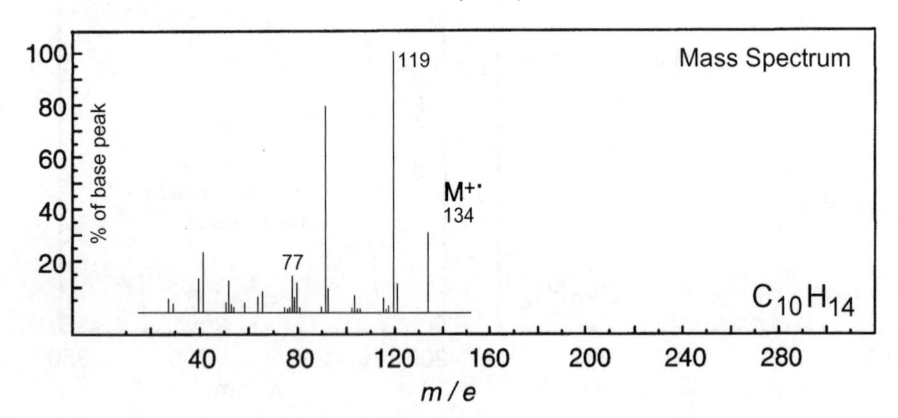

UV Spectrum

λ_{max} 262 nm ($\log_{10}\varepsilon$ 2.5)

solvent : ethanol

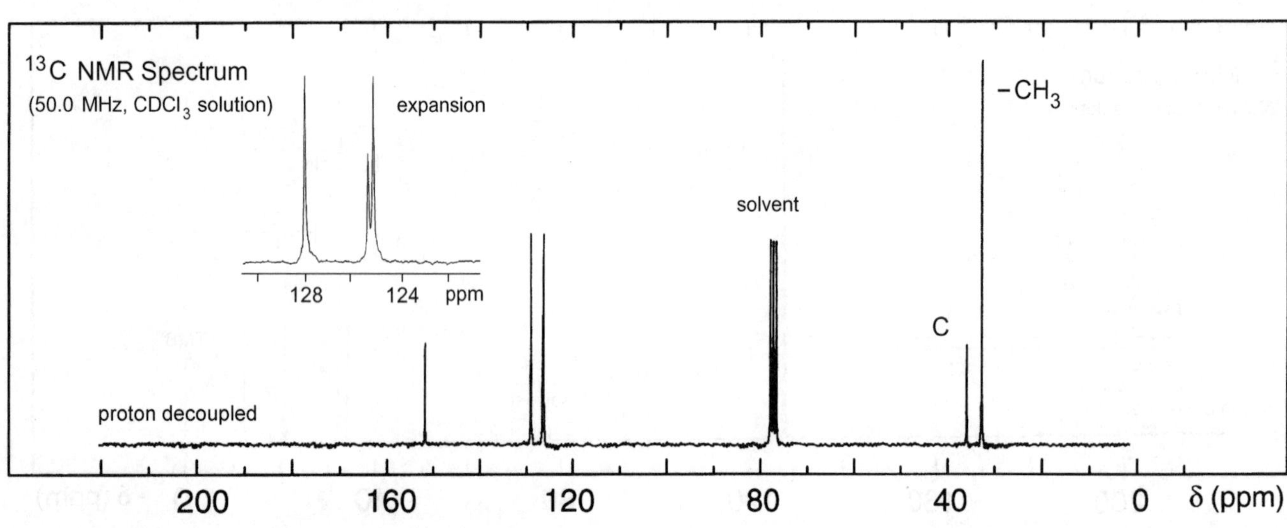

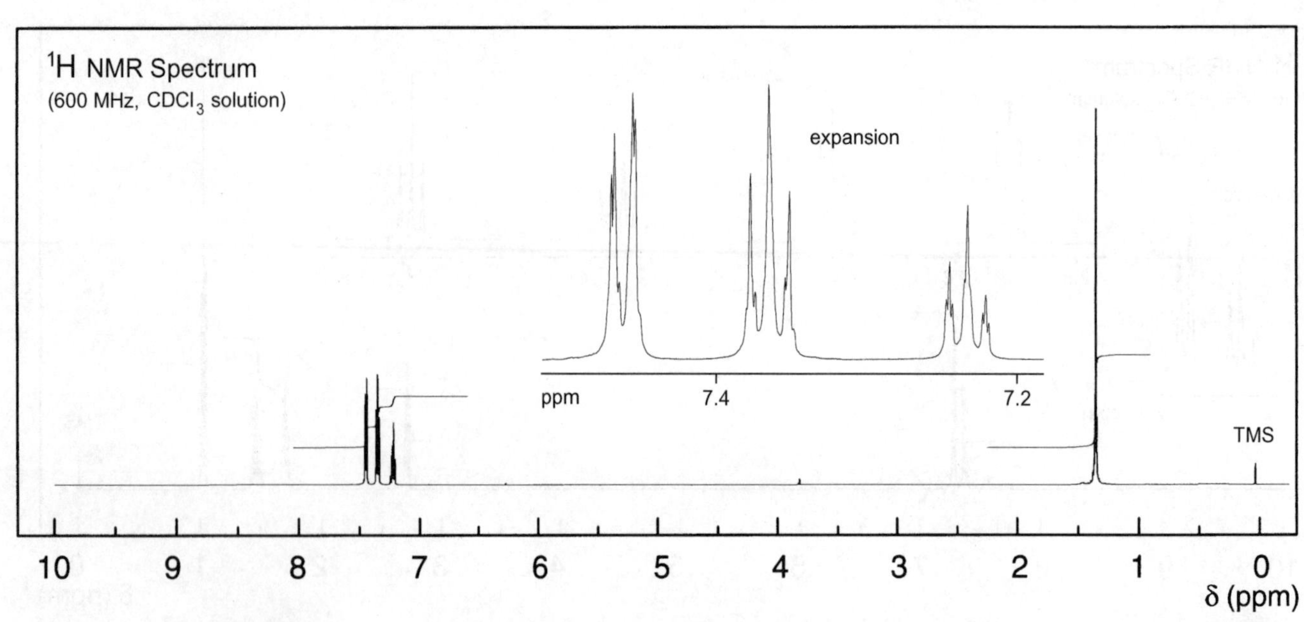

Problem 78

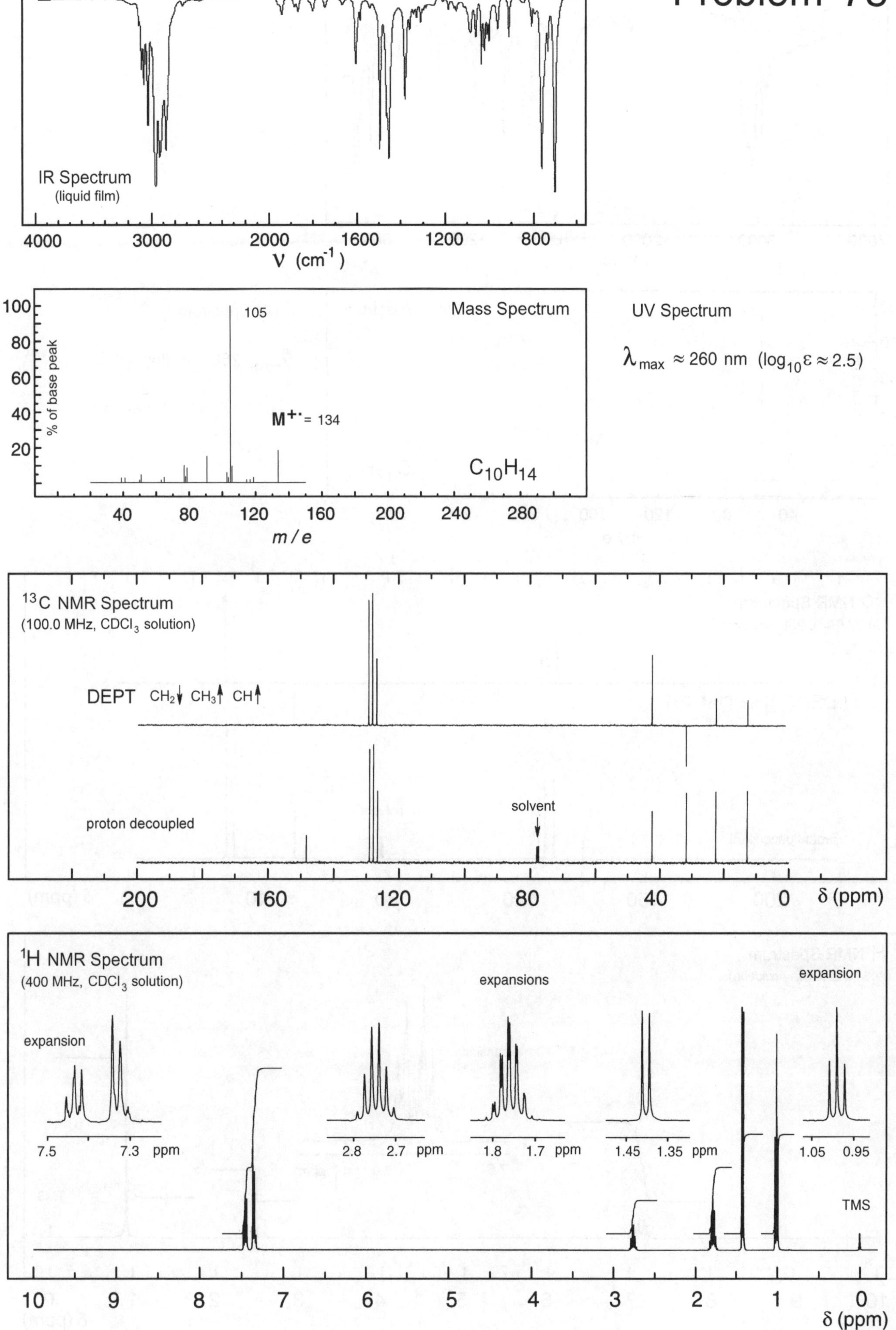

IR Spectrum
(liquid film)
4000
3000
2000
1600
1200
800
ν (cm-1)
Mass Spectrum
% of base peak
105
M+· = 134
C10H14
m / e
UV Spectrum
λmax ≈ 260 nm (log10ε ≈ 2.5)
13C NMR Spectrum
(100.0 MHz, CDCl3 solution)
DEPT CH2↓ CH3↑ CH↑
proton decoupled
solvent
δ (ppm)
1H NMR Spectrum
(400 MHz, CDCl3 solution)
expansion
expansions
TMS
δ (ppm)

Problem 79

4000 3000 2000 1600 1200 800

ν (cm^{-1})

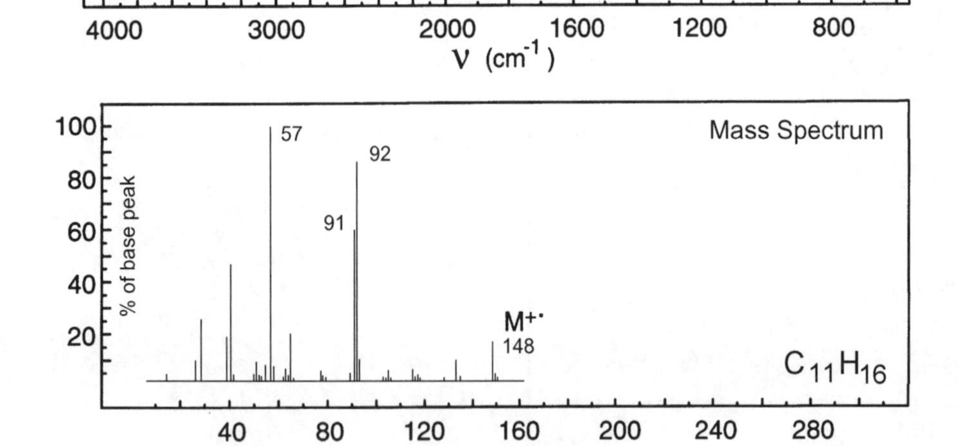

UV Spectrum

λ_{max} 260 nm ($\log_{10}\varepsilon$ 2.5)

solvent : methanol

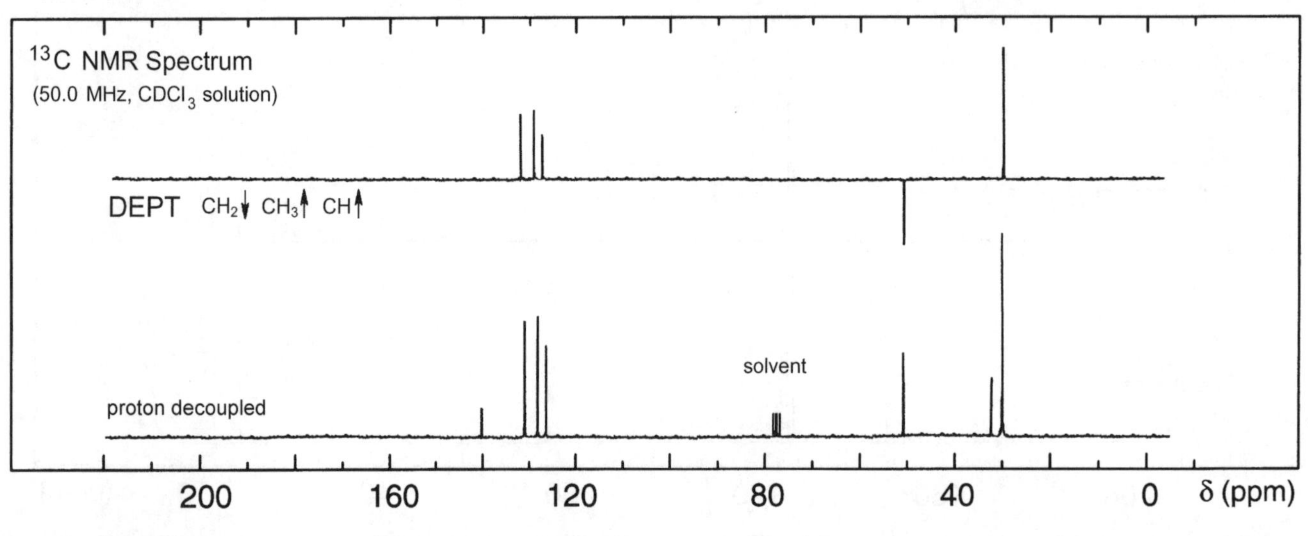

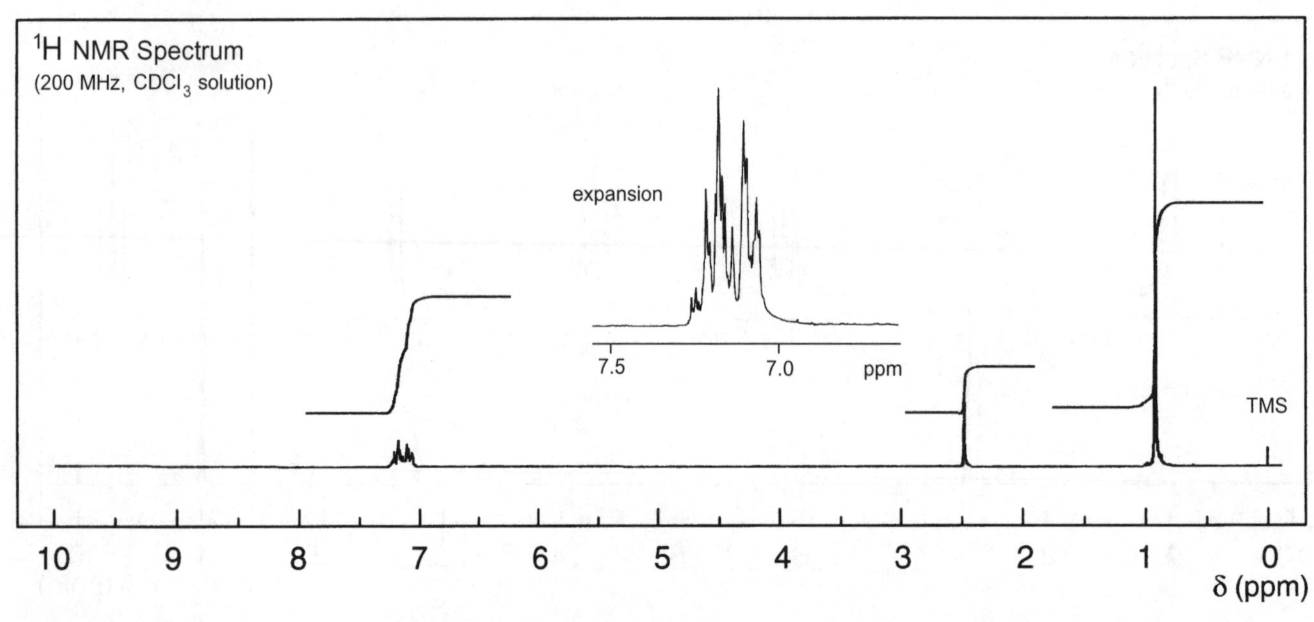

Problem 80

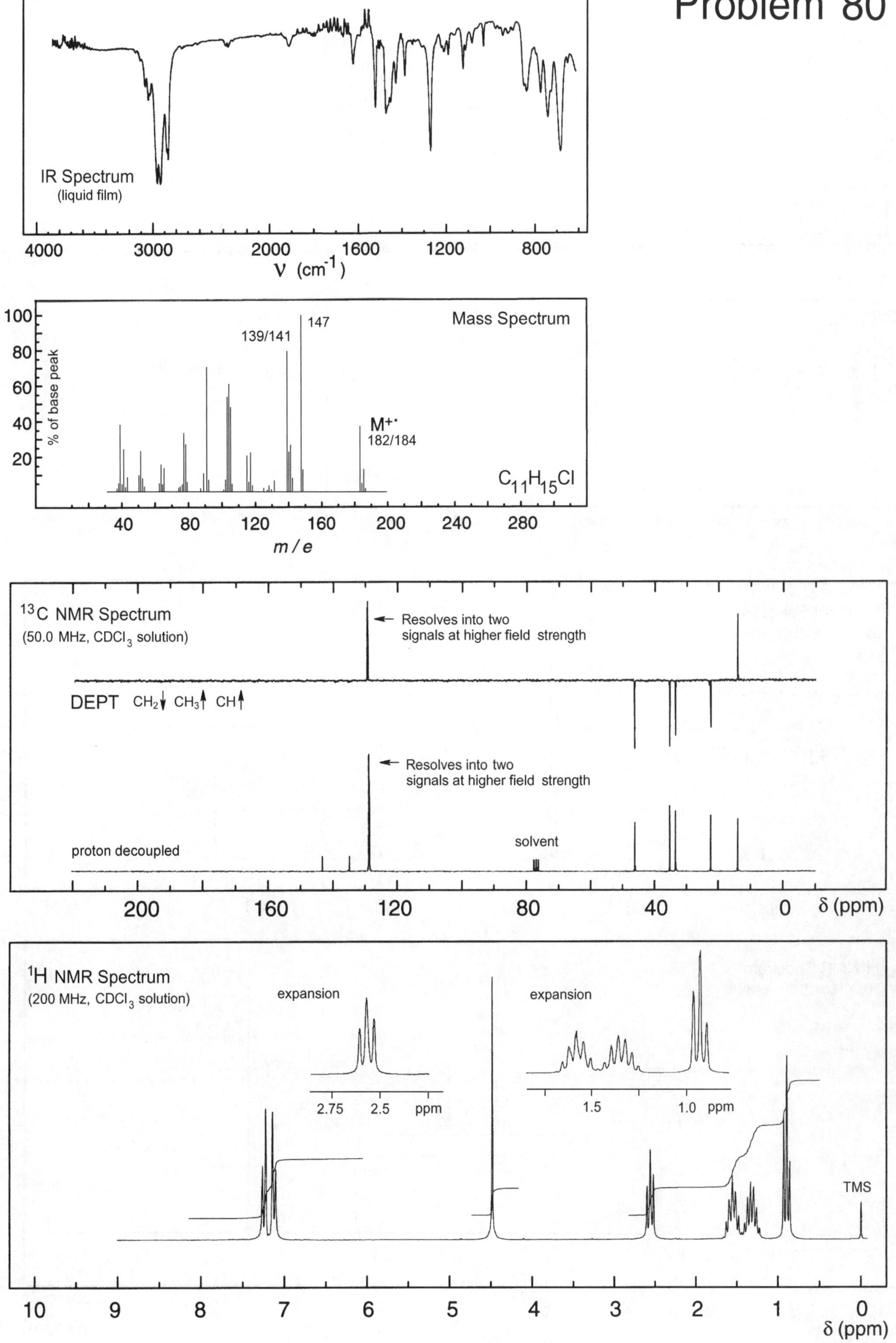
IR Spectrum
(liquid film)
4000
3000
2000
1600
1200
800
ν (cm^{-1})
Mass Spectrum
% of base peak
100
80
60
40
20
147
139/141
M$^{+\cdot}$
182/184
$C_{11}H_{15}Cl$
40
80
120
160
200
240
280
m / e
^{13}C NMR Spectrum
(50.0 MHz, $CDCl_3$ solution)
Resolves into two signals at higher field strength
DEPT CH_2↓ CH_3↑ CH↑
Resolves into two signals at higher field strength
solvent
proton decoupled
200
160
120
80
40
0
δ (ppm)
^{1}H NMR Spectrum
(200 MHz, $CDCl_3$ solution)
expansion
2.75
2.5
ppm
expansion
1.5
1.0
ppm
TMS
10
9
8
7
6
5
4
3
2
1
0
δ (ppm)

Problem 81

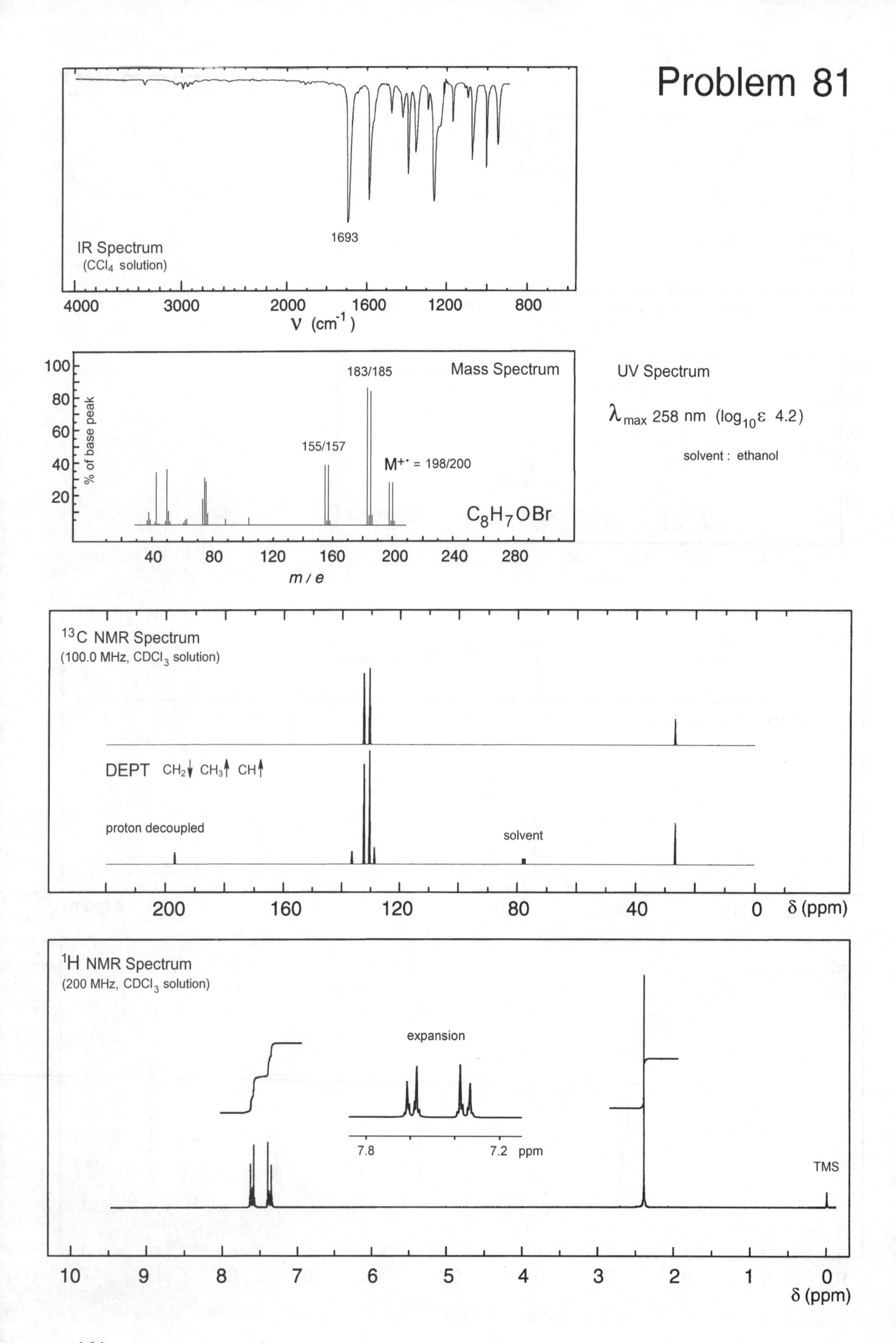

IR Spectrum
(CCl4 solution)
1693
4000
3000
2000
1600
1200
800
ν (cm-1)
Mass Spectrum
183/185
155/157
M+· = 198/200
C8H7OBr
% of base peak
m / e
UV Spectrum
λmax 258 nm (log10ε 4.2)
solvent : ethanol
13C NMR Spectrum
(100.0 MHz, CDCl3 solution)
DEPT CH2↓ CH3↑ CH↑
proton decoupled
solvent
δ (ppm)
1H NMR Spectrum
(200 MHz, CDCl3 solution)
expansion
7.8
7.2 ppm
TMS
δ (ppm)

Problem 82

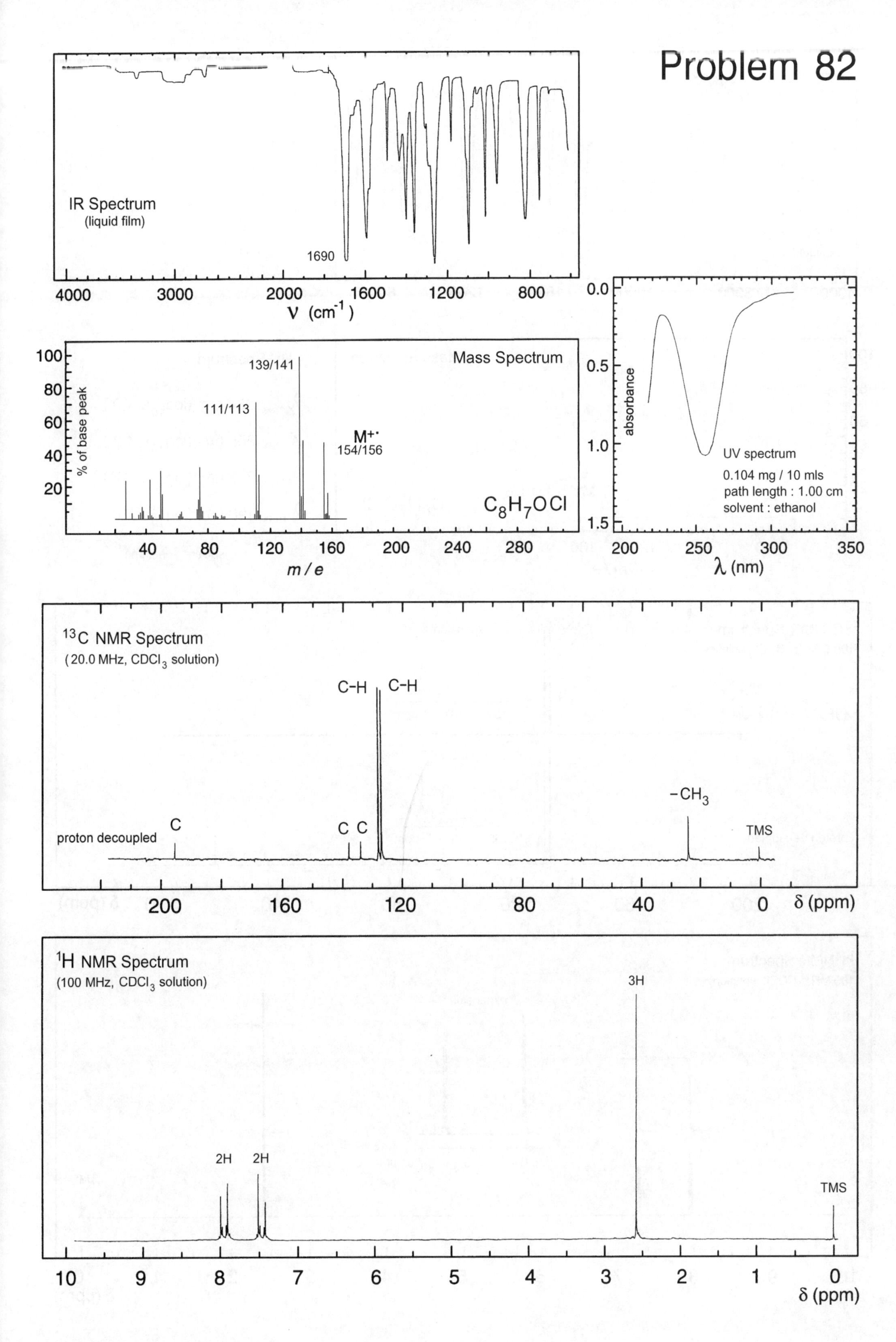

IR Spectrum
(liquid film)
1690
4000
3000
2000
1600
1200
800
ν (cm^{-1})
Mass Spectrum
100
80
60
40
20
% of base peak
139/141
111/113
M+·
154/156
C_8H_7OCl
40
80
120
160
200
240
280
m / e
0.0
0.5
1.0
1.5
absorbance
UV spectrum
0.104 mg / 10 mls
path length : 1.00 cm
solvent : ethanol
200
250
300
350
λ (nm)
^{13}C NMR Spectrum
(20.0 MHz, $CDCl_3$ solution)
C-H
C-H
-CH_3
C
C C
proton decoupled
TMS
200
160
120
80
40
0
δ (ppm)
^{1}H NMR Spectrum
(100 MHz, $CDCl_3$ solution)
3H
2H
2H
TMS
10
9
8
7
6
5
4
3
2
1
0
δ (ppm)

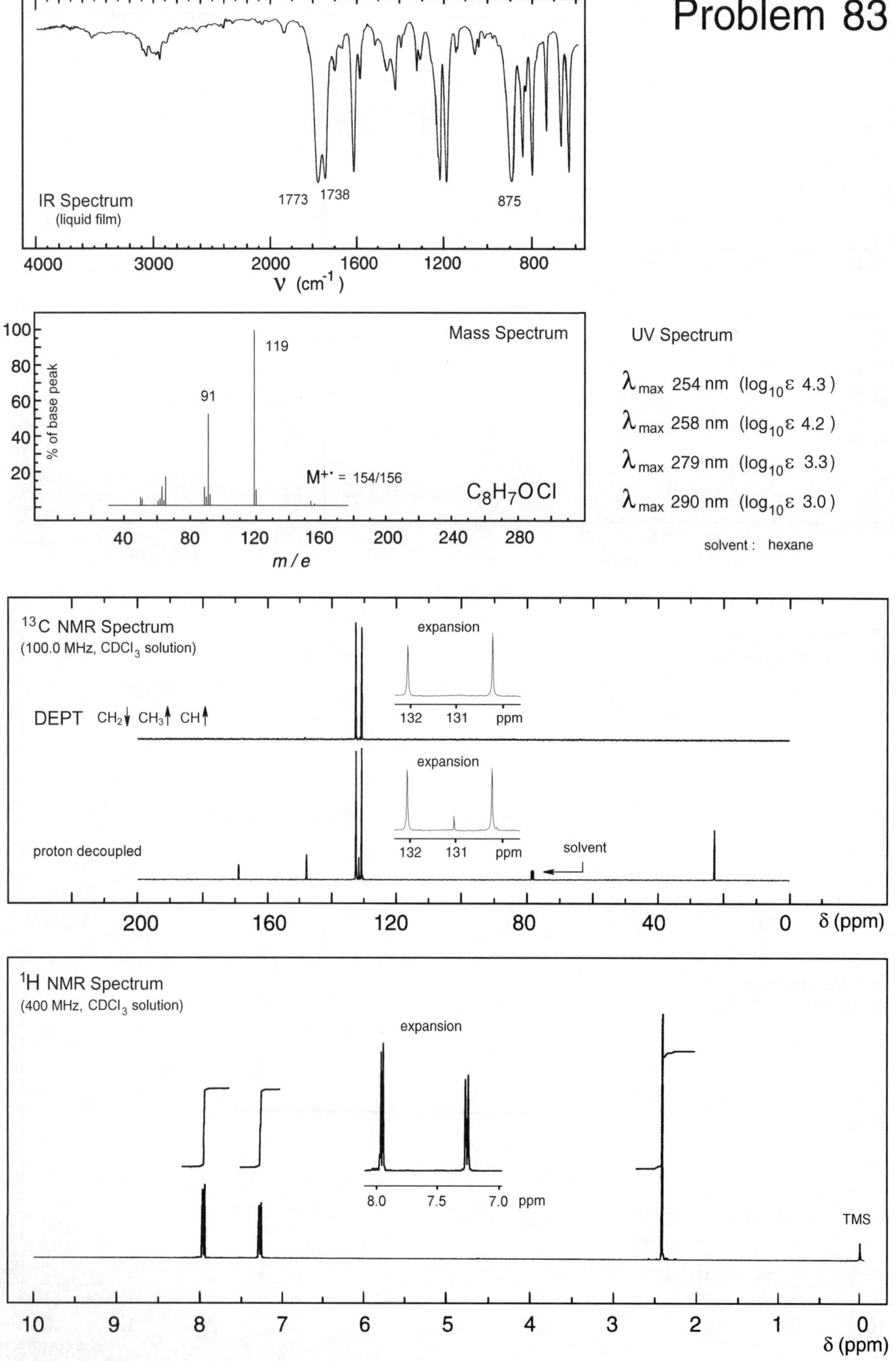
Problem 83
IR Spectrum
(liquid film)
1773
1738
875
4000
3000
2000
1600
1200
800
ν (cm-1)
Mass Spectrum
100
80
60
40
20
% of base peak
119
91
M+· = 154/156
C8H7OCl
40
80
120
160
200
240
280
m/e
UV Spectrum
λmax 254 nm (log10ε 4.3)
λmax 258 nm (log10ε 4.2)
λmax 279 nm (log10ε 3.3)
λmax 290 nm (log10ε 3.0)
solvent : hexane
13C NMR Spectrum
(100.0 MHz, CDCl3 solution)
expansion
DEPT CH2↓ CH3↑ CH↑
132
131
ppm
expansion
proton decoupled
132
131
ppm
solvent
200
160
120
80
40
0
δ (ppm)
1H NMR Spectrum
(400 MHz, CDCl3 solution)
expansion
8.0
7.5
7.0
ppm
TMS
10
9
8
7
6
5
4
3
2
1
0
δ (ppm)

Problem 84

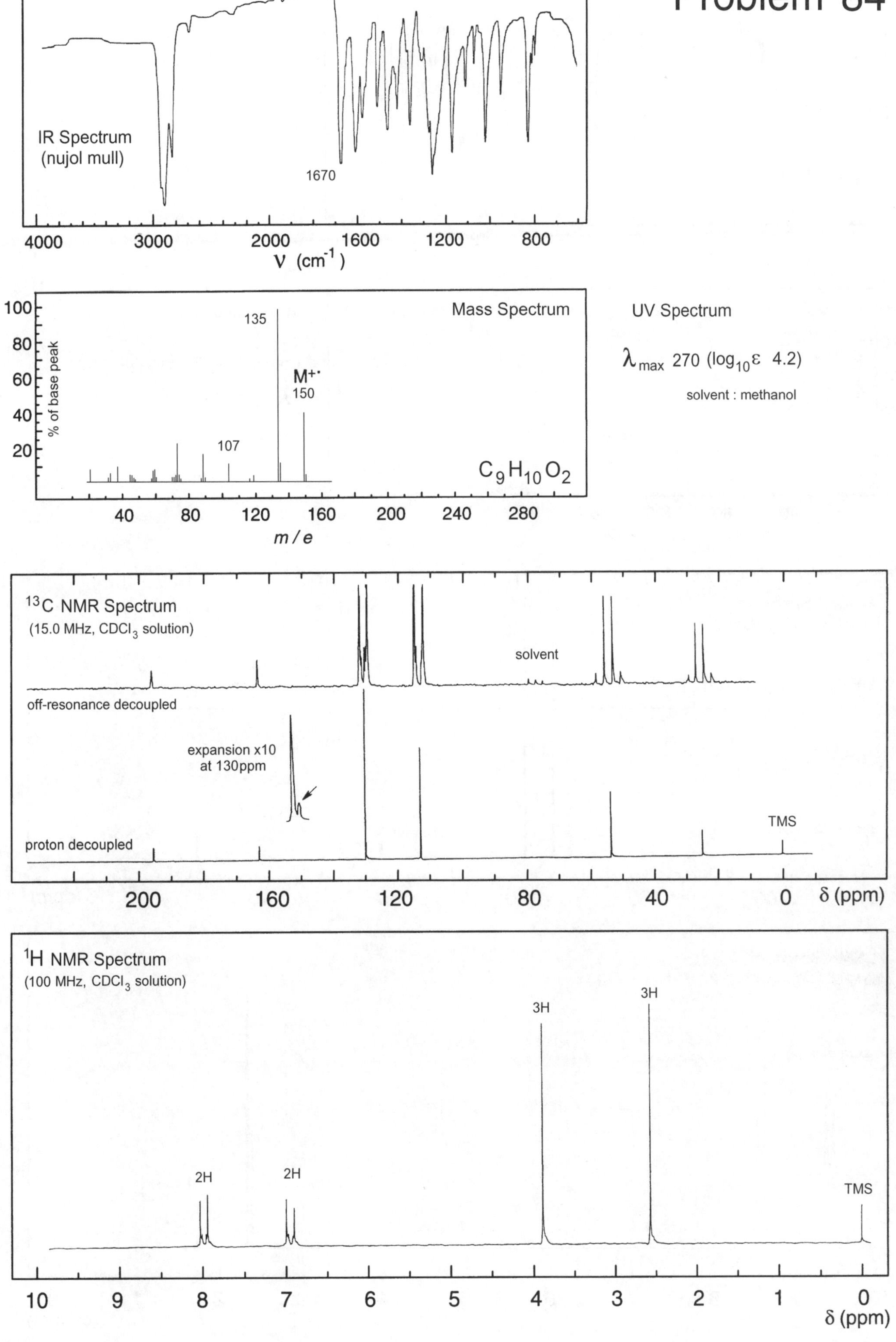

IR Spectrum
(nujol mull)
1670
4000
3000
2000
1600
1200
800
ν (cm-1)
Mass Spectrum
% of base peak
135
M+·
150
107
C9H10O2
m / e
UV Spectrum
λmax 270 (log10ε 4.2)
solvent : methanol
13C NMR Spectrum
(15.0 MHz, CDCl3 solution)
solvent
off-resonance decoupled
expansion x10
at 130ppm
TMS
proton decoupled
δ (ppm)
1H NMR Spectrum
(100 MHz, CDCl3 solution)
3H
3H
2H
2H
TMS
δ (ppm)

Problem 85

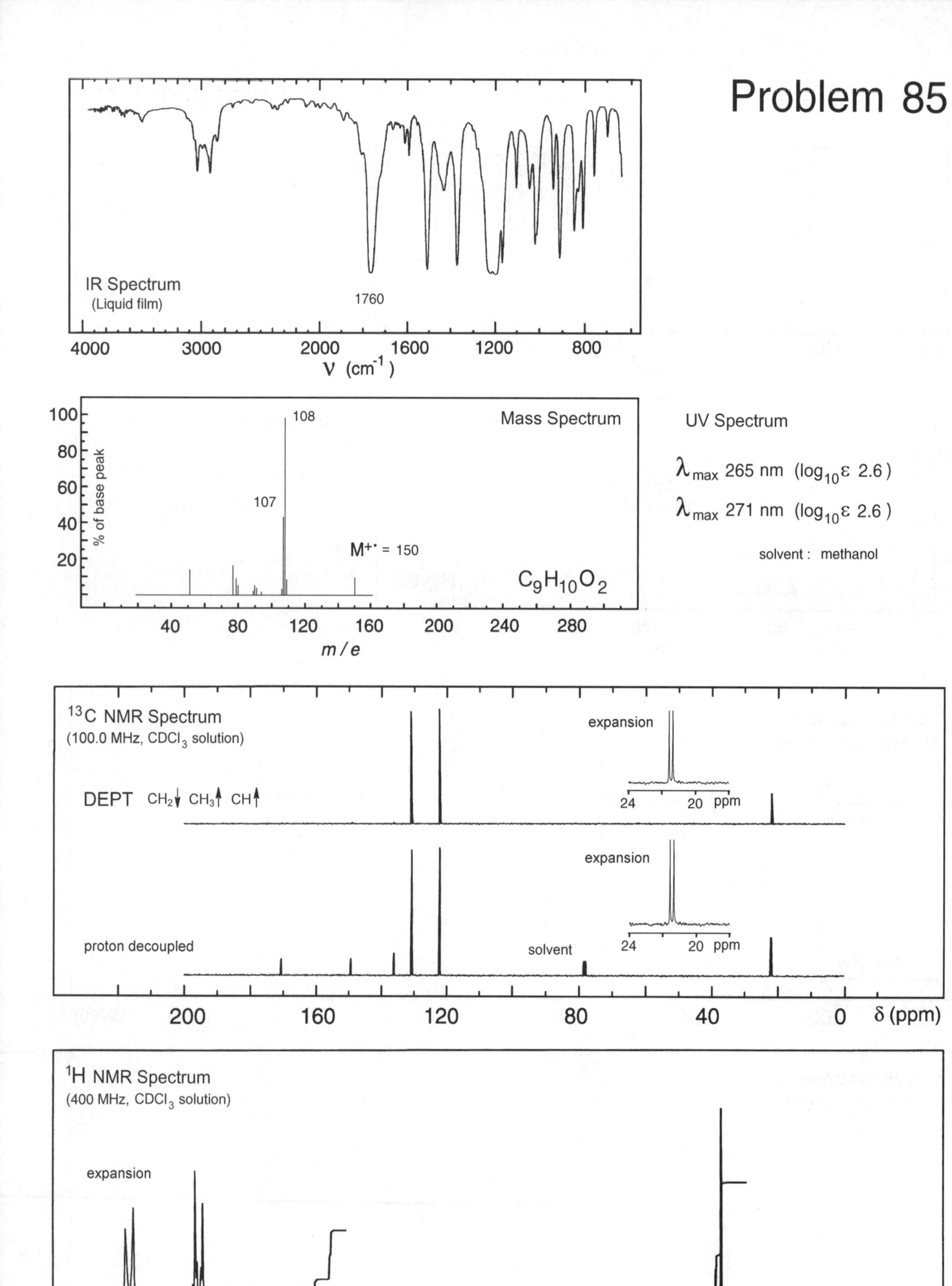

UV Spectrum

λ_{max} 265 nm ($\log_{10}\varepsilon$ 2.6)

λ_{max} 271 nm ($\log_{10}\varepsilon$ 2.6)

solvent : methanol

Problem 86

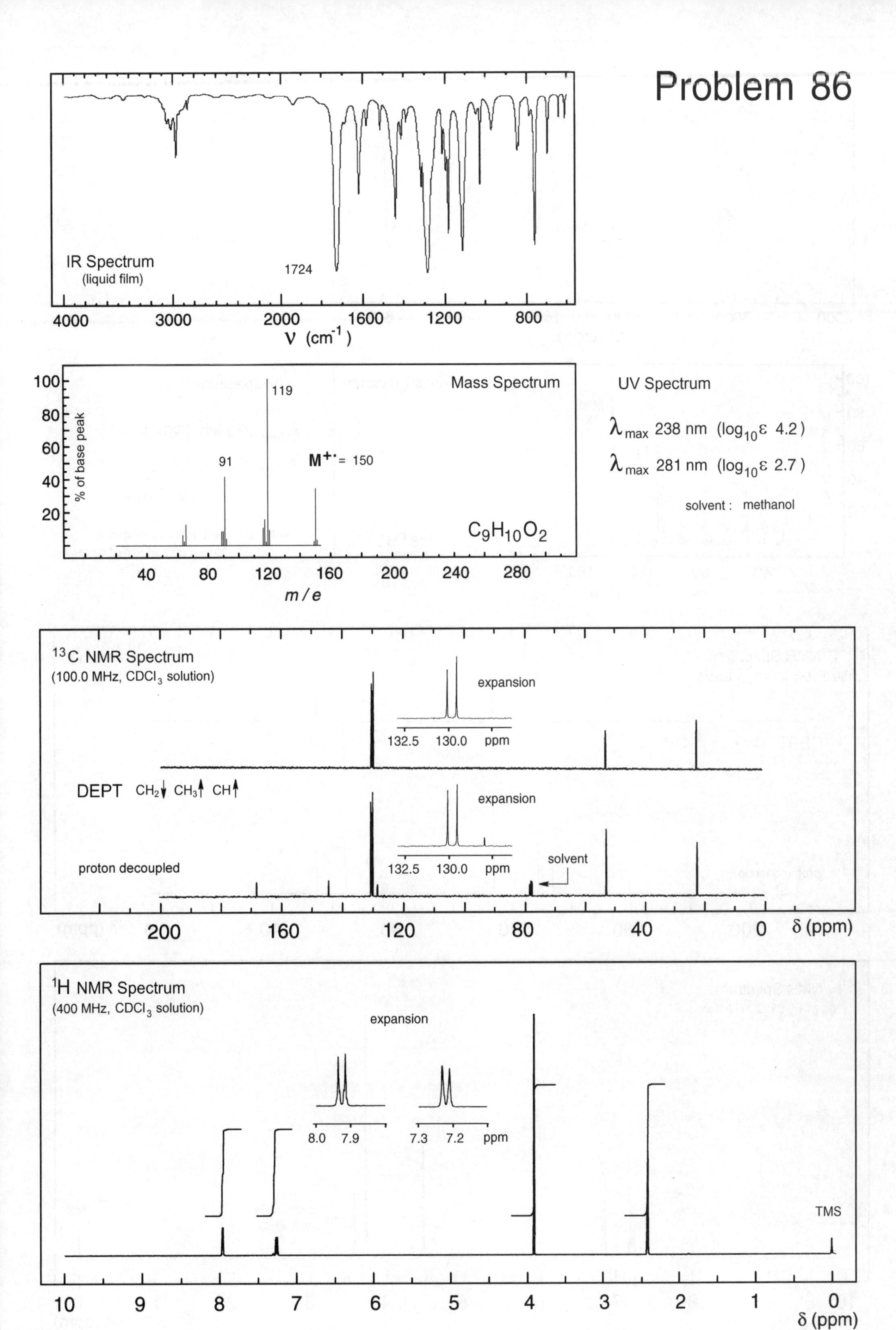
IR Spectrum
(liquid film)
1724
4000
3000
2000
1600
1200
800
ν (cm⁻¹)
Mass Spectrum
% of base peak
100
80
60
40
20
119
91
M⁺· = 150
C9H10O2
40
80
120
160
200
240
280
m / e
UV Spectrum
λmax 238 nm (log10ε 4.2)
λmax 281 nm (log10ε 2.7)
solvent : methanol
13C NMR Spectrum
(100.0 MHz, CDCl3 solution)
expansion
132.5
130.0
ppm
DEPT CH2↓ CH3↑ CH↑
proton decoupled
solvent
200
160
120
80
40
0
δ (ppm)
1H NMR Spectrum
(400 MHz, CDCl3 solution)
expansion
8.0
7.9
7.3
7.2
ppm
TMS
10
9
8
7
6
5
4
3
2
1
0
δ (ppm)

Problem 87

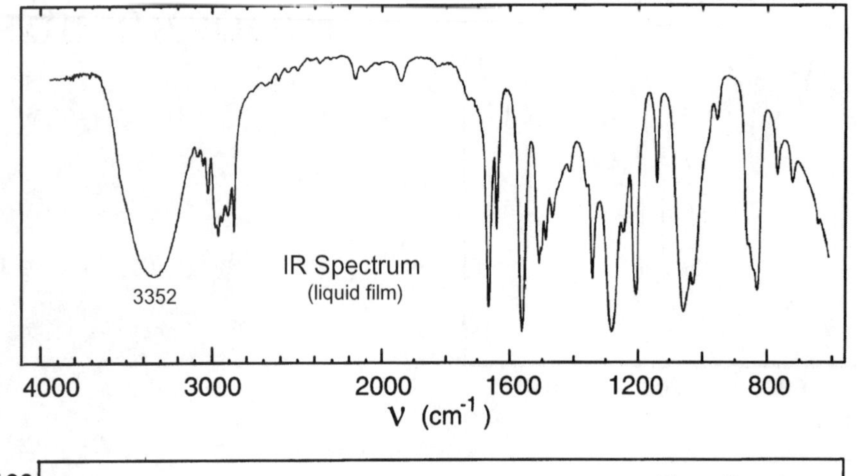

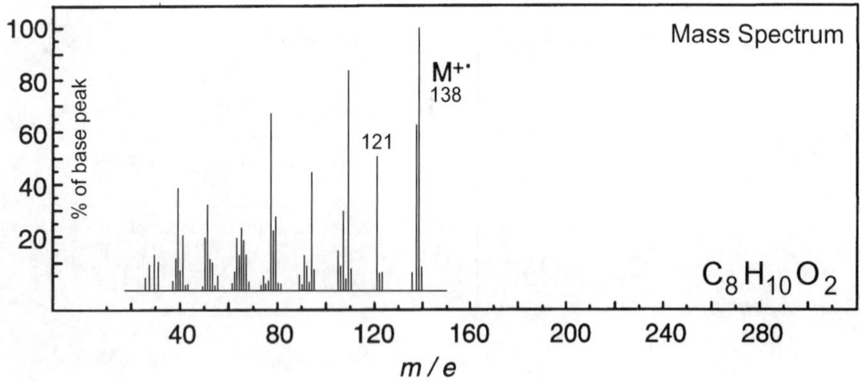

UV Spectrum

λ_{max} 270 nm ($\log_{10}\varepsilon$ 3.1)

λ_{max} 282 nm ($\log_{10}\varepsilon$ 3.1)

solvent : hexane

Note: UV spectrum not changed significantly on addition of base

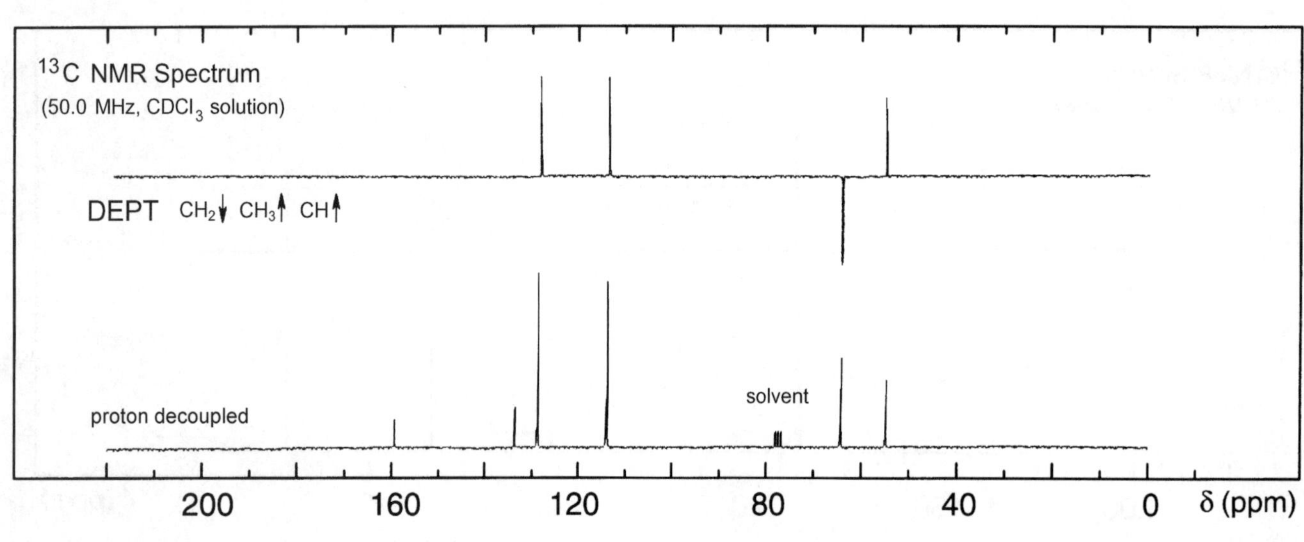

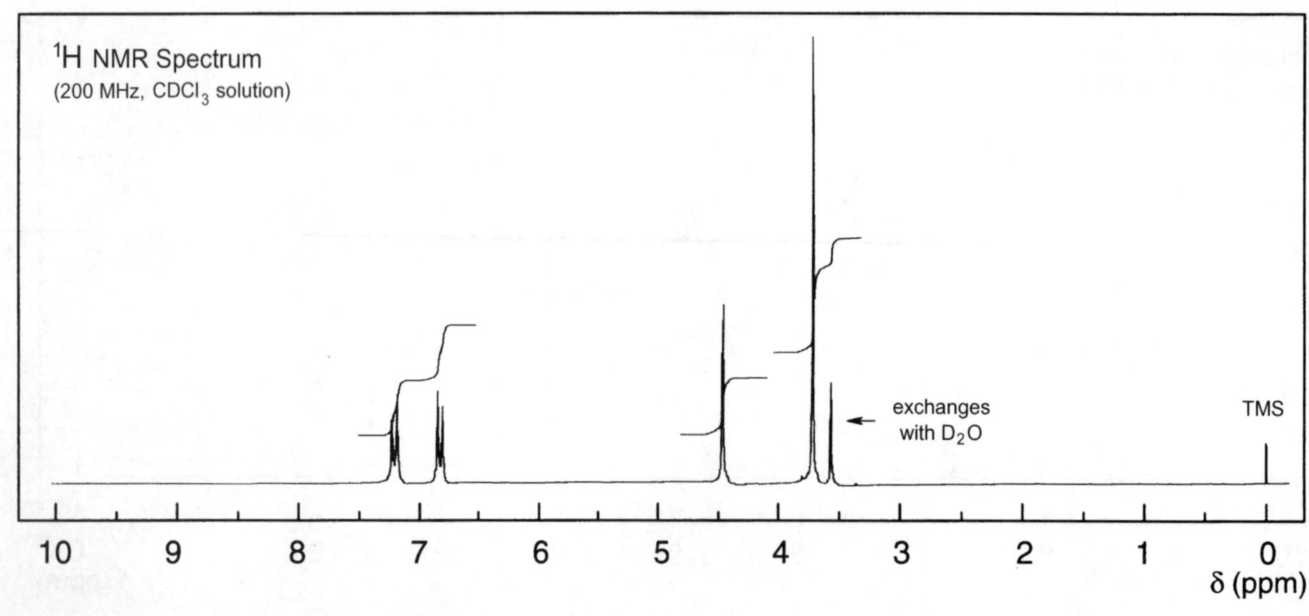

Problem 88

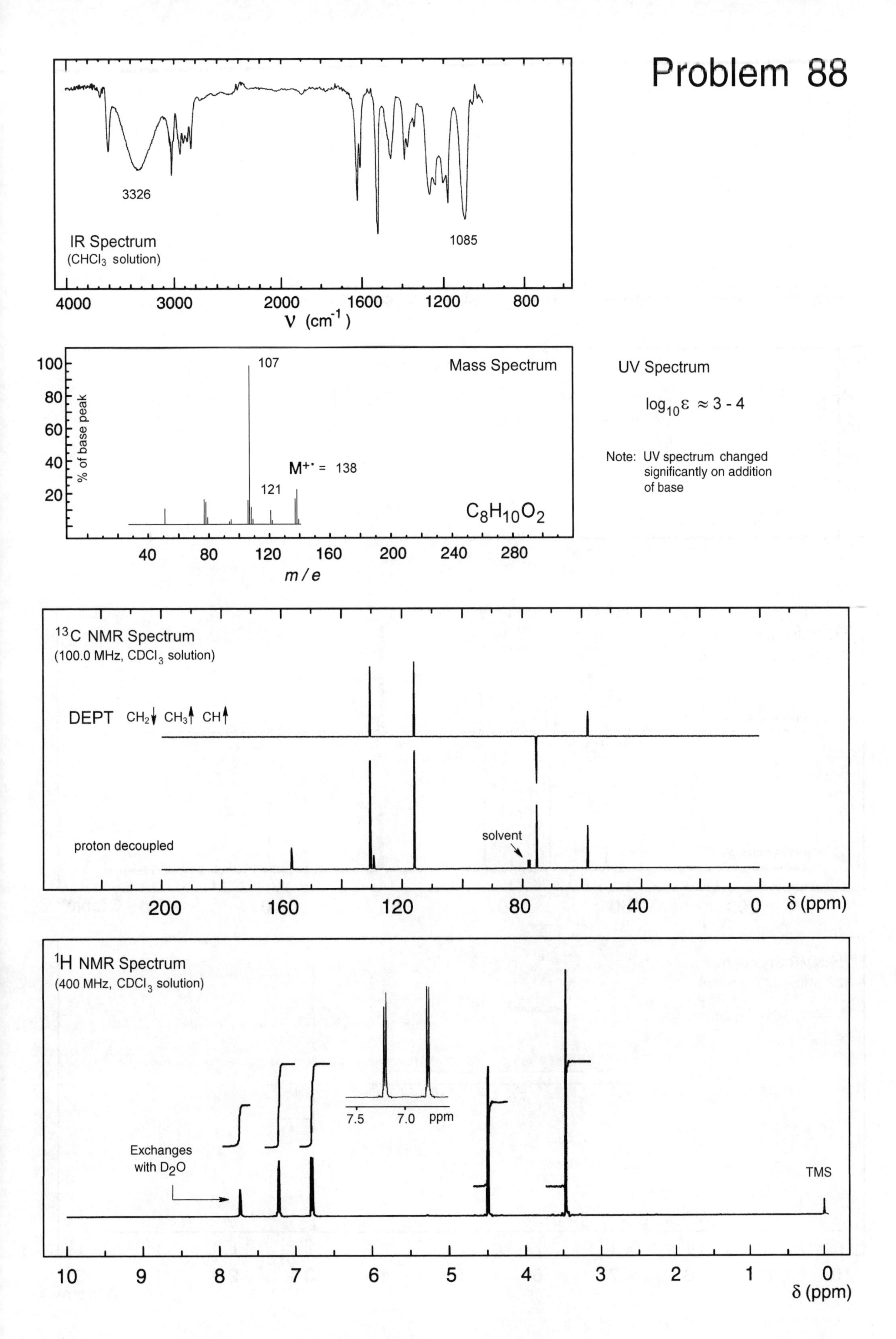

3326
1085
IR Spectrum
(CHCl3 solution)
4000
3000
2000
1600
1200
800
ν (cm-1)
Mass Spectrum
107
M+· = 138
121
C8H10O2
% of base peak
100
80
60
40
20
40
80
120
160
200
240
280
m / e
UV Spectrum
log10ε ≈ 3 - 4
Note: UV spectrum changed significantly on addition of base
13C NMR Spectrum
(100.0 MHz, CDCl3 solution)
DEPT CH2↓ CH3↑ CH↑
proton decoupled
solvent
200
160
120
80
40
0
δ (ppm)
1H NMR Spectrum
(400 MHz, CDCl3 solution)
7.5
7.0
ppm
Exchanges with D2O
TMS
10
9
8
7
6
5
4
3
2
1
0
δ (ppm)

Problem 89

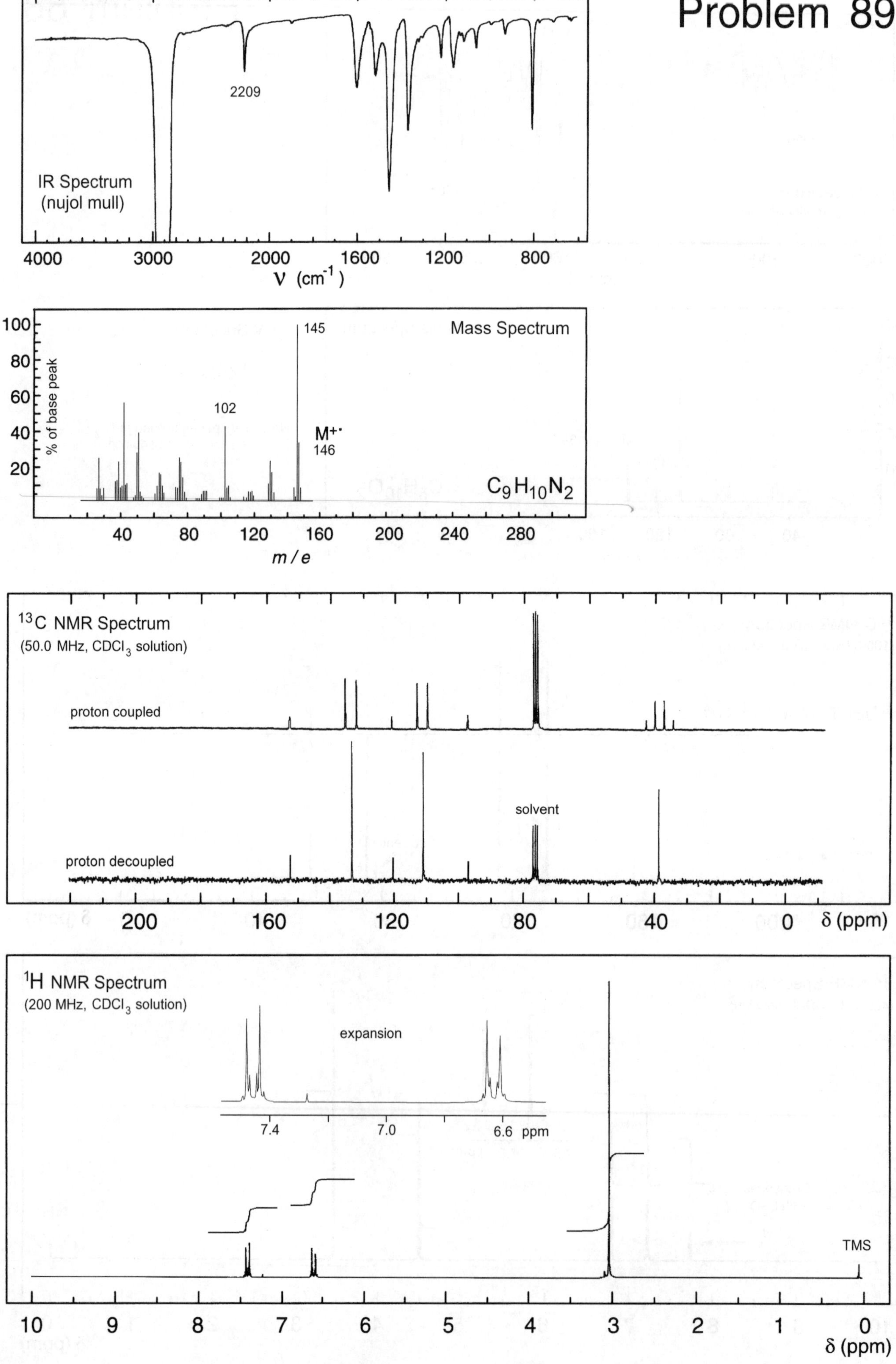

IR Spectrum
(nujol mull)
2209
4000
3000
2000
1600
1200
800
ν (cm^{-1})
Mass Spectrum
% of base peak
145
102
M+·
146
$C_9H_{10}N_2$
m / e
^{13}C NMR Spectrum
(50.0 MHz, $CDCl_3$ solution)
proton coupled
proton decoupled
solvent
δ (ppm)
^{1}H NMR Spectrum
(200 MHz, $CDCl_3$ solution)
expansion
7.4
7.0
6.6 ppm
TMS
δ (ppm)

Problem 90

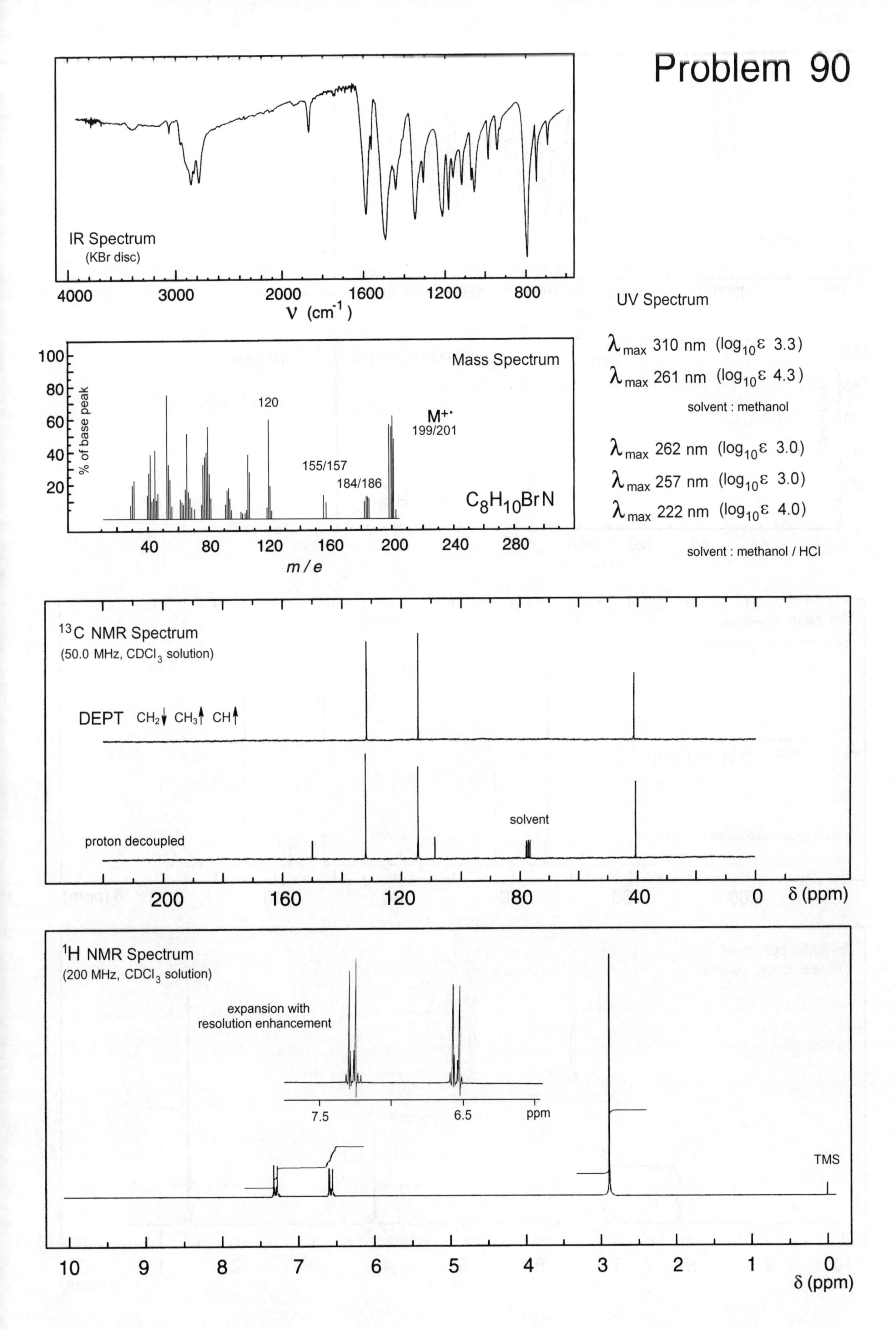
IR Spectrum
(KBr disc)
4000
3000
2000
1600
1200
800
ν (cm-1)
Mass Spectrum
% of base peak
120
155/157
184/186
M+·
199/201
C8H10BrN
m / e
UV Spectrum
λmax 310 nm (log10ε 3.3)
λmax 261 nm (log10ε 4.3)
solvent : methanol
λmax 262 nm (log10ε 3.0)
λmax 257 nm (log10ε 3.0)
λmax 222 nm (log10ε 4.0)
solvent : methanol / HCl
13C NMR Spectrum
(50.0 MHz, CDCl3 solution)
DEPT CH2↓ CH3↑ CH↑
proton decoupled
solvent
δ (ppm)
1H NMR Spectrum
(200 MHz, CDCl3 solution)
expansion with
resolution enhancement
7.5
6.5
ppm
TMS
δ (ppm)

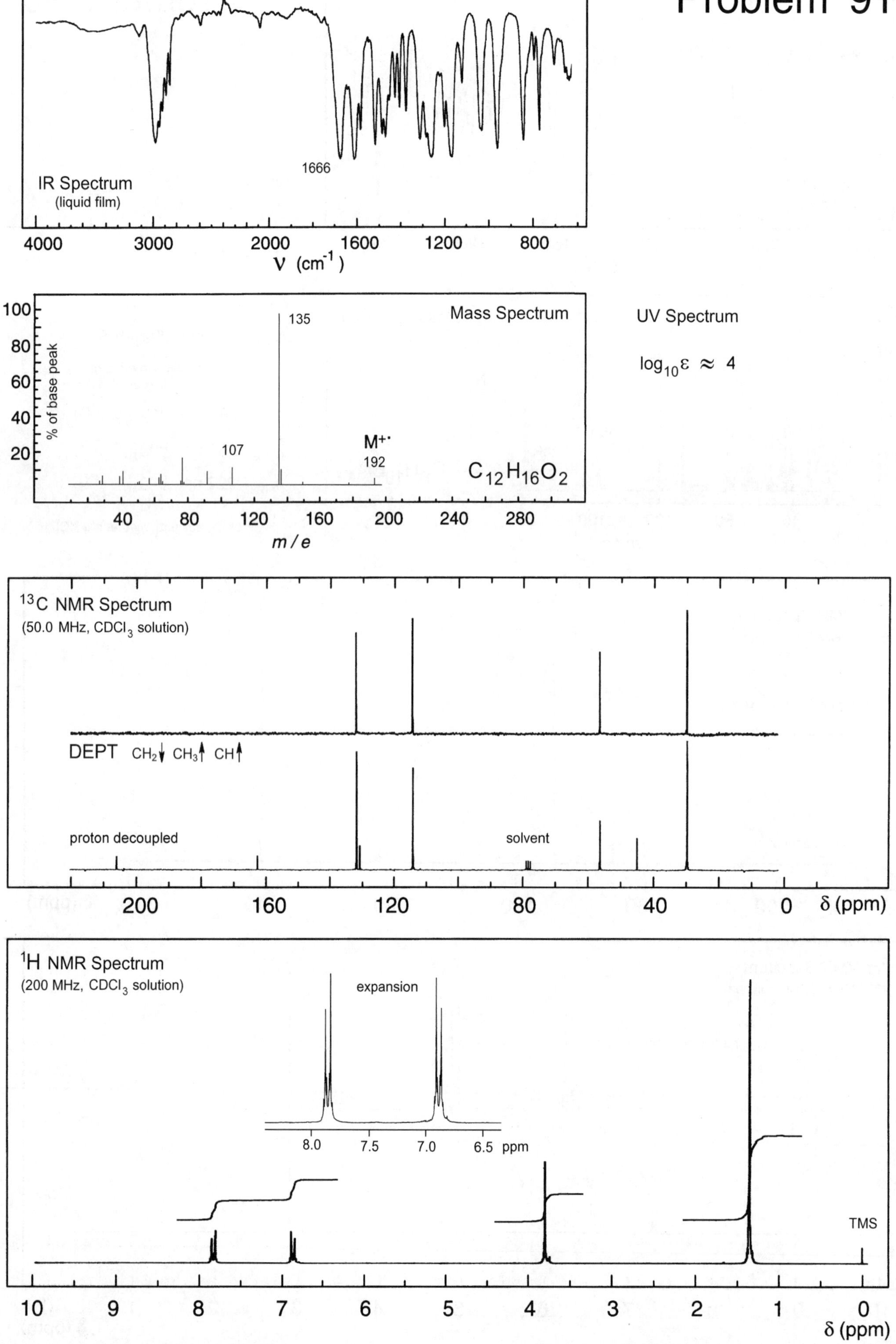
Problem 91
IR Spectrum
(liquid film)
1666
4000
3000
2000
1600
1200
800
ν (cm^{-1})
Mass Spectrum
% of base peak
100
80
60
40
20
135
107
M+·
192
$C_{12}H_{16}O_2$
40
80
120
160
200
240
280
m / e
UV Spectrum
$\log_{10}\varepsilon \approx 4$
^{13}C NMR Spectrum
(50.0 MHz, $CDCl_3$ solution)
DEPT CH_2↓ CH_3↑ CH↑
proton decoupled
solvent
200
160
120
80
40
0
δ (ppm)
^{1}H NMR Spectrum
(200 MHz, $CDCl_3$ solution)
expansion
8.0
7.5
7.0
6.5
ppm
TMS
10
9
8
7
6
5
4
3
2
1
0
δ (ppm)

Problem 92

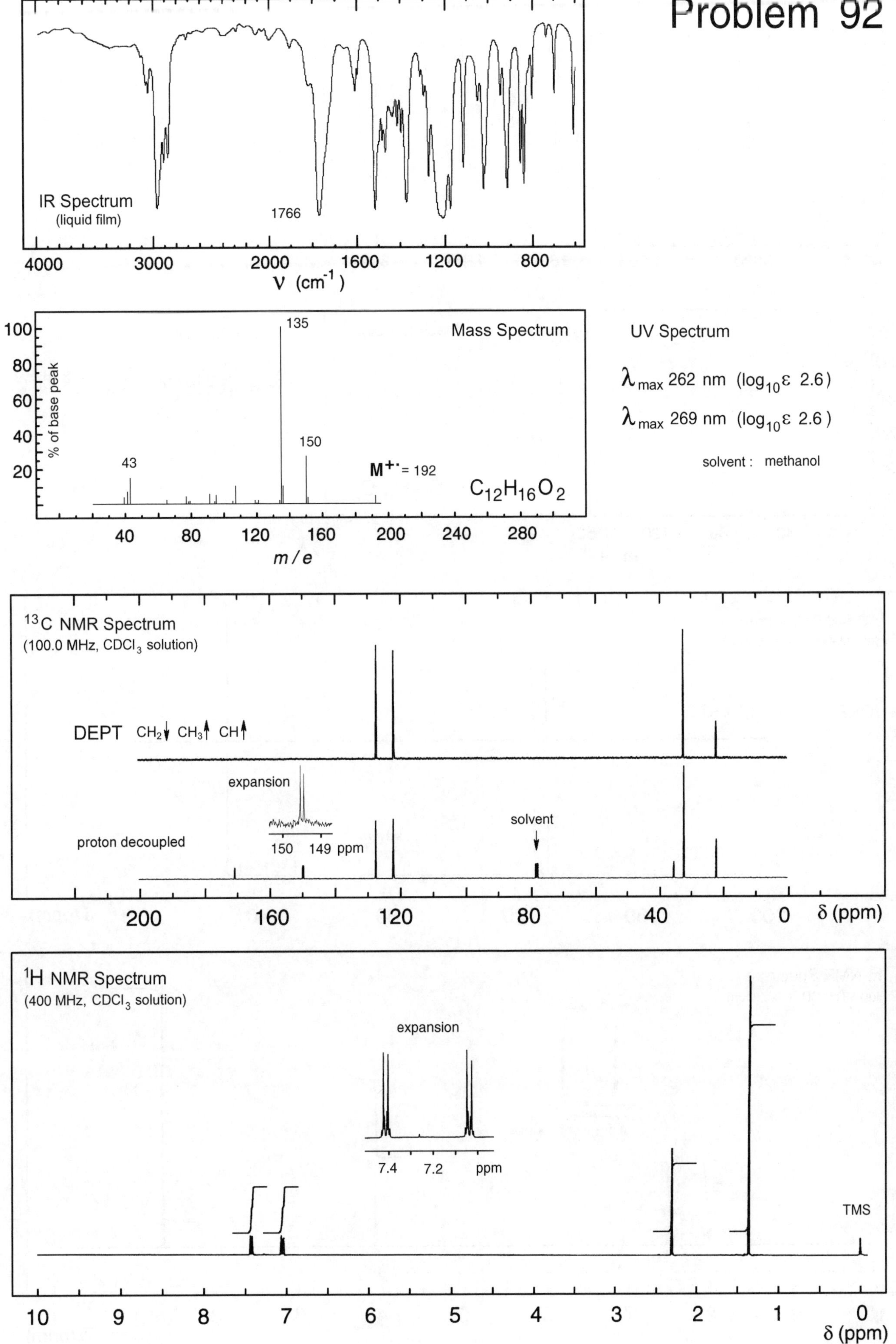

IR Spectrum
(liquid film)
1766
4000
3000
2000
1600
1200
800
ν (cm⁻¹)
Mass Spectrum
% of base peak
100
80
60
40
20
135
150
43
M⁺· = 192
C12H16O2
40
80
120
160
200
240
280
m / e
UV Spectrum
λmax 262 nm (log10ε 2.6)
λmax 269 nm (log10ε 2.6)
solvent : methanol
13C NMR Spectrum
(100.0 MHz, CDCl3 solution)
DEPT CH2↓ CH3↑ CH↑
expansion
150 149 ppm
solvent
proton decoupled
200
160
120
80
40
0
δ (ppm)
1H NMR Spectrum
(400 MHz, CDCl3 solution)
expansion
7.4 7.2 ppm
TMS
10
9
8
7
6
5
4
3
2
1
0
δ (ppm)

Problem 93

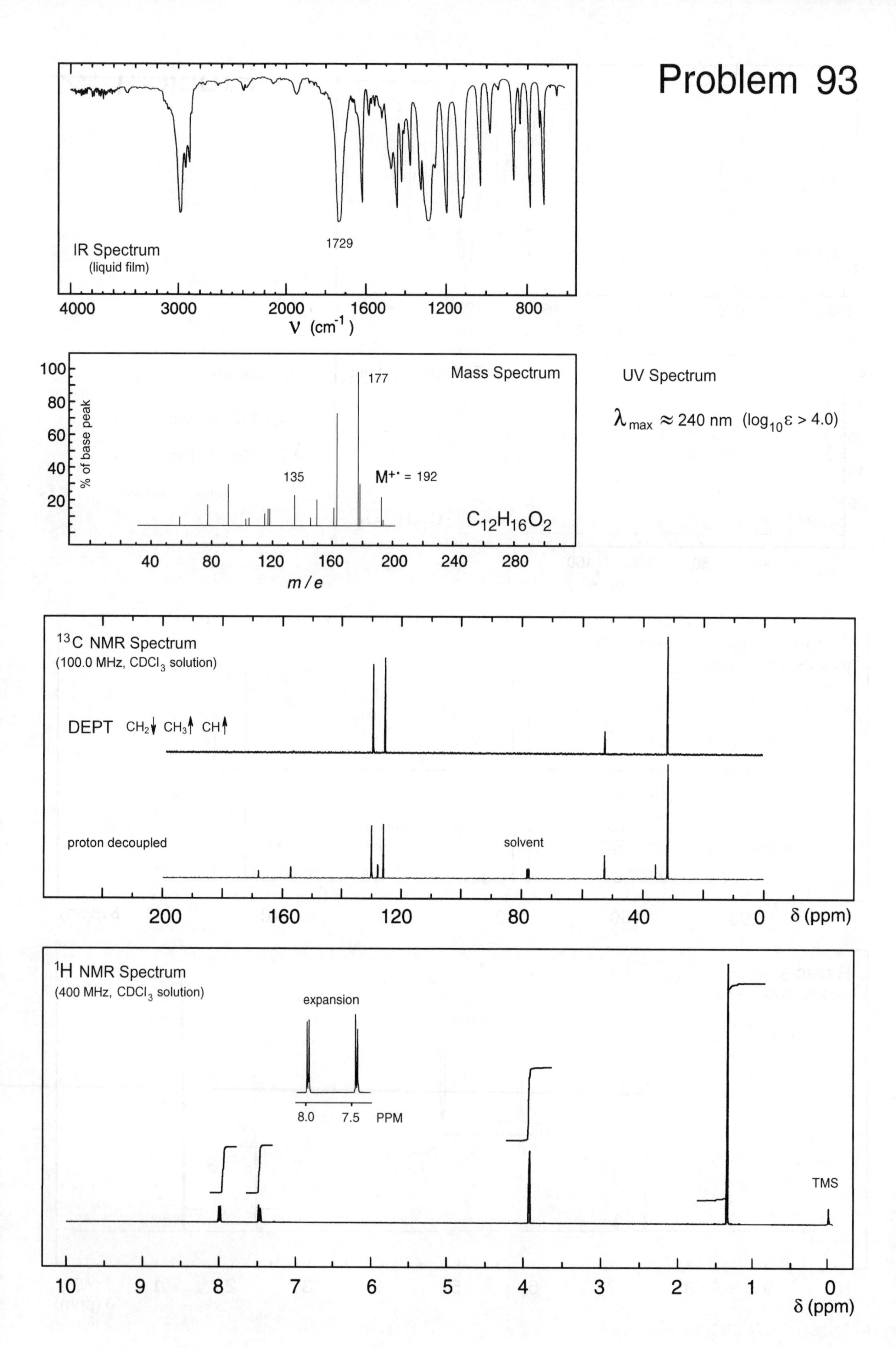

IR Spectrum
(liquid film)
1729
4000
3000
2000
1600
1200
800
ν (cm-1)
Mass Spectrum
100
80
60
40
20
% of base peak
177
135
M+• = 192
C12H16O2
40
80
120
160
200
240
280
m / e
UV Spectrum
λmax ≈ 240 nm (log10ε > 4.0)
13C NMR Spectrum
(100.0 MHz, CDCl3 solution)
DEPT CH2↓ CH3↑ CH↑
proton decoupled
solvent
200
160
120
80
40
0
δ (ppm)
1H NMR Spectrum
(400 MHz, CDCl3 solution)
expansion
8.0
7.5
PPM
TMS
10
9
8
7
6
5
4
3
2
1
0
δ (ppm)

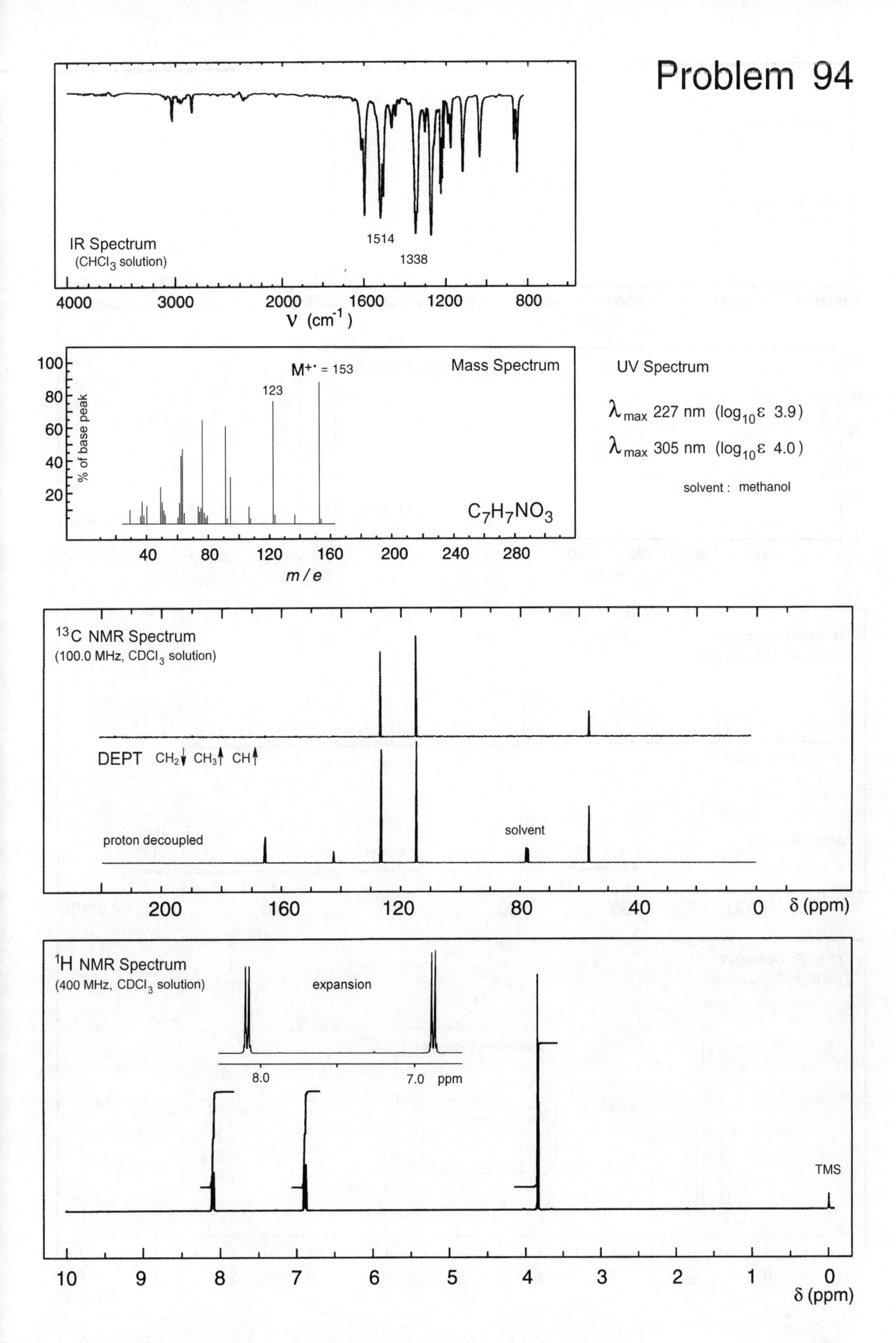

Problem 94
IR Spectrum
(CHCl3 solution)
1514
1338
4000 3000 2000 1600 1200 800
ν (cm-1)
Mass Spectrum
M+· = 153
123
% of base peak
100 80 60 40 20
C7H7NO3
40 80 120 160 200 240 280
m / e
UV Spectrum
λmax 227 nm (log10ε 3.9)
λmax 305 nm (log10ε 4.0)
solvent : methanol
13C NMR Spectrum
(100.0 MHz, CDCl3 solution)
DEPT CH2↓ CH3↑ CH↑
proton decoupled
solvent
200 160 120 80 40 0 δ (ppm)
1H NMR Spectrum
(400 MHz, CDCl3 solution)
expansion
8.0
7.0 ppm
TMS
10 9 8 7 6 5 4 3 2 1 0
δ (ppm)

Problem 95

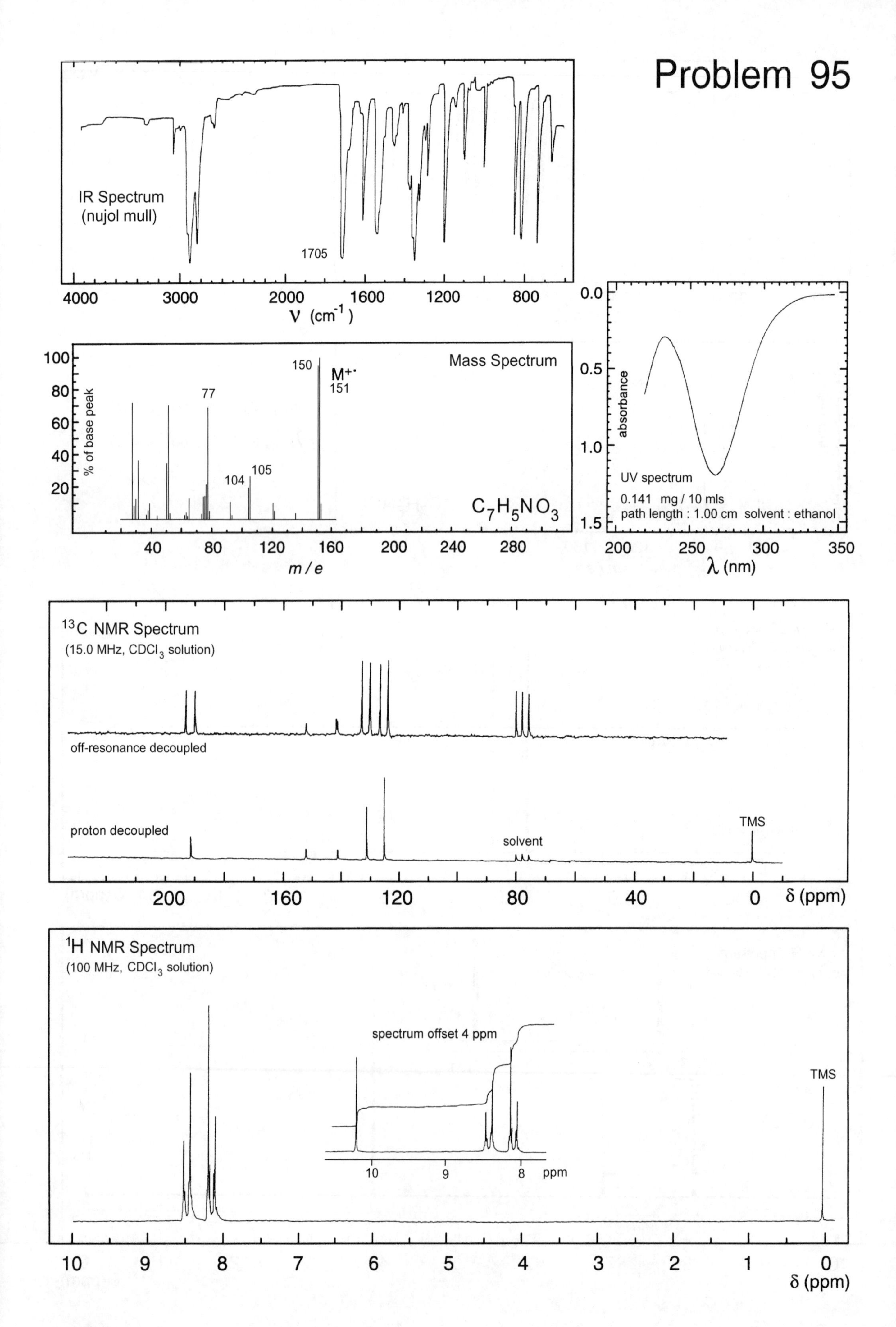

IR Spectrum
(nujol mull)
1705
4000
3000
2000
1600
1200
800
ν (cm⁻¹)
Mass Spectrum
% of base peak
150
M+·
151
77
104
105
$C_7H_5NO_3$
m / e
UV spectrum
0.141 mg / 10 mls
path length : 1.00 cm solvent : ethanol
absorbance
λ (nm)
¹³C NMR Spectrum
(15.0 MHz, CDCl₃ solution)
off-resonance decoupled
proton decoupled
solvent
TMS
δ (ppm)
¹H NMR Spectrum
(100 MHz, CDCl₃ solution)
spectrum offset 4 ppm
ppm
TMS
δ (ppm)

Problem 96

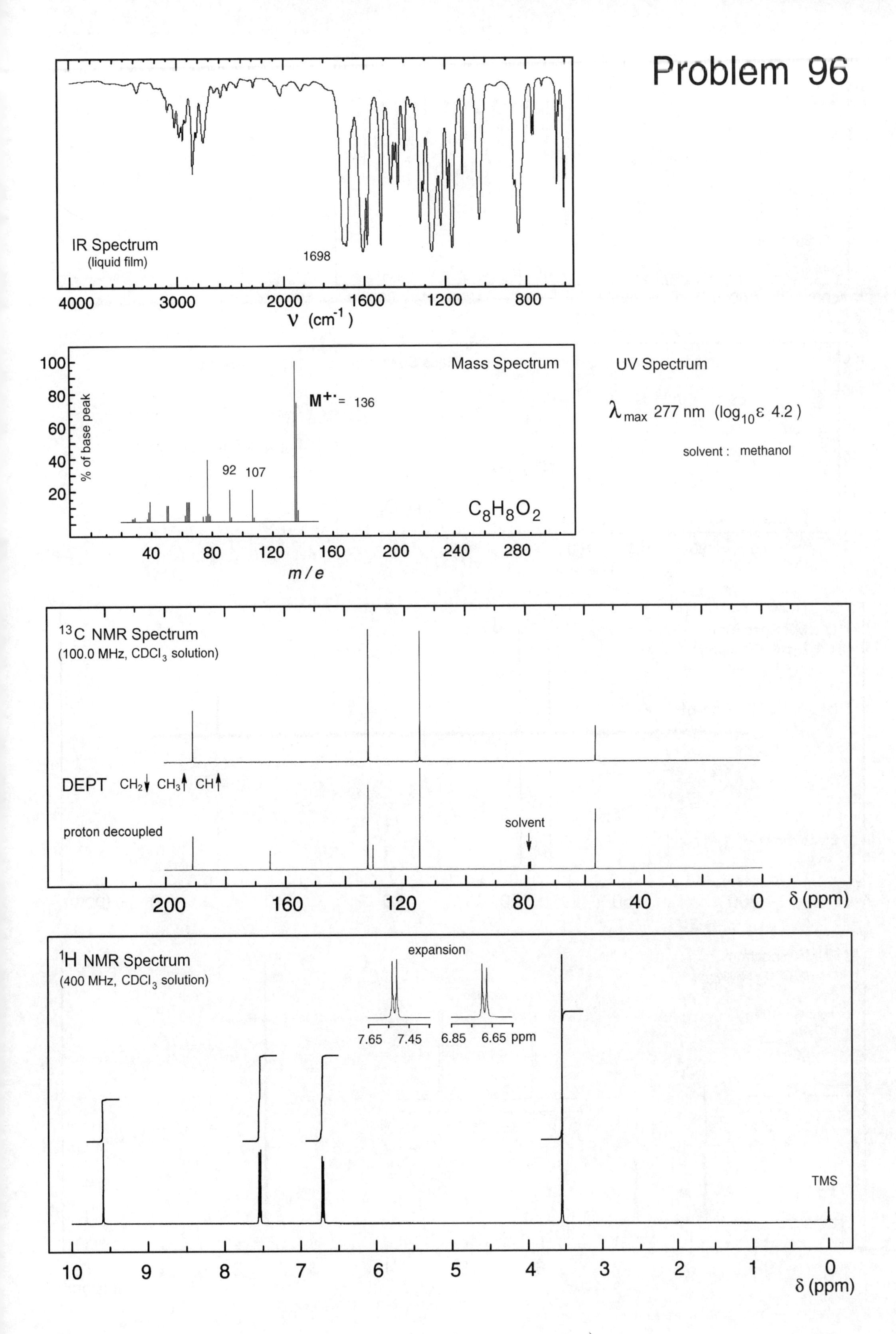

IR Spectrum
(liquid film)
1698
4000
3000
2000
1600
1200
800
ν (cm⁻¹)
Mass Spectrum
% of base peak
M⁺˙= 136
92
107
C₈H₈O₂
m / e
UV Spectrum
λmax 277 nm (log₁₀ε 4.2)
solvent : methanol
¹³C NMR Spectrum
(100.0 MHz, CDCl₃ solution)
DEPT CH₂↓ CH₃↑ CH↑
proton decoupled
solvent
δ (ppm)
¹H NMR Spectrum
(400 MHz, CDCl₃ solution)
expansion
7.65 7.45
6.85 6.65 ppm
TMS
δ (ppm)

Problem 97

IR Spectrum
(KBr disc)
1766
1685
4000 3000 2000 1600 1200 800
ν (cm^{-1})

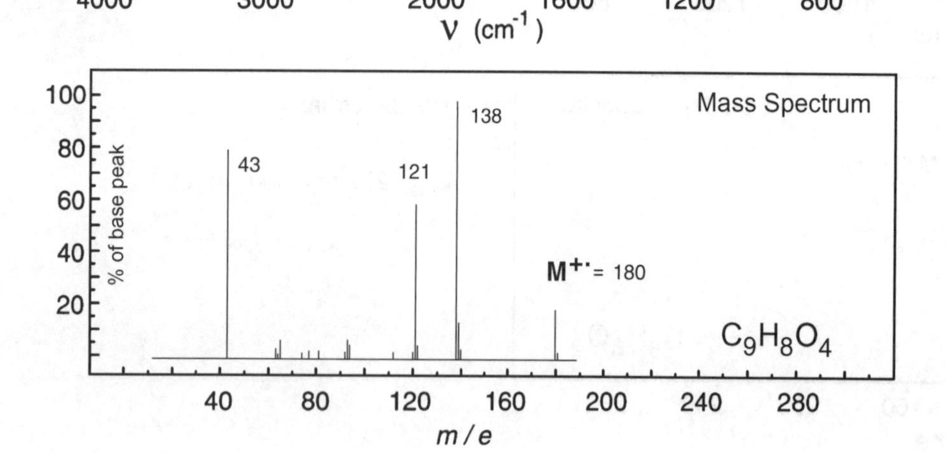

UV Spectrum

λ_{max} 235 nm ($\log_{10}\varepsilon$ 4.1)

λ_{max} 265 nm ($\log_{10}\varepsilon$ 3.0)

solvent : ethanol

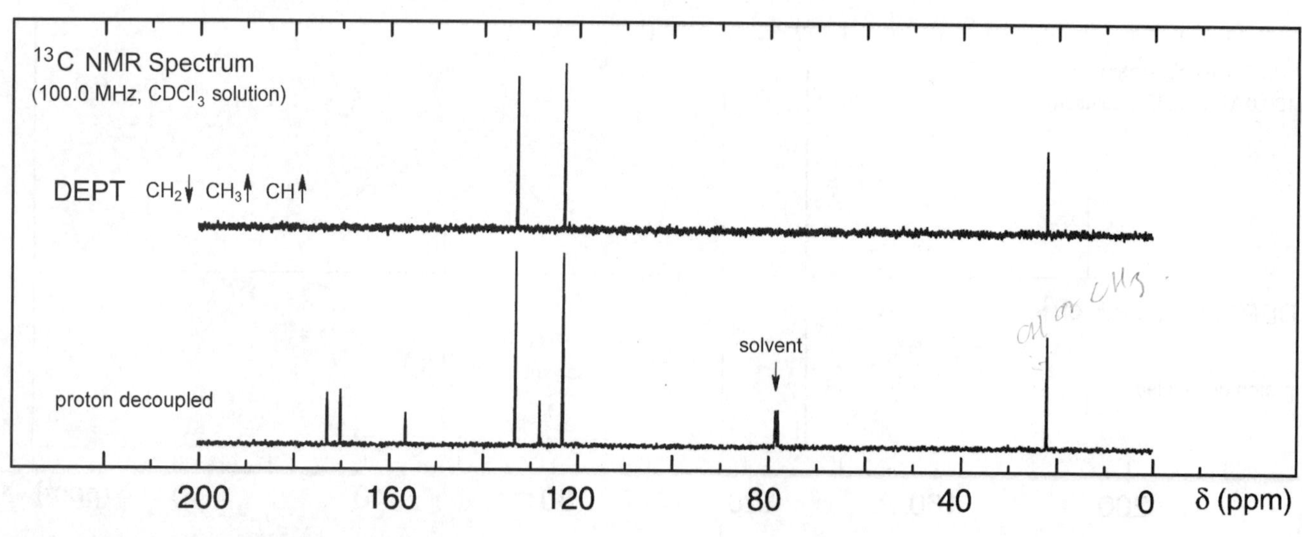

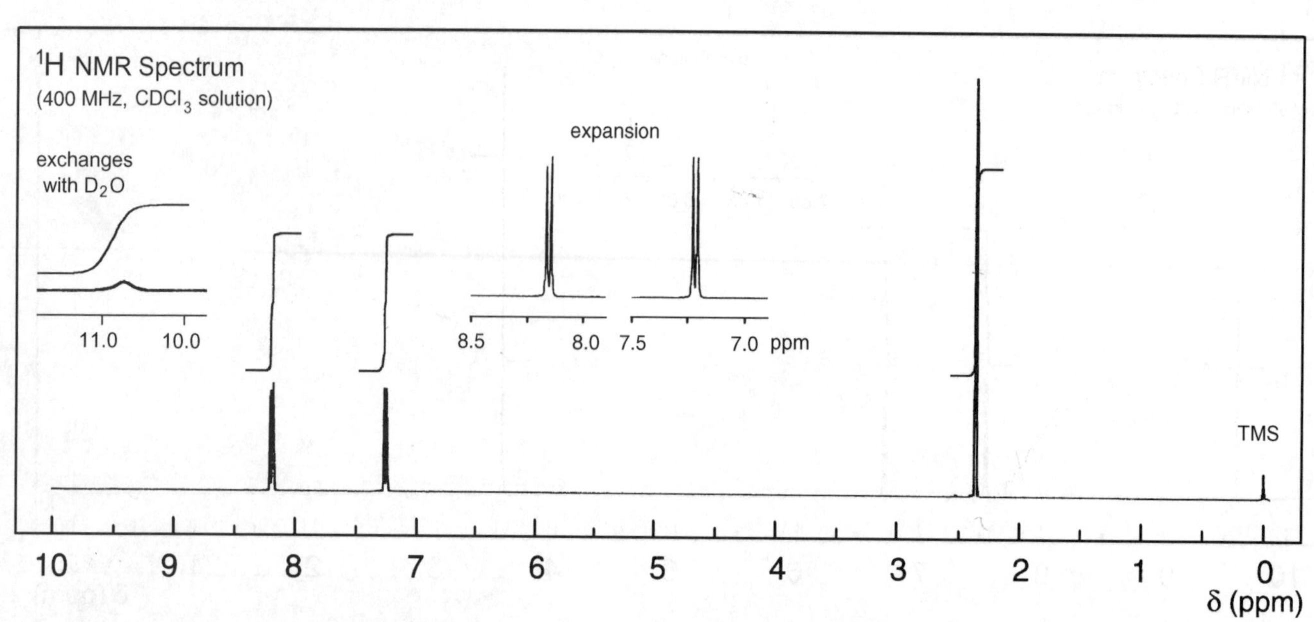

Problem 98

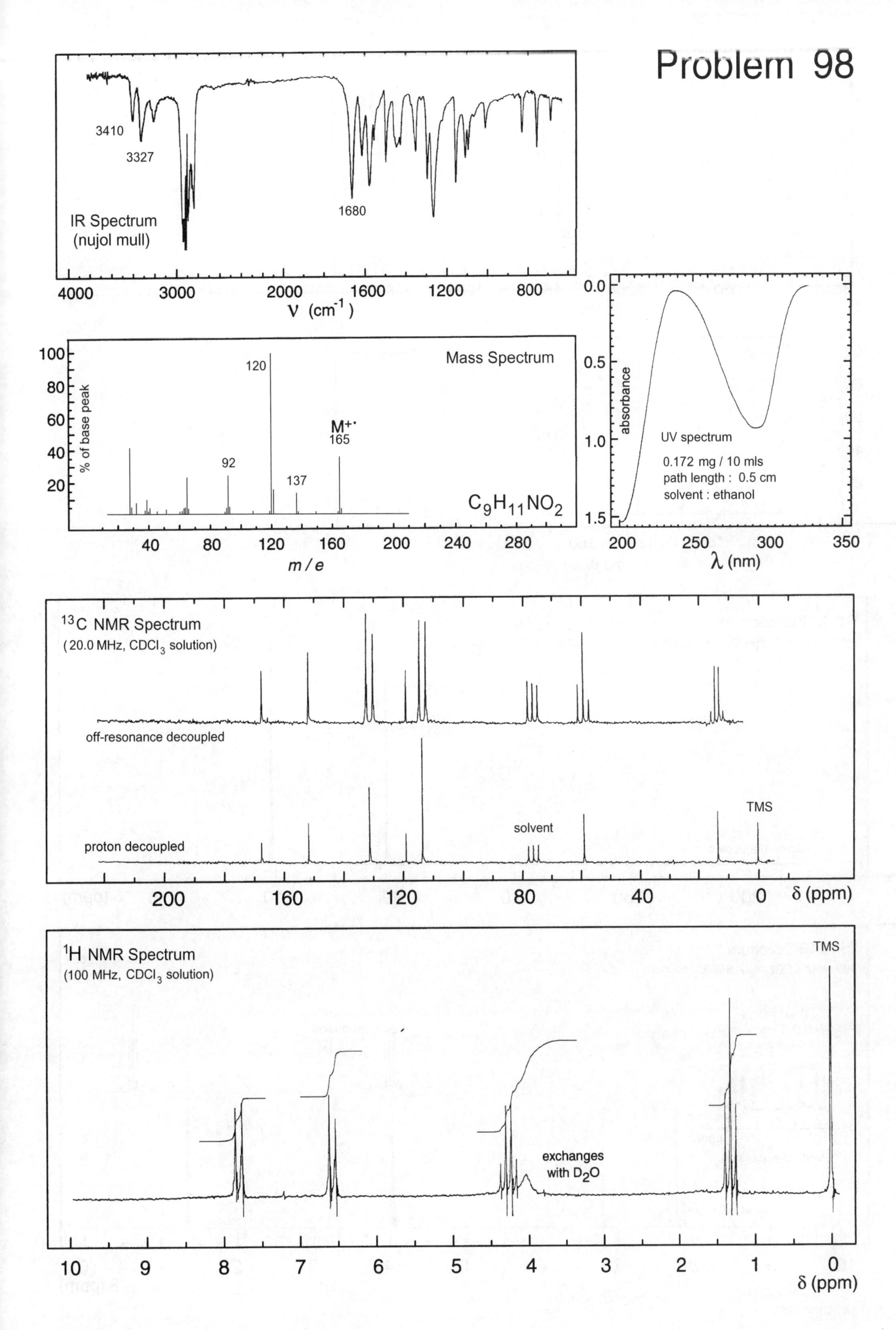

3410
3327
1680
IR Spectrum
(nujol mull)
4000
3000
2000
1600
1200
800
ν (cm⁻¹)
Mass Spectrum
100
80
60
40
20
% of base peak
120
92
137
M+·
165
C9H11NO2
40
80
120
160
200
240
280
m / e
0.0
0.5
1.0
1.5
absorbance
UV spectrum
0.172 mg / 10 mls
path length : 0.5 cm
solvent : ethanol
200
250
300
350
λ (nm)
13C NMR Spectrum
(20.0 MHz, CDCl3 solution)
off-resonance decoupled
proton decoupled
solvent
TMS
200
160
120
80
40
0
δ (ppm)
1H NMR Spectrum
(100 MHz, CDCl3 solution)
TMS
exchanges
with D2O
10
9
8
7
6
5
4
3
2
1
0
δ (ppm)

Problem 99

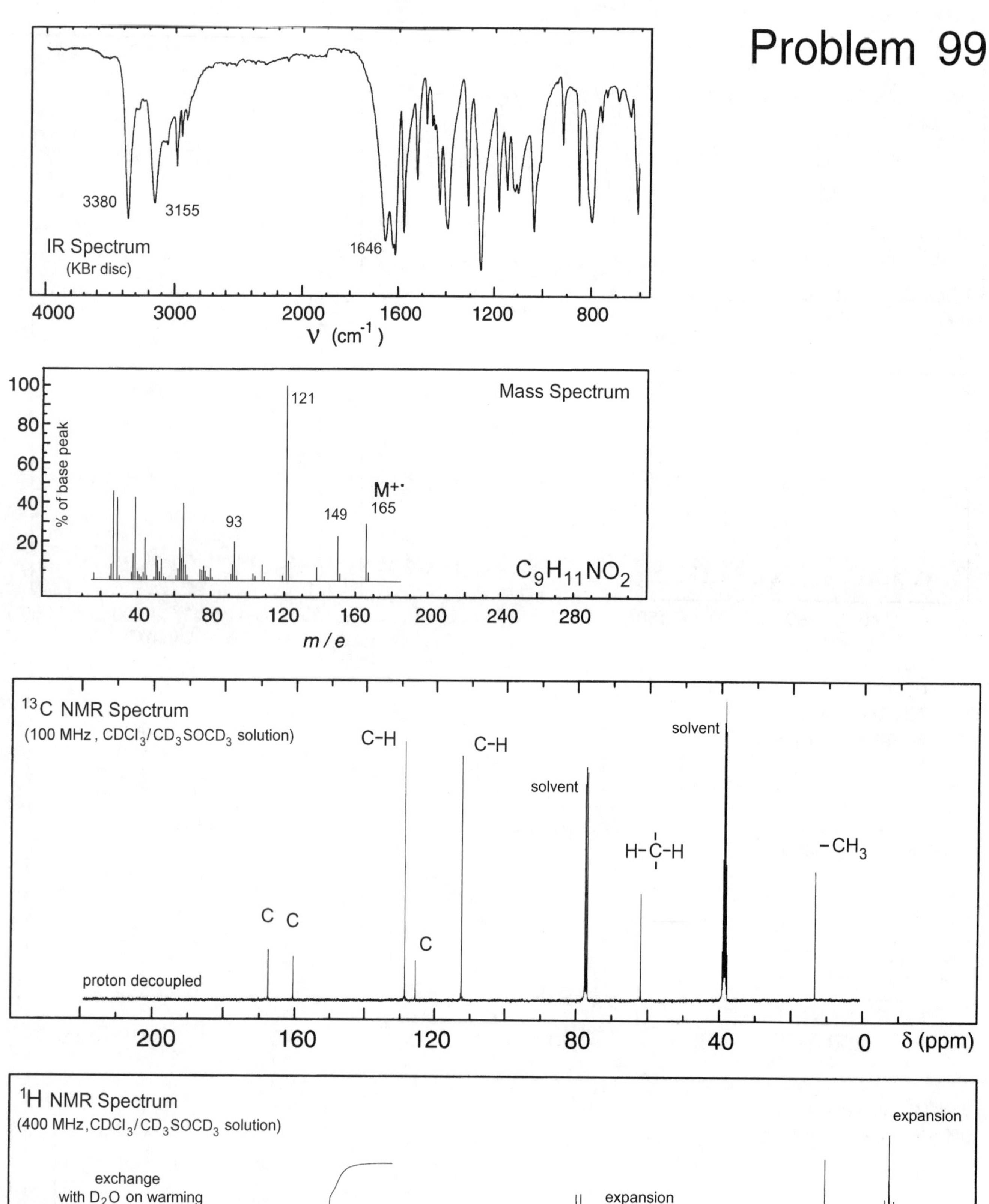

3380
3155
IR Spectrum
(KBr disc)
1646
4000
3000
2000
1600
1200
800
ν (cm^{-1})
100
80
60
40
20
% of base peak
121
Mass Spectrum
M+·
165
149
93
$C_9H_{11}NO_2$
40
80
120
160
200
240
280
m / e
^{13}C NMR Spectrum
(100 MHz, $CDCl_3/CD_3SOCD_3$ solution)
C-H
C-H
solvent
solvent
H-C-H
$-CH_3$
C C
C
proton decoupled
200
160
120
80
40
0
δ (ppm)

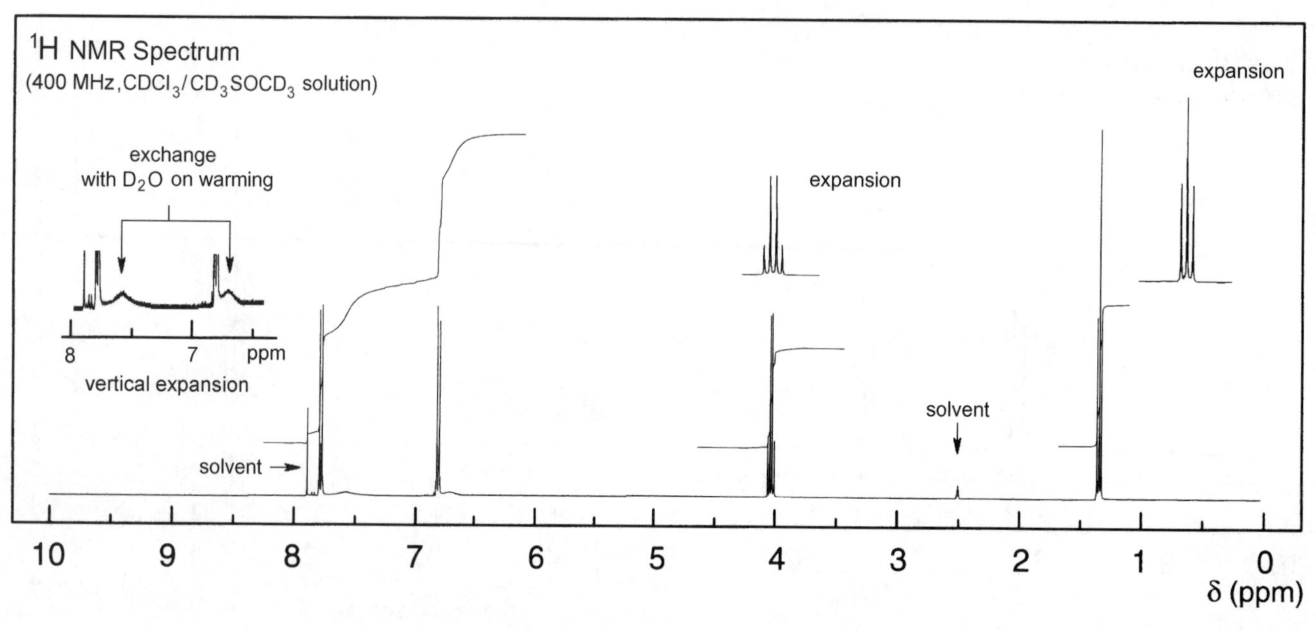

^{1}H NMR Spectrum
(400 MHz, $CDCl_3/CD_3SOCD_3$ solution)
expansion
exchange
with D_2O on warming
expansion
8
7
ppm
vertical expansion
solvent
solvent
10
9
8
7
6
5
4
3
2
1
0
δ (ppm)

Problem 100

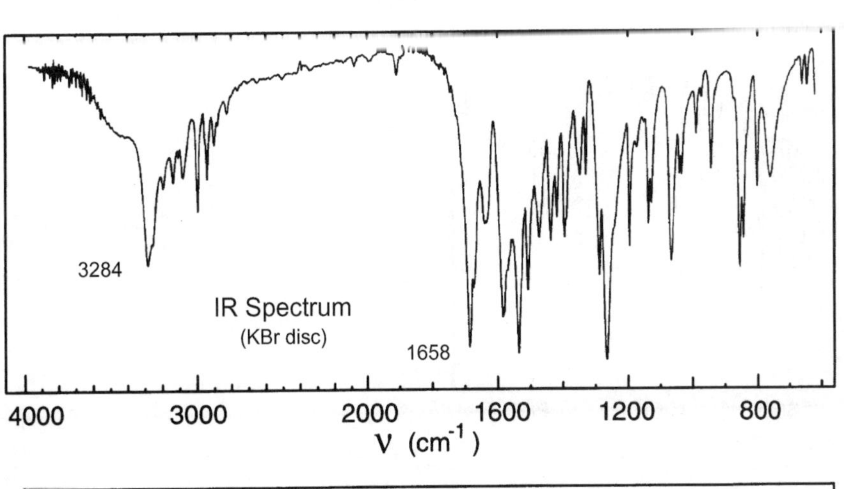

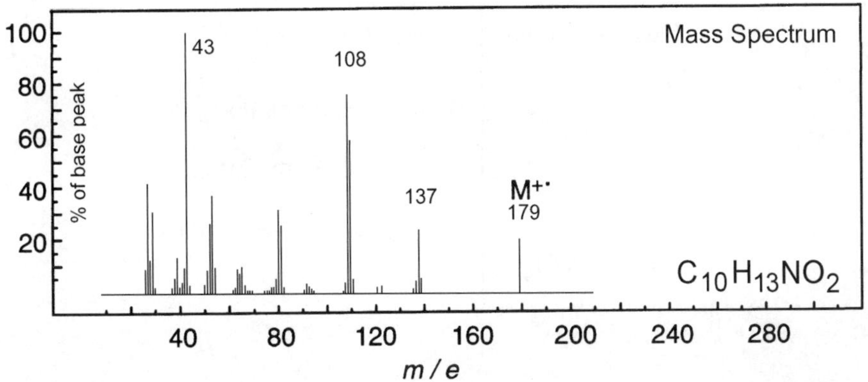

UV Spectrum

λ_{max} 250 nm ($\log_{10}\varepsilon$ 3.1)

λ_{max} 287 nm ($\log_{10}\varepsilon$ 2.2)

solvent : chloroform

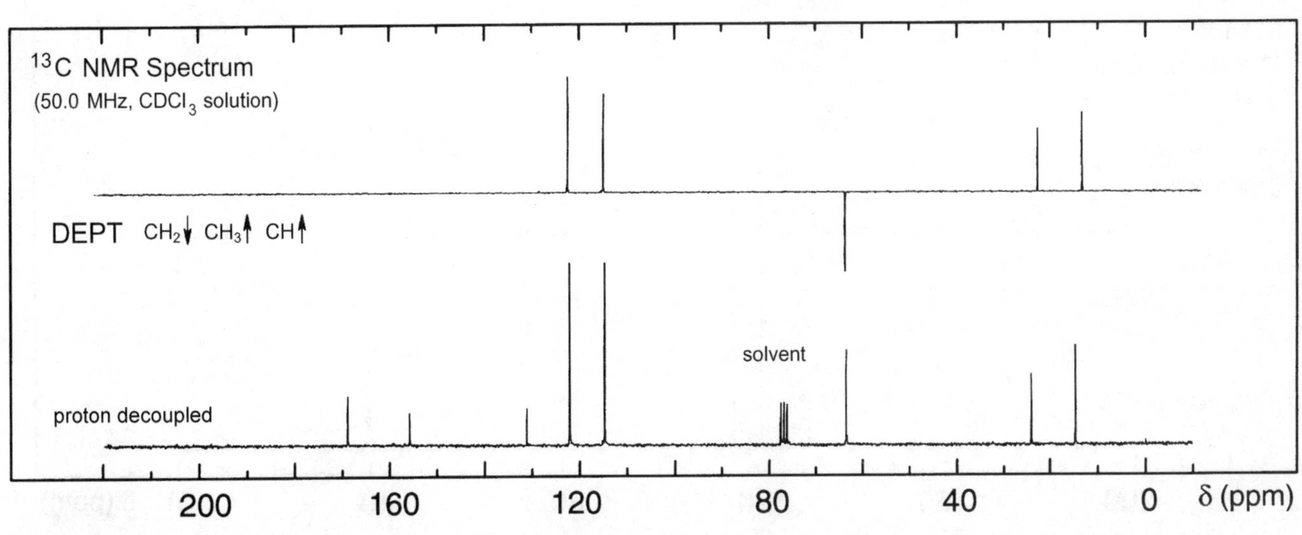

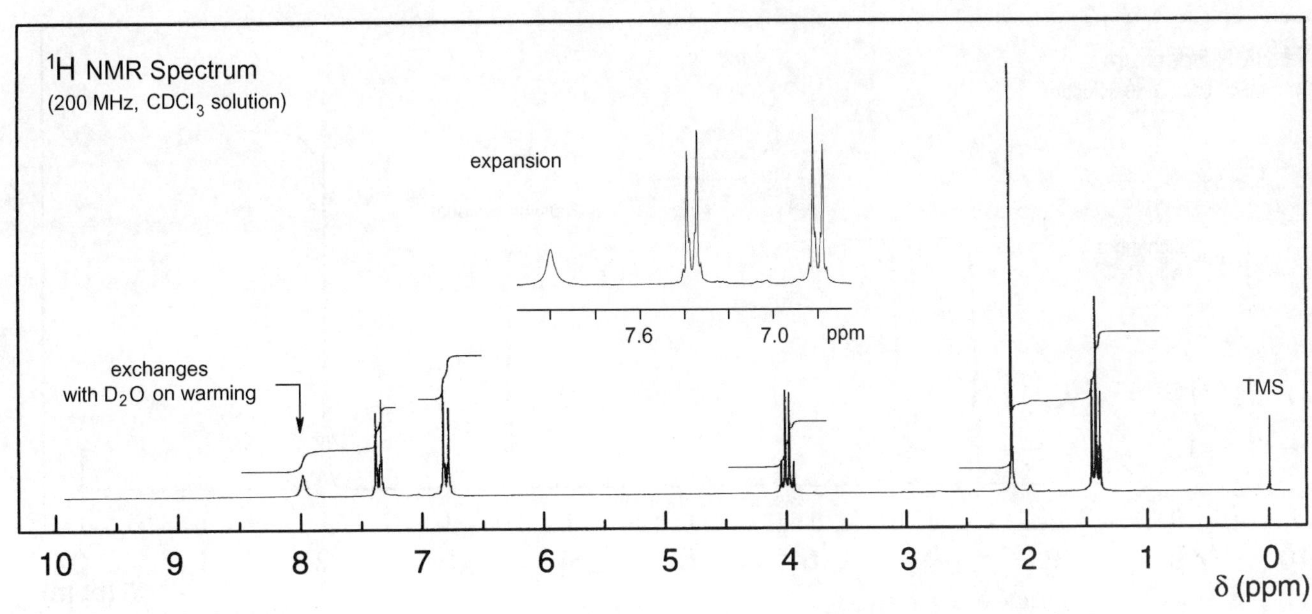

Problem 101

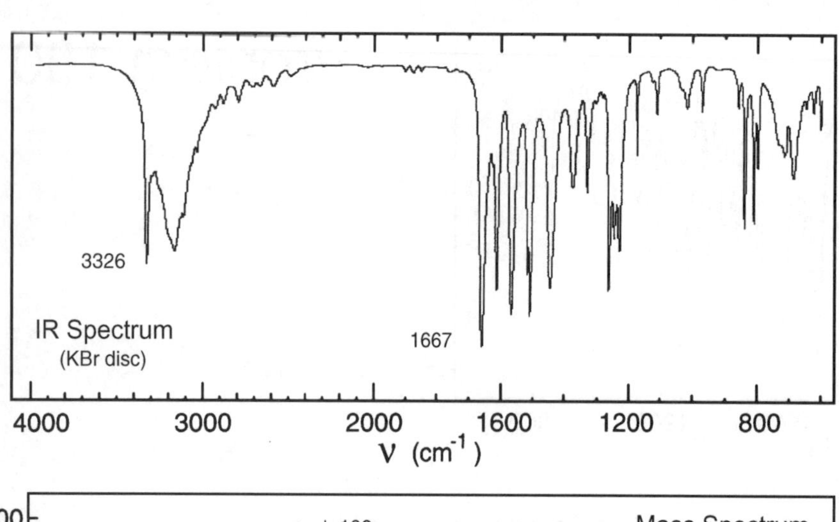

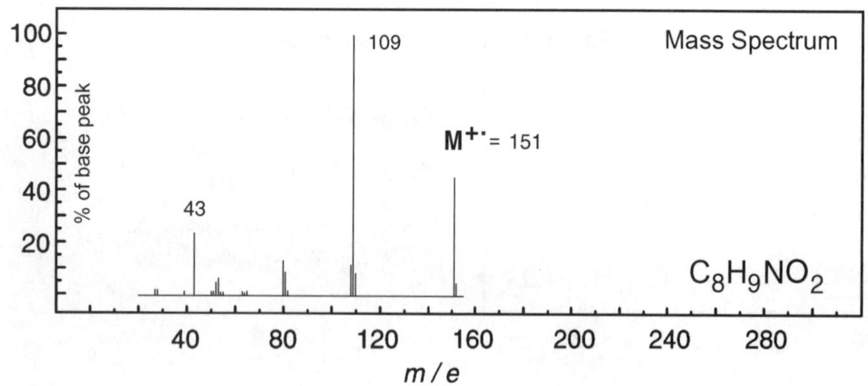

UV Spectrum

λ_{max} 250 nm ($\log_{10}\varepsilon$ 4.1)

λ_{max} 285 nm ($\log_{10}\varepsilon$ 3.6)

solvent : ethanol

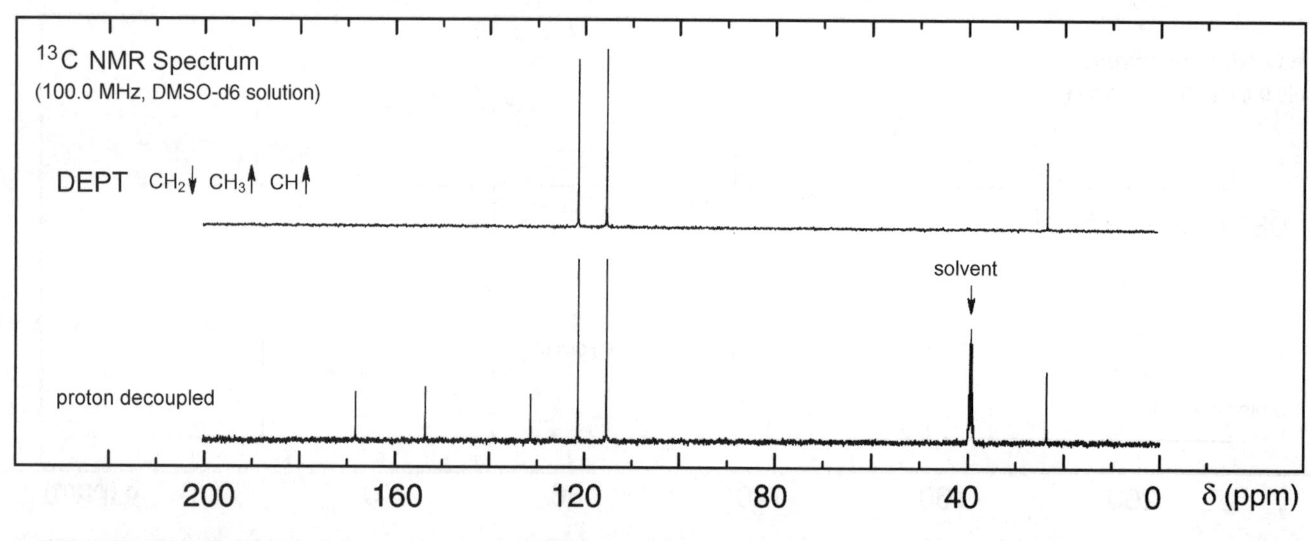

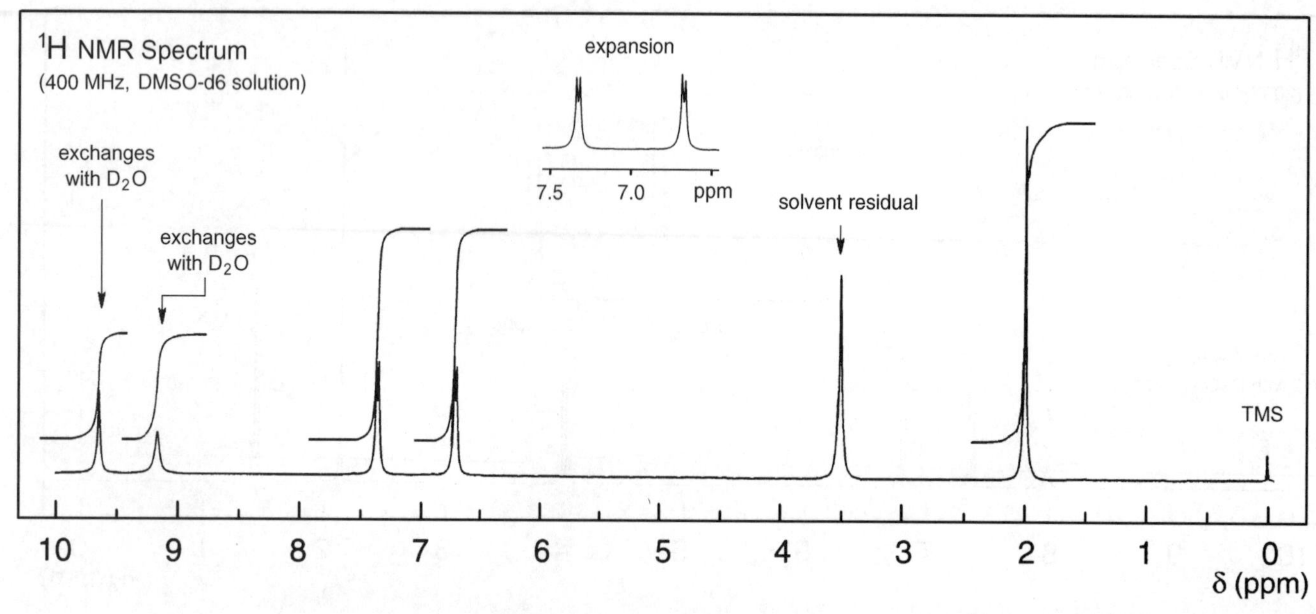

Problem 102

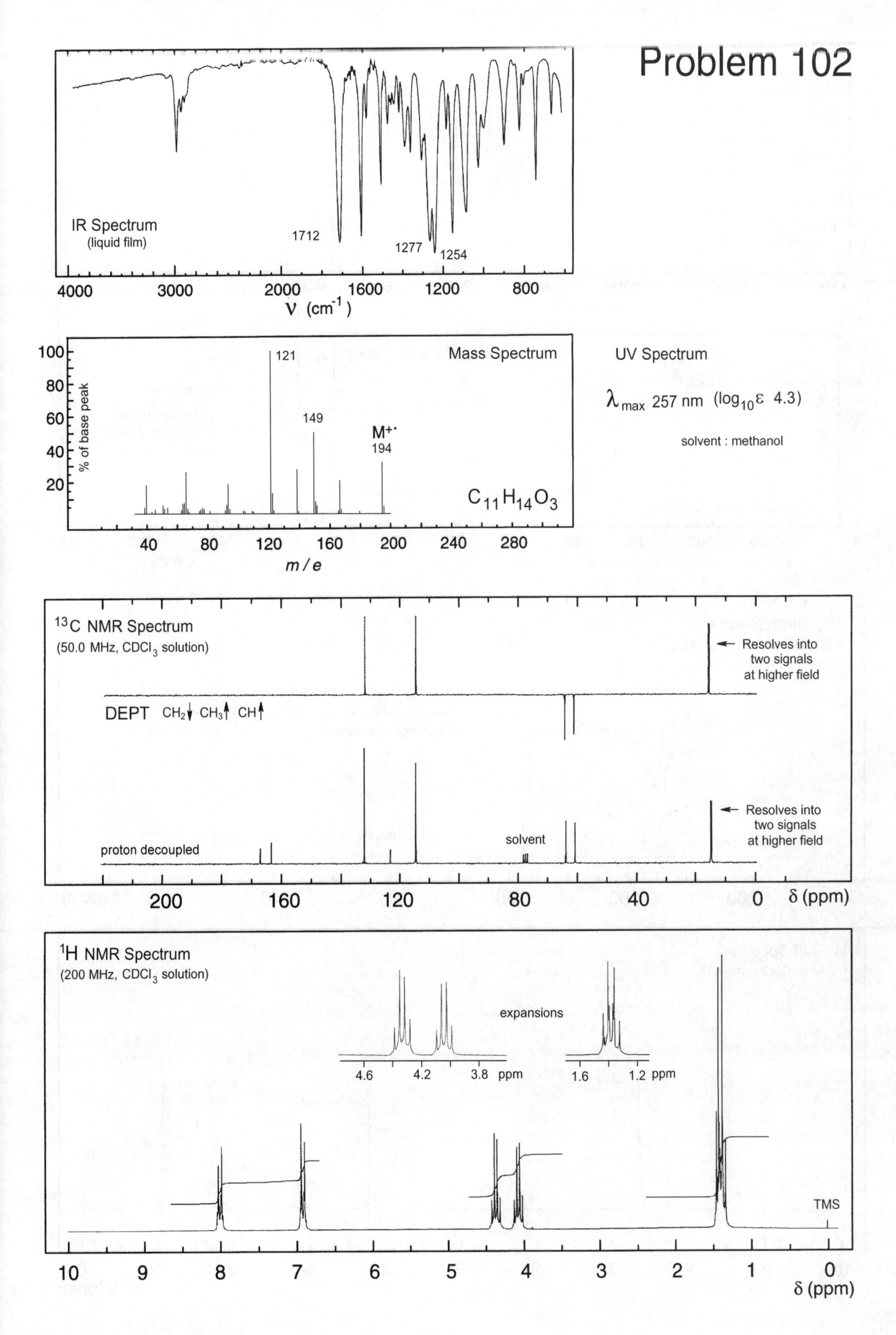

IR Spectrum
(liquid film)
1712
1277
1254
4000
3000
2000
1600
1200
800
ν (cm⁻¹)
Mass Spectrum
% of base peak
100
80
60
40
20
121
149
M⁺˙
194
C11H14O3
40
80
120
160
200
240
280
m / e
UV Spectrum
λmax 257 nm (log10ε 4.3)
solvent : methanol
13C NMR Spectrum
(50.0 MHz, CDCl3 solution)
Resolves into two signals at higher field
DEPT CH2↓ CH3↑ CH↑
solvent
proton decoupled
Resolves into two signals at higher field
200
160
120
80
40
0
δ (ppm)
1H NMR Spectrum
(200 MHz, CDCl3 solution)
expansions
4.6
4.2
3.8 ppm
1.6
1.2 ppm
TMS
10
9
8
7
6
5
4
3
2
1
0
δ (ppm)

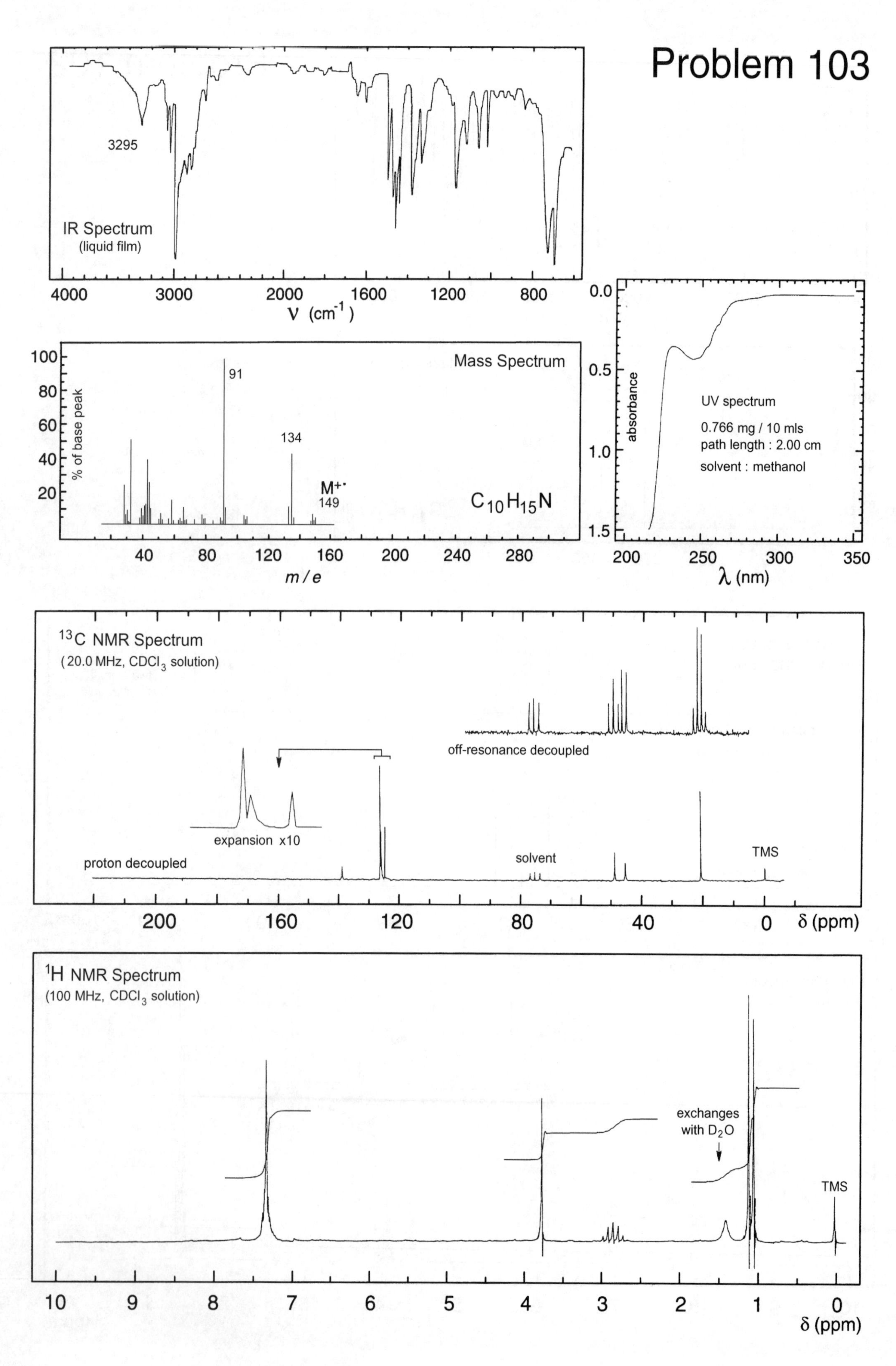
Problem 103
3295
IR Spectrum
(liquid film)
4000
3000
2000
1600
1200
800
ν (cm⁻¹)
Mass Spectrum
% of base peak
91
134
M+·
149
$C_{10}H_{15}N$
m / e
UV spectrum
0.766 mg / 10 mls
path length : 2.00 cm
solvent : methanol
absorbance
λ (nm)
13C NMR Spectrum
(20.0 MHz, CDCl3 solution)
off-resonance decoupled
expansion x10
proton decoupled
solvent
TMS
δ (ppm)
1H NMR Spectrum
(100 MHz, CDCl3 solution)
exchanges
with D2O
TMS
δ (ppm)

Problem 104

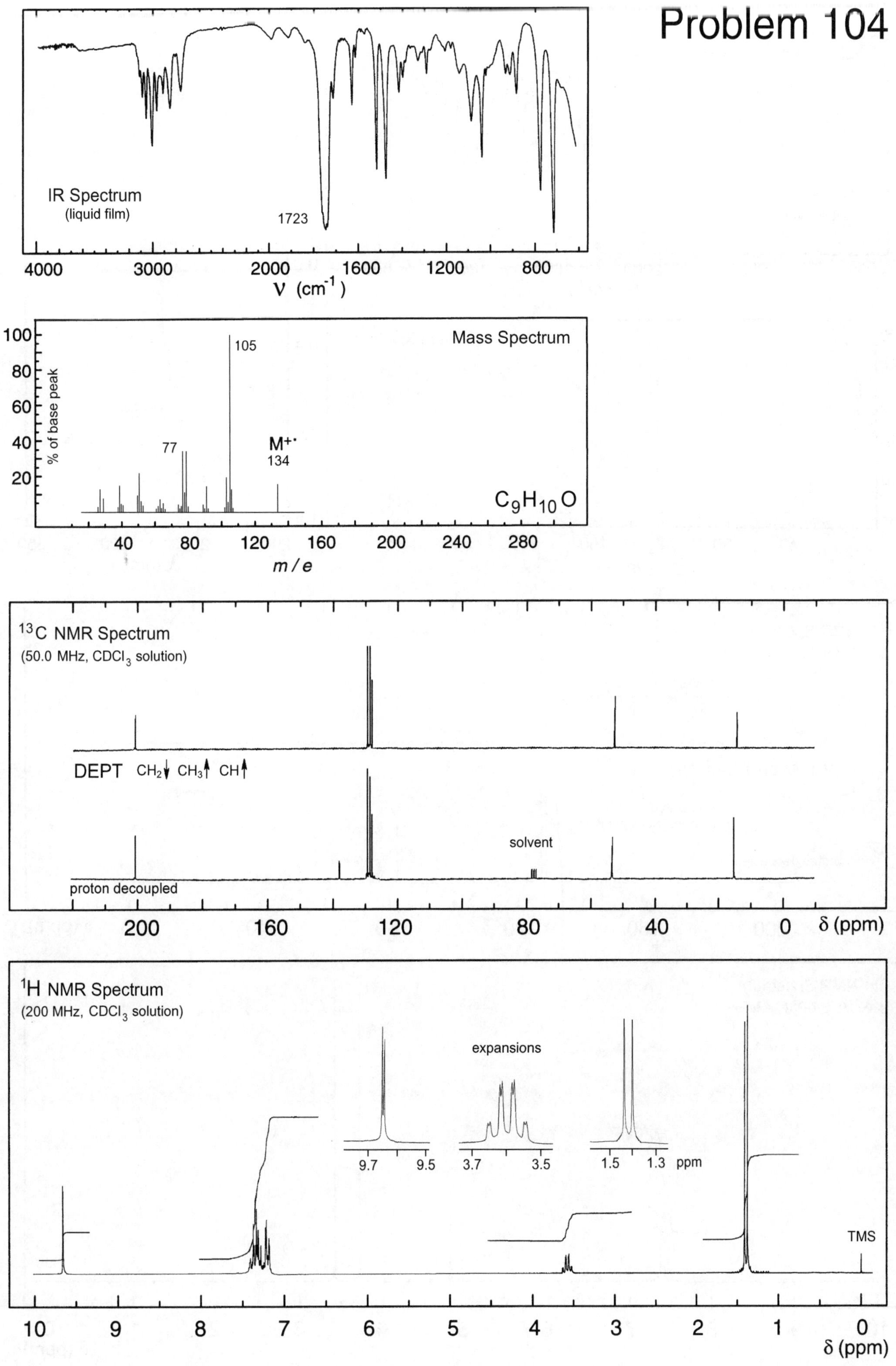

IR Spectrum
(liquid film)
1723
4000
3000
2000
1600
1200
800
ν (cm⁻¹)
Mass Spectrum
% of base peak
105
77
M+·
134
$C_9H_{10}O$
m / e
13C NMR Spectrum
(50.0 MHz, CDCl3 solution)
DEPT CH2↓ CH3↑ CH↑
solvent
proton decoupled
δ (ppm)
1H NMR Spectrum
(200 MHz, CDCl3 solution)
expansions
ppm
TMS
δ (ppm)

Problem 105

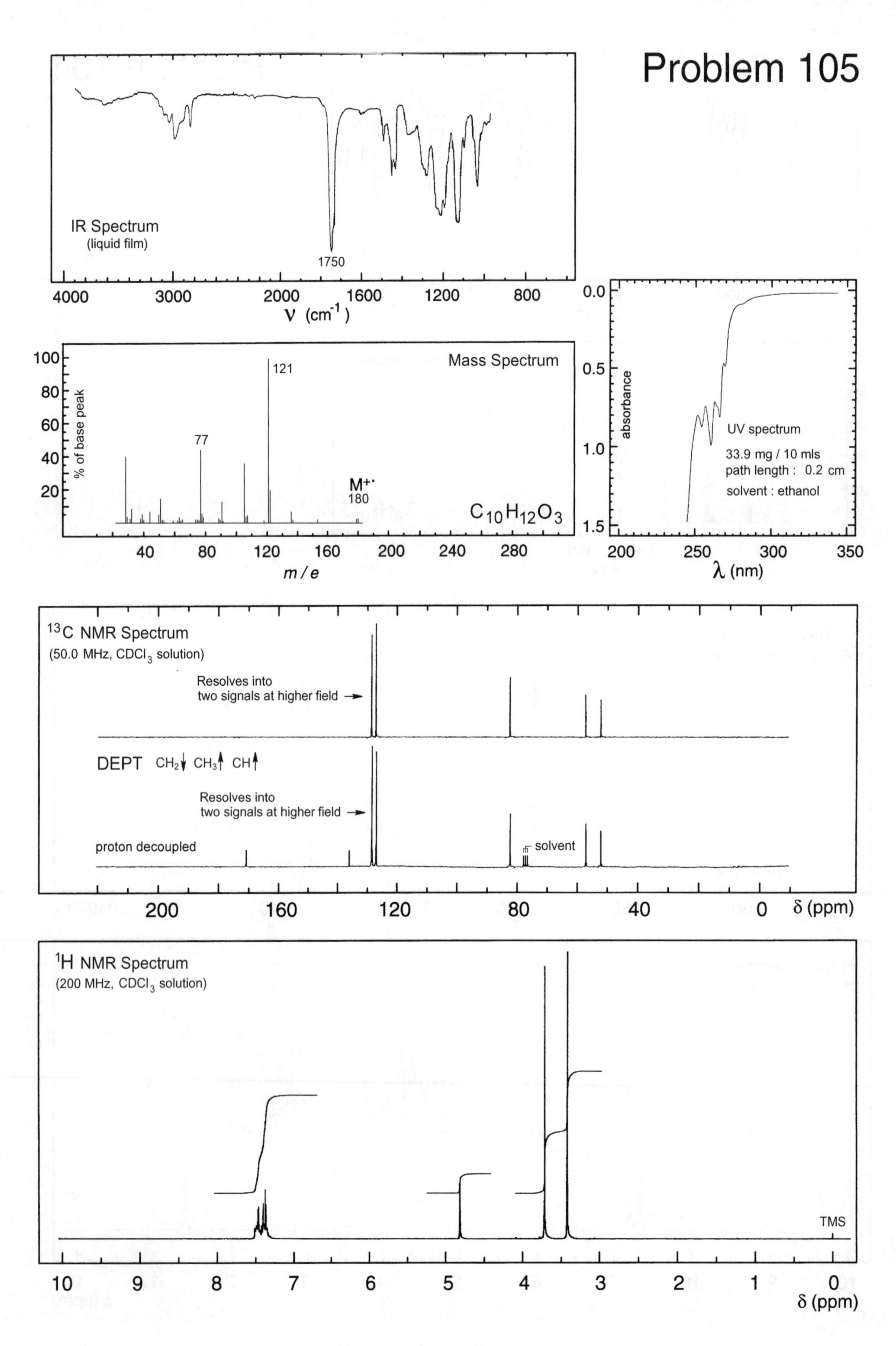

IR Spectrum
(liquid film)
1750
4000
3000
2000
1600
1200
800
ν (cm⁻¹)
Mass Spectrum
100
80
60
40
20
% of base peak
121
77
M+·
180
C10H12O3
40
80
120
160
200
240
280
m / e
0.0
0.5
1.0
1.5
absorbance
UV spectrum
33.9 mg / 10 mls
path length : 0.2 cm
solvent : ethanol
200
250
300
350
λ (nm)
13C NMR Spectrum
(50.0 MHz, CDCl3 solution)
Resolves into
two signals at higher field
DEPT CH2 CH3 CH
Resolves into
two signals at higher field
proton decoupled
solvent
200
160
120
80
40
0
δ (ppm)
1H NMR Spectrum
(200 MHz, CDCl3 solution)
TMS
10
9
8
7
6
5
4
3
2
1
0
δ (ppm)

Problem 106

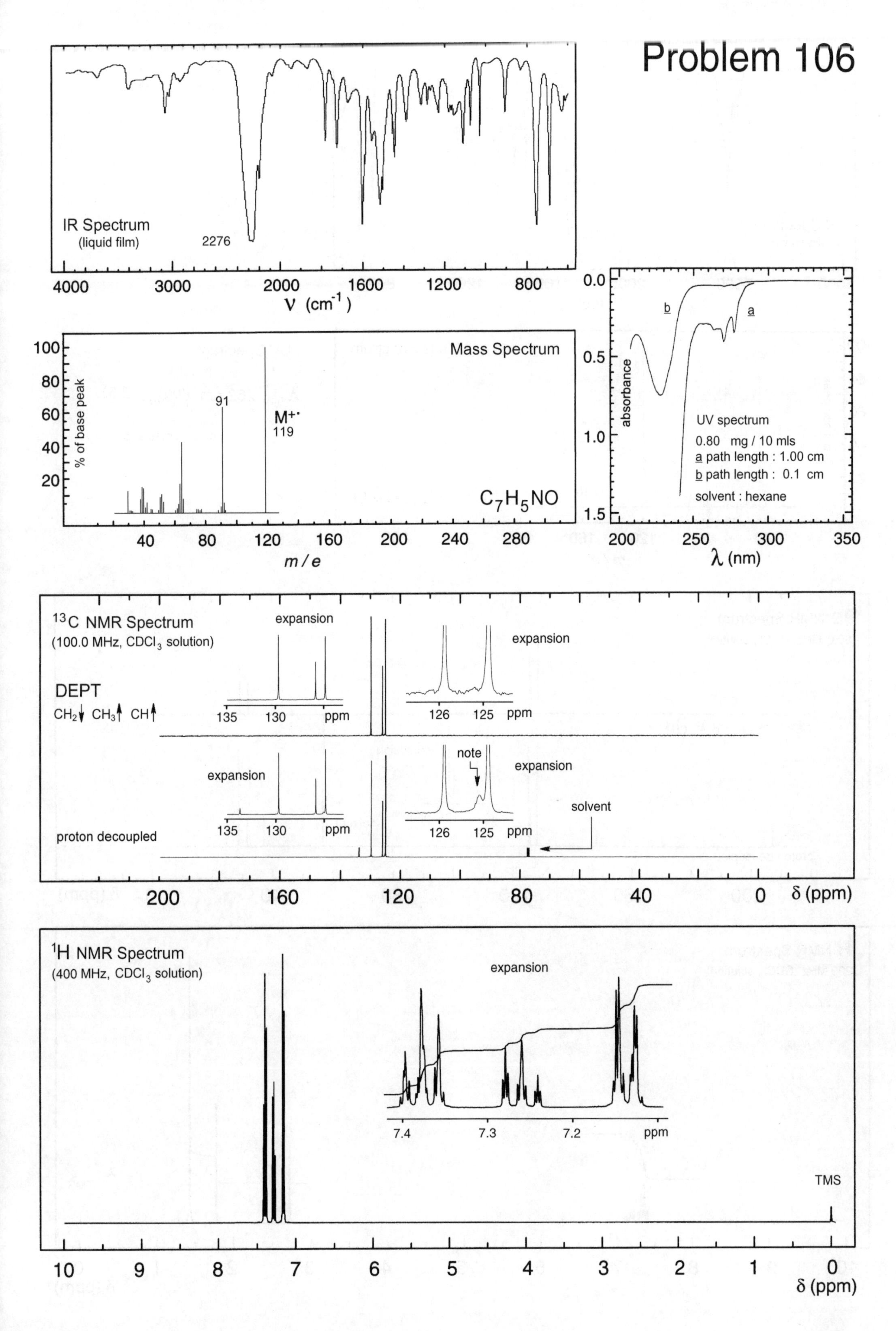

IR Spectrum
(liquid film)
2276
4000
3000
2000
1600
1200
800
ν (cm⁻¹)
Mass Spectrum
100
80
60
40
20
% of base peak
91
M+·
119
C7H5NO
40
80
120
160
200
240
280
m / e
0.0
0.5
1.0
1.5
absorbance
b
a
UV spectrum
0.80 mg / 10 mls
a path length : 1.00 cm
b path length : 0.1 cm
solvent : hexane
200
250
300
350
λ (nm)
13C NMR Spectrum
(100.0 MHz, CDCl3 solution)
DEPT
CH2↓ CH3↑ CH↑
expansion
135
130
ppm
126
125
ppm
note
solvent
proton decoupled
200
160
120
80
40
0
δ (ppm)
1H NMR Spectrum
(400 MHz, CDCl3 solution)
7.4
7.3
7.2
TMS
10
9
8
7
6
5
4
3
2
1
0

Problem 107

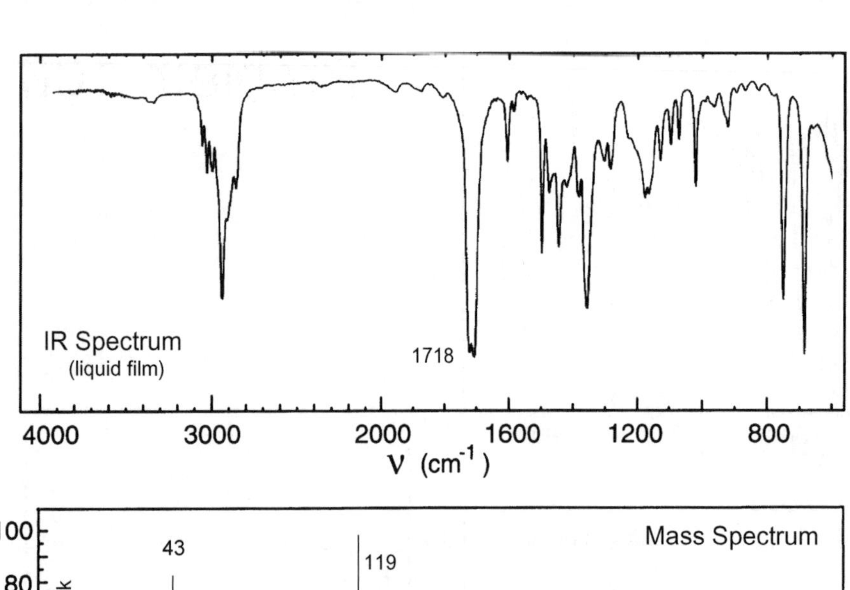

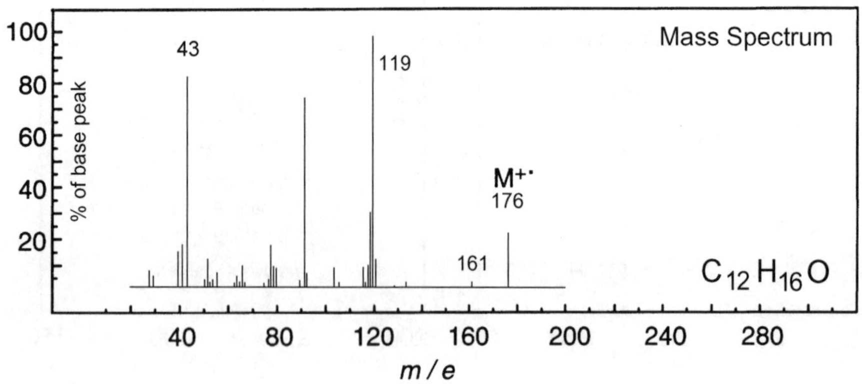

UV Spectrum

λ_{max} 262 nm ($\log_{10}\varepsilon$ 2.5)

solvent : methanol

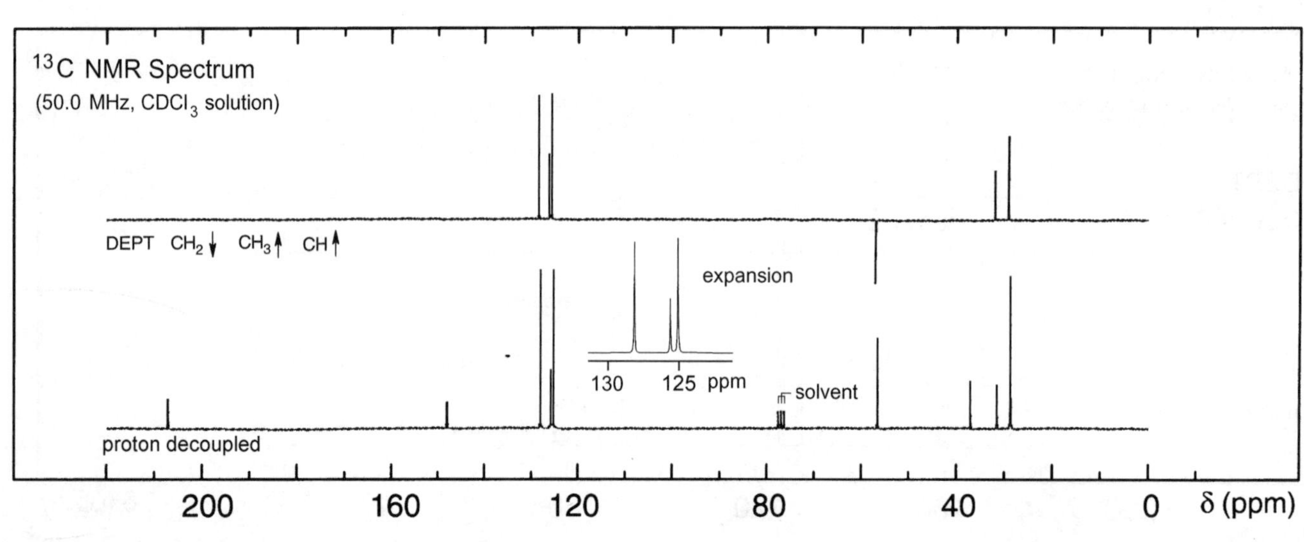

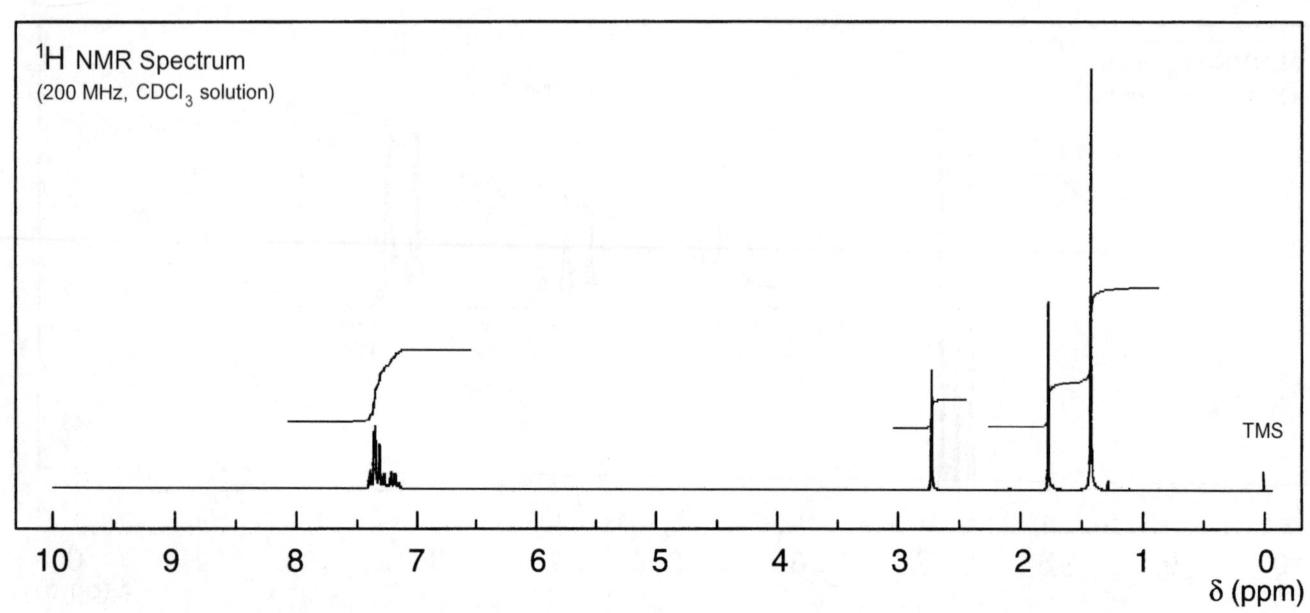

Problem 108

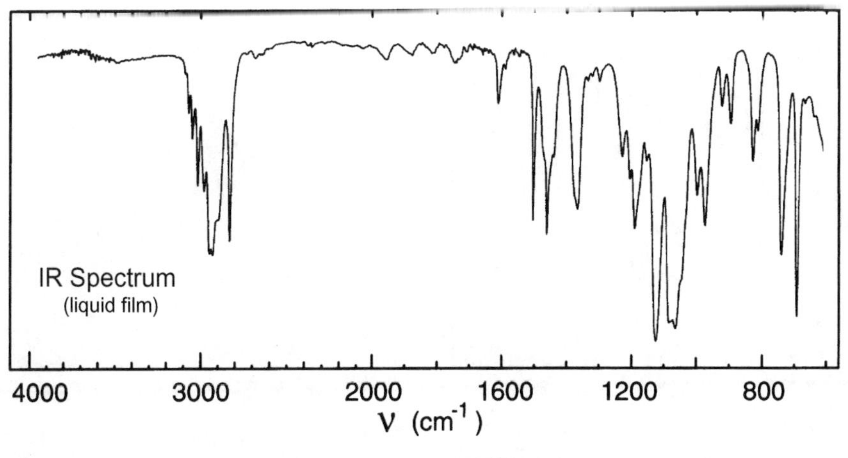

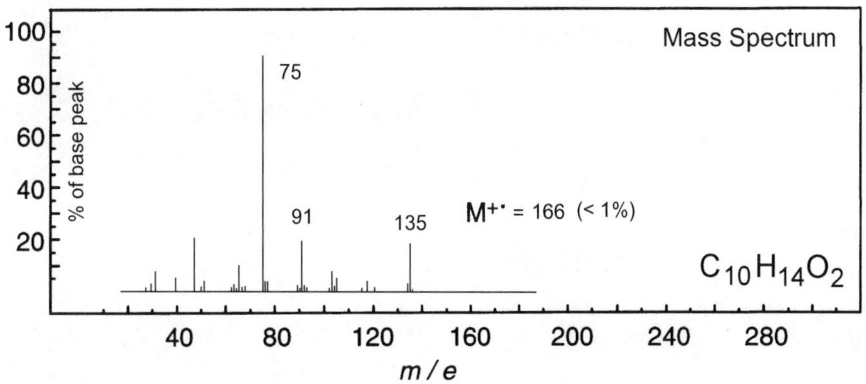

UV Spectrum

λ_{max} 262 nm ($\log_{10}\varepsilon$ 2.3)

solvent : methanol

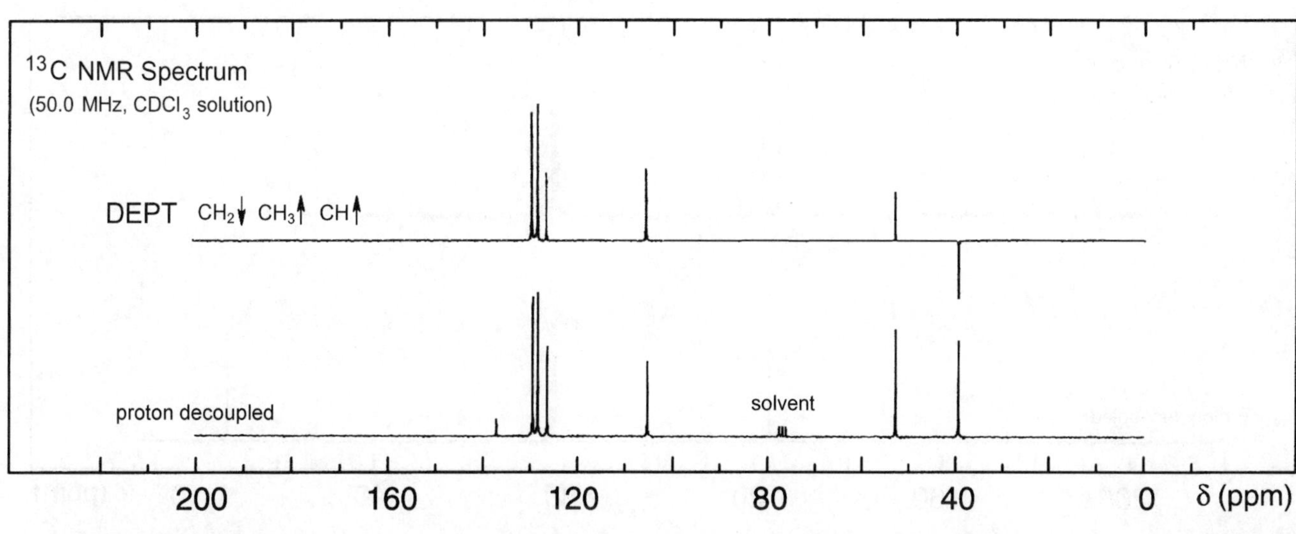

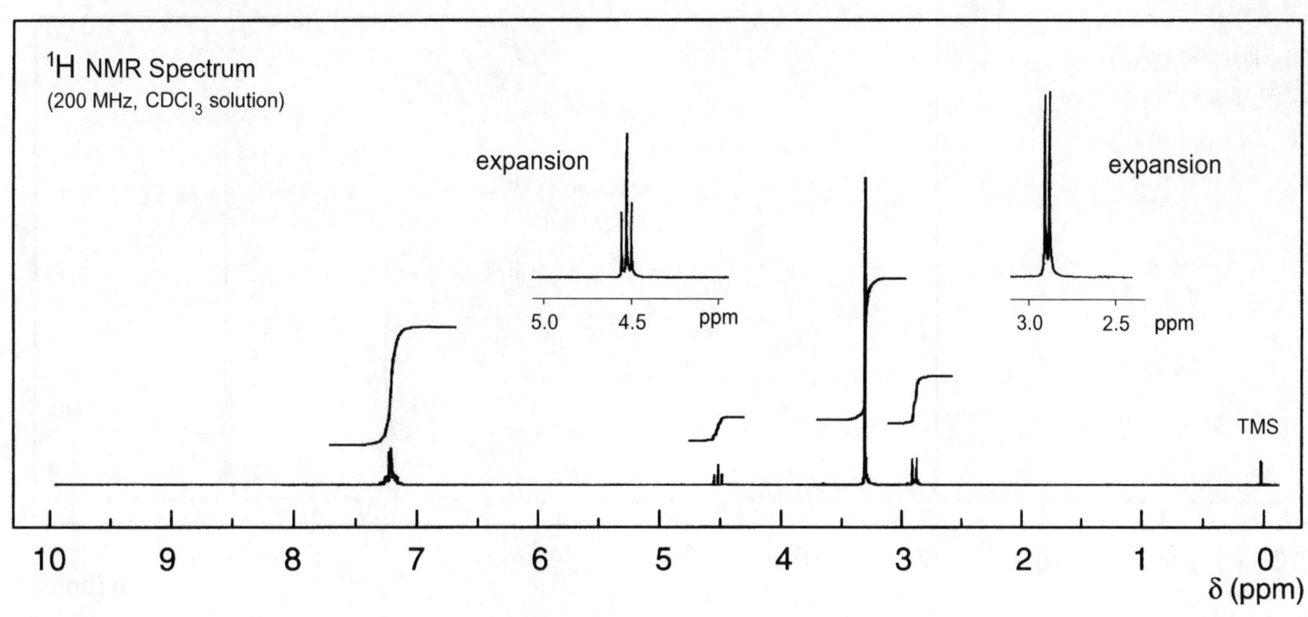

Problem 109

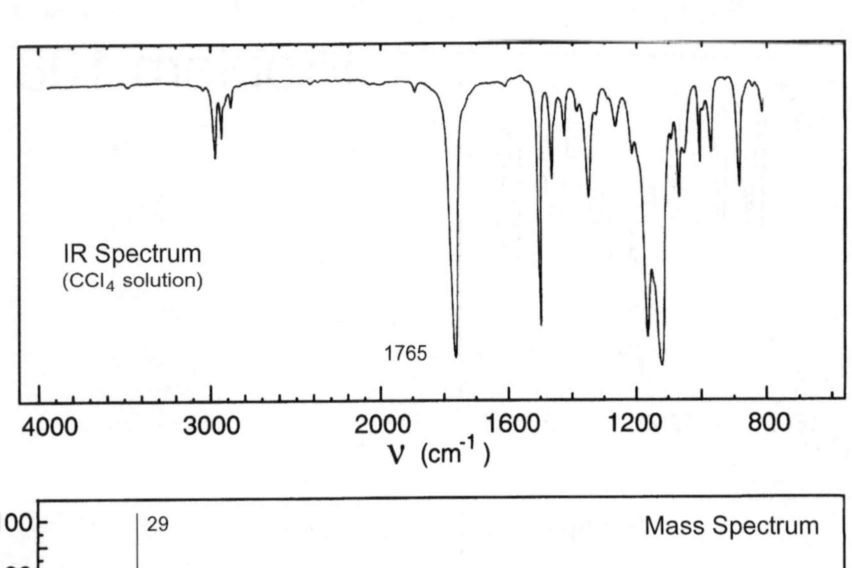

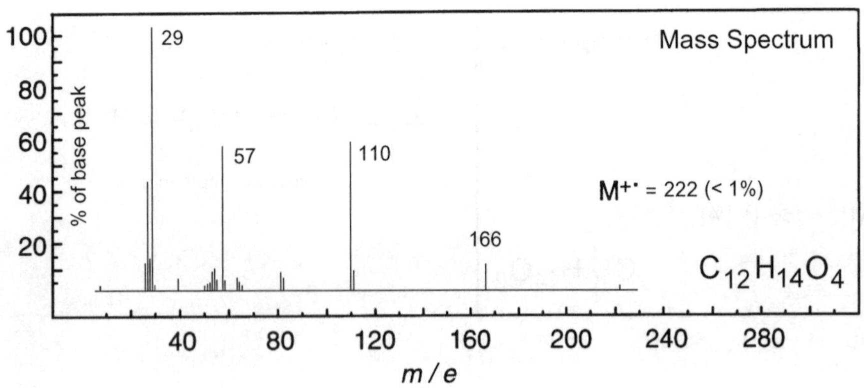

UV Spectrum

λ_{max} 269 nm ($\log_{10}\varepsilon$ 2.7)

λ_{max} 263 nm ($\log_{10}\varepsilon$ 2.7)

solvent : methanol

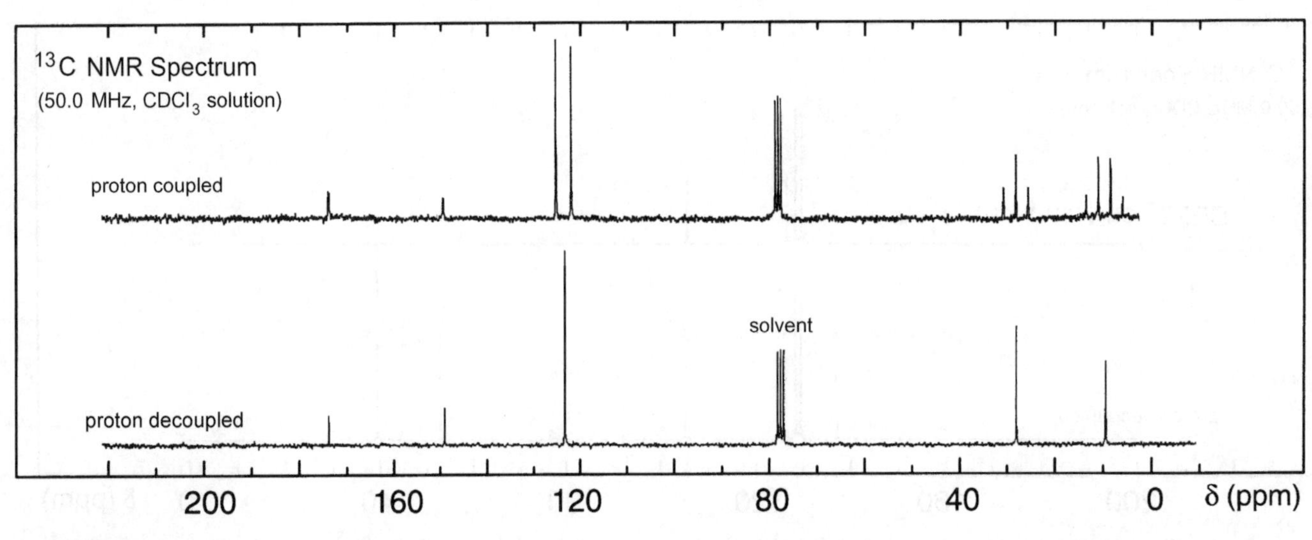

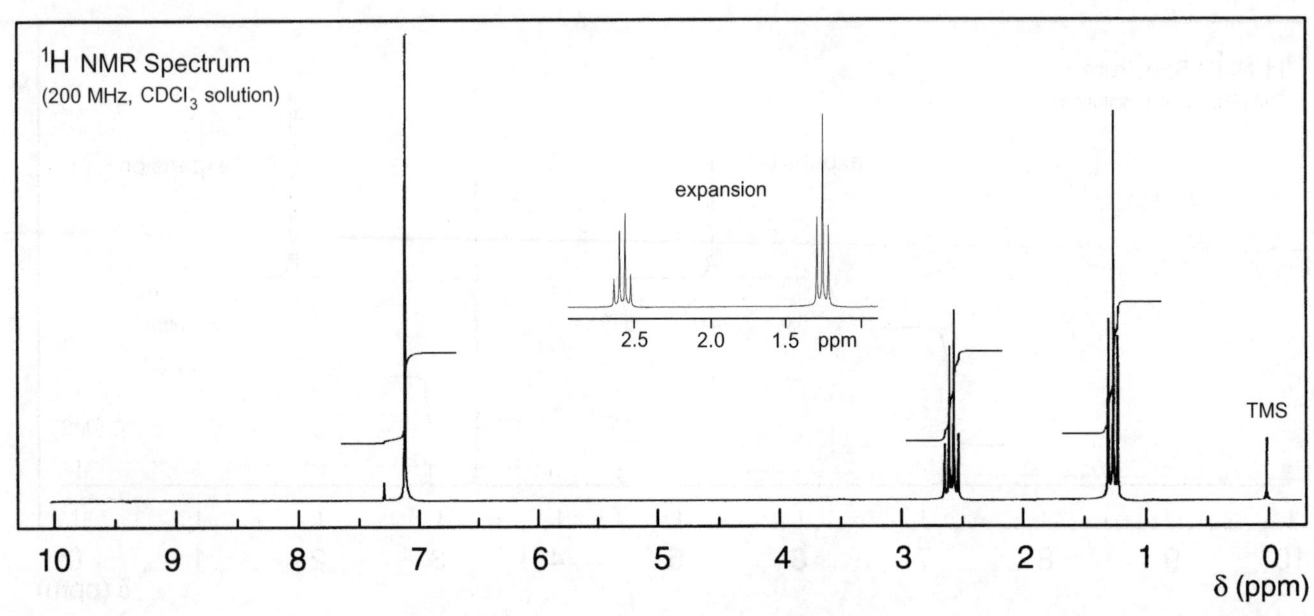

Problem 110

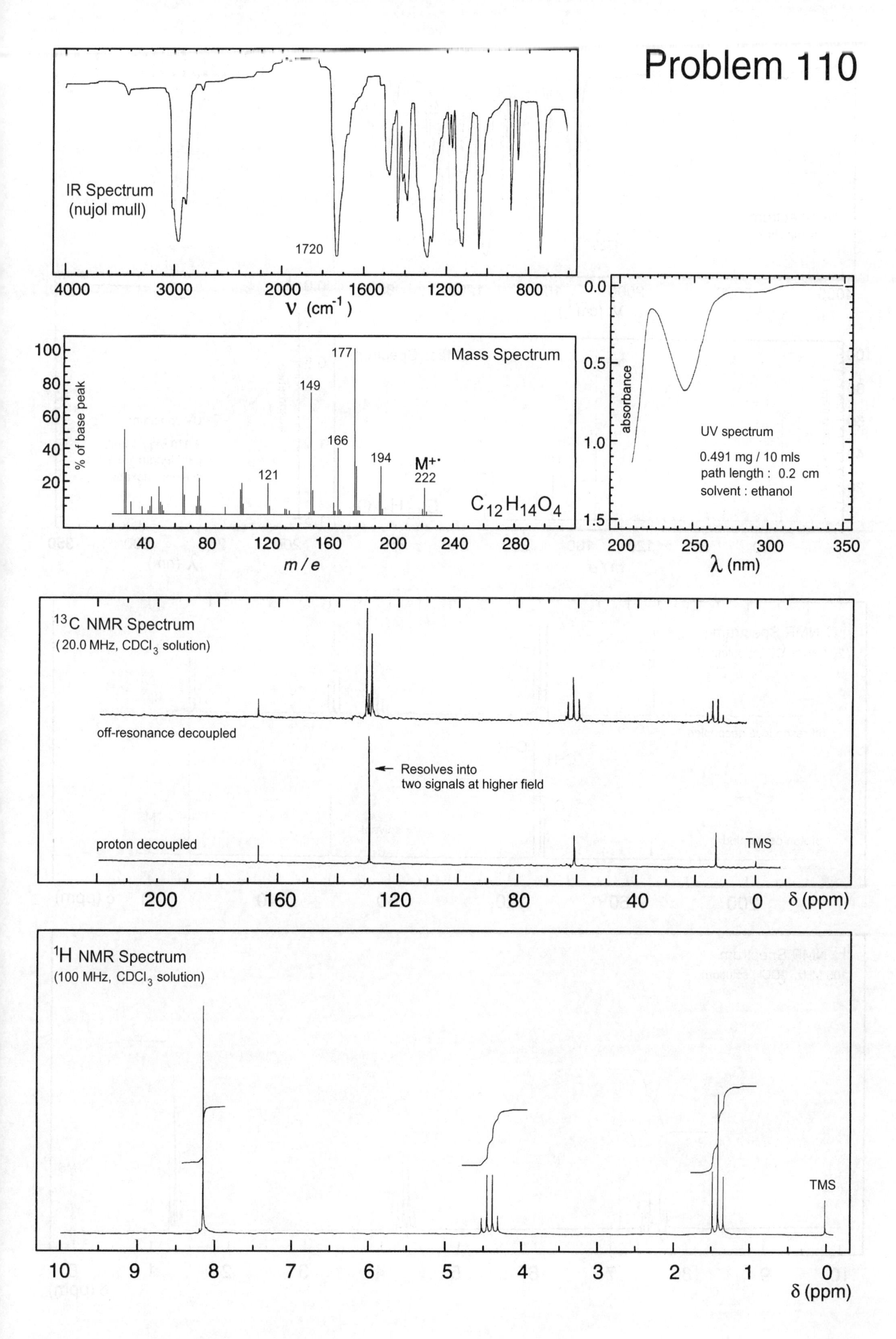

IR Spectrum
(nujol mull)
1720
4000
3000
2000
1600
1200
800
ν (cm⁻¹)
Mass Spectrum
% of base peak
177
149
166
194
121
M+·
222
C12H14O4
m / e
UV spectrum
0.491 mg / 10 mls
path length : 0.2 cm
solvent : ethanol
absorbance
λ (nm)
13C NMR Spectrum
(20.0 MHz, CDCl3 solution)
off-resonance decoupled
Resolves into
two signals at higher field
proton decoupled
TMS
δ (ppm)
1H NMR Spectrum
(100 MHz, CDCl3 solution)
TMS
δ (ppm)

Problem 111

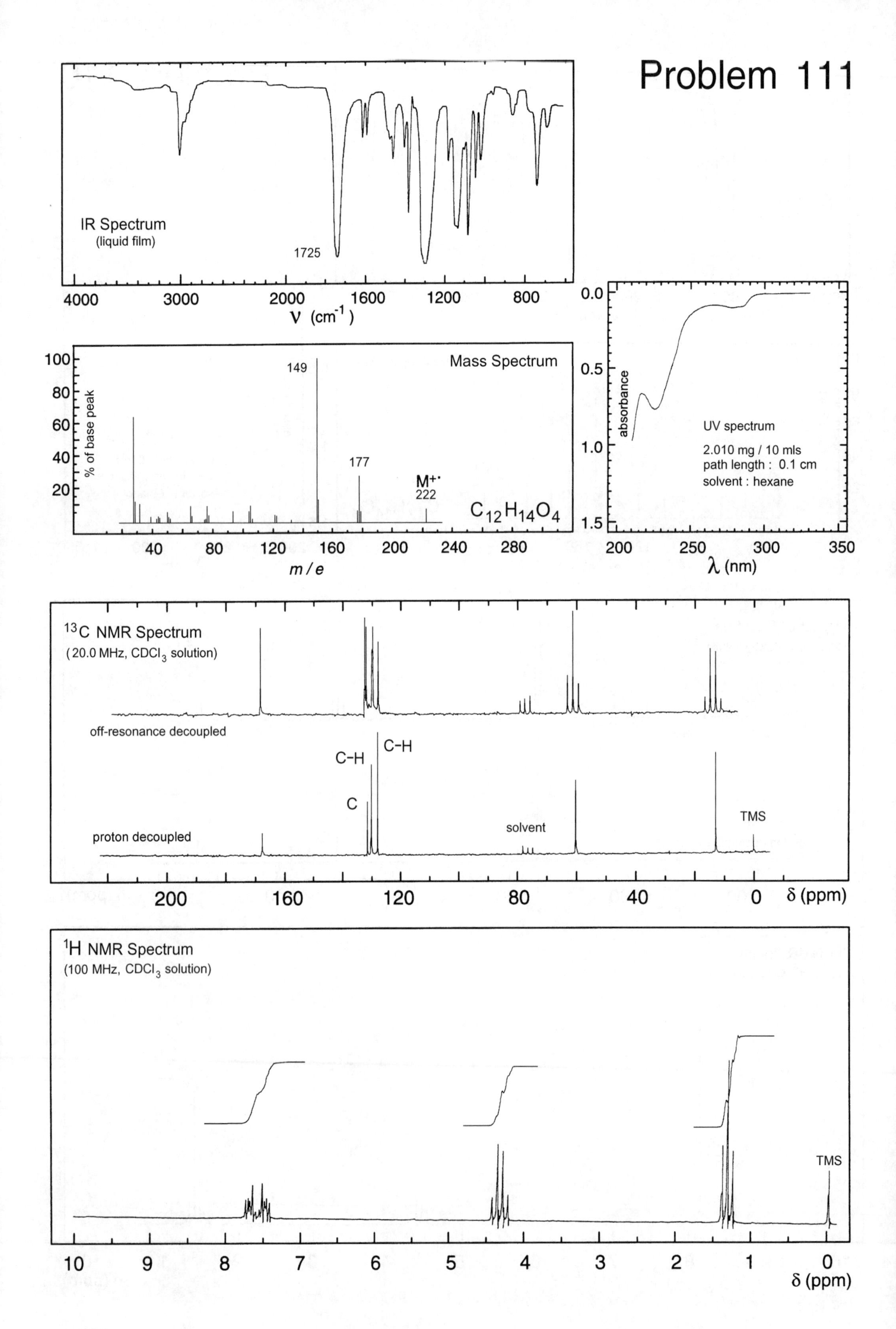

IR Spectrum
(liquid film)
1725
4000
3000
2000
1600
1200
800
ν (cm^{-1})
Mass Spectrum
% of base peak
149
177
M+·
222
$C_{12}H_{14}O_4$
m / e
absorbance
UV spectrum
2.010 mg / 10 mls
path length : 0.1 cm
solvent : hexane
λ (nm)
^{13}C NMR Spectrum
(20.0 MHz, $CDCl_3$ solution)
off-resonance decoupled
C-H
C-H
C
solvent
TMS
proton decoupled
δ (ppm)
^{1}H NMR Spectrum
(100 MHz, $CDCl_3$ solution)
TMS
δ (ppm)

Problem 112

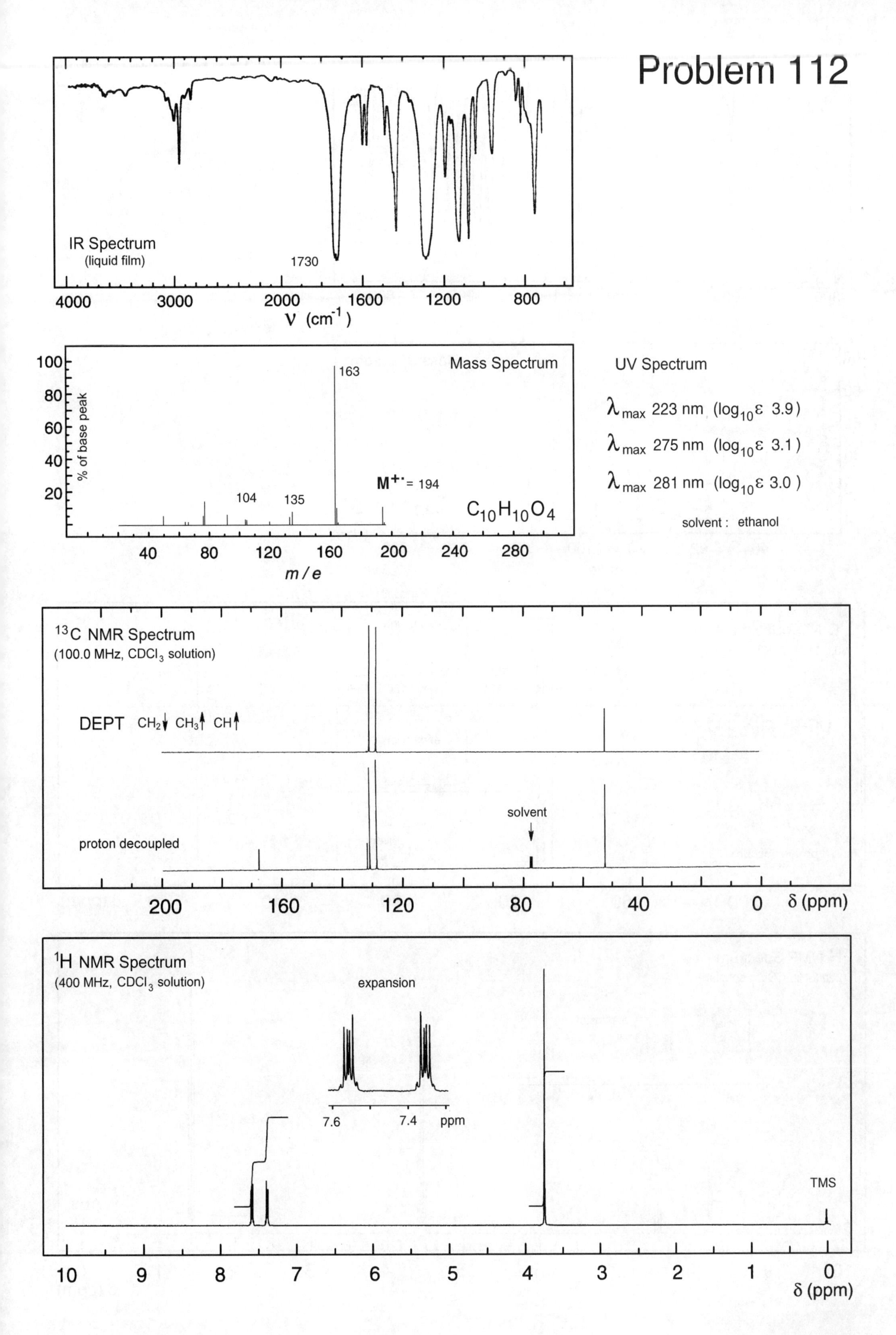

IR Spectrum
(liquid film)
1730
4000
3000
2000
1600
1200
800
ν (cm⁻¹)
Mass Spectrum
% of base peak
163
104
135
M⁺· = 194
C10H10O4
m / e
UV Spectrum
λmax 223 nm (log10ε 3.9)
λmax 275 nm (log10ε 3.1)
λmax 281 nm (log10ε 3.0)
solvent : ethanol
13C NMR Spectrum
(100.0 MHz, CDCl3 solution)
DEPT CH2↓ CH3↑ CH↑
solvent
proton decoupled
δ (ppm)
1H NMR Spectrum
(400 MHz, CDCl3 solution)
expansion
7.6
7.4
ppm
TMS
δ (ppm)

Problem 113

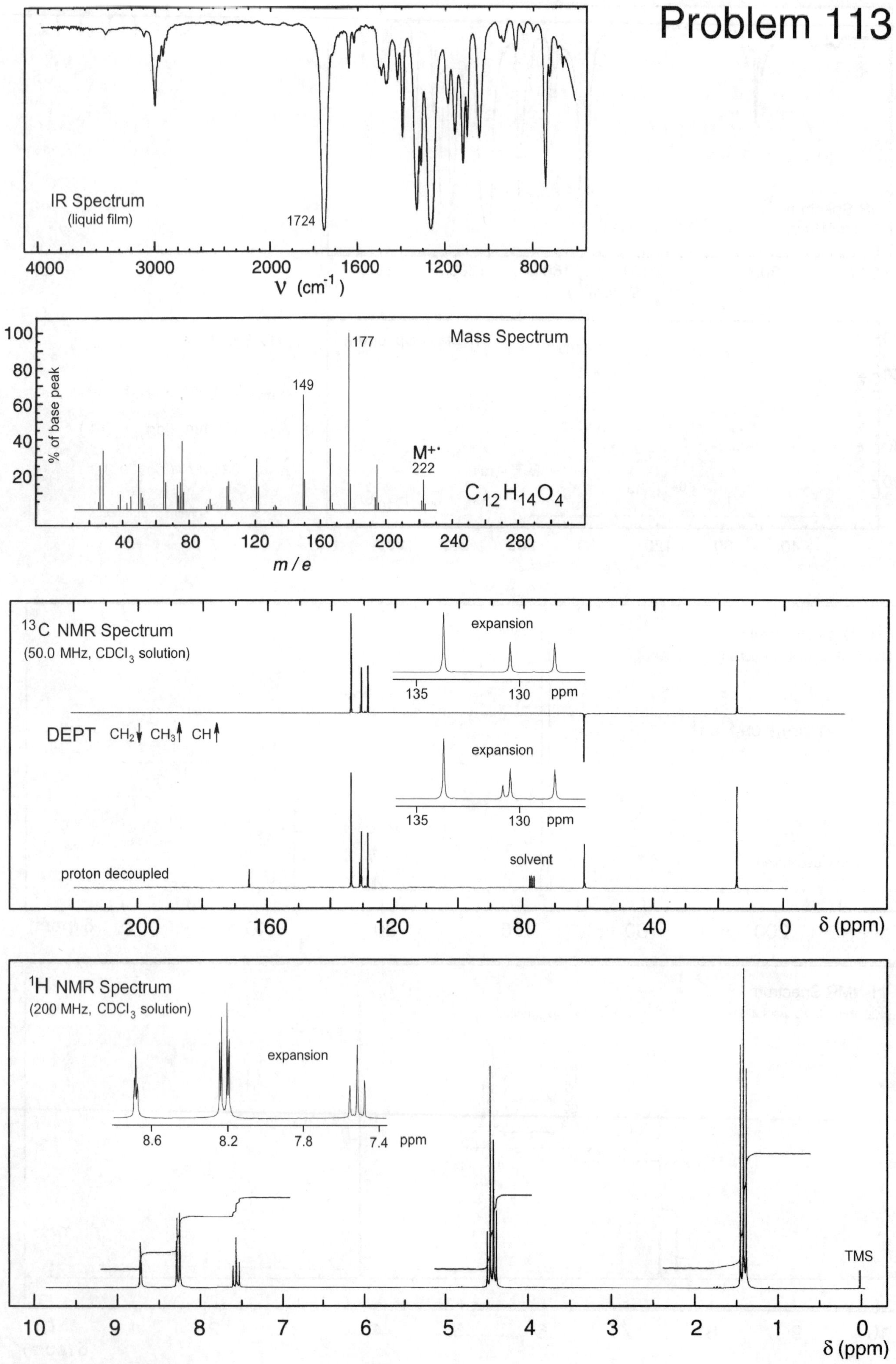

IR Spectrum
(liquid film)
1724
4000
3000
2000
1600
1200
800
ν (cm⁻¹)
Mass Spectrum
% of base peak
100
80
60
40
20
177
149
M+·
222
C12H14O4
40
80
120
160
200
240
280
m / e
13C NMR Spectrum
(50.0 MHz, CDCl3 solution)
expansion
135
130
ppm
DEPT CH2↓ CH3↑ CH↑
expansion
135
130
ppm
solvent
proton decoupled
200
160
120
80
40
0
δ (ppm)
1H NMR Spectrum
(200 MHz, CDCl3 solution)
expansion
8.6
8.2
7.8
7.4
ppm
TMS
10
9
8
7
6
5
4
3
2
1
0
δ (ppm)

Problem 114

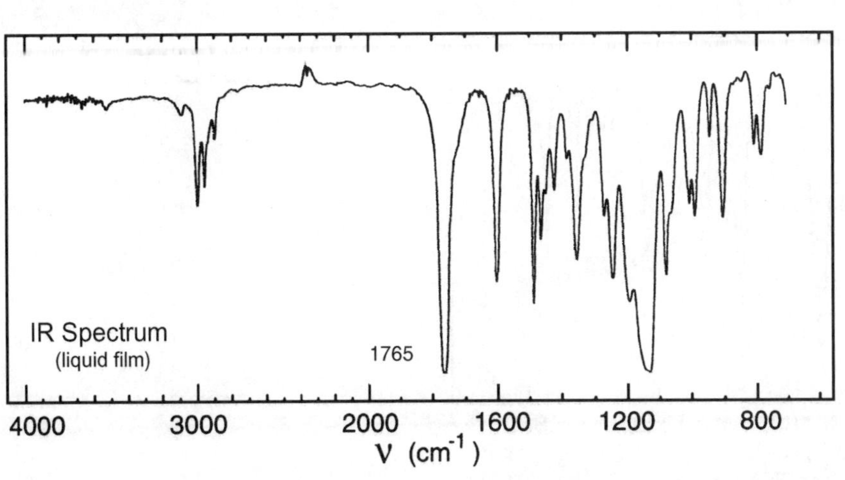

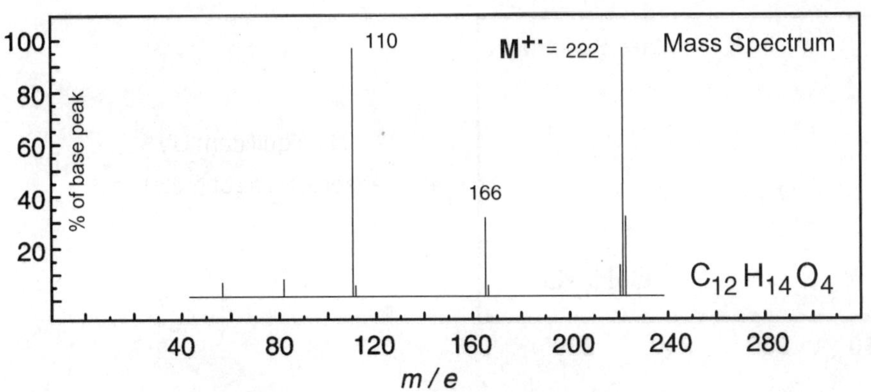

UV Spectrum

λ_{max} 220 nm ($log_{10}\varepsilon$ 3.7)

λ_{max} 274 nm ($log_{10}\varepsilon$ 3.3)

solvent : ethanol

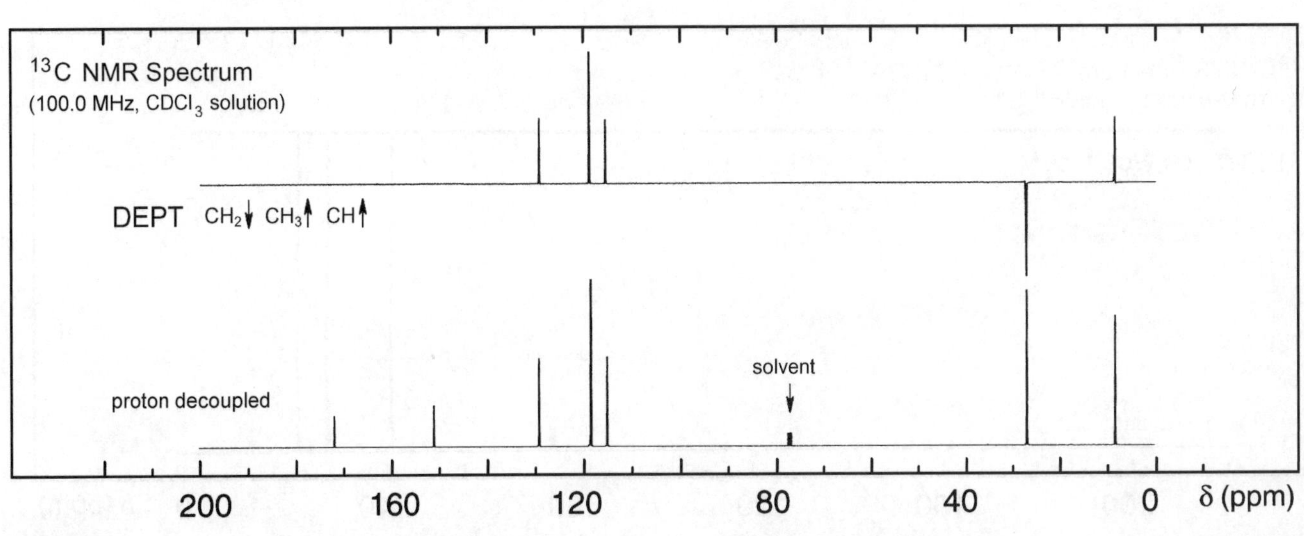

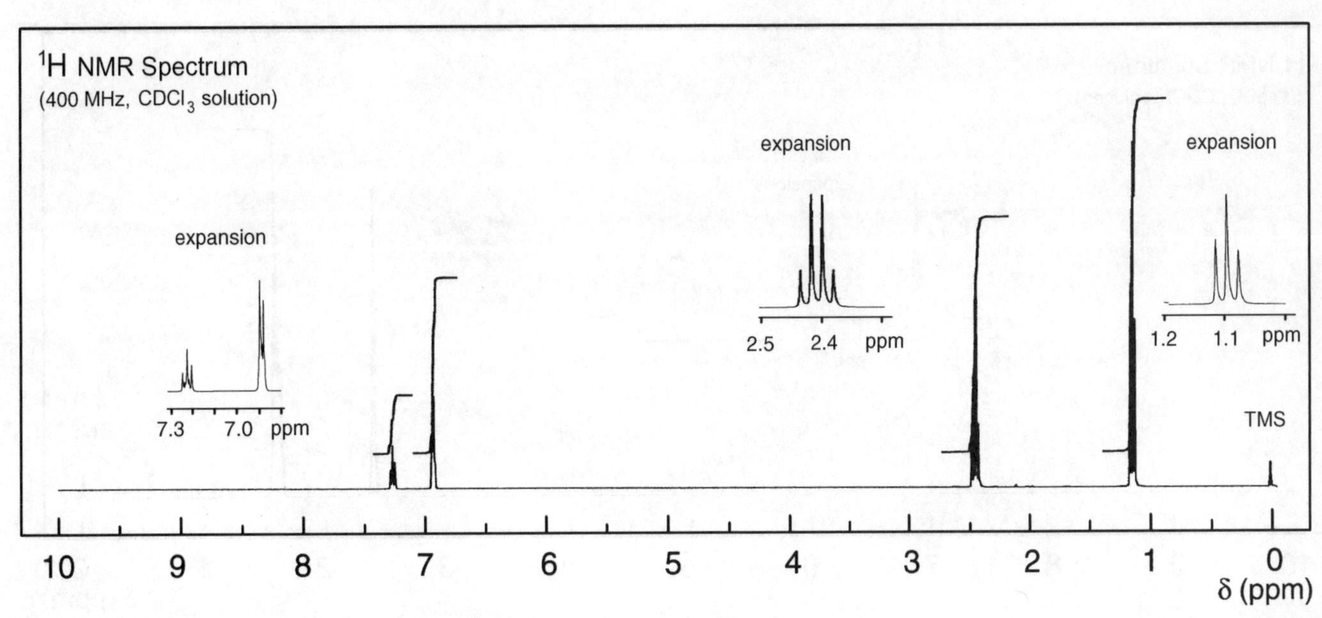

Problem 115

IR Spectrum
(liquid film)
1702
4000 3000 2000 1600 1200 800
ν (cm⁻¹)

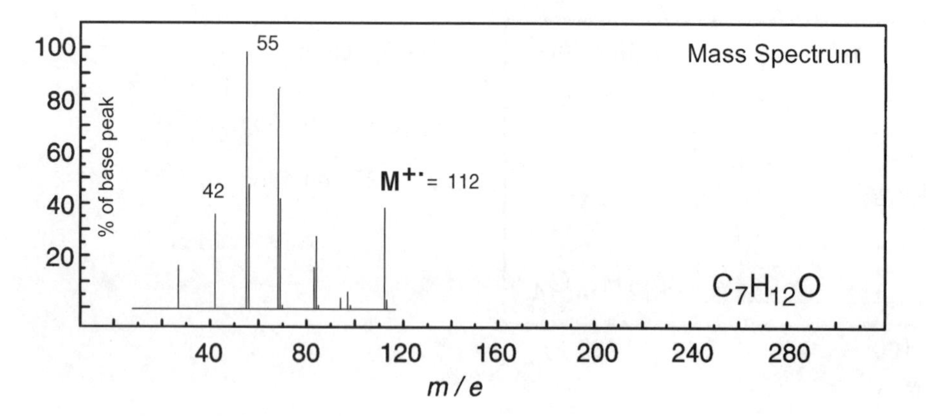

No significant UV absorption above 220 nm

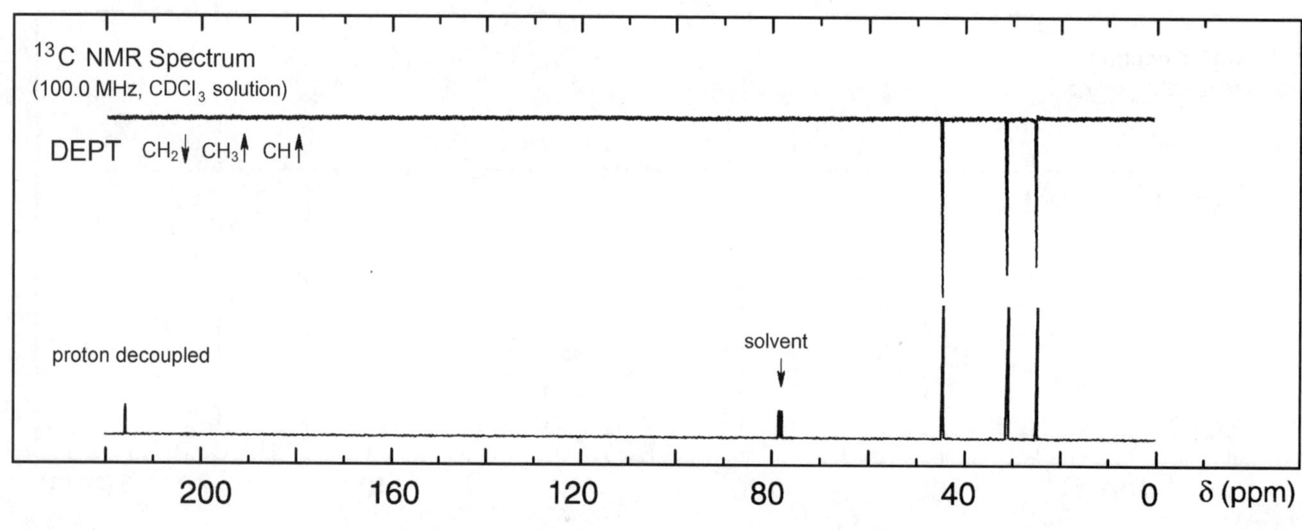

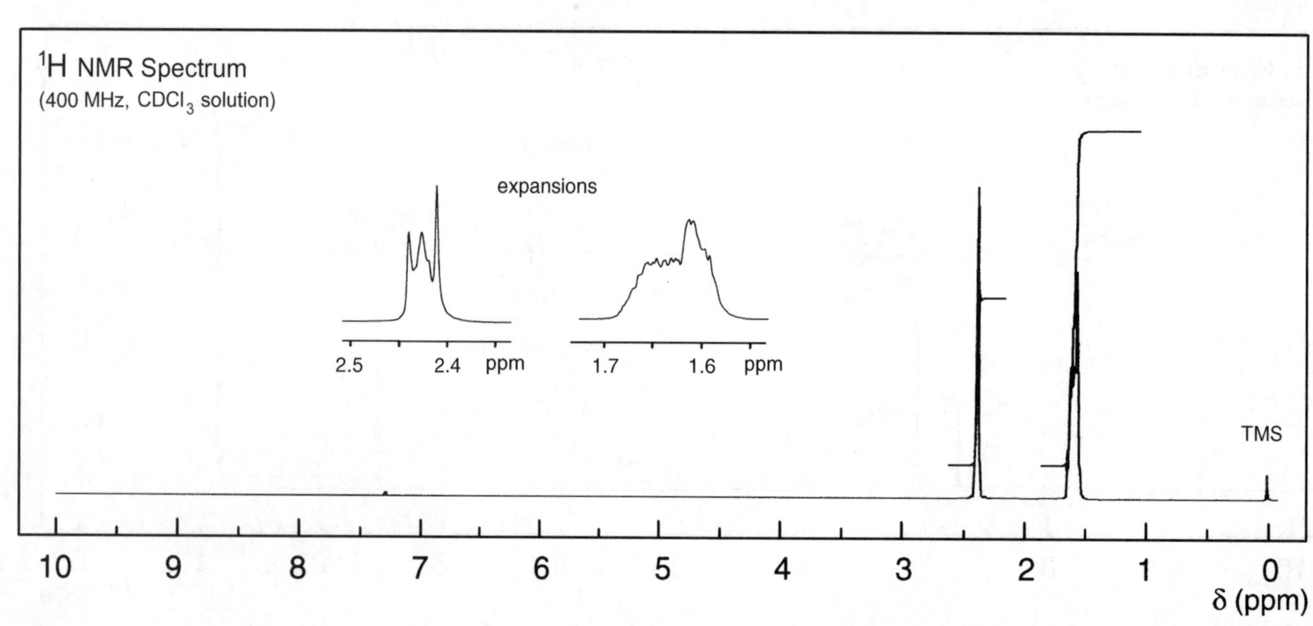

Problem 116

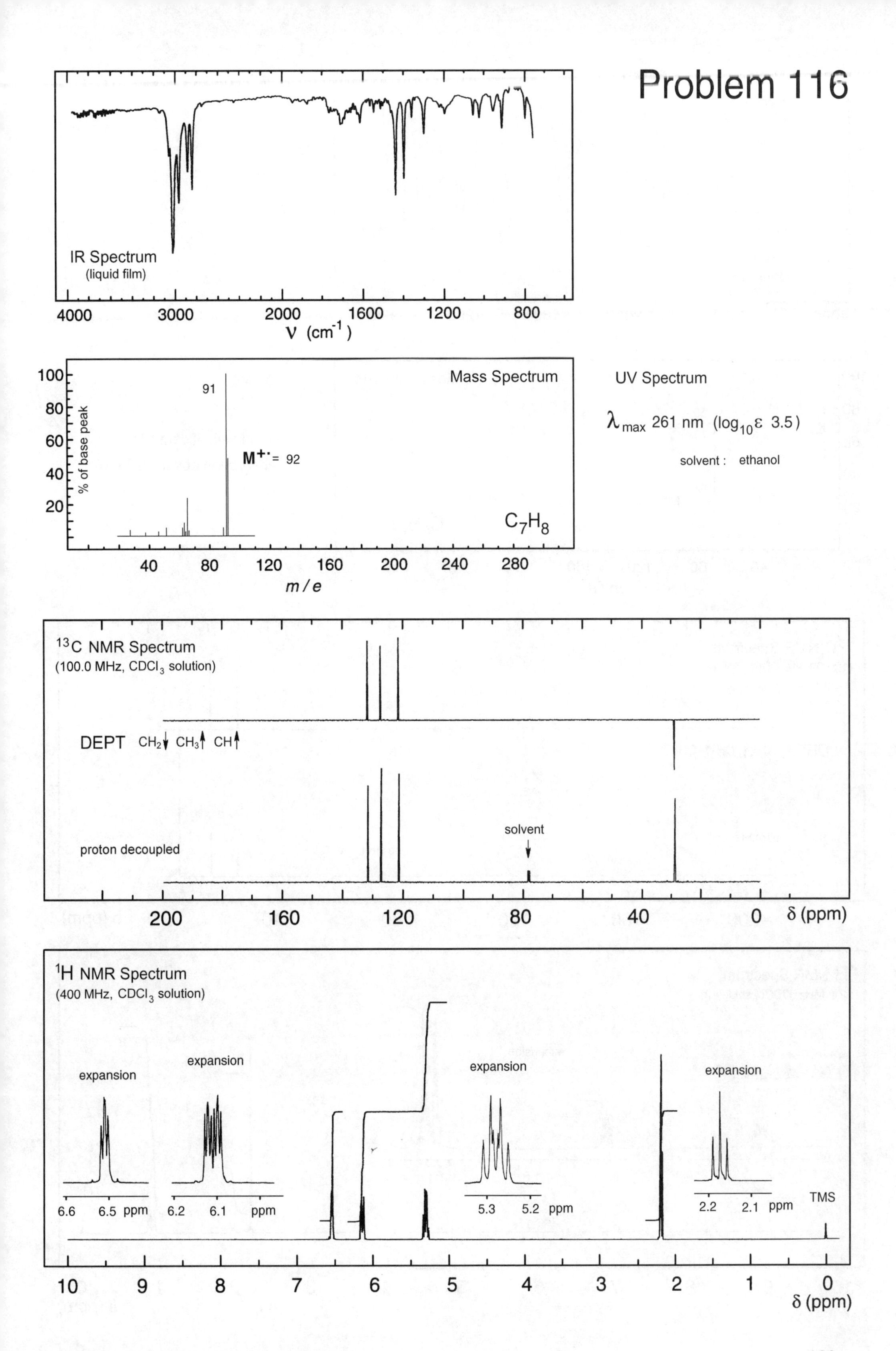

IR Spectrum
(liquid film)
4000
3000
2000
1600
1200
800
ν (cm⁻¹)
Mass Spectrum
% of base peak
100
80
60
40
20
91
M⁺˙= 92
C7H8
40
80
120
160
200
240
280
m / e
UV Spectrum
λmax 261 nm (log10ε 3.5)
solvent : ethanol
13C NMR Spectrum
(100.0 MHz, CDCl3 solution)
DEPT CH2↓ CH3↑ CH↑
solvent
proton decoupled
200
160
120
80
40
0
δ (ppm)
1H NMR Spectrum
(400 MHz, CDCl3 solution)
expansion
expansion
expansion
expansion
6.6 6.5 ppm
6.2 6.1 ppm
5.3 5.2 ppm
2.2 2.1 ppm
TMS
10 9 8 7 6 5 4 3 2 1 0
δ (ppm)

Problem 117

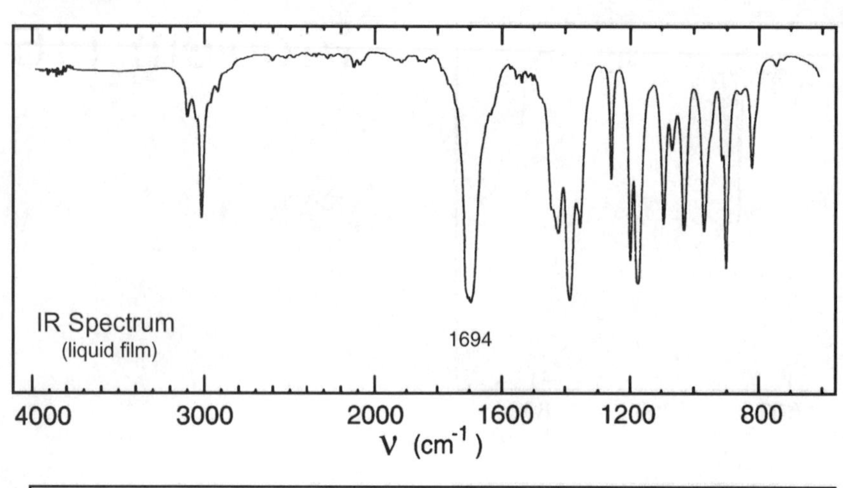

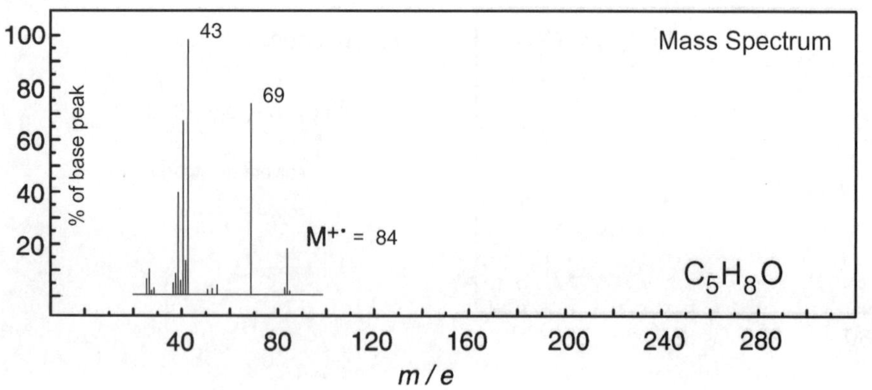

No significant UV absorption above 220 nm

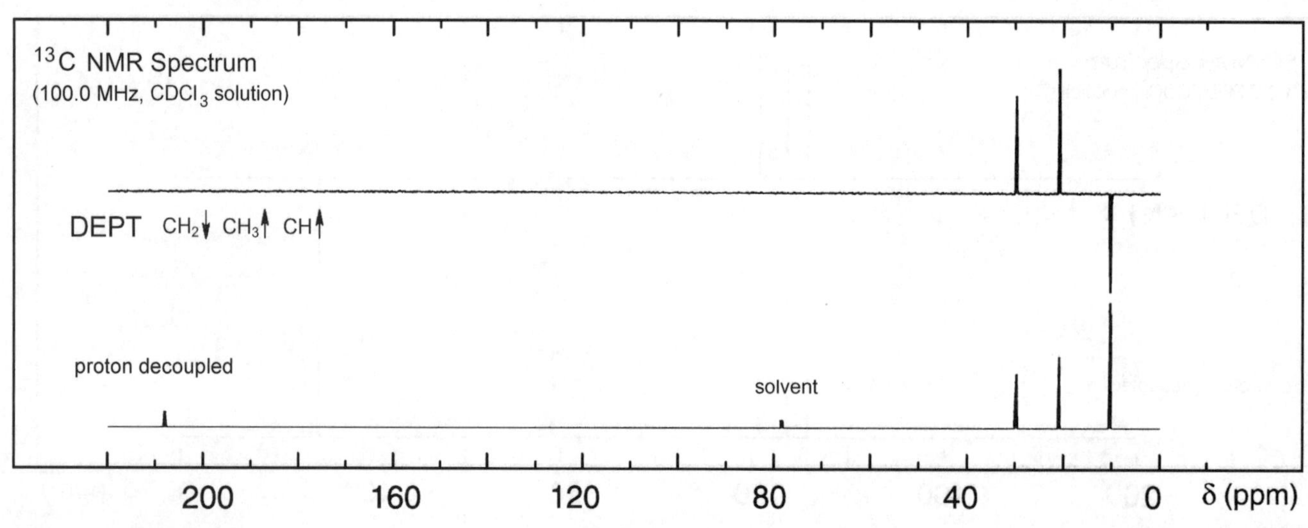

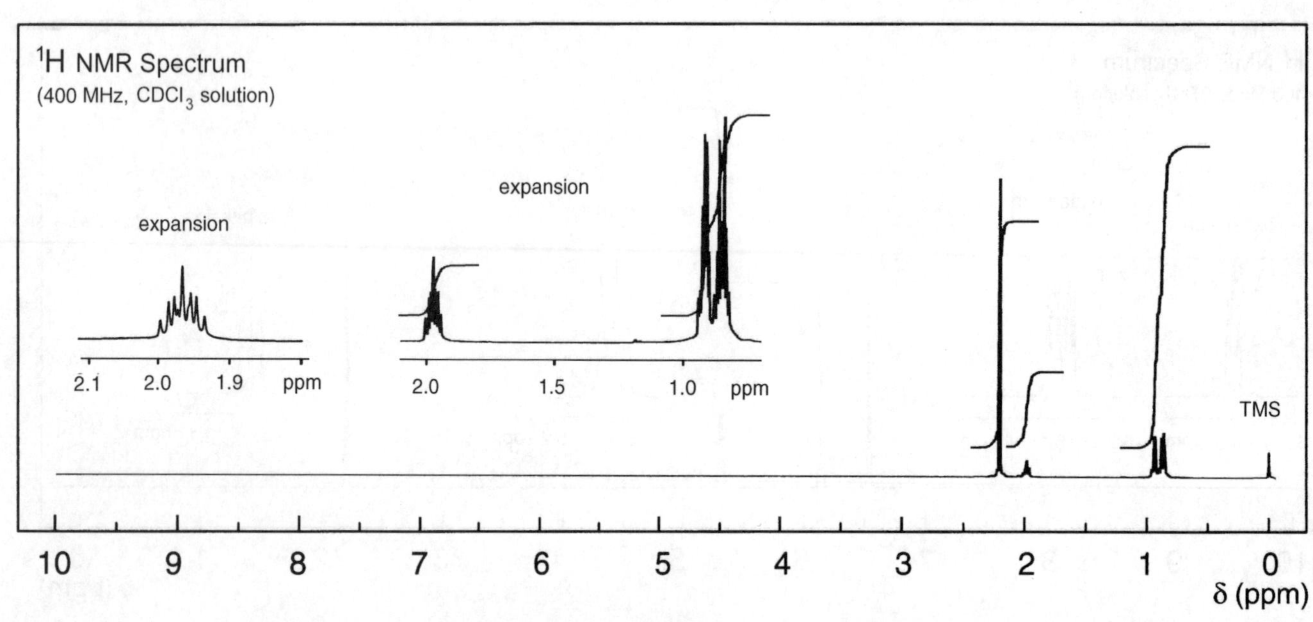

Problem 118

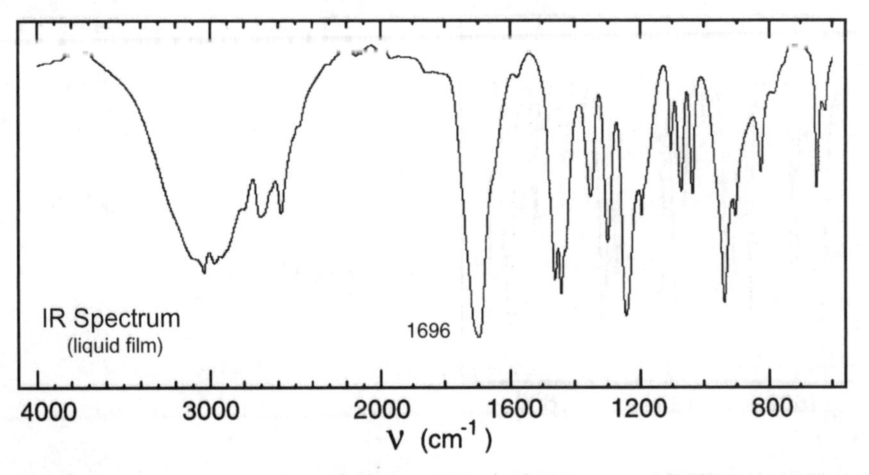

IR Spectrum
(liquid film)
1696
4000
3000
2000
1600
1200
800
ν (cm⁻¹)

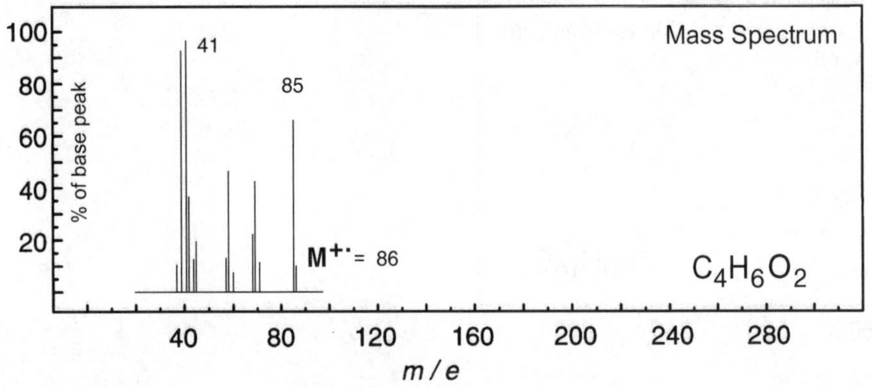

Mass Spectrum
100
80
60
40
20
% of base peak
41
85
M+· = 86
$C_4H_6O_2$
40
80
120
160
200
240
280
m / e

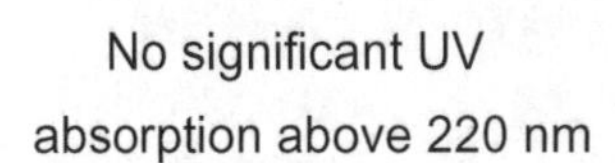

No significant UV
absorption above 220 nm

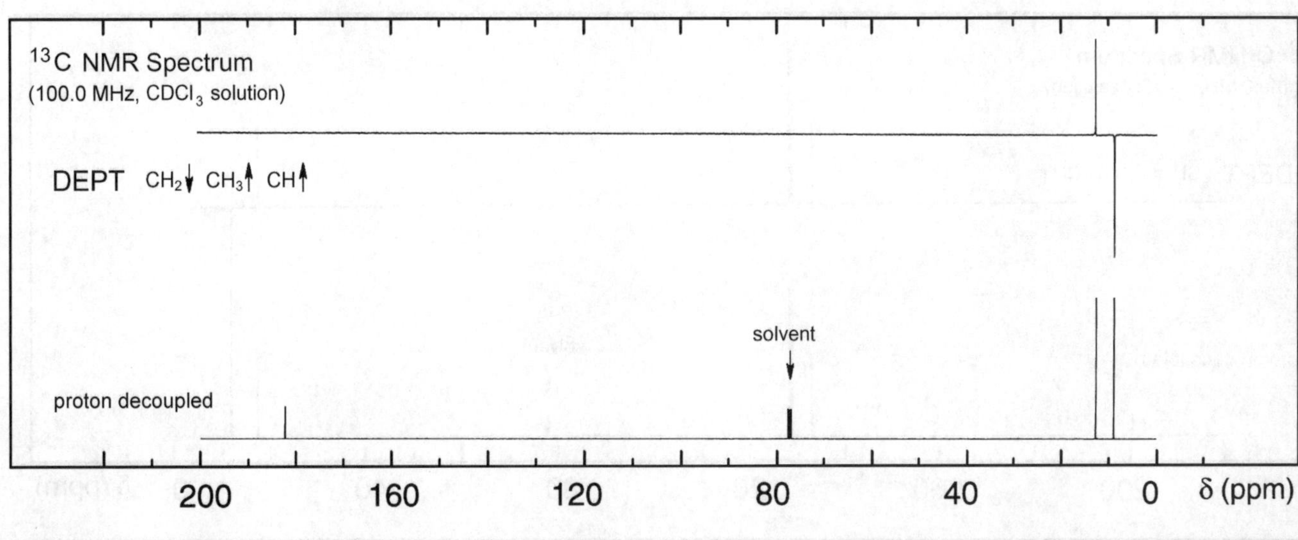

13C NMR Spectrum
(100.0 MHz, CDCl3 solution)
DEPT CH2↓ CH3↑ CH↑
solvent
proton decoupled
200
160
120
80
40
0
δ (ppm)

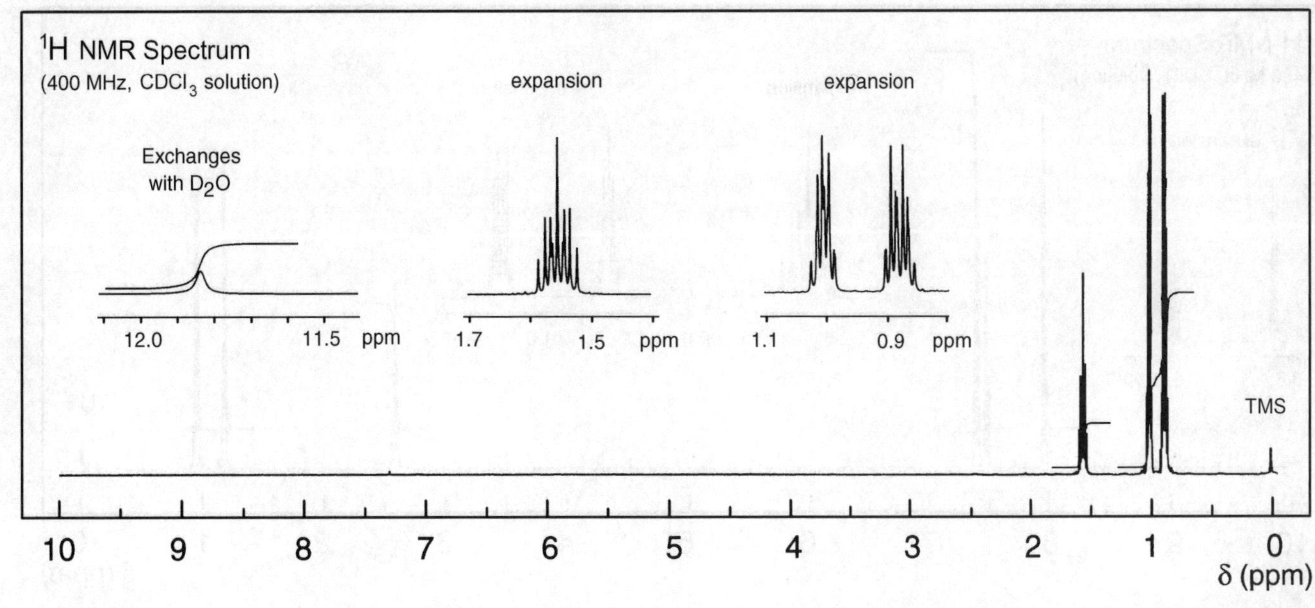

1H NMR Spectrum
(400 MHz, CDCl3 solution)
Exchanges
with D2O
expansion
expansion
12.0
11.5
ppm
1.7
1.5
ppm
1.1
0.9
ppm
TMS
10
9
8
7
6
5
4
3
2
1
0
δ (ppm)

Problem 119

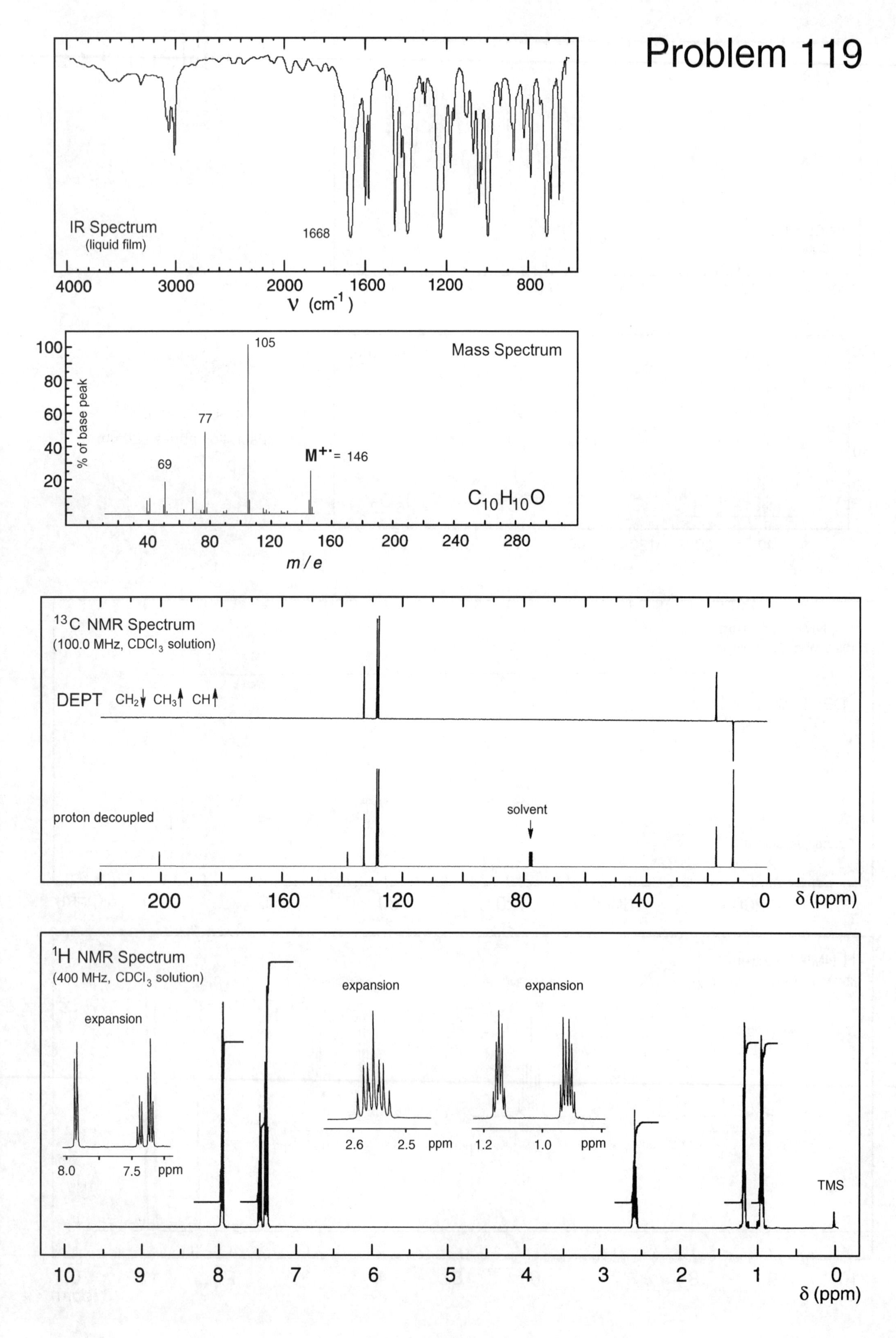

IR Spectrum
(liquid film)
1668
4000
3000
2000
1600
1200
800
ν (cm-1)
100
80
60
40
20
% of base peak
105
77
69
M+· = 146
Mass Spectrum
C10H10O
40
80
120
160
200
240
280
m / e
13C NMR Spectrum
(100.0 MHz, CDCl3 solution)
DEPT CH2 CH3 CH
proton decoupled
solvent
200
160
120
80
40
0
δ (ppm)
1H NMR Spectrum
(400 MHz, CDCl3 solution)
expansion
8.0
7.5
ppm
expansion
2.6
2.5
ppm
expansion
1.2
1.0
ppm
TMS
10
9
8
7
6
5
4
3
2
1
0
δ (ppm)

Problem 120

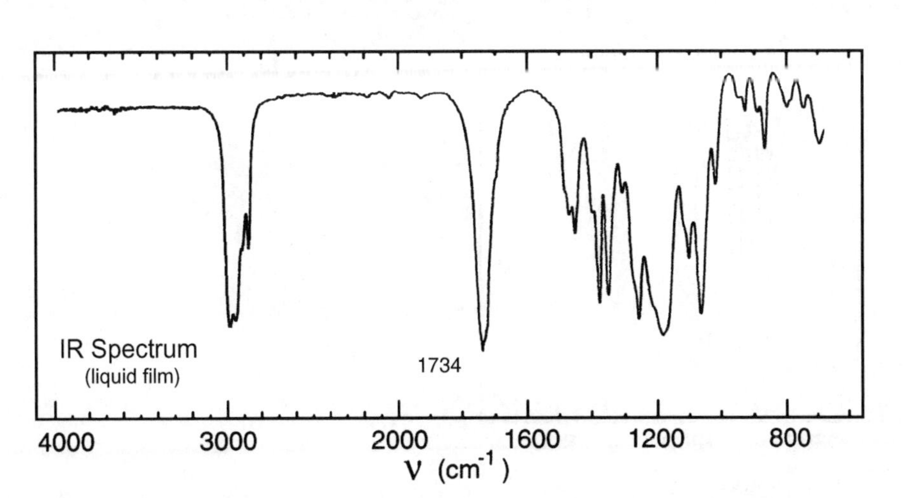

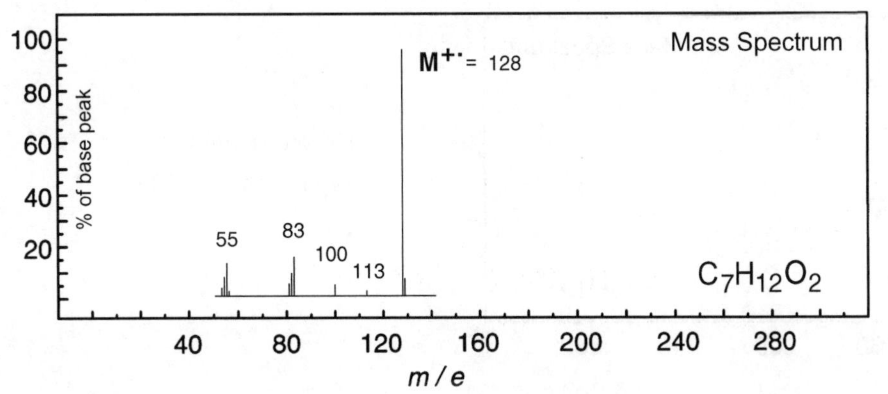

No significant UV absorption above 220 nm

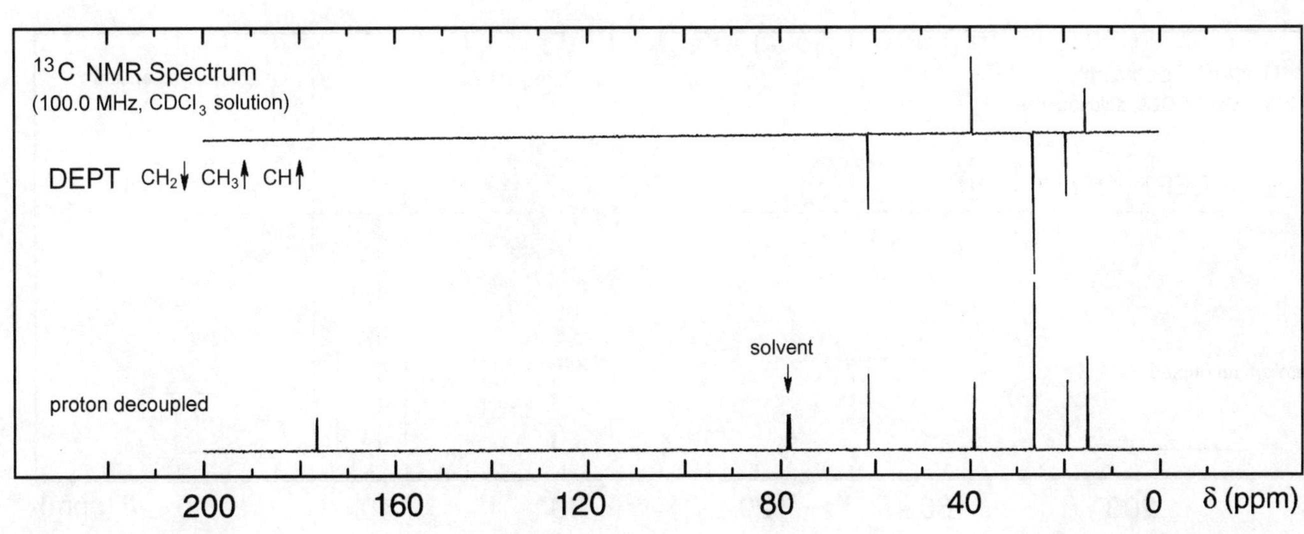

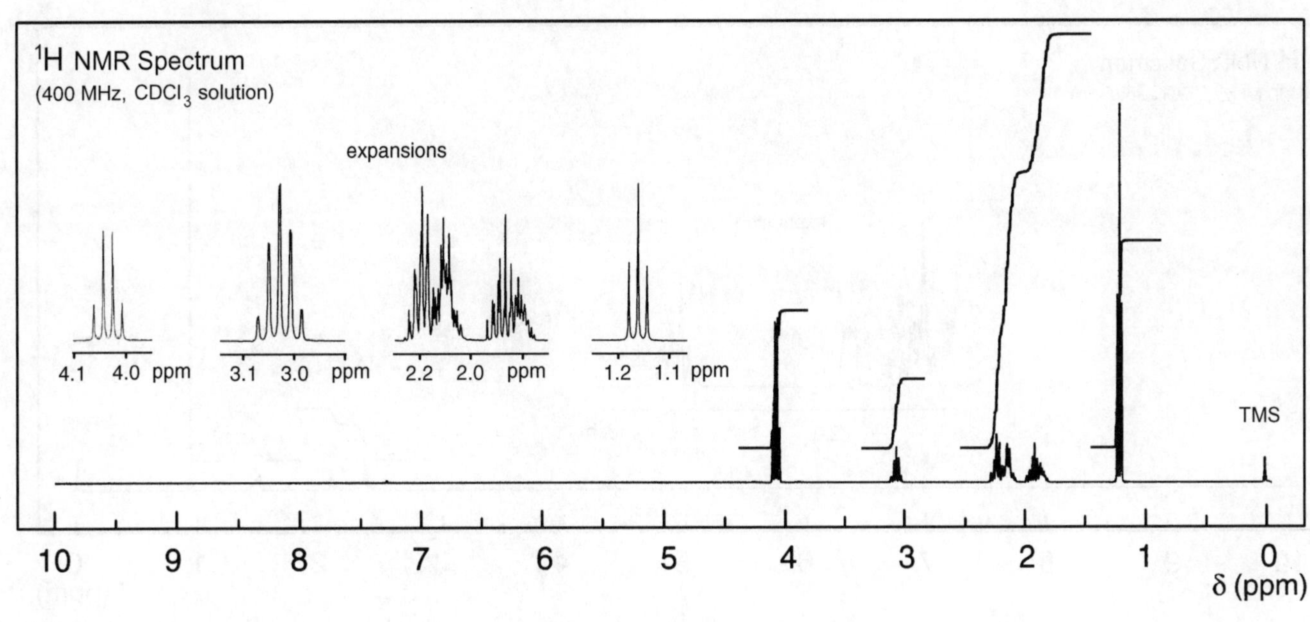

Problem 121

IR Spectrum
(KBr disc)
1728
4000 3000 2000 1600 1200 800
ν (cm^{-1})

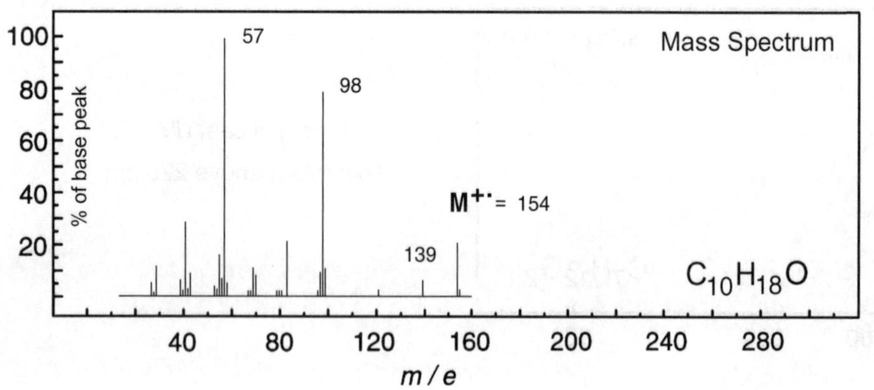

No significant UV absorption above 220 nm

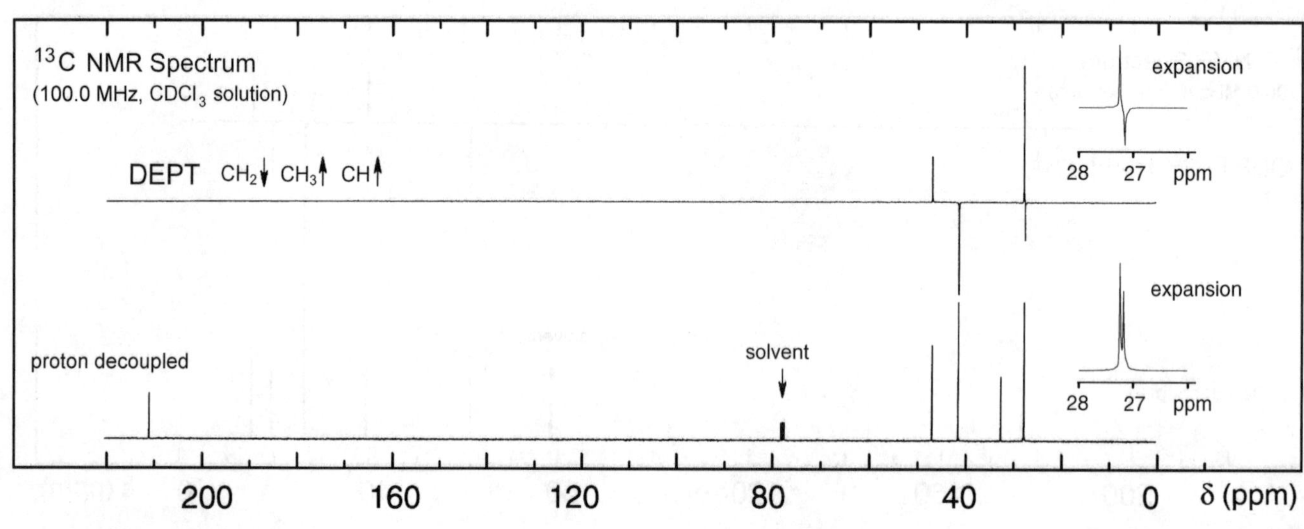

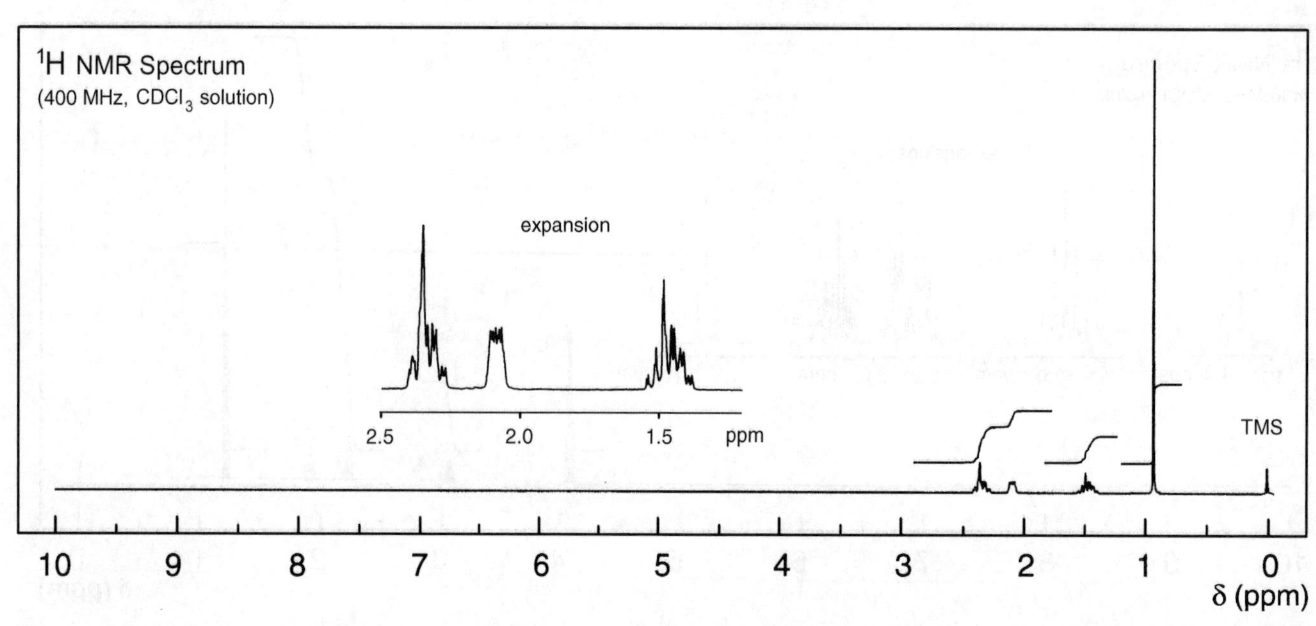

Problem 122

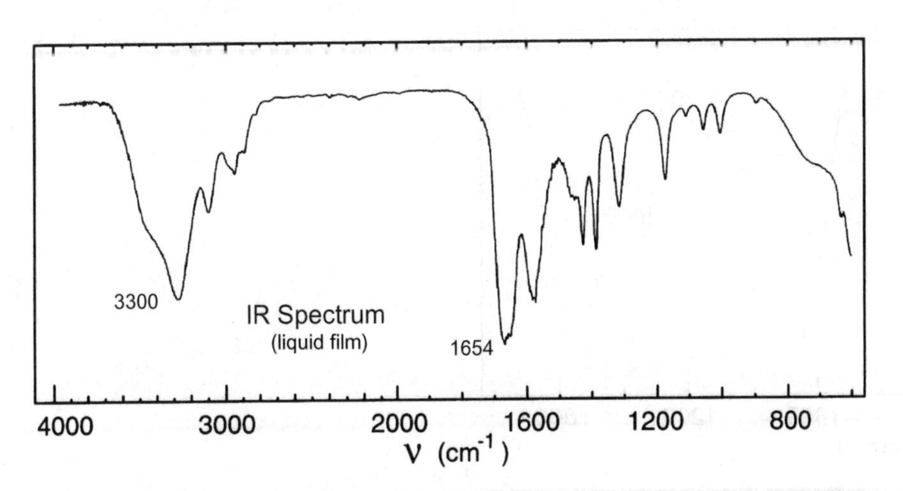

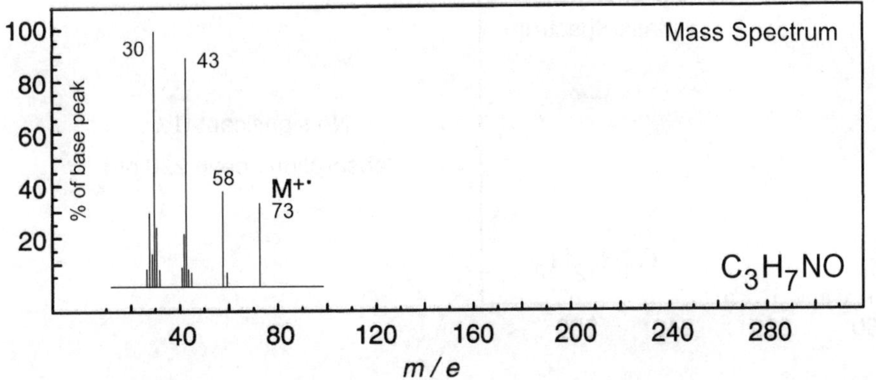

No significant UV absorption above 220 nm

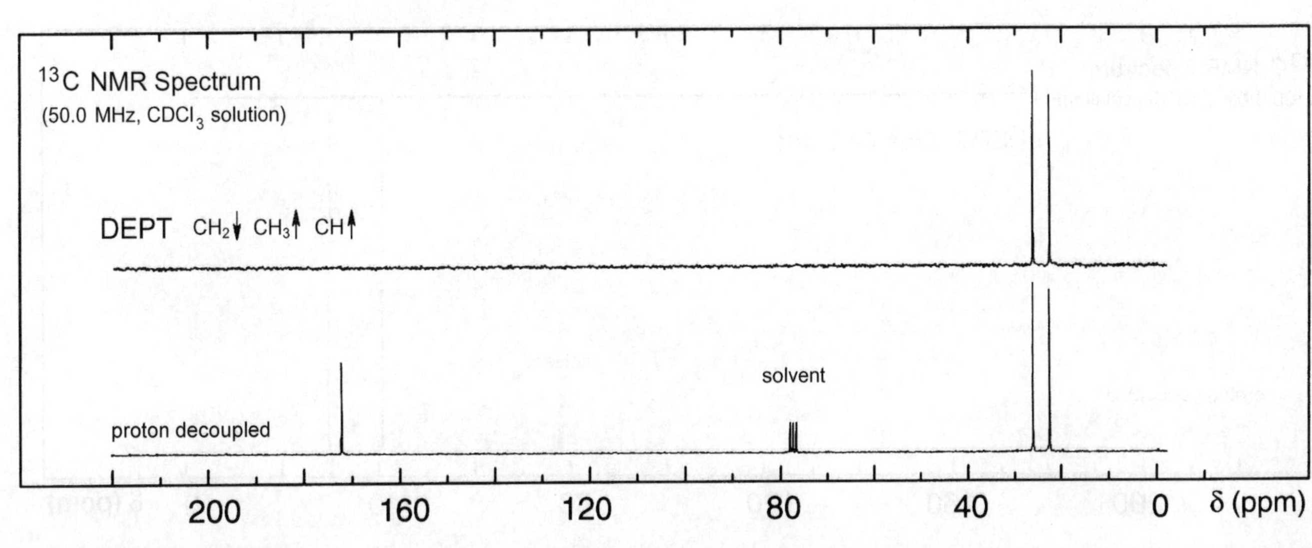

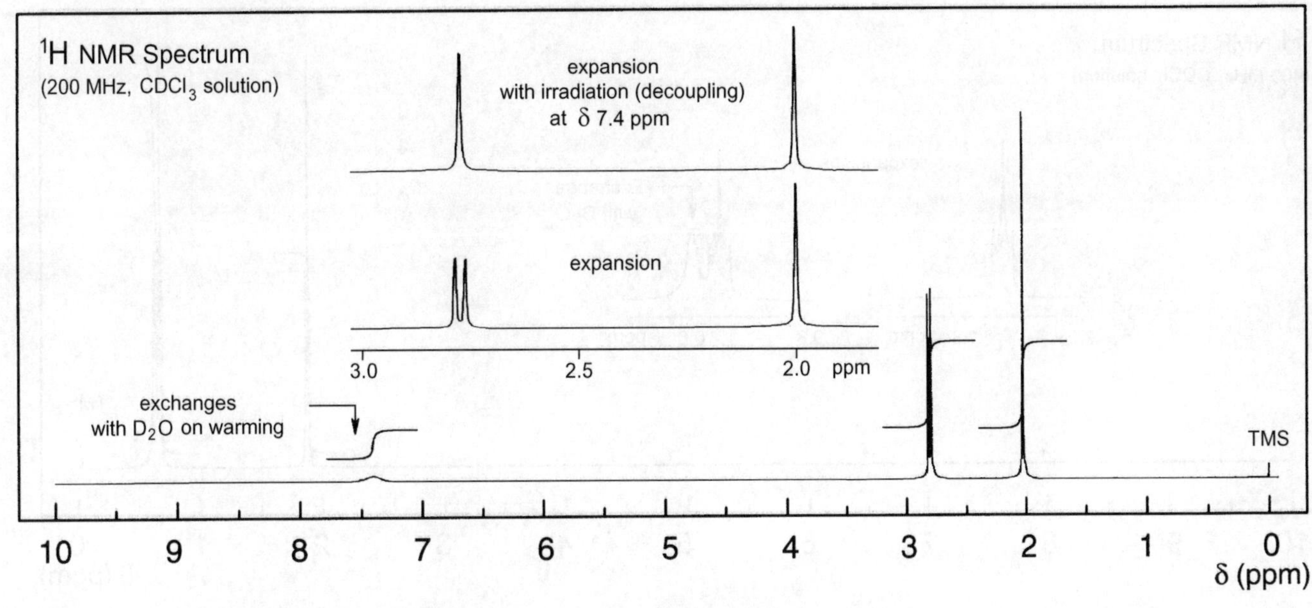

Problem 123

3360
3283
IR Spectrum
(liquid film)
4000 3000 2000 1600 1200 800
ν (cm⁻¹)

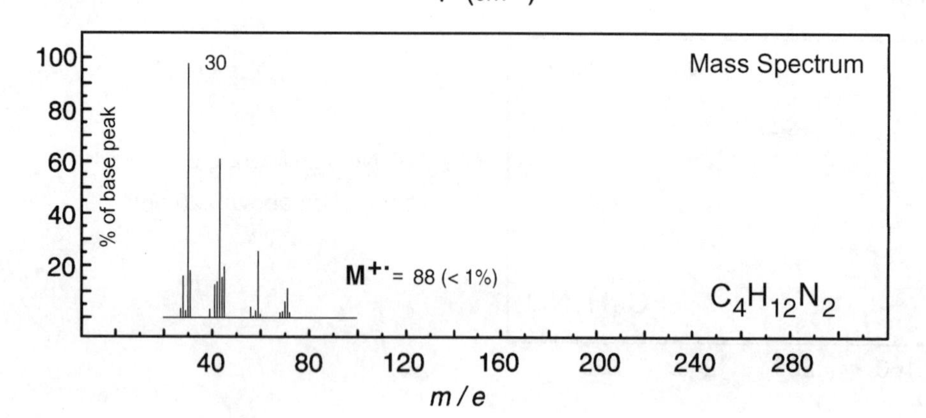

No significant UV absorption above 220 nm

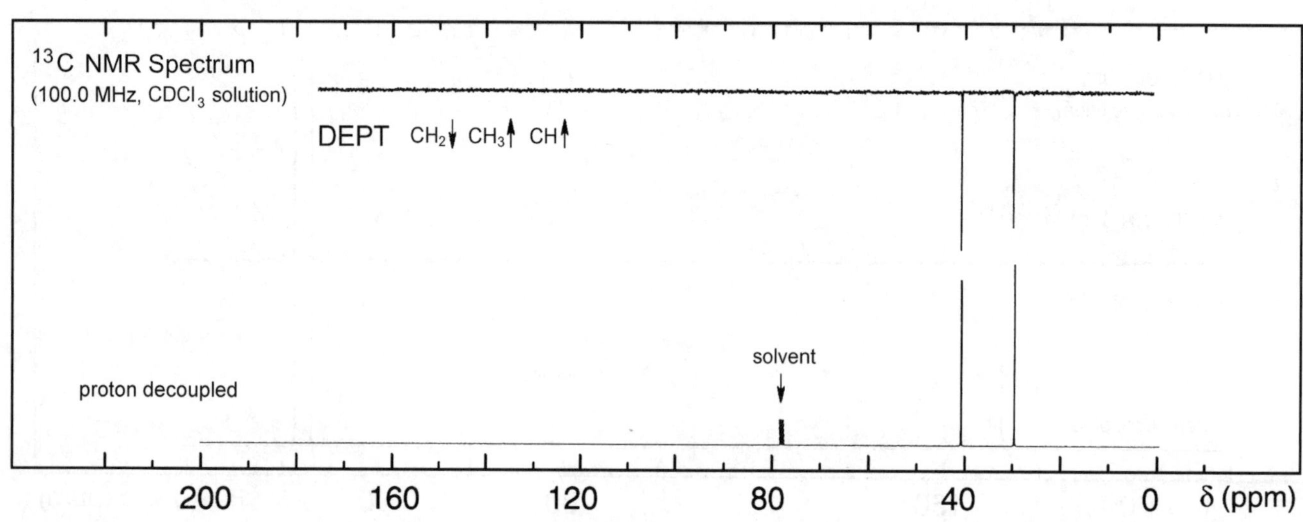

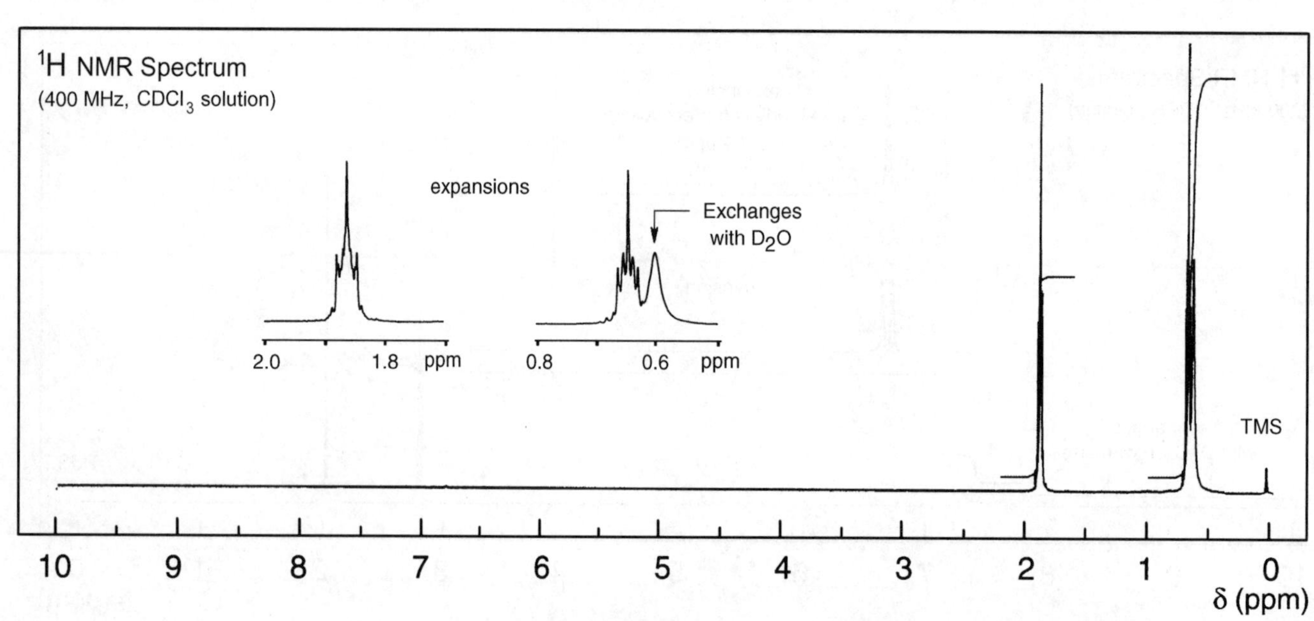

Problem 124

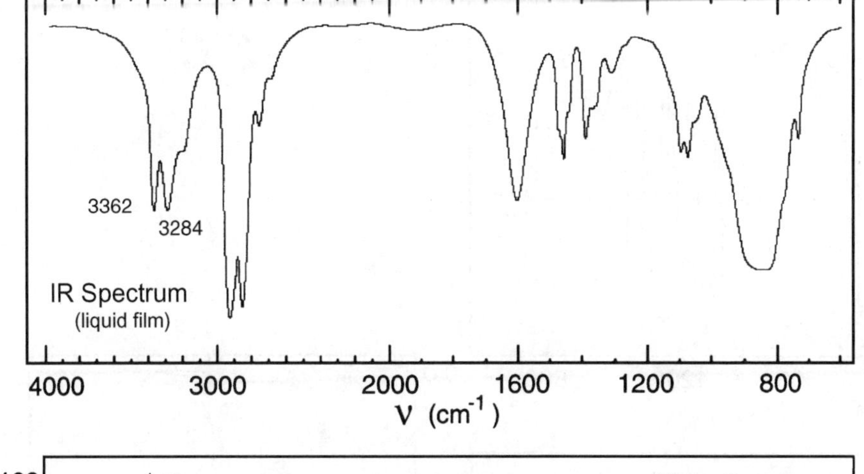

3362
3284
IR Spectrum
(liquid film)
4000
3000
2000
1600
1200
800
ν (cm-1)

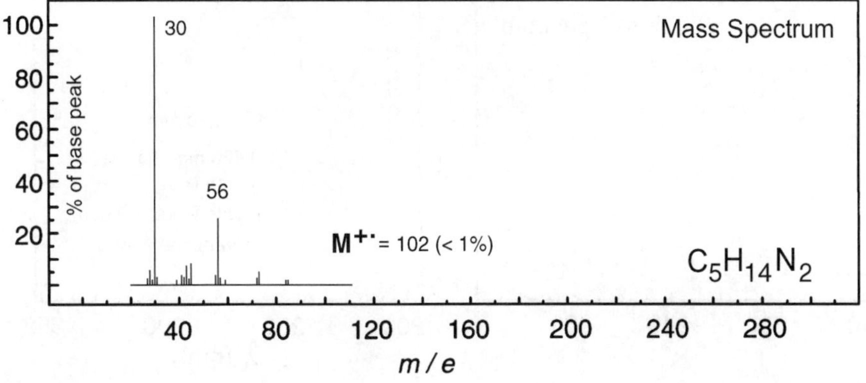

100
80
60
40
20
% of base peak
30
56
Mass Spectrum
M+· = 102 (< 1%)
C5H14N2
40
80
120
160
200
240
280
m / e

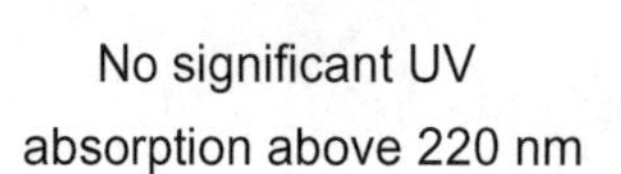

No significant UV absorption above 220 nm

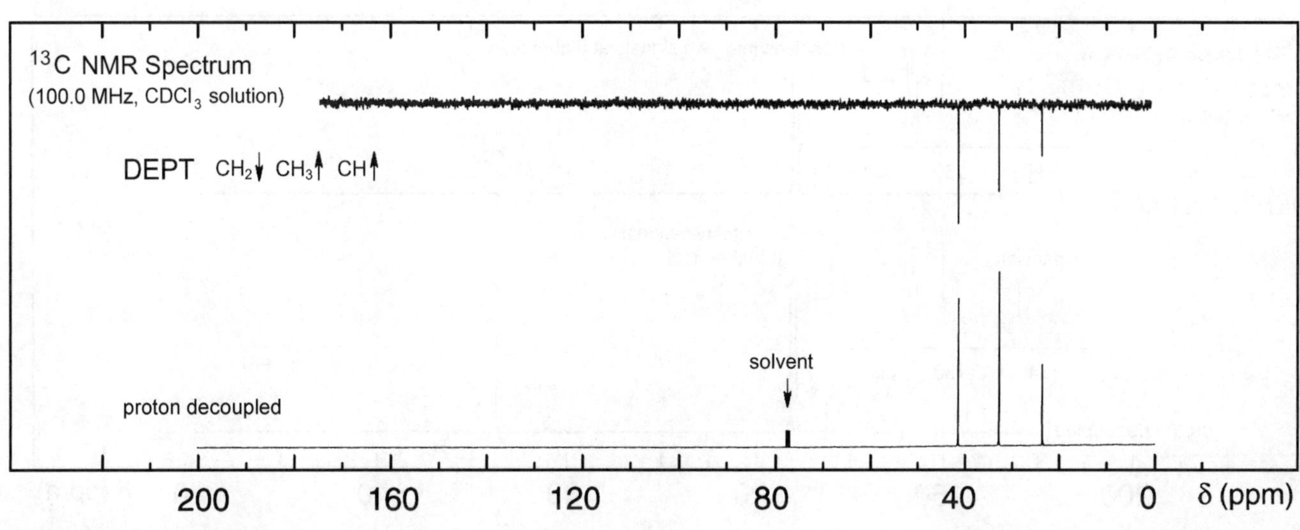

13C NMR Spectrum
(100.0 MHz, CDCl3 solution)
DEPT CH2↓ CH3↑ CH↑
solvent
proton decoupled
200
160
120
80
40
0
δ (ppm)

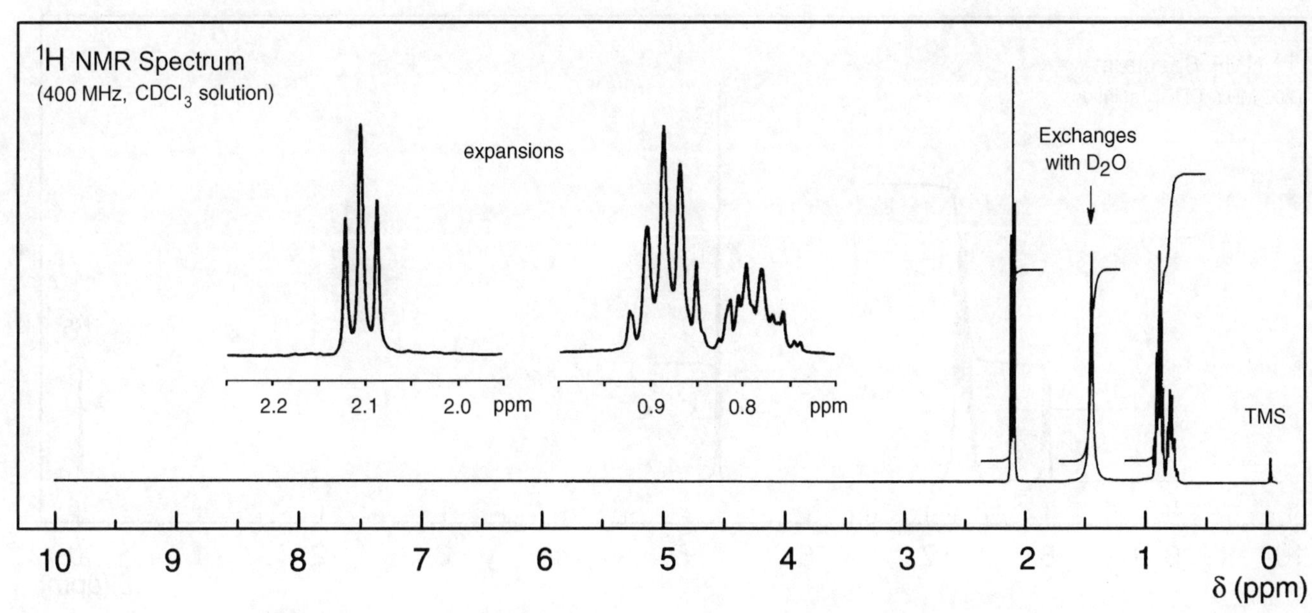

1H NMR Spectrum
(400 MHz, CDCl3 solution)
expansions
2.2
2.1
2.0
ppm
0.9
0.8
ppm
Exchanges with D2O
TMS
10
9
8
7
6
5
4
3
2
1
0
δ (ppm)

Problem 125

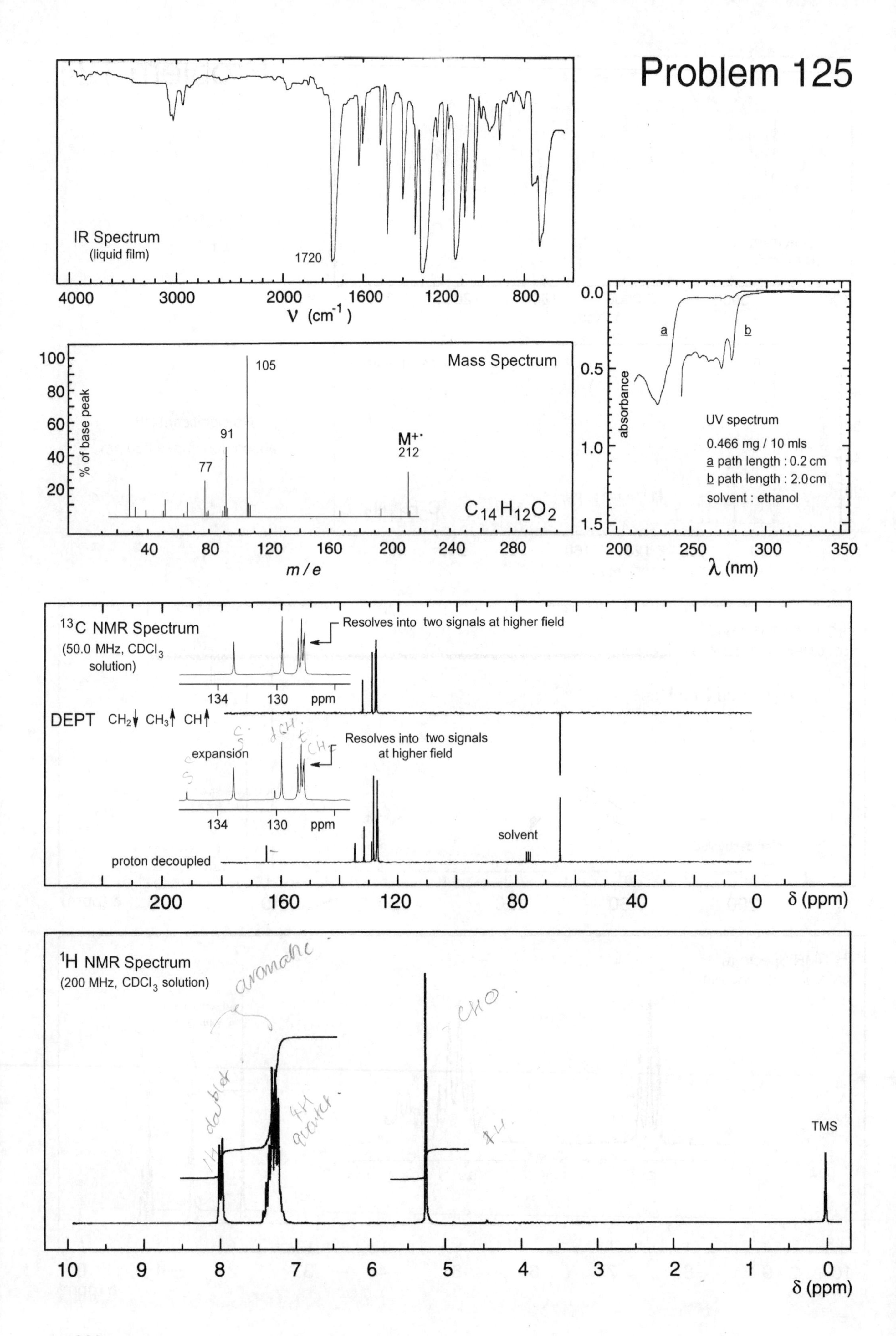

IR Spectrum
(liquid film)
1720
4000
3000
2000
1600
1200
800
ν (cm⁻¹)
Mass Spectrum
% of base peak
105
91
77
M+·
212
C14H12O2
m / e
UV spectrum
0.466 mg / 10 mls
a path length : 0.2 cm
b path length : 2.0 cm
solvent : ethanol
absorbance
λ (nm)
13C NMR Spectrum
(50.0 MHz, CDCl3 solution)
Resolves into two signals at higher field
DEPT CH2 CH3 CH
expansion
Resolves into two signals at higher field
134
130
ppm
solvent
proton decoupled
δ (ppm)
1H NMR Spectrum
(200 MHz, CDCl3 solution)
TMS
δ (ppm)

Problem 126

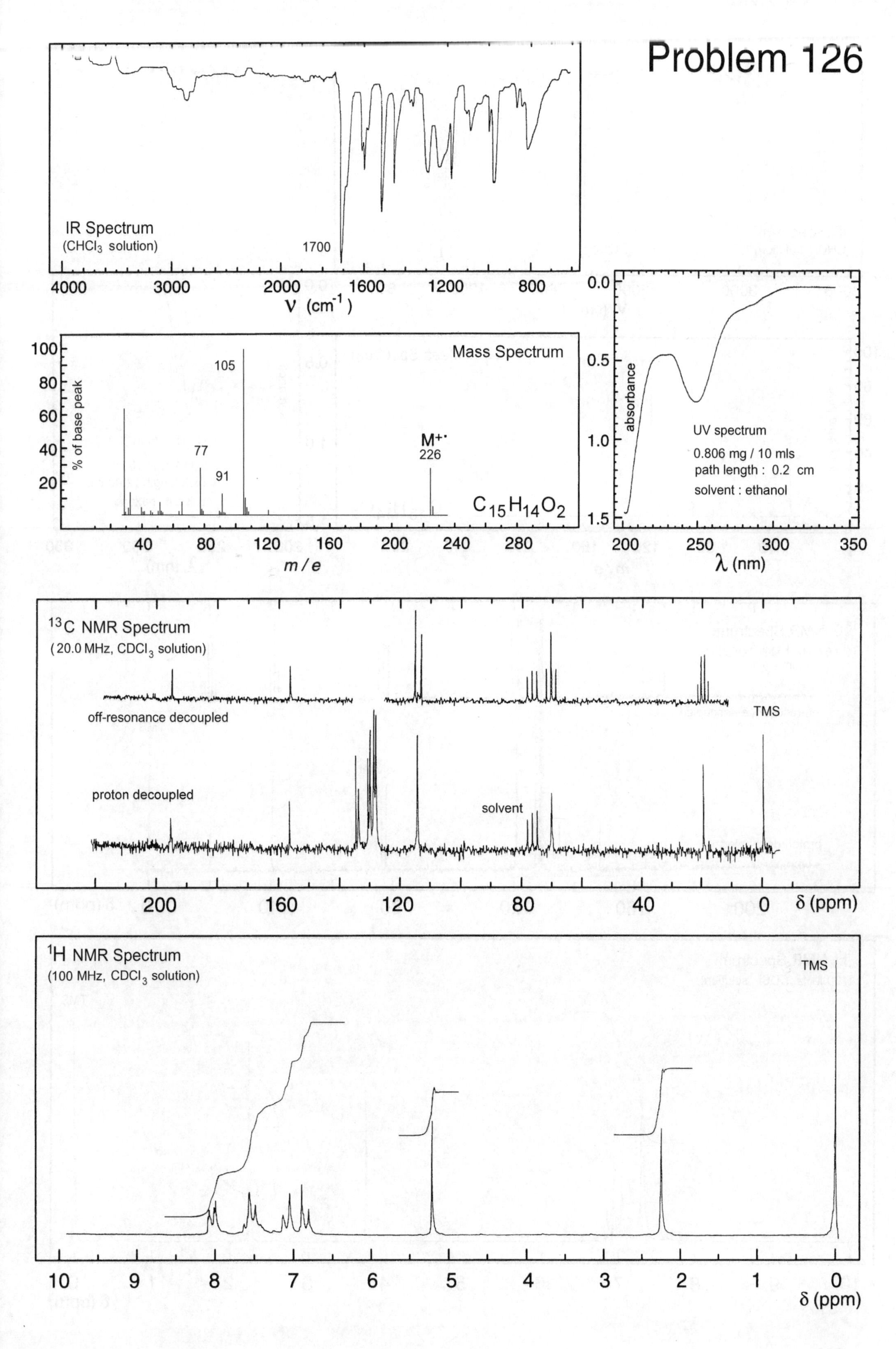

IR Spectrum
(CHCl3 solution)
1700
4000
3000
2000
1600
1200
800
ν (cm-1)
Mass Spectrum
100
80
60
40
20
% of base peak
105
77
91
M+·
226
C15H14O2
40
80
120
160
200
240
280
m / e
0.0
0.5
1.0
1.5
absorbance
UV spectrum
0.806 mg / 10 mls
path length : 0.2 cm
solvent : ethanol
200
250
300
350
λ (nm)
13C NMR Spectrum
(20.0 MHz, CDCl3 solution)
off-resonance decoupled
TMS
proton decoupled
solvent
200
160
120
80
40
0
δ (ppm)
1H NMR Spectrum
(100 MHz, CDCl3 solution)
TMS
10
9
8
7
6
5
4
3
2
1
0
δ (ppm)

Problem 127

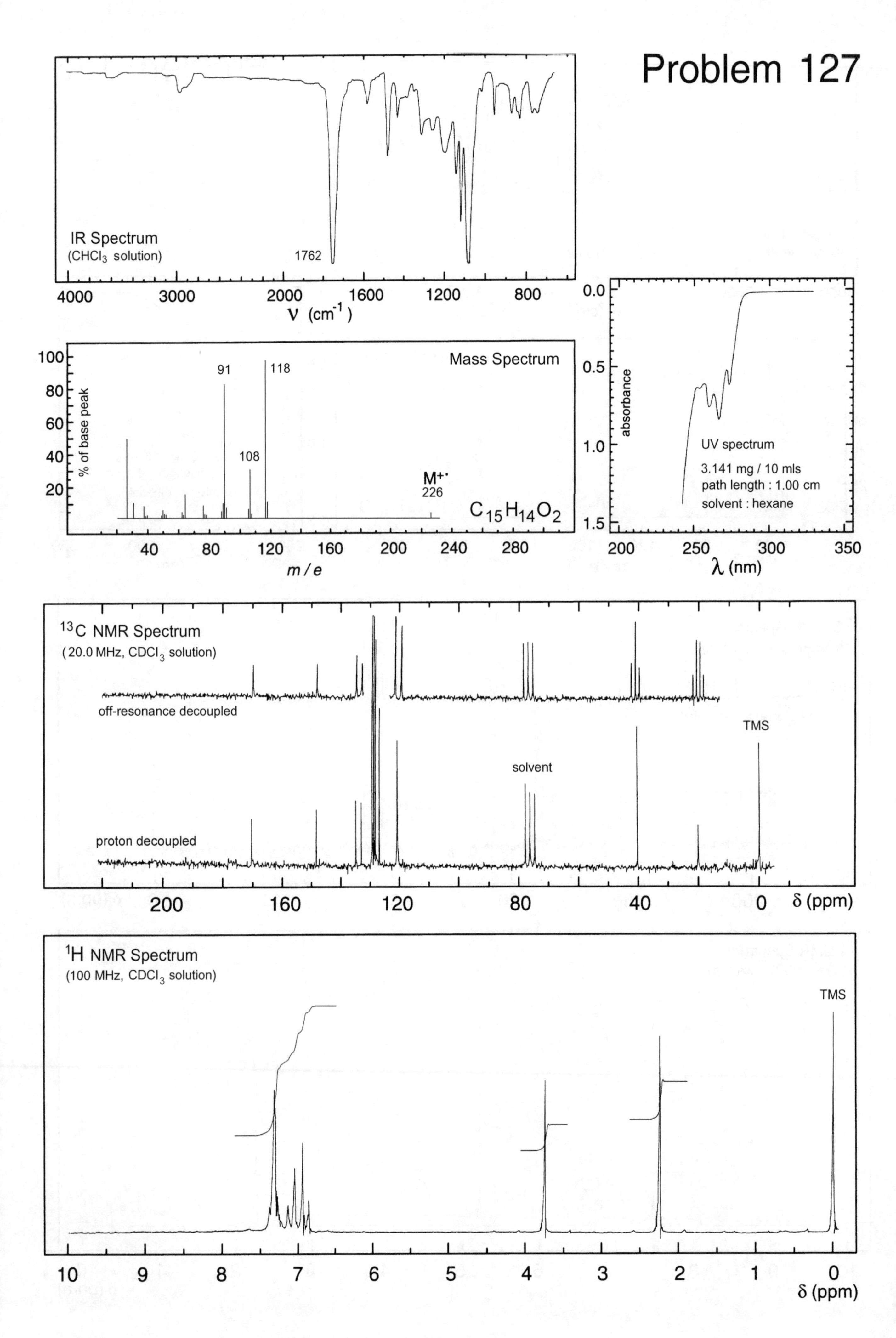

IR Spectrum
(CHCl3 solution)
1762
4000
3000
2000
1600
1200
800
ν (cm-1)
Mass Spectrum
% of base peak
100
80
60
40
20
91
118
108
M+·
226
C15H14O2
40
80
120
160
200
240
280
m / e
absorbance
0.0
0.5
1.0
1.5
UV spectrum
3.141 mg / 10 mls
path length : 1.00 cm
solvent : hexane
200
250
300
350
λ (nm)
13C NMR Spectrum
(20.0 MHz, CDCl3 solution)
off-resonance decoupled
proton decoupled
solvent
TMS
200
160
120
80
40
0
δ (ppm)
1H NMR Spectrum
(100 MHz, CDCl3 solution)
TMS
10
9
8
7
6
5
4
3
2
1
0
δ (ppm)

Problem 128

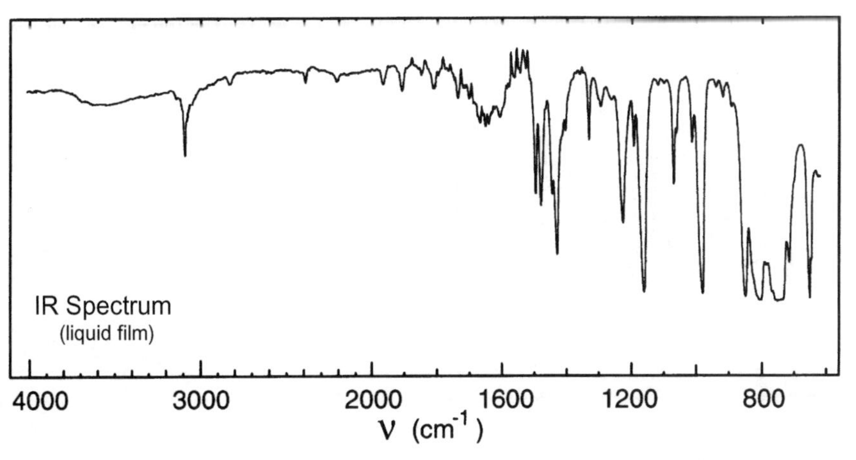

IR Spectrum
(liquid film)
4000
3000
2000
1600
1200
800
ν (cm⁻¹)

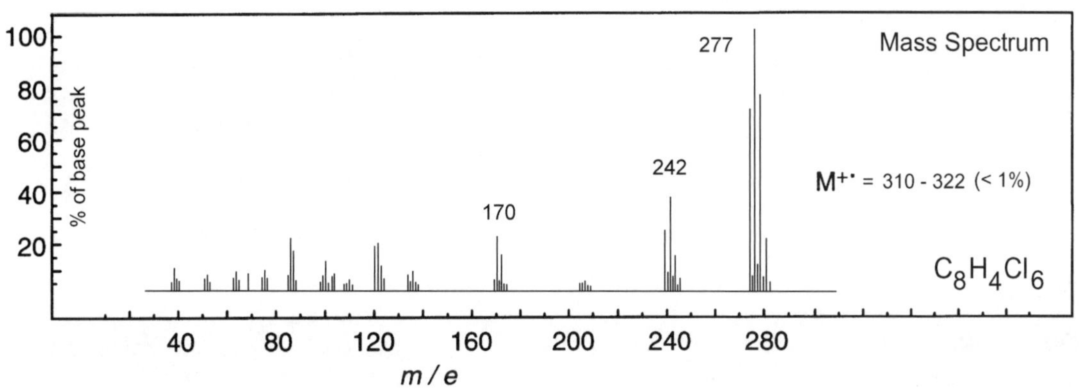

Mass Spectrum
% of base peak
100
80
60
40
20
277
242
170
M+• = 310 - 322 (< 1%)
C8H4Cl6
40
80
120
160
200
240
280
m / e

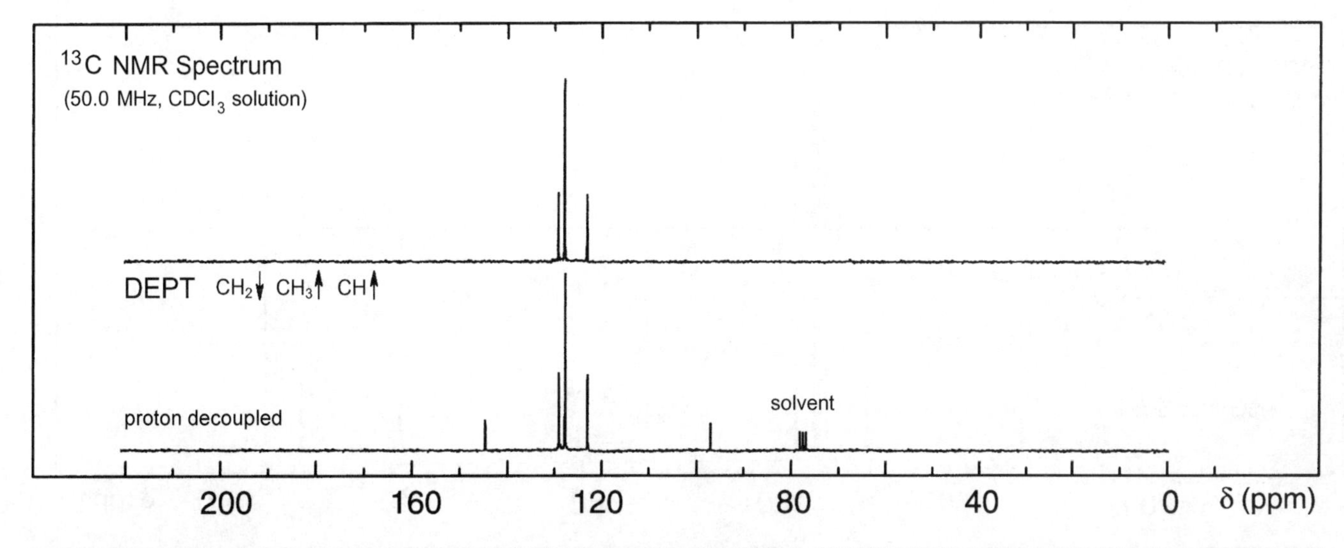

13C NMR Spectrum
(50.0 MHz, CDCl3 solution)
DEPT CH2 CH3 CH
proton decoupled
solvent
200
160
120
80
40
0
δ (ppm)

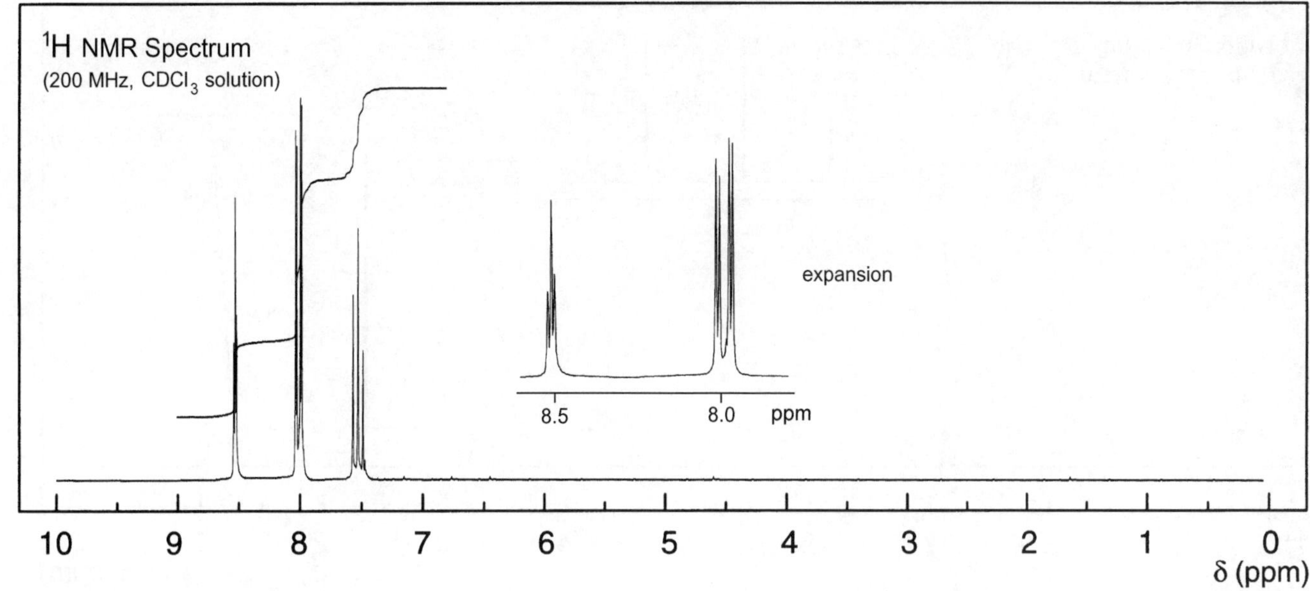

1H NMR Spectrum
(200 MHz, CDCl3 solution)
expansion
8.5
8.0
ppm
10
9
8
7
6
5
4
3
2
1
0
δ (ppm)

Problem 129

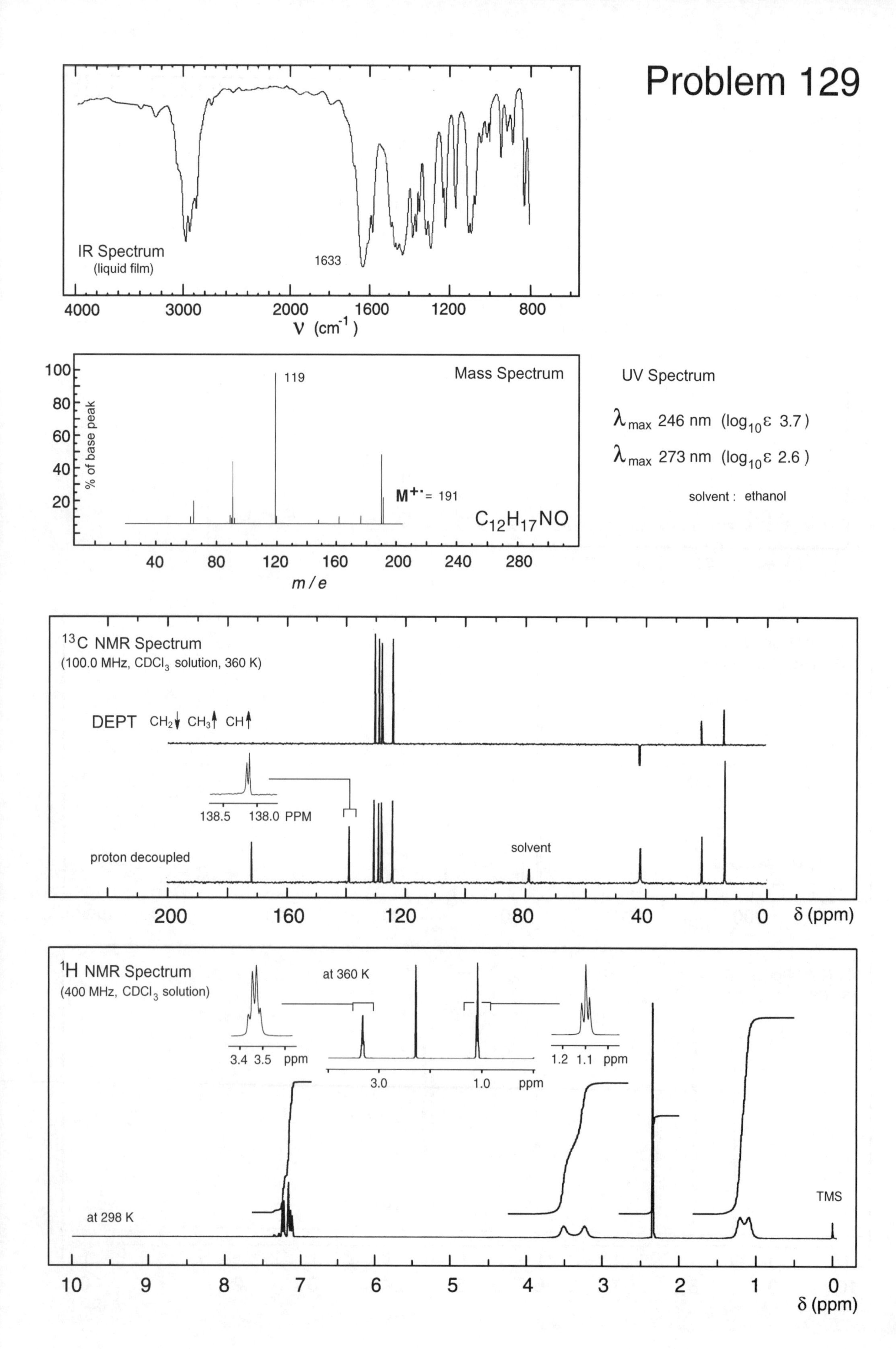

IR Spectrum
(liquid film)
1633
4000
3000
2000
1600
1200
800
ν (cm⁻¹)
Mass Spectrum
119
M⁺˙ = 191
C12H17NO
% of base peak
100
80
60
40
20
40
80
120
160
200
240
280
m / e
UV Spectrum
λmax 246 nm (log10ε 3.7)
λmax 273 nm (log10ε 2.6)
solvent : ethanol
13C NMR Spectrum
(100.0 MHz, CDCl3 solution, 360 K)
DEPT CH2↓ CH3↑ CH↑
138.5 138.0 PPM
proton decoupled
solvent
200
160
120
80
40
0
δ (ppm)
1H NMR Spectrum
(400 MHz, CDCl3 solution)
at 360 K
3.4 3.5 ppm
3.0
1.0
ppm
1.2 1.1 ppm
at 298 K
TMS
10
9
8
7
6
5
4
3
2
1
0
δ (ppm)

Problem 130

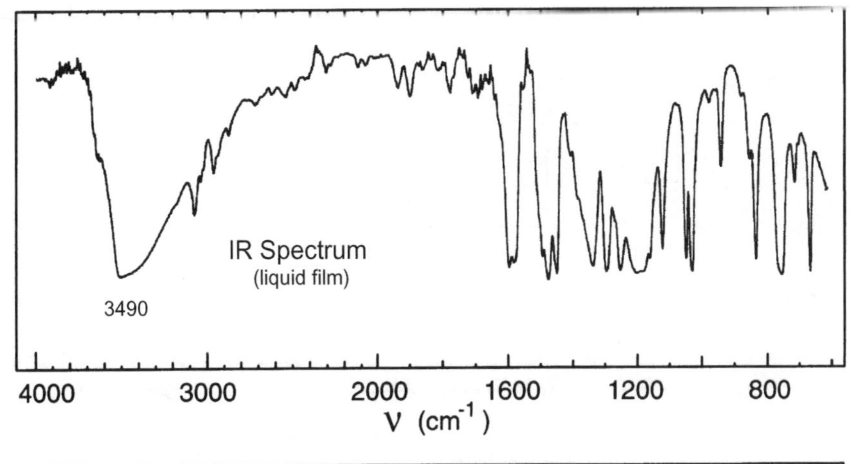

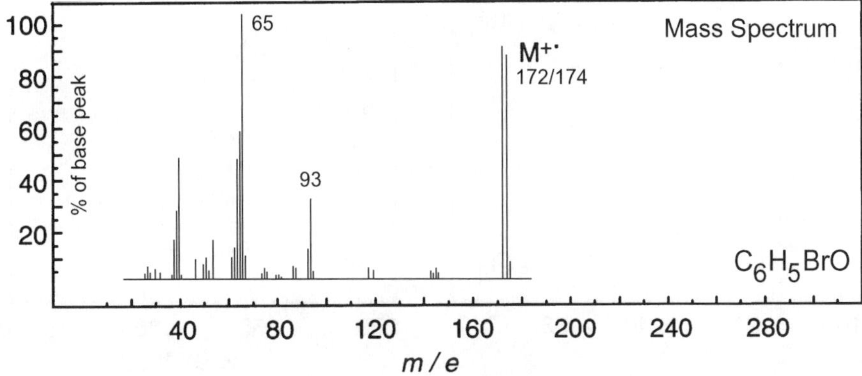

UV Spectrum

λ_{max} 277 nm ($\log_{10}\varepsilon$ 3.4)

solvent : methanol

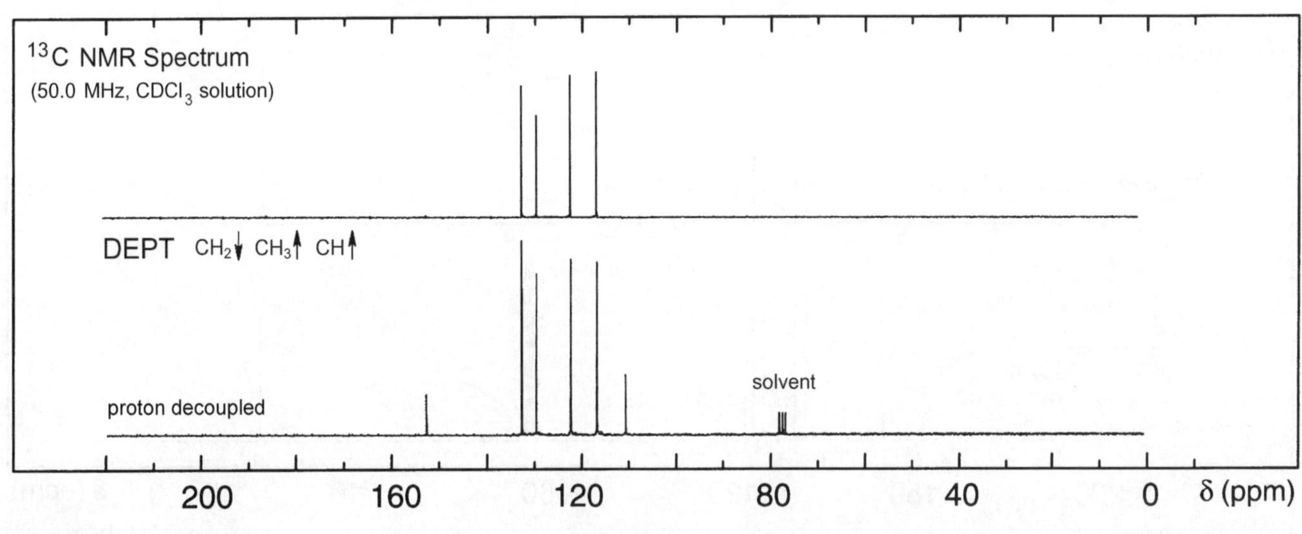

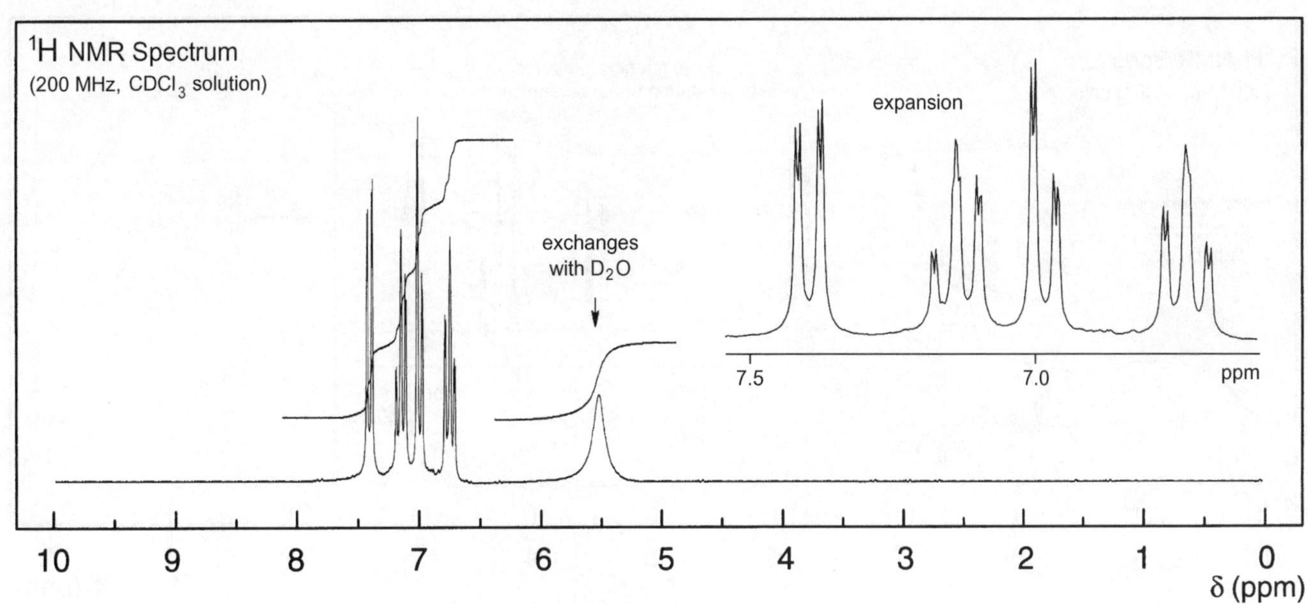

Problem 131

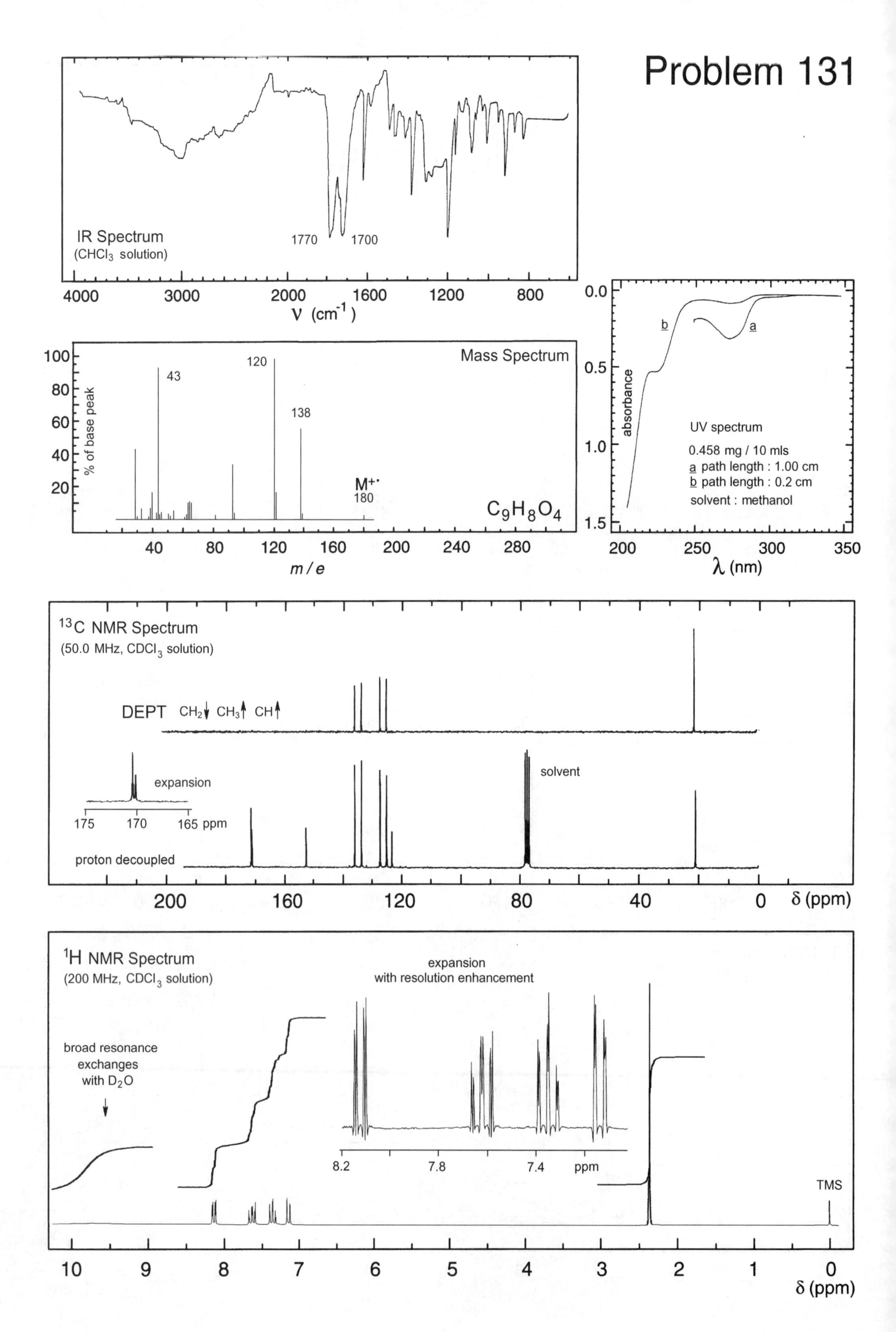

IR Spectrum
(CHCl3 solution)
1770
1700
4000
3000
2000
1600
1200
800
ν (cm-1)
Mass Spectrum
100
80
60
40
20
% of base peak
43
120
138
M+·
180
C9H8O4
40
80
120
160
200
240
280
m / e
0.0
0.5
1.0
1.5
absorbance
b
a
UV spectrum
0.458 mg / 10 mls
a path length : 1.00 cm
b path length : 0.2 cm
solvent : methanol
200
250
300
350
λ (nm)
13C NMR Spectrum
(50.0 MHz, CDCl3 solution)
DEPT CH2↓ CH3↑ CH↑
expansion
175
170
165 ppm
solvent
proton decoupled
200
160
120
80
40
0
δ (ppm)
1H NMR Spectrum
(200 MHz, CDCl3 solution)
expansion
with resolution enhancement
broad resonance
exchanges
with D2O
8.2
7.8
7.4
ppm
TMS
10
9
8
7
6
5
4
3
2
1
0
δ (ppm)

Problem 132

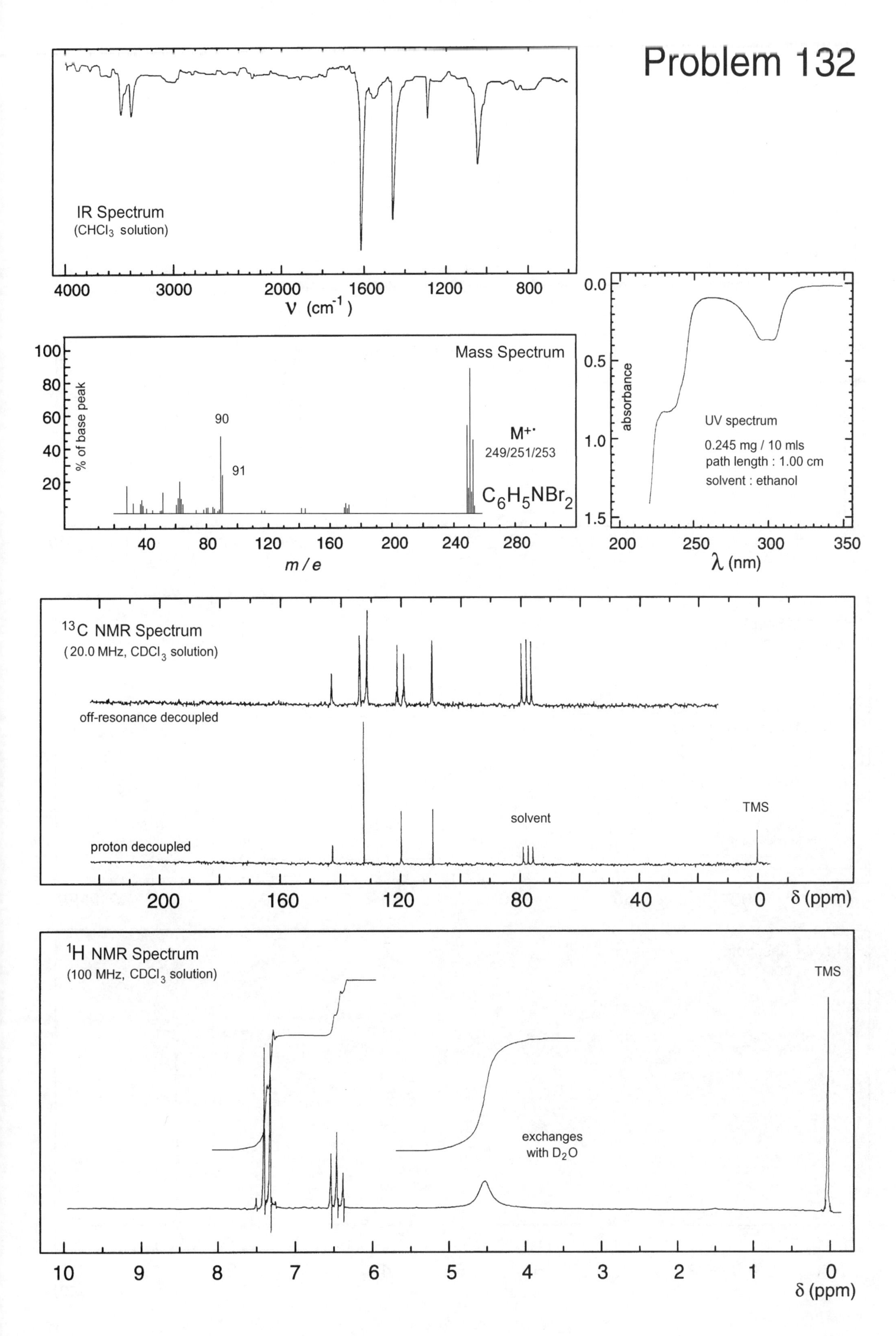

IR Spectrum
(CHCl3 solution)
4000
3000
2000
1600
1200
800
ν (cm-1)
Mass Spectrum
% of base peak
100
80
60
40
20
90
91
M+·
249/251/253
C6H5NBr2
40
80
120
160
200
240
280
m / e
0.0
0.5
1.0
1.5
absorbance
UV spectrum
0.245 mg / 10 mls
path length : 1.00 cm
solvent : ethanol
200
250
300
350
λ (nm)
13C NMR Spectrum
(20.0 MHz, CDCl3 solution)
off-resonance decoupled
proton decoupled
solvent
TMS
200
160
120
80
40
0
δ (ppm)
1H NMR Spectrum
(100 MHz, CDCl3 solution)
TMS
exchanges
with D2O
10
9
8
7
6
5
4
3
2
1
0
δ (ppm)

Problem 133

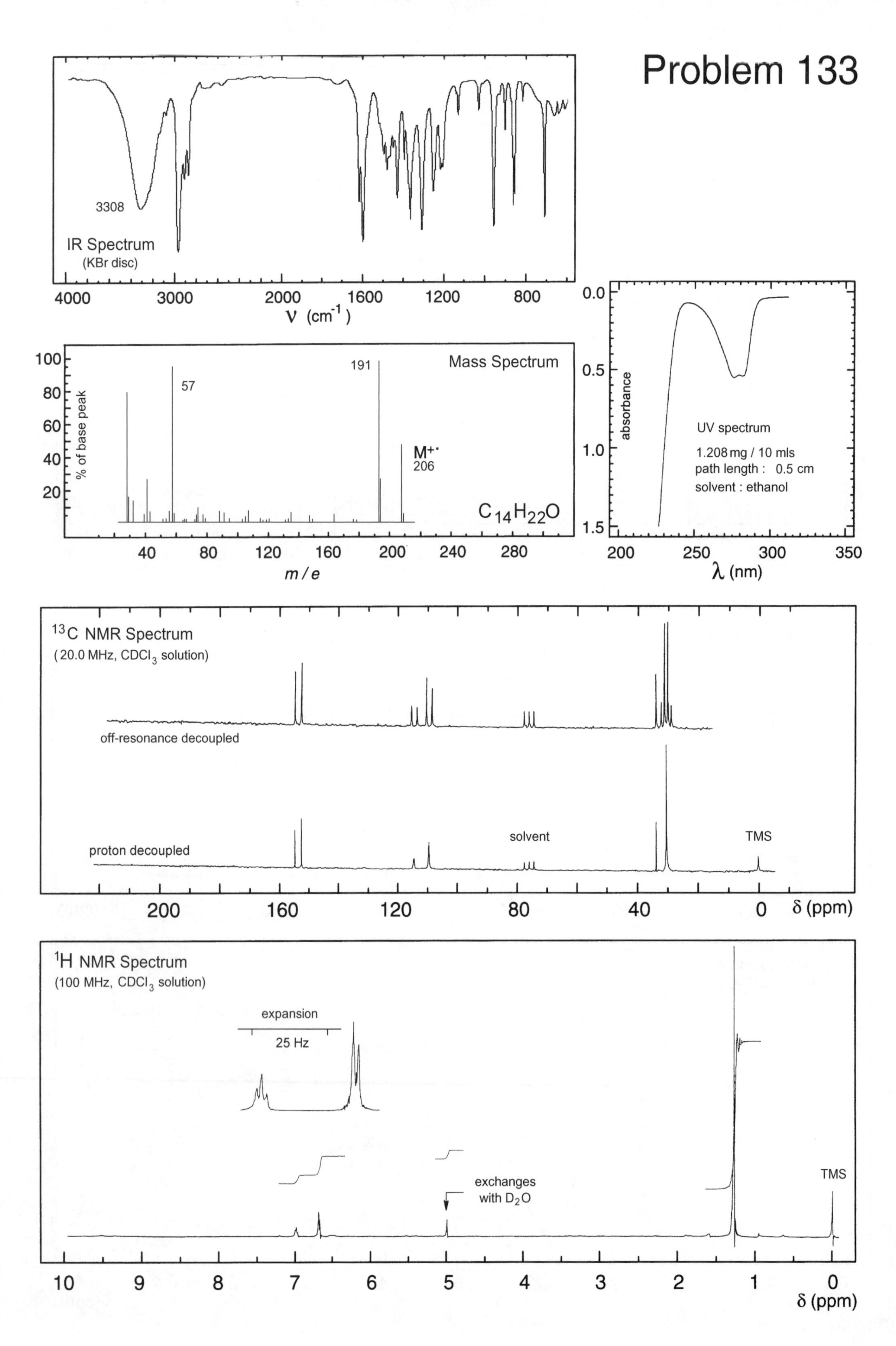

3308
IR Spectrum
(KBr disc)
4000
3000
2000
1600
1200
800
ν (cm⁻¹)
100
80
60
40
20
% of base peak
57
191
Mass Spectrum
M+·
206
C14H22O
40
80
120
160
200
240
280
m / e
0.0
0.5
1.0
1.5
absorbance
UV spectrum
1.208 mg / 10 mls
path length : 0.5 cm
solvent : ethanol
200
250
300
350
λ (nm)
13C NMR Spectrum
(20.0 MHz, CDCl3 solution)
off-resonance decoupled
proton decoupled
solvent
TMS
200
160
120
80
40
0
δ (ppm)
1H NMR Spectrum
(100 MHz, CDCl3 solution)
expansion
25 Hz
exchanges with D2O
TMS
10
9
8
7
6
5
4
3
2
1
0
δ (ppm)

Problem 134

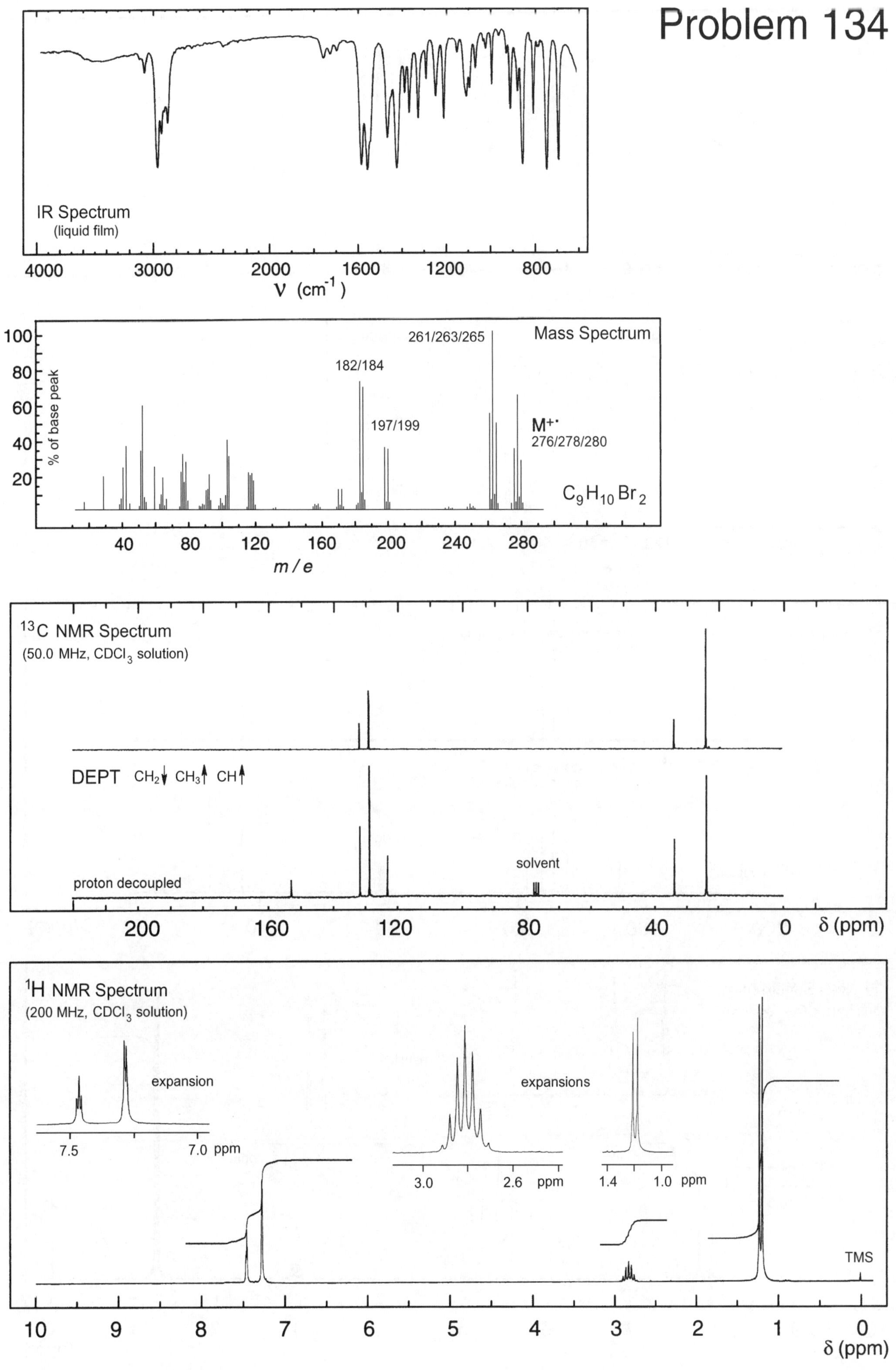

IR Spectrum
(liquid film)
4000
3000
2000
1600
1200
800
ν (cm^{-1})
Mass Spectrum
100
80
60
40
20
% of base peak
261/263/265
182/184
197/199
M$^{+\cdot}$
276/278/280
$C_9H_{10}Br_2$
40
80
120
160
200
240
280
m / e
^{13}C NMR Spectrum
(50.0 MHz, $CDCl_3$ solution)
DEPT CH$_2$↓ CH$_3$↑ CH↑
solvent
proton decoupled
200
160
120
80
40
0
δ (ppm)
^{1}H NMR Spectrum
(200 MHz, $CDCl_3$ solution)
expansion
7.5
7.0 ppm
expansions
3.0
2.6 ppm
1.4
1.0 ppm
TMS
10
9
8
7
6
5
4
3
2
1
0
δ (ppm)

Problem 135

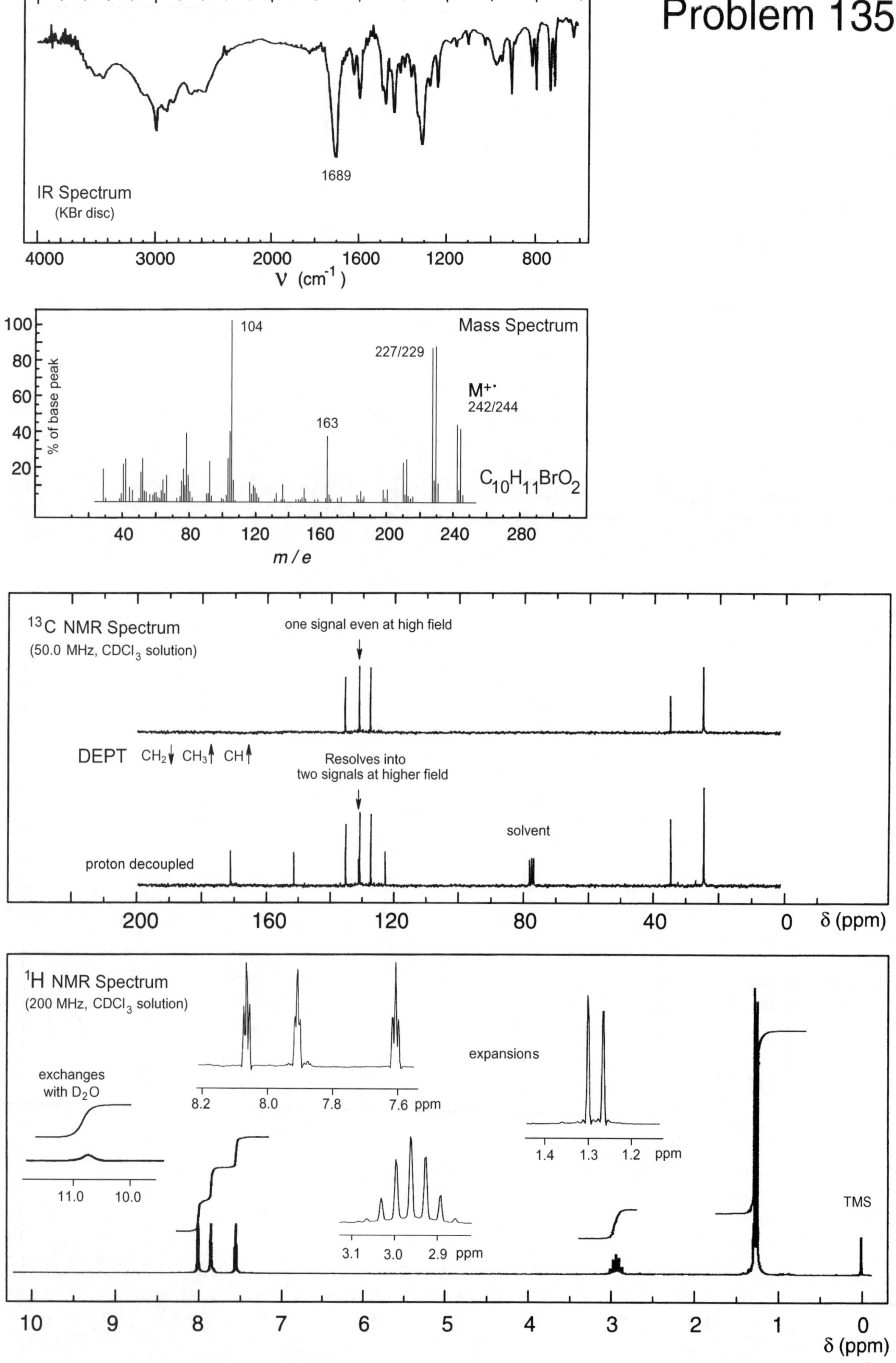

IR Spectrum
(KBr disc)
1689
4000
3000
2000
1600
1200
800
ν (cm⁻¹)
100
80
60
40
20
% of base peak
104
Mass Spectrum
227/229
M+·
242/244
163
C10H11BrO2
40
80
120
160
200
240
280
m / e
13C NMR Spectrum
(50.0 MHz, CDCl3 solution)
one signal even at high field
DEPT CH2↓ CH3↑ CH↑
Resolves into
two signals at higher field
solvent
proton decoupled
200
160
120
80
40
0
δ (ppm)
1H NMR Spectrum
(200 MHz, CDCl3 solution)
exchanges
with D2O
11.0
10.0
8.2
8.0
7.8
7.6 ppm
expansions
1.4
1.3
1.2
ppm
3.1
3.0
2.9 ppm
TMS
10
9
8
7
6
5
4
3
2
1
0
δ (ppm)

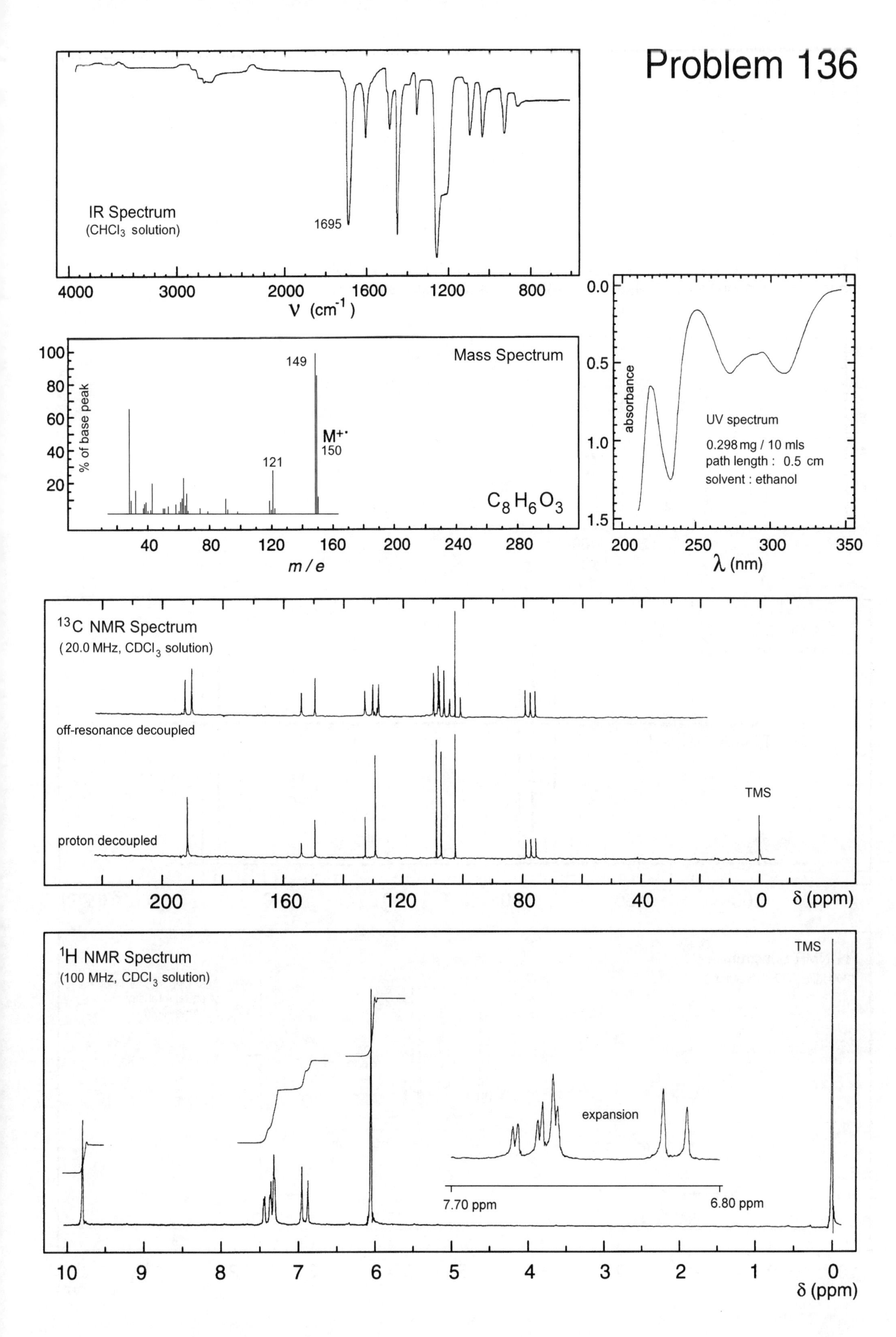
Problem 136
IR Spectrum
(CHCl3 solution)
1695
4000
3000
2000
1600
1200
800
ν (cm-1)
Mass Spectrum
100
80
60
40
20
% of base peak
149
M+·
150
121
C8 H6 O3
40
80
120
160
200
240
280
m / e
0.0
0.5
1.0
1.5
absorbance
UV spectrum
0.298 mg / 10 mls
path length : 0.5 cm
solvent : ethanol
200
250
300
350
λ (nm)
13C NMR Spectrum
(20.0 MHz, CDCl3 solution)
off-resonance decoupled
proton decoupled
TMS
200
160
120
80
40
0
δ (ppm)
1H NMR Spectrum
(100 MHz, CDCl3 solution)
TMS
expansion
7.70 ppm
6.80 ppm
10
9
8
7
6
5
4
3
2
1
0
δ (ppm)

Problem 137

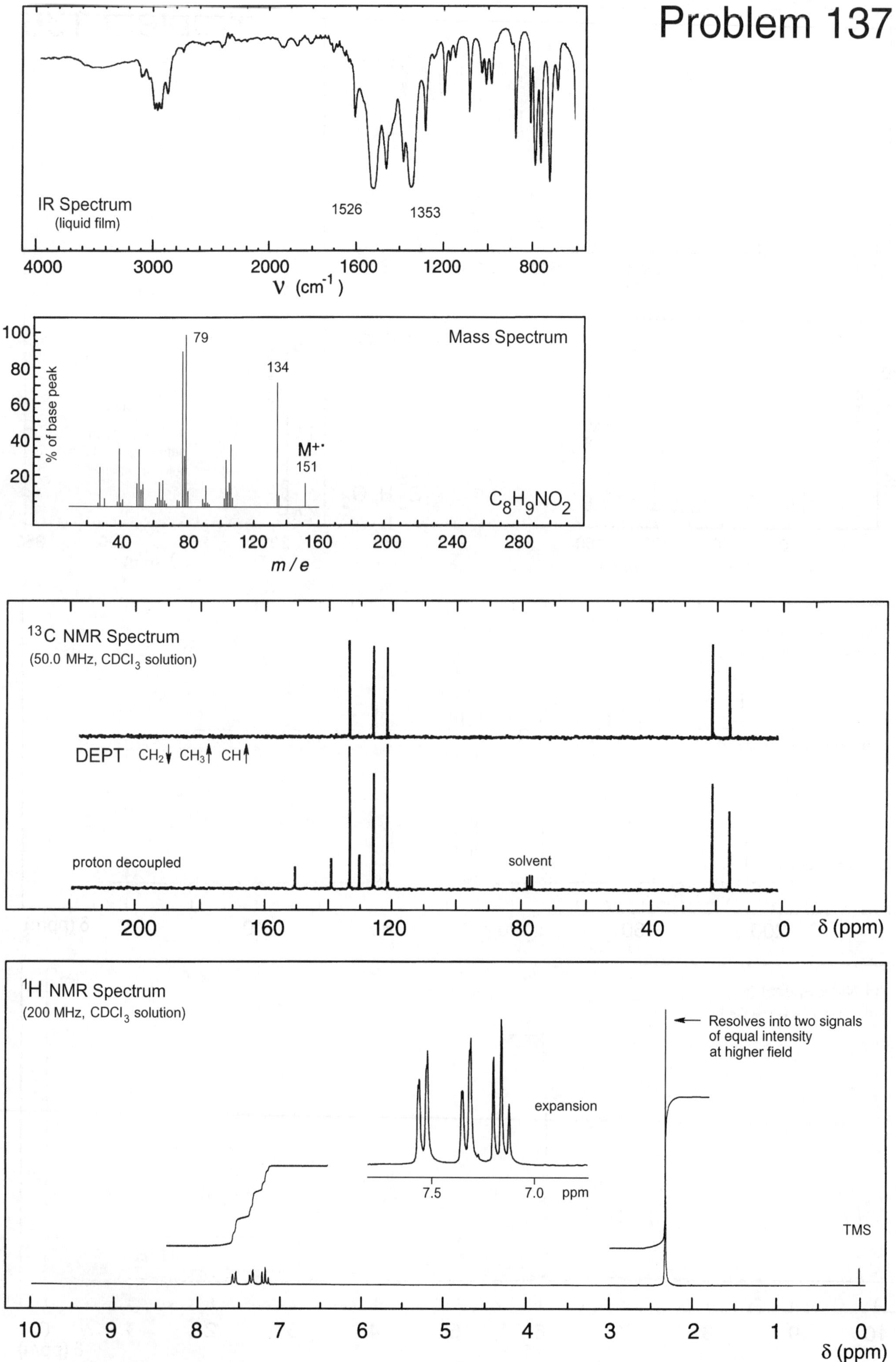

IR Spectrum
(liquid film)
1526
1353
4000
3000
2000
1600
1200
800
ν (cm⁻¹)
Mass Spectrum
% of base peak
100
80
60
40
20
79
134
M+·
151
C8H9NO2
40
80
120
160
200
240
280
m / e
13C NMR Spectrum
(50.0 MHz, CDCl3 solution)
DEPT CH2↓ CH3↑ CH↑
proton decoupled
solvent
200
160
120
80
40
0
δ (ppm)
1H NMR Spectrum
(200 MHz, CDCl3 solution)
Resolves into two signals of equal intensity at higher field
expansion
7.5
7.0
ppm
TMS
10
9
8
7
6
5
4
3
2
1
0
δ (ppm)

Problem 138

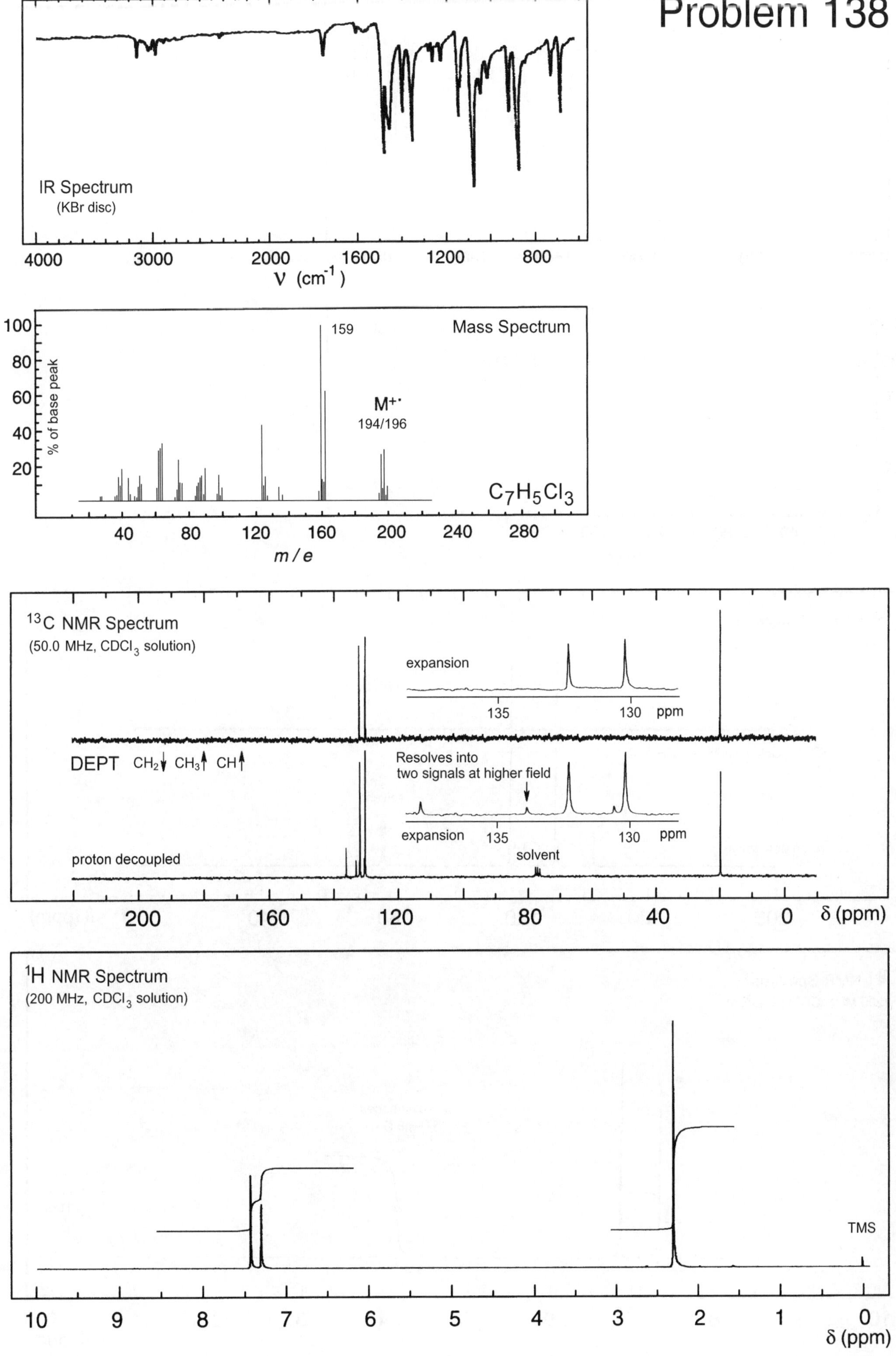

IR Spectrum
(KBr disc)
4000
3000
2000
1600
1200
800
ν (cm⁻¹)
Mass Spectrum
100
80
60
40
20
% of base peak
159
M+·
194/196
C7H5Cl3
40
80
120
160
200
240
280
m / e
13C NMR Spectrum
(50.0 MHz, CDCl3 solution)
expansion
135
130
ppm
DEPT CH2↓ CH3↑ CH↑
Resolves into
two signals at higher field
expansion
135
130
ppm
solvent
proton decoupled
200
160
120
80
40
0
δ (ppm)
1H NMR Spectrum
(200 MHz, CDCl3 solution)
TMS
10
9
8
7
6
5
4
3
2
1
0
δ (ppm)

Problem 139

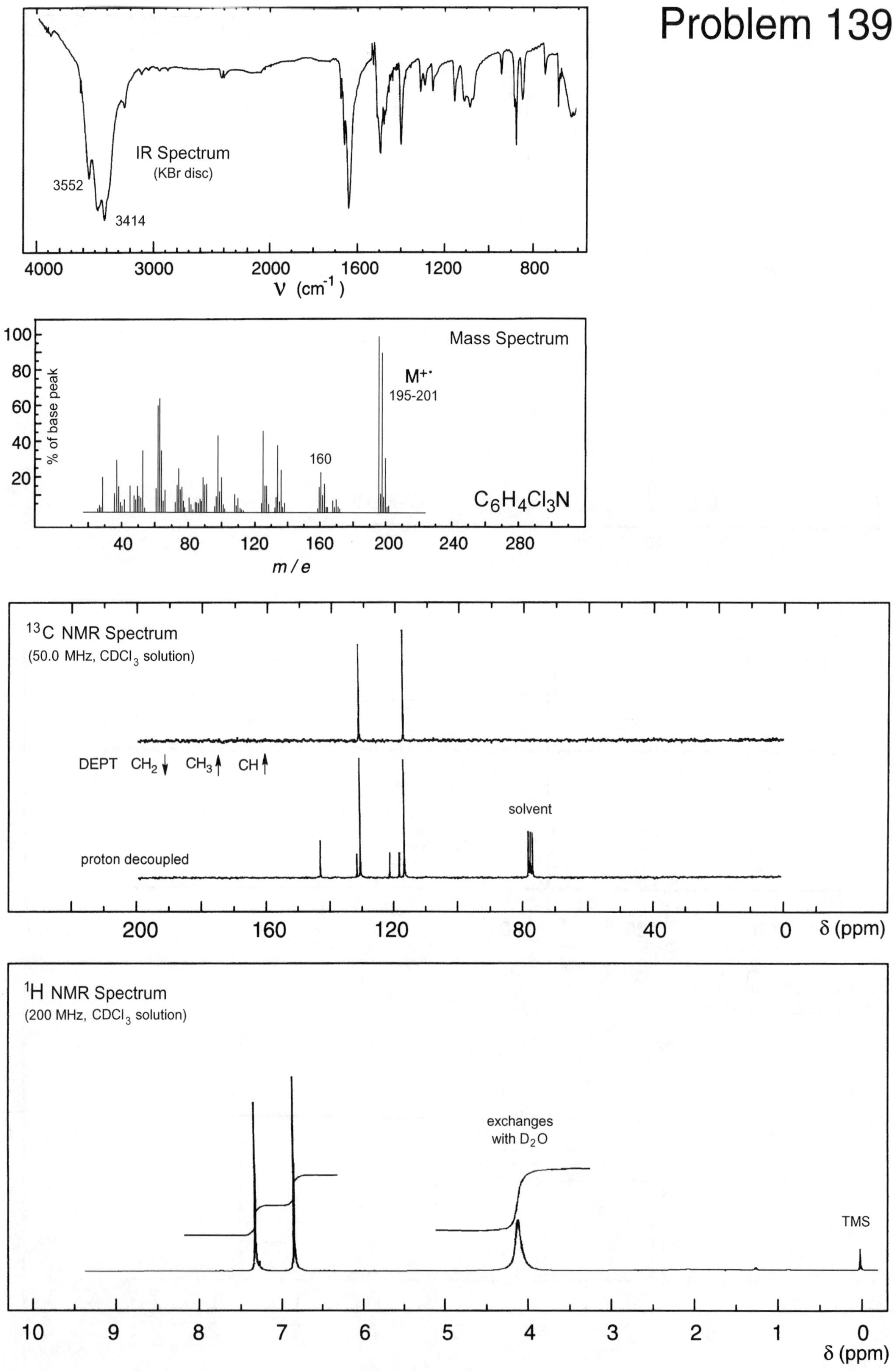

IR Spectrum
(KBr disc)
3552
3414
4000
3000
2000
1600
1200
800
ν (cm⁻¹)
Mass Spectrum
100
80
60
40
20
% of base peak
M+·
195-201
160
$C_6H_4Cl_3N$
40
80
120
160
200
240
280
m / e
13C NMR Spectrum
(50.0 MHz, CDCl3 solution)
DEPT CH2 ↓ CH3 ↑ CH ↑
solvent
proton decoupled
200
160
120
80
40
0
δ (ppm)
1H NMR Spectrum
(200 MHz, CDCl3 solution)
exchanges
with D2O
TMS
10
9
8
7
6
5
4
3
2
1
0
δ (ppm)

Problem 140

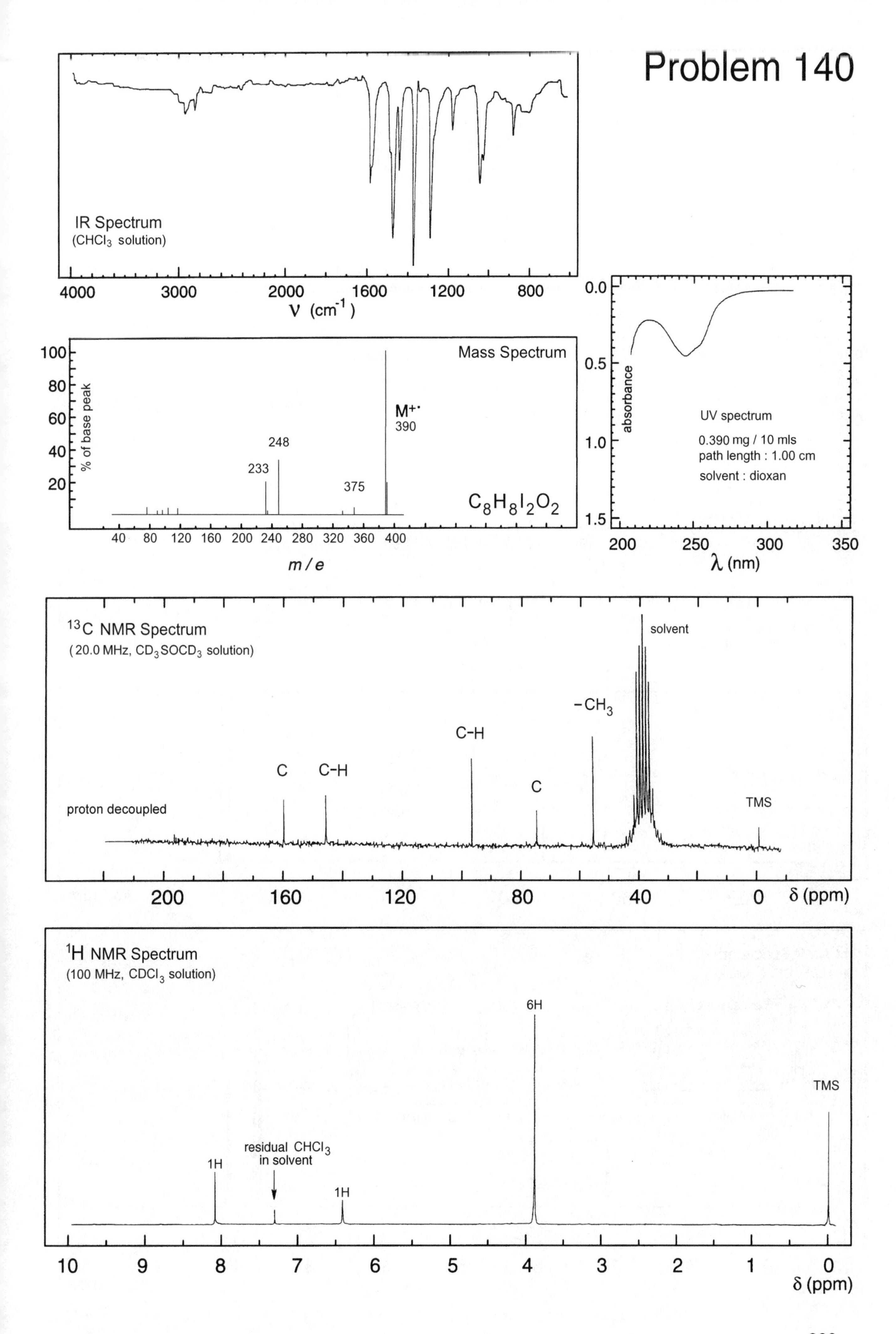

IR Spectrum
(CHCl3 solution)
4000
3000
2000
1600
1200
800
ν (cm-1)
Mass Spectrum
100
80
60
40
20
% of base peak
M+·
390
248
233
375
C8H8I2O2
40
80
120
160
200
240
280
320
360
400
m / e
0.0
0.5
1.0
1.5
absorbance
UV spectrum
0.390 mg / 10 mls
path length : 1.00 cm
solvent : dioxan
200
250
300
350
λ (nm)
13C NMR Spectrum
(20.0 MHz, CD3SOCD3 solution)
solvent
-CH3
C-H
C
C-H
C
proton decoupled
TMS
200
160
120
80
40
0
δ (ppm)
1H NMR Spectrum
(100 MHz, CDCl3 solution)
6H
TMS
residual CHCl3
in solvent
1H
1H
10
9
8
7
6
5
4
3
2
1
0
δ (ppm)

Problem 141

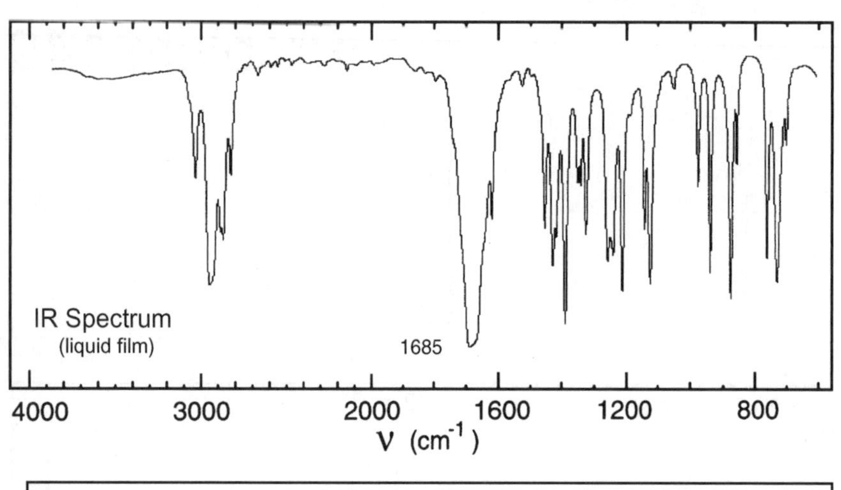

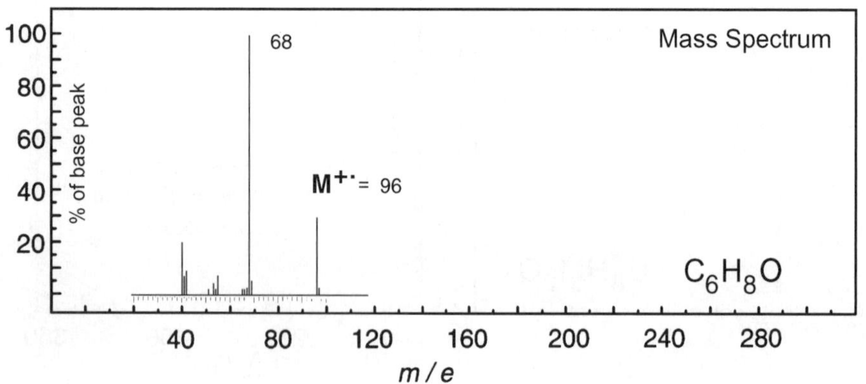

UV Spectrum

λ_{max} 225 nm ($\log_{10}\varepsilon$ 3.9)

solvent : methanol

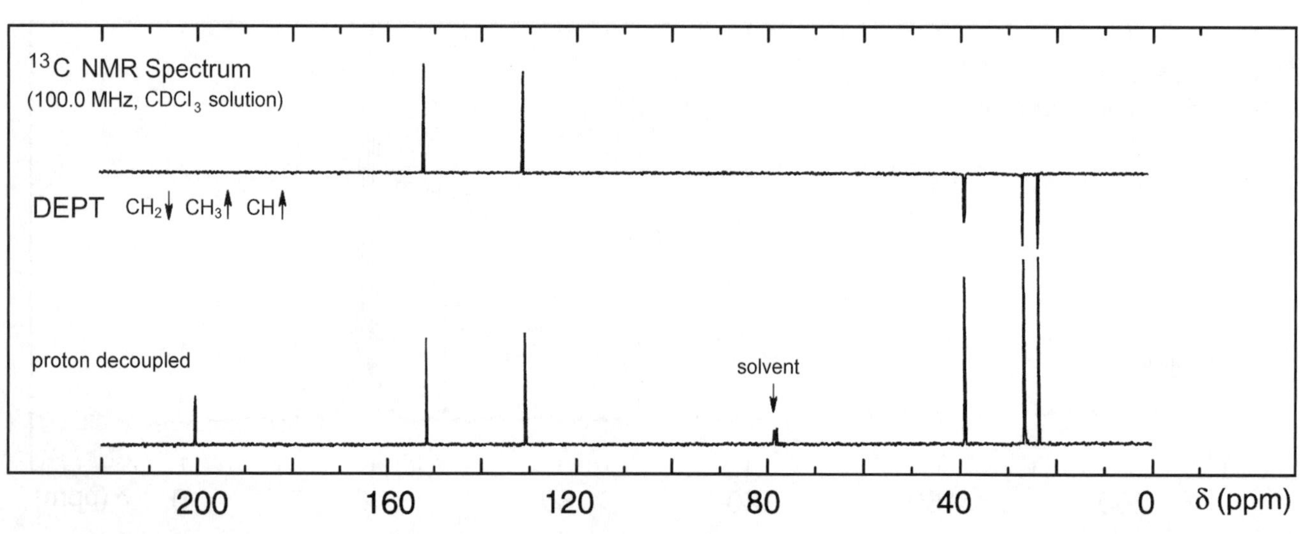

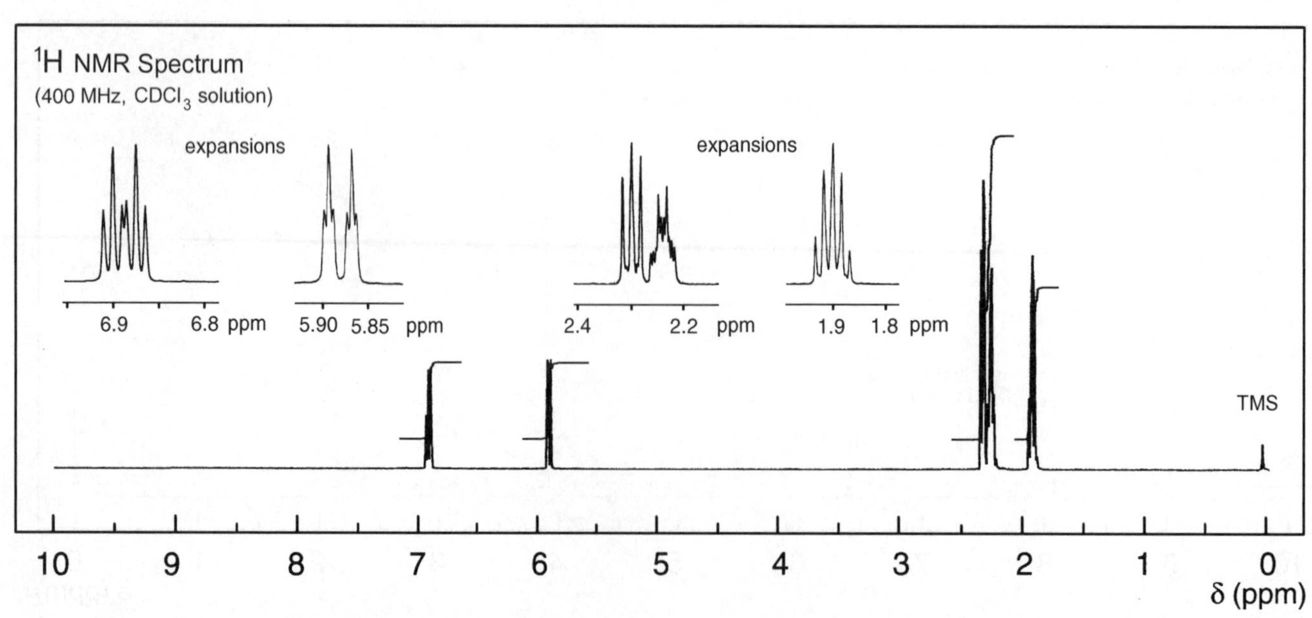

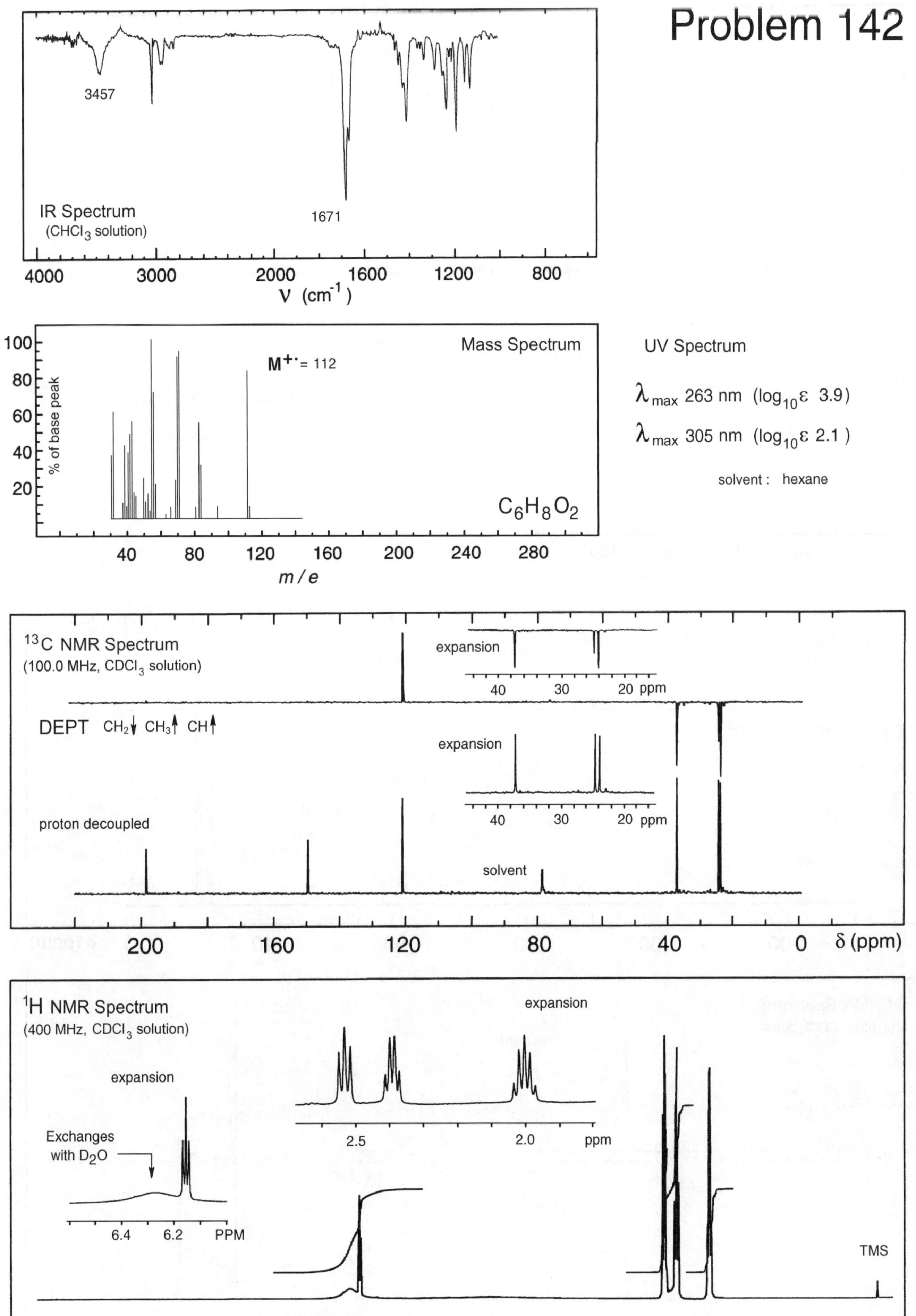
Problem 142
IR Spectrum
(CHCl3 solution)
3457
1671
4000 3000 2000 1600 1200 800
ν (cm-1)
Mass Spectrum
M+· = 112
% of base peak
100 80 60 40 20
40 80 120 160 200 240 280
m / e
C6H8O2
UV Spectrum
λmax 263 nm (log10ε 3.9)
λmax 305 nm (log10ε 2.1)
solvent : hexane
13C NMR Spectrum
(100.0 MHz, CDCl3 solution)
expansion
40 30 20 ppm
DEPT CH2↓ CH3↑ CH↑
expansion
40 30 20 ppm
proton decoupled
solvent
200 160 120 80 40 0 δ (ppm)
1H NMR Spectrum
(400 MHz, CDCl3 solution)
expansion
Exchanges with D2O
6.4 6.2 PPM
expansion
2.5 2.0 ppm
TMS
10 9 8 7 6 5 4 3 2 1 0
δ (ppm)

Problem 143

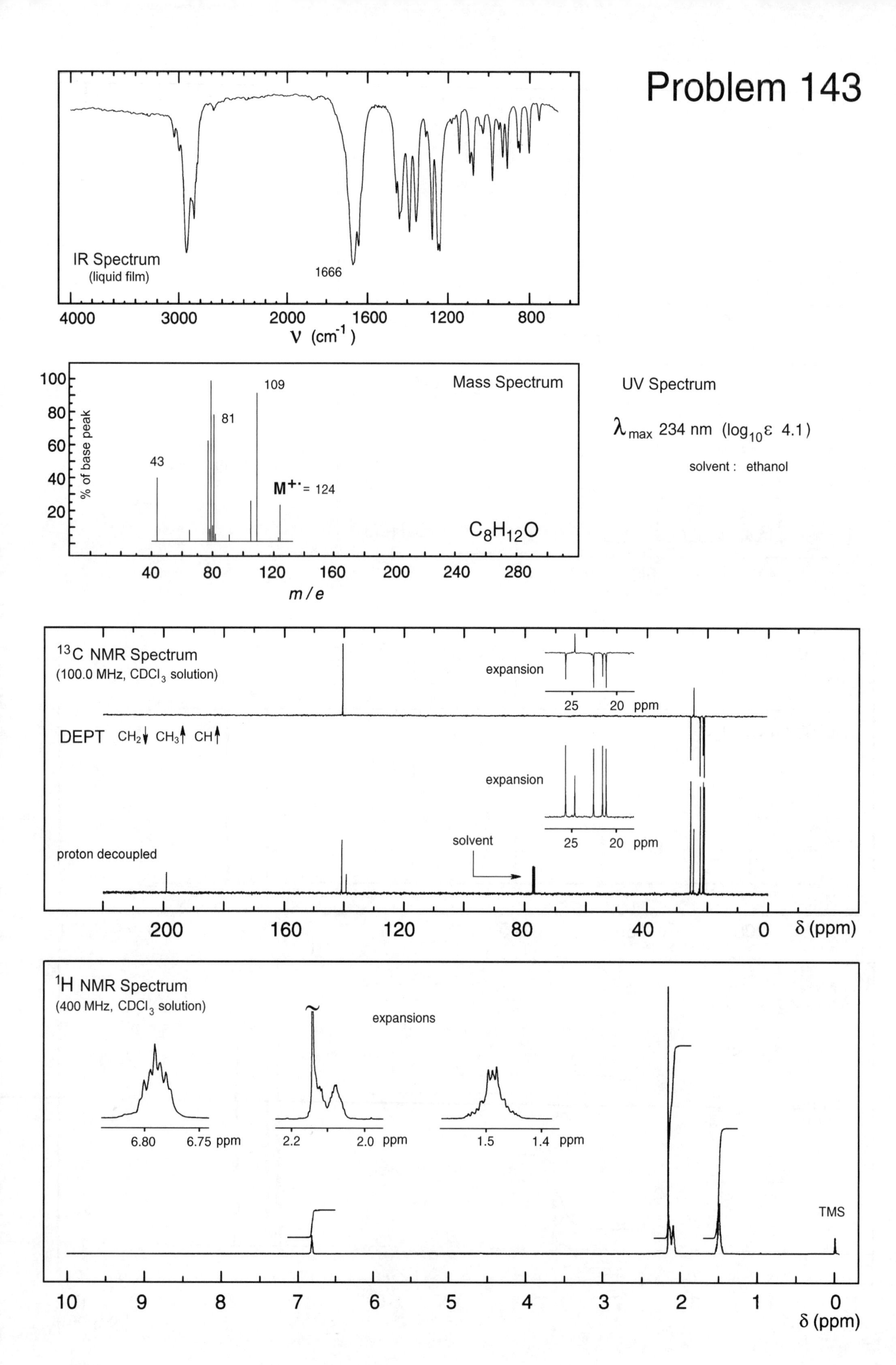

IR Spectrum
(liquid film)
1666
4000
3000
2000
1600
1200
800
ν (cm⁻¹)
100
80
60
40
20
% of base peak
109
81
43
M⁺· = 124
Mass Spectrum
C8H12O
40
80
120
160
200
240
280
m / e
UV Spectrum
λmax 234 nm (log10ε 4.1)
solvent : ethanol
13C NMR Spectrum
(100.0 MHz, CDCl3 solution)
expansion
25
20
ppm
DEPT CH2↓ CH3↑ CH↑
expansion
25
20
ppm
solvent
proton decoupled
200
160
120
80
40
0
δ (ppm)
1H NMR Spectrum
(400 MHz, CDCl3 solution)
expansions
6.80
6.75
ppm
2.2
2.0
ppm
1.5
1.4
ppm
TMS
10
9
8
7
6
5
4
3
2
1
0
δ (ppm)

Problem 144

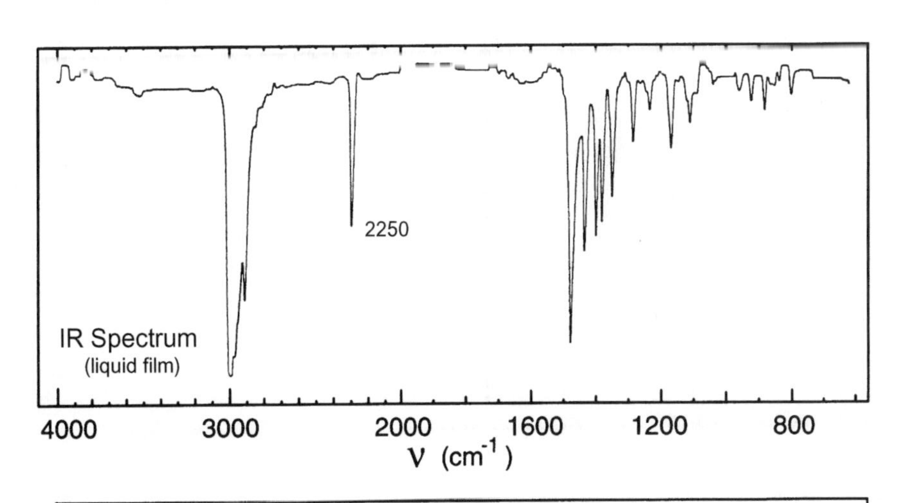

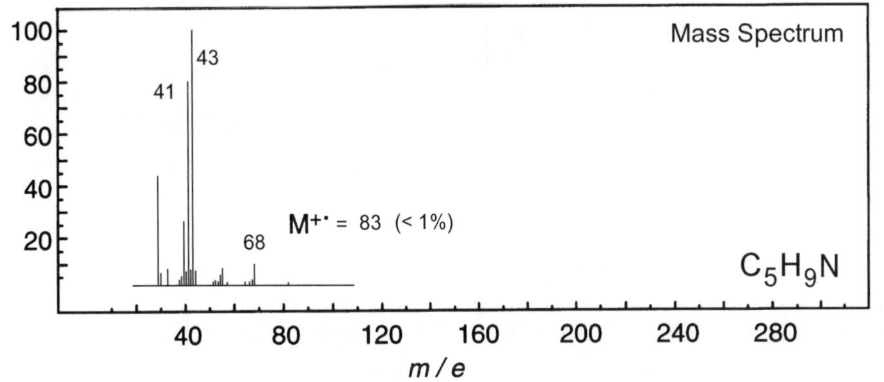

No significant UV absorption above 220 nm

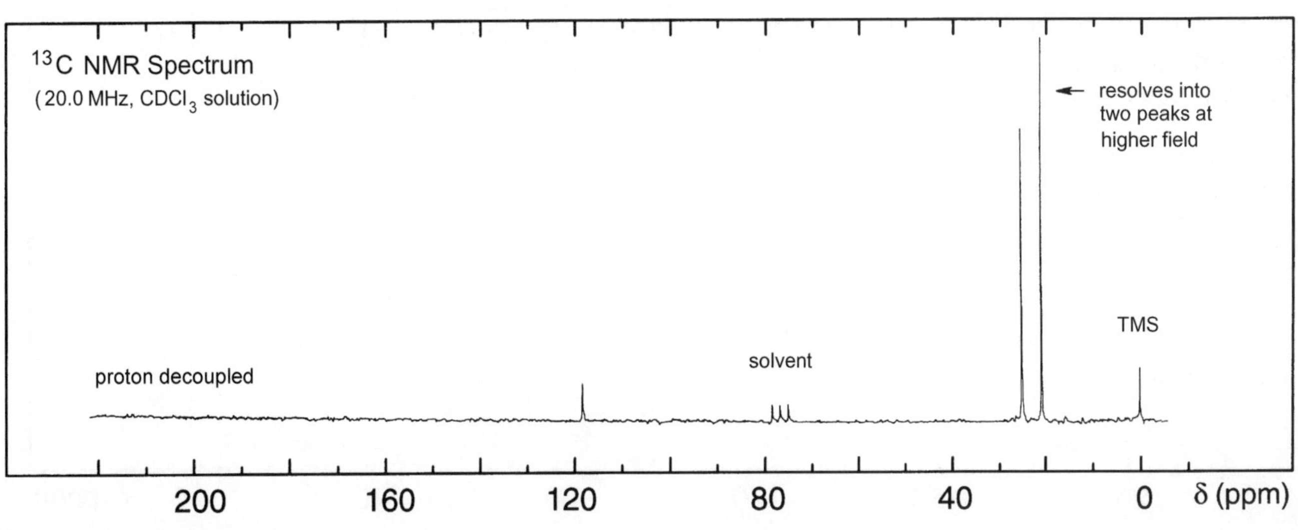

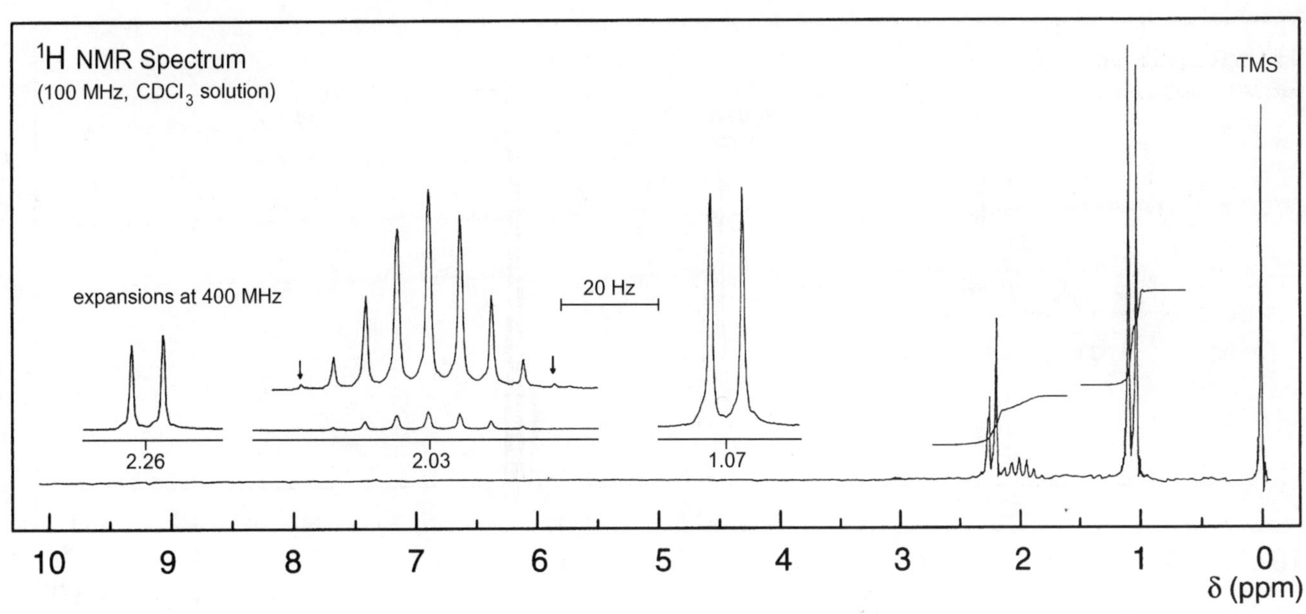

Problem 145

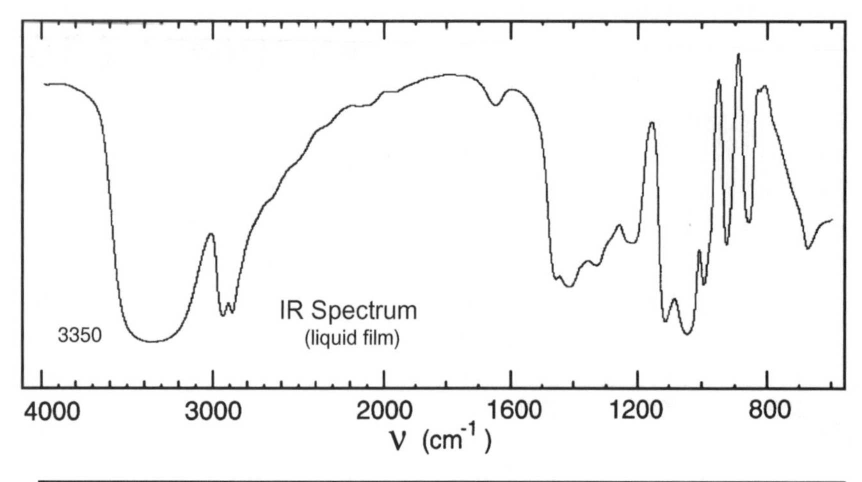

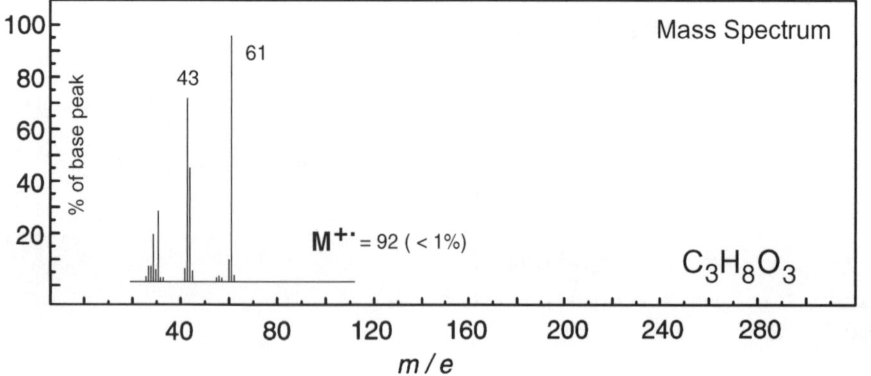

No significant UV absorption above 220 nm

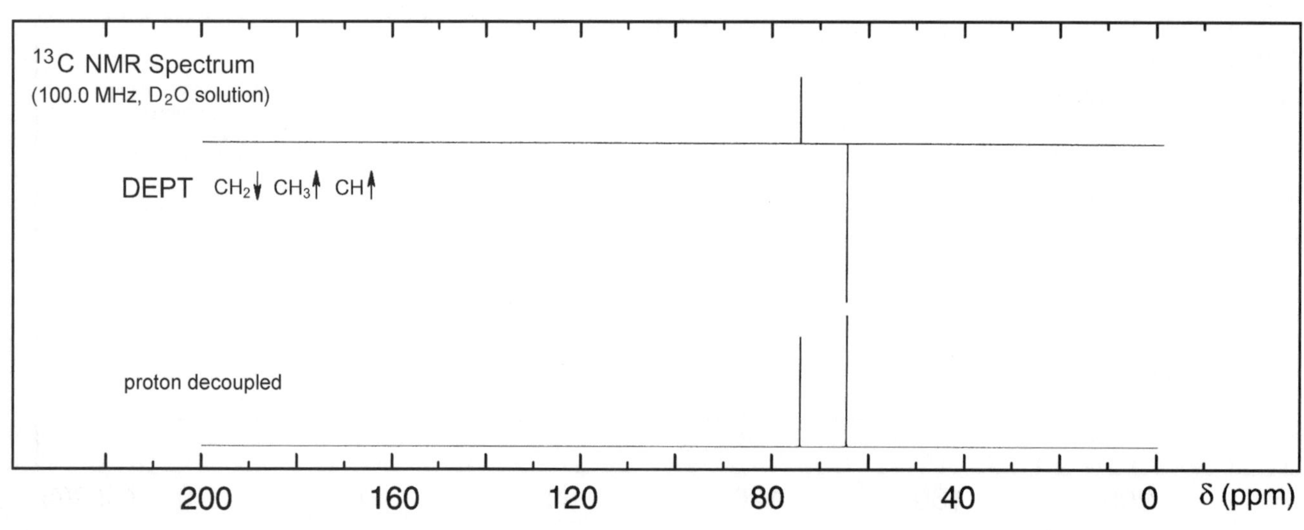

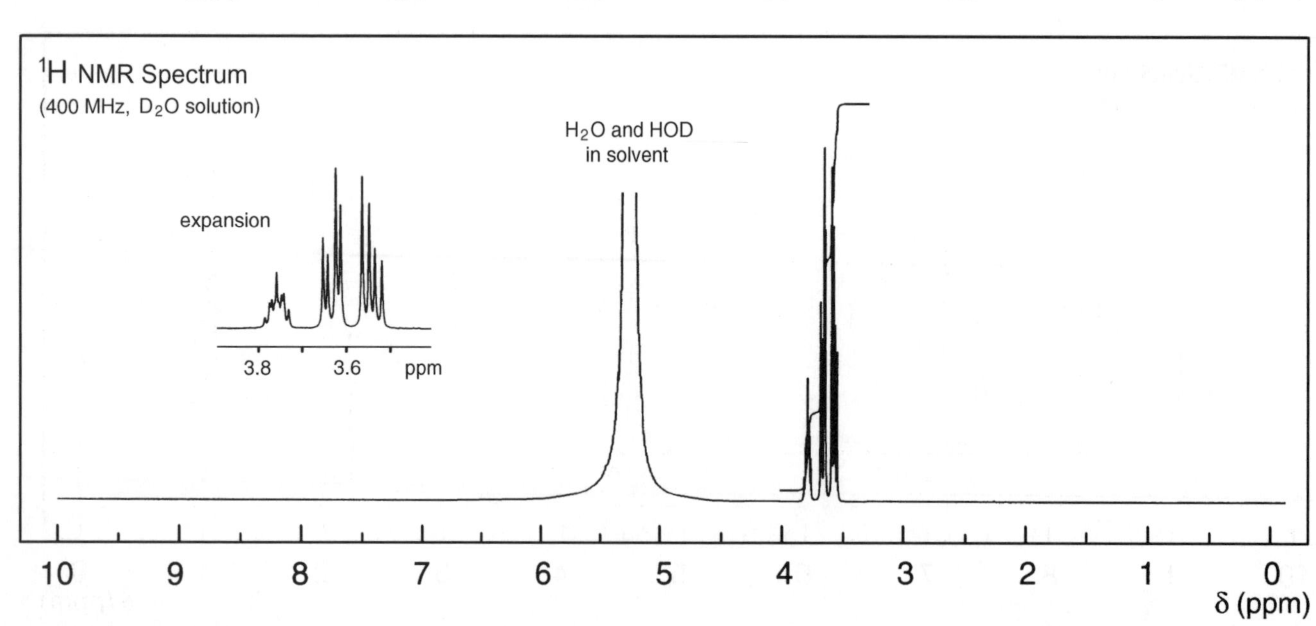

Problem 146

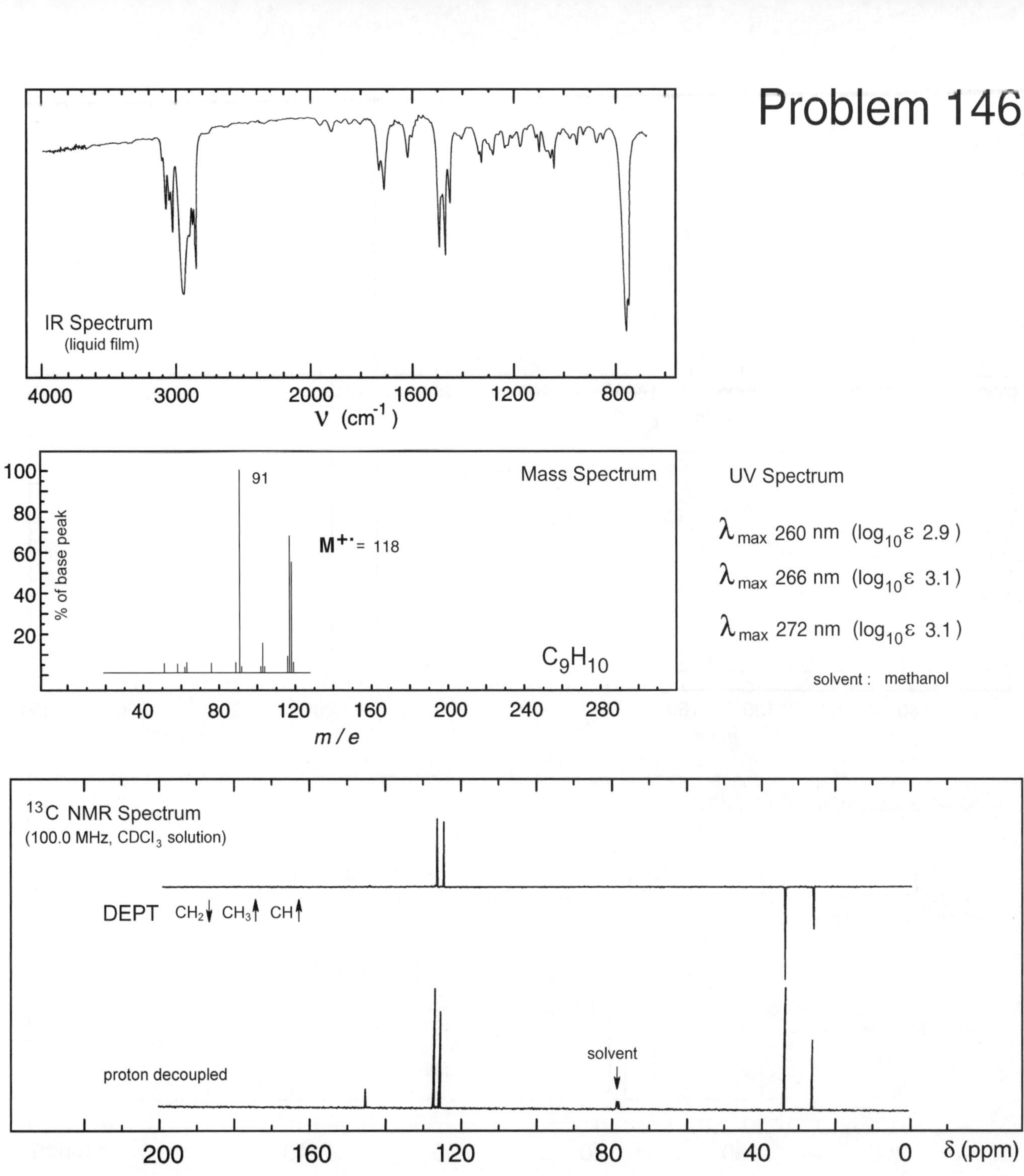

IR Spectrum
(liquid film)
4000
3000
2000
1600
1200
800
ν (cm⁻¹)
100
80
60
40
20
% of base peak
91
Mass Spectrum
M⁺· = 118
C9H10
40
80
120
160
200
240
280
m / e
UV Spectrum
λmax 260 nm (log10ε 2.9)
λmax 266 nm (log10ε 3.1)
λmax 272 nm (log10ε 3.1)
solvent : methanol
13C NMR Spectrum
(100.0 MHz, CDCl3 solution)
DEPT CH2↓ CH3↑ CH↑
proton decoupled
solvent
200
160
120
80
40
0
δ (ppm)

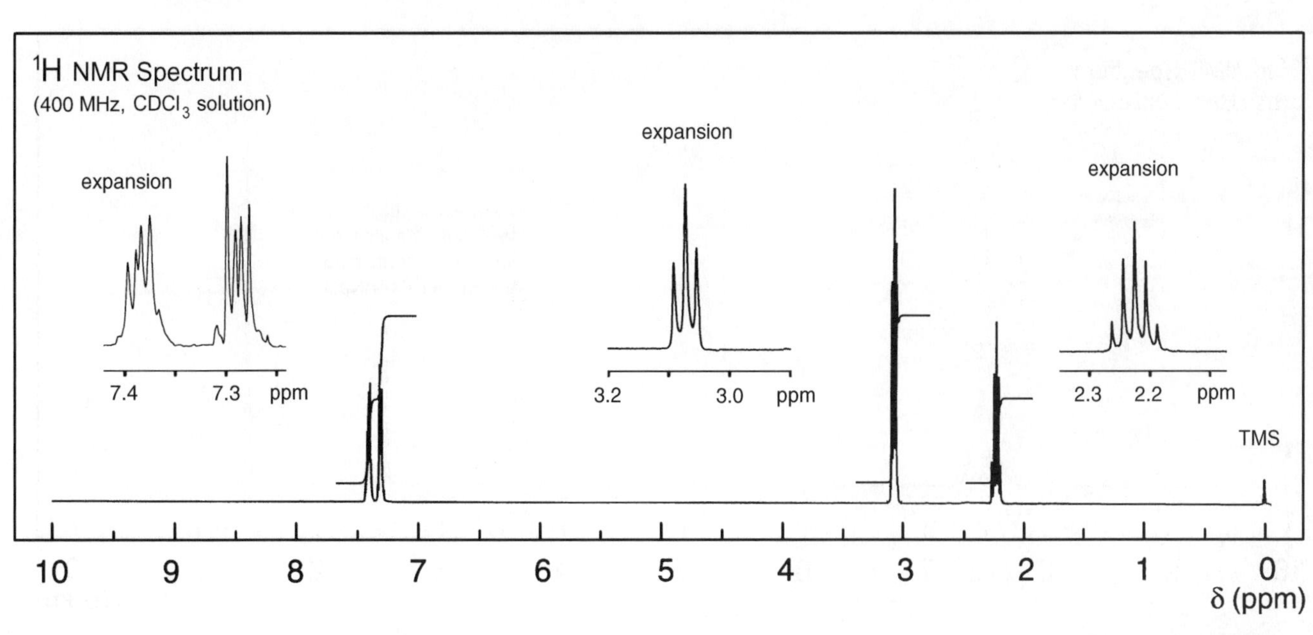

1H NMR Spectrum
(400 MHz, CDCl3 solution)
expansion
7.4
7.3
ppm
expansion
3.2
3.0
ppm
expansion
2.3
2.2
ppm
TMS
10
9
8
7
6
5
4
3
2
1
0
δ (ppm)

Problem 147

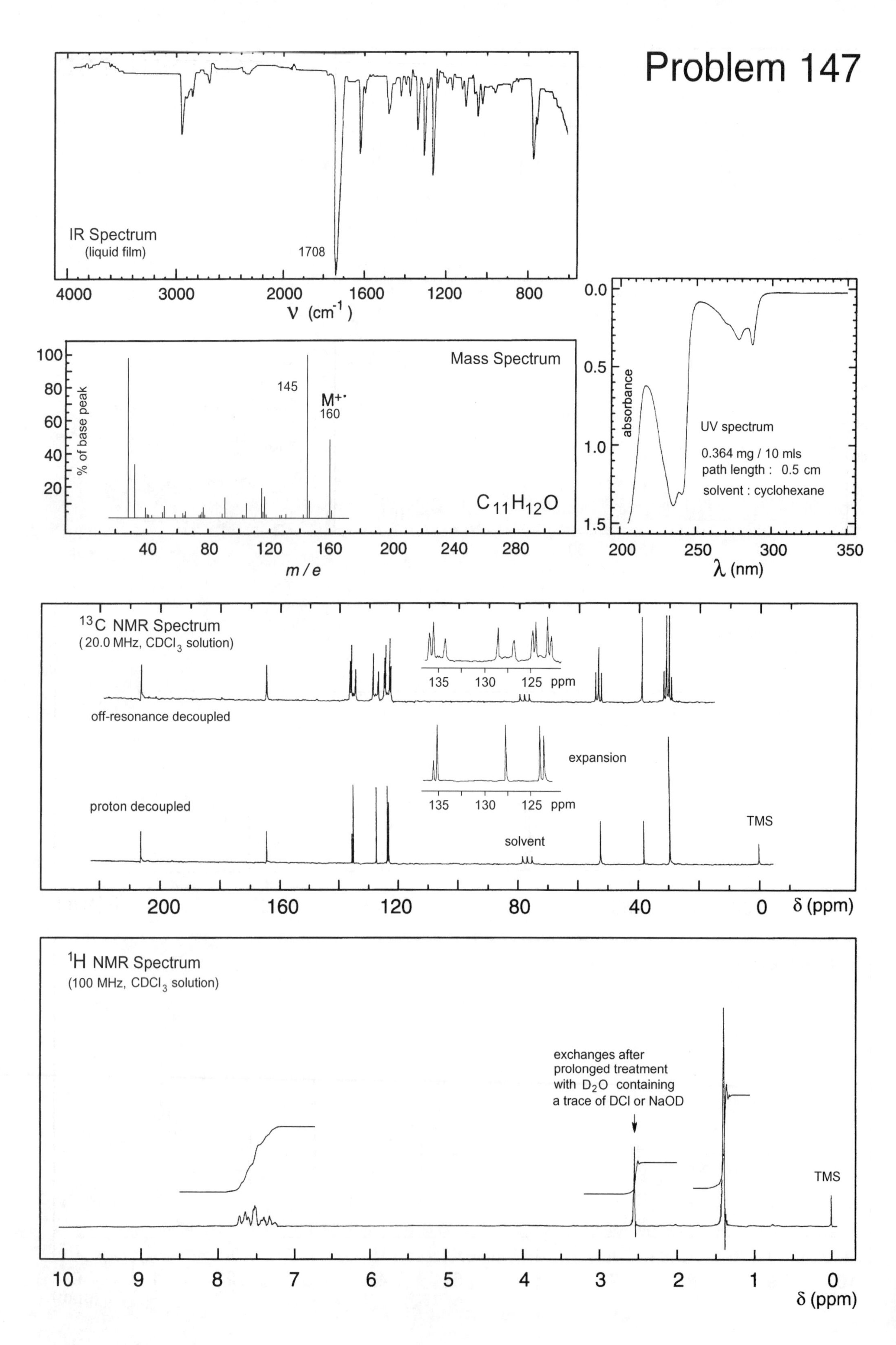

IR Spectrum
(liquid film)
1708
4000
3000
2000
1600
1200
800
ν (cm^{-1})
Mass Spectrum
100
80
60
40
20
% of base peak
145
M+·
160
$C_{11}H_{12}O$
40
80
120
160
200
240
280
m / e
0.0
0.5
1.0
1.5
absorbance
UV spectrum
0.364 mg / 10 mls
path length : 0.5 cm
solvent : cyclohexane
200
250
300
350
λ (nm)
^{13}C NMR Spectrum
(20.0 MHz, $CDCl_3$ solution)
135
130
125 ppm
off-resonance decoupled
expansion
135
130
125 ppm
proton decoupled
solvent
TMS
200
160
120
80
40
0
δ (ppm)
^{1}H NMR Spectrum
(100 MHz, $CDCl_3$ solution)
exchanges after
prolonged treatment
with D_2O containing
a trace of DCl or NaOD
TMS
10
9
8
7
6
5
4
3
2
1
0
δ (ppm)

Problem 148

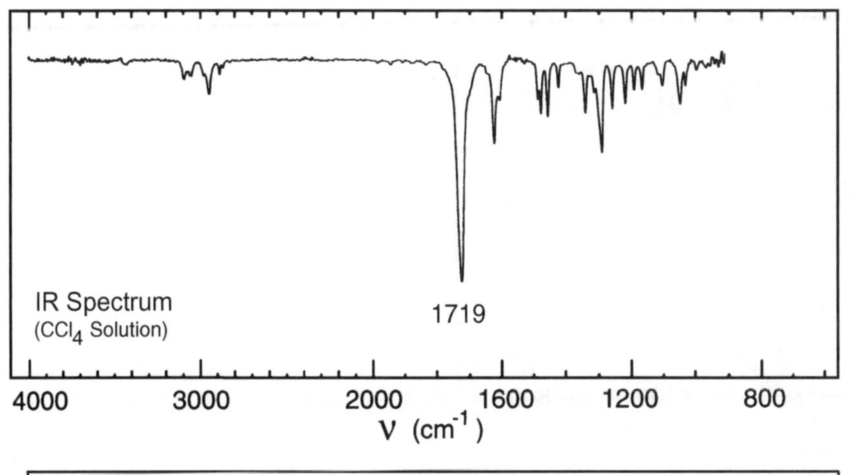

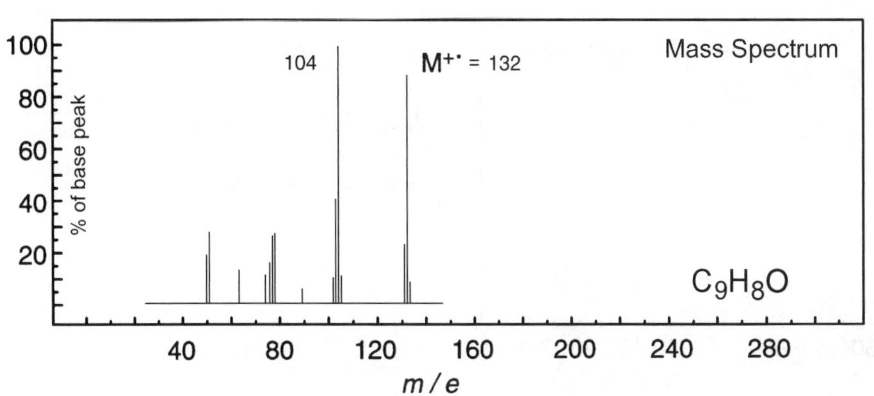

UV Spectrum

λ_{max} = 243 nm ($\log_{10}\varepsilon$ 4.1)

λ_{max} = 291 nm ($\log_{10}\varepsilon$ 3.4)

solvent : ethanol

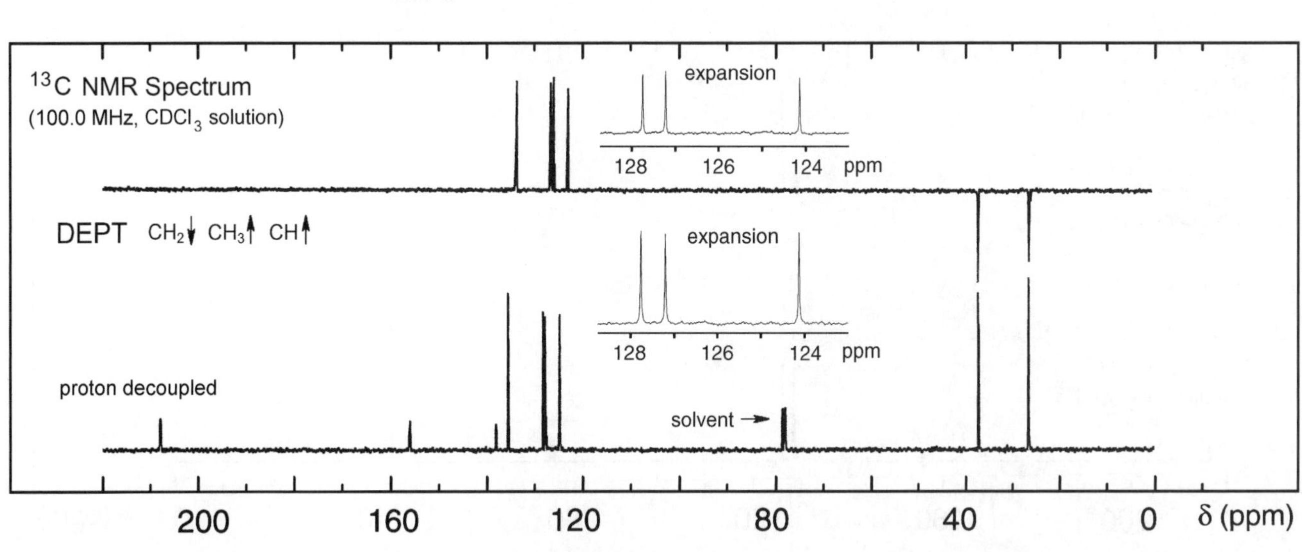

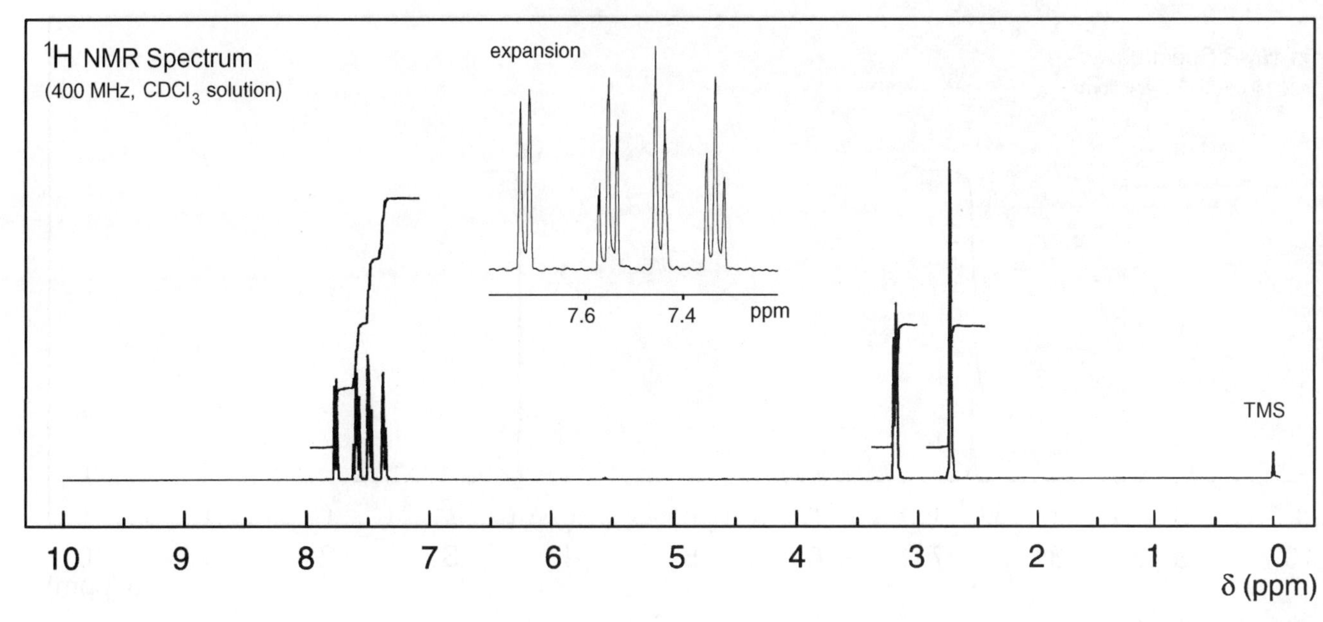

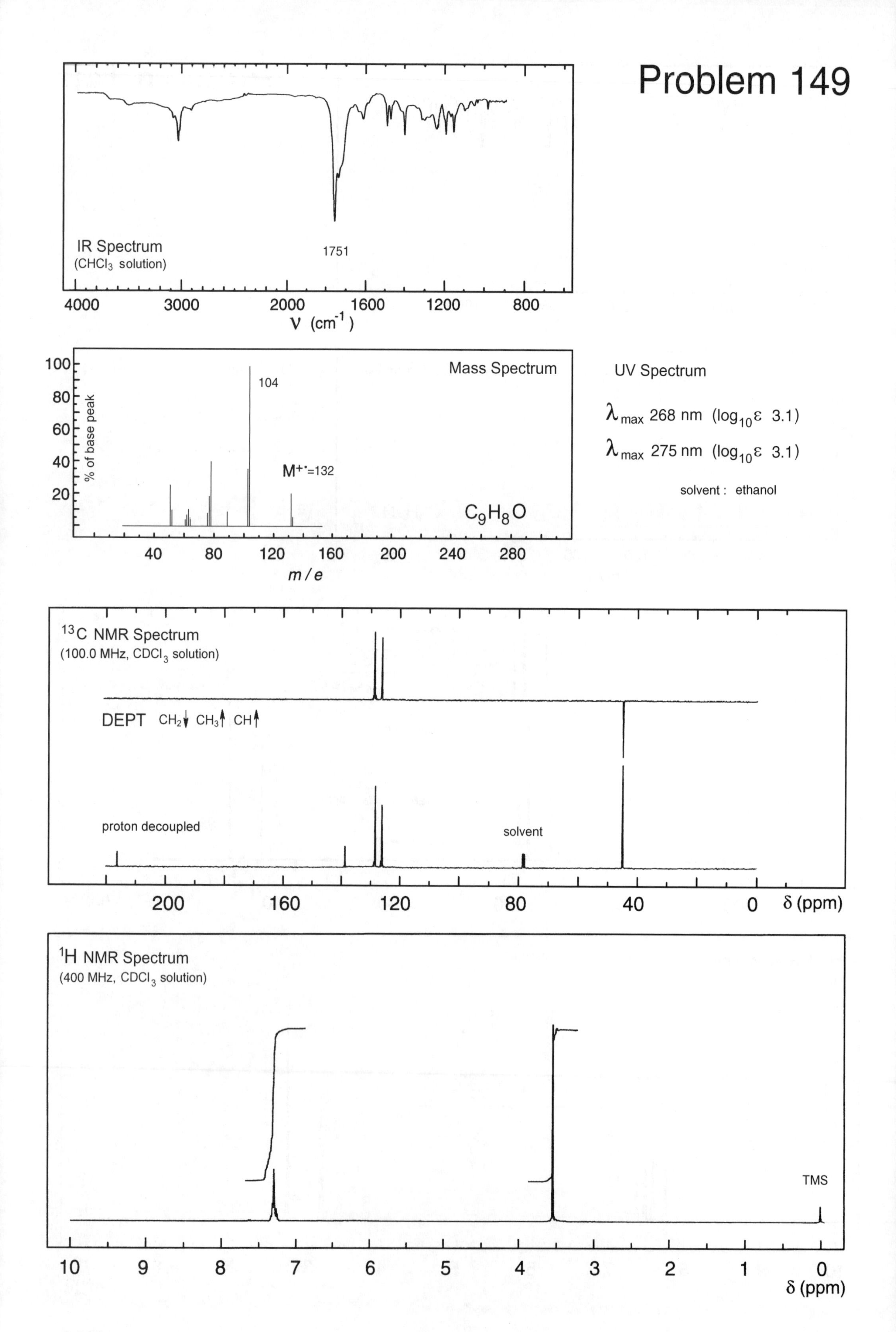

Problem 149
IR Spectrum
(CHCl3 solution)
1751
4000
3000
2000
1600
1200
800
ν (cm-1)
Mass Spectrum
104
M+•=132
C9H8O
% of base peak
m / e
UV Spectrum
λmax 268 nm (log10ε 3.1)
λmax 275 nm (log10ε 3.1)
solvent : ethanol
13C NMR Spectrum
(100.0 MHz, CDCl3 solution)
DEPT CH2↓ CH3↑ CH↑
proton decoupled
solvent
δ (ppm)
1H NMR Spectrum
(400 MHz, CDCl3 solution)
TMS
δ (ppm)

Problem 150

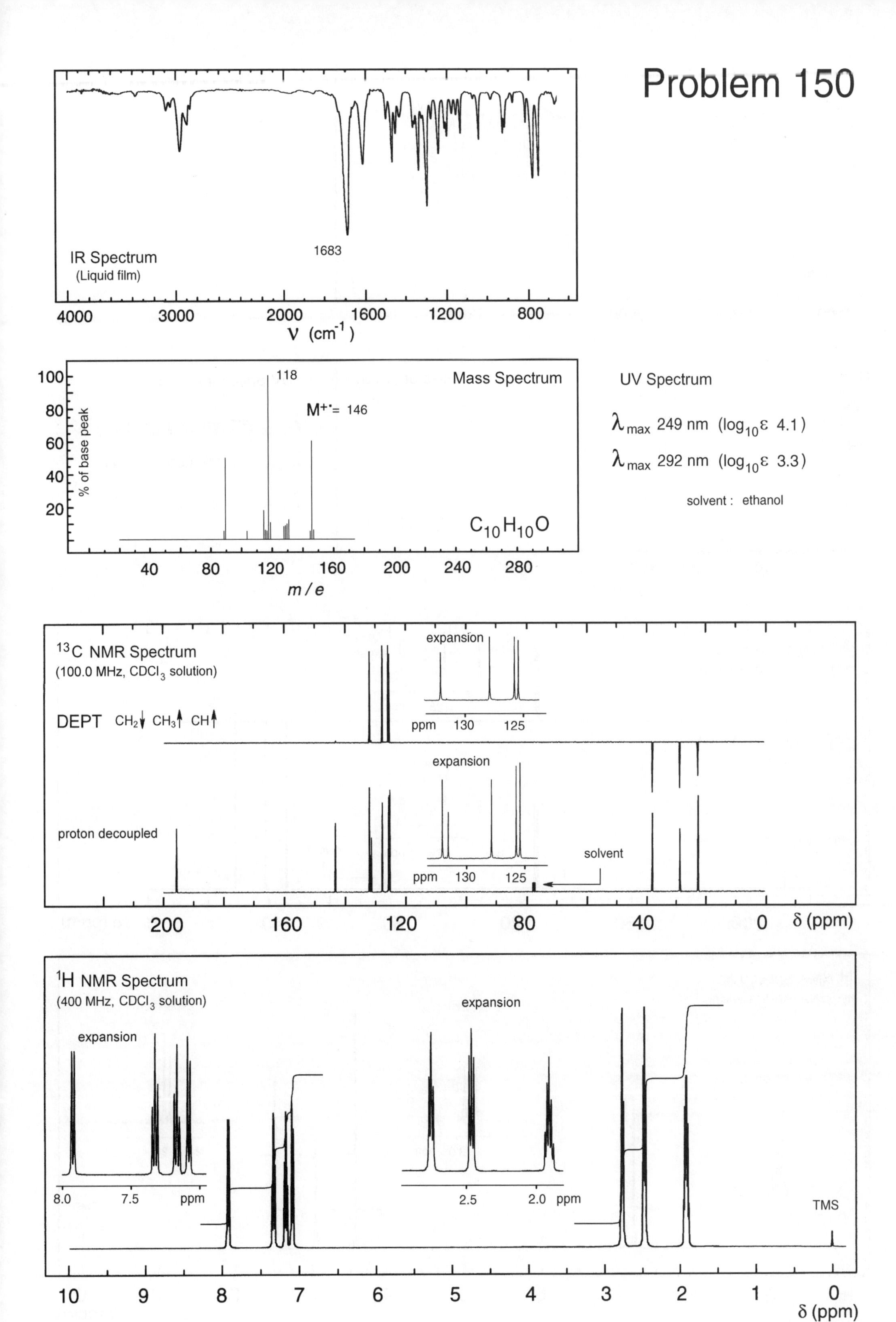

UV Spectrum

λ_{max} 249 nm ($\log_{10}\varepsilon$ 4.1)

λ_{max} 292 nm ($\log_{10}\varepsilon$ 3.3)

solvent : ethanol

Problem 151

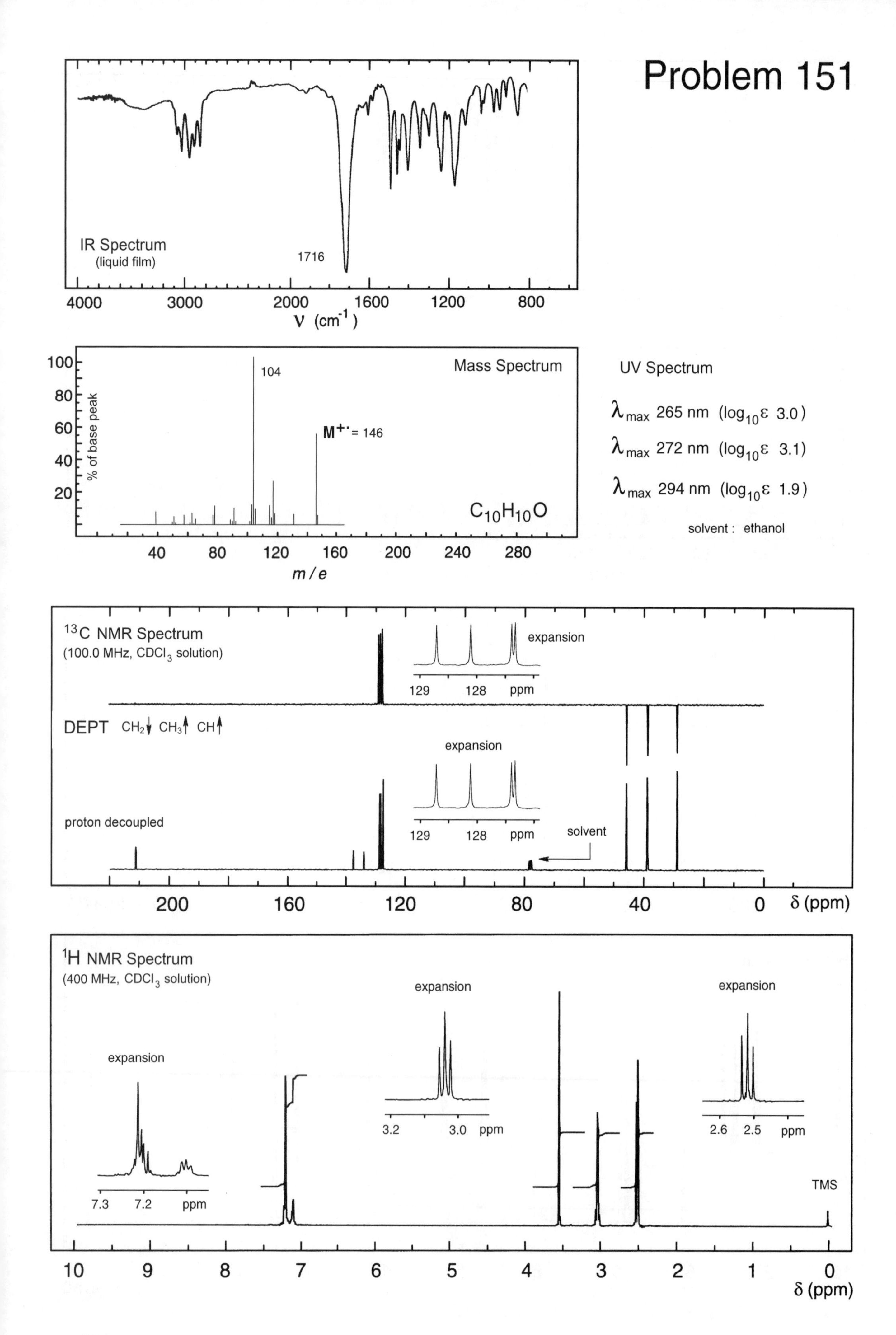

IR Spectrum
(liquid film)
1716
4000
3000
2000
1600
1200
800
ν (cm⁻¹)
Mass Spectrum
% of base peak
104
M⁺˙ = 146
C₁₀H₁₀O
m / e
UV Spectrum
λmax 265 nm (log10ε 3.0)
λmax 272 nm (log10ε 3.1)
λmax 294 nm (log10ε 1.9)
solvent : ethanol
13C NMR Spectrum
(100.0 MHz, CDCl3 solution)
expansion
129
128
ppm
DEPT CH2 CH3 CH
proton decoupled
solvent
200
160
120
80
40
0
δ (ppm)
1H NMR Spectrum
(400 MHz, CDCl3 solution)
expansion
7.3
7.2
3.2
3.0
2.6
2.5
TMS

Problem 152

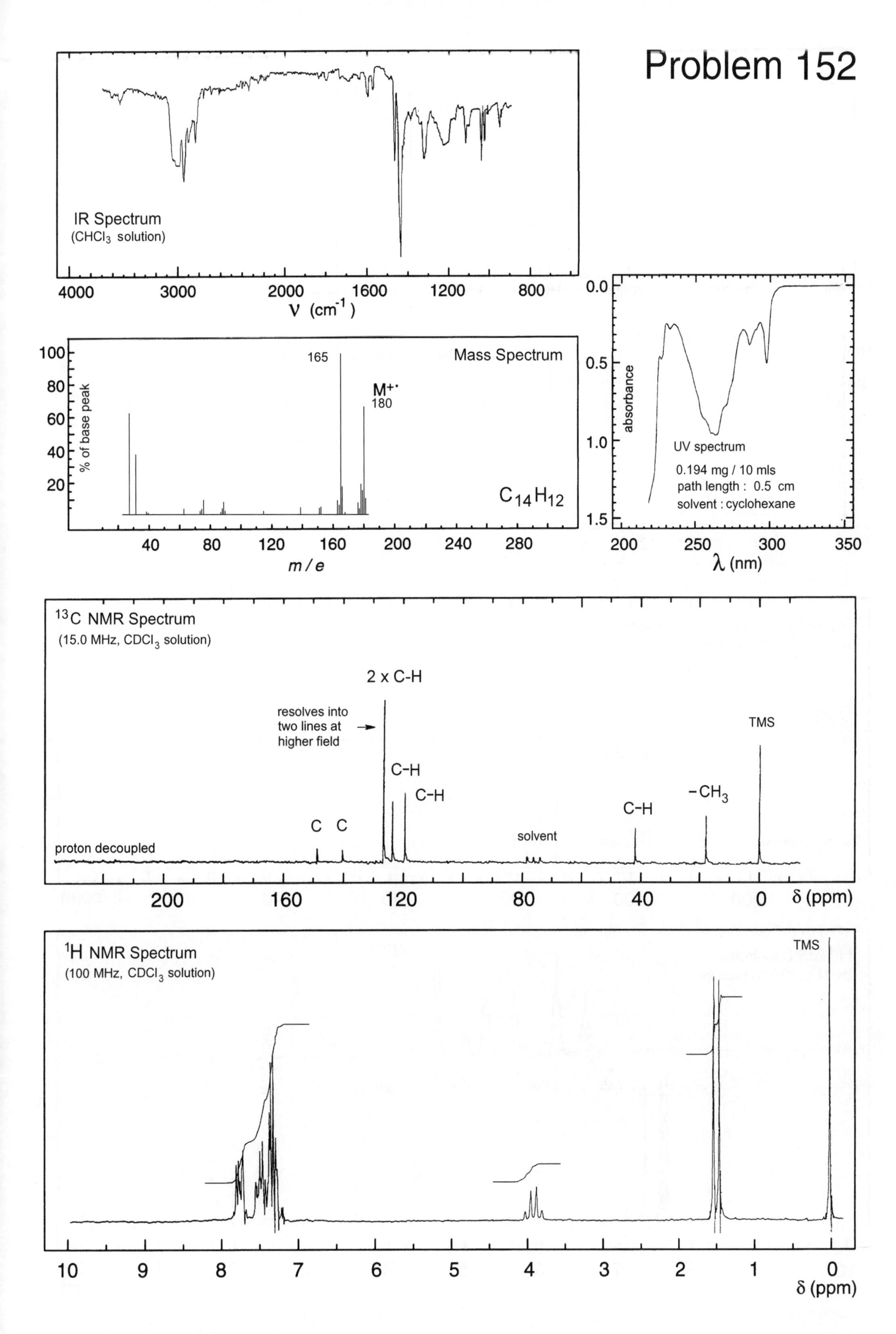

IR Spectrum
(CHCl3 solution)
4000
3000
2000
1600
1200
800
ν (cm-1)
Mass Spectrum
% of base peak
165
M+·
180
C14H12
m / e
absorbance
UV spectrum
0.194 mg / 10 mls
path length : 0.5 cm
solvent : cyclohexane
λ (nm)
13C NMR Spectrum
(15.0 MHz, CDCl3 solution)
2 x C-H
resolves into
two lines at
higher field
TMS
C-H
C-H
-CH3
C-H
C C
solvent
proton decoupled
δ (ppm)
1H NMR Spectrum
(100 MHz, CDCl3 solution)
TMS
δ (ppm)

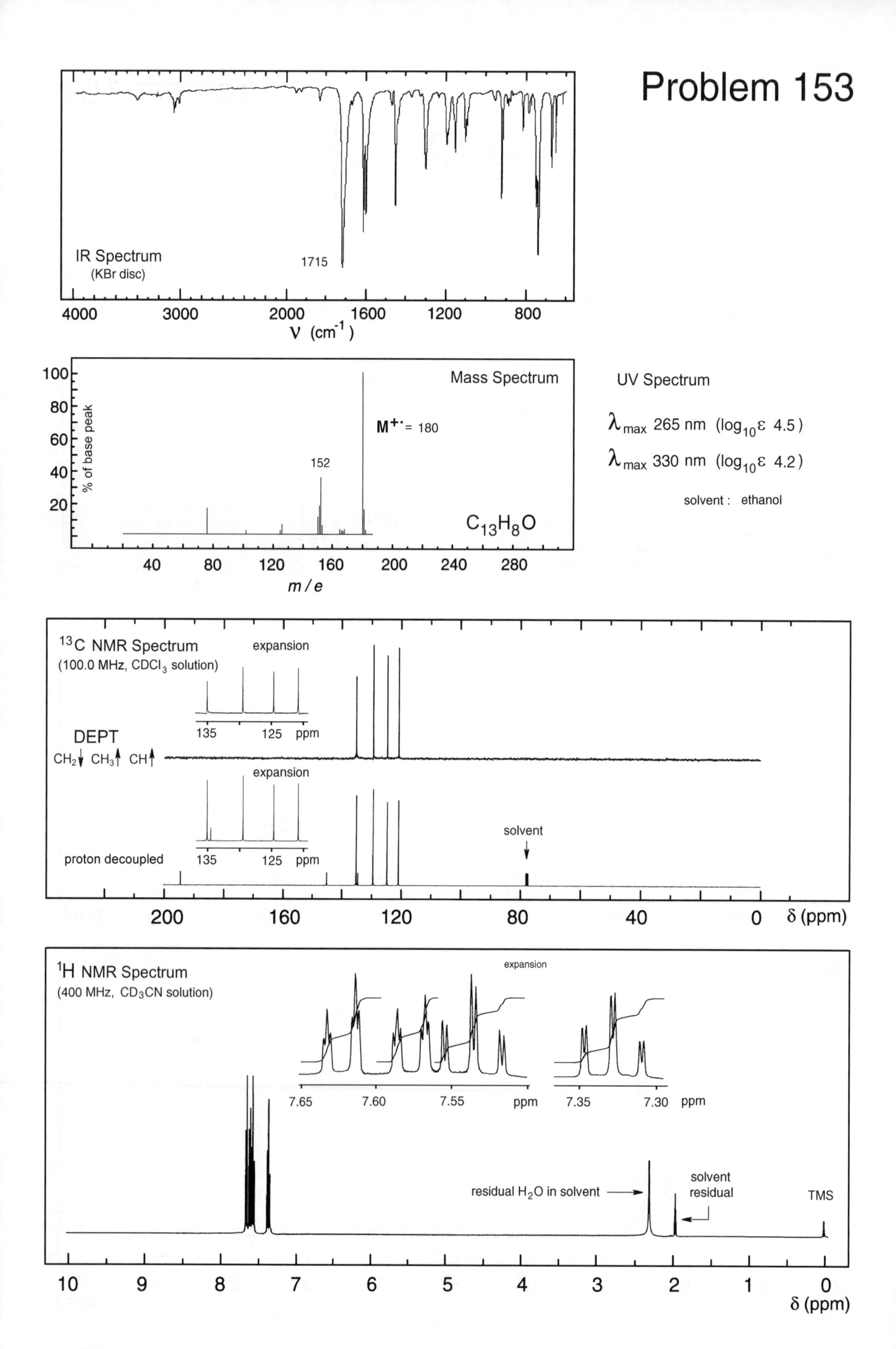

Problem 153
IR Spectrum
(KBr disc)
1715
4000
3000
2000
1600
1200
800
ν (cm⁻¹)
Mass Spectrum
% of base peak
M+· = 180
152
C13H8O
m / e
UV Spectrum
λmax 265 nm (log10ε 4.5)
λmax 330 nm (log10ε 4.2)
solvent : ethanol
13C NMR Spectrum
(100.0 MHz, CDCl3 solution)
expansion
DEPT
CH2↓ CH3↑ CH↑
proton decoupled
solvent
135
125
ppm
δ (ppm)
1H NMR Spectrum
(400 MHz, CD3CN solution)
expansion
7.65
7.60
7.55
7.35
7.30
residual H2O in solvent
solvent residual
TMS

Problem 154

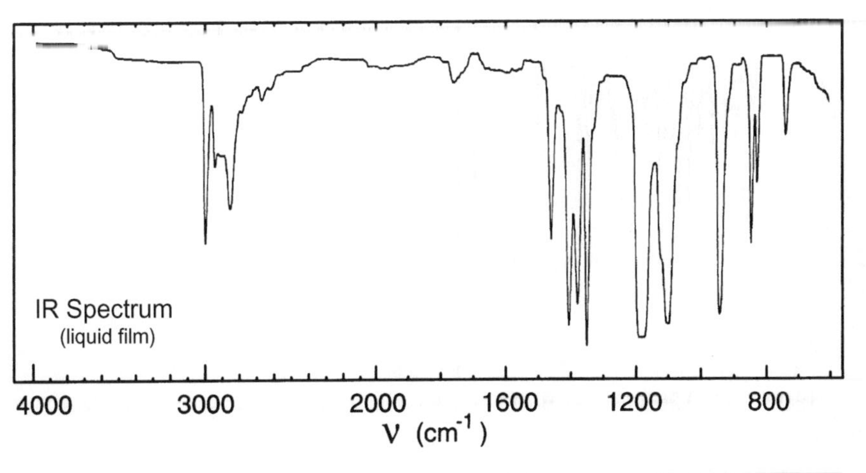

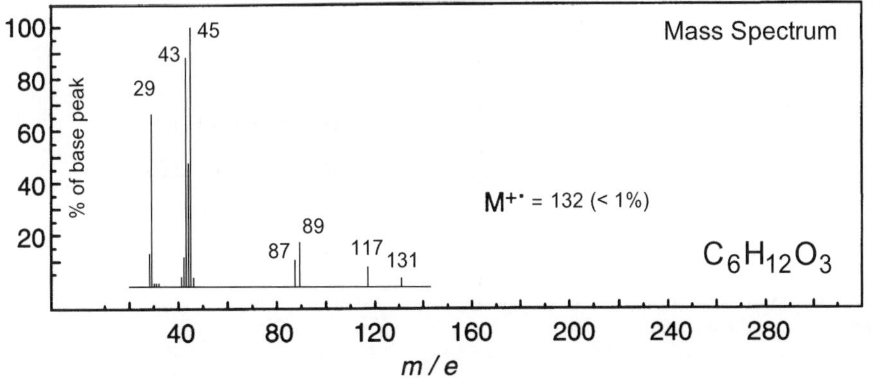

No significant UV absorption above 220 nm

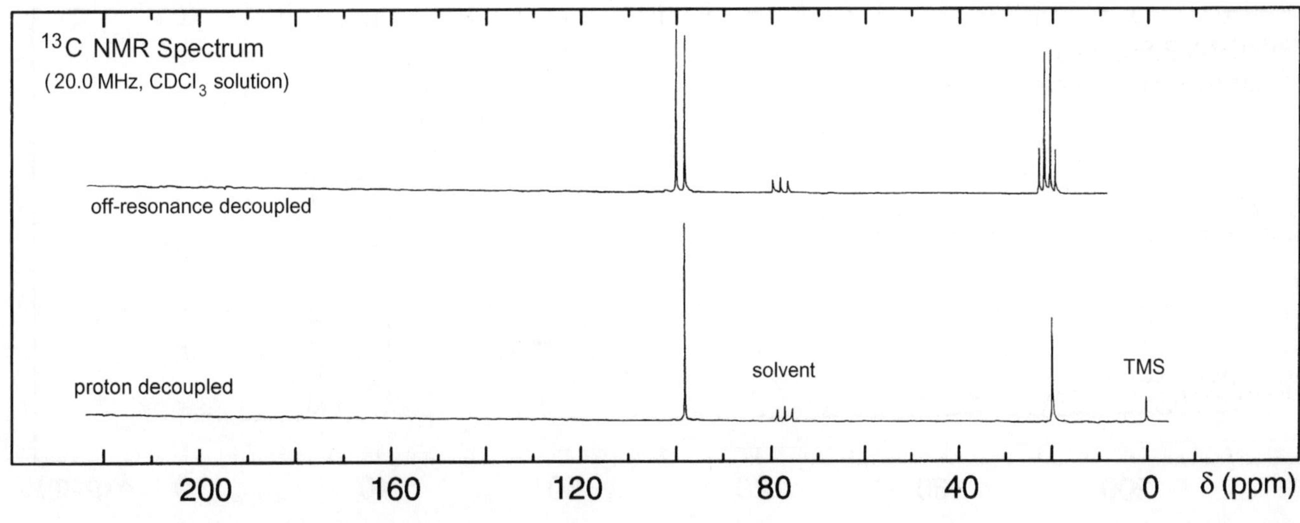

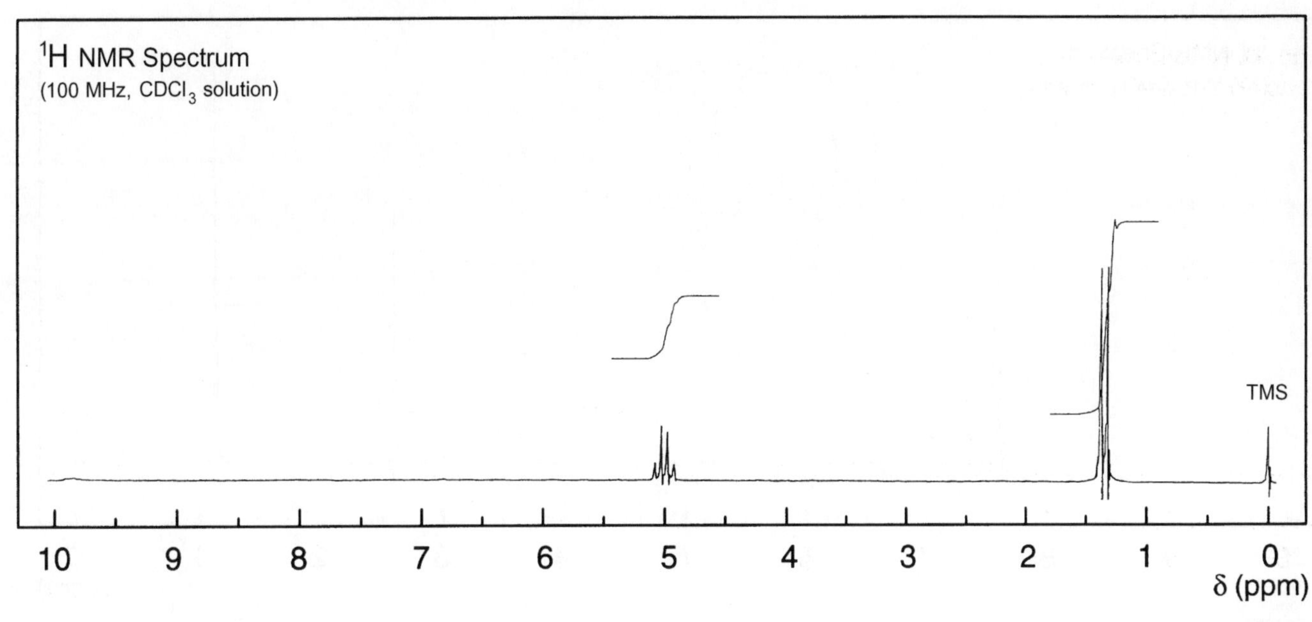

Problem 155

IR Spectrum
(KBr disc)
1811
1777
4000 3000 2000 1600 1200 800
ν (cm^{-1})

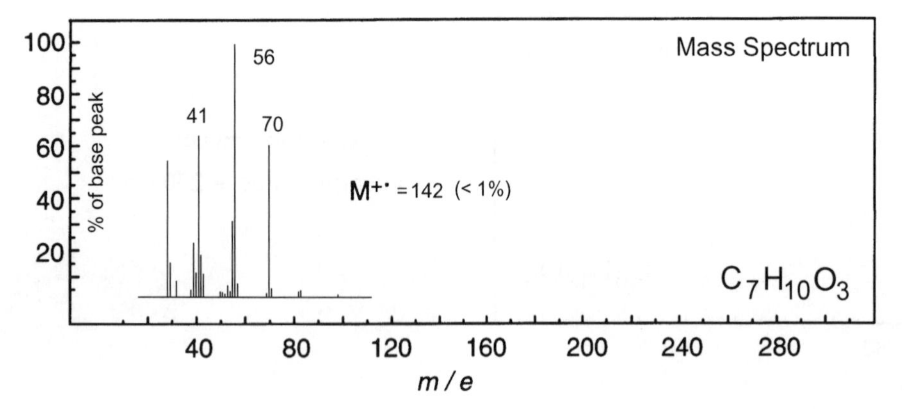

No significant UV absorption above 220 nm

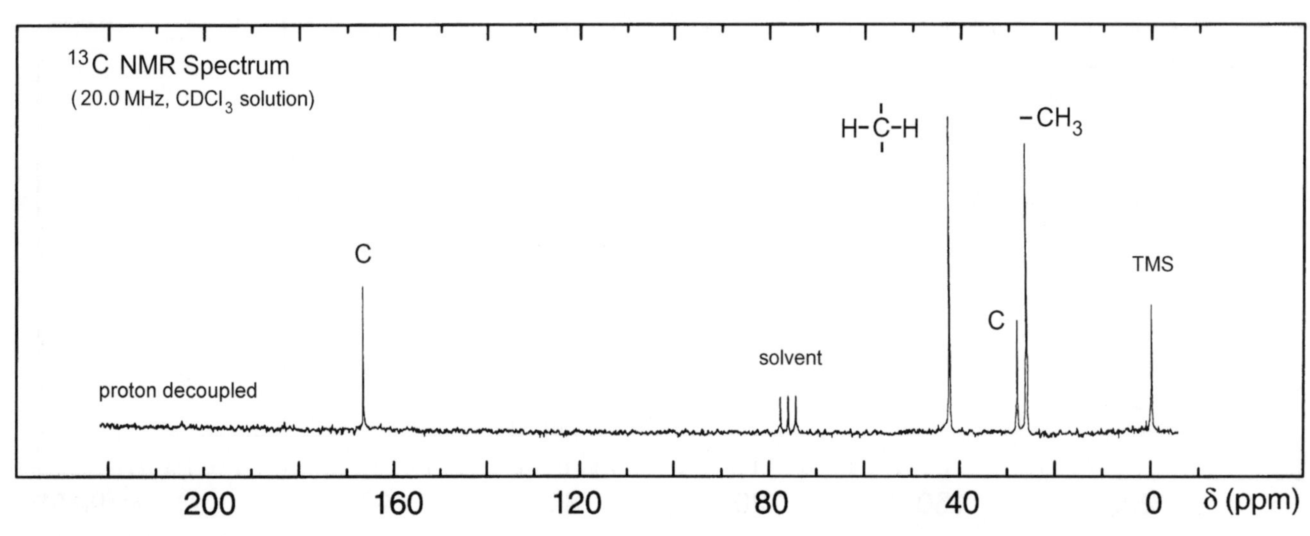

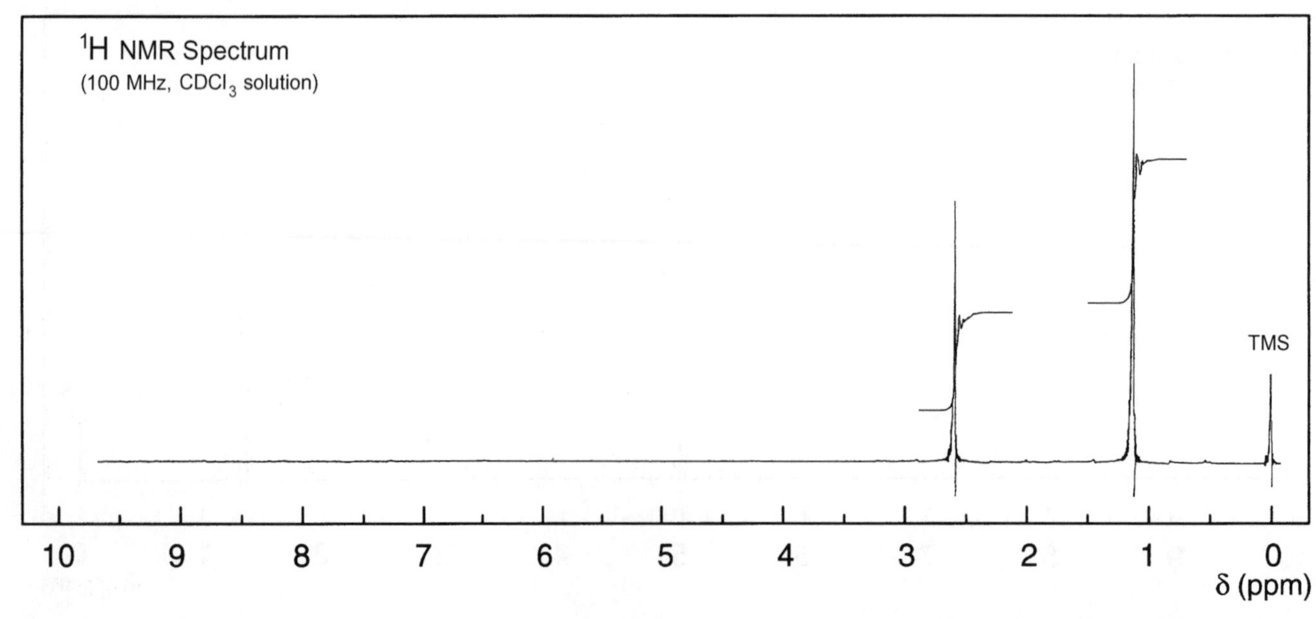

Problem 156

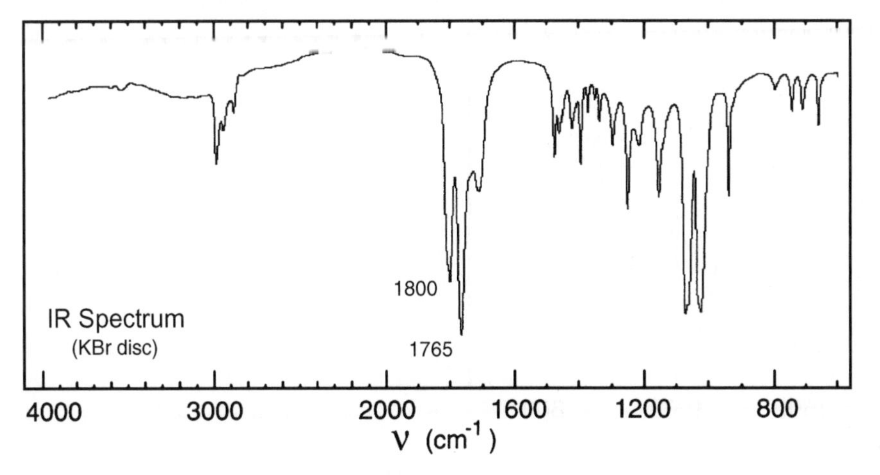

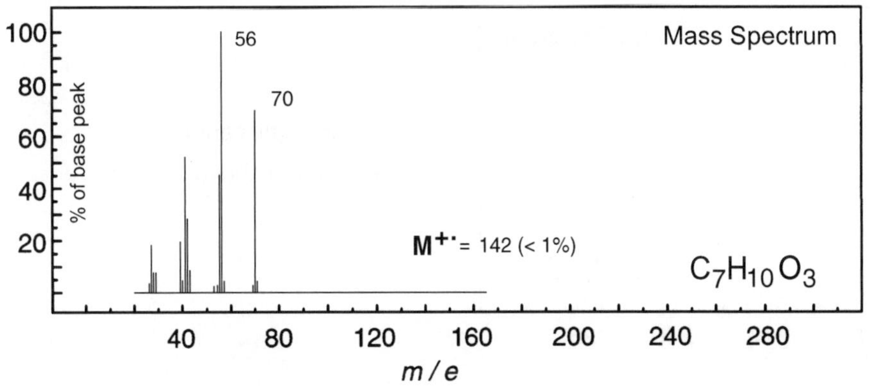

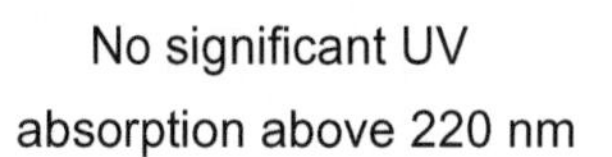

$C_7H_{10}O_3$

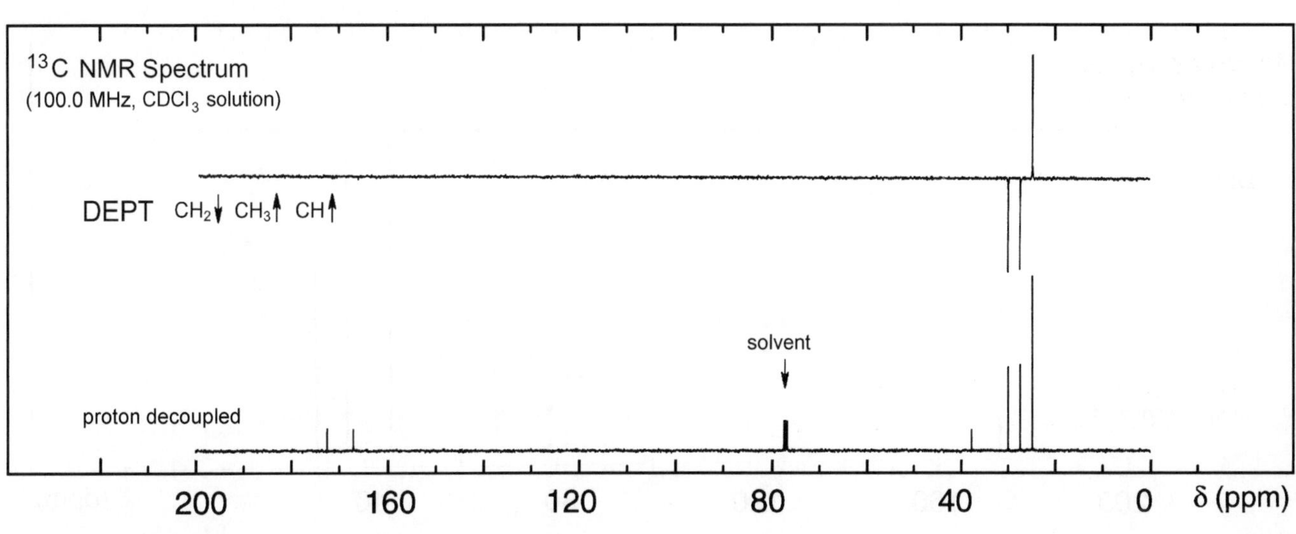

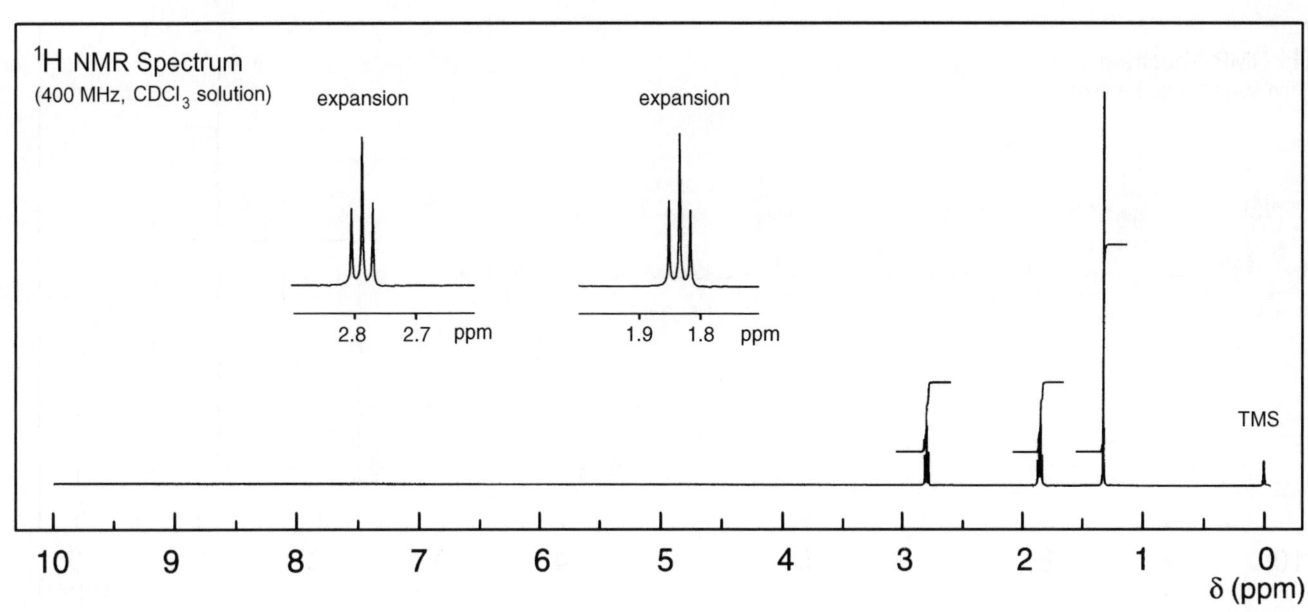

Problem 157

IR Spectrum
(liquid film)
3405
1721
4000 3000 2000 1600 1200 800
ν (cm^{-1})

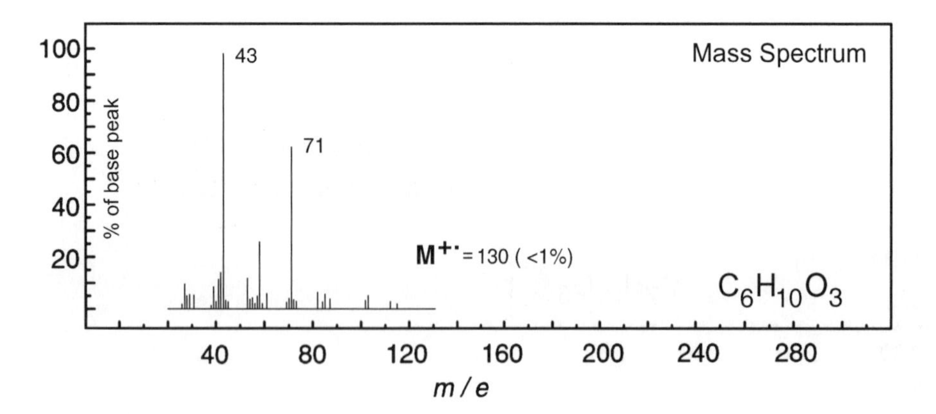

No significant UV absorption above 220 nm

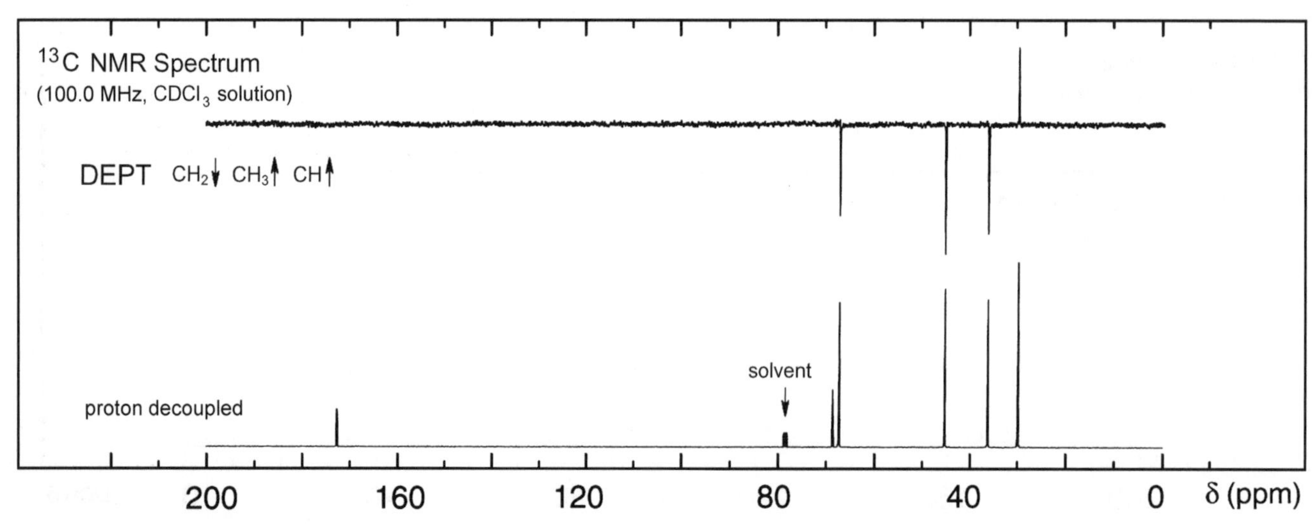

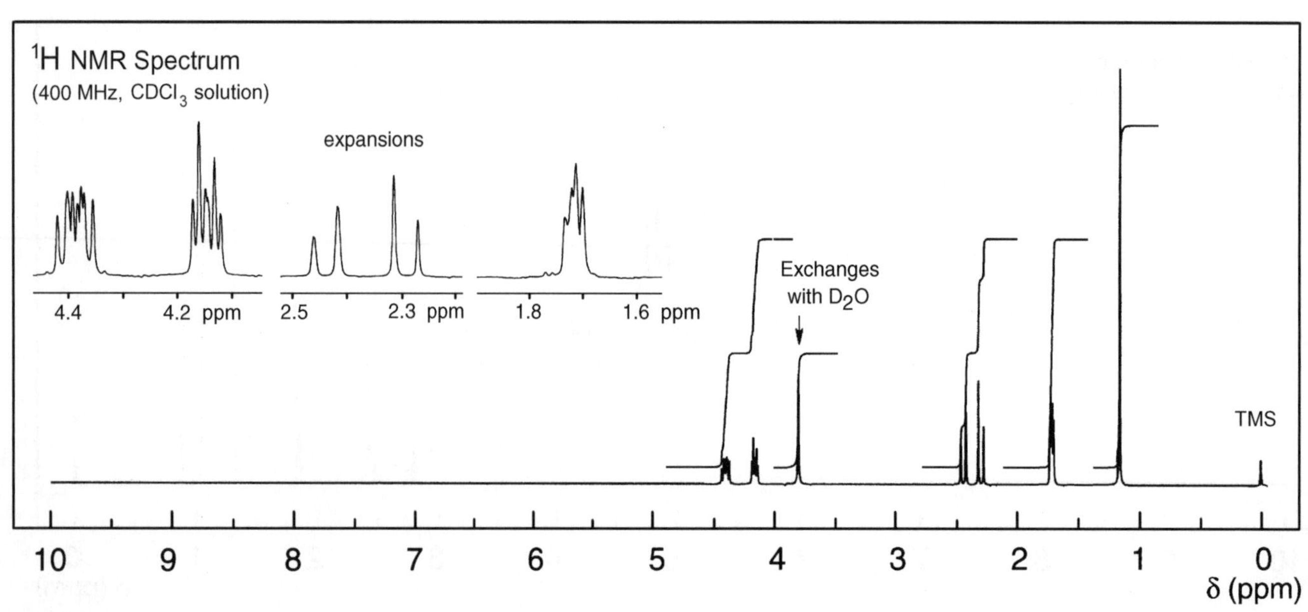

Problem 158

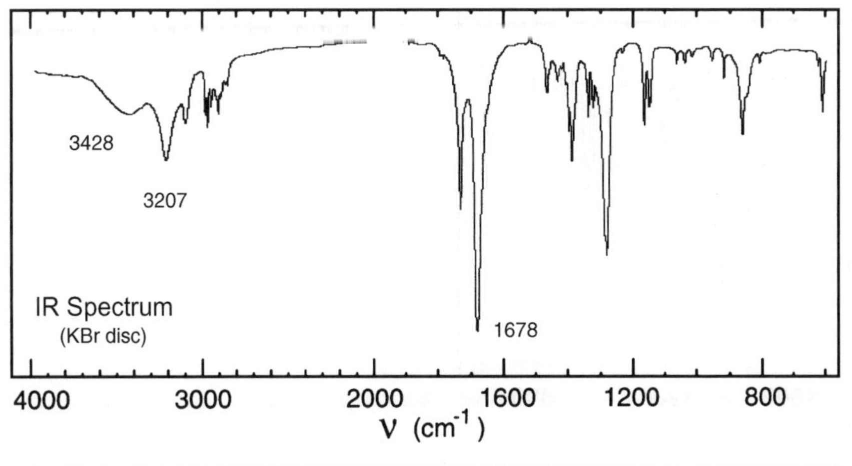

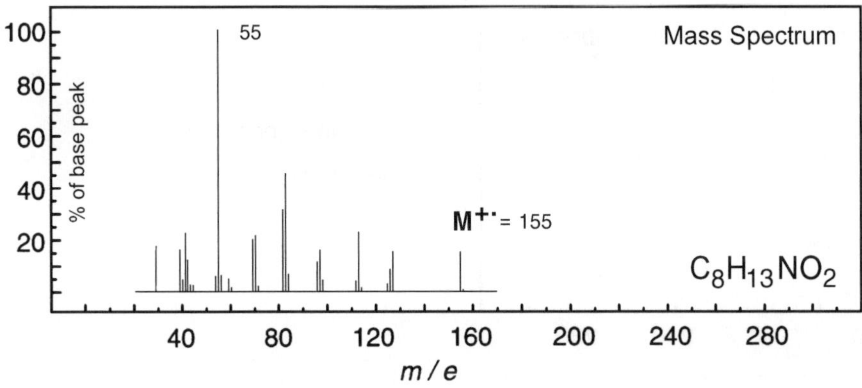

UV Spectrum

λ_{max} 230 nm ($\log_{10}\varepsilon$ 3.6) pH 11

λ_{max} 230 nm ($\log_{10}\varepsilon$ 4.4) pH 13

solvent : water

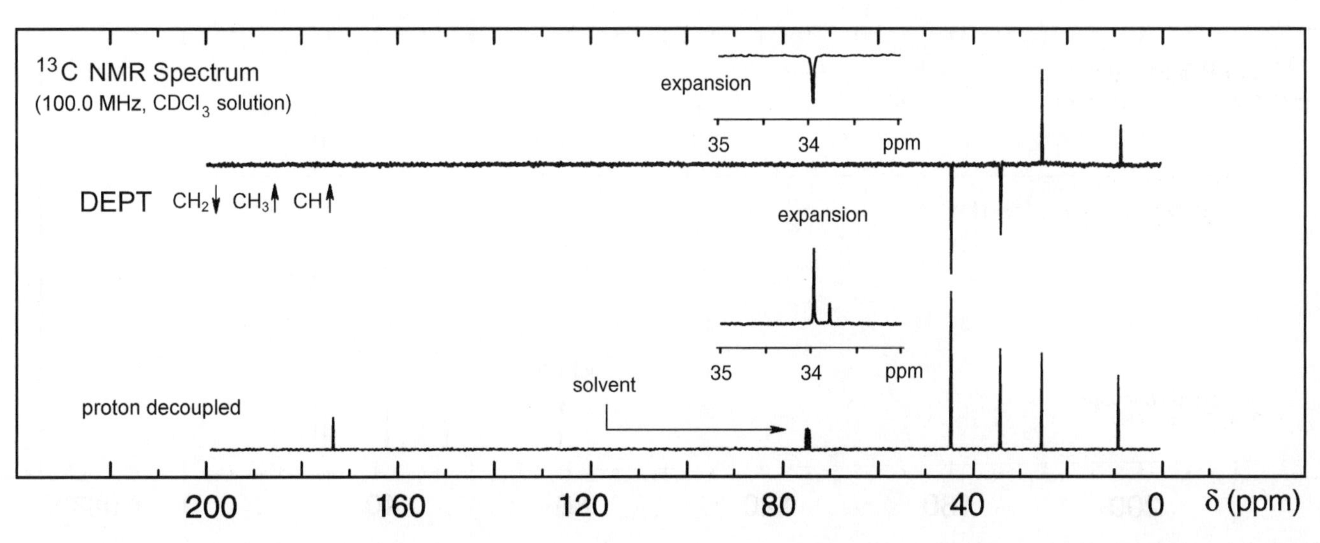

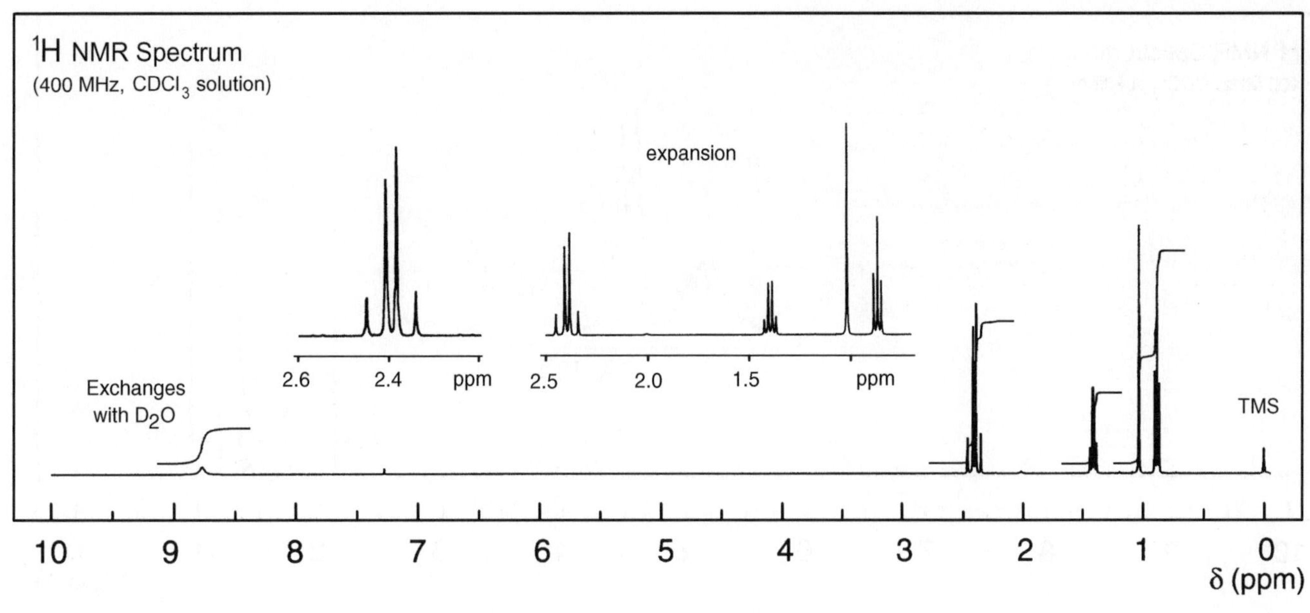

Problem 159

IR Spectrum (liquid film)

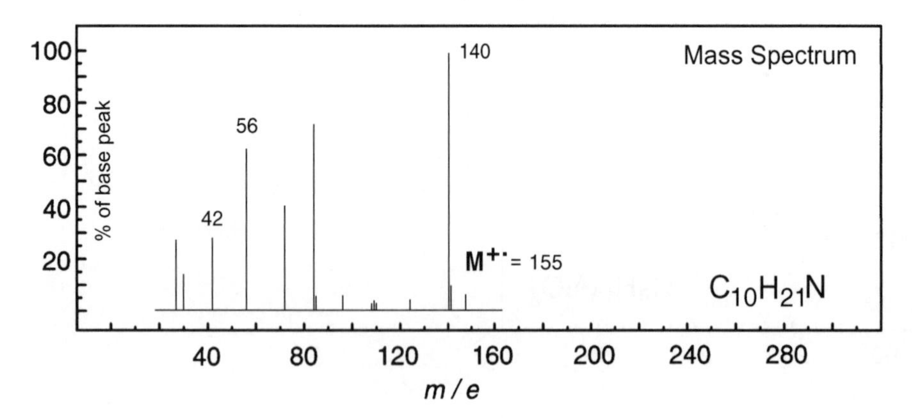

No significant UV absorption above 220 nm

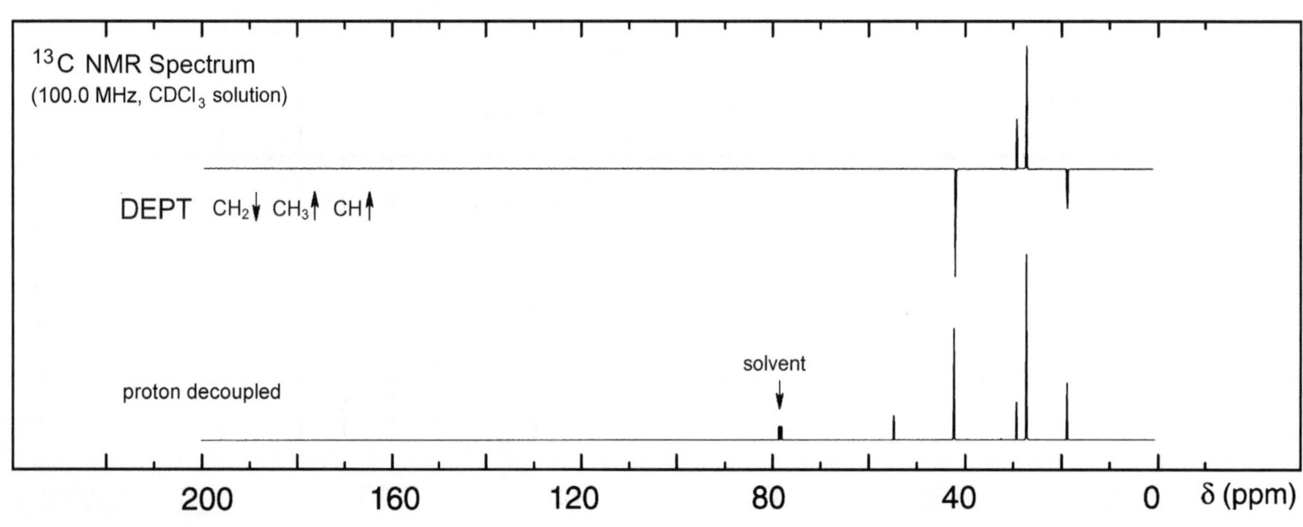

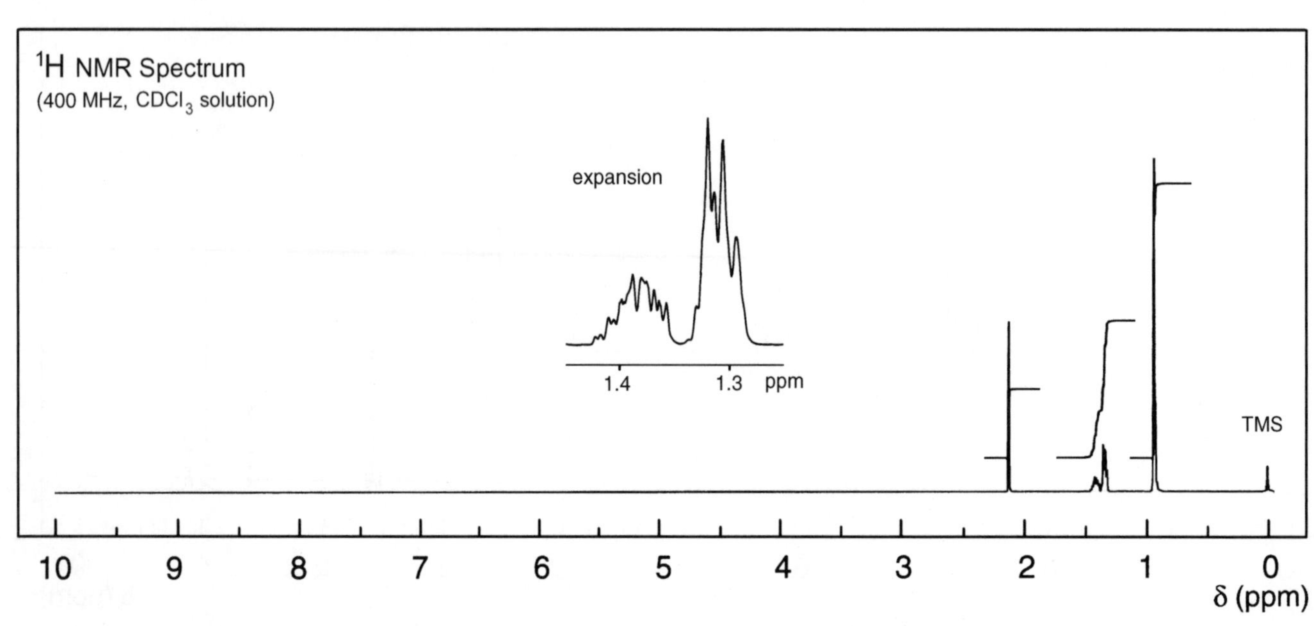

Problem 160

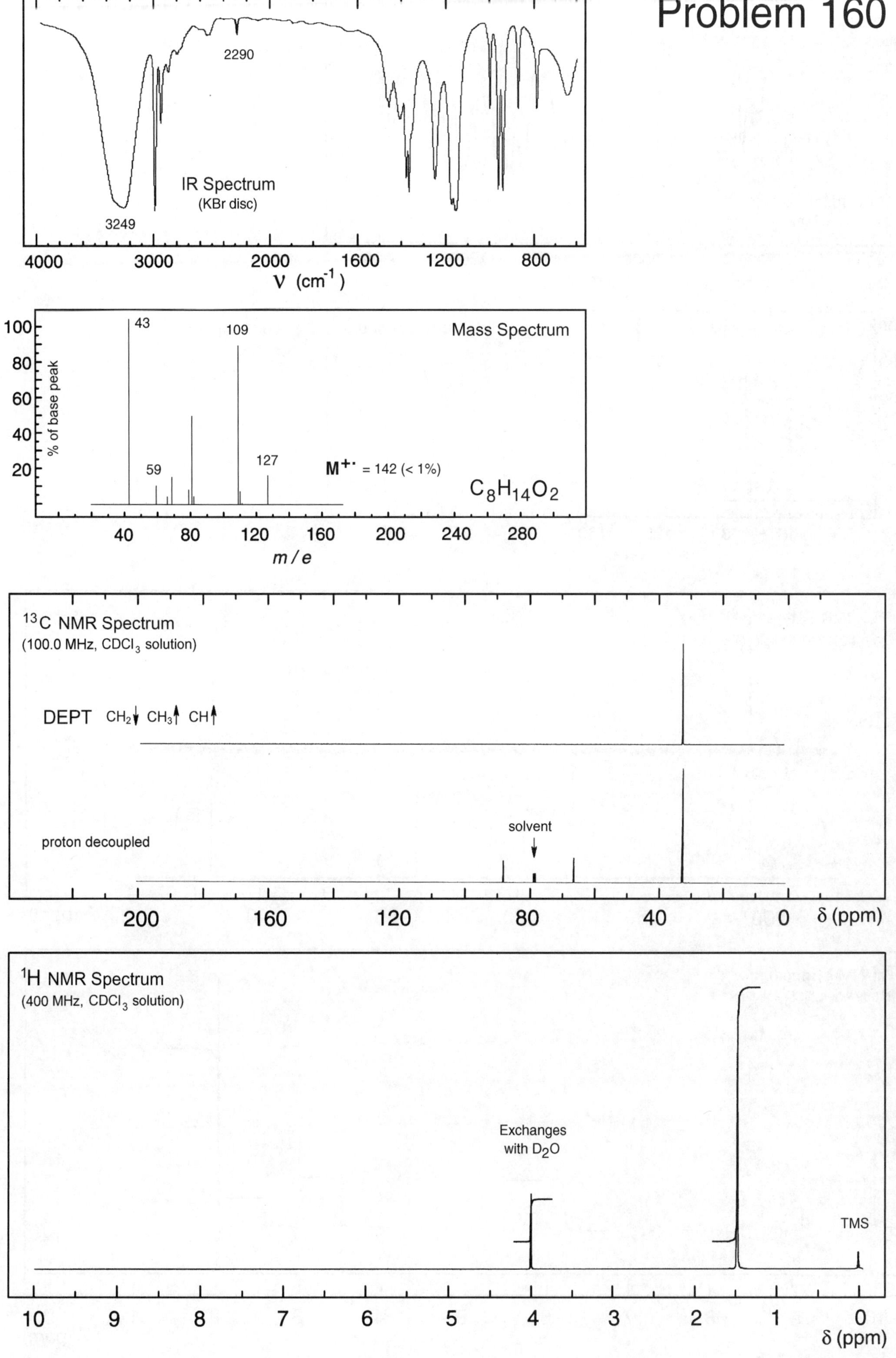

IR Spectrum
(KBr disc)
3249
2290
4000
3000
2000
1600
1200
800
ν (cm⁻¹)
Mass Spectrum
% of base peak
100
80
60
40
20
43
109
59
127
M+· = 142 (< 1%)
C8H14O2
40
80
120
160
200
240
280
m / e
13C NMR Spectrum
(100.0 MHz, CDCl3 solution)
DEPT CH2↓ CH3↑ CH↑
proton decoupled
solvent
200
160
120
80
40
0
δ (ppm)
1H NMR Spectrum
(400 MHz, CDCl3 solution)
Exchanges with D2O
TMS
10
9
8
7
6
5
4
3
2
1
0
δ (ppm)

Problem 161

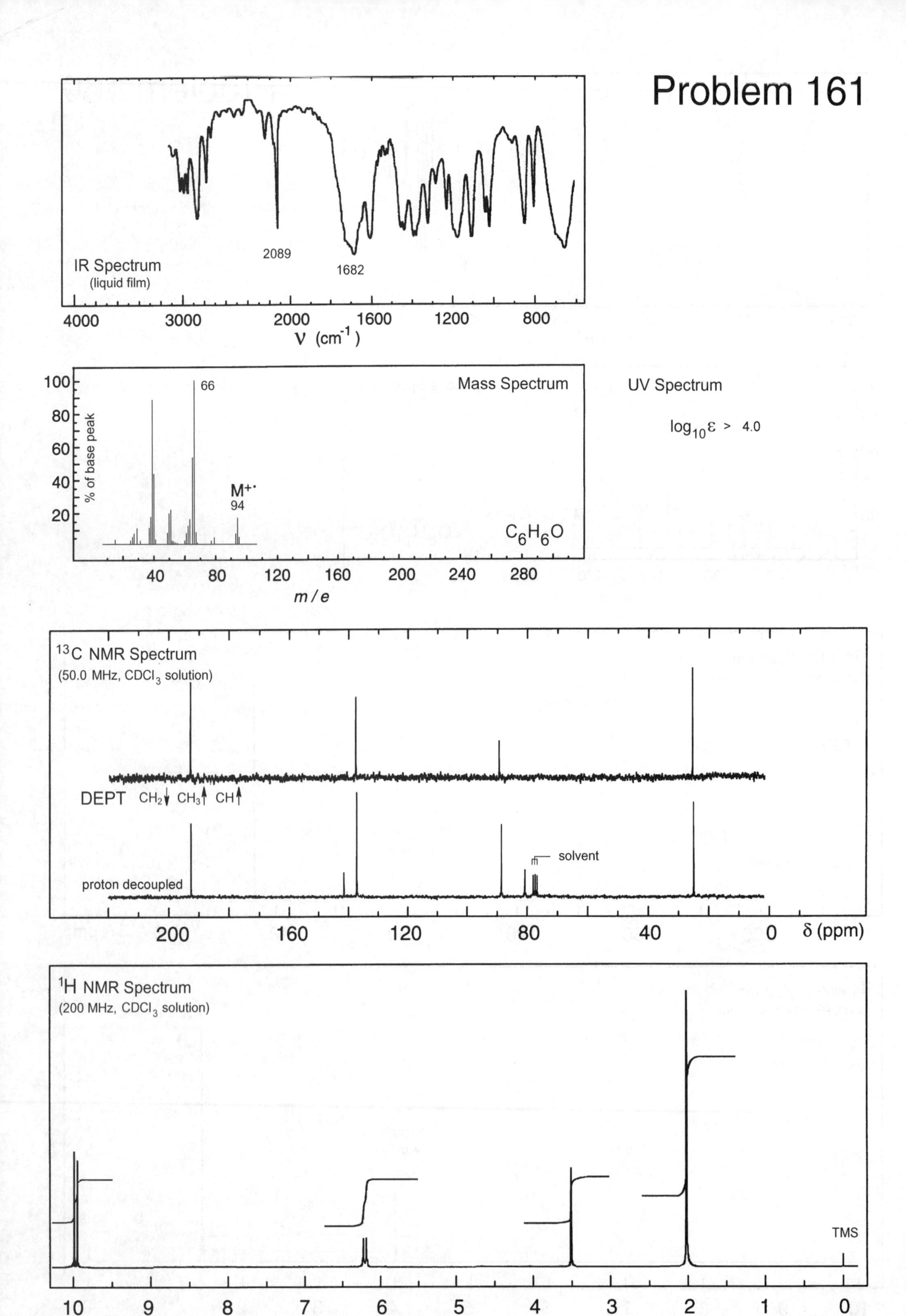

IR Spectrum
(liquid film)
2089
1682
4000
3000
2000
1600
1200
800
ν (cm⁻¹)
Mass Spectrum
66
M+·
94
C_6H_6O
% of base peak
100
80
60
40
20
40
80
120
160
200
240
280
m / e
UV Spectrum
log₁₀ε > 4.0
13C NMR Spectrum
(50.0 MHz, CDCl3 solution)
DEPT CH2↓ CH3↑ CH↑
solvent
proton decoupled
200
160
120
80
40
0
δ (ppm)
1H NMR Spectrum
(200 MHz, CDCl3 solution)
TMS
10
9
8
7
6
5
4
3
2
1
0
δ (ppm)

Problem 162

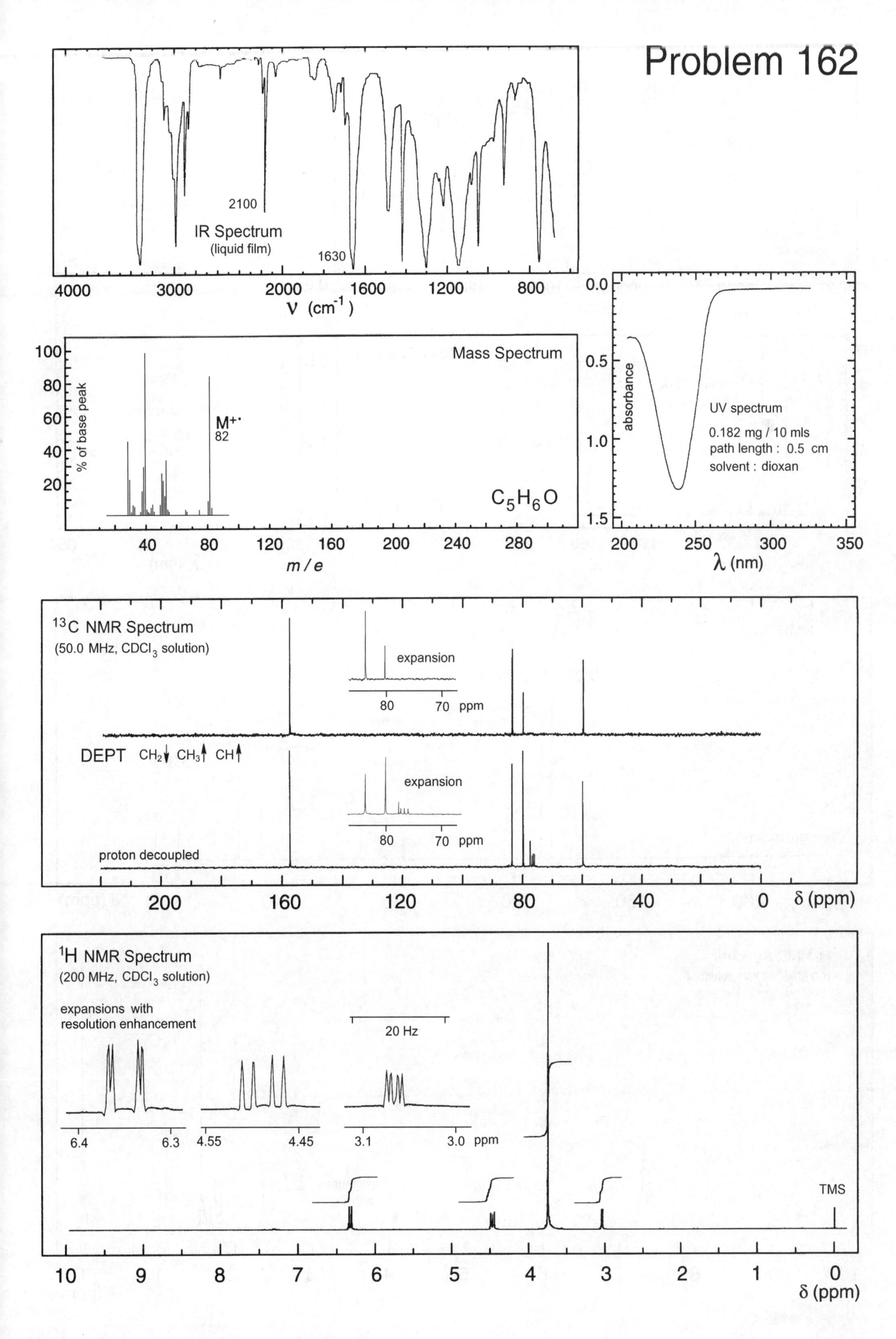

IR Spectrum
(liquid film)
2100
1630
4000
3000
2000
1600
1200
800
ν (cm⁻¹)
Mass Spectrum
% of base peak
M+·
82
C5H6O
m / e
UV spectrum
0.182 mg / 10 mls
path length : 0.5 cm
solvent : dioxan
absorbance
λ (nm)
13C NMR Spectrum
(50.0 MHz, CDCl3 solution)
expansion
ppm
DEPT CH2↓ CH3↑ CH↑
proton decoupled
δ (ppm)
1H NMR Spectrum
(200 MHz, CDCl3 solution)
expansions with
resolution enhancement
20 Hz
TMS

Problem 163

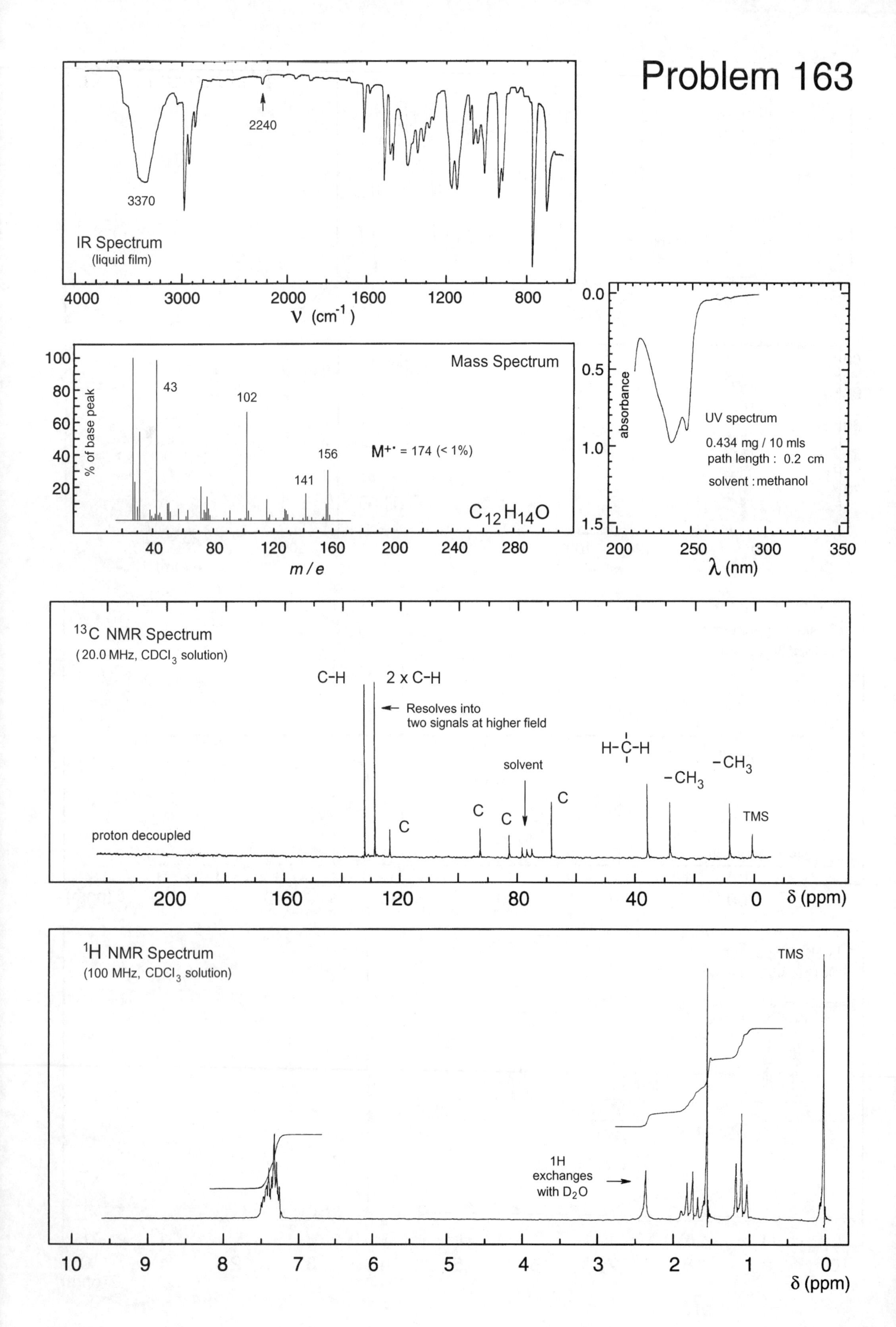

IR Spectrum
(liquid film)
3370
2240
4000
3000
2000
1600
1200
800
ν (cm⁻¹)
Mass Spectrum
% of base peak
43
102
156
141
M⁺˙ = 174 (< 1%)
C12H14O
m / e
UV spectrum
0.434 mg / 10 mls
path length : 0.2 cm
solvent : methanol
absorbance
λ (nm)
13C NMR Spectrum
(20.0 MHz, CDCl3 solution)
C-H
2 x C-H
Resolves into
two signals at higher field
C
C
C
C
solvent
H-C-H
-CH3
-CH3
TMS
proton decoupled
δ (ppm)
1H NMR Spectrum
(100 MHz, CDCl3 solution)
TMS
1H
exchanges
with D2O
δ (ppm)

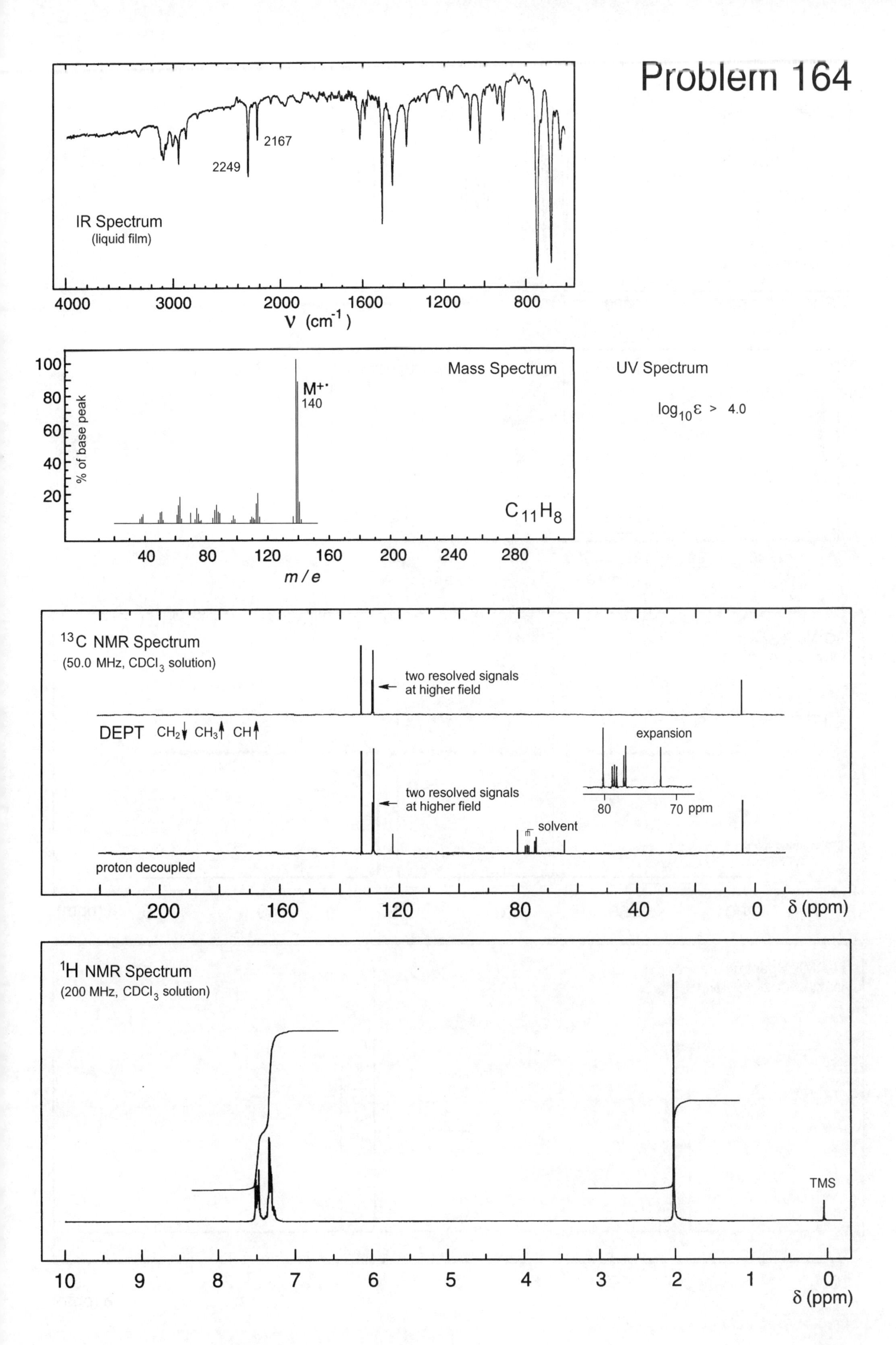
Problem 164
2167
2249
IR Spectrum
(liquid film)
4000
3000
2000
1600
1200
800
ν (cm⁻¹)
Mass Spectrum
UV Spectrum
log10ε > 4.0
100
80
60
40
20
% of base peak
M+·
140
C11H8
40
80
120
160
200
240
280
m / e
13C NMR Spectrum
(50.0 MHz, CDCl3 solution)
two resolved signals
at higher field
DEPT CH2 CH3 CH
expansion
80
70 ppm
two resolved signals
at higher field
solvent
proton decoupled
200
160
120
80
40
0
δ (ppm)
1H NMR Spectrum
(200 MHz, CDCl3 solution)
TMS
10
9
8
7
6
5
4
3
2
1
0
δ (ppm)

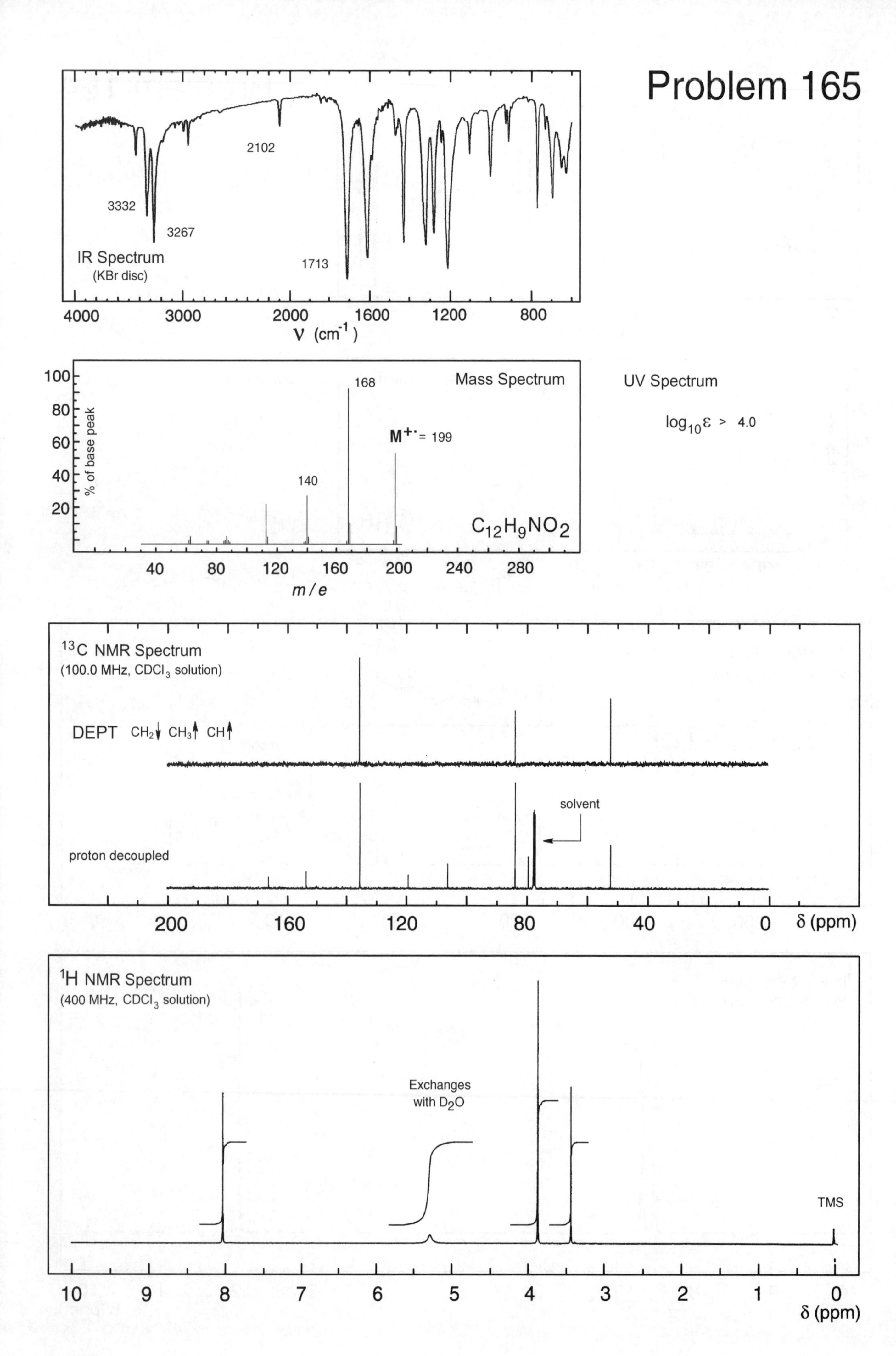
Problem 165
IR Spectrum
(KBr disc)
3332
3267
2102
1713
4000
3000
2000
1600
1200
800
ν (cm⁻¹)
Mass Spectrum
% of base peak
168
M⁺· = 199
140
C12H9NO2
m / e
UV Spectrum
log10ε > 4.0
13C NMR Spectrum
(100.0 MHz, CDCl3 solution)
DEPT CH2 CH3 CH
proton decoupled
solvent
δ (ppm)
1H NMR Spectrum
(400 MHz, CDCl3 solution)
Exchanges
with D2O
TMS
δ (ppm)

Problem 166

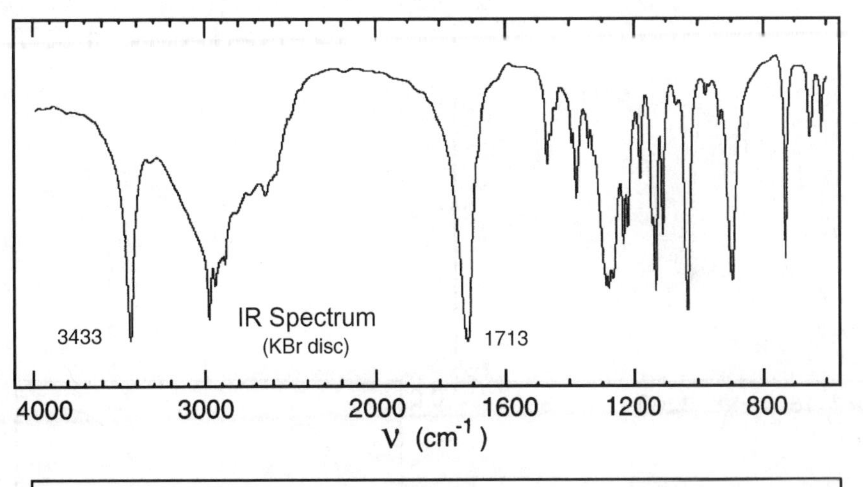

3433
IR Spectrum
(KBr disc)
1713
4000
3000
2000
1600
1200
800
ν (cm⁻¹)

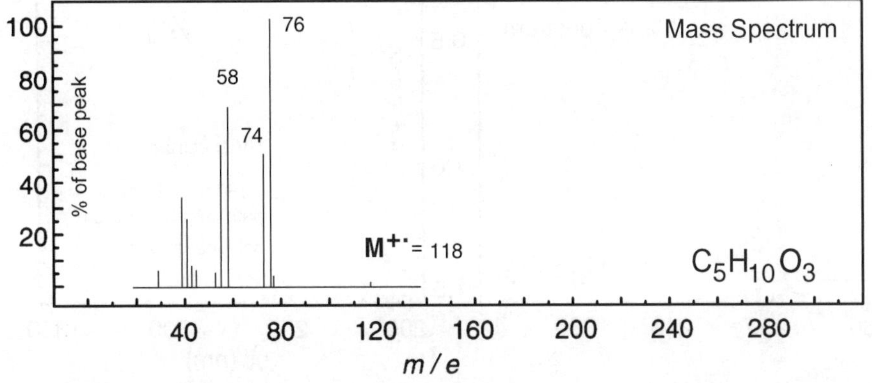

Mass Spectrum
% of base peak
76
58
74
M⁺˙ = 118
C5H10O3
m / e

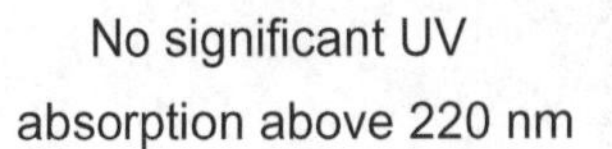

No significant UV
absorption above 220 nm

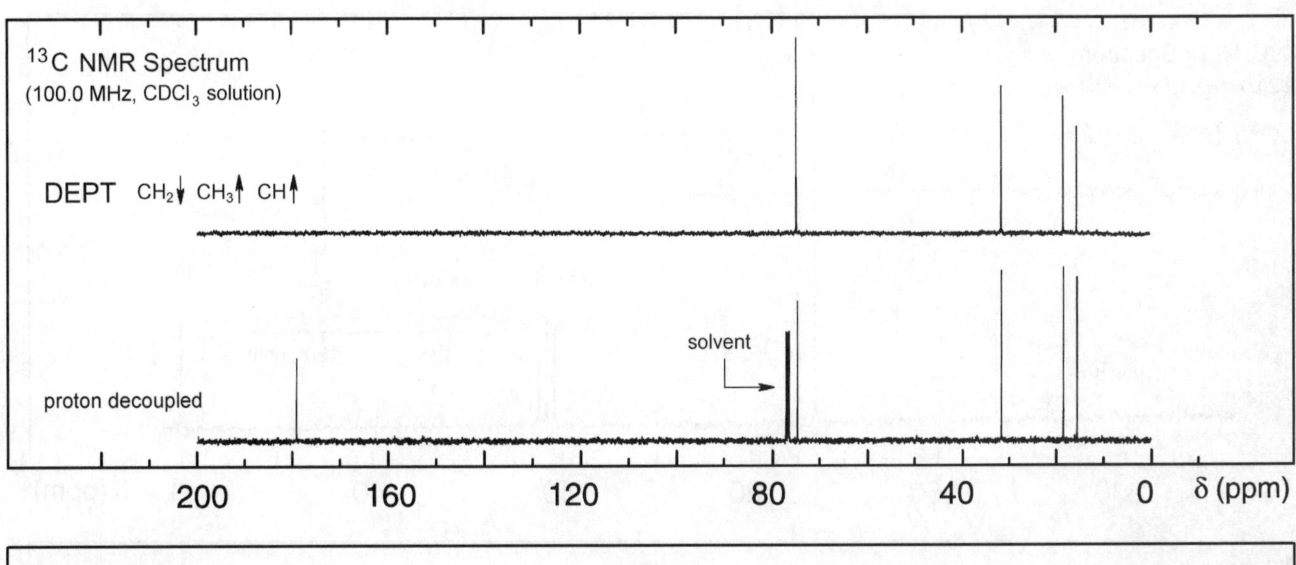

13C NMR Spectrum
(100.0 MHz, CDCl3 solution)
DEPT CH2↓ CH3↑ CH↑
solvent
proton decoupled
δ (ppm)

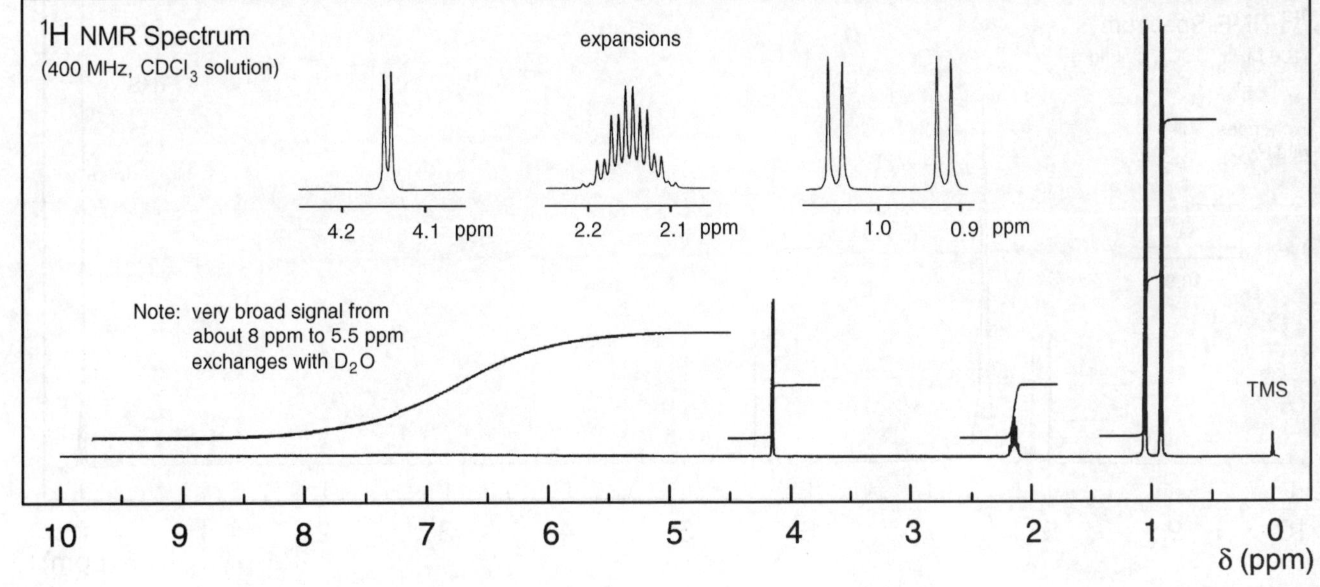

1H NMR Spectrum
(400 MHz, CDCl3 solution)
expansions
ppm
Note: very broad signal from
about 8 ppm to 5.5 ppm
exchanges with D2O
TMS
δ (ppm)

Problem 167

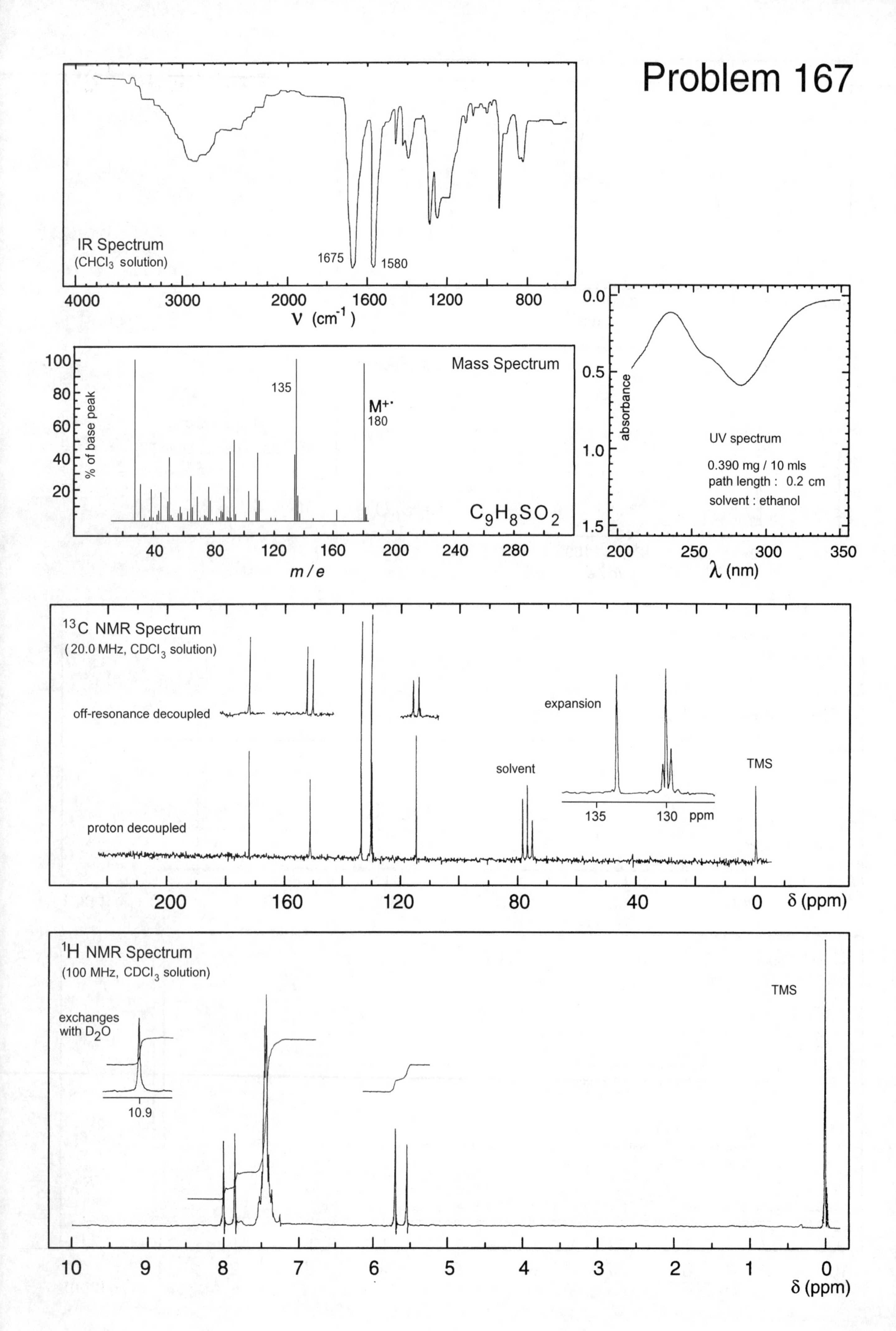

IR Spectrum
(CHCl3 solution)
1675
1580
4000
3000
2000
1600
1200
800
ν (cm-1)
Mass Spectrum
% of base peak
135
M+·
180
C9H8SO2
m / e
absorbance
UV spectrum
0.390 mg / 10 mls
path length : 0.2 cm
solvent : ethanol
λ (nm)
13C NMR Spectrum
(20.0 MHz, CDCl3 solution)
off-resonance decoupled
expansion
solvent
TMS
proton decoupled
135
130
ppm
δ (ppm)
1H NMR Spectrum
(100 MHz, CDCl3 solution)
TMS
exchanges
with D2O
10.9
δ (ppm)

Problem 168

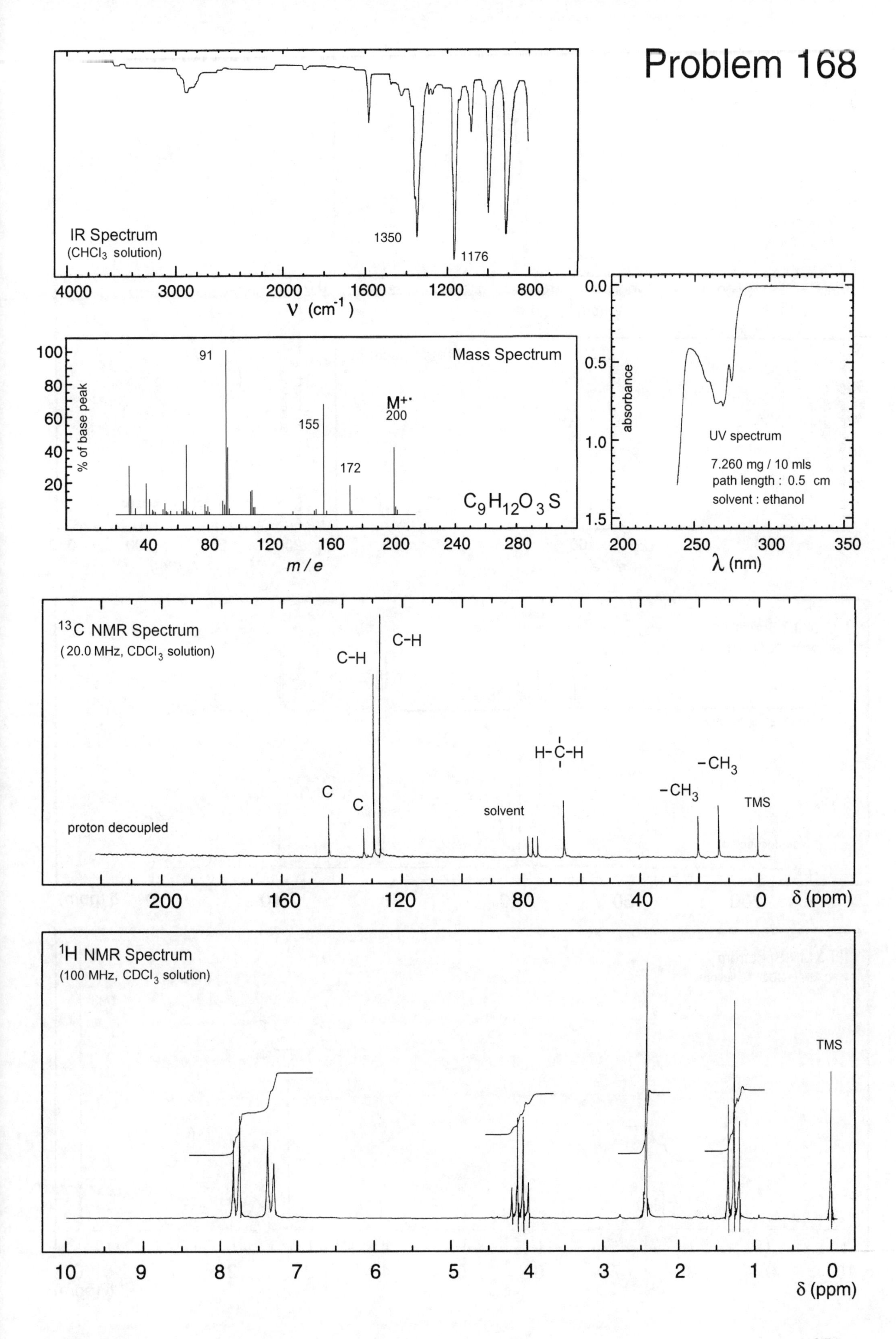

IR Spectrum
($CHCl_3$ solution)
1350
1176
4000
3000
2000
1600
1200
800
ν (cm⁻¹)
Mass Spectrum
91
155
172
M+·
200
% of base peak
$C_9H_{12}O_3S$
40
80
120
160
200
240
280
m / e
absorbance
UV spectrum
7.260 mg / 10 mls
path length : 0.5 cm
solvent : ethanol
λ (nm)
¹³C NMR Spectrum
(20.0 MHz, $CDCl_3$ solution)
proton decoupled
C
C
C-H
C-H
solvent
H-C-H
-CH3
-CH3
TMS
δ (ppm)
¹H NMR Spectrum
(100 MHz, $CDCl_3$ solution)
TMS
δ (ppm)

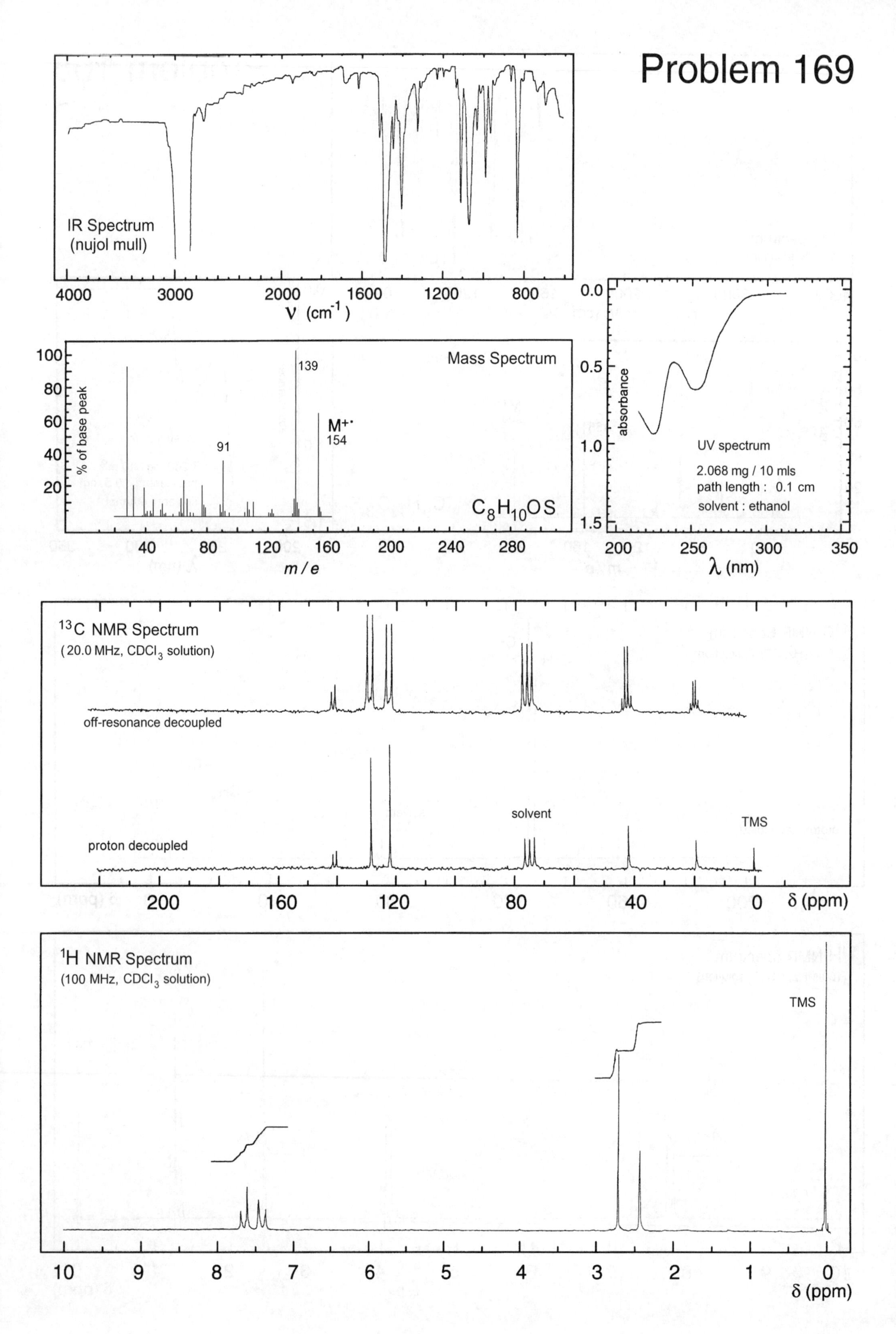
Problem 169
IR Spectrum
(nujol mull)
4000
3000
2000
1600
1200
800
ν (cm-1)
Mass Spectrum
% of base peak
100
80
60
40
20
139
91
M+·
154
C8H10OS
40
80
120
160
200
240
280
m / e
UV spectrum
2.068 mg / 10 mls
path length : 0.1 cm
solvent : ethanol
absorbance
0.0
0.5
1.0
1.5
200
250
300
350
λ (nm)
13C NMR Spectrum
(20.0 MHz, CDCl3 solution)
off-resonance decoupled
proton decoupled
solvent
TMS
200
160
120
80
40
0
δ (ppm)
1H NMR Spectrum
(100 MHz, CDCl3 solution)
TMS
10
9
8
7
6
5
4
3
2
1
0
δ (ppm)

Proble

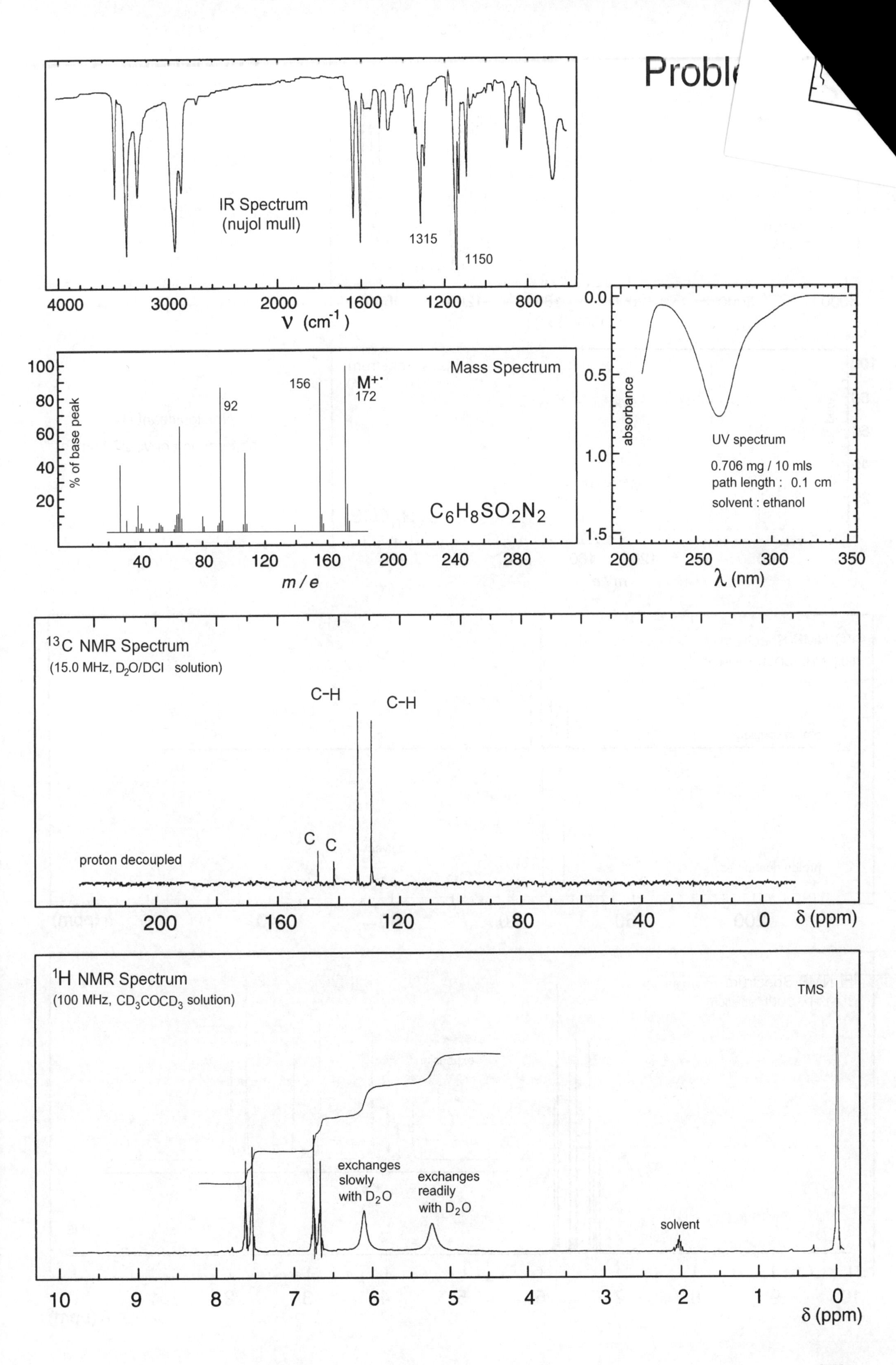

IR Spectrum
(nujol mull)
1315
1150
4000
3000
2000
1600
1200
800
ν (cm⁻¹)
Mass Spectrum
156
M+·
172
92
% of base peak
C6H8SO2N2
m / e
absorbance
UV spectrum
0.706 mg / 10 mls
path length : 0.1 cm
solvent : ethanol
λ (nm)
13C NMR Spectrum
(15.0 MHz, D2O/DCl solution)
C-H
C-H
C
C
proton decoupled
δ (ppm)
1H NMR Spectrum
(100 MHz, CD3COCD3 solution)
TMS
exchanges
slowly
with D2O
exchanges
readily
with D2O
solvent
δ (ppm)

Problem 171

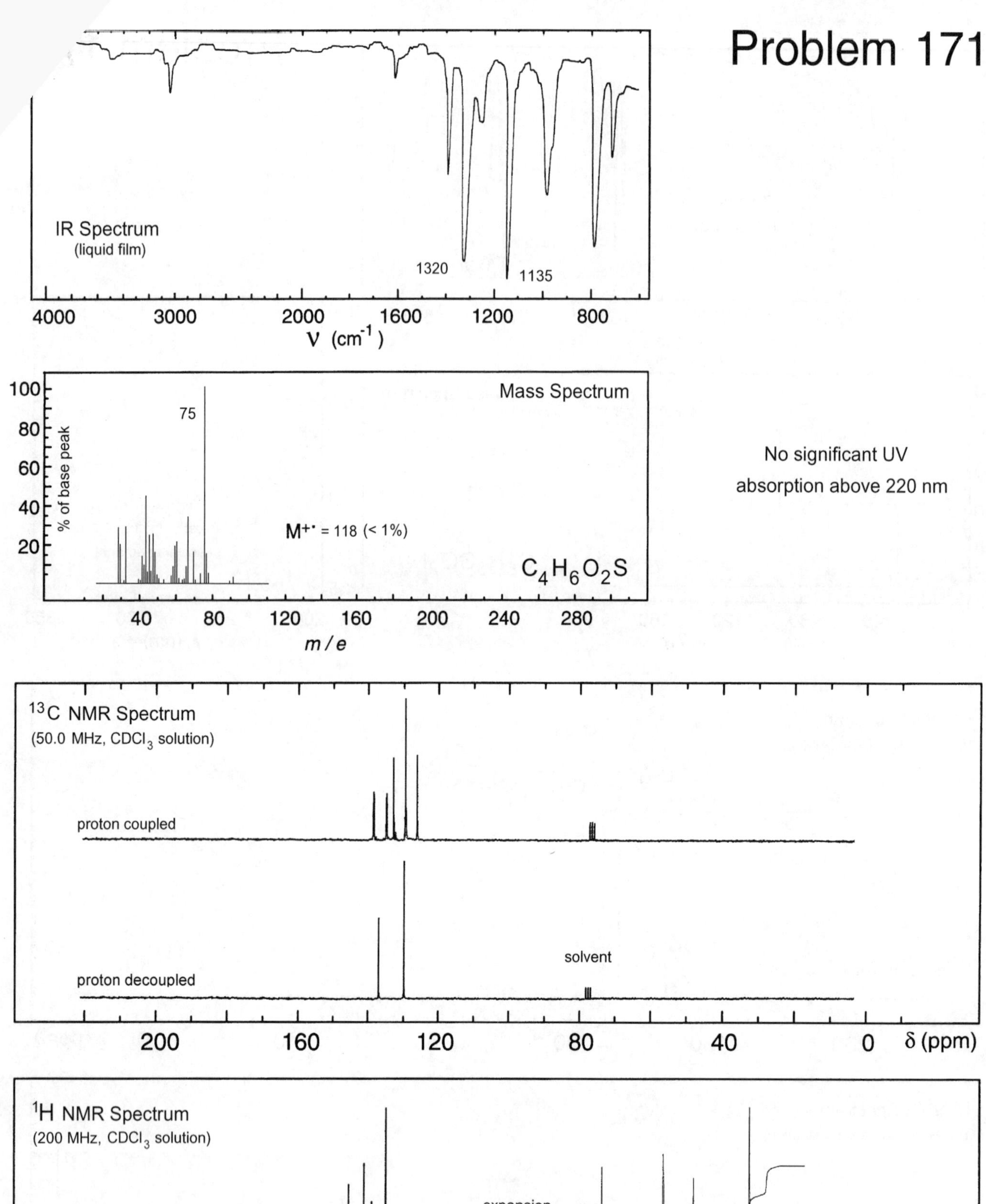

No significant UV absorption above 220 nm

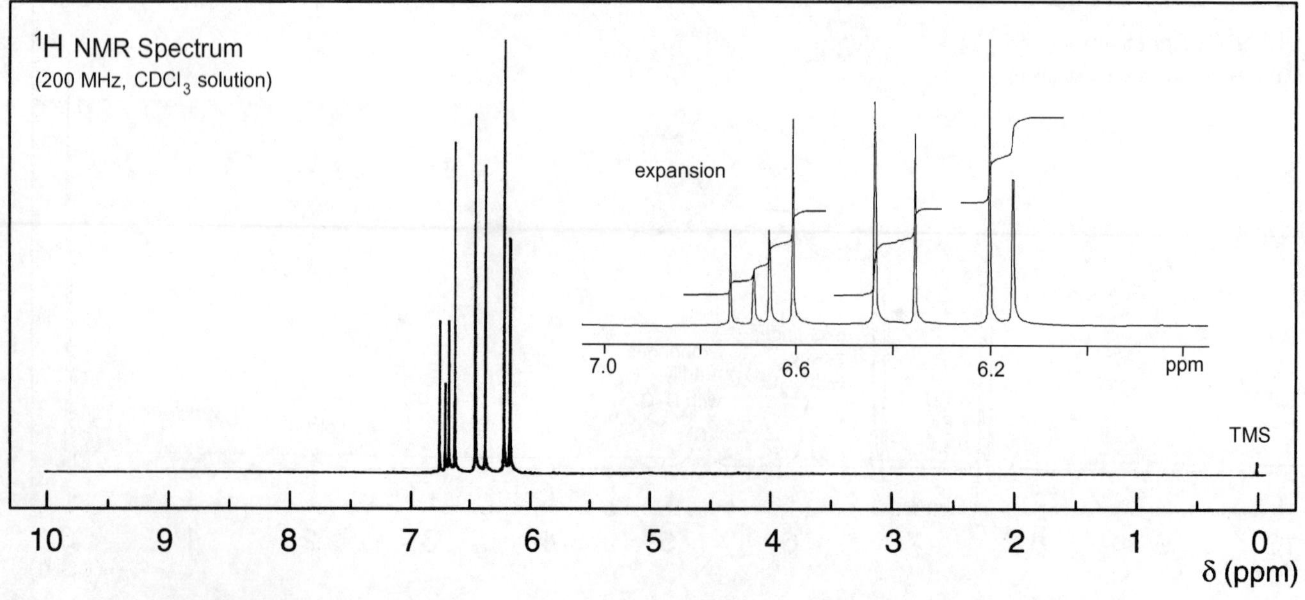

Problem 172

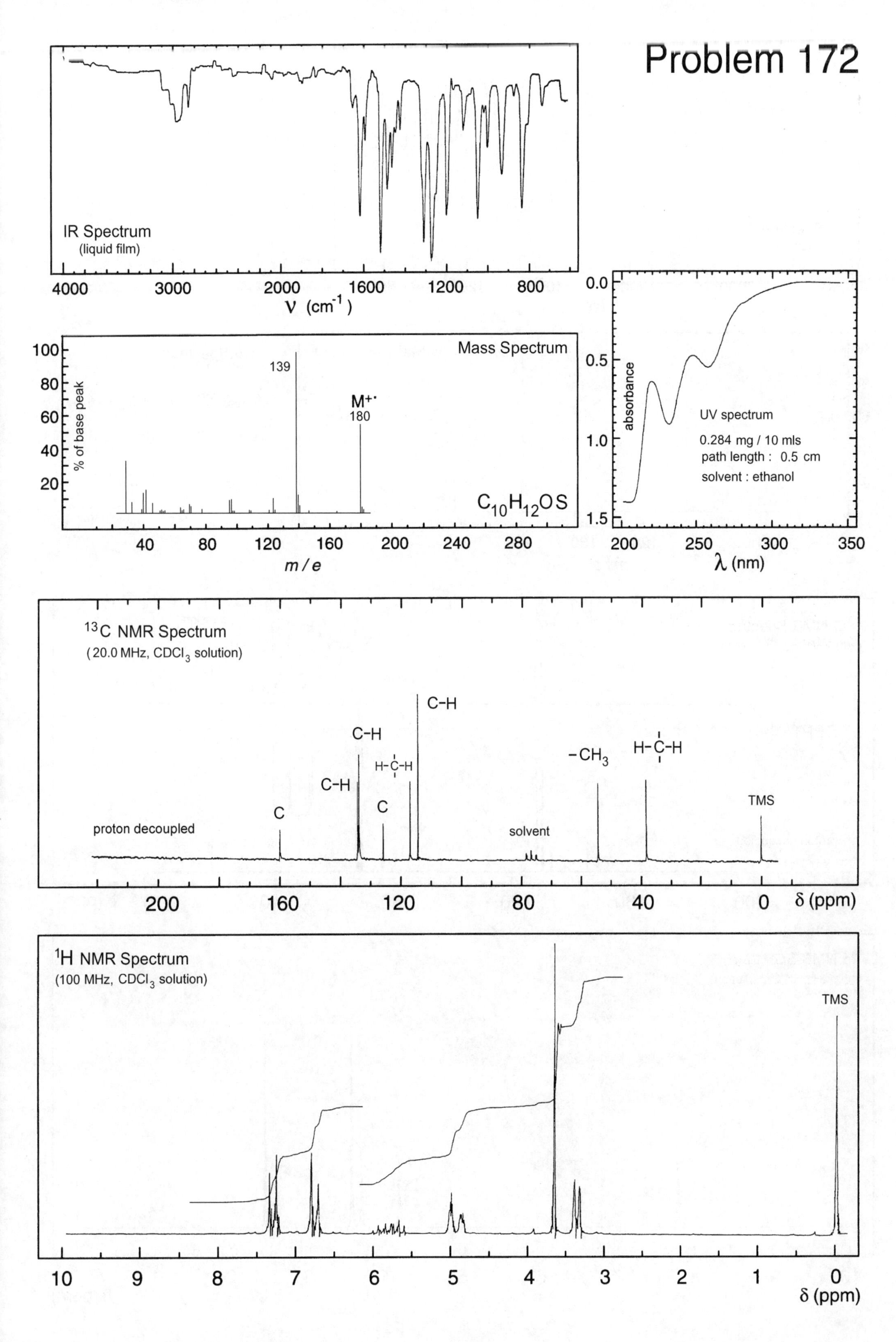

IR Spectrum
(liquid film)
4000
3000
2000
1600
1200
800
ν (cm⁻¹)
Mass Spectrum
100
80
60
40
20
% of base peak
139
M+·
180
$C_{10}H_{12}OS$
40
80
120
160
200
240
280
m / e
0.0
0.5
1.0
1.5
absorbance
UV spectrum
0.284 mg / 10 mls
path length : 0.5 cm
solvent : ethanol
200
250
300
350
λ (nm)
^{13}C NMR Spectrum
(20.0 MHz, $CDCl_3$ solution)
C-H
C-H
H-C-H
C-H
C
C
proton decoupled
solvent
$-CH_3$
H-C-H
TMS
200
160
120
80
40
0
δ (ppm)
^{1}H NMR Spectrum
(100 MHz, $CDCl_3$ solution)
TMS
10
9
8
7
6
5
4
3
2
1
0
δ (ppm)

Problem 173

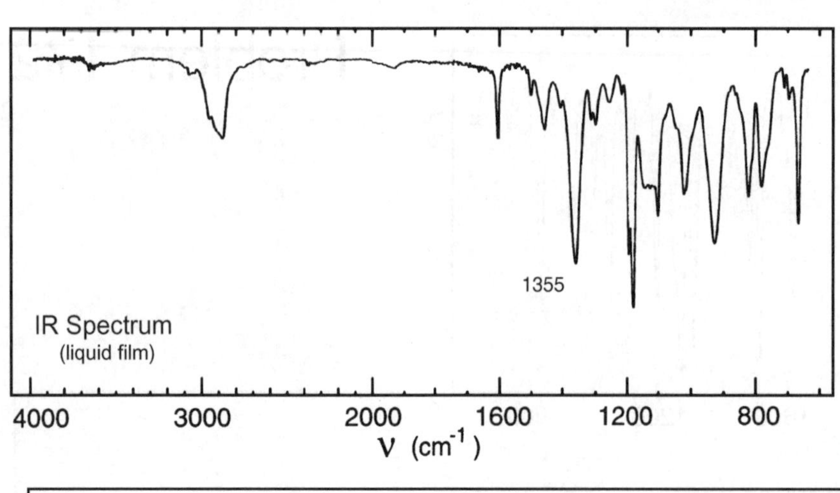

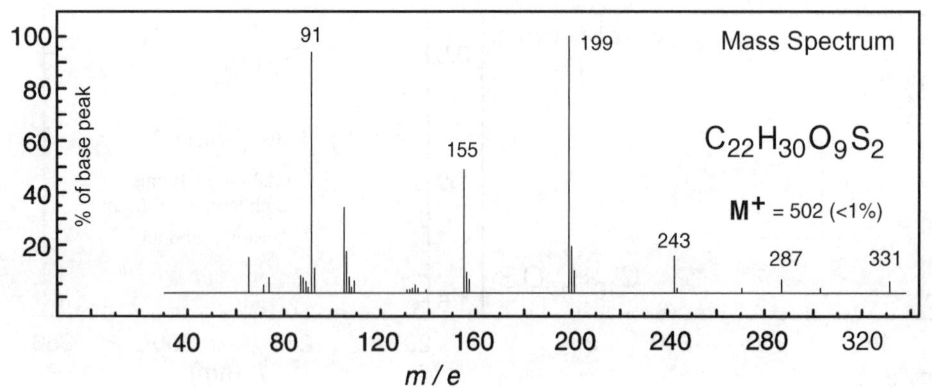

UV Spectrum

λ_{max} 225 nm ($\log_{10}\varepsilon$ 4.6)

λ_{max} 260 nm ($\log_{10}\varepsilon$ 3.3)

λ_{max} 272 nm ($\log_{10}\varepsilon$ 3.1)

solvent : methanol

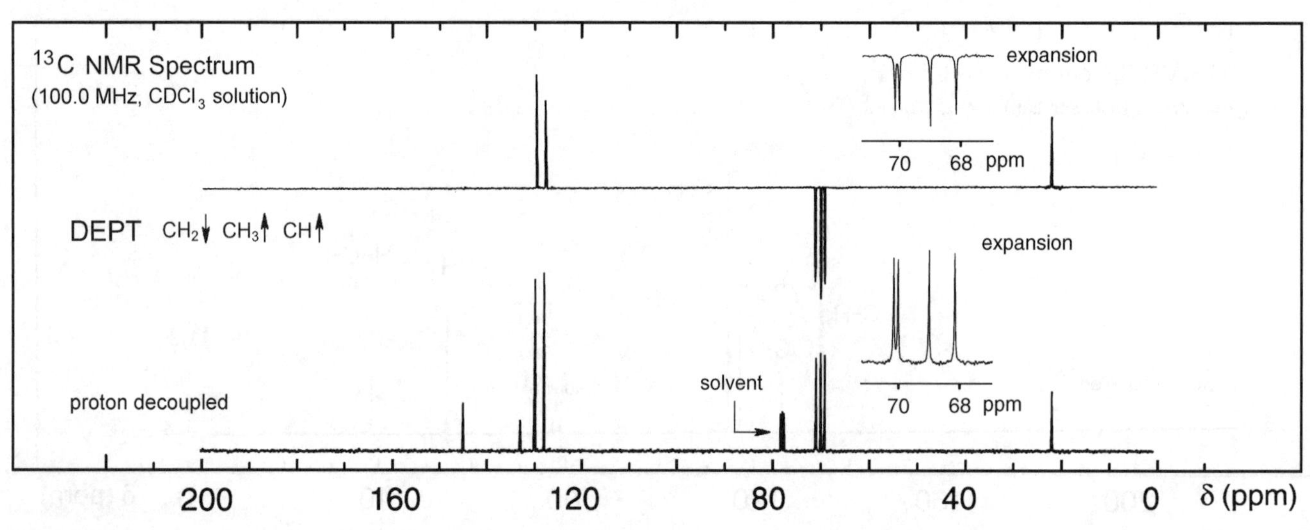

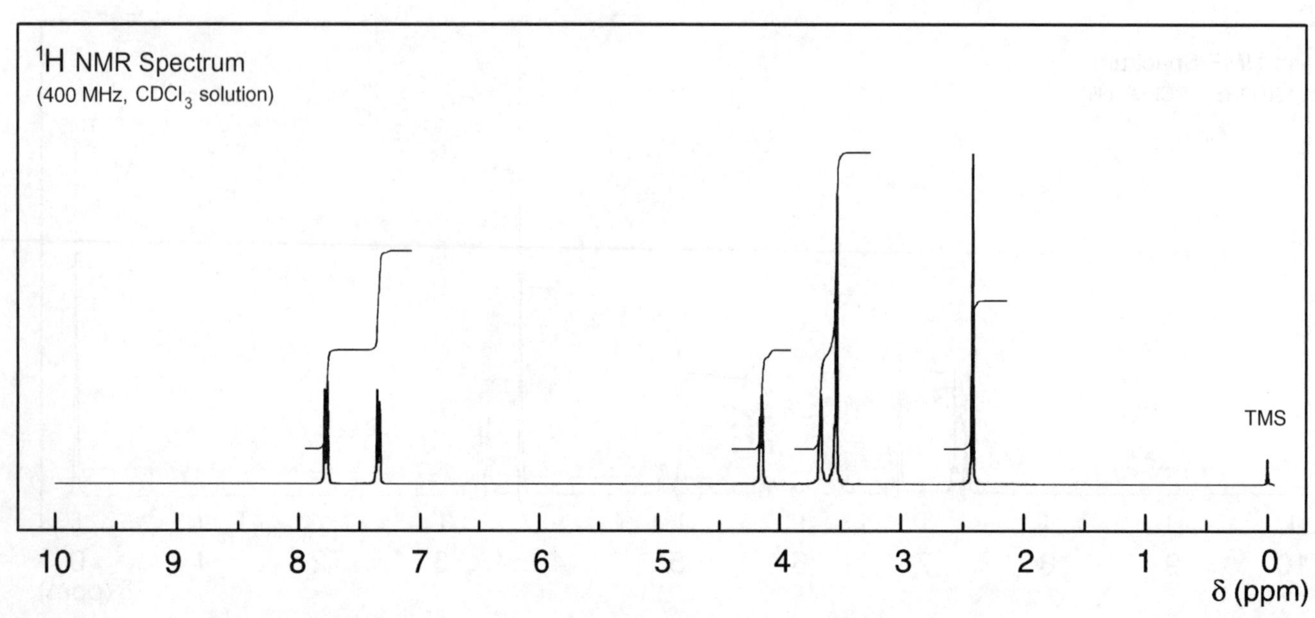

Problem 174

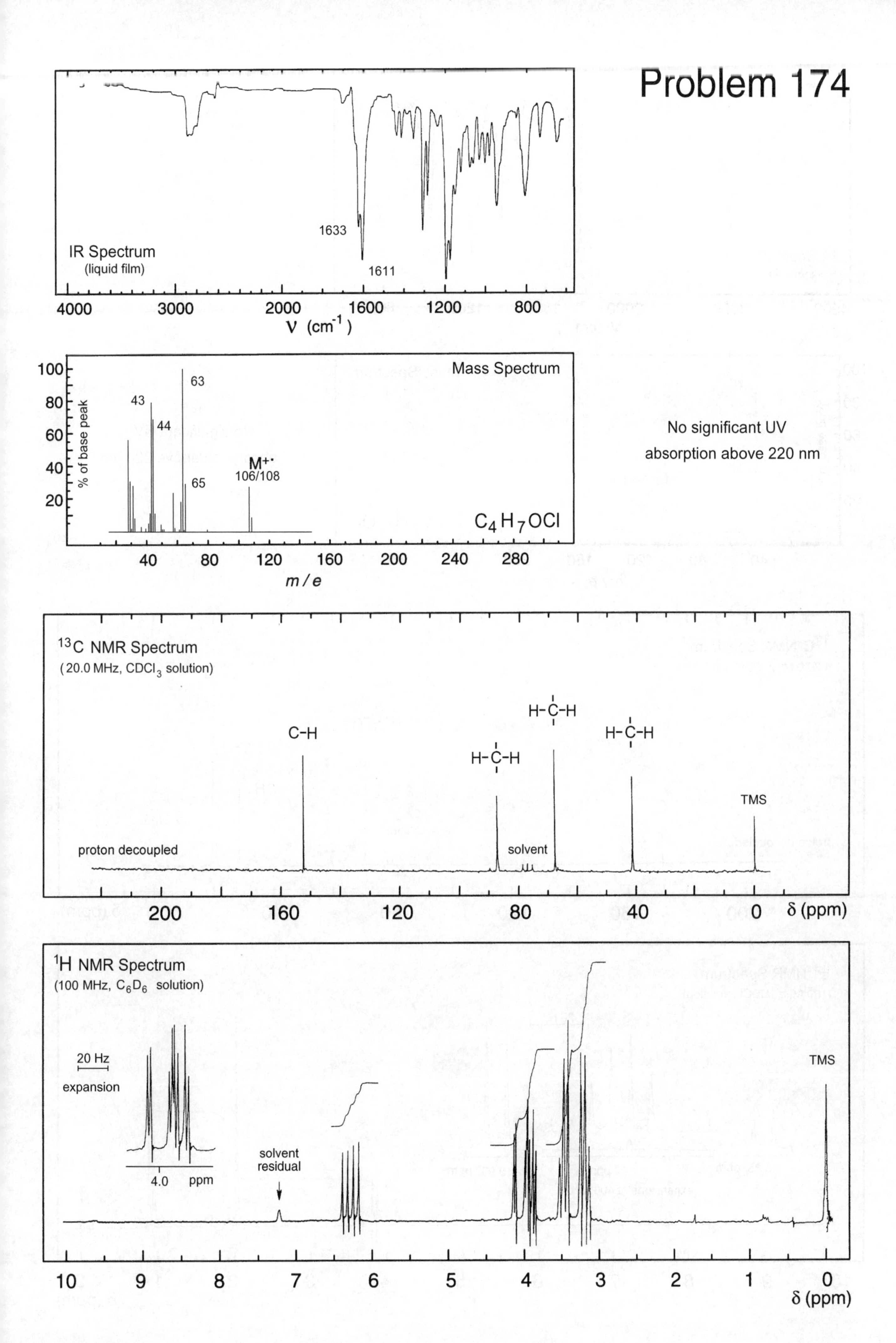

IR Spectrum
(liquid film)
1633
1611
4000
3000
2000
1600
1200
800
ν (cm-1)
Mass Spectrum
% of base peak
43
44
63
65
M+·
106/108
C4H7OCl
m/e
No significant UV
absorption above 220 nm
13C NMR Spectrum
(20.0 MHz, CDCl3 solution)
C-H
H-C-H
H-C-H
H-C-H
TMS
proton decoupled
solvent
δ (ppm)
1H NMR Spectrum
(100 MHz, C6D6 solution)
20 Hz
expansion
4.0
ppm
solvent
residual
TMS
δ (ppm)

Problem 175

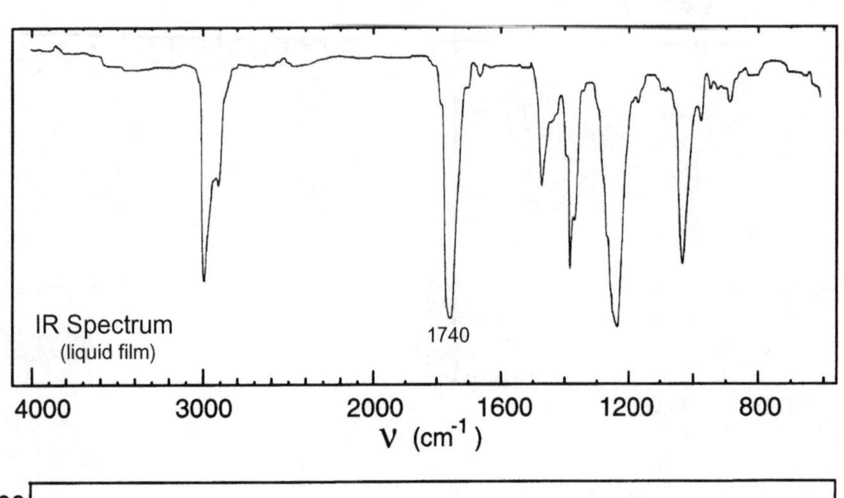

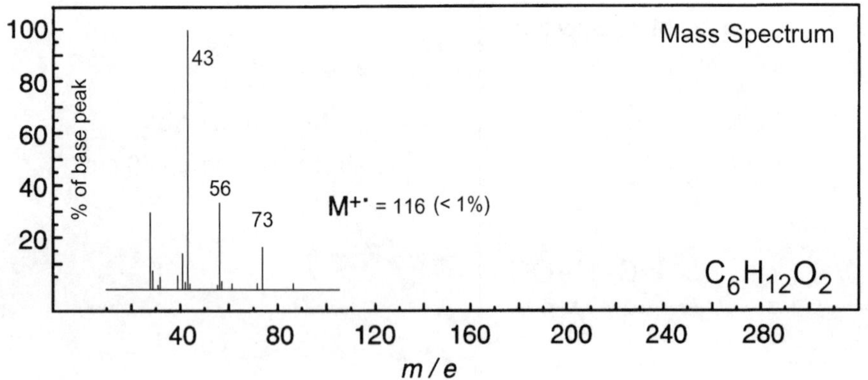

No significant UV absorption above 220 nm

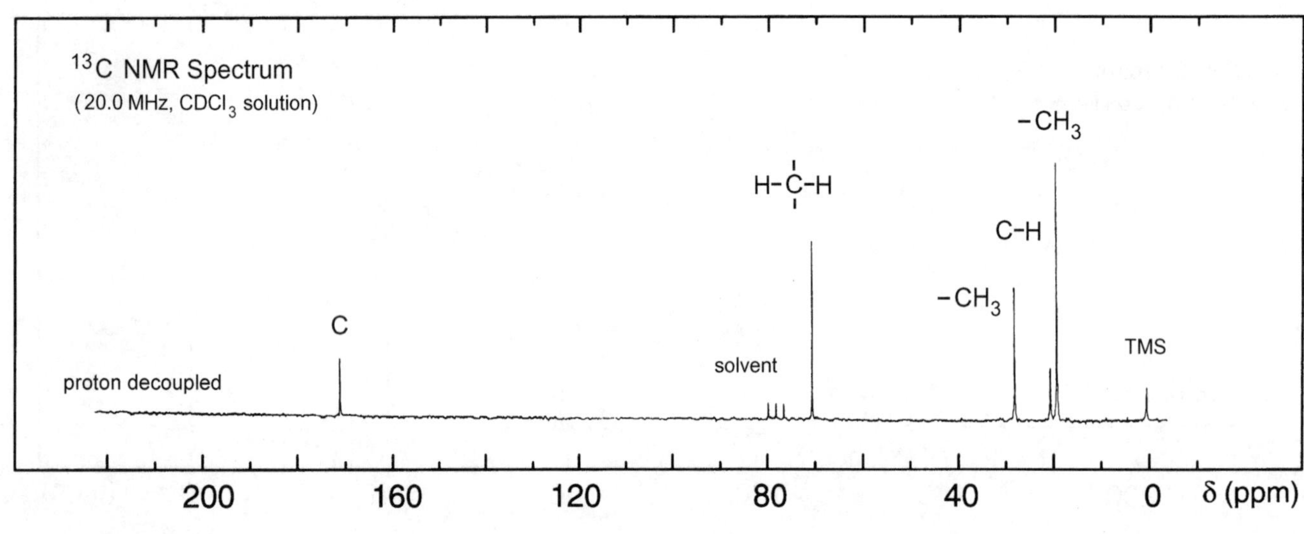

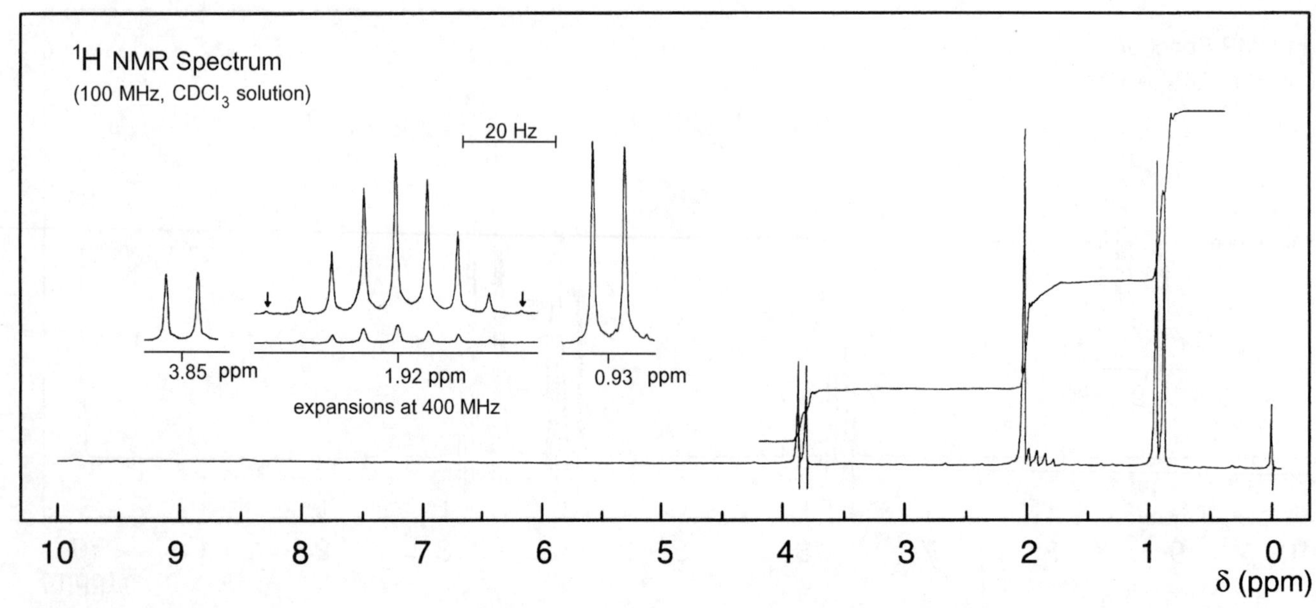

Problem 176

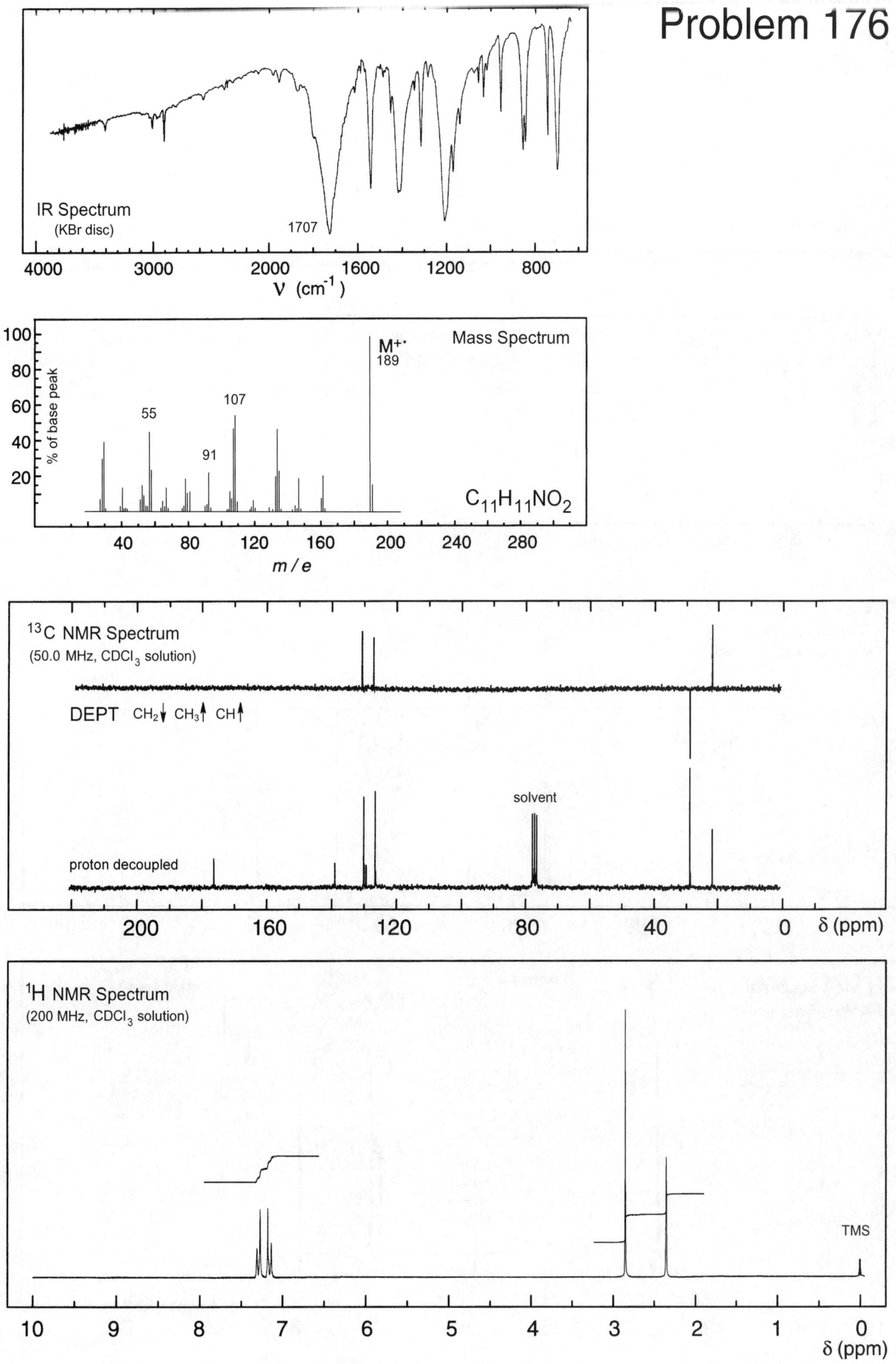

IR Spectrum
(KBr disc)
1707
4000
3000
2000
1600
1200
800
ν (cm⁻¹)
Mass Spectrum
M+·
189
107
55
91
% of base peak
100
80
60
40
20
40
80
120
160
200
240
280
m / e
C11H11NO2
13C NMR Spectrum
(50.0 MHz, CDCl3 solution)
DEPT CH2↓ CH3↑ CH↑
solvent
proton decoupled
200
160
120
80
40
0
δ (ppm)
1H NMR Spectrum
(200 MHz, CDCl3 solution)
TMS
10
9
8
7
6
5
4
3
2
1
0
δ (ppm)

Problem 177

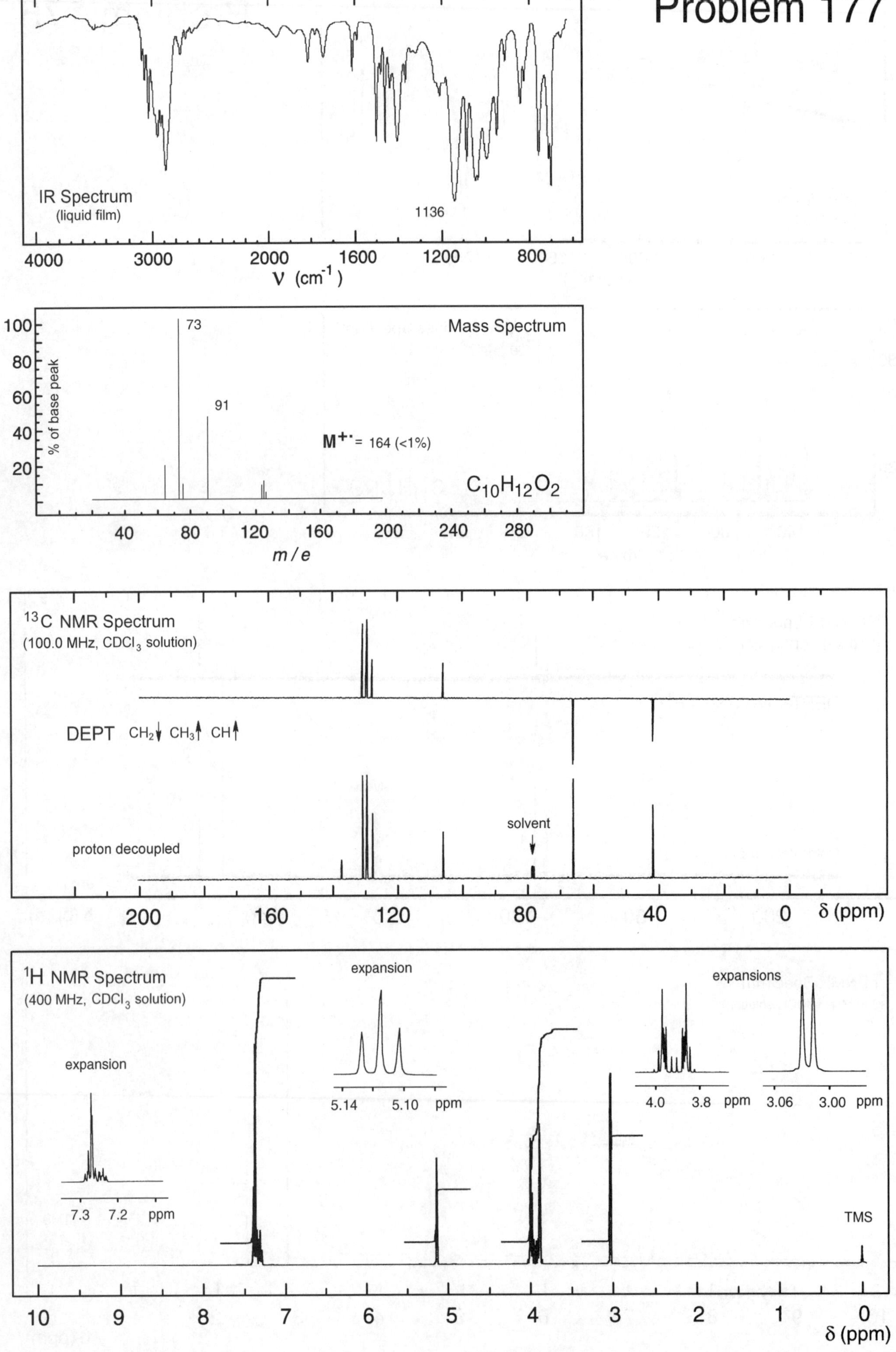

IR Spectrum
(liquid film)
1136
4000
3000
2000
1600
1200
800
ν (cm-1)
Mass Spectrum
% of base peak
100
80
60
40
20
73
91
M+· = 164 (<1%)
C10H12O2
40
80
120
160
200
240
280
m / e
13C NMR Spectrum
(100.0 MHz, CDCl3 solution)
DEPT CH2↓ CH3↑ CH↑
solvent
proton decoupled
200
160
120
80
40
0
δ (ppm)
1H NMR Spectrum
(400 MHz, CDCl3 solution)
expansion
7.3 7.2 ppm
expansion
5.14 5.10 ppm
expansions
4.0 3.8 ppm
3.06 3.00 ppm
TMS
10
9
8
7
6
5
4
3
2
1
0
δ (ppm)

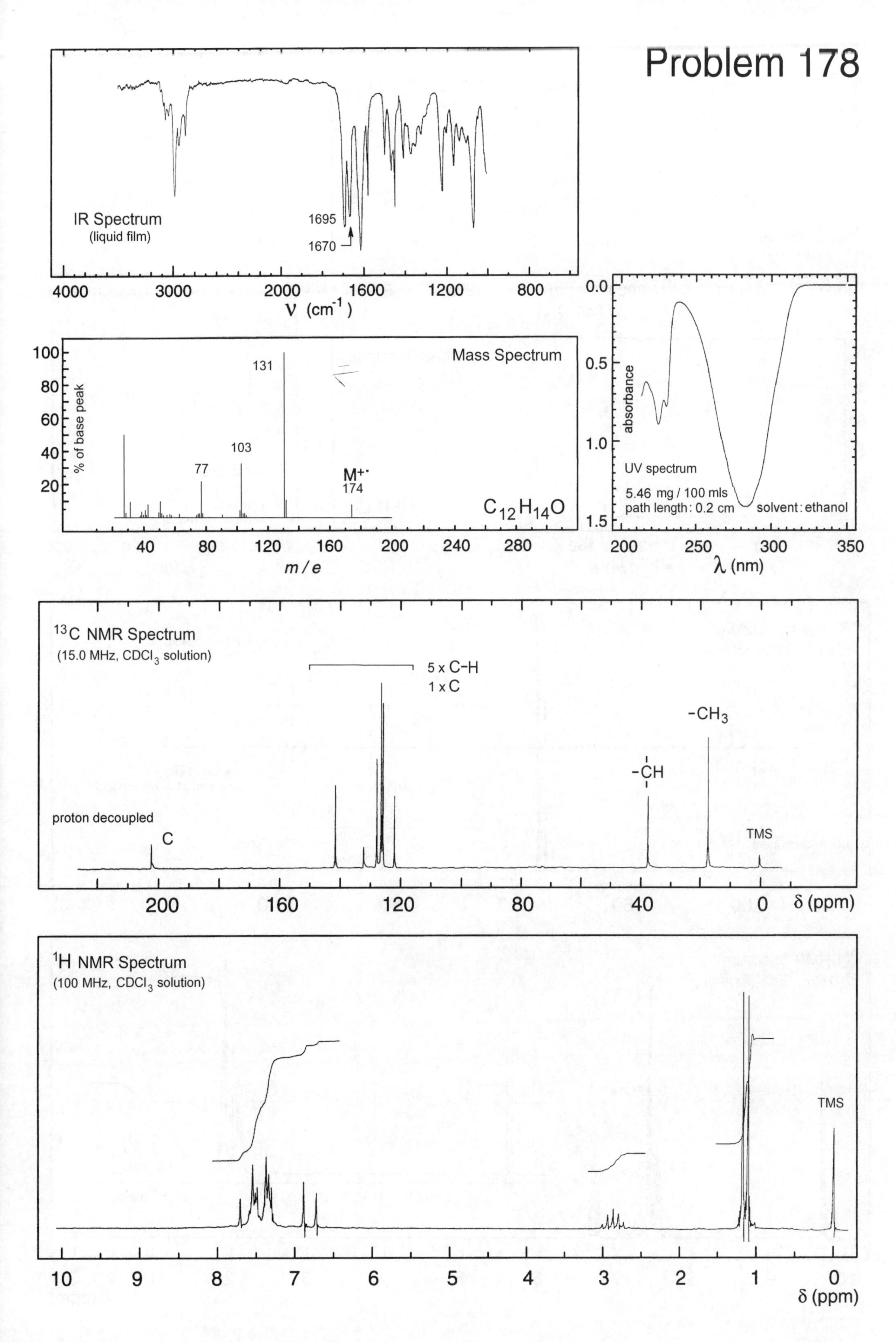

Problem 178
IR Spectrum
(liquid film)
1695
1670
4000
3000
2000
1600
1200
800
ν (cm⁻¹)
Mass Spectrum
% of base peak
131
103
77
M+·
174
C12H14O
m / e
absorbance
UV spectrum
5.46 mg / 100 mls
path length: 0.2 cm
solvent: ethanol
λ (nm)
13C NMR Spectrum
(15.0 MHz, CDCl3 solution)
5 x C-H
1 x C
-CH3
-CH
proton decoupled
C
TMS
δ (ppm)
1H NMR Spectrum
(100 MHz, CDCl3 solution)
TMS
δ (ppm)

Problem 179

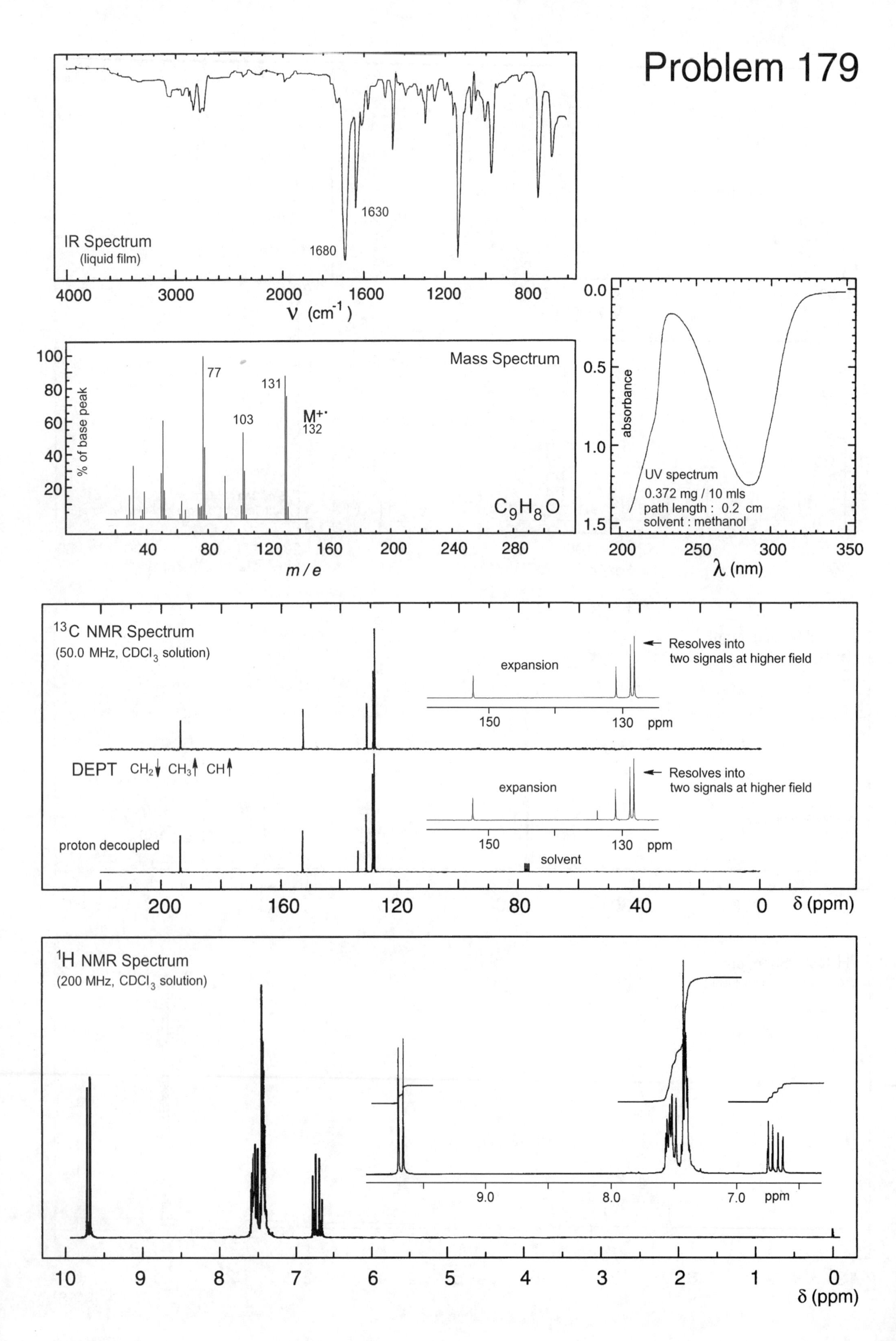

IR Spectrum
(liquid film)
1680
1630
4000
3000
2000
1600
1200
800
ν (cm-1)
Mass Spectrum
% of base peak
100
80
60
40
20
77
131
103
M+·
132
C9H8O
40
80
120
160
200
240
280
m / e
UV spectrum
0.372 mg / 10 mls
path length : 0.2 cm
solvent : methanol
absorbance
0.0
0.5
1.0
1.5
200
250
300
350
λ (nm)
13C NMR Spectrum
(50.0 MHz, CDCl3 solution)
expansion
Resolves into
two signals at higher field
150
130
ppm
DEPT CH2↓ CH3↑ CH↑
proton decoupled
solvent
200
160
120
80
40
0
δ (ppm)
1H NMR Spectrum
(200 MHz, CDCl3 solution)
9.0
8.0
7.0
ppm
10
9
8
7
6
5
4
3
2
1
0
δ (ppm)

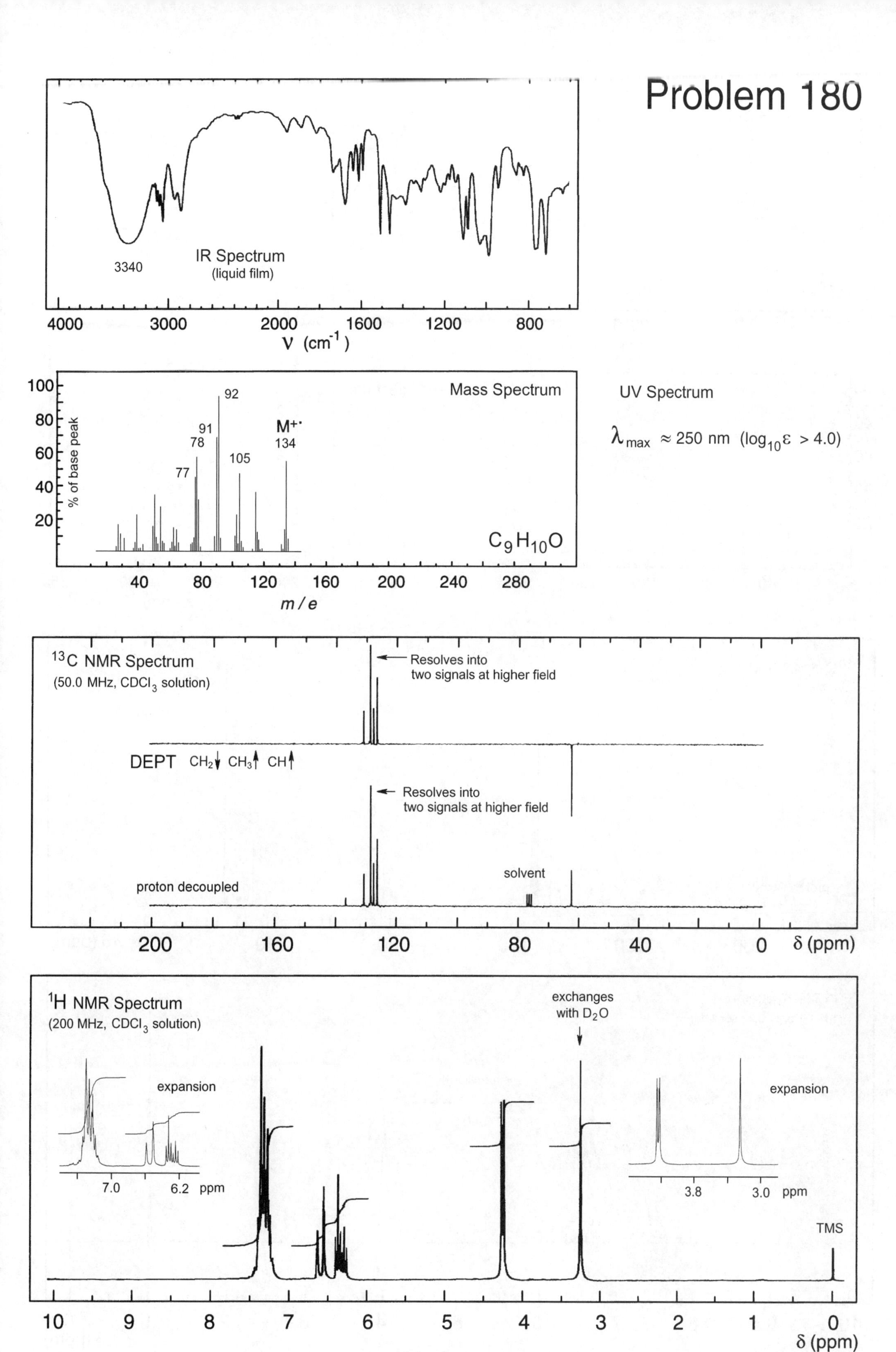
Problem 180
IR Spectrum
(liquid film)
3340
4000
3000
2000
1600
1200
800
ν (cm⁻¹)
Mass Spectrum
% of base peak
100
80
60
40
20
92
91
78
77
105
M+·
134
C9H10O
40
80
120
160
200
240
280
m / e
UV Spectrum
λmax ≈ 250 nm (log10ε > 4.0)
13C NMR Spectrum
(50.0 MHz, CDCl3 solution)
Resolves into
two signals at higher field
DEPT CH2↓ CH3↑ CH↑
Resolves into
two signals at higher field
solvent
proton decoupled
200
160
120
80
40
0
δ (ppm)
1H NMR Spectrum
(200 MHz, CDCl3 solution)
expansion
7.0
6.2 ppm
exchanges
with D2O
expansion
3.8
3.0 ppm
TMS
10
9
8
7
6
5
4
3
2
1
0
δ (ppm)

Problem 181

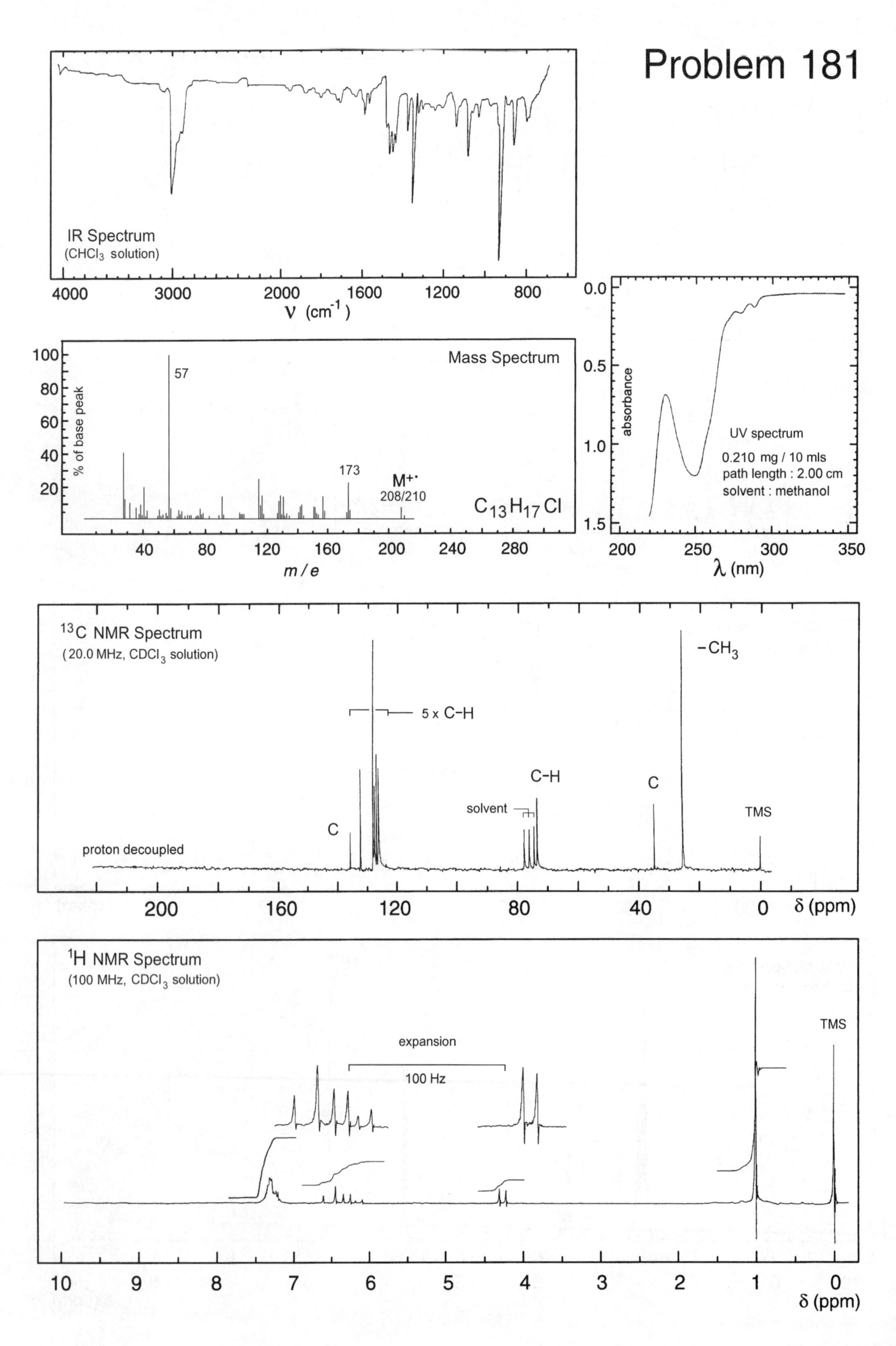

IR Spectrum
(CHCl3 solution)
4000
3000
2000
1600
1200
800
ν (cm-1)
Mass Spectrum
% of base peak
57
173
M+·
208/210
C13H17Cl
m / e
absorbance
UV spectrum
0.210 mg / 10 mls
path length : 2.00 cm
solvent : methanol
λ (nm)
13C NMR Spectrum
(20.0 MHz, CDCl3 solution)
5 x C-H
-CH3
C-H
C
solvent
C
TMS
proton decoupled
δ (ppm)
1H NMR Spectrum
(100 MHz, CDCl3 solution)
expansion
100 Hz
TMS
δ (ppm)

Problem 182

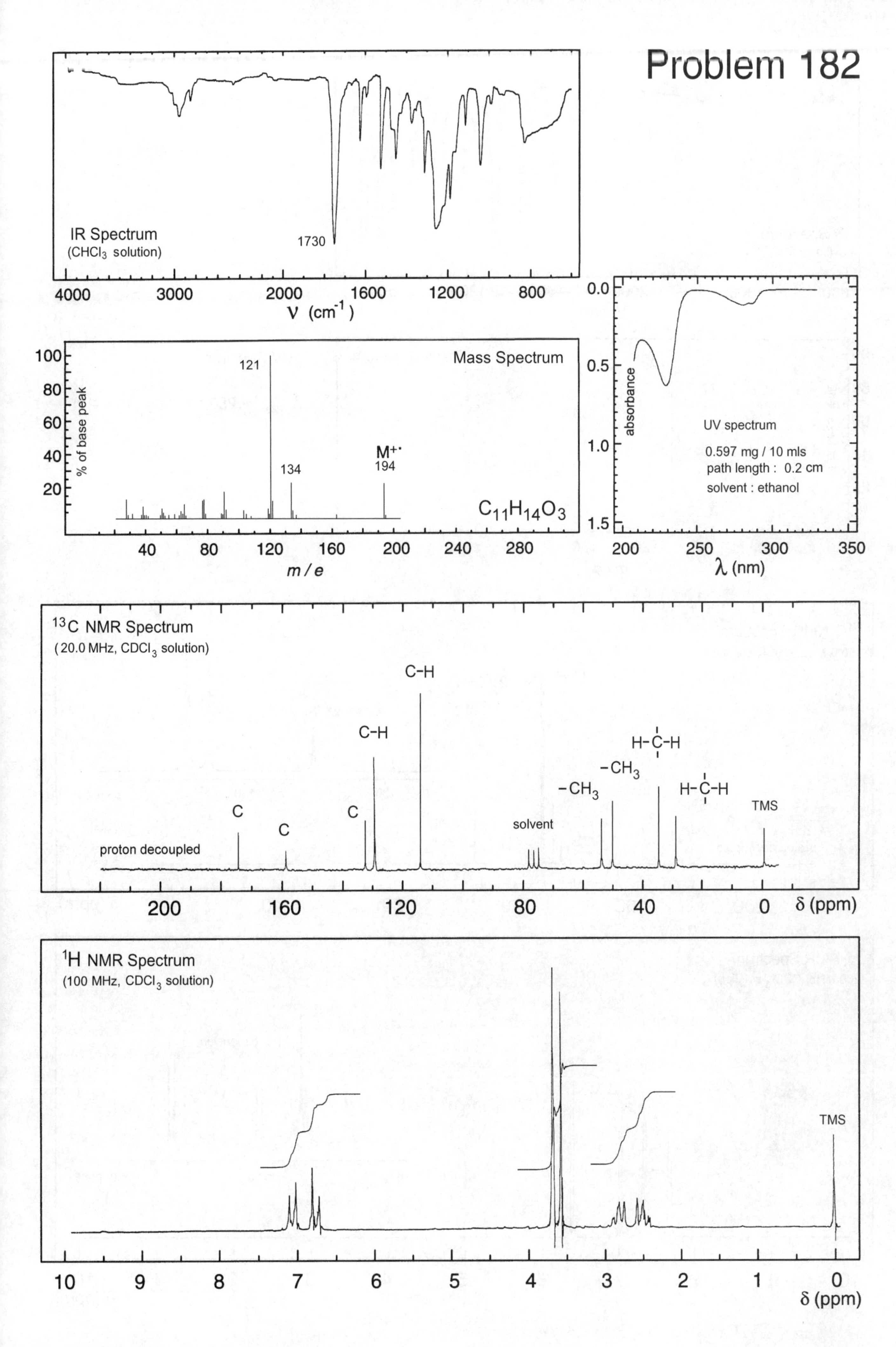

IR Spectrum
(CHCl3 solution)
1730
4000
3000
2000
1600
1200
800
ν (cm-1)
Mass Spectrum
% of base peak
121
134
M+·
194
C11H14O3
m / e
absorbance
UV spectrum
0.597 mg / 10 mls
path length : 0.2 cm
solvent : ethanol
λ (nm)
13C NMR Spectrum
(20.0 MHz, CDCl3 solution)
proton decoupled
C
C
C
C-H
C-H
solvent
-CH3
-CH3
H-C-H
H-C-H
TMS
δ (ppm)
1H NMR Spectrum
(100 MHz, CDCl3 solution)
TMS
δ (ppm)

Problem 183

IR Spectrum
(liquid film)

4000 3000 2000 1600 1200 800
ν (cm^{-1})

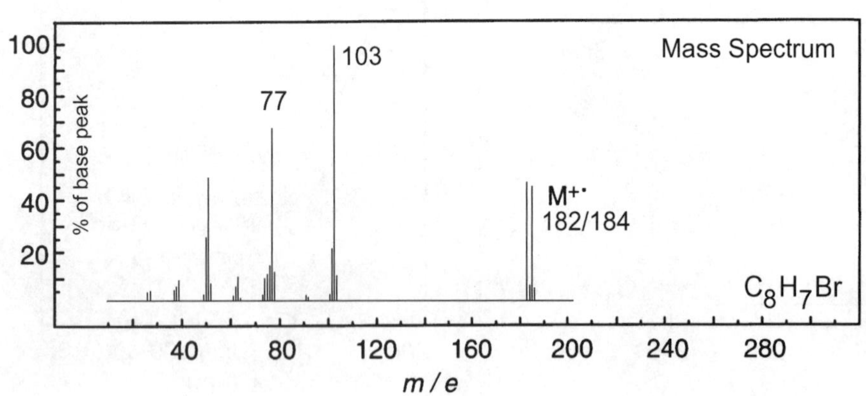

UV Spectrum

$\lambda_{max} \approx 250$ nm ($\log_{10}\varepsilon > 4.0$)

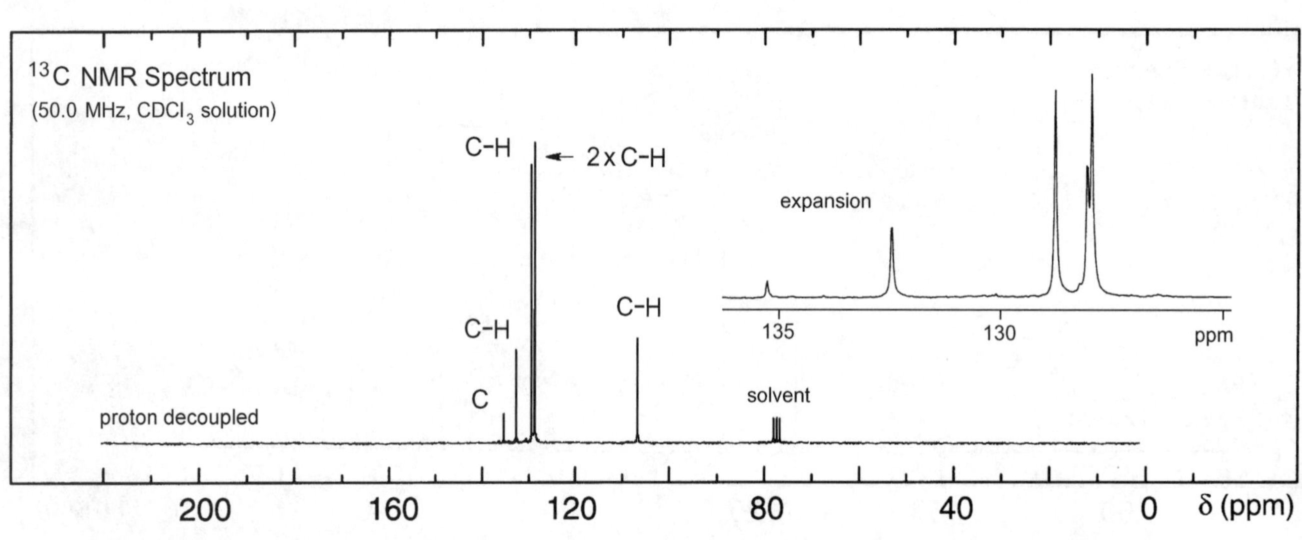

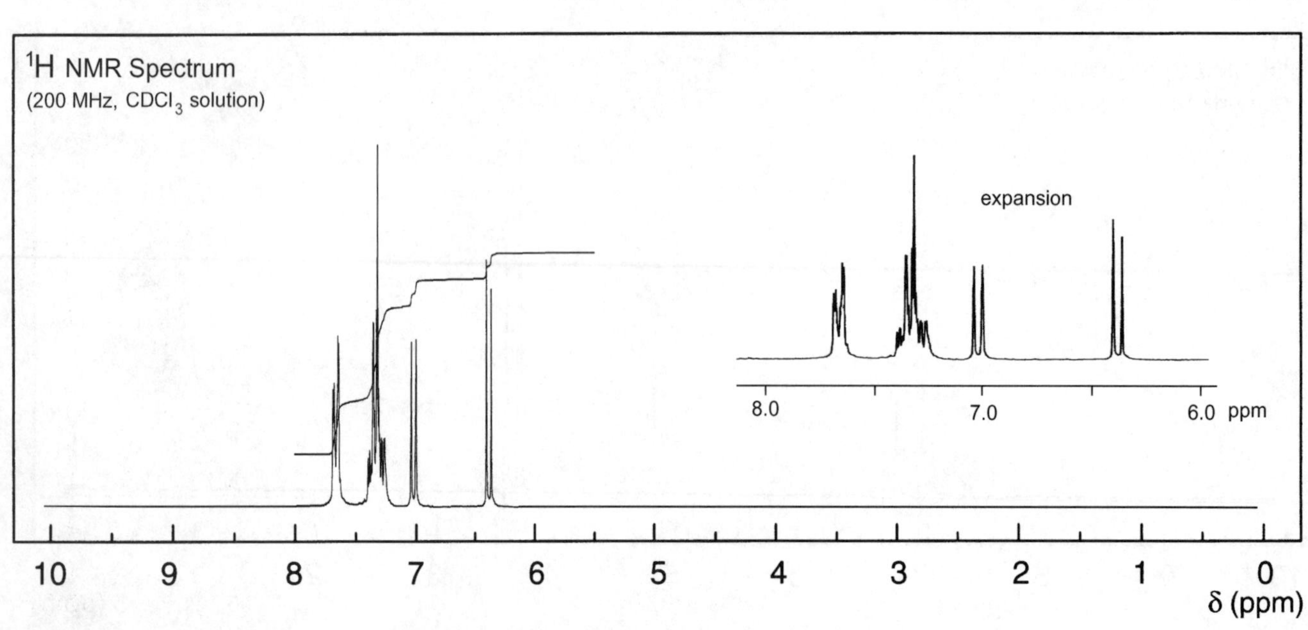

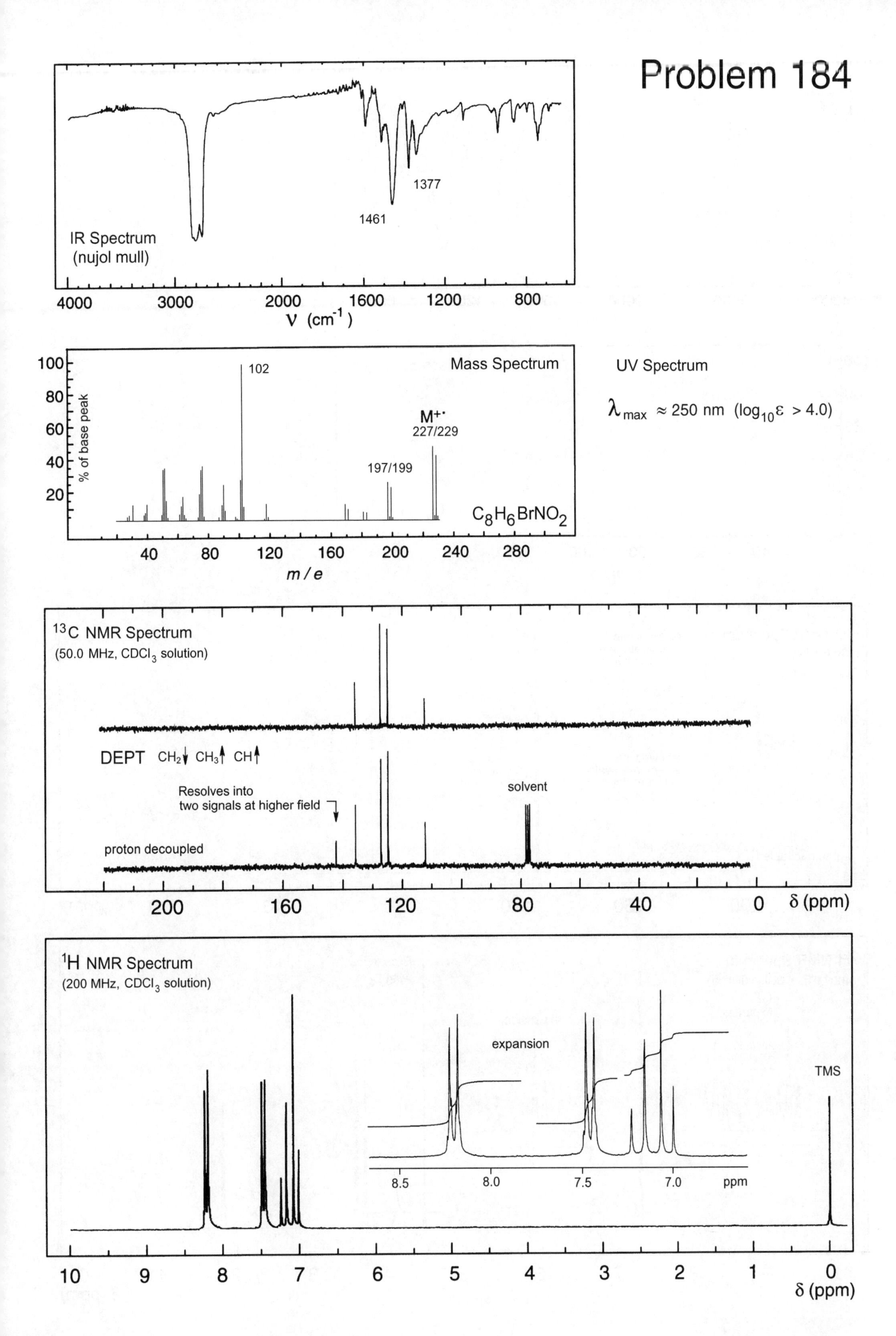

Problem 184
IR Spectrum
(nujol mull)
1461
1377
4000
3000
2000
1600
1200
800
ν (cm⁻¹)
Mass Spectrum
% of base peak
102
M+·
227/229
197/199
C8H6BrNO2
m / e
UV Spectrum
λmax ≈ 250 nm (log10ε > 4.0)
13C NMR Spectrum
(50.0 MHz, CDCl3 solution)
DEPT CH2↓ CH3↑ CH↑
Resolves into
two signals at higher field
solvent
proton decoupled
δ (ppm)
1H NMR Spectrum
(200 MHz, CDCl3 solution)
expansion
TMS
ppm
δ (ppm)

Problem 185

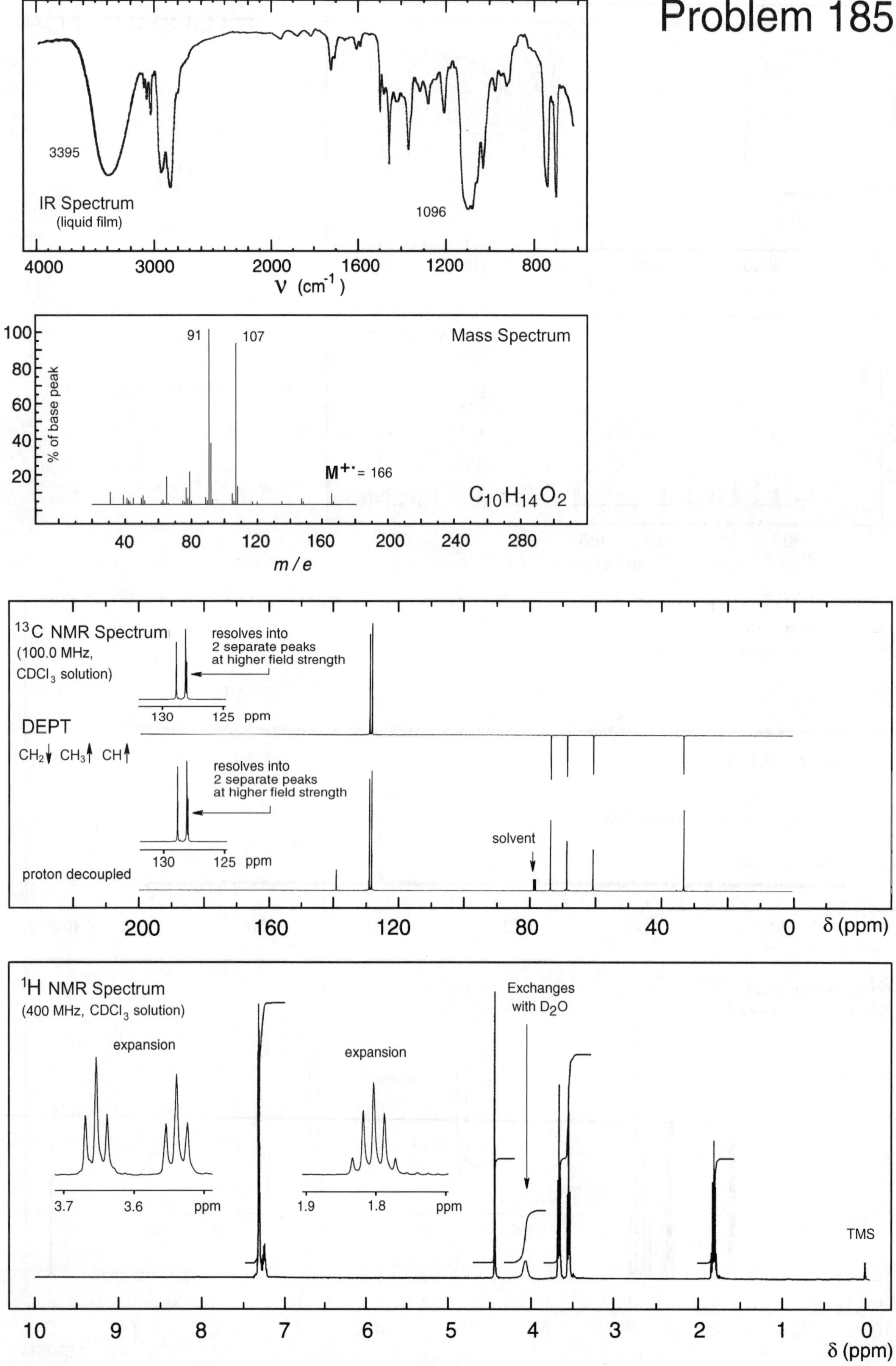

3395
IR Spectrum
(liquid film)
1096
4000
3000
2000
1600
1200
800
ν (cm^{-1})
100
80
60
40
20
% of base peak
91
107
Mass Spectrum
M+· = 166
$C_{10}H_{14}O_2$
40
80
120
160
200
240
280
m / e
^{13}C NMR Spectrum
(100.0 MHz,
$CDCl_3$ solution)
resolves into
2 separate peaks
at higher field strength
130
125 ppm
DEPT
CH2↓ CH3↑ CH↑
solvent
proton decoupled
200
160
120
80
40
0
δ (ppm)
^{1}H NMR Spectrum
(400 MHz, $CDCl_3$ solution)
expansion
3.7
3.6
ppm
1.9
1.8
Exchanges
with D_2O
TMS
10
9
8
7
6
5
4
3
2
1
0

Problem 186

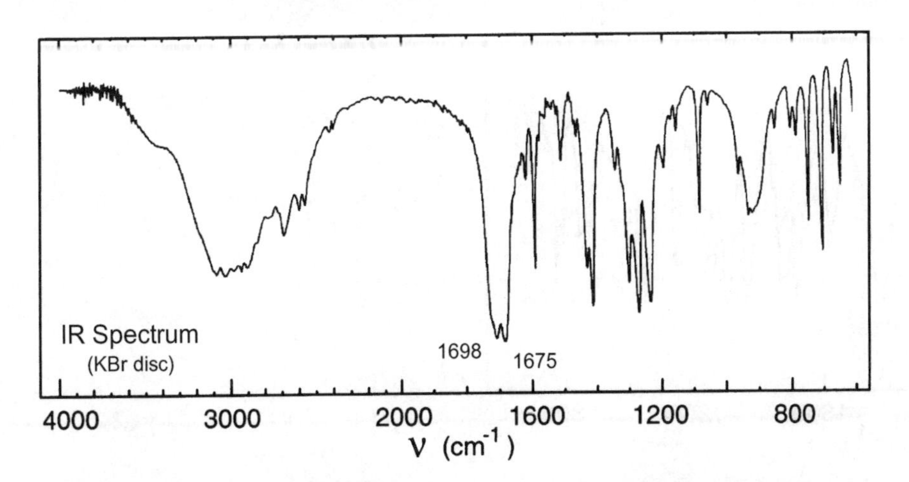

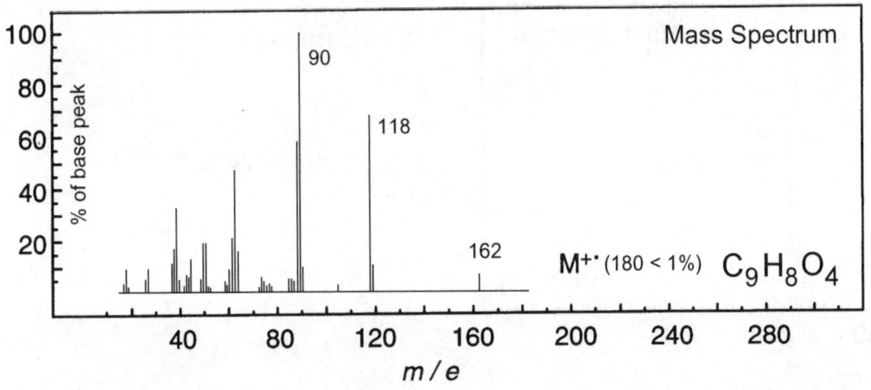

UV Spectrum

λ_{max} 276 nm ($\log_{10}\varepsilon$ 3.1)

λ_{max} 228 nm ($\log_{10}\varepsilon$ 3.9)

solvent : methanol

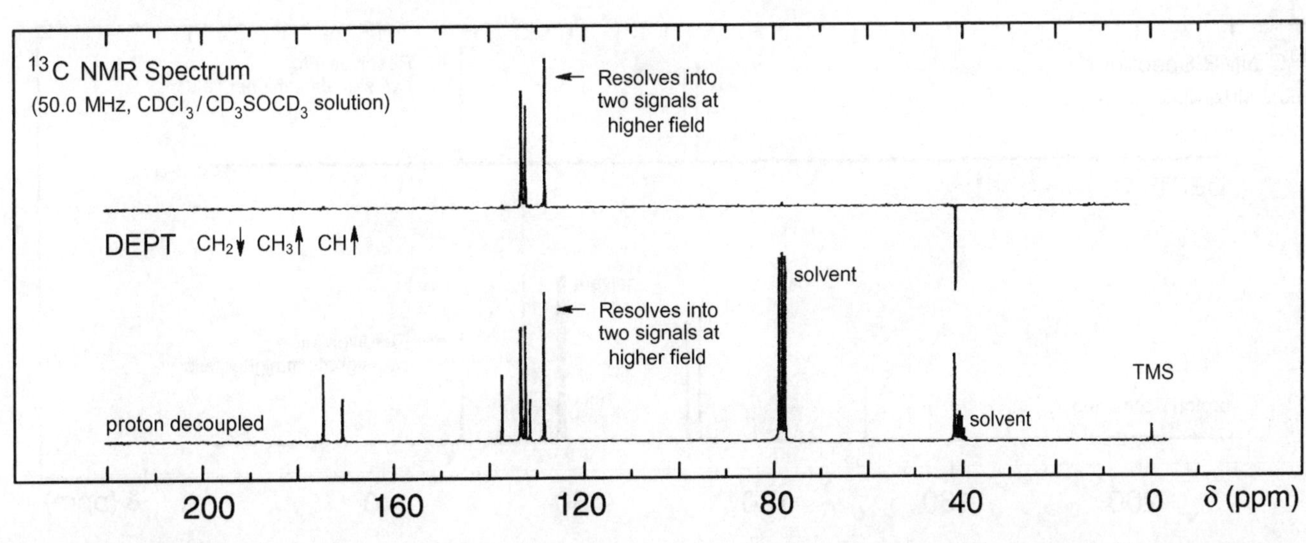

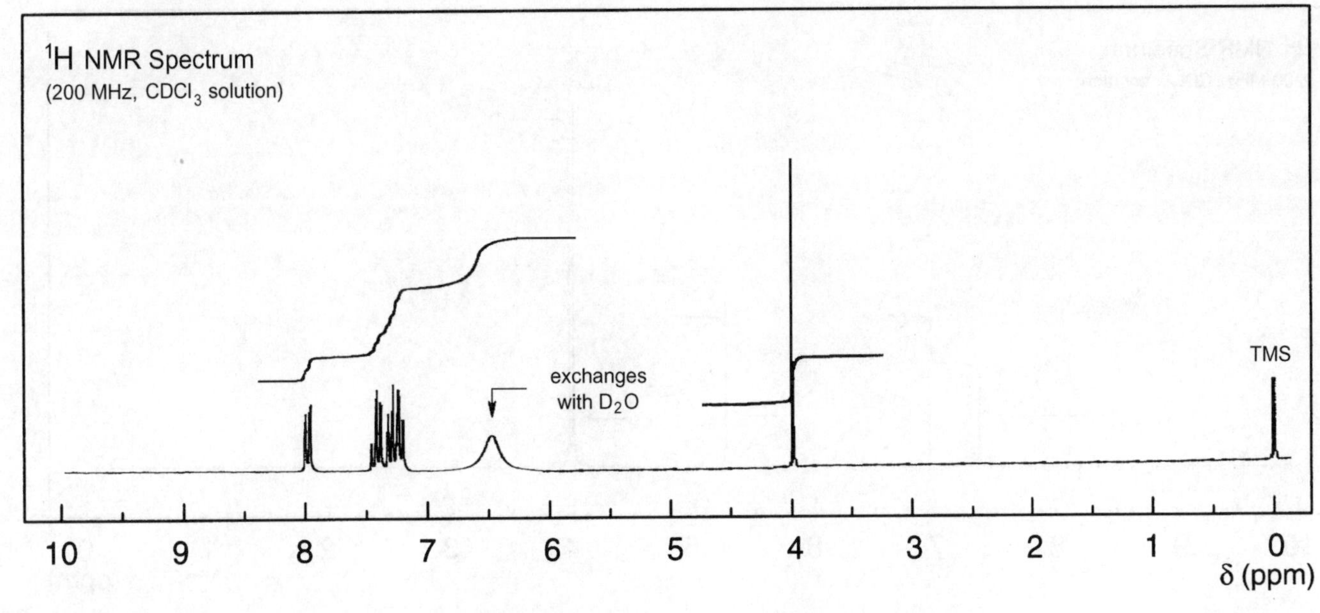

Problem 187

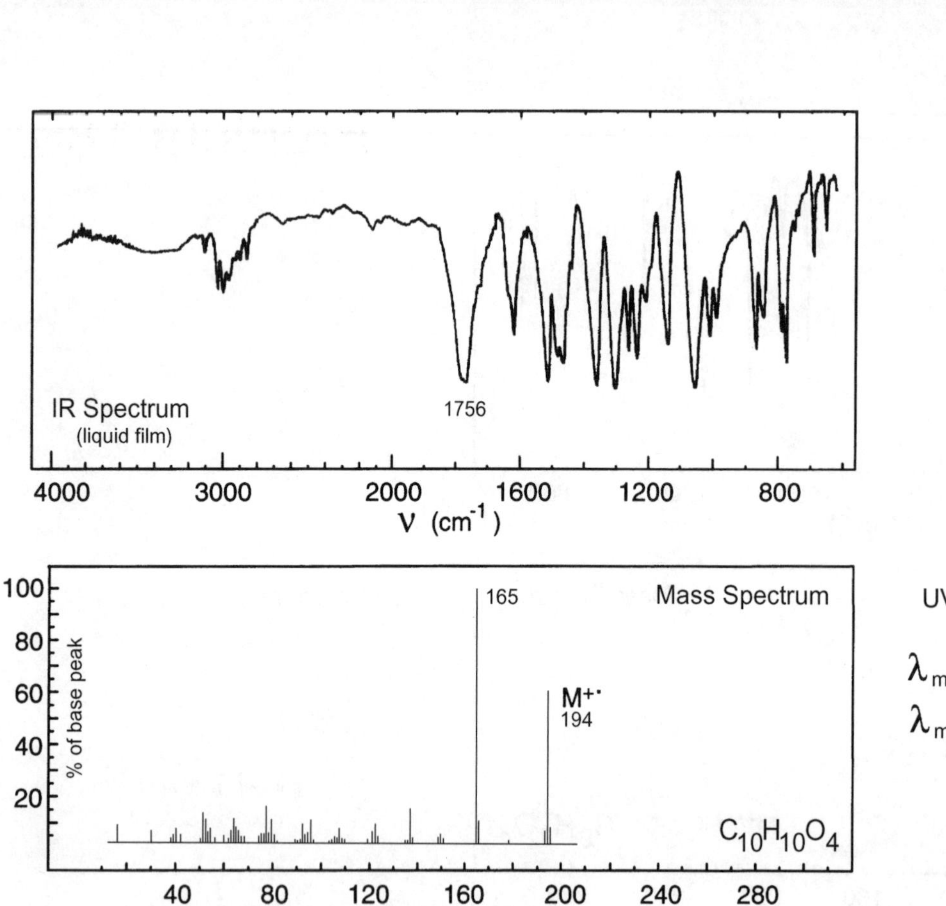

UV Spectrum

λ_{max} 232 nm ($\log_{10}\varepsilon$ 3.8)

λ_{max} 300 nm ($\log_{10}\varepsilon$ 3.6)

solvent : methanol

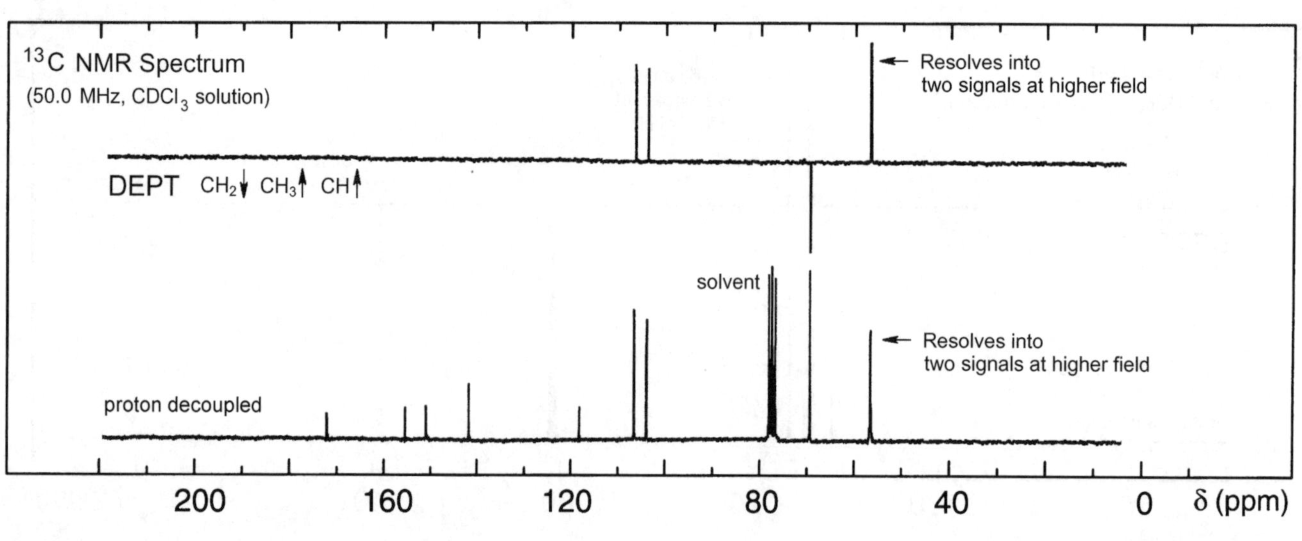

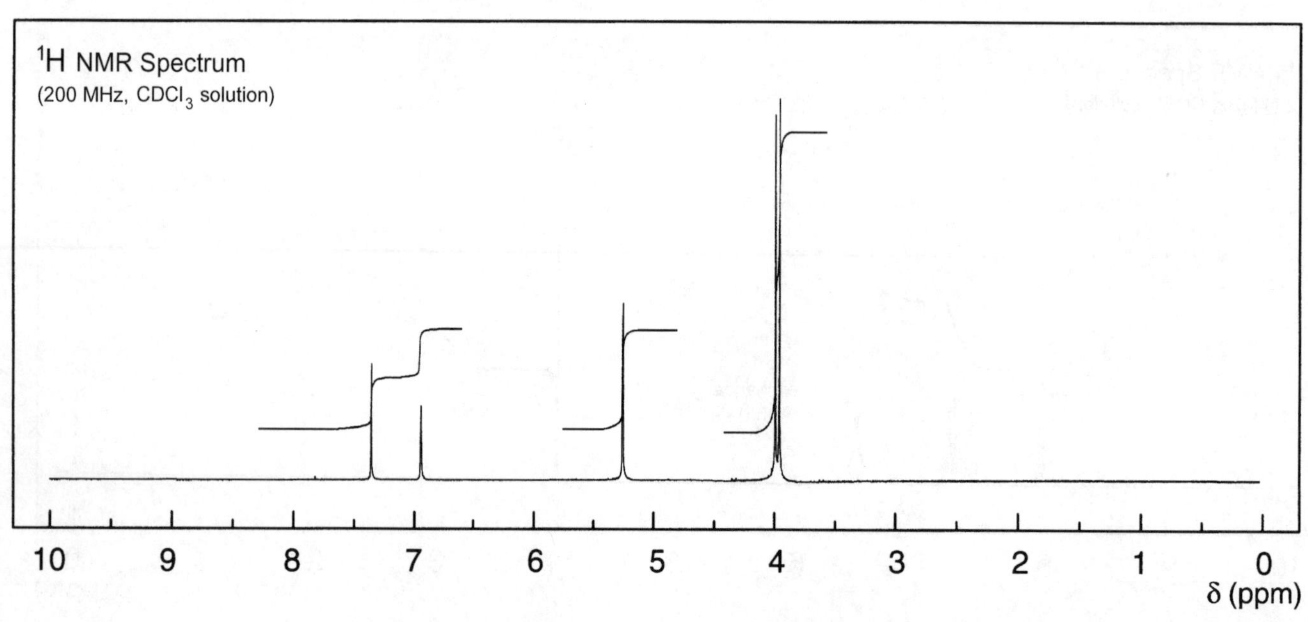

Problem 188

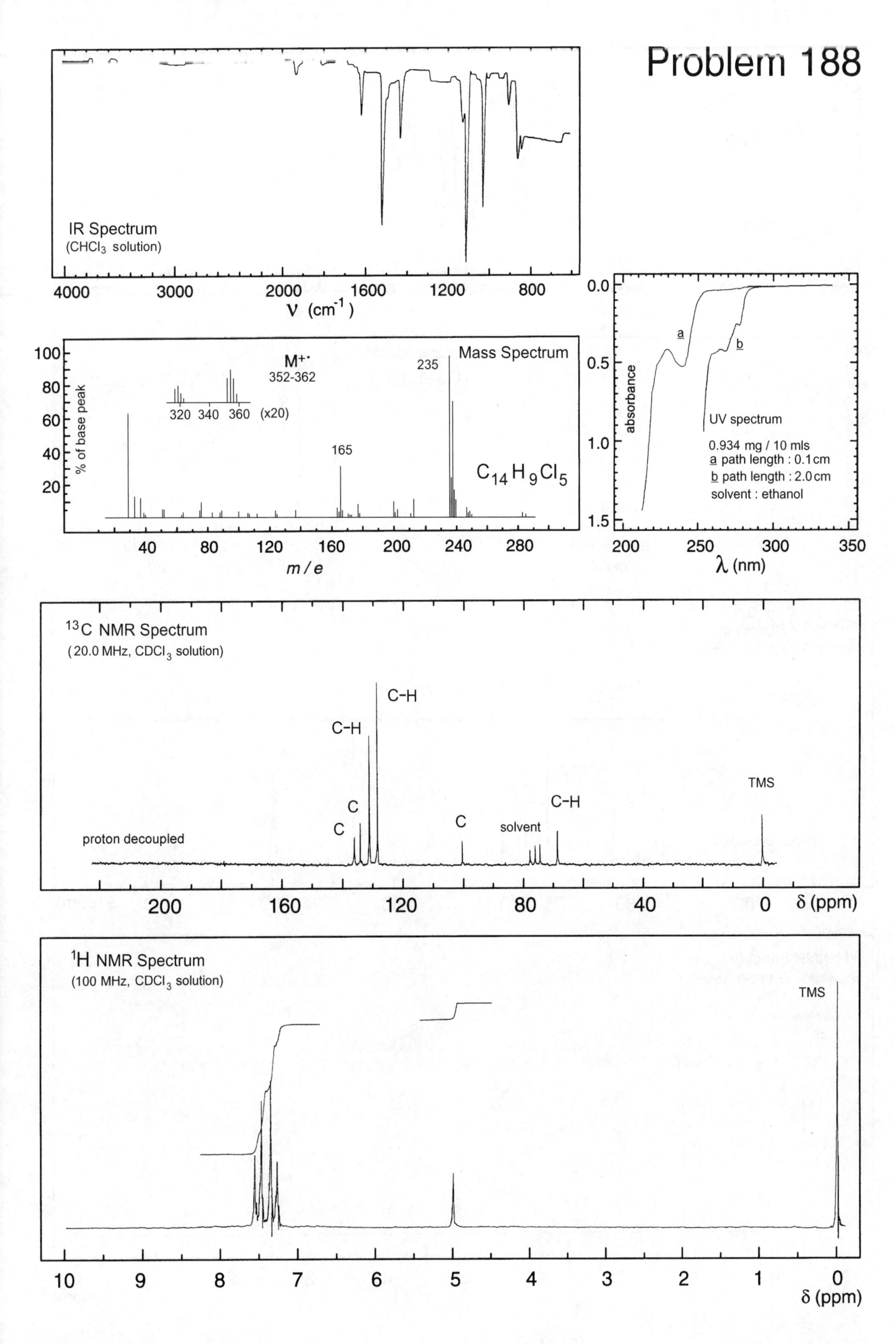

IR Spectrum
(CHCl3 solution)
4000
3000
2000
1600
1200
800
ν (cm-1)
Mass Spectrum
M+·
352-362
235
165
320 340 360 (x20)
% of base peak
100
80
60
40
20
40
80
120
160
200
240
280
m / e
C14 H9 Cl5
0.0
0.5
1.0
1.5
absorbance
a
b
UV spectrum
0.934 mg / 10 mls
a path length : 0.1 cm
b path length : 2.0 cm
solvent : ethanol
200
250
300
350
λ (nm)
13C NMR Spectrum
(20.0 MHz, CDCl3 solution)
C-H
C-H
C
C
C
solvent
C-H
TMS
proton decoupled
200
160
120
80
40
0
δ (ppm)
1H NMR Spectrum
(100 MHz, CDCl3 solution)
TMS
10
9
8
7
6
5
4
3
2
1
0
δ (ppm)

Problem 189

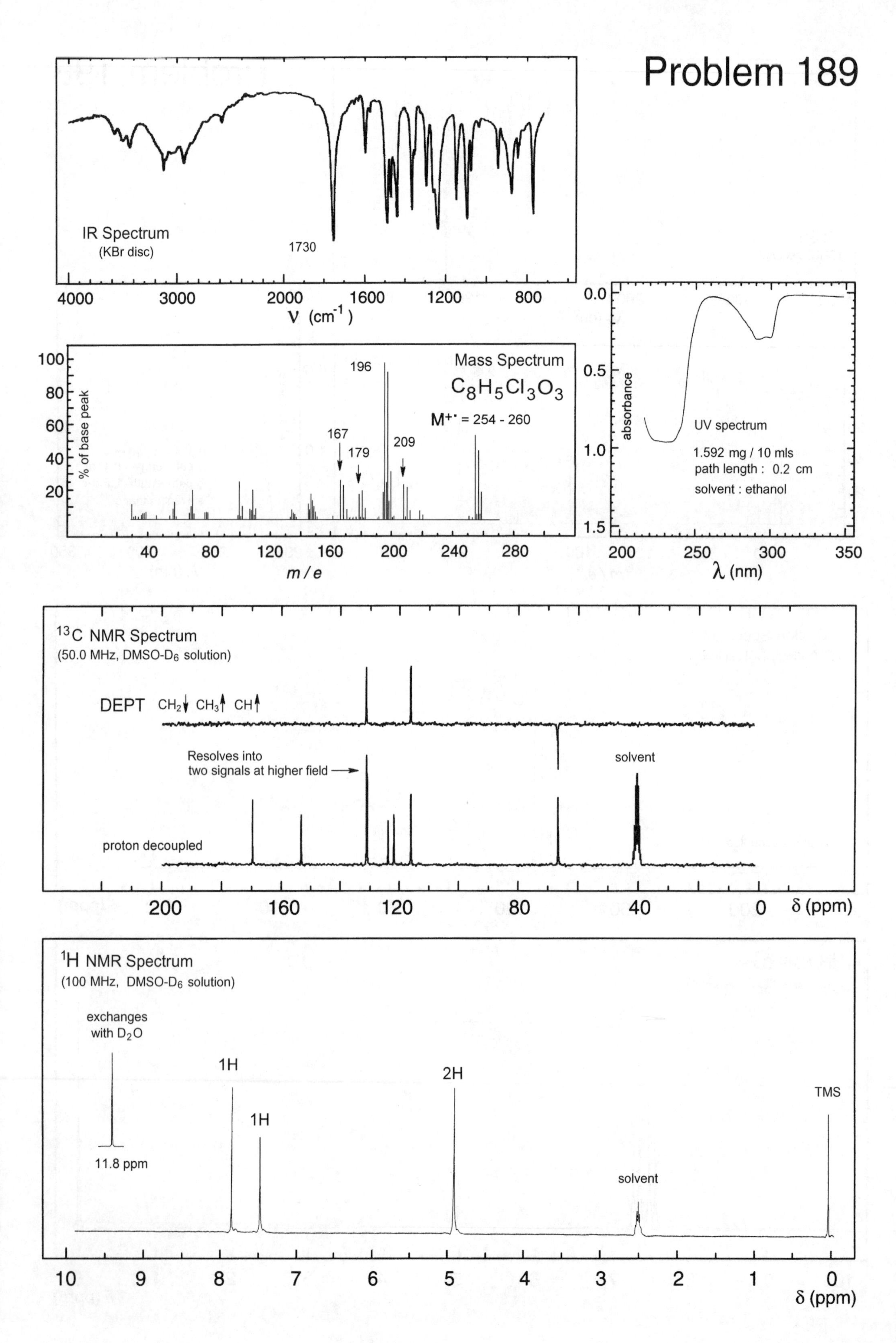

IR Spectrum
(KBr disc)
1730
4000
3000
2000
1600
1200
800
ν (cm⁻¹)
Mass Spectrum
$C_8H_5Cl_3O_3$
$M^{+\bullet}$ = 254 - 260
196
167
179
209
% of base peak
m / e
UV spectrum
1.592 mg / 10 mls
path length : 0.2 cm
solvent : ethanol
absorbance
λ (nm)
^{13}C NMR Spectrum
(50.0 MHz, DMSO-D6 solution)
DEPT CH2↓ CH3↑ CH↑
Resolves into
two signals at higher field
solvent
proton decoupled
δ (ppm)
^{1}H NMR Spectrum
(100 MHz, DMSO-D6 solution)
exchanges
with D2O
11.8 ppm
1H
1H
2H
solvent
TMS
δ (ppm)

Problem 190

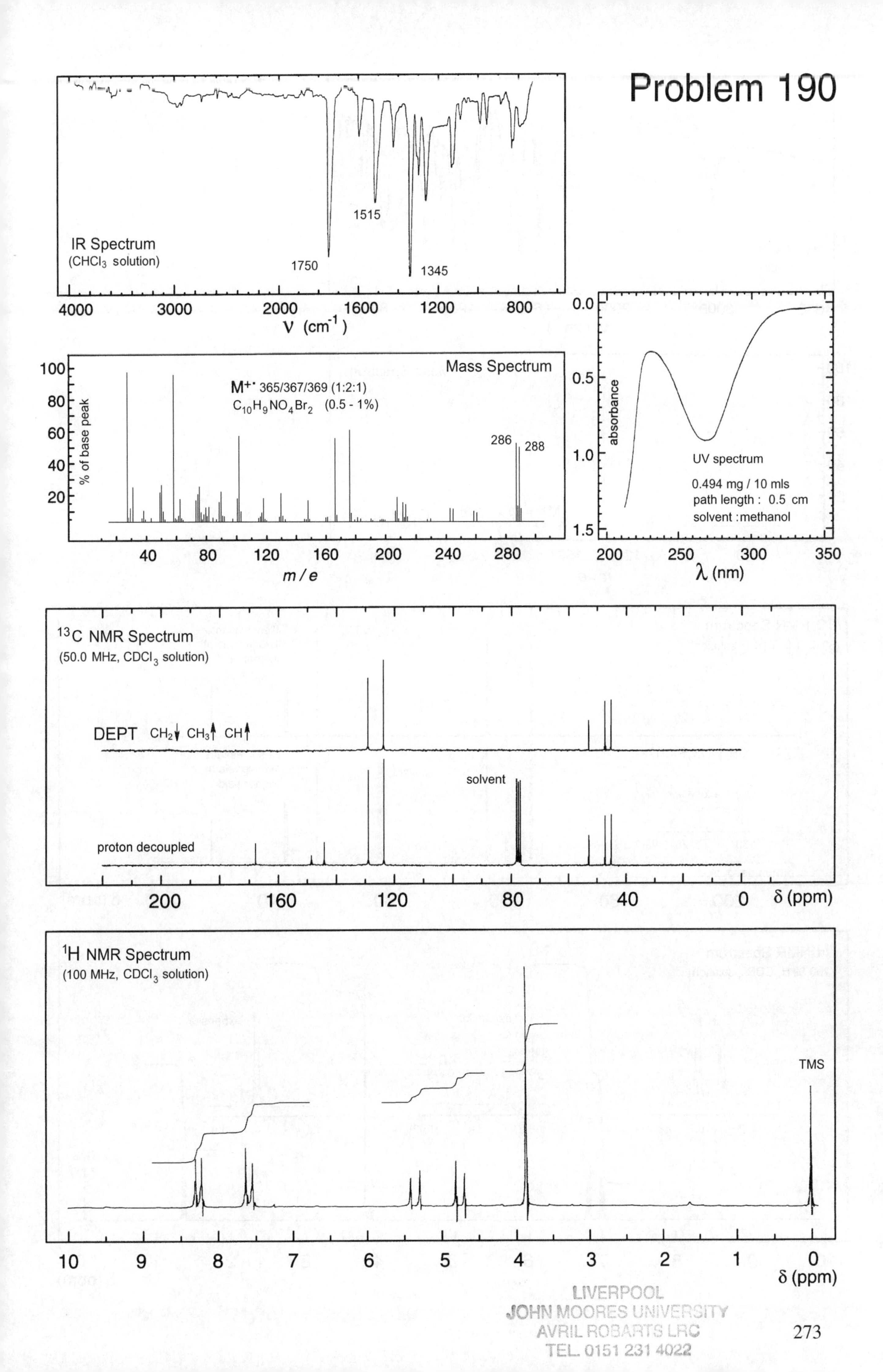

IR Spectrum
(CHCl3 solution)
1750
1515
1345
4000
3000
2000
1600
1200
800
ν (cm-1)
Mass Spectrum
M+· 365/367/369 (1:2:1)
C10H9NO4Br2 (0.5 - 1%)
% of base peak
286
288
m / e
UV spectrum
0.494 mg / 10 mls
path length : 0.5 cm
solvent : methanol
absorbance
λ (nm)
13C NMR Spectrum
(50.0 MHz, CDCl3 solution)
DEPT CH2↓ CH3↑ CH↑
solvent
proton decoupled
δ (ppm)
1H NMR Spectrum
(100 MHz, CDCl3 solution)
TMS
δ (ppm)

Problem 191

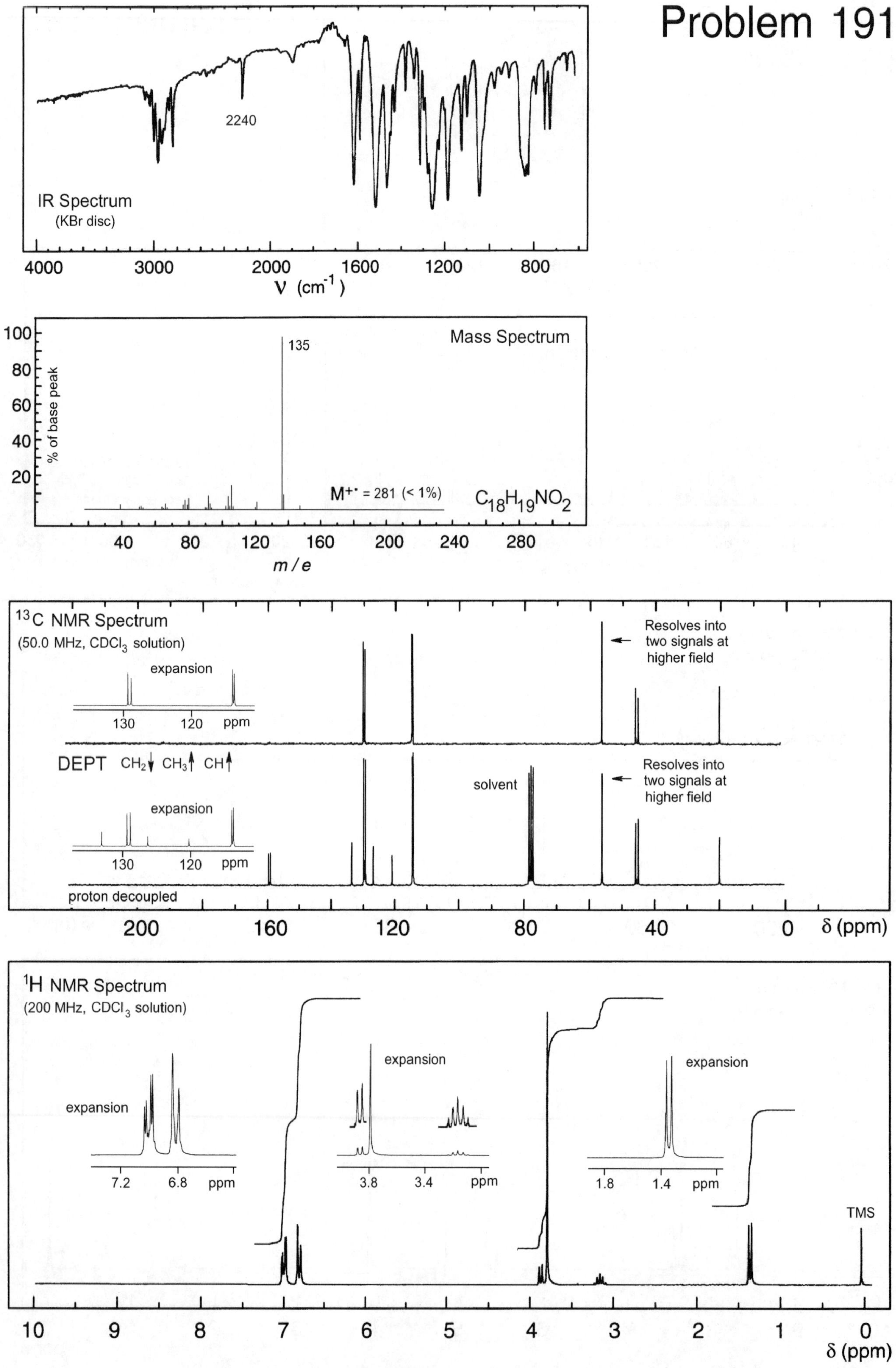

IR Spectrum
(KBr disc)
2240
4000
3000
2000
1600
1200
800
ν (cm^{-1})
Mass Spectrum
% of base peak
135
M$^{+\cdot}$ = 281 (< 1%)
$C_{18}H_{19}NO_2$
m / e
^{13}C NMR Spectrum
(50.0 MHz, $CDCl_3$ solution)
expansion
Resolves into two signals at higher field
DEPT CH_2↓ CH_3↑ CH↑
solvent
proton decoupled
δ (ppm)
^{1}H NMR Spectrum
(200 MHz, $CDCl_3$ solution)
expansion
TMS
ppm

Problem 192

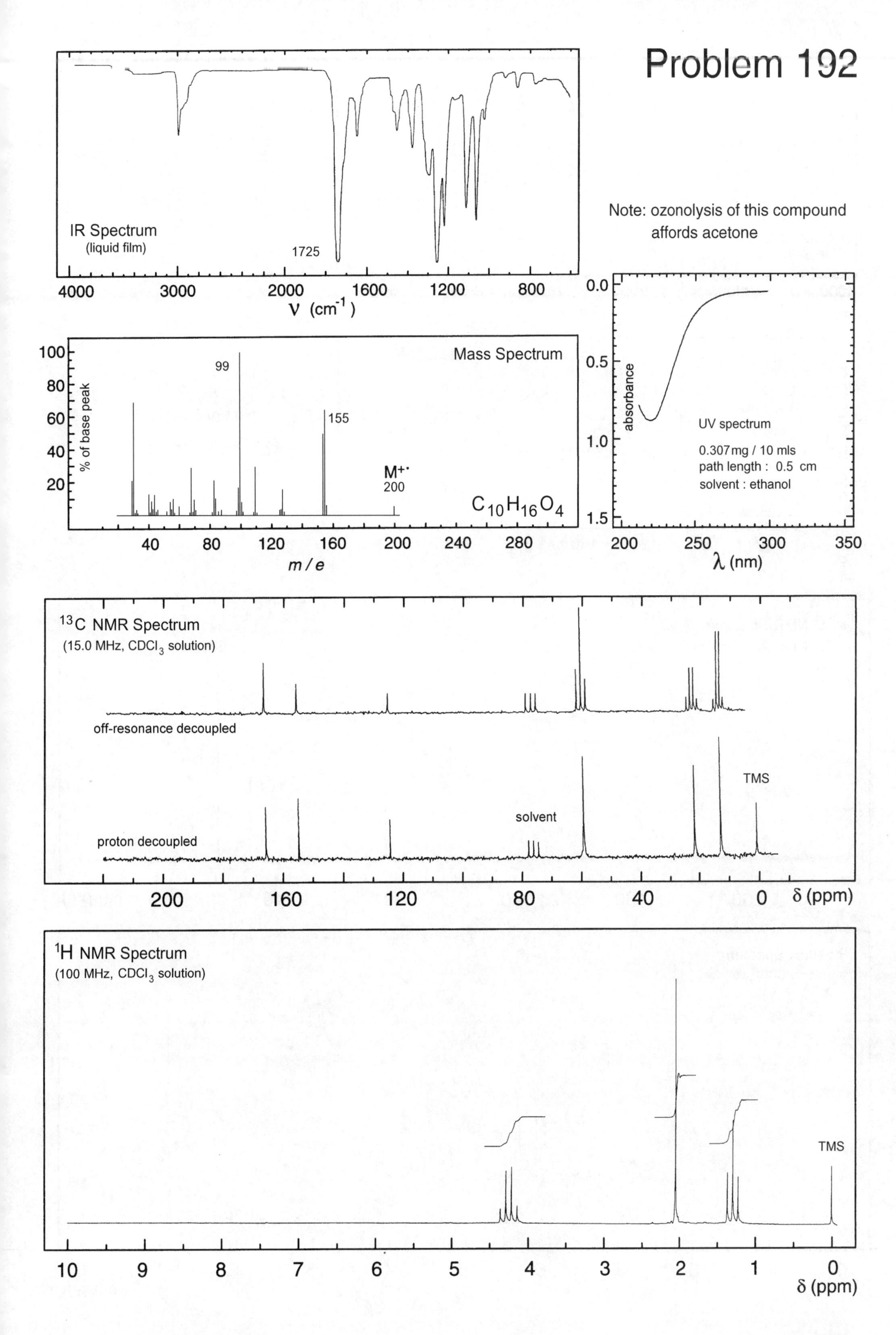

IR Spectrum
(liquid film)
1725
4000
3000
2000
1600
1200
800
ν (cm⁻¹)
Note: ozonolysis of this compound affords acetone
Mass Spectrum
% of base peak
99
155
M+·
200
$C_{10}H_{16}O_4$
m / e
absorbance
UV spectrum
0.307 mg / 10 mls
path length : 0.5 cm
solvent : ethanol
λ (nm)
13C NMR Spectrum
(15.0 MHz, CDCl3 solution)
off-resonance decoupled
proton decoupled
solvent
TMS
δ (ppm)
1H NMR Spectrum
(100 MHz, CDCl3 solution)
TMS
δ (ppm)

Problem 193

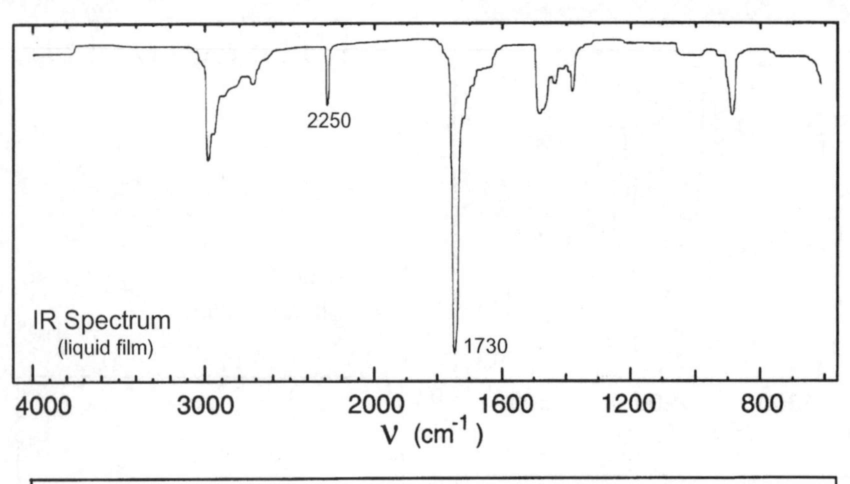

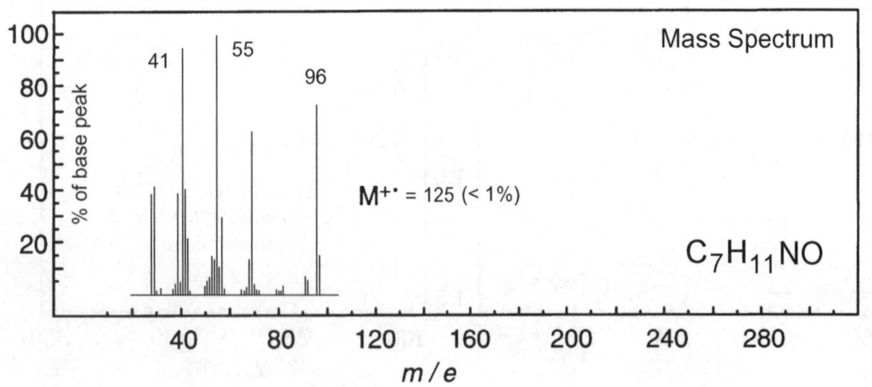

No significant UV absorption above 220 nm

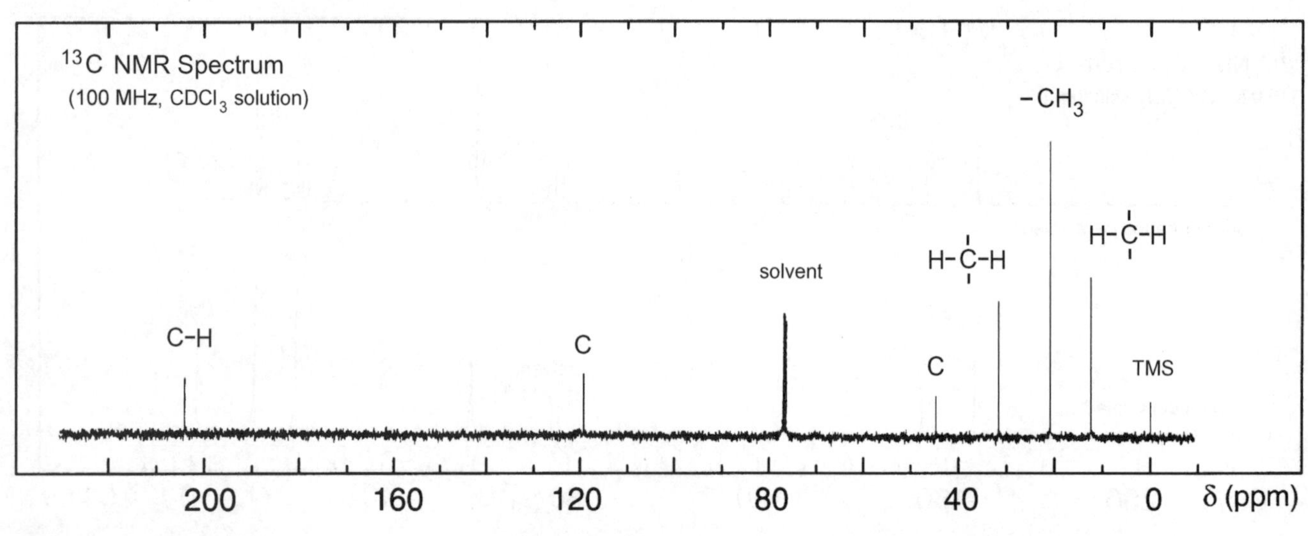

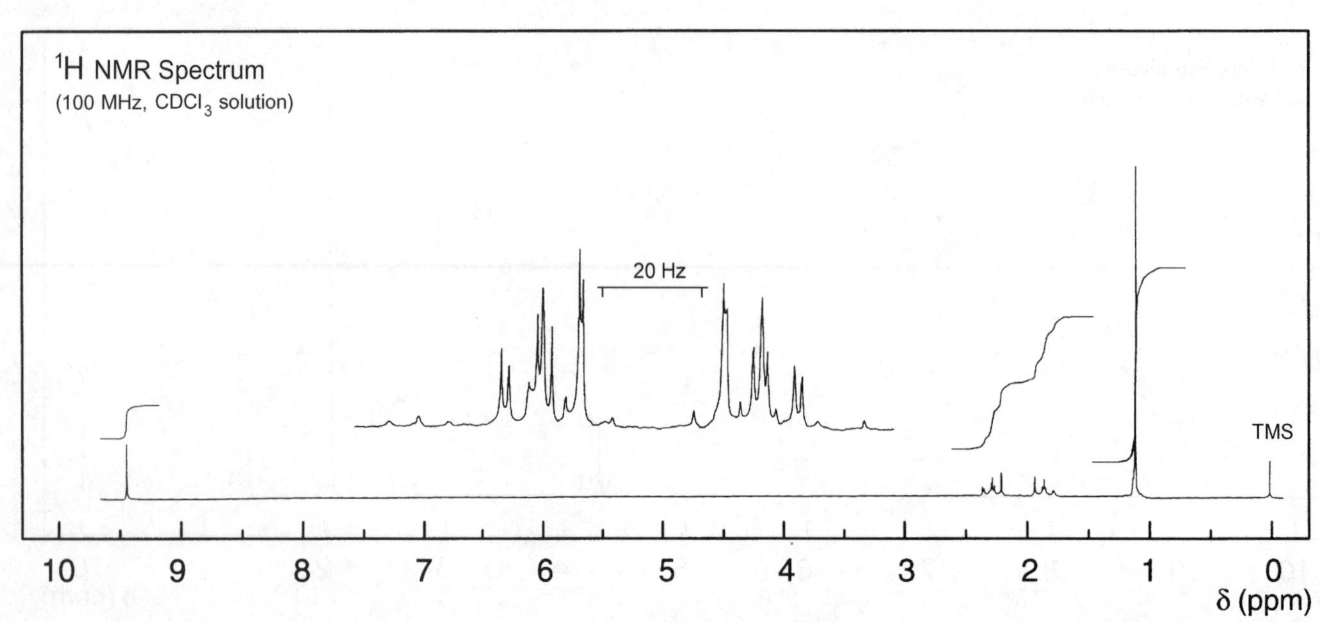

Problem 194

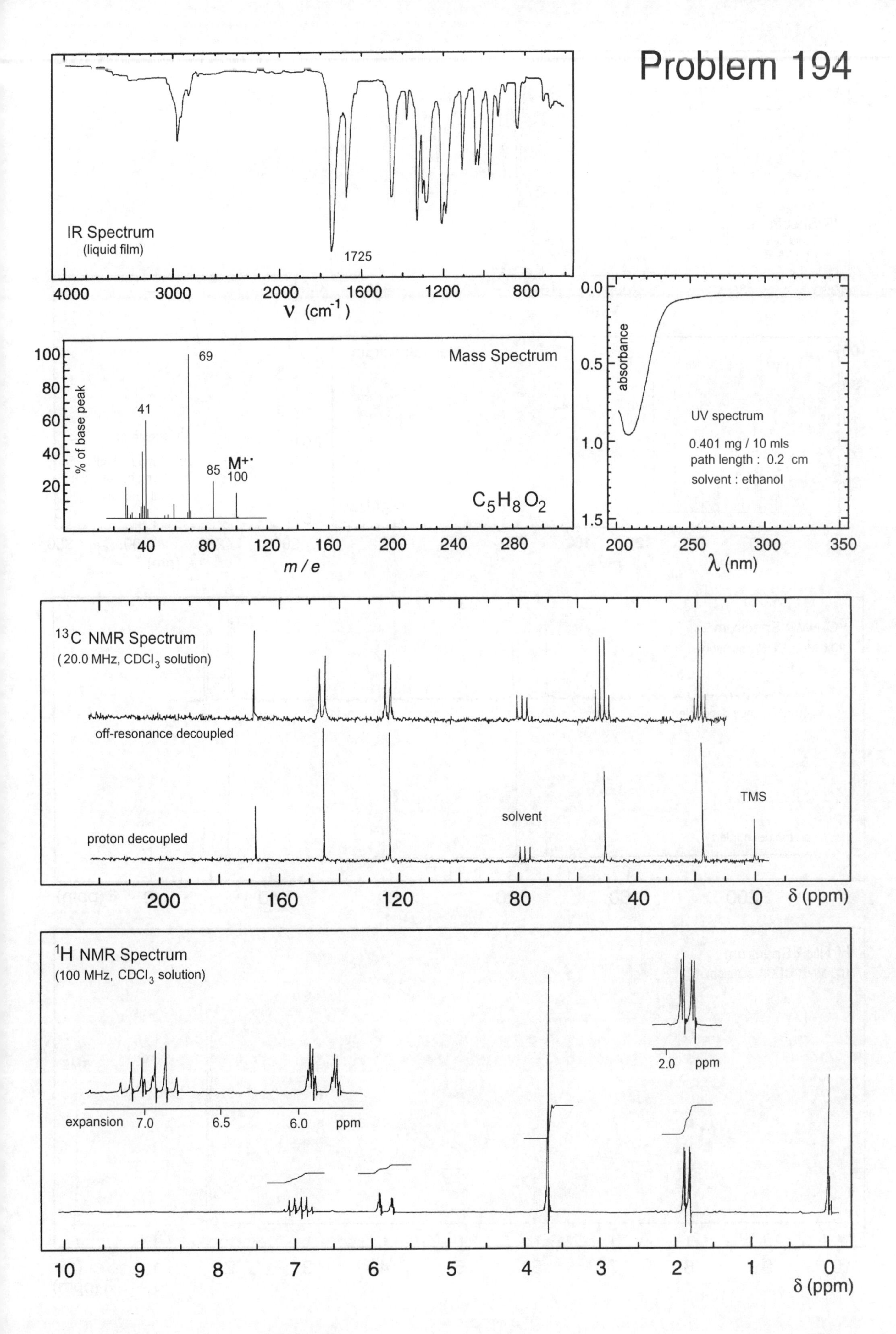

IR Spectrum
(liquid film)
1725
4000
3000
2000
1600
1200
800
ν (cm⁻¹)
Mass Spectrum
% of base peak
100
80
60
40
20
69
41
85
M⁺·
100
$C_5H_8O_2$
40
80
120
160
200
240
280
m / e
absorbance
0.0
0.5
1.0
1.5
UV spectrum
0.401 mg / 10 mls
path length : 0.2 cm
solvent : ethanol
200
250
300
350
λ (nm)
¹³C NMR Spectrum
(20.0 MHz, CDCl₃ solution)
off-resonance decoupled
proton decoupled
solvent
TMS
200
160
120
80
40
0
δ (ppm)
¹H NMR Spectrum
(100 MHz, CDCl₃ solution)
expansion
7.0
6.5
6.0
ppm
2.0
ppm
10
9
8
7
6
5
4
3
2
1
0
δ (ppm)

Problem 195

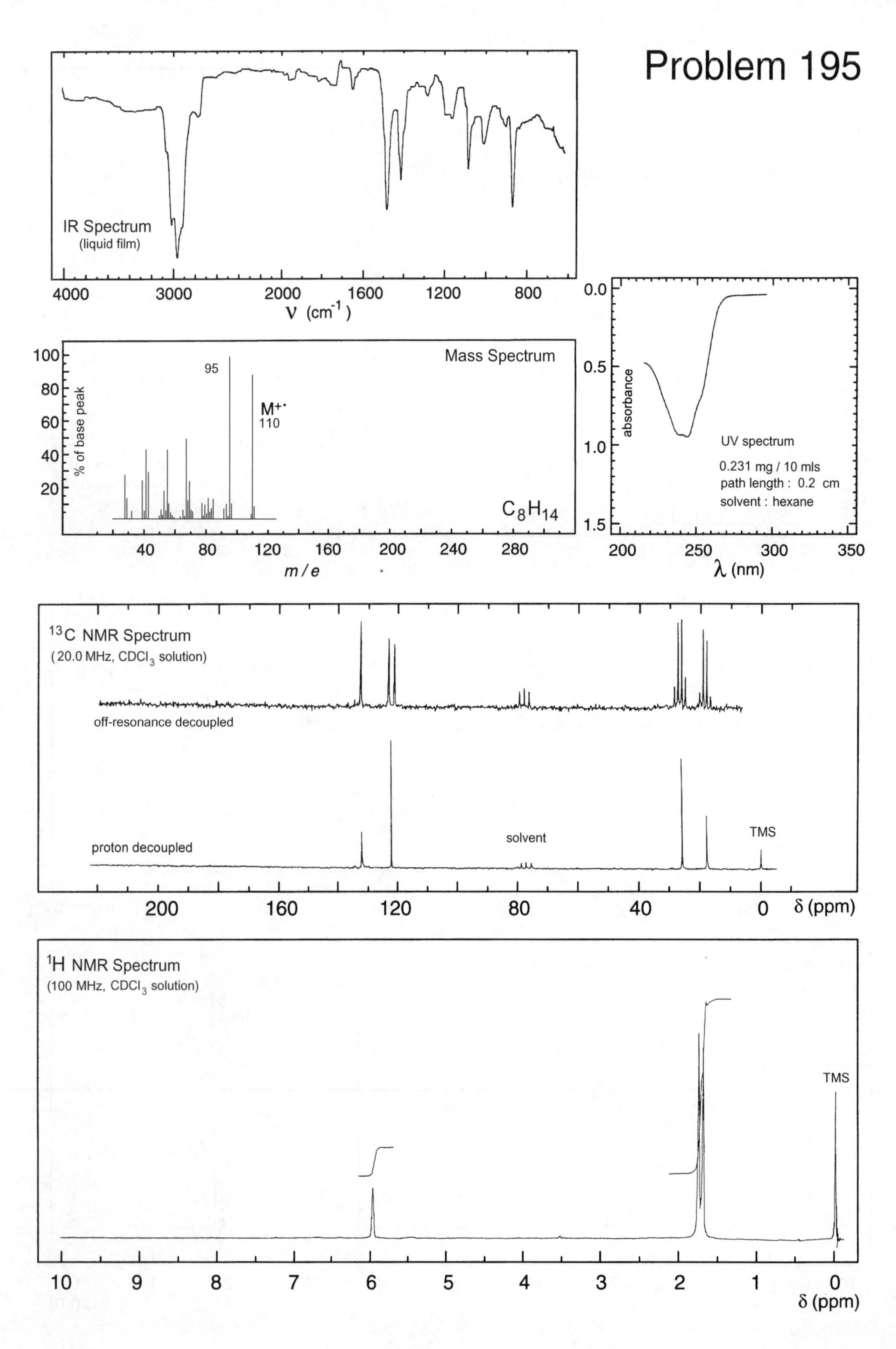

IR Spectrum
(liquid film)
4000
3000
2000
1600
1200
800
ν (cm⁻¹)
Mass Spectrum
% of base peak
95
M+·
110
C8H14
m / e
UV spectrum
0.231 mg / 10 mls
path length : 0.2 cm
solvent : hexane
absorbance
λ (nm)
13C NMR Spectrum
(20.0 MHz, CDCl3 solution)
off-resonance decoupled
proton decoupled
solvent
TMS
δ (ppm)
1H NMR Spectrum
(100 MHz, CDCl3 solution)
TMS
δ (ppm)

Problem 196

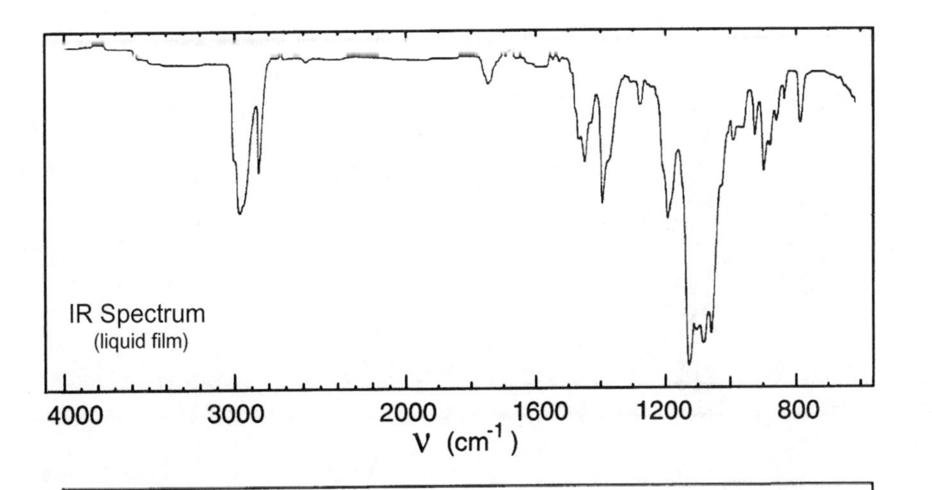

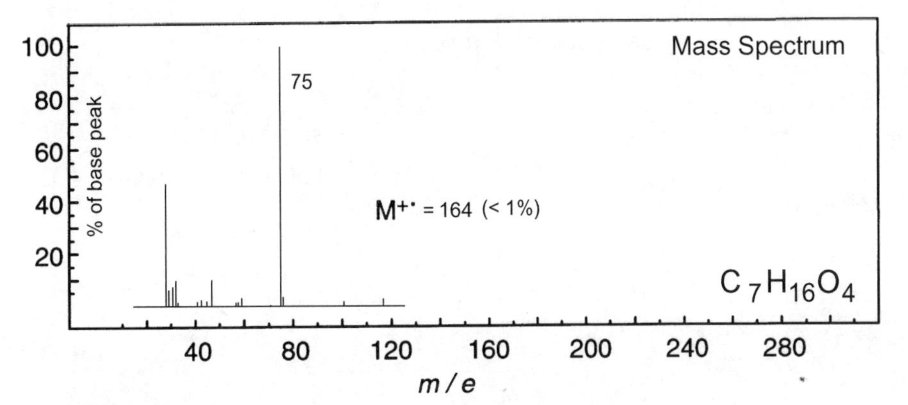

No significant UV absorption above 220 nm

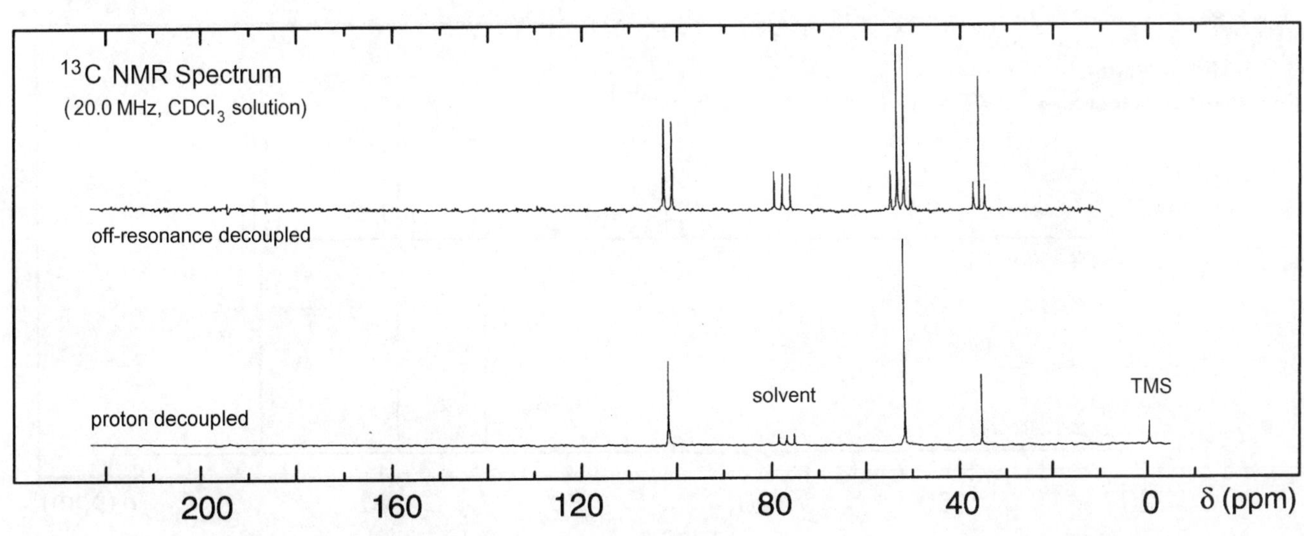

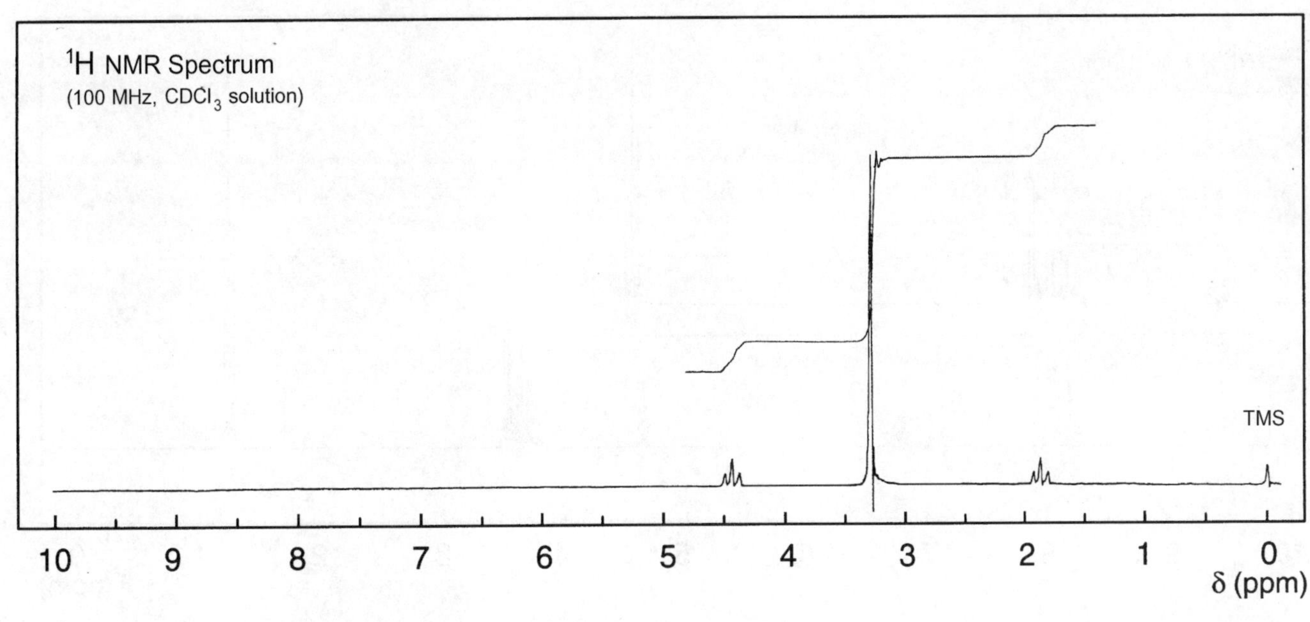

Problem 197

IR Spectrum
(liquid film)

4000 3000 2000 1600 1200 800
ν (cm^{-1})

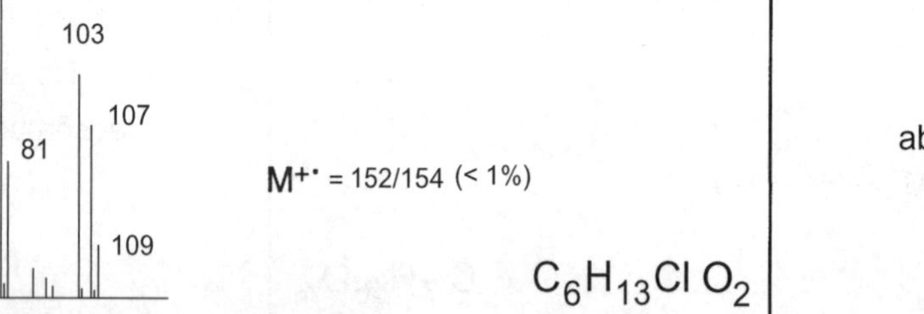

No significant UV absorption above 220 nm

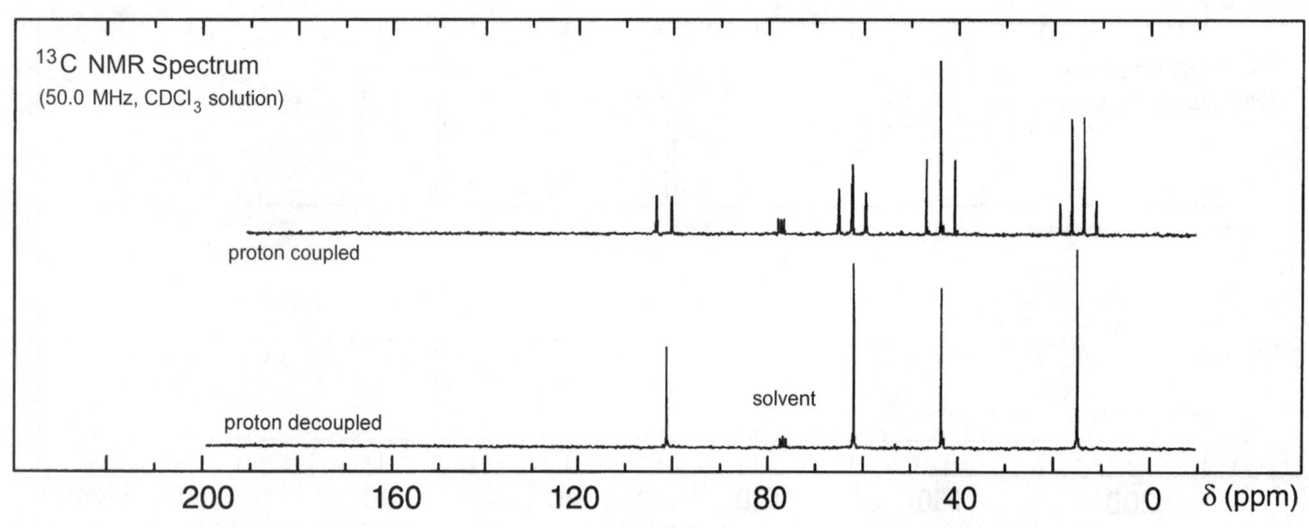

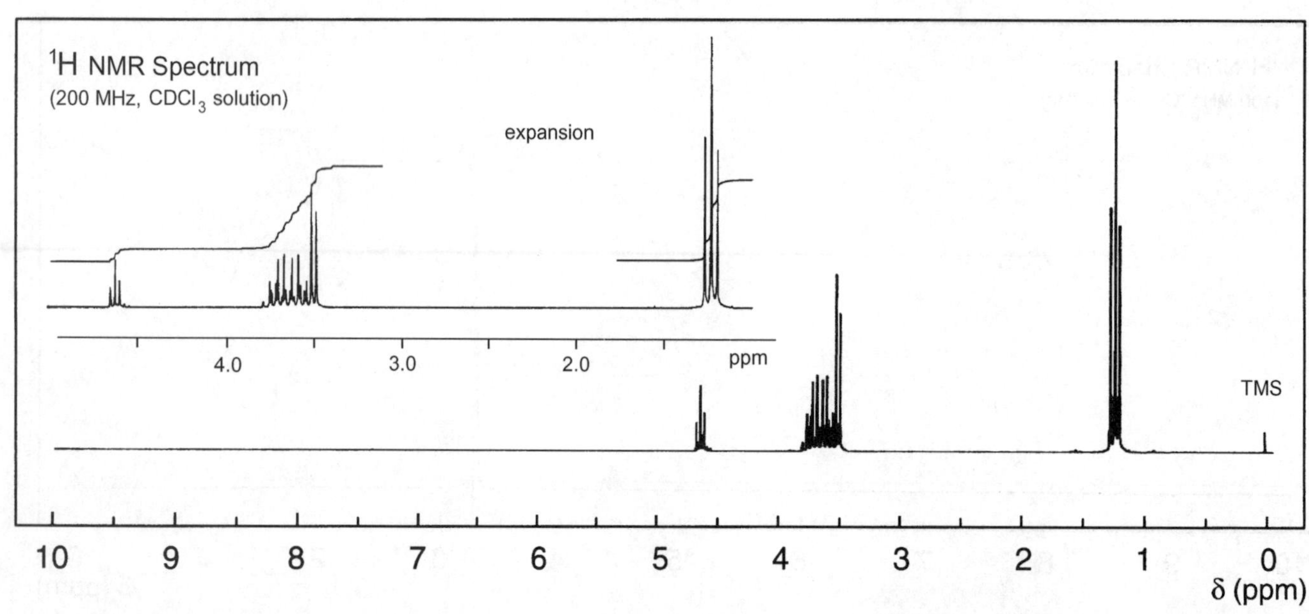

Problem 198

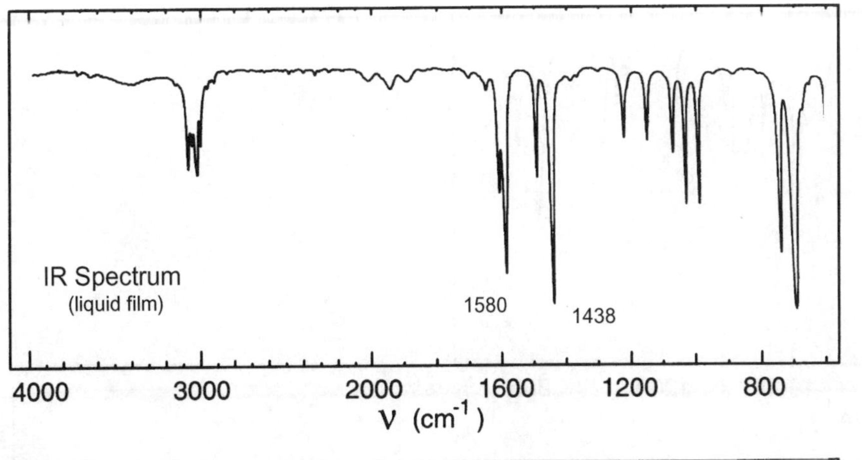

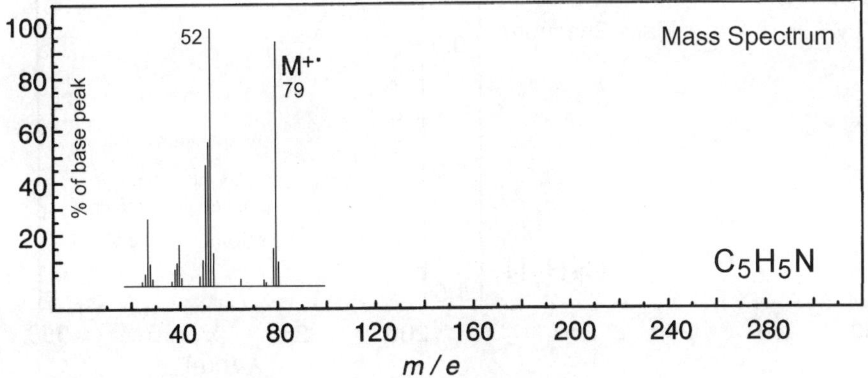

UV Spectrum

λ_{max} 257 nm ($\log_{10}\varepsilon$ 3.4)

λ_{max} 270 nm ($\log_{10}\varepsilon$ 2.6)

solvent : methanol

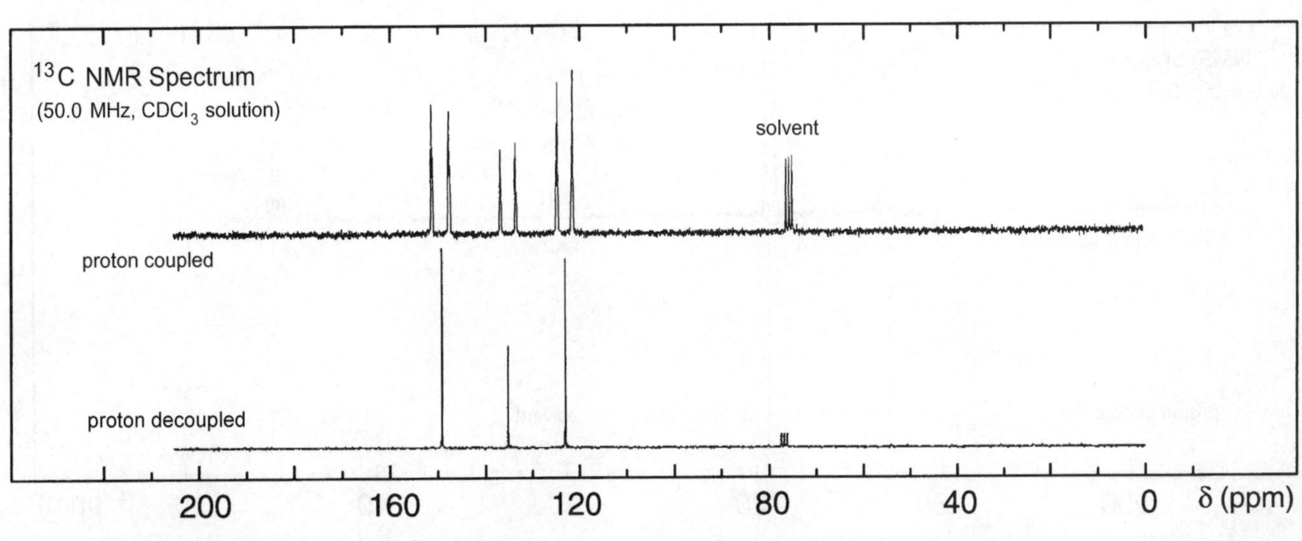

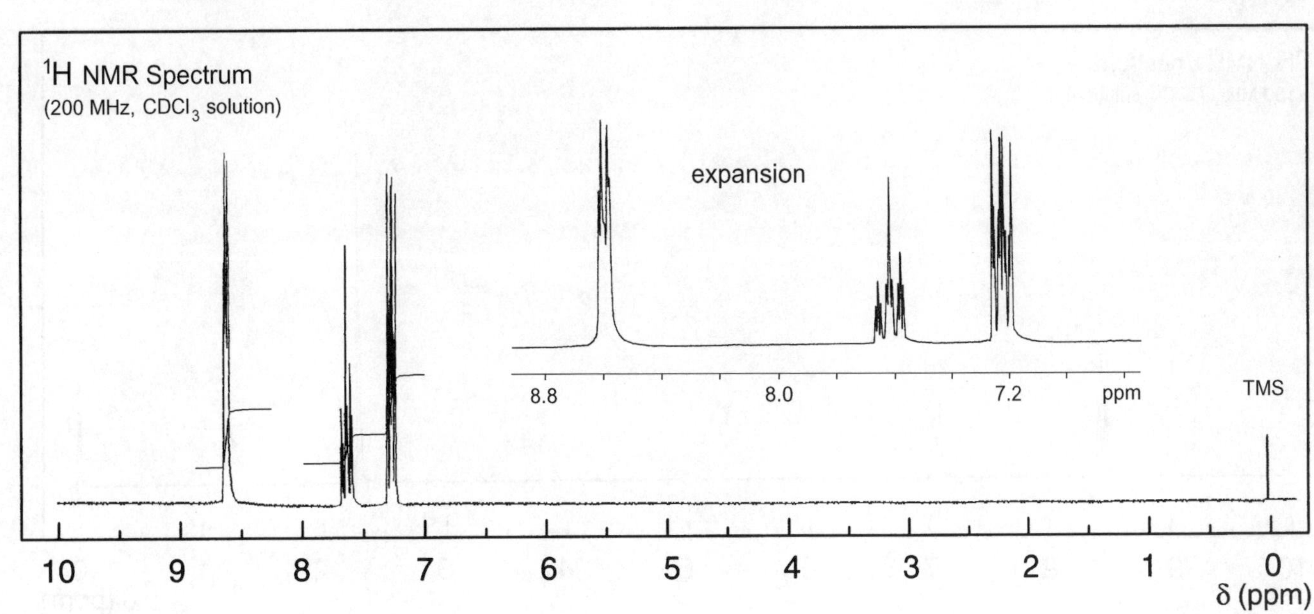

Problem 199

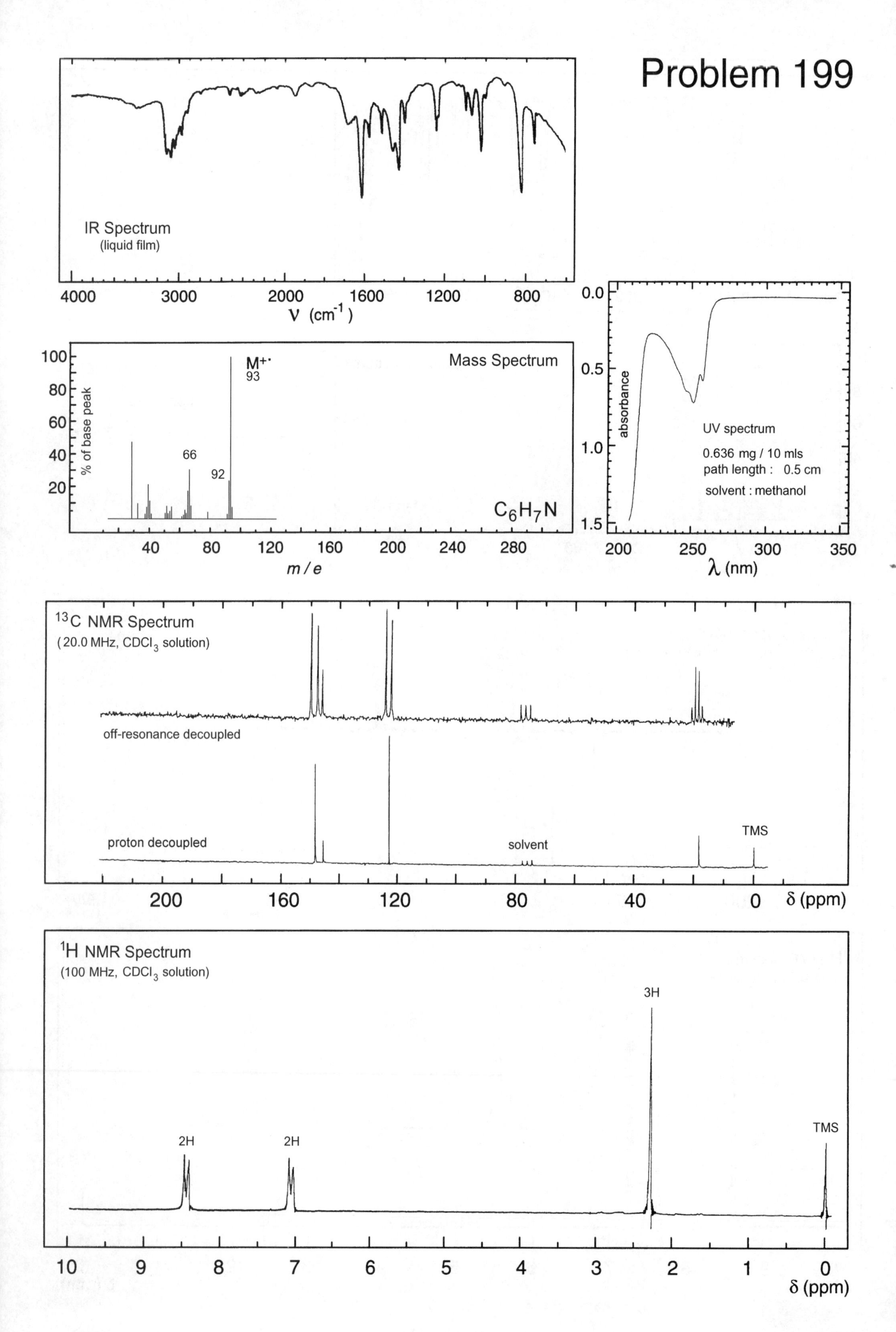

IR Spectrum
(liquid film)
4000
3000
2000
1600
1200
800
ν (cm-1)
Mass Spectrum
100
80
60
40
20
% of base peak
M+·
93
66
92
C6H7N
40
80
120
160
200
240
280
m / e
UV spectrum
0.636 mg / 10 mls
path length : 0.5 cm
solvent : methanol
absorbance
0.0
0.5
1.0
1.5
200
250
300
350
λ (nm)
13C NMR Spectrum
(20.0 MHz, CDCl3 solution)
off-resonance decoupled
proton decoupled
solvent
TMS
200
160
120
80
40
0
δ (ppm)
1H NMR Spectrum
(100 MHz, CDCl3 solution)
3H
2H
2H
TMS
10
9
8
7
6
5
4
3
2
1
0
δ (ppm)

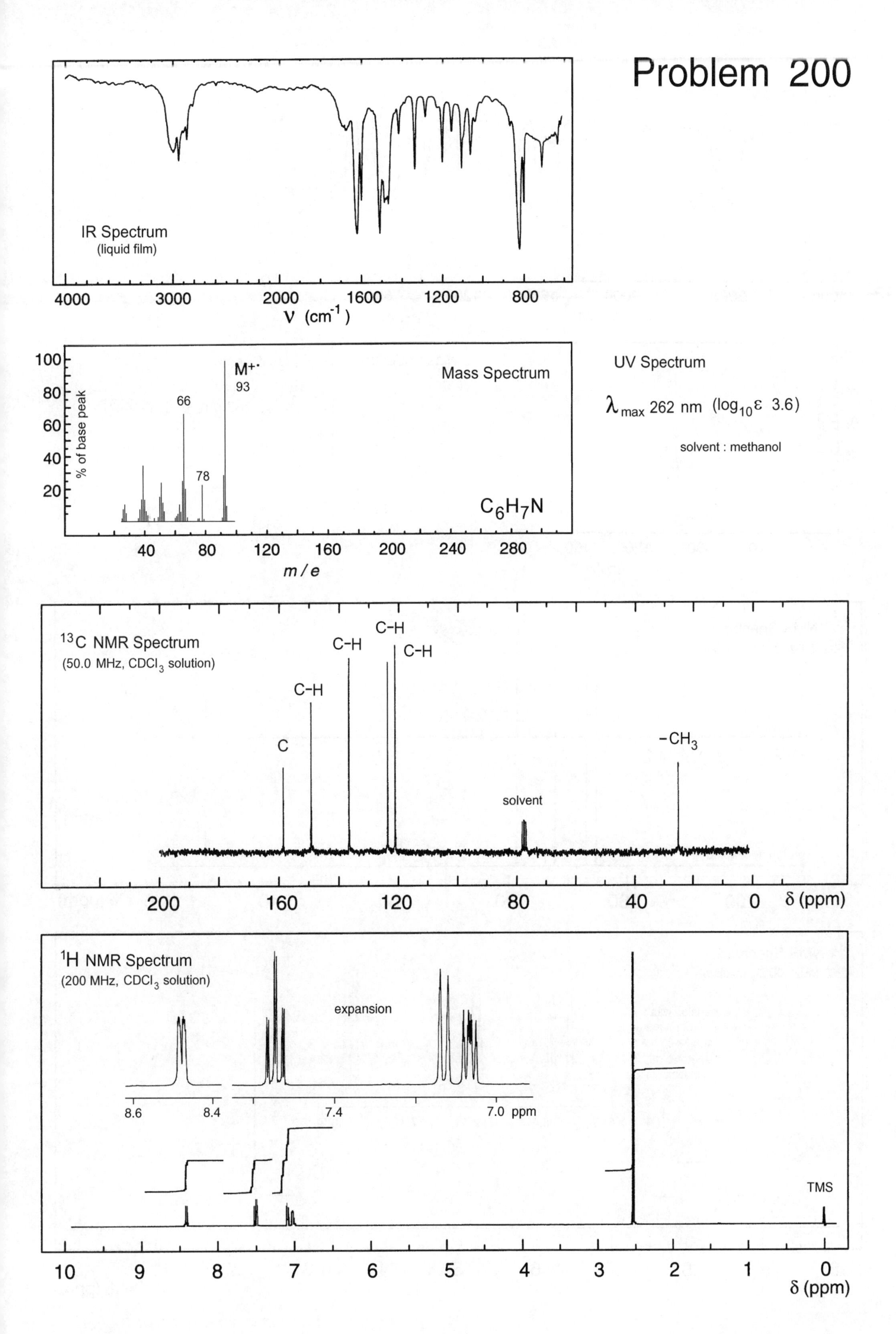
Problem 200
IR Spectrum
(liquid film)
4000
3000
2000
1600
1200
800
ν (cm⁻¹)
Mass Spectrum
% of base peak
M+·
93
66
78
C6H7N
m / e
UV Spectrum
λmax 262 nm (log10ε 3.6)
solvent : methanol
13C NMR Spectrum
(50.0 MHz, CDCl3 solution)
C-H
C
-CH3
solvent
δ (ppm)
1H NMR Spectrum
(200 MHz, CDCl3 solution)
expansion
7.0 ppm
TMS

Problem 201

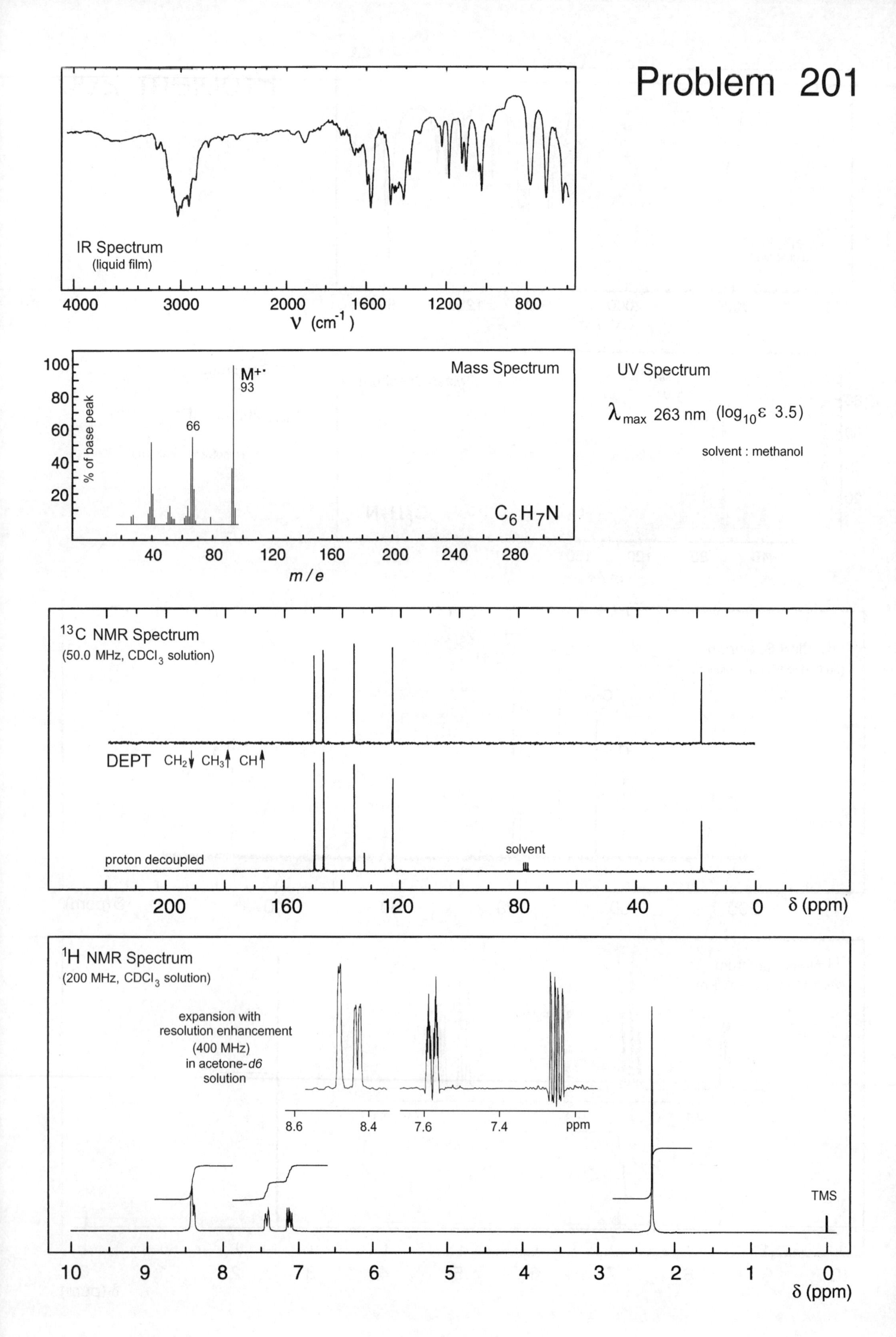

IR Spectrum
(liquid film)
4000
3000
2000
1600
1200
800
ν (cm-1)
Mass Spectrum
% of base peak
M+·
93
66
C6H7N
m / e
UV Spectrum
λmax 263 nm (log10ε 3.5)
solvent : methanol
13C NMR Spectrum
(50.0 MHz, CDCl3 solution)
DEPT CH2↓ CH3↑ CH↑
proton decoupled
solvent
δ (ppm)
1H NMR Spectrum
(200 MHz, CDCl3 solution)
expansion with
resolution enhancement
(400 MHz)
in acetone-d6
solution
ppm
TMS
δ (ppm)

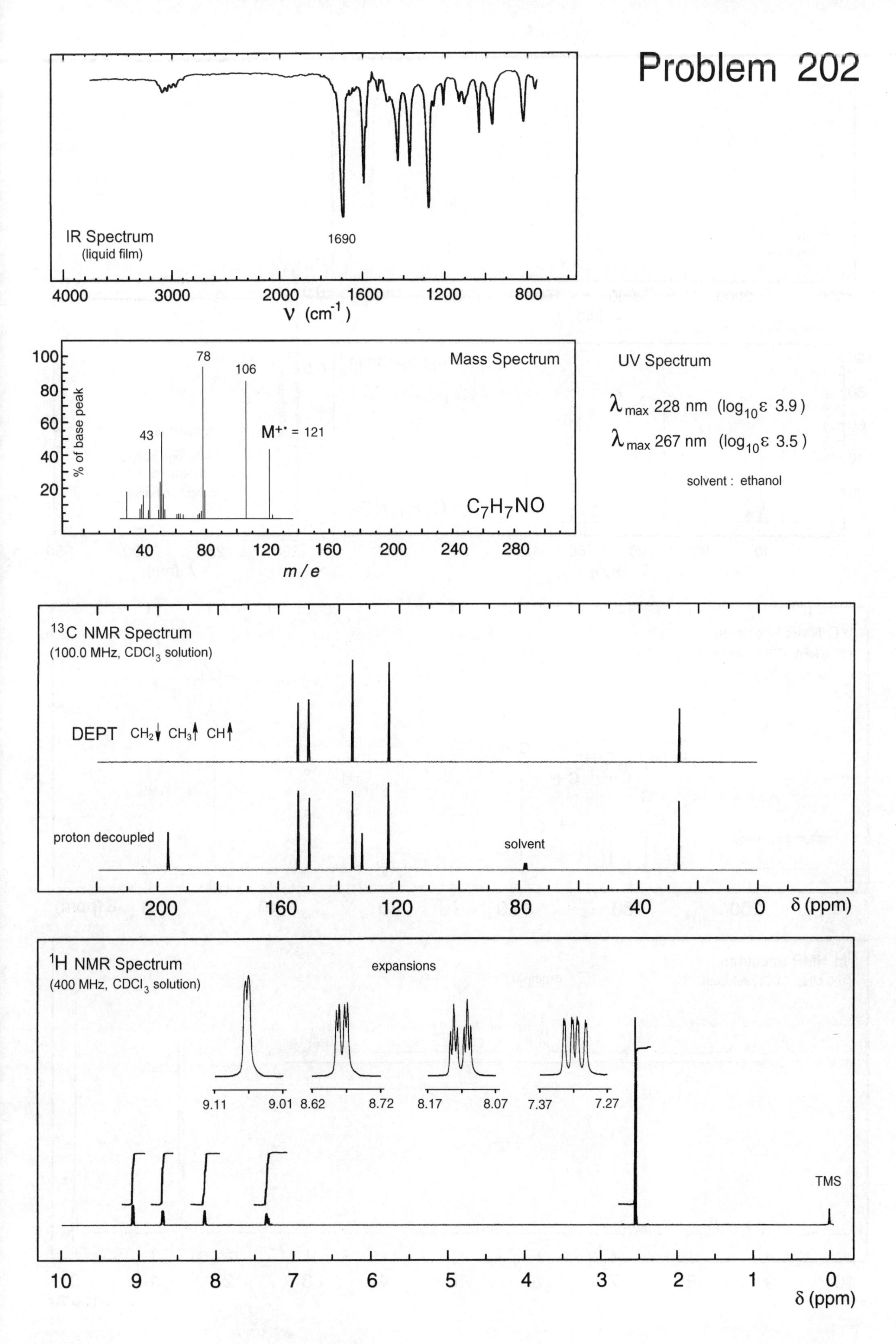
Problem 202
IR Spectrum
(liquid film)
1690
4000
3000
2000
1600
1200
800
ν (cm⁻¹)
Mass Spectrum
% of base peak
100
80
60
40
20
43
78
106
M⁺˙ = 121
C7H7NO
40
80
120
160
200
240
280
m / e
UV Spectrum
λmax 228 nm (log10ε 3.9)
λmax 267 nm (log10ε 3.5)
solvent : ethanol
13C NMR Spectrum
(100.0 MHz, CDCl3 solution)
DEPT CH2↓ CH3↑ CH↑
proton decoupled
solvent
200
160
120
80
40
0
δ (ppm)
1H NMR Spectrum
(400 MHz, CDCl3 solution)
expansions
9.11
9.01
8.62
8.72
8.17
8.07
7.37
7.27
TMS
10
9
8
7
6
5
4
3
2
1
0
δ (ppm)

Problem 203

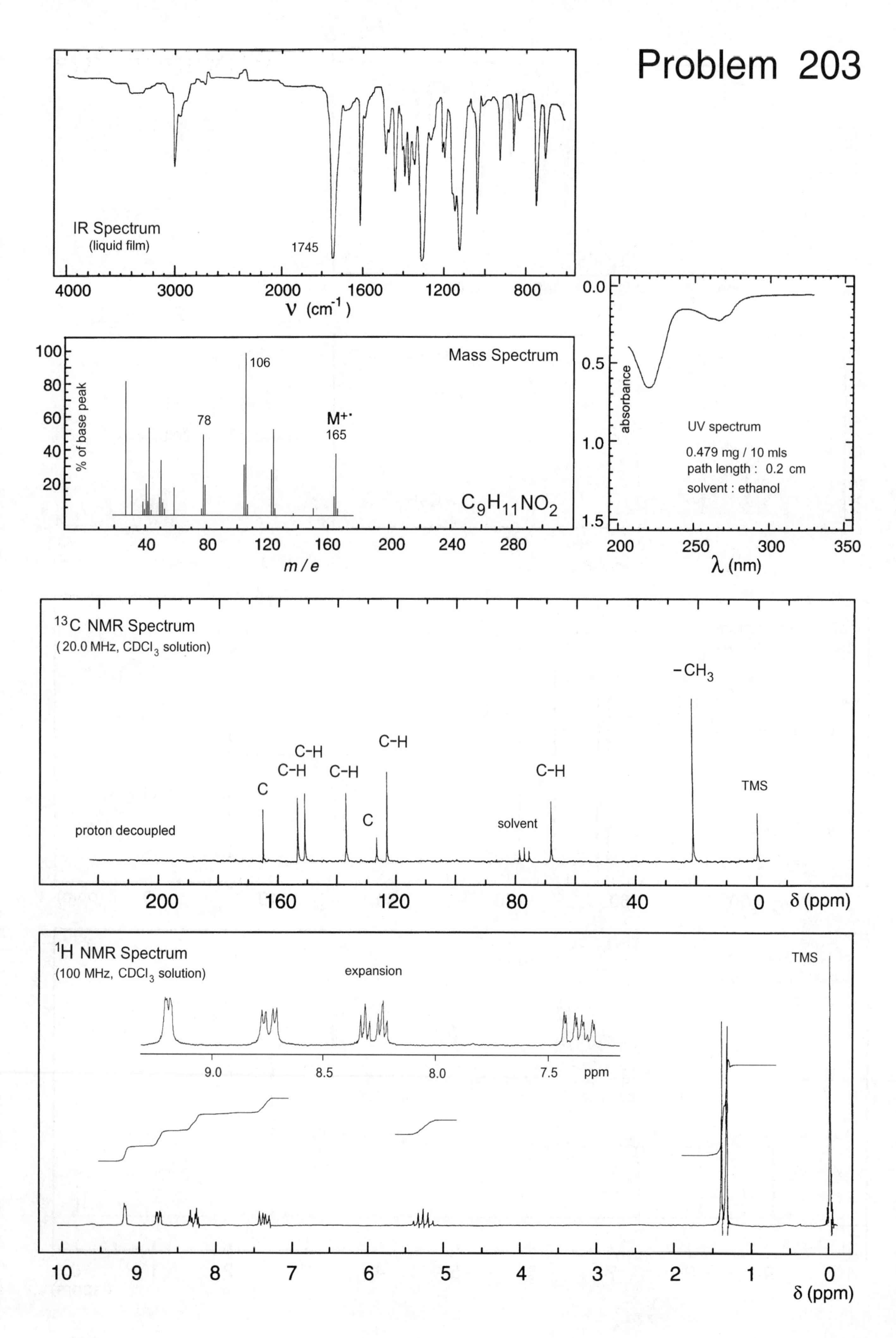

IR Spectrum
(liquid film)
1745
4000
3000
2000
1600
1200
800
ν (cm⁻¹)
Mass Spectrum
100
80
60
40
20
% of base peak
106
78
M+·
165
C9H11NO2
m / e
0.0
0.5
1.0
1.5
absorbance
UV spectrum
0.479 mg / 10 mls
path length : 0.2 cm
solvent : ethanol
200
250
300
350
λ (nm)
13C NMR Spectrum
(20.0 MHz, CDCl3 solution)
-CH3
C-H
C
TMS
solvent
proton decoupled
δ (ppm)
1H NMR Spectrum
(100 MHz, CDCl3 solution)
expansion
9.0
8.5
8.0
7.5
ppm

Problem 204

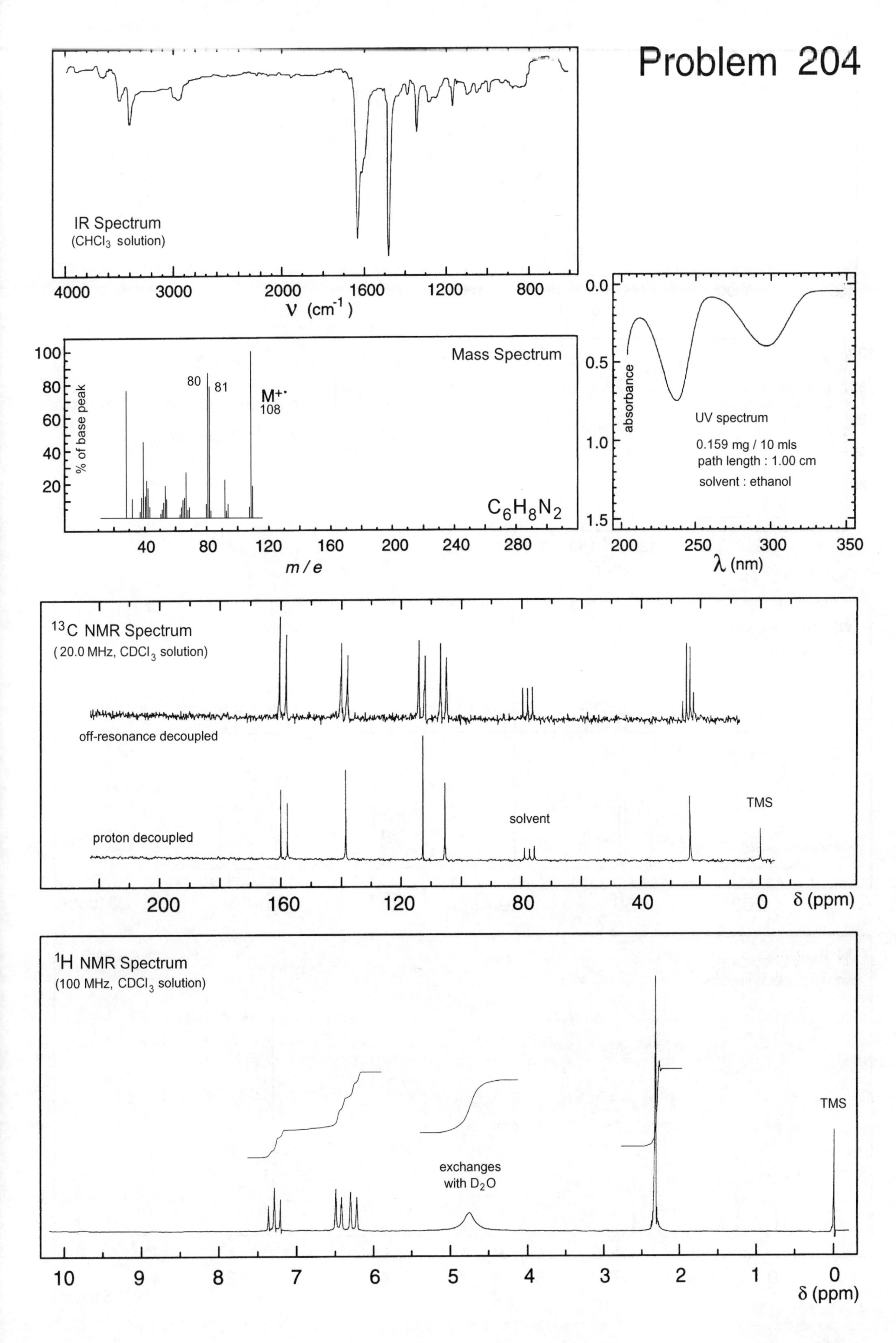

IR Spectrum
(CHCl3 solution)
4000
3000
2000
1600
1200
800
ν (cm-1)
Mass Spectrum
100
80
60
40
20
% of base peak
80
81
M+·
108
C6H8N2
40
80
120
160
200
240
280
m / e
0.0
0.5
1.0
1.5
absorbance
UV spectrum
0.159 mg / 10 mls
path length : 1.00 cm
solvent : ethanol
200
250
300
350
λ (nm)
13C NMR Spectrum
(20.0 MHz, CDCl3 solution)
off-resonance decoupled
proton decoupled
solvent
TMS
200
160
120
80
40
0
δ (ppm)
1H NMR Spectrum
(100 MHz, CDCl3 solution)
TMS
exchanges
with D2O
10
9
8
7
6
5
4
3
2
1
0
δ (ppm)

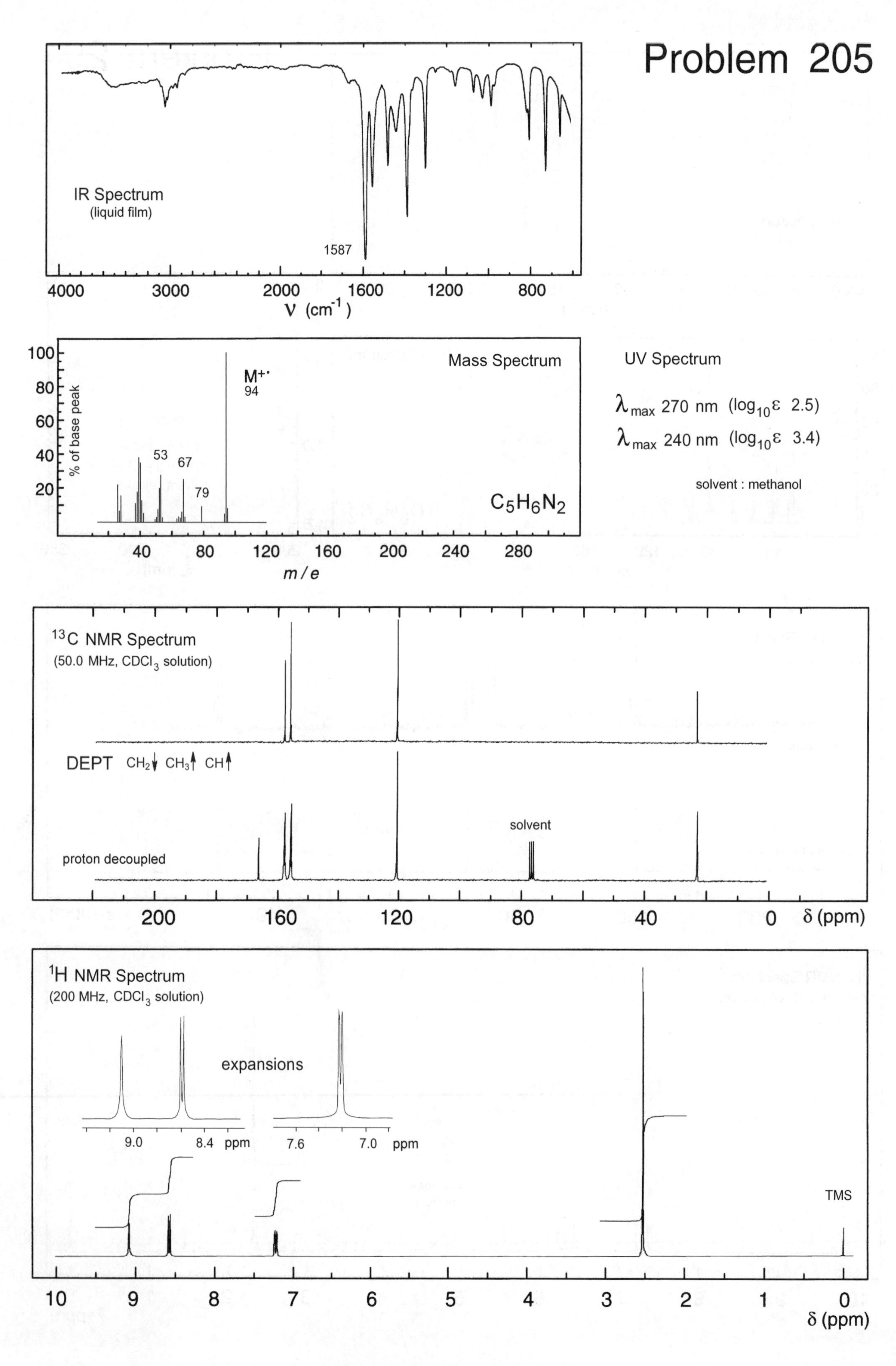

Problem 205
IR Spectrum
(liquid film)
1587
4000
3000
2000
1600
1200
800
ν (cm-1)
Mass Spectrum
100
80
60
40
20
% of base peak
M+·
94
53
67
79
C5H6N2
40
80
120
160
200
240
280
m / e
UV Spectrum
λmax 270 nm (log10ε 2.5)
λmax 240 nm (log10ε 3.4)
solvent : methanol
13C NMR Spectrum
(50.0 MHz, CDCl3 solution)
DEPT CH2↓ CH3↑ CH↑
solvent
proton decoupled
200
160
120
80
40
0
δ (ppm)
1H NMR Spectrum
(200 MHz, CDCl3 solution)
expansions
9.0
8.4
ppm
7.6
7.0
ppm
TMS
10
9
8
7
6
5
4
3
2
1
0
δ (ppm)

Problem 206

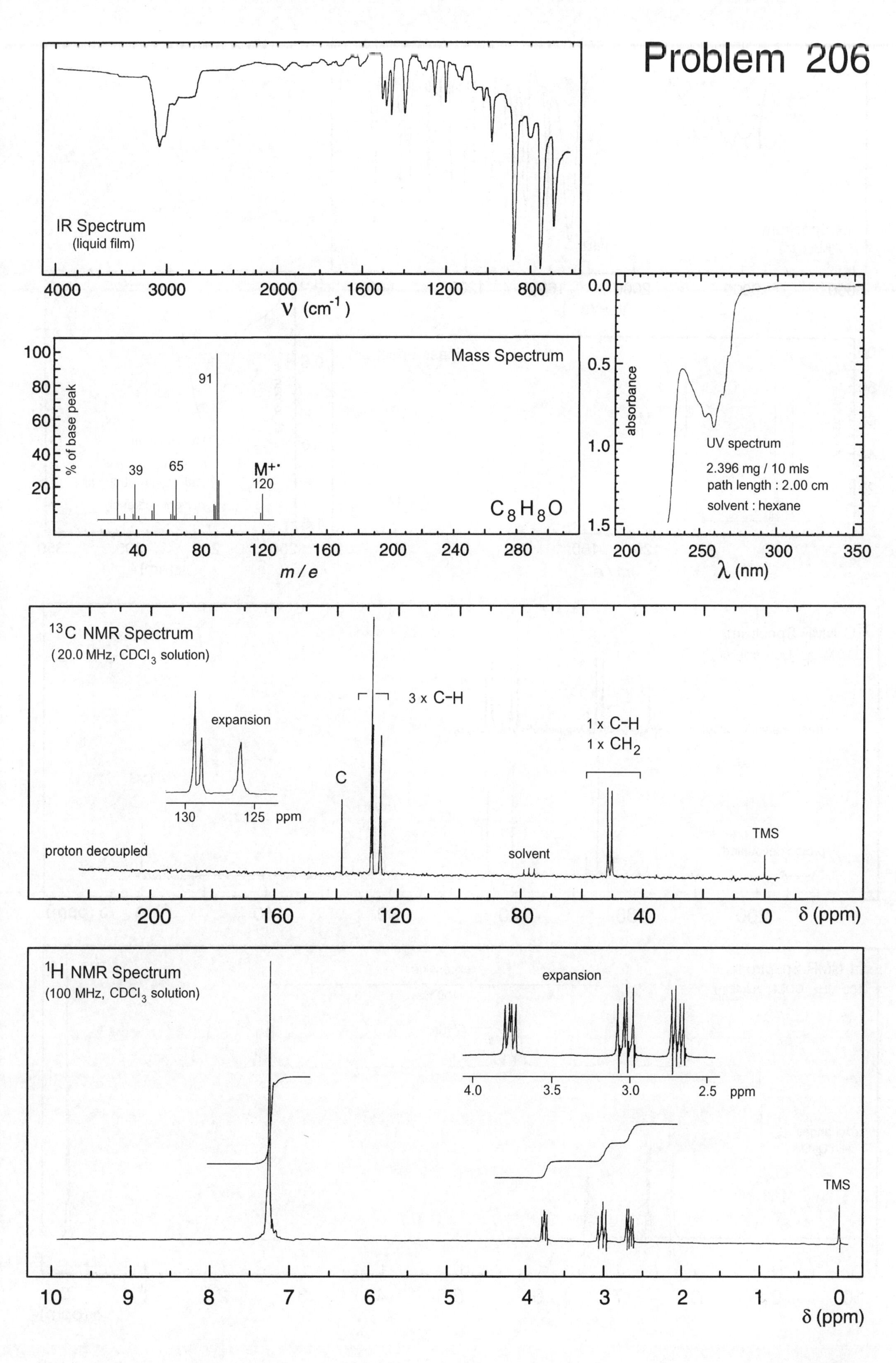

IR Spectrum
(liquid film)
4000
3000
2000
1600
1200
800
ν (cm⁻¹)
Mass Spectrum
% of base peak
100
80
60
40
20
39
65
91
M⁺˙
120
C8H8O
40
80
120
160
200
240
280
m / e
0.0
0.5
1.0
1.5
absorbance
UV spectrum
2.396 mg / 10 mls
path length : 2.00 cm
solvent : hexane
200
250
300
350
λ (nm)
13C NMR Spectrum
(20.0 MHz, CDCl3 solution)
expansion
130
125
ppm
3 x C–H
C
1 x C–H
1 x CH2
proton decoupled
solvent
TMS
200
160
120
80
40
0
δ (ppm)
1H NMR Spectrum
(100 MHz, CDCl3 solution)
expansion
4.0
3.5
3.0
2.5
ppm
TMS
10
9
8
7
6
5
4
3
2
1
0
δ (ppm)

Problem 207

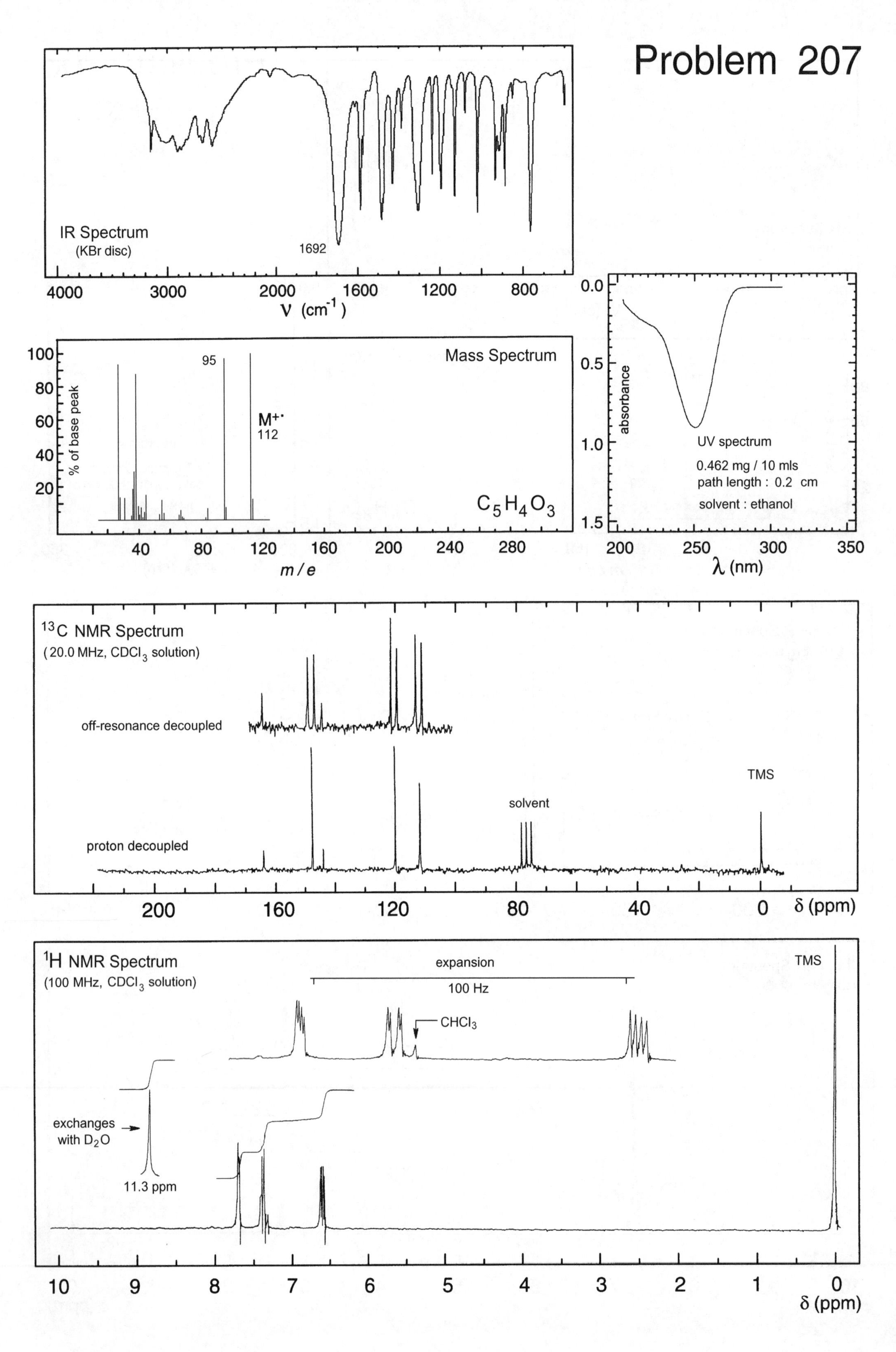

IR Spectrum
(KBr disc)
1692
4000
3000
2000
1600
1200
800
ν (cm⁻¹)
Mass Spectrum
100
80
60
40
20
% of base peak
95
M+·
112
C5H4O3
40
80
120
160
200
240
280
m / e
0.0
0.5
1.0
1.5
absorbance
UV spectrum
0.462 mg / 10 mls
path length : 0.2 cm
solvent : ethanol
200
250
300
350
λ (nm)
13C NMR Spectrum
(20.0 MHz, CDCl3 solution)
off-resonance decoupled
proton decoupled
solvent
TMS
200
160
120
80
40
0
δ (ppm)
1H NMR Spectrum
(100 MHz, CDCl3 solution)
expansion
100 Hz
CHCl3
TMS
exchanges with D2O
11.3 ppm
10
9
8
7
6
5
4
3
2
1
0
δ (ppm)

Problem 208

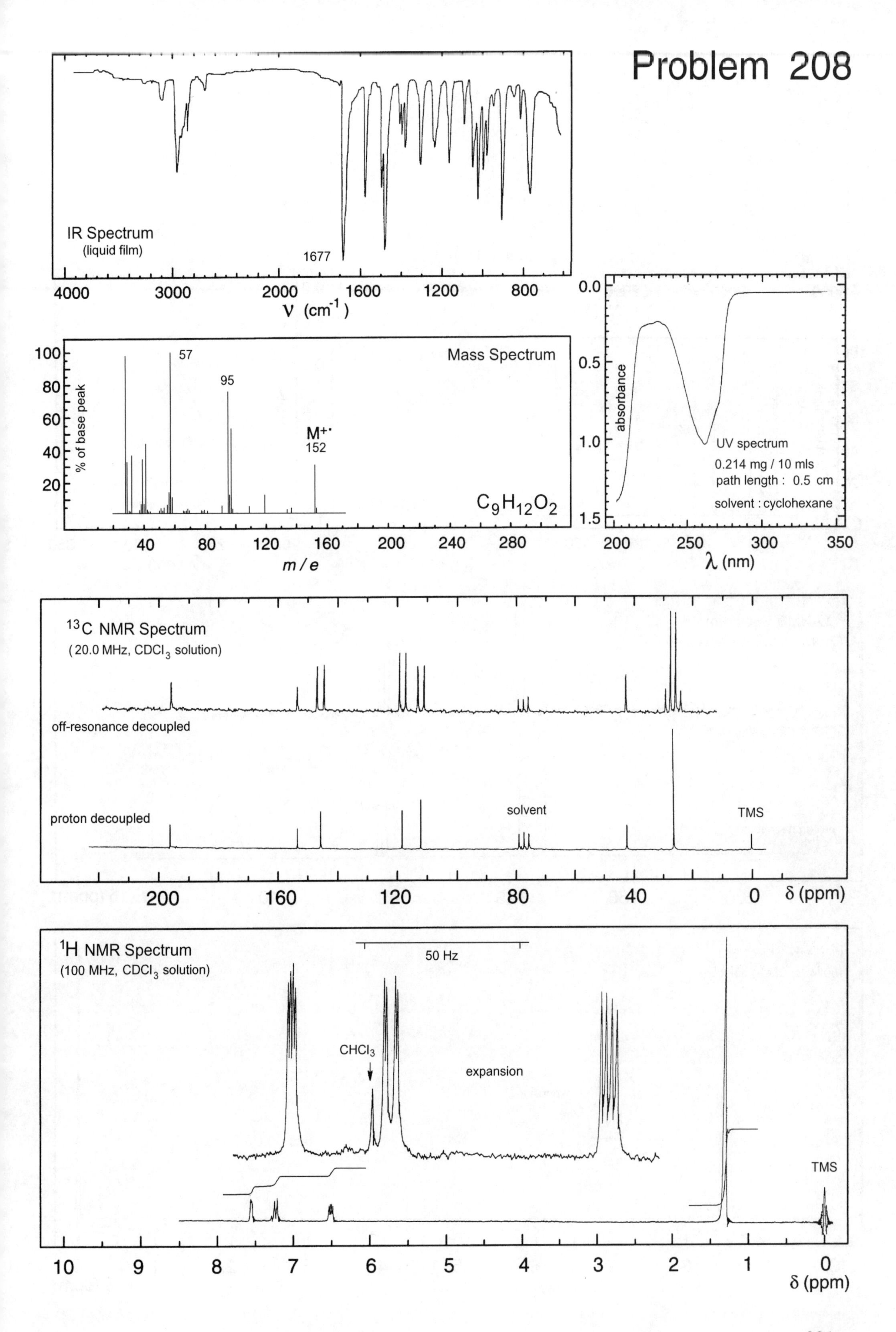

IR Spectrum
(liquid film)
1677
4000
3000
2000
1600
1200
800
ν (cm⁻¹)
Mass Spectrum
% of base peak
57
95
M⁺˙
152
$C_9H_{12}O_2$
m / e
UV spectrum
0.214 mg / 10 mls
path length : 0.5 cm
solvent : cyclohexane
absorbance
λ (nm)
¹³C NMR Spectrum
(20.0 MHz, CDCl₃ solution)
off-resonance decoupled
proton decoupled
solvent
TMS
δ (ppm)
¹H NMR Spectrum
(100 MHz, CDCl₃ solution)
50 Hz
CHCl₃
expansion
TMS
δ (ppm)

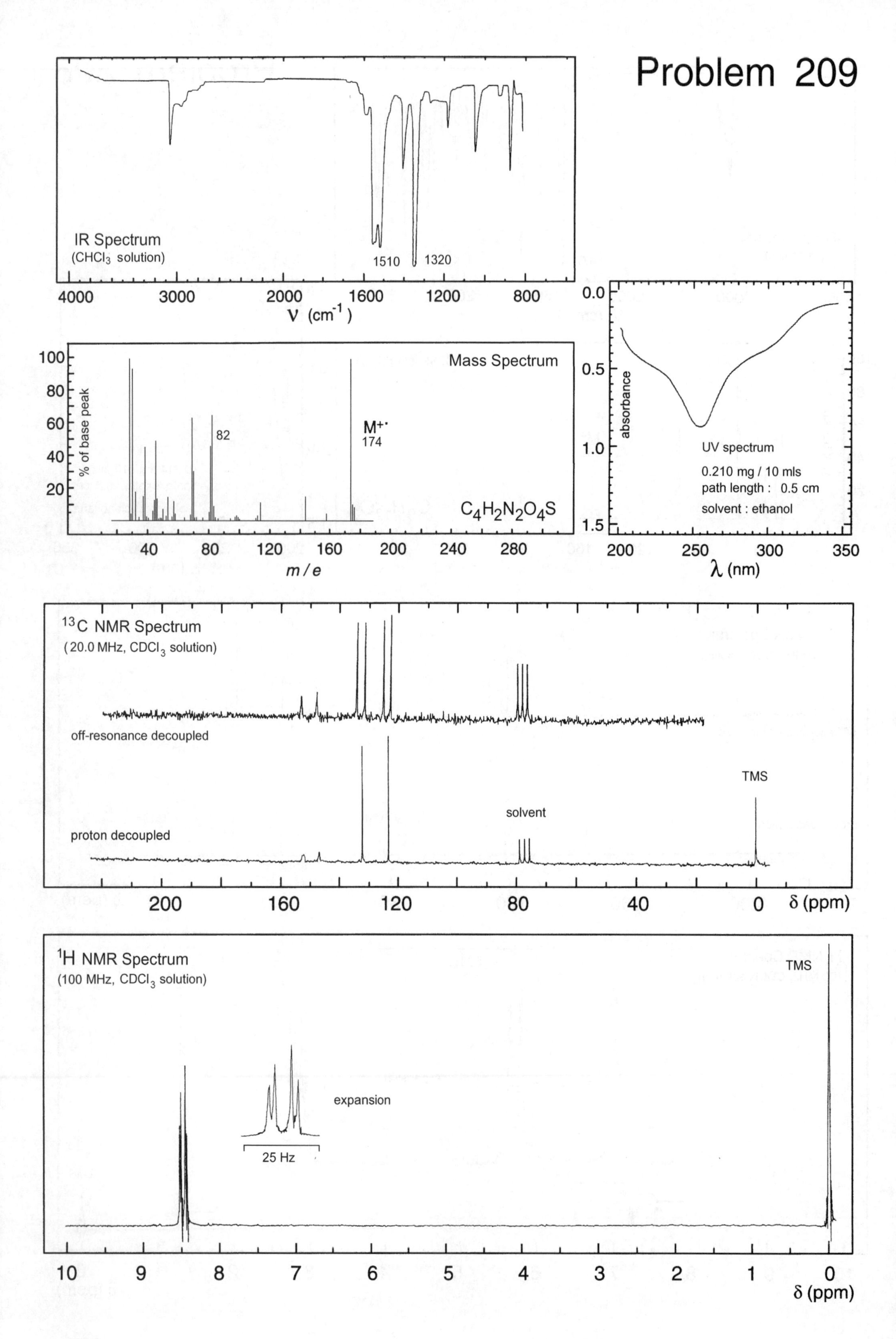

Problem 209
IR Spectrum
(CHCl3 solution)
1510
1320
4000
3000
2000
1600
1200
800
ν (cm-1)
Mass Spectrum
% of base peak
100
80
60
40
20
82
M+·
174
C4H2N2O4S
40
80
120
160
200
240
280
m / e
UV spectrum
0.210 mg / 10 mls
path length : 0.5 cm
solvent : ethanol
absorbance
0.0
0.5
1.0
1.5
200
250
300
350
λ (nm)
13C NMR Spectrum
(20.0 MHz, CDCl3 solution)
off-resonance decoupled
proton decoupled
solvent
TMS
200
160
120
80
40
0
δ (ppm)
1H NMR Spectrum
(100 MHz, CDCl3 solution)
expansion
25 Hz
TMS
10
9
8
7
6
5
4
3
2
1
0
δ (ppm)

Problem 210

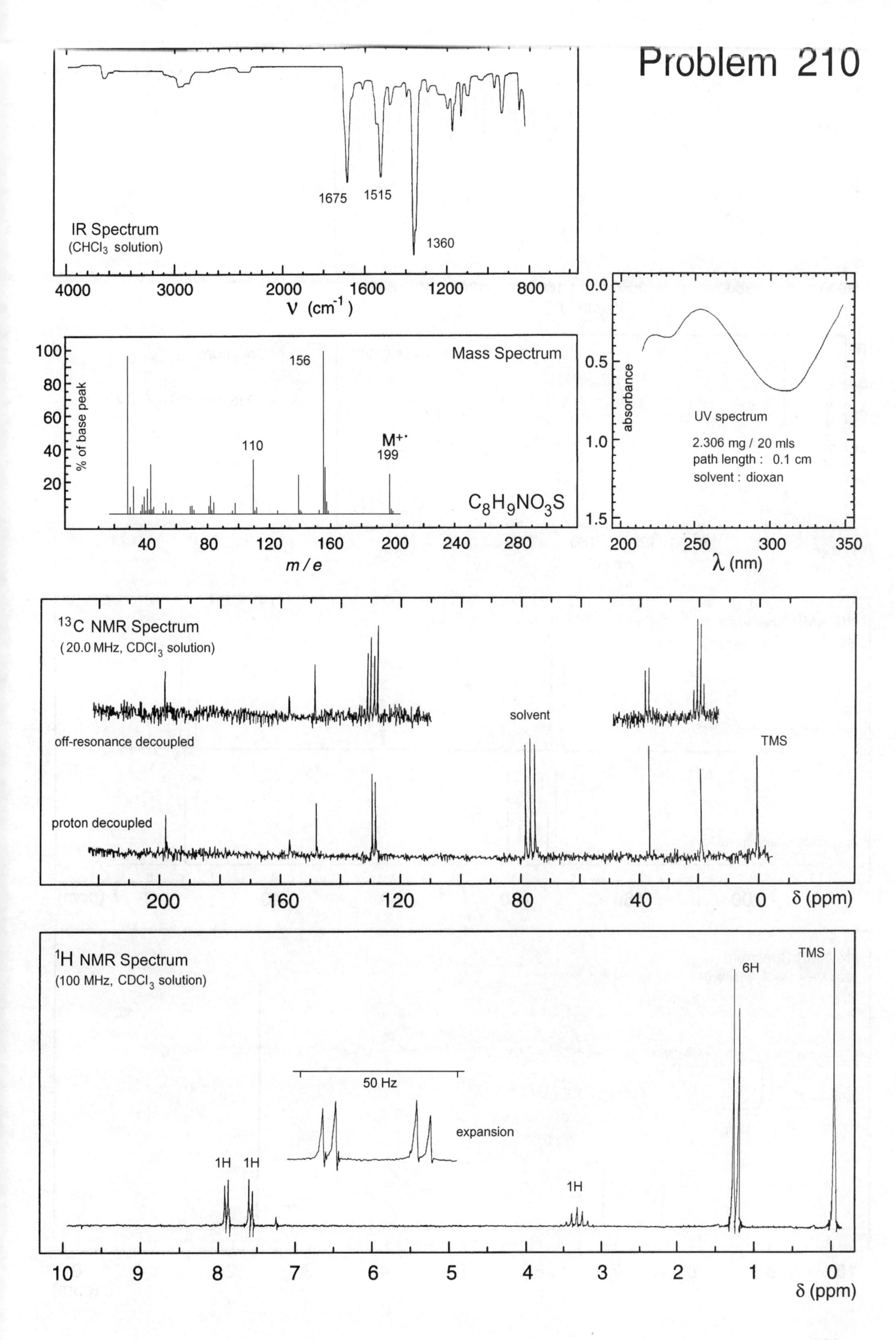

IR Spectrum
(CHCl3 solution)
1675
1515
1360
4000
3000
2000
1600
1200
800
ν (cm-1)
Mass Spectrum
% of base peak
156
110
M+·
199
C8H9NO3S
m / e
UV spectrum
2.306 mg / 20 mls
path length : 0.1 cm
solvent : dioxan
absorbance
λ (nm)
13C NMR Spectrum
(20.0 MHz, CDCl3 solution)
off-resonance decoupled
proton decoupled
solvent
TMS
δ (ppm)
1H NMR Spectrum
(100 MHz, CDCl3 solution)
50 Hz
expansion
1H
1H
1H
6H
TMS
δ (ppm)

Problem 211

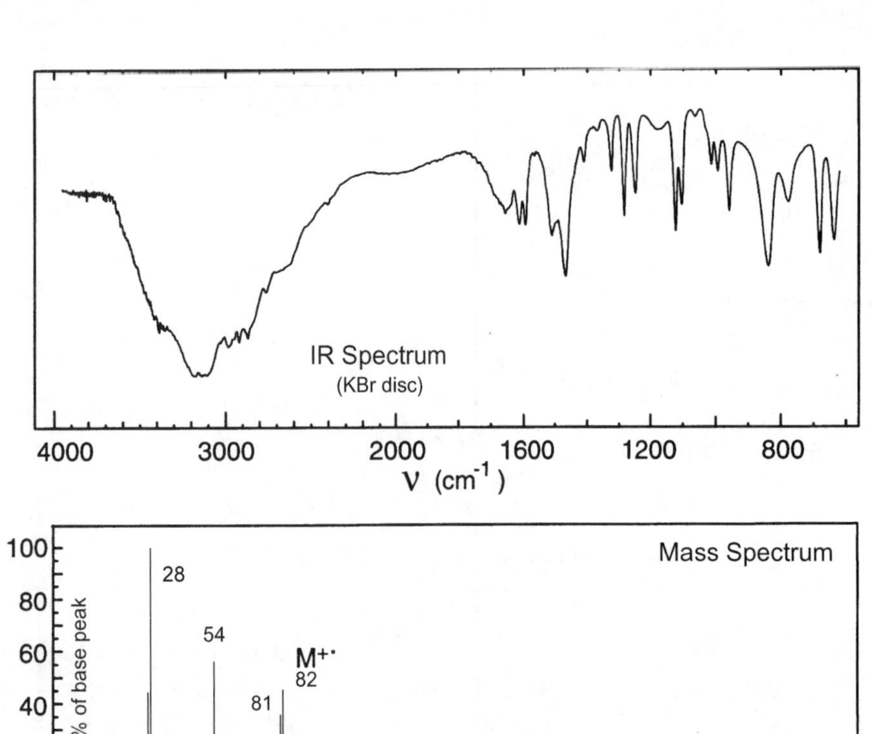

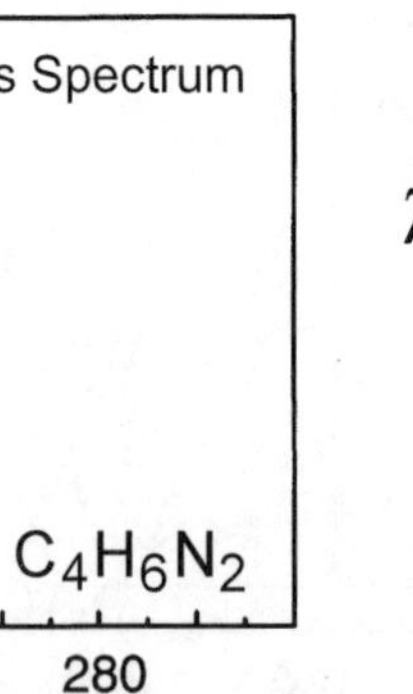

UV Spectrum

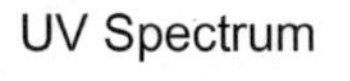

solvent : water

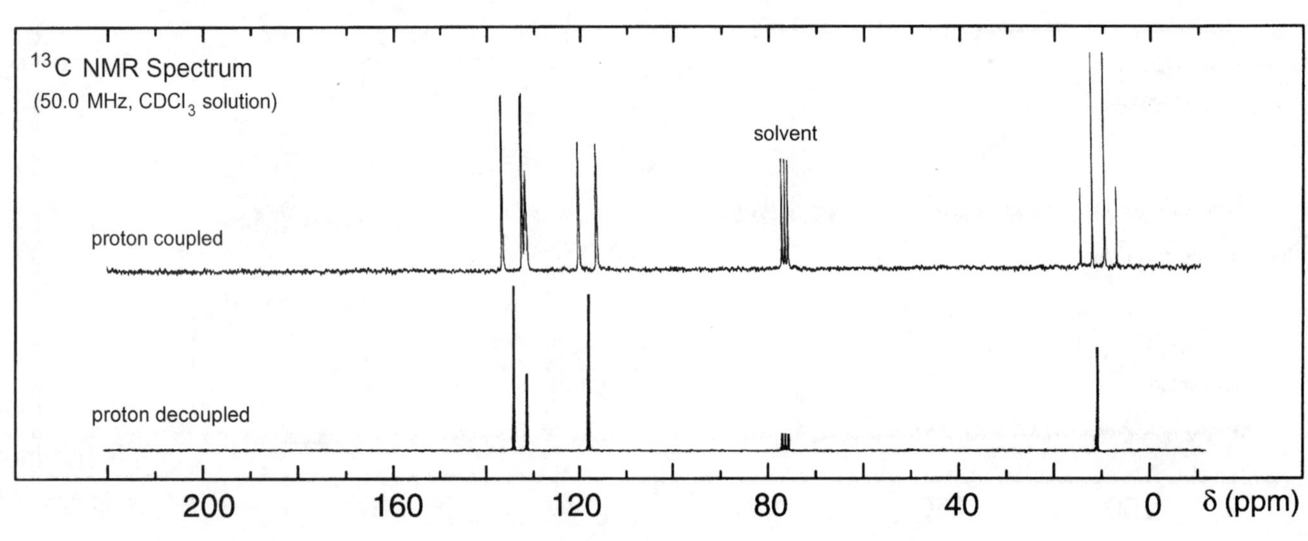

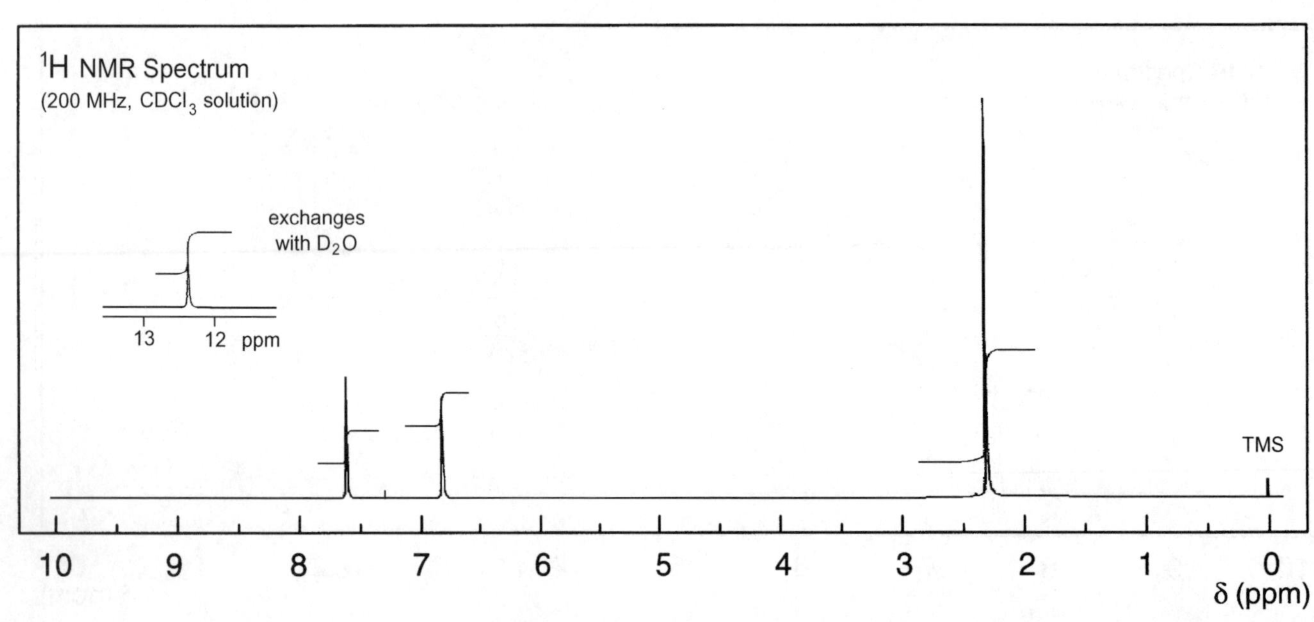

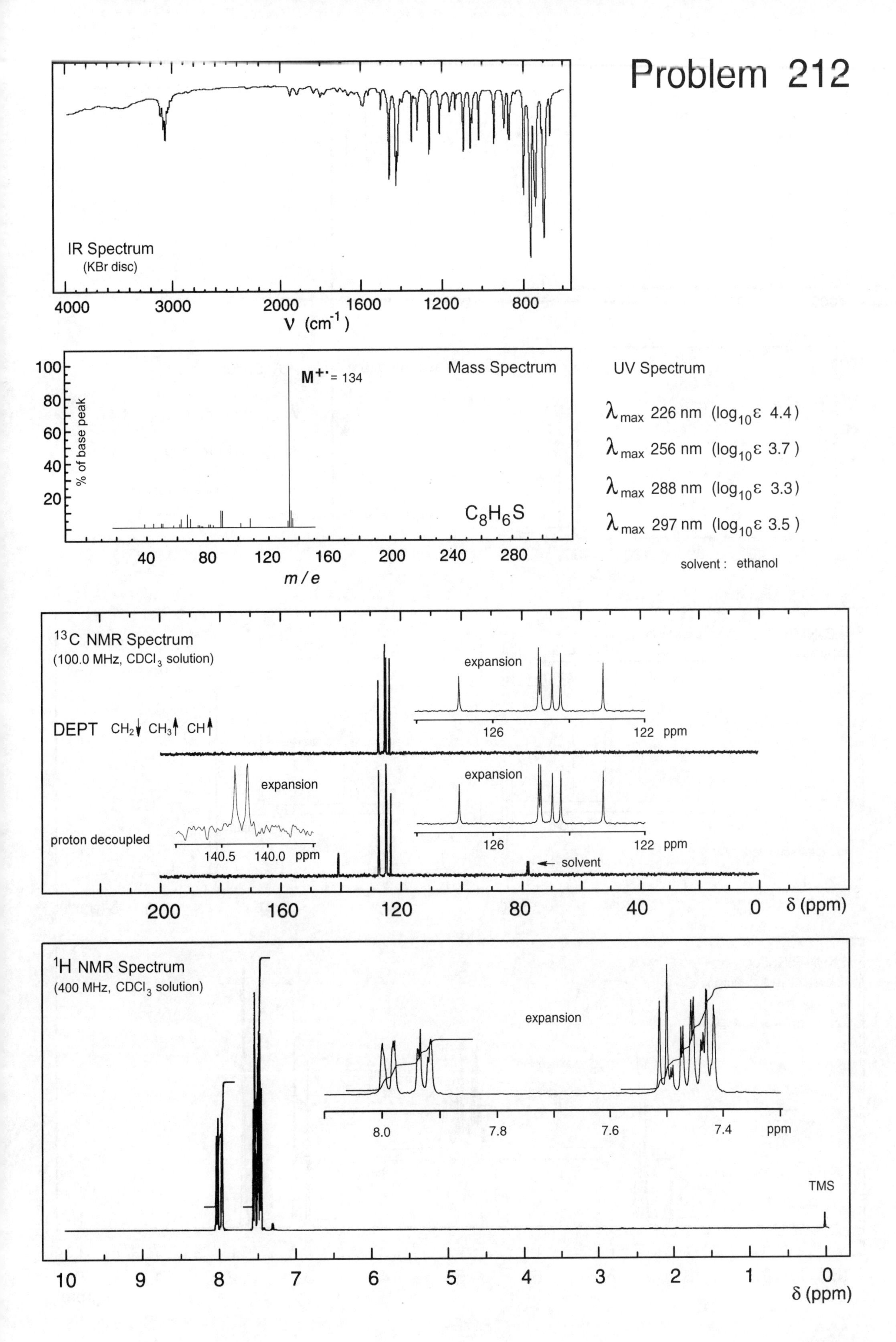
Problem 212
IR Spectrum
(KBr disc)
4000
3000
2000
1600
1200
800
ν (cm⁻¹)
Mass Spectrum
M+· = 134
% of base peak
C8H6S
m / e
UV Spectrum
λmax 226 nm (log10ε 4.4)
λmax 256 nm (log10ε 3.7)
λmax 288 nm (log10ε 3.3)
λmax 297 nm (log10ε 3.5)
solvent : ethanol
13C NMR Spectrum
(100.0 MHz, CDCl3 solution)
DEPT CH2↓ CH3↑ CH↑
expansion
126
122 ppm
proton decoupled
140.5
140.0 ppm
solvent
δ (ppm)
1H NMR Spectrum
(400 MHz, CDCl3 solution)
expansion
8.0
7.8
7.6
7.4
ppm
TMS
δ (ppm)

Problem 213

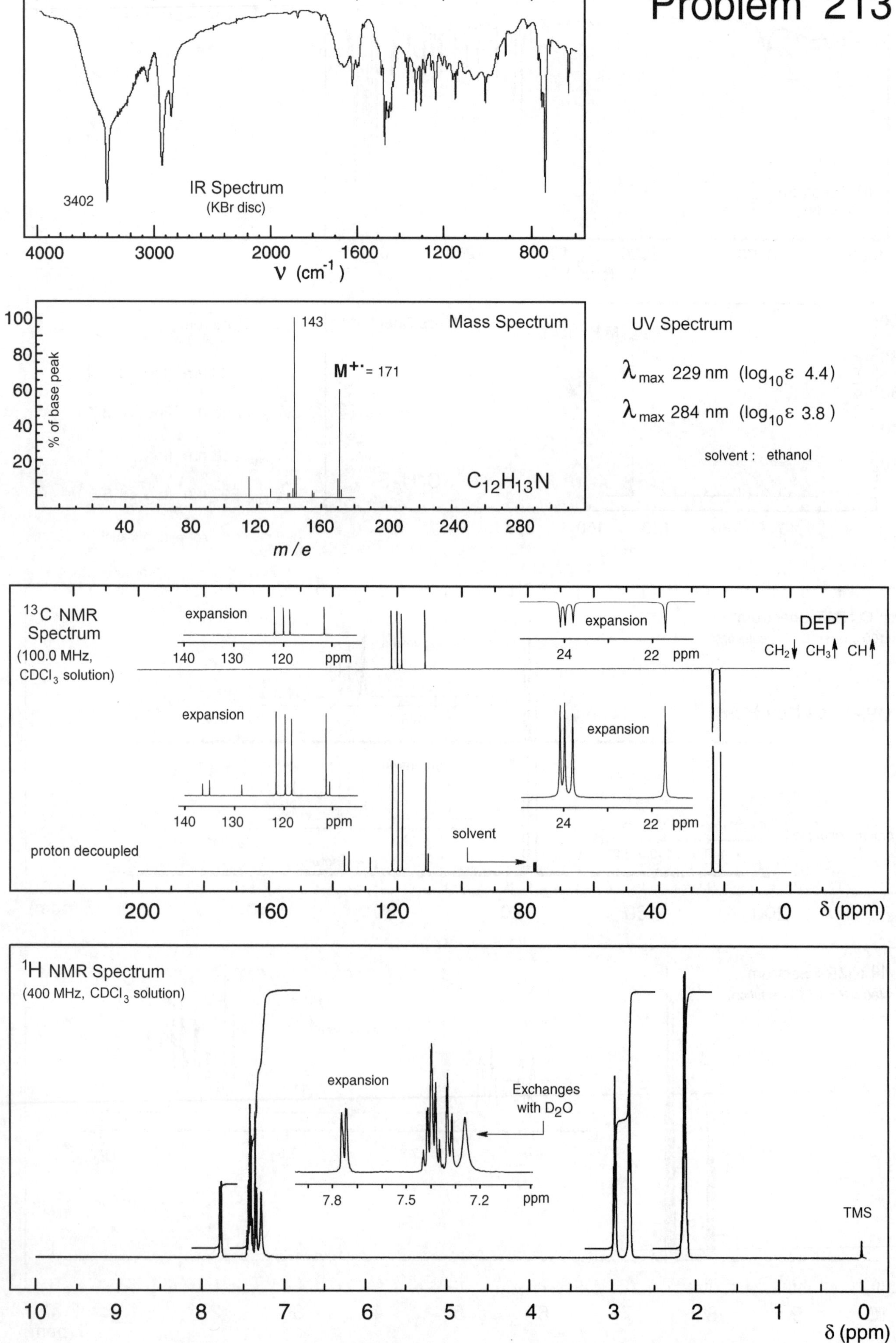

UV Spectrum

λ_{max} 229 nm ($\log_{10}\varepsilon$ 4.4)

λ_{max} 284 nm ($\log_{10}\varepsilon$ 3.8)

solvent : ethanol

Problem 214

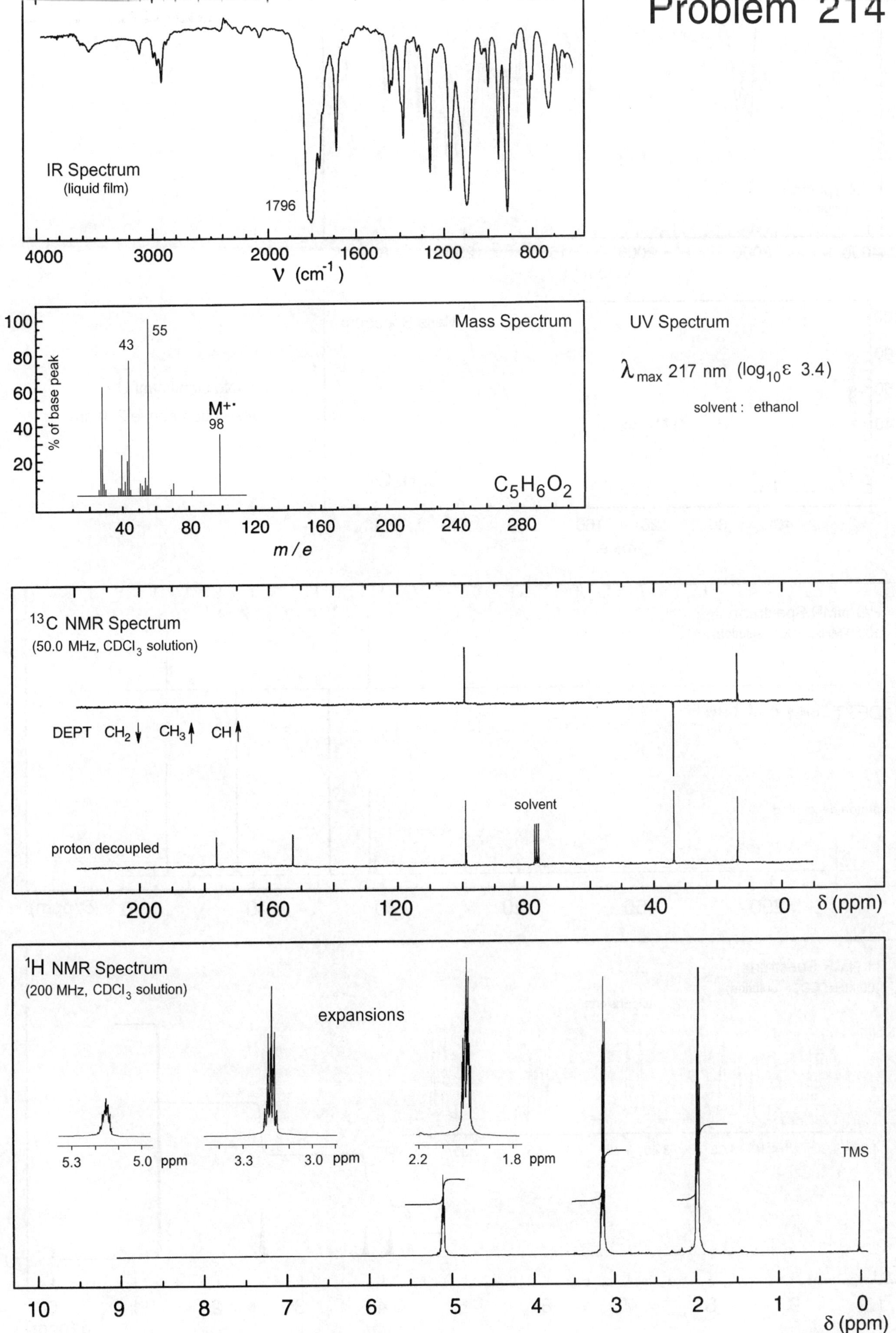

IR Spectrum
(liquid film)
1796
4000
3000
2000
1600
1200
800
ν (cm⁻¹)
Mass Spectrum
% of base peak
100
80
60
40
20
43
55
M+·
98
C5H6O2
40
80
120
160
200
240
280
m / e
UV Spectrum
λmax 217 nm (log10ε 3.4)
solvent : ethanol
13C NMR Spectrum
(50.0 MHz, CDCl3 solution)
DEPT CH2 ↓ CH3 ↑ CH ↑
solvent
proton decoupled
200
160
120
80
40
0
δ (ppm)
1H NMR Spectrum
(200 MHz, CDCl3 solution)
expansions
5.3
5.0 ppm
3.3
3.0 ppm
2.2
1.8 ppm
TMS
10
9
8
7
6
5
4
3
2
1
0
δ (ppm)

Problem 215

IR Spectrum
(liquid film)

1760

4000 3000 2000 1600 1200 800

ν (cm^{-1})

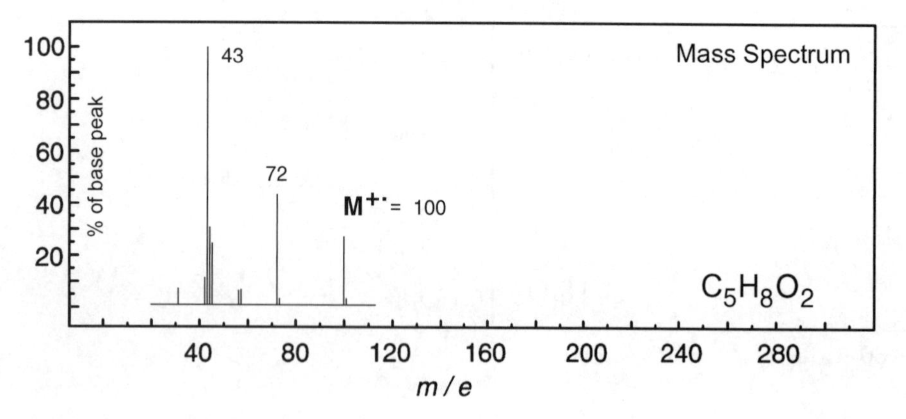

No significant UV absorption above 220 nm

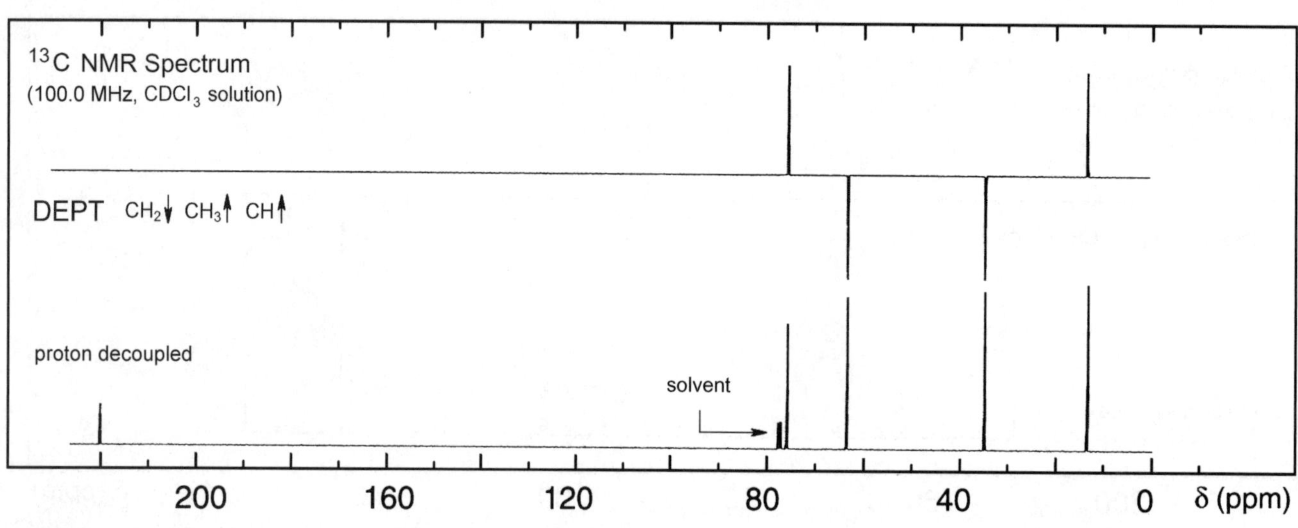

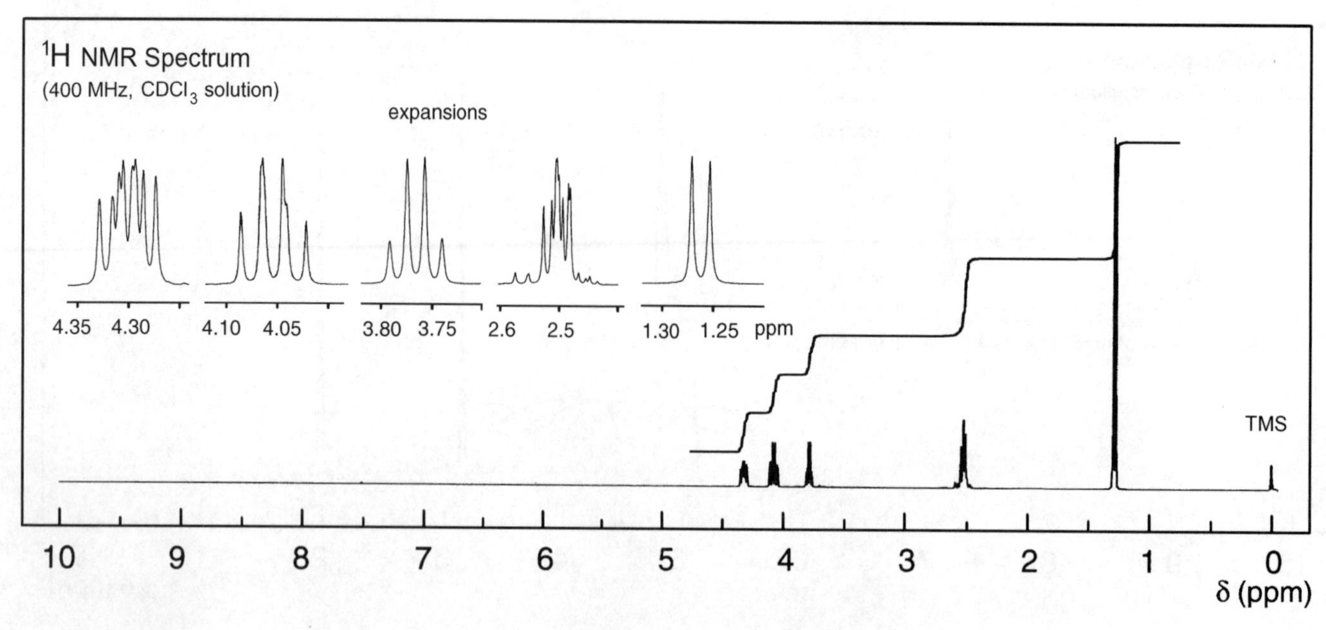

Problem 216

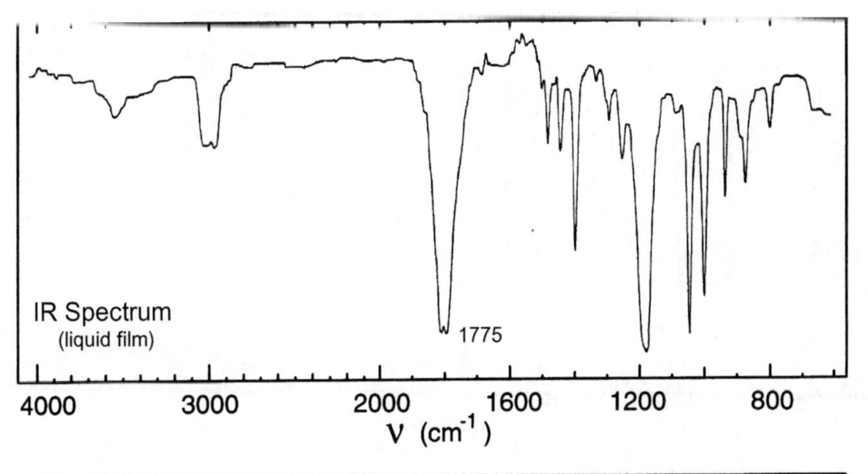

IR Spectrum
(liquid film)
1775
4000
3000
2000
1600
1200
800
ν (cm⁻¹)

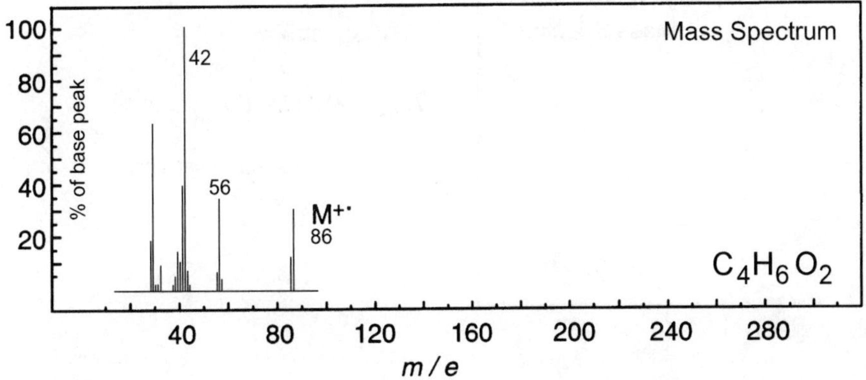

Mass Spectrum
% of base peak
100
80
60
40
20
42
56
M+·
86
$C_4H_6O_2$
40
80
120
160
200
240
280
m / e

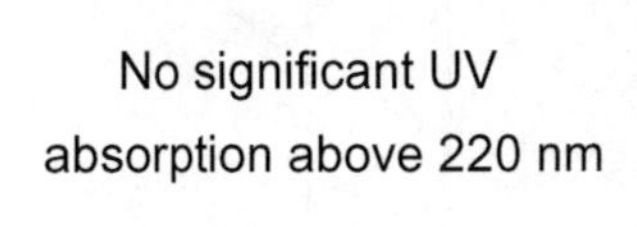

No significant UV
absorption above 220 nm

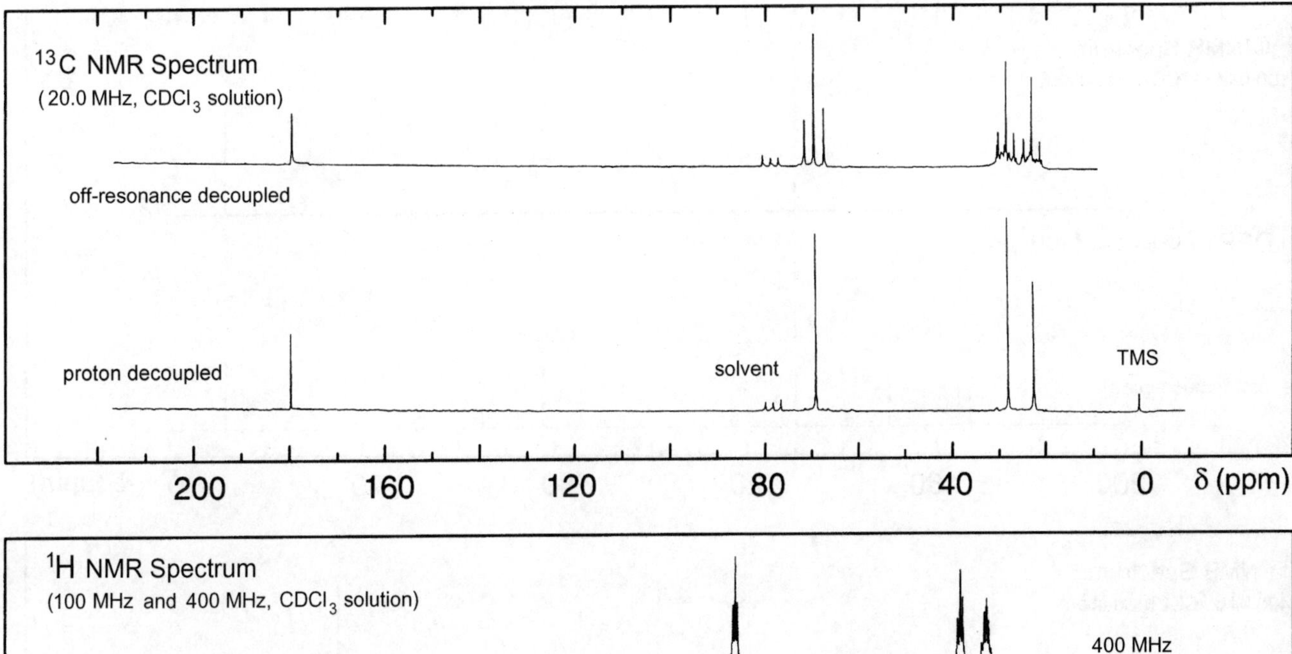

13C NMR Spectrum
(20.0 MHz, CDCl3 solution)
off-resonance decoupled
proton decoupled
solvent
TMS
200
160
120
80
40
0
δ (ppm)

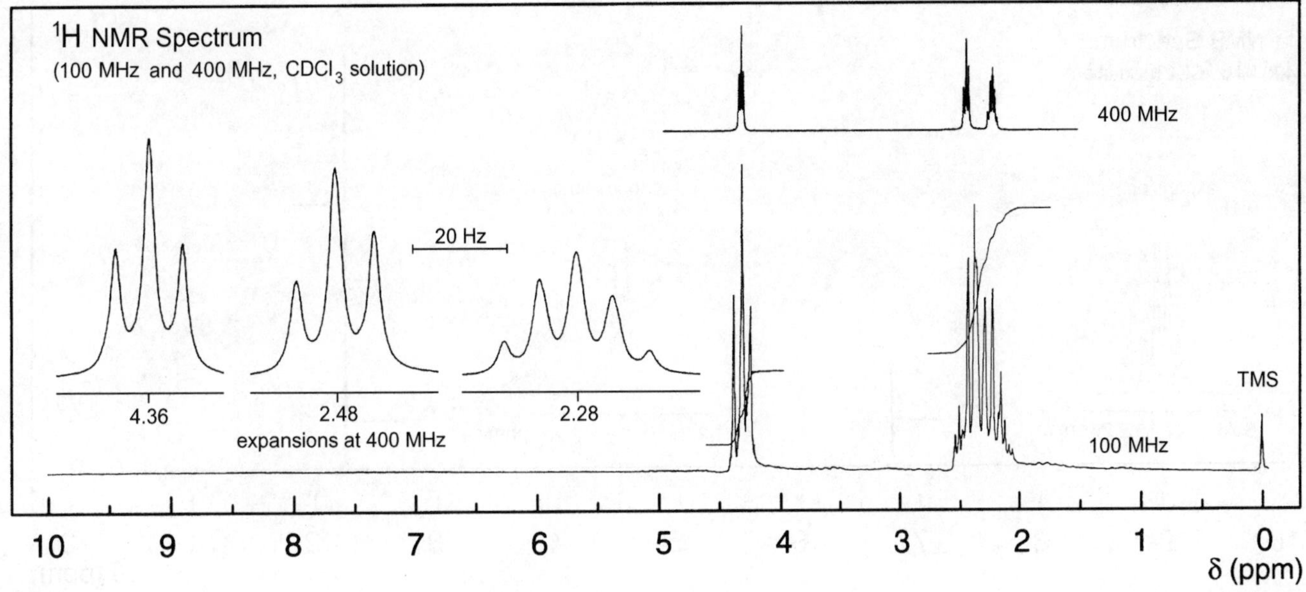

1H NMR Spectrum
(100 MHz and 400 MHz, CDCl3 solution)
400 MHz
20 Hz
4.36
2.48
2.28
expansions at 400 MHz
TMS
100 MHz
10
9
8
7
6
5
4
3
2
1
0
δ (ppm)

Problem 217

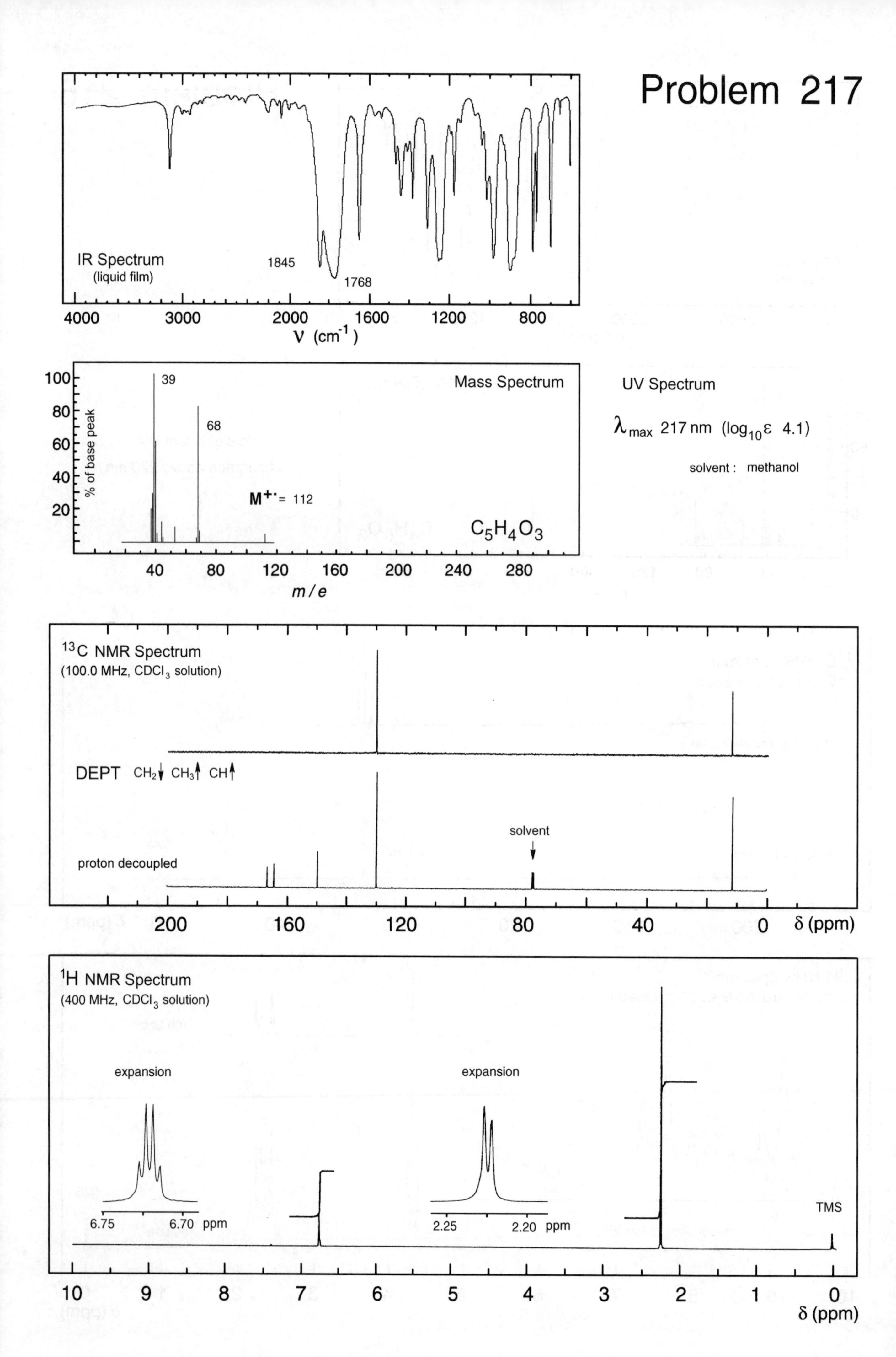

IR Spectrum
(liquid film)
1845
1768
4000
3000
2000
1600
1200
800
ν (cm⁻¹)
Mass Spectrum
% of base peak
39
68
M+· = 112
$C_5H_4O_3$
m/e
UV Spectrum
λmax 217 nm (log10ε 4.1)
solvent : methanol
13C NMR Spectrum
(100.0 MHz, CDCl3 solution)
DEPT CH2↓ CH3↑ CH↑
solvent
proton decoupled
δ (ppm)
1H NMR Spectrum
(400 MHz, CDCl3 solution)
expansion
6.75
6.70 ppm
expansion
2.25
2.20 ppm
TMS
δ (ppm)

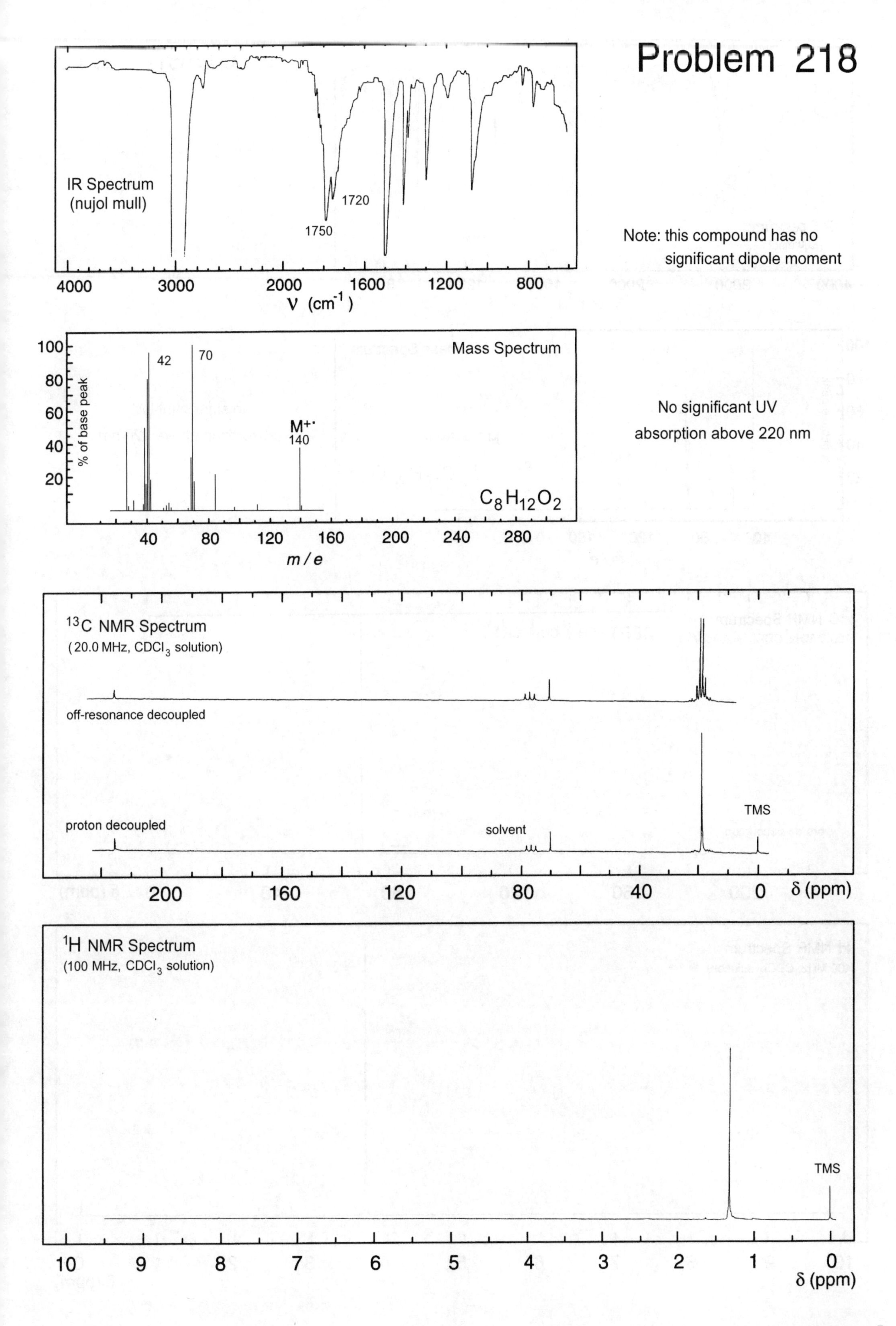

Problem 218
IR Spectrum
(nujol mull)
1720
1750
4000
3000
2000
1600
1200
800
ν (cm⁻¹)
Note: this compound has no significant dipole moment
Mass Spectrum
% of base peak
100
80
60
40
20
42
70
M+·
140
C8H12O2
40
80
120
160
200
240
280
m / e
No significant UV absorption above 220 nm
13C NMR Spectrum
(20.0 MHz, CDCl3 solution)
off-resonance decoupled
proton decoupled
solvent
TMS
200
160
120
80
40
0
δ (ppm)
1H NMR Spectrum
(100 MHz, CDCl3 solution)
TMS
10
9
8
7
6
5
4
3
2
1
0
δ (ppm)

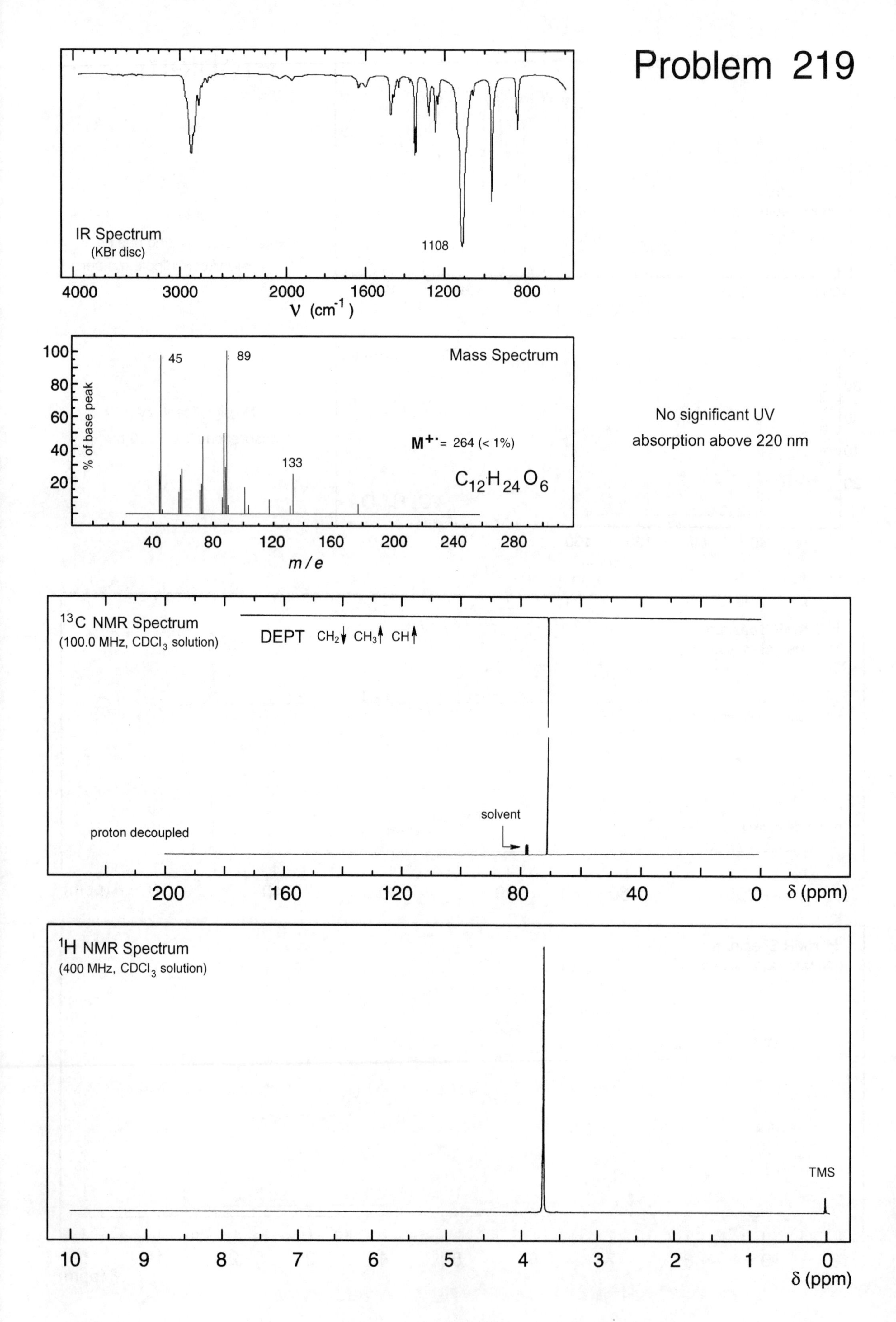
Problem 219
IR Spectrum
(KBr disc)
1108
4000
3000
2000
1600
1200
800
ν (cm-1)
Mass Spectrum
100
80
60
40
20
% of base peak
45
89
133
M+· = 264 (< 1%)
C12H24O6
40
80
120
160
200
240
280
m / e
No significant UV
absorption above 220 nm
13C NMR Spectrum
(100.0 MHz, CDCl3 solution)
DEPT CH2↓ CH3↑ CH↑
solvent
proton decoupled
200
160
120
80
40
0
δ (ppm)
1H NMR Spectrum
(400 MHz, CDCl3 solution)
TMS
10
9
8
7
6
5
4
3
2
1
0
δ (ppm)

Problem 220

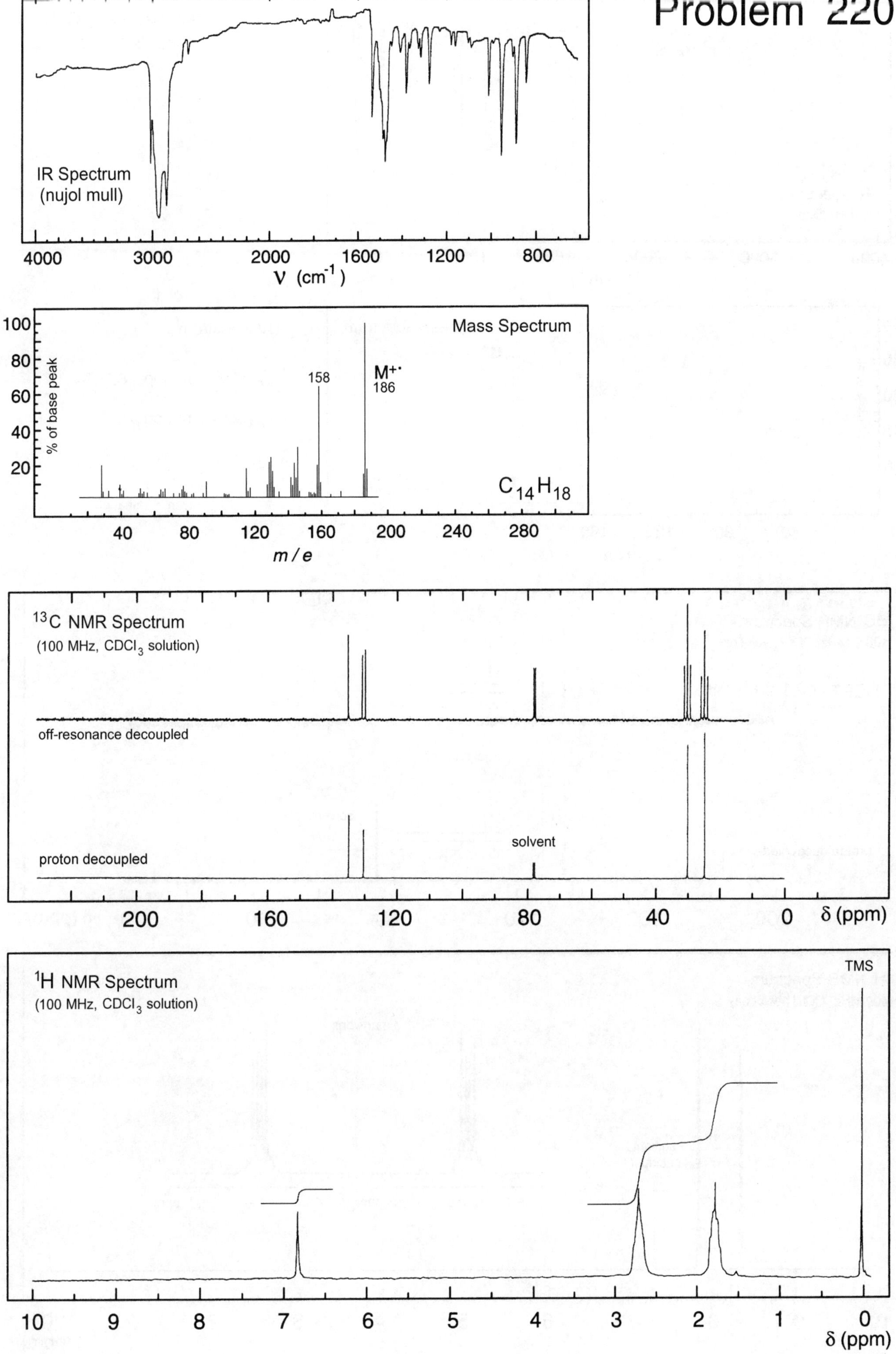

IR Spectrum
(nujol mull)
4000
3000
2000
1600
1200
800
ν (cm^{-1})
Mass Spectrum
100
80
60
40
20
% of base peak
158
M+·
186
C14H18
40
80
120
160
200
240
280
m / e
13C NMR Spectrum
(100 MHz, CDCl3 solution)
off-resonance decoupled
proton decoupled
solvent
200
160
120
80
40
0
δ (ppm)
1H NMR Spectrum
(100 MHz, CDCl3 solution)
TMS
10
9
8
7
6
5
4
3
2
1
0
δ (ppm)

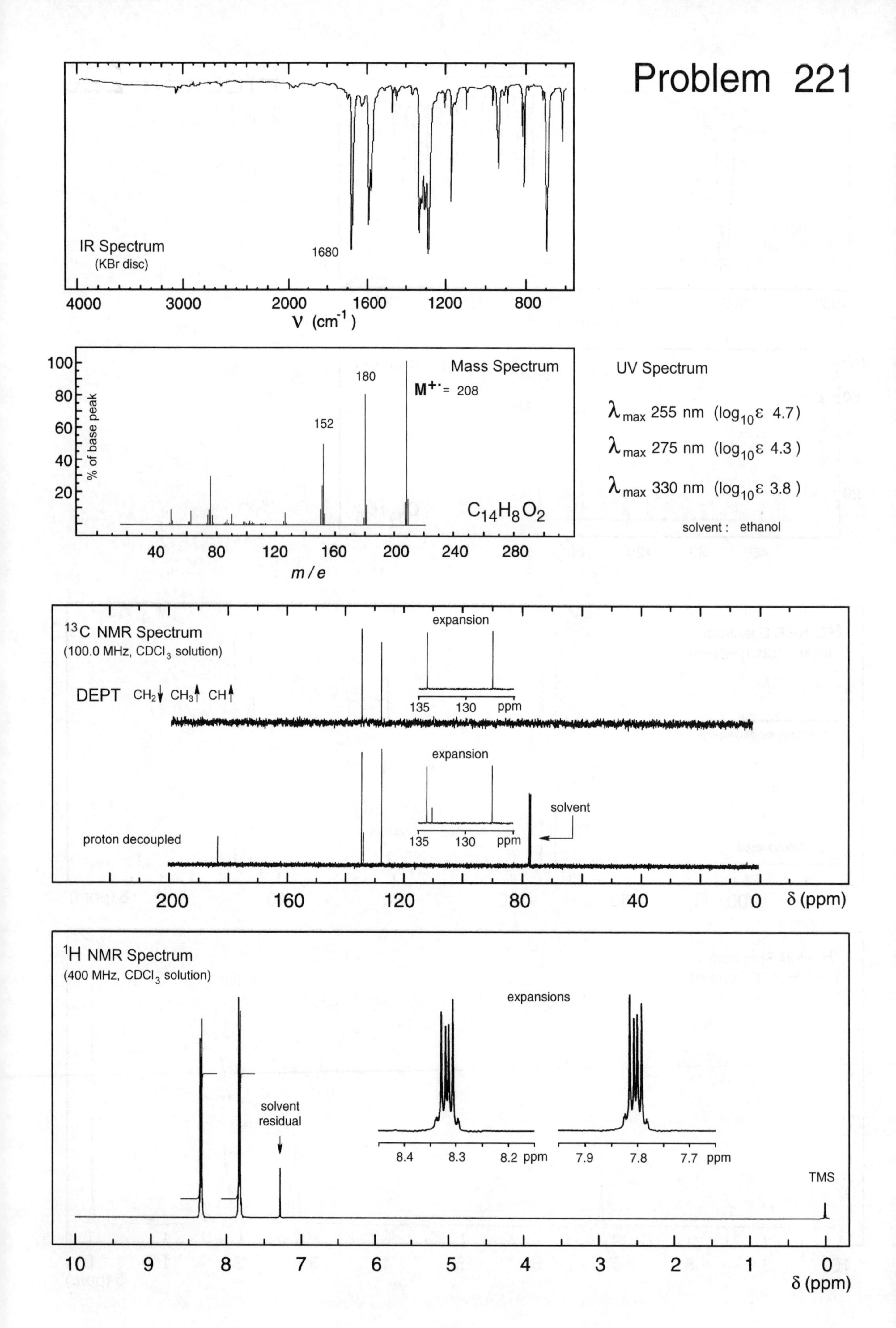
Problem 221
IR Spectrum
(KBr disc)
1680
4000
3000
2000
1600
1200
800
ν (cm⁻¹)
Mass Spectrum
M⁺˙ = 208
180
152
100
80
60
40
20
% of base peak
40
80
120
160
200
240
280
m / e
C14H8O2
UV Spectrum
λmax 255 nm (log10ε 4.7)
λmax 275 nm (log10ε 4.3)
λmax 330 nm (log10ε 3.8)
solvent : ethanol
13C NMR Spectrum
(100.0 MHz, CDCl3 solution)
DEPT CH2↓ CH3↑ CH↑
expansion
135
130
ppm
proton decoupled
solvent
200
160
120
80
40
0
δ (ppm)
1H NMR Spectrum
(400 MHz, CDCl3 solution)
expansions
8.4
8.3
8.2 ppm
7.9
7.8
7.7 ppm
solvent residual
TMS
10
9
8
7
6
5
4
3
2
1
0
δ (ppm)

Problem 222

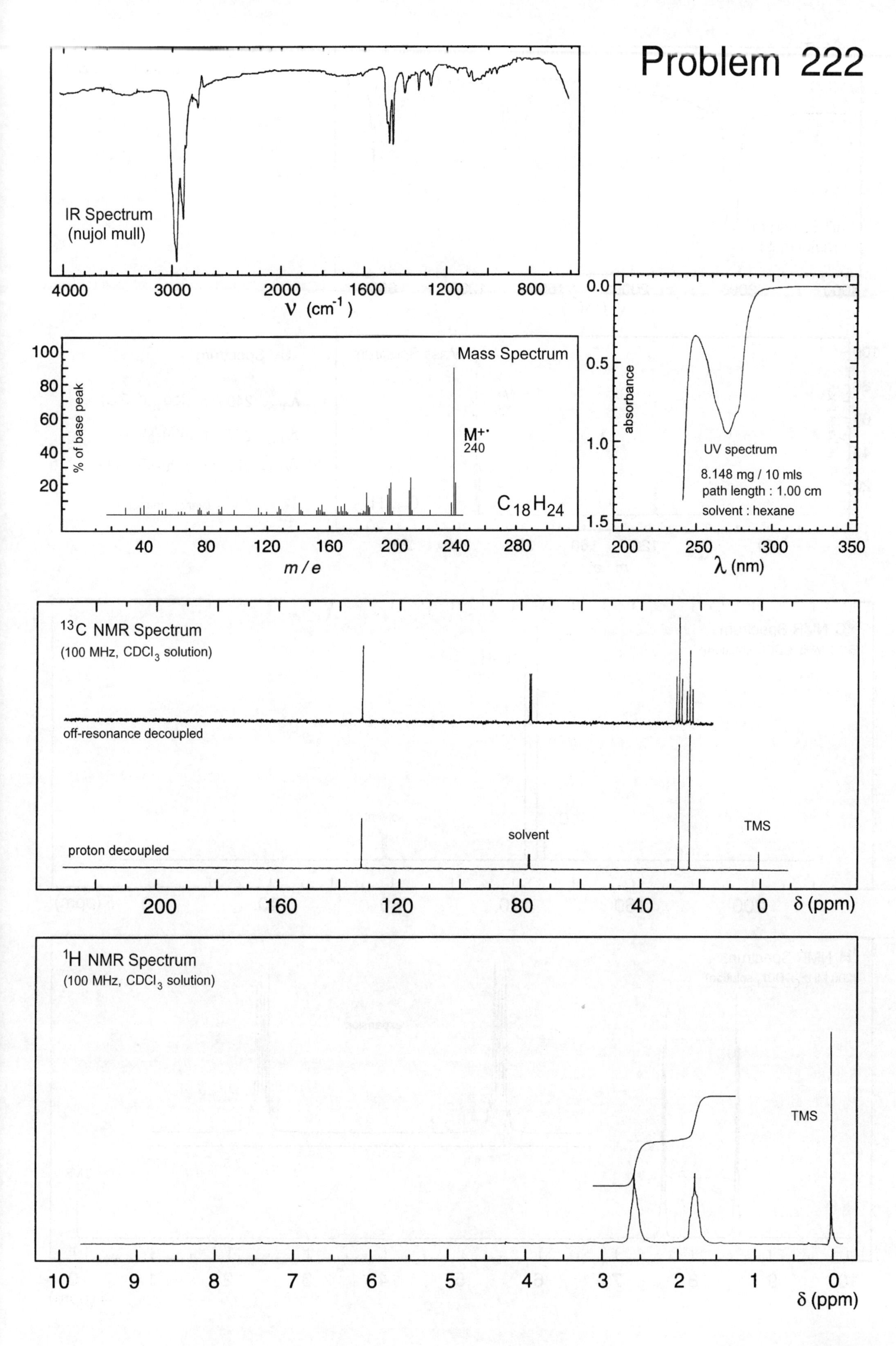

IR Spectrum
(nujol mull)
4000
3000
2000
1600
1200
800
ν (cm-1)
Mass Spectrum
% of base peak
100
80
60
40
20
M+·
240
C18H24
40
80
120
160
200
240
280
m / e
0.0
0.5
1.0
1.5
absorbance
UV spectrum
8.148 mg / 10 mls
path length : 1.00 cm
solvent : hexane
200
250
300
350
λ (nm)
13C NMR Spectrum
(100 MHz, CDCl3 solution)
off-resonance decoupled
proton decoupled
solvent
TMS
200
160
120
80
40
0
δ (ppm)
1H NMR Spectrum
(100 MHz, CDCl3 solution)
TMS
10
9
8
7
6
5
4
3
2
1
0
δ (ppm)

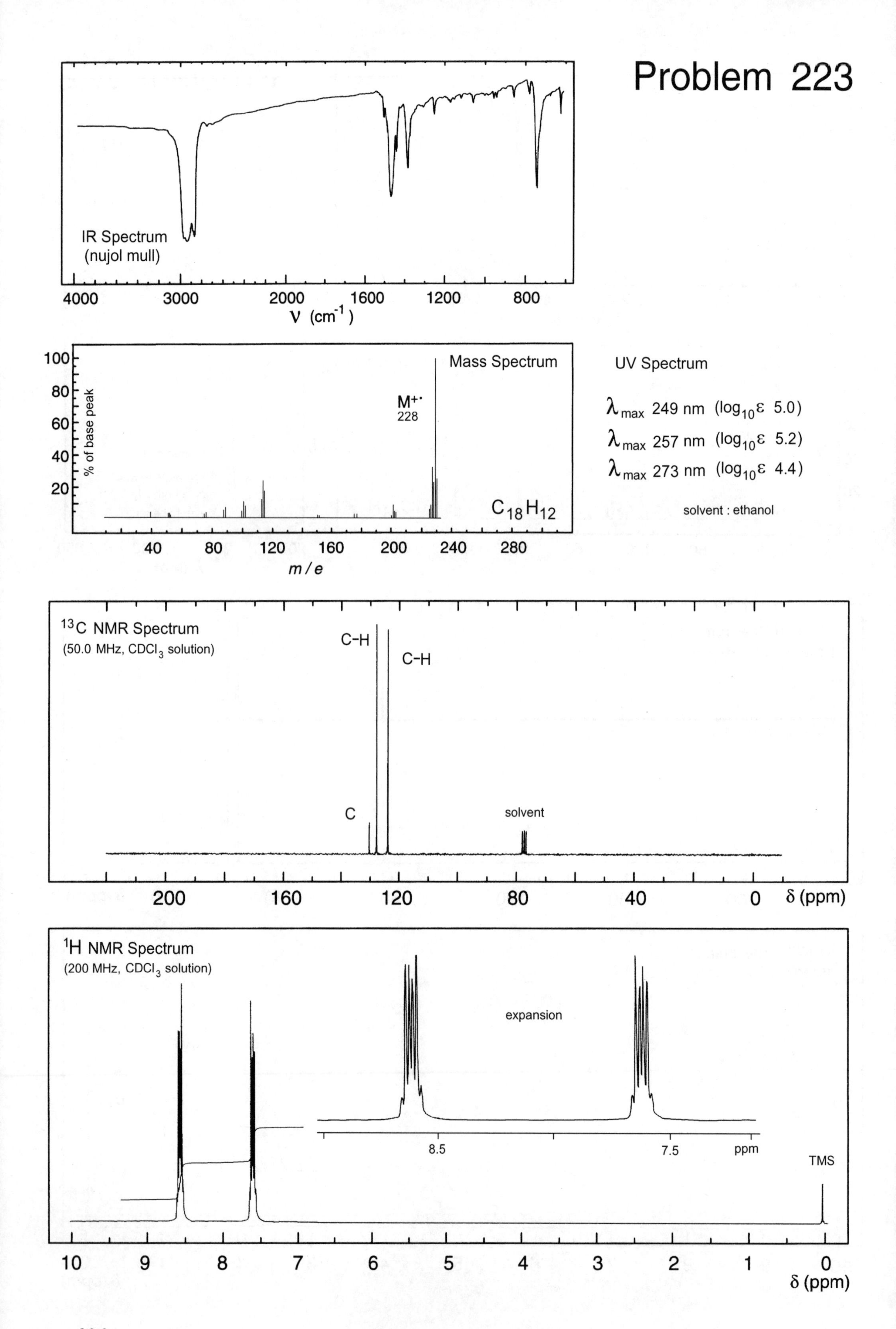
Problem 223
IR Spectrum
(nujol mull)
4000
3000
2000
1600
1200
800
ν (cm^{-1})
Mass Spectrum
M+·
228
% of base peak
100
80
60
40
20
40
80
120
160
200
240
280
m / e
$C_{18}H_{12}$
UV Spectrum
λ_{max} 249 nm ($\log_{10}\varepsilon$ 5.0)
λ_{max} 257 nm ($\log_{10}\varepsilon$ 5.2)
λ_{max} 273 nm ($\log_{10}\varepsilon$ 4.4)
solvent : ethanol
^{13}C NMR Spectrum
(50.0 MHz, $CDCl_3$ solution)
C-H
C-H
C
solvent
200
160
120
80
40
0
δ (ppm)
^{1}H NMR Spectrum
(200 MHz, $CDCl_3$ solution)
expansion
8.5
7.5
ppm
TMS
10
9
8
7
6
5
4
3
2
1
0
δ (ppm)

Problem 224

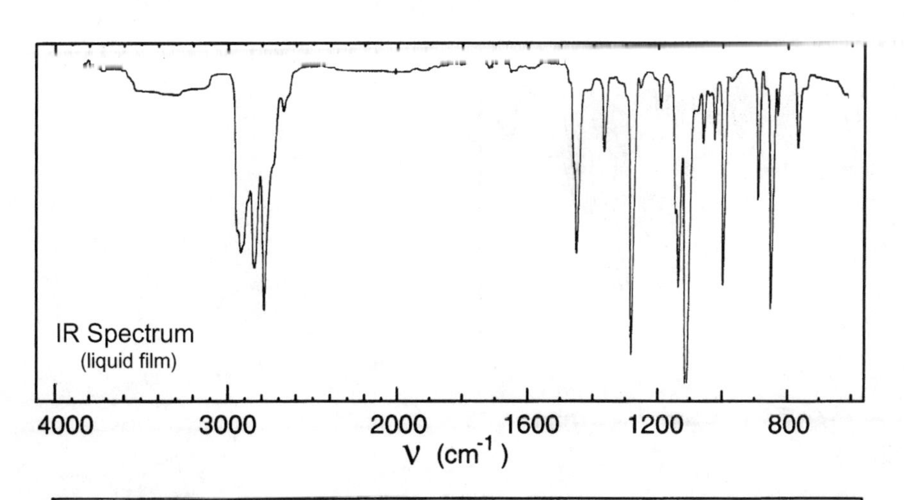

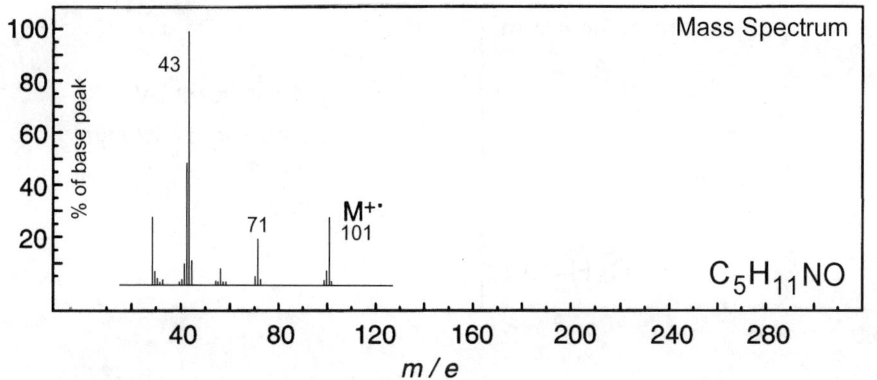

No significant UV absorption above 220 nm

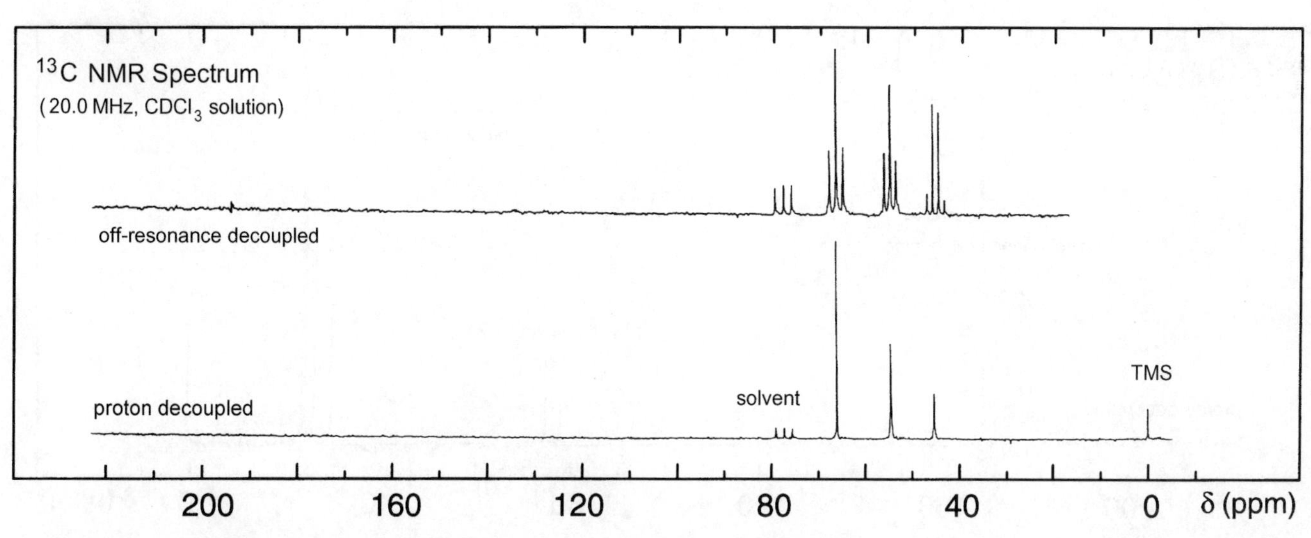

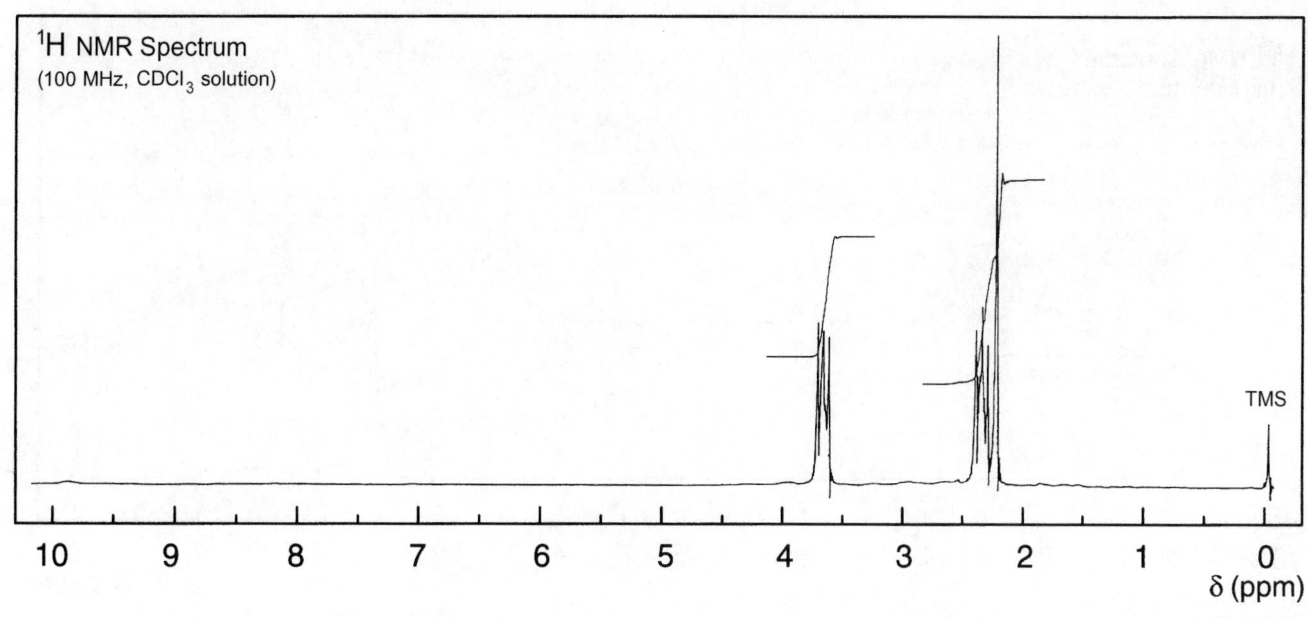

Problem 225

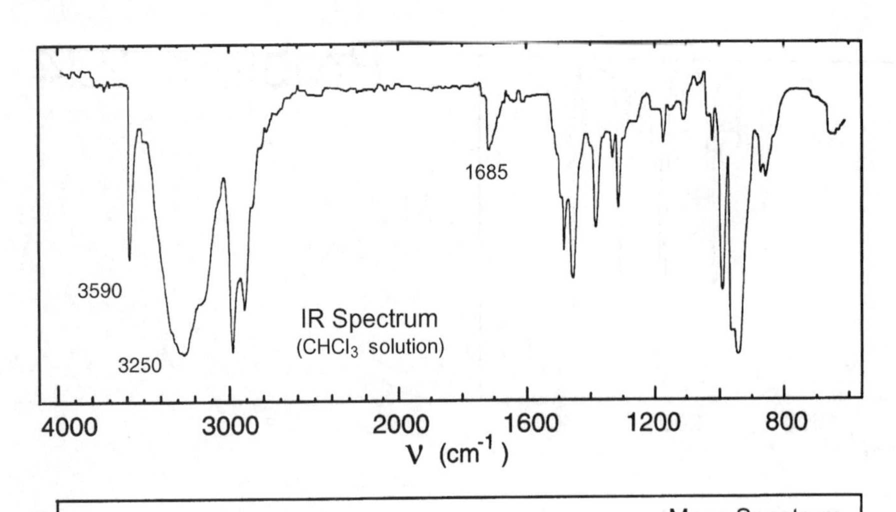

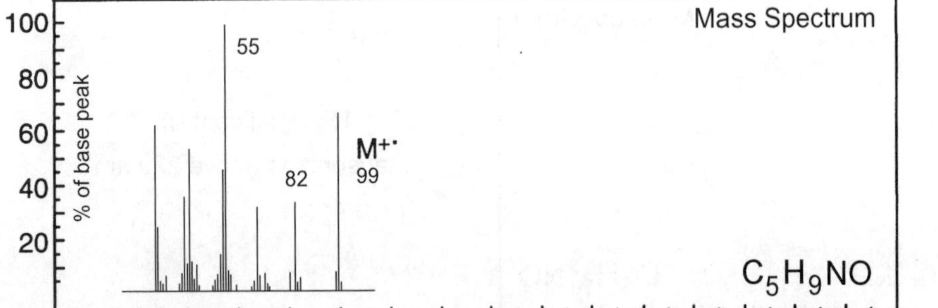

No significant UV absorption above 220 nm

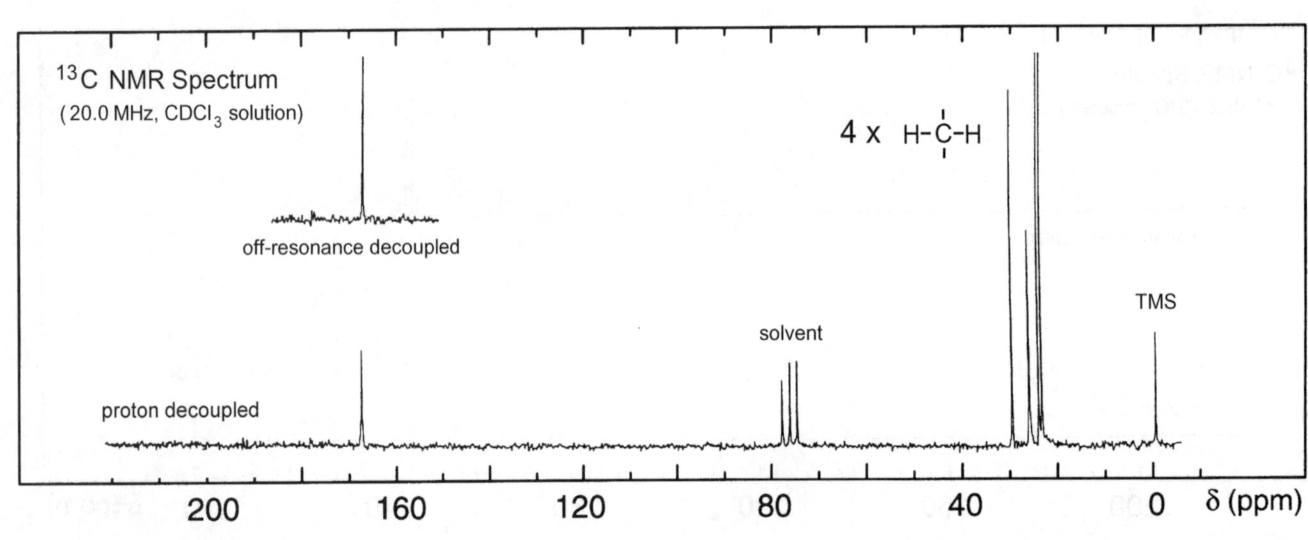

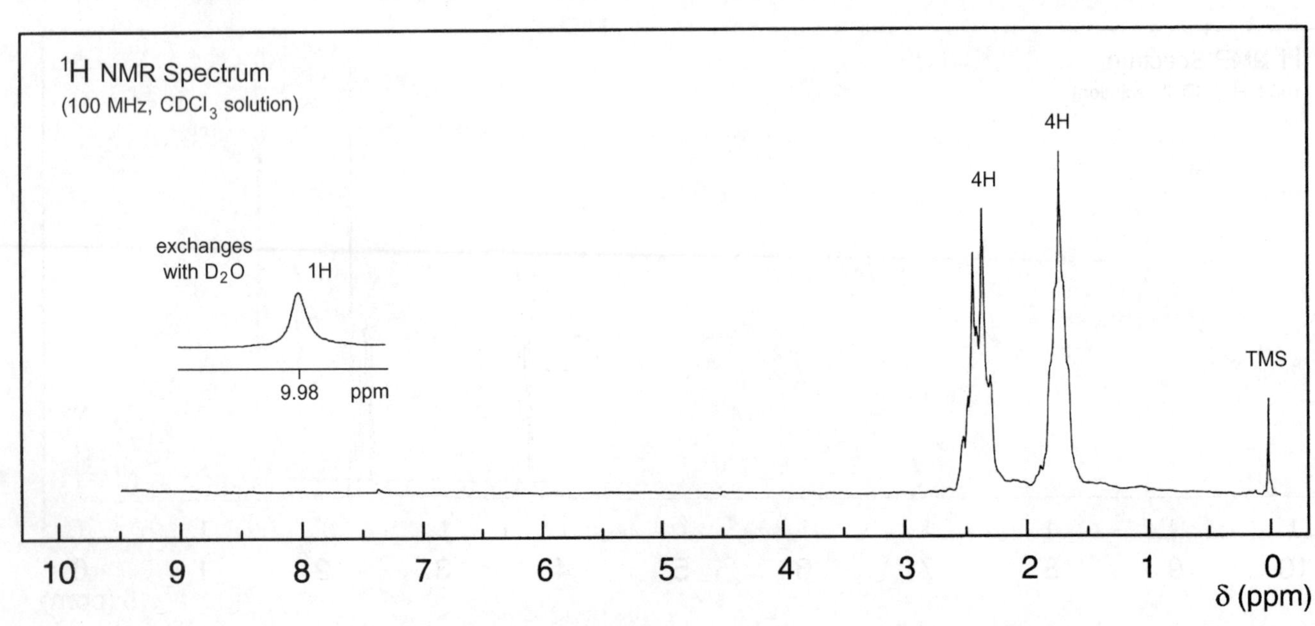

Problem 226

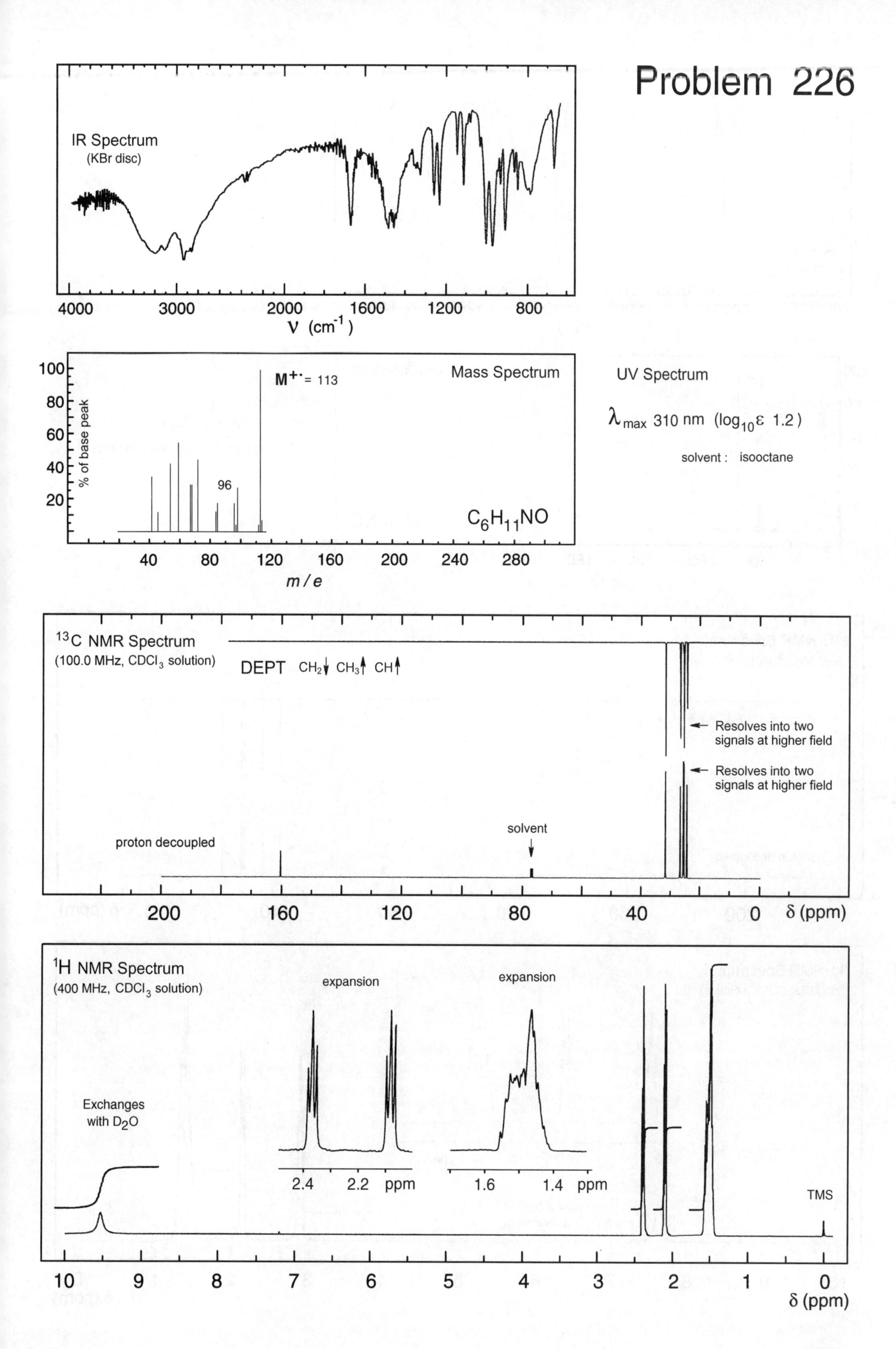

IR Spectrum
(KBr disc)
4000
3000
2000
1600
1200
800
ν (cm⁻¹)
Mass Spectrum
M⁺˙= 113
96
% of base peak
100
80
60
40
20
40
80
120
160
200
240
280
m / e
C6H11NO
UV Spectrum
λmax 310 nm (log10ε 1.2)
solvent : isooctane
13C NMR Spectrum
(100.0 MHz, CDCl3 solution)
DEPT CH2↓ CH3↑ CH↑
Resolves into two signals at higher field
Resolves into two signals at higher field
solvent
proton decoupled
200
160
120
80
40
0
δ (ppm)
1H NMR Spectrum
(400 MHz, CDCl3 solution)
expansion
expansion
2.4
2.2
ppm
1.6
1.4
ppm
Exchanges with D2O
TMS
10
9
8
7
6
5
4
3
2
1
0
δ (ppm)

Problem 227

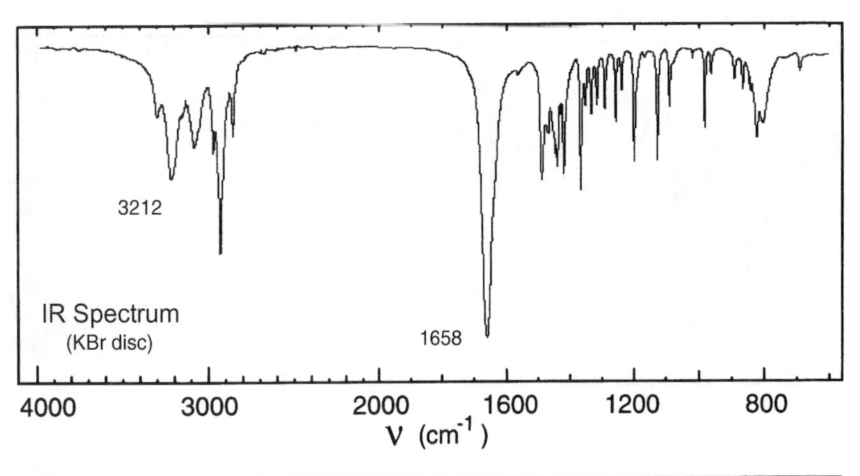

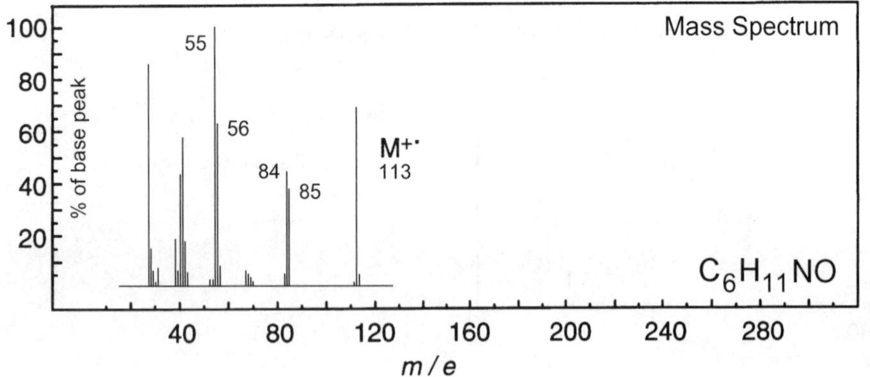

No significant UV absorption above 220 nm

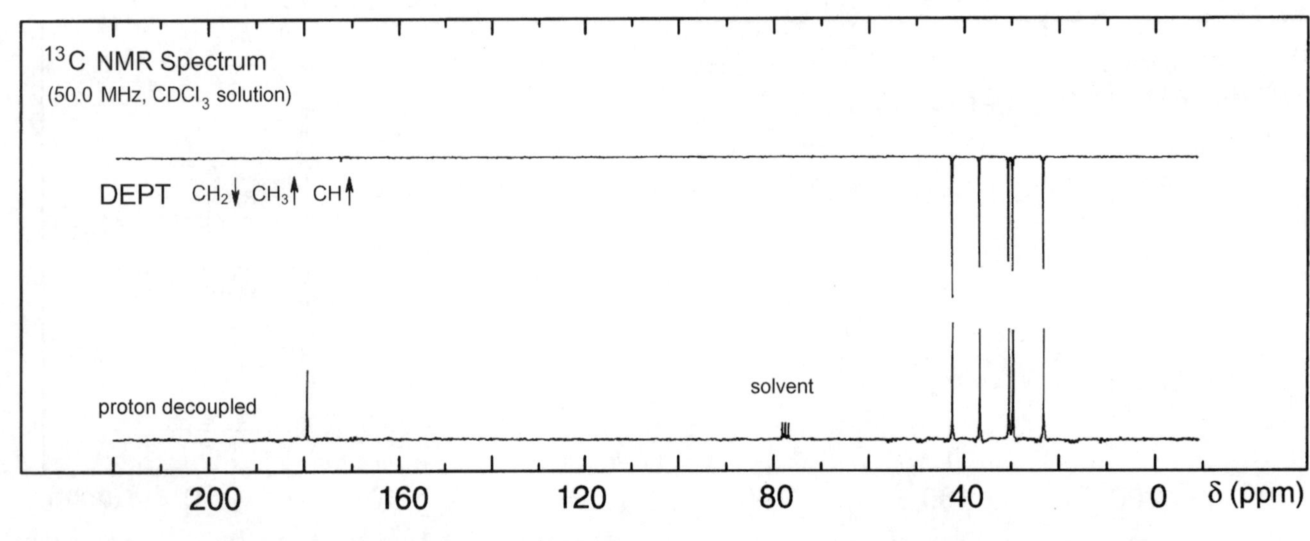

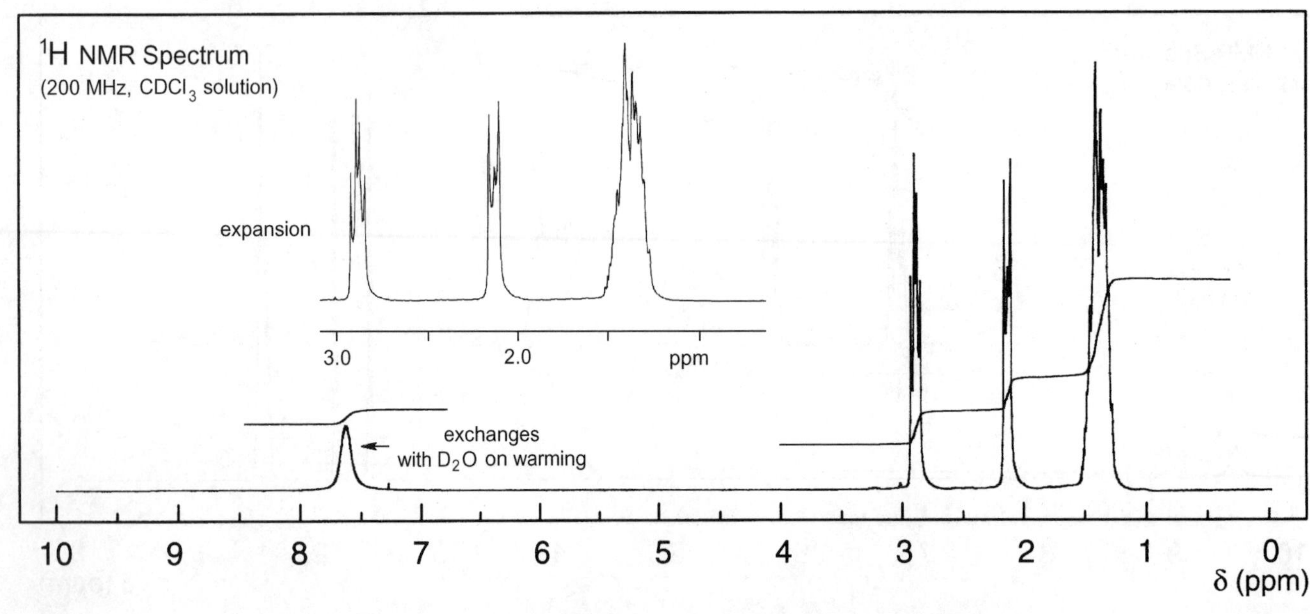

Problem 228

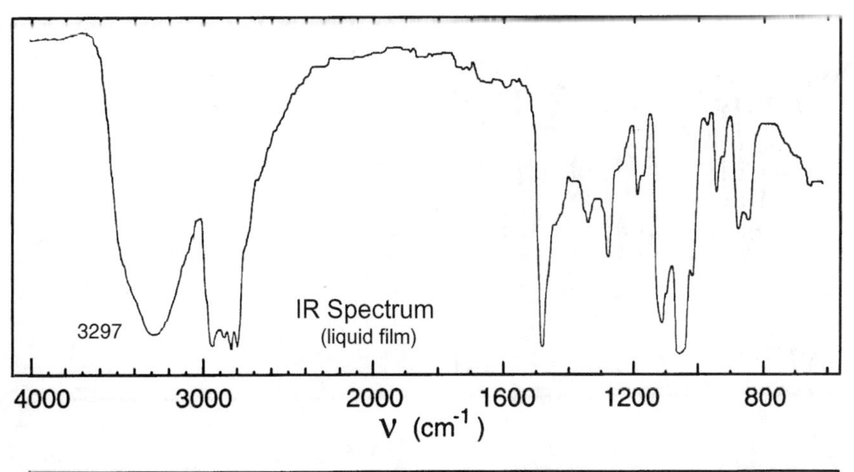

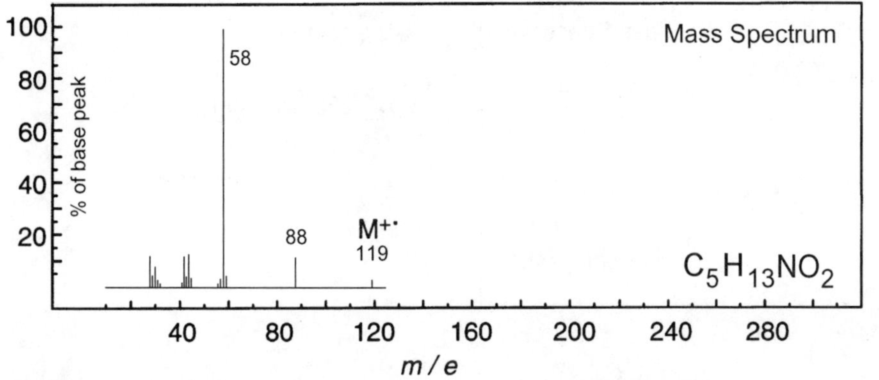

No significant UV absorption above 220 nm

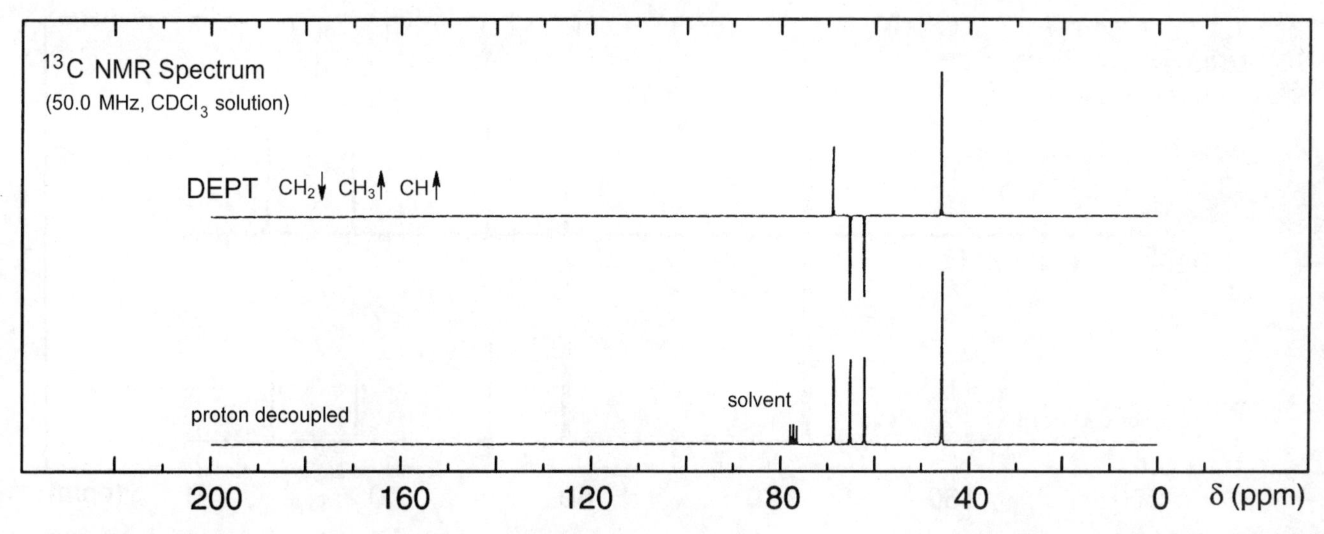

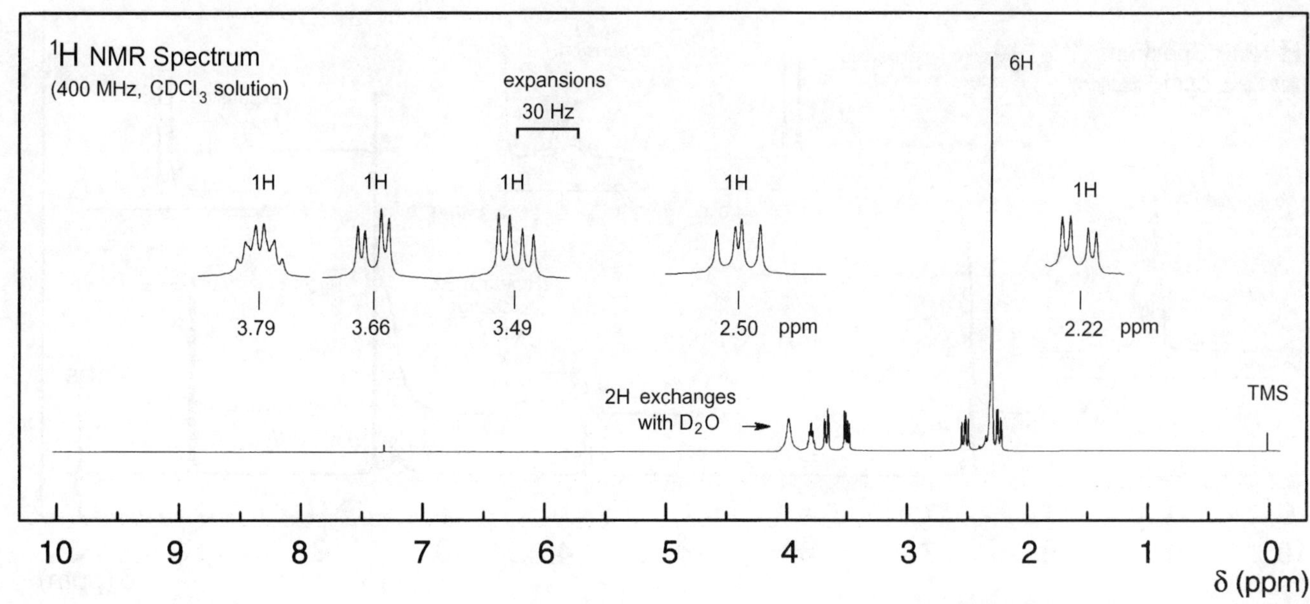

Problem 229

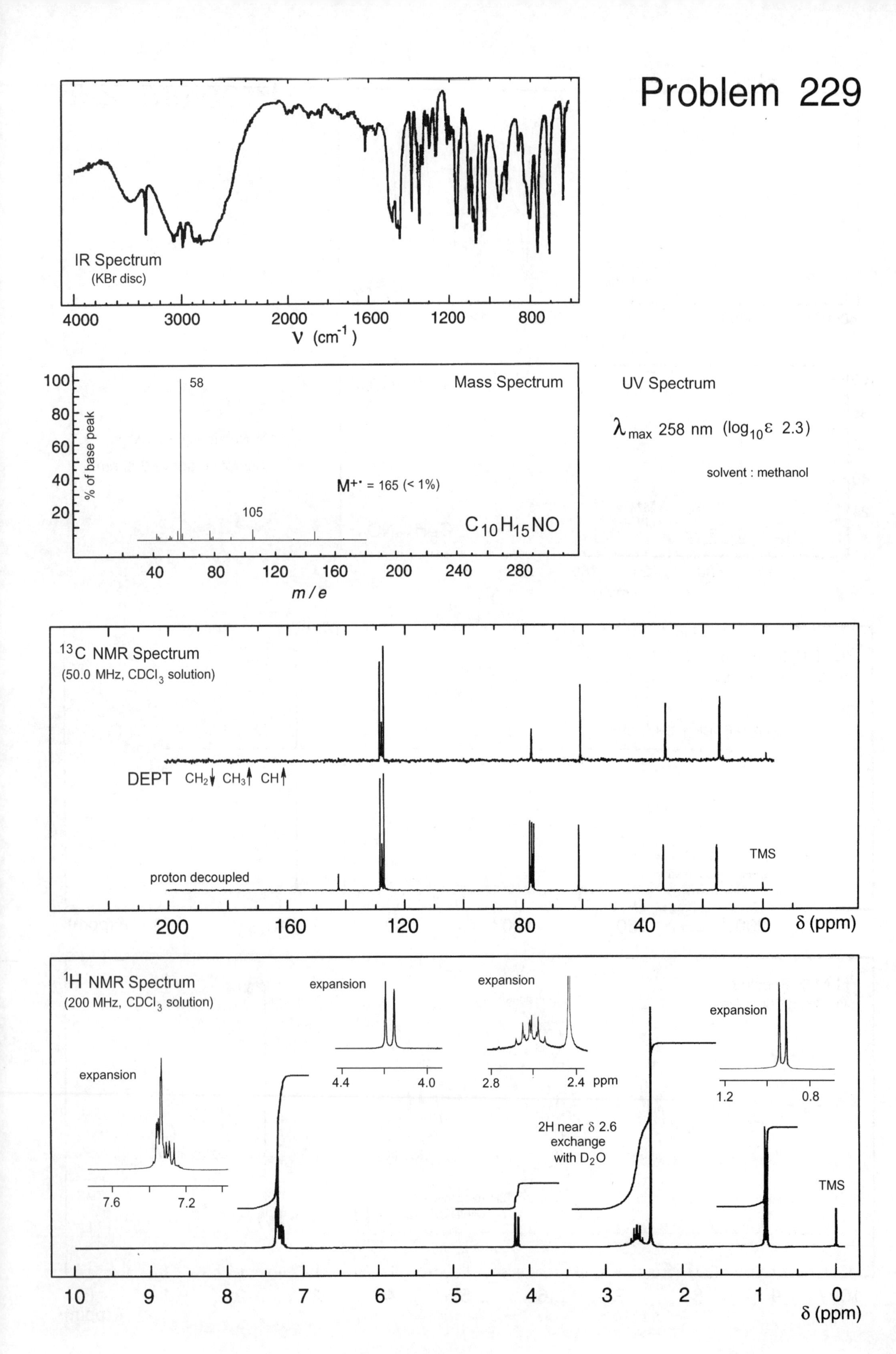

IR Spectrum
(KBr disc)
4000
3000
2000
1600
1200
800
ν (cm⁻¹)
100
80
60
40
20
% of base peak
58
105
Mass Spectrum
M+• = 165 (< 1%)
C10H15NO
40
80
120
160
200
240
280
m / e
UV Spectrum
λmax 258 nm (log10ε 2.3)
solvent : methanol
13C NMR Spectrum
(50.0 MHz, CDCl3 solution)
DEPT CH2↓ CH3↑ CH↑
proton decoupled
TMS
200
160
120
80
40
0
δ (ppm)
1H NMR Spectrum
(200 MHz, CDCl3 solution)
expansion
7.6
7.2
expansion
4.4
4.0
expansion
2.8
2.4 ppm
expansion
1.2
0.8
2H near δ 2.6
exchange
with D2O
TMS
10
9
8
7
6
5
4
3
2
1
0
δ (ppm)

Problem 230

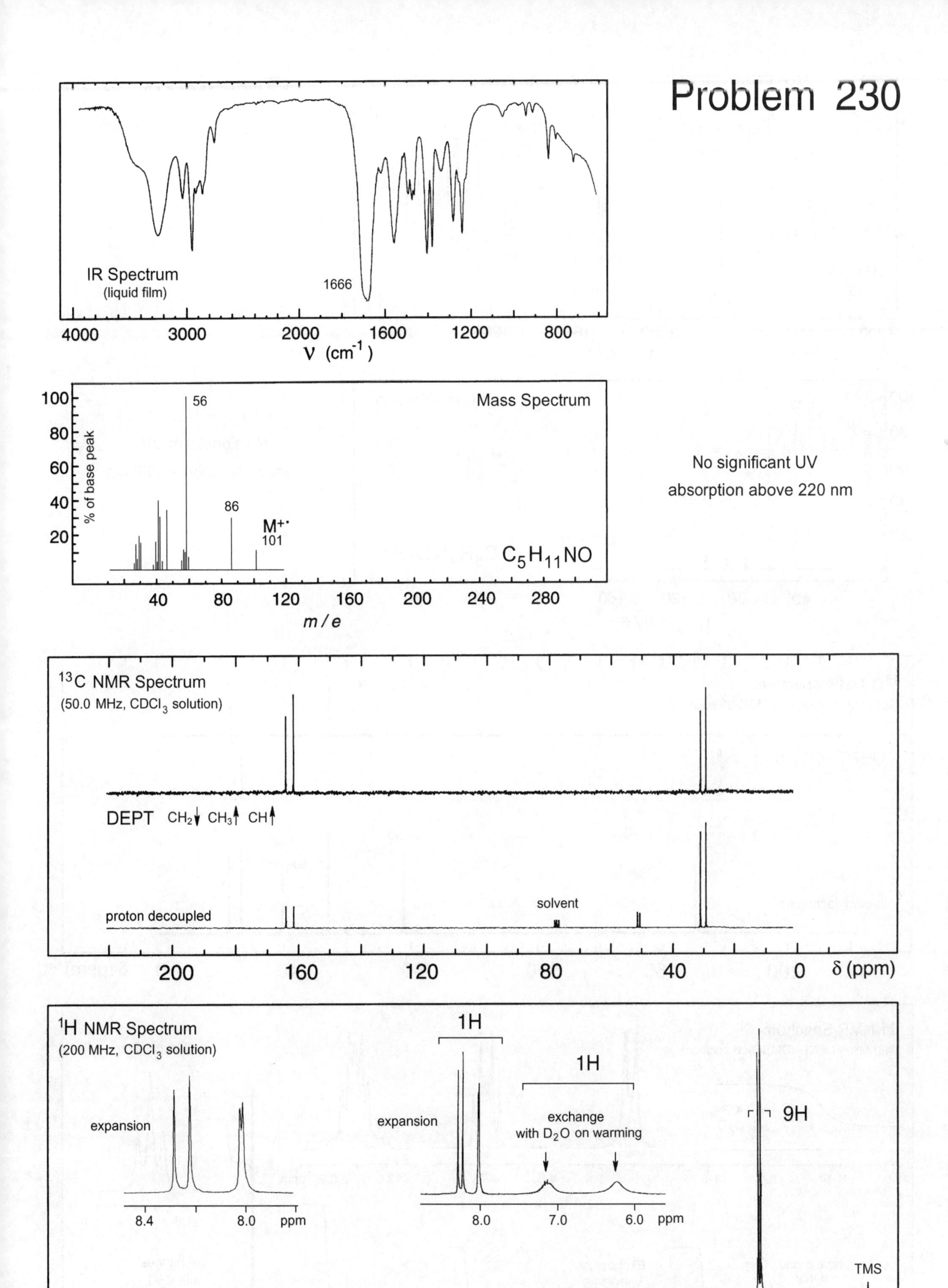

No significant UV absorption above 220 nm

Problem 231

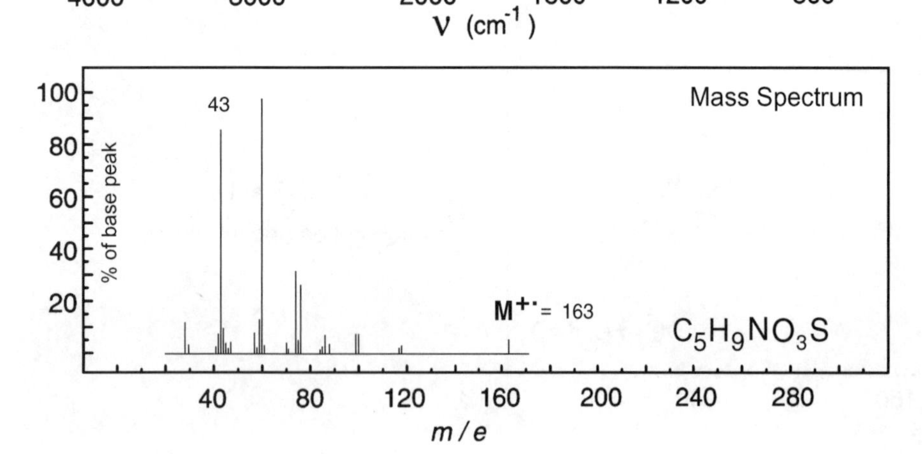

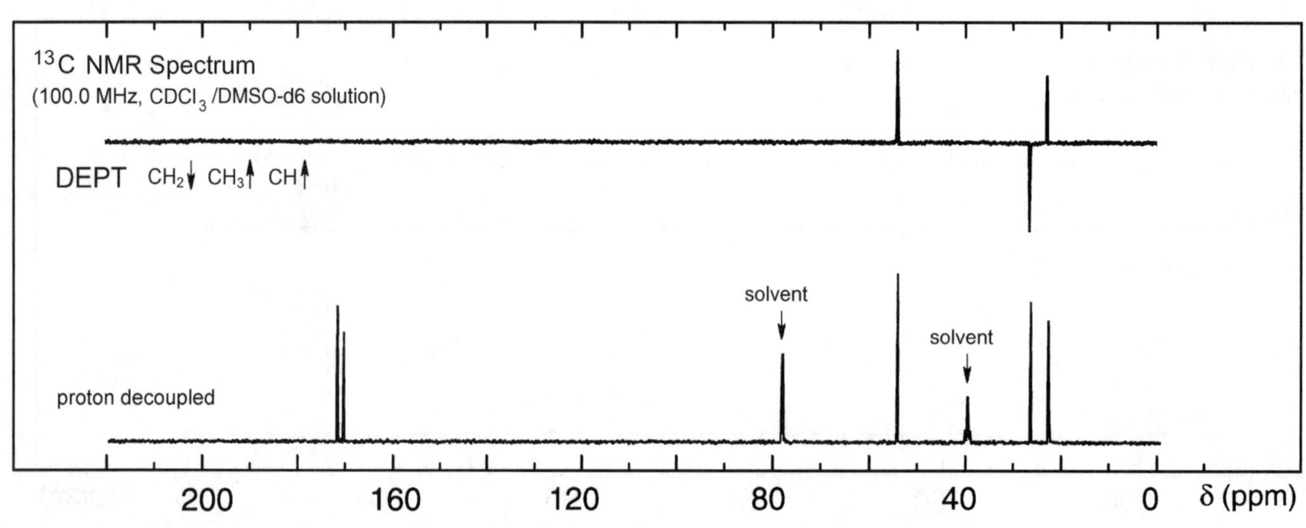

No significant UV absorption above 220 nm

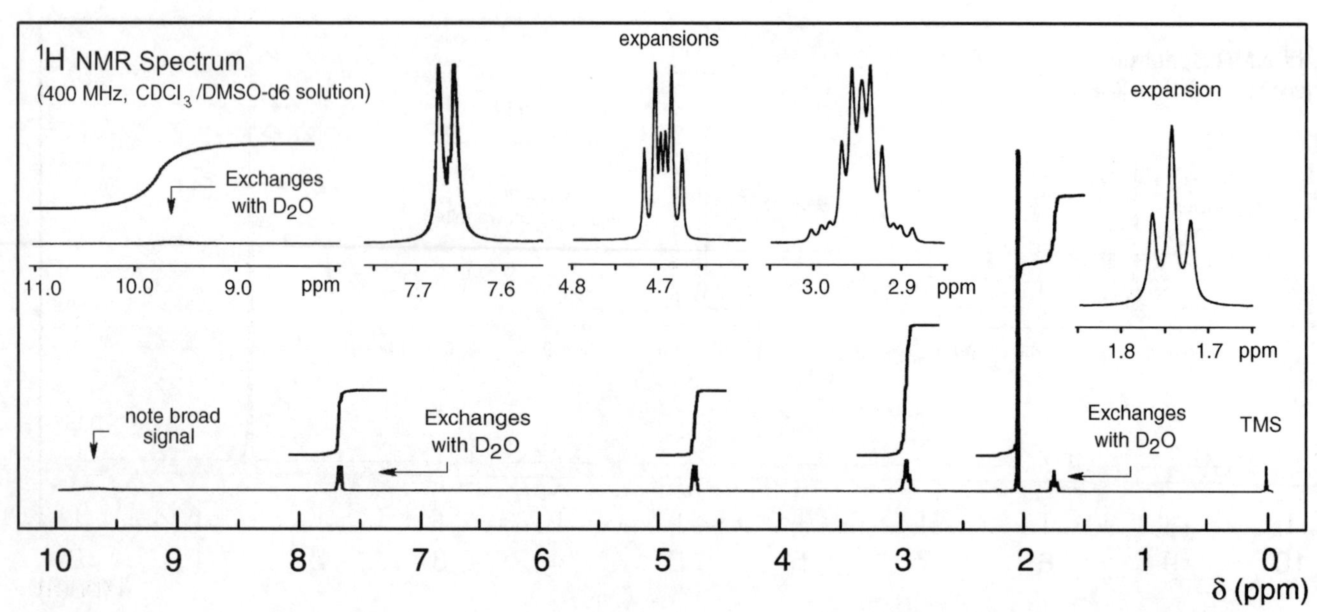

Problem 232

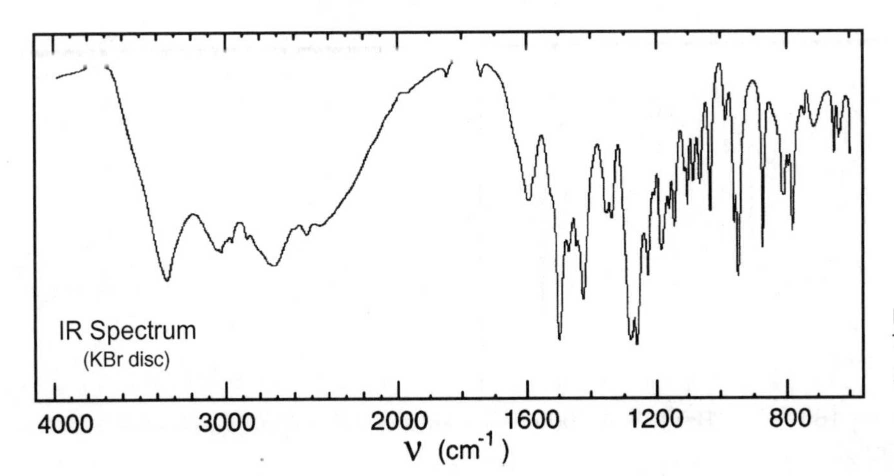

Note: irradiation of the signal at δ 6.58 in the ^1H NMR produces an enhancement of the signals at δ 6.66 and 4.45 ppm via the NOE

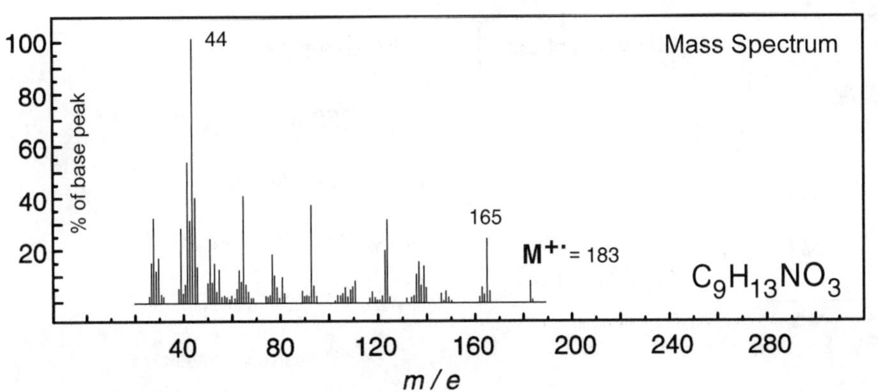

UV Spectrum

λ_{max} 220 nm ($\log_{10}\varepsilon$ 3.8)

λ_{max} 280 nm ($\log_{10}\varepsilon$ 3.4)

solvent : H_2O pH 7

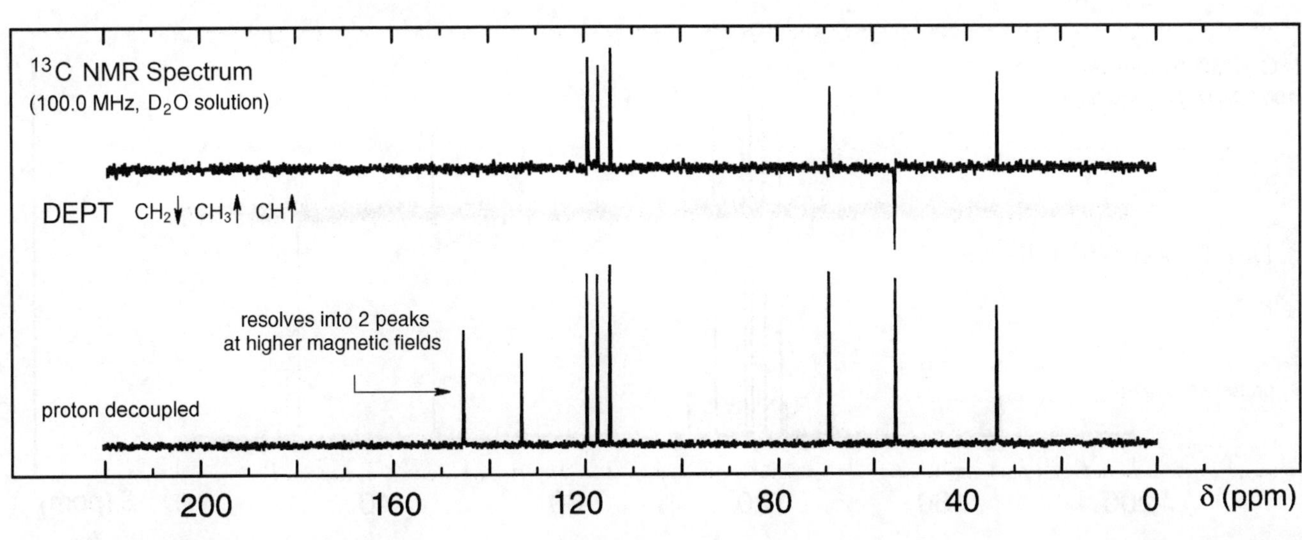

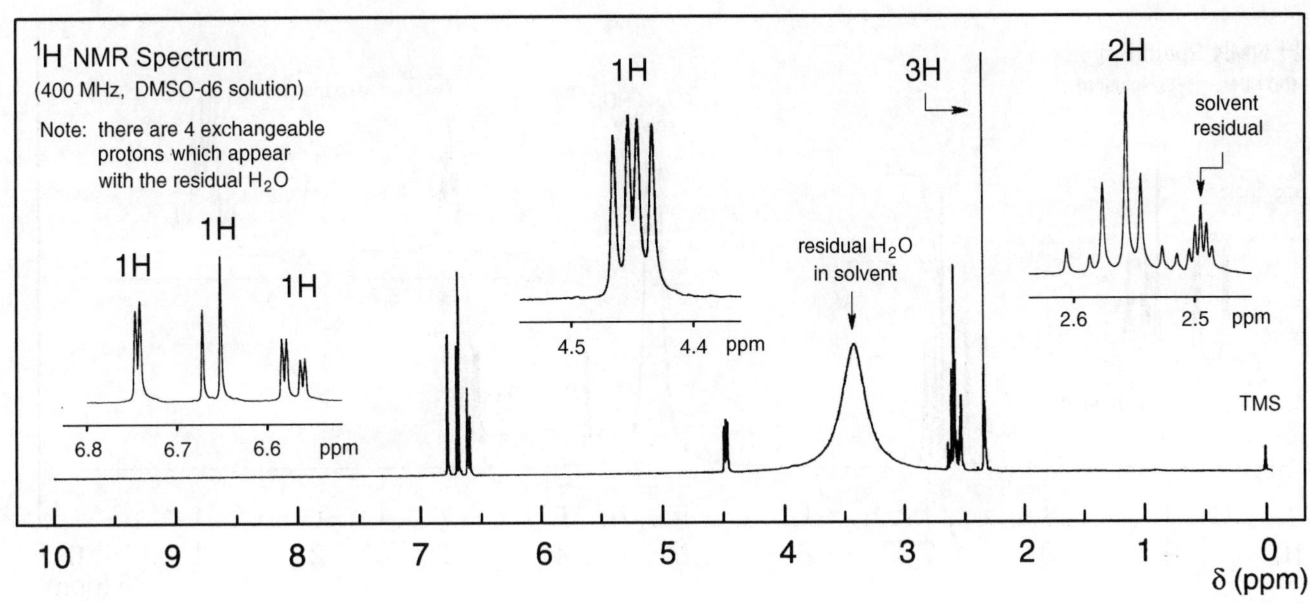

Problem 233

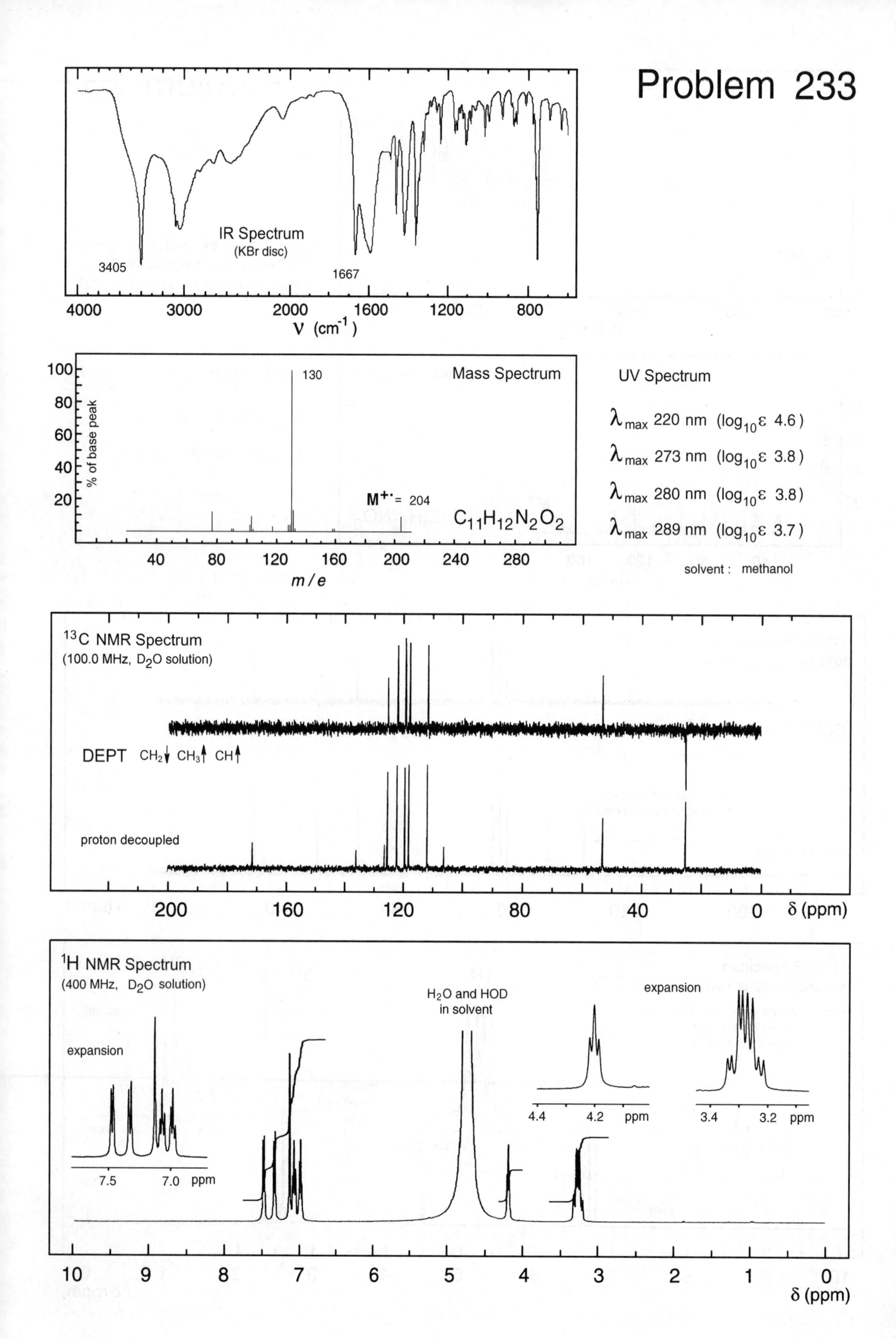

IR Spectrum
(KBr disc)
3405
1667
4000
3000
2000
1600
1200
800
ν (cm⁻¹)
Mass Spectrum
% of base peak
130
M⁺˙ = 204
$C_{11}H_{12}N_2O_2$
m / e
UV Spectrum
λmax 220 nm (log10ε 4.6)
λmax 273 nm (log10ε 3.8)
λmax 280 nm (log10ε 3.8)
λmax 289 nm (log10ε 3.7)
solvent : methanol
13C NMR Spectrum
(100.0 MHz, D2O solution)
DEPT CH2↓ CH3↑ CH↑
proton decoupled
δ (ppm)
1H NMR Spectrum
(400 MHz, D2O solution)
expansion
H2O and HOD in solvent
ppm
δ (ppm)

Problem 234

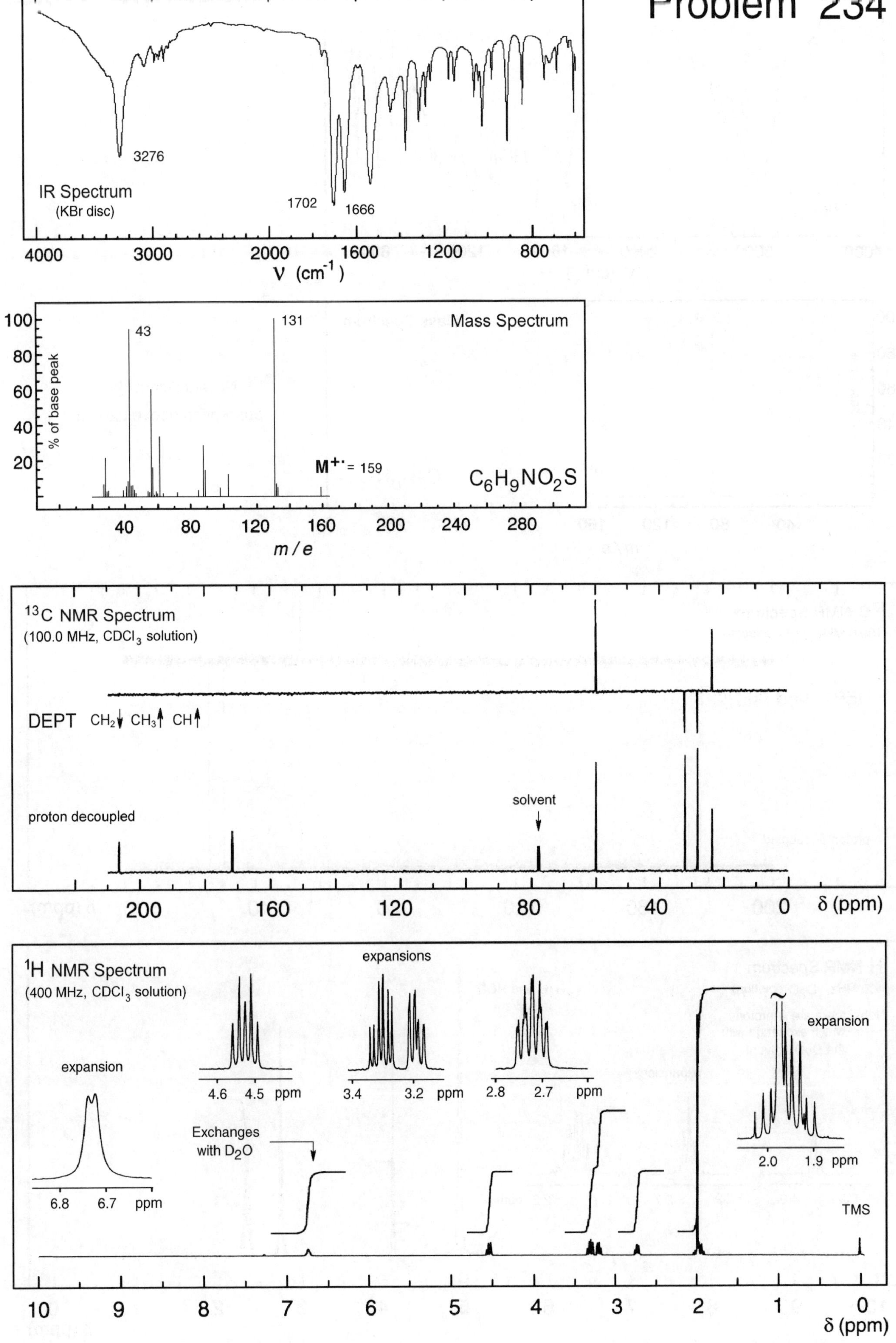

3276
IR Spectrum
(KBr disc)
1702
1666
4000
3000
2000
1600
1200
800
ν (cm^{-1})
Mass Spectrum
43
131
% of base peak
$M^{+\cdot}$ = 159
$C_6H_9NO_2S$
m / e
^{13}C NMR Spectrum
(100.0 MHz, $CDCl_3$ solution)
DEPT CH_2↓ CH_3↑ CH↑
solvent
proton decoupled
δ (ppm)
^{1}H NMR Spectrum
(400 MHz, $CDCl_3$ solution)
expansions
expansion
Exchanges with D_2O
TMS
ppm
δ (ppm)

Problem 235

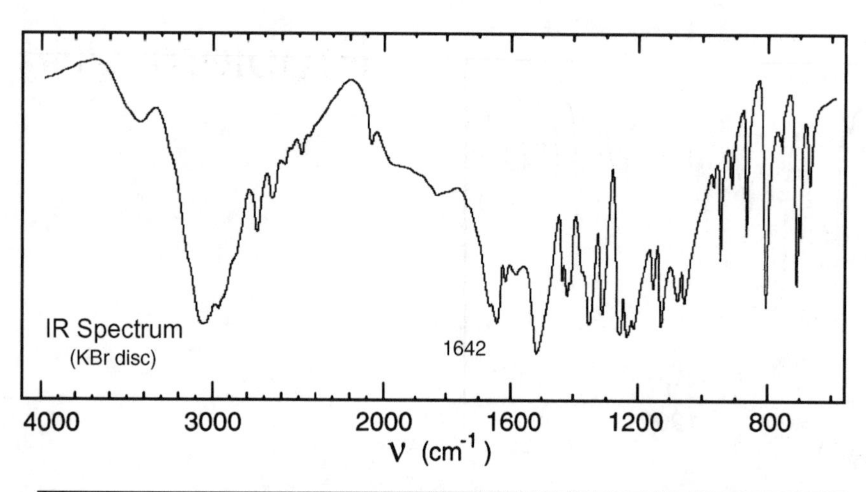

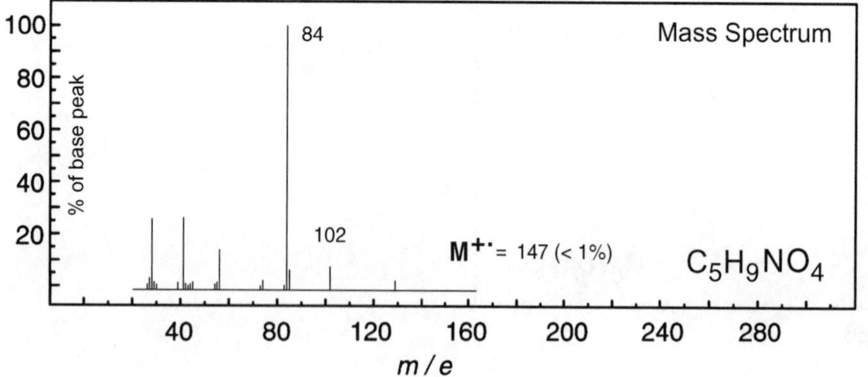

No significant UV absorption above 220 nm

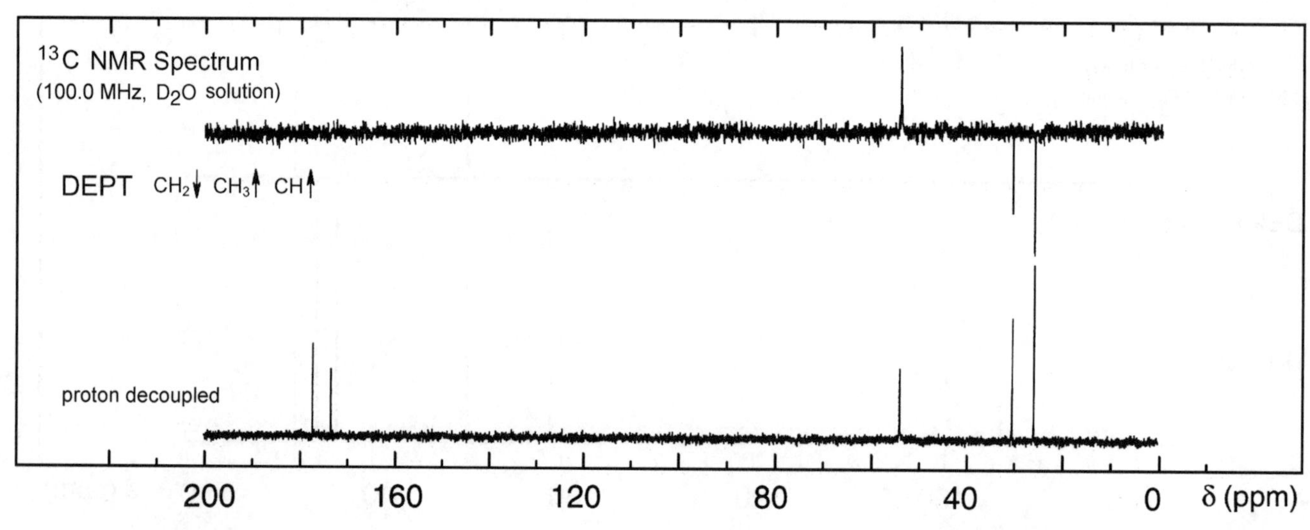

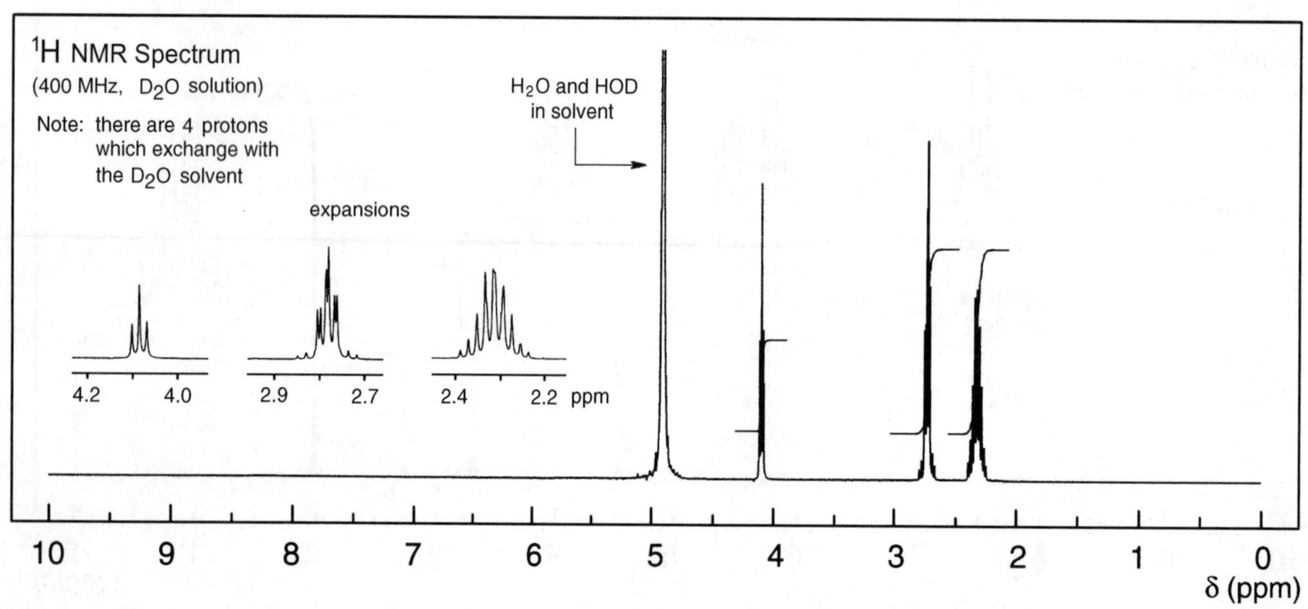

Problem 236

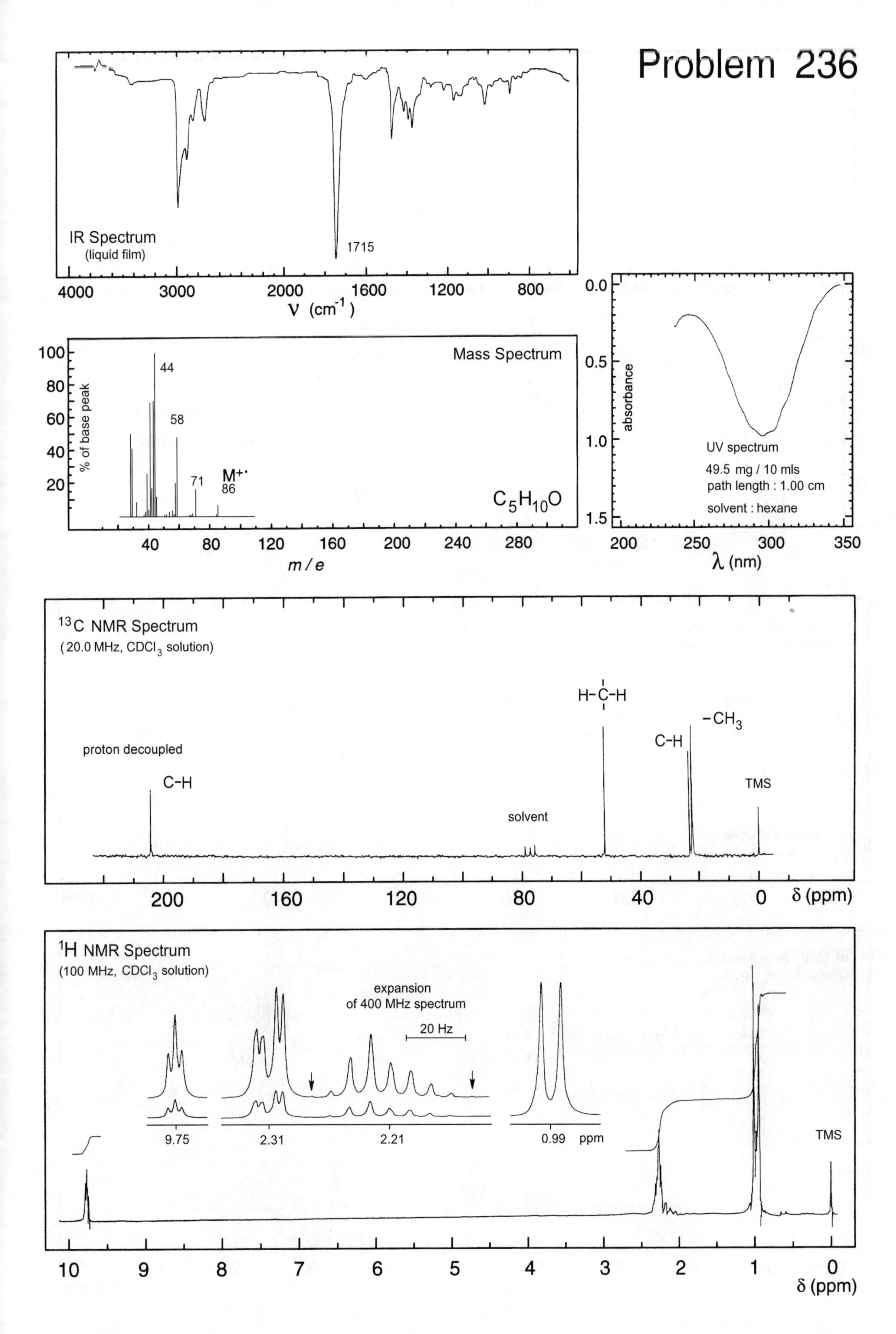

IR Spectrum
(liquid film)
1715
4000
3000
2000
1600
1200
800
ν (cm⁻¹)
Mass Spectrum
% of base peak
100
80
60
40
20
44
58
71
M+·
86
C5H10O
40
80
120
160
200
240
280
m / e
UV spectrum
49.5 mg / 10 mls
path length : 1.00 cm
solvent : hexane
absorbance
0.0
0.5
1.0
1.5
200
250
300
350
λ (nm)
13C NMR Spectrum
(20.0 MHz, CDCl3 solution)
proton decoupled
C-H
solvent
H-C-H
C-H
-CH3
TMS
200
160
120
80
40
0
δ (ppm)
1H NMR Spectrum
(100 MHz, CDCl3 solution)
expansion
of 400 MHz spectrum
20 Hz
9.75
2.31
2.21
0.99
ppm
TMS
10
9
8
7
6
5
4
3
2
1
0
δ (ppm)

Problem 237a

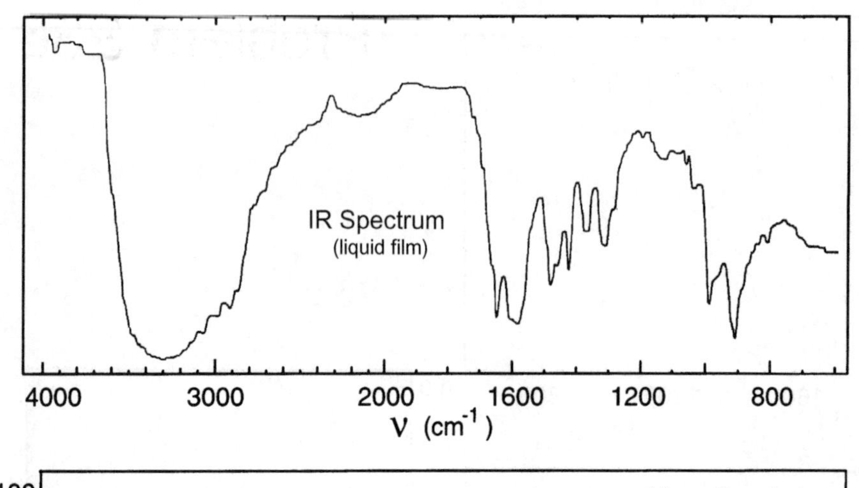

A 400 MHz ^1H NMR spectrum (with expansions) is given on the following page

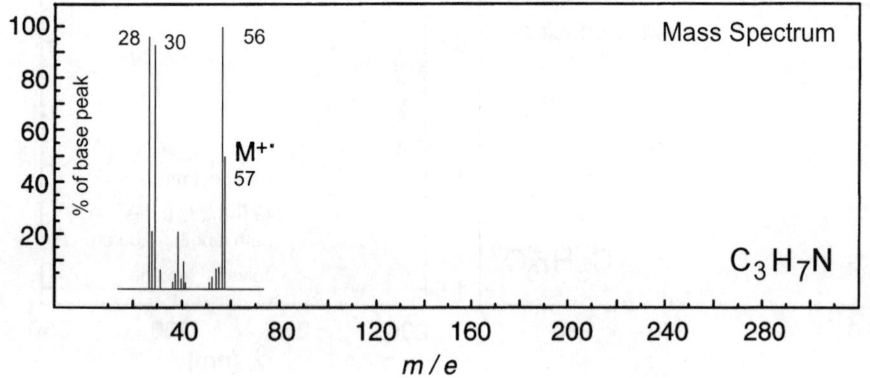

No significant UV absorption above 220 nm

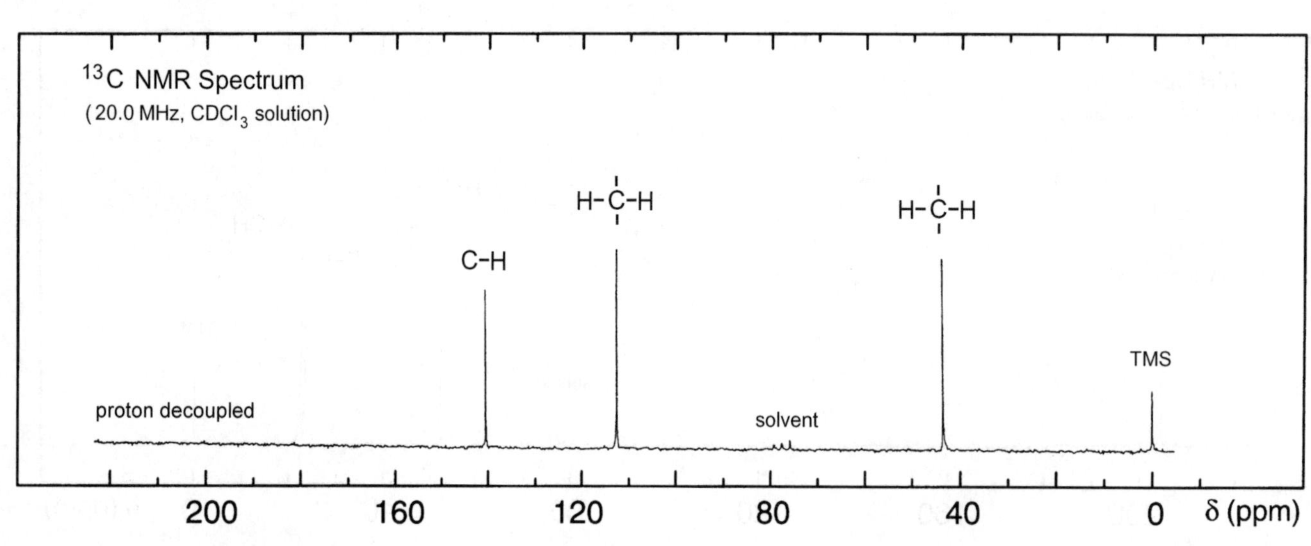

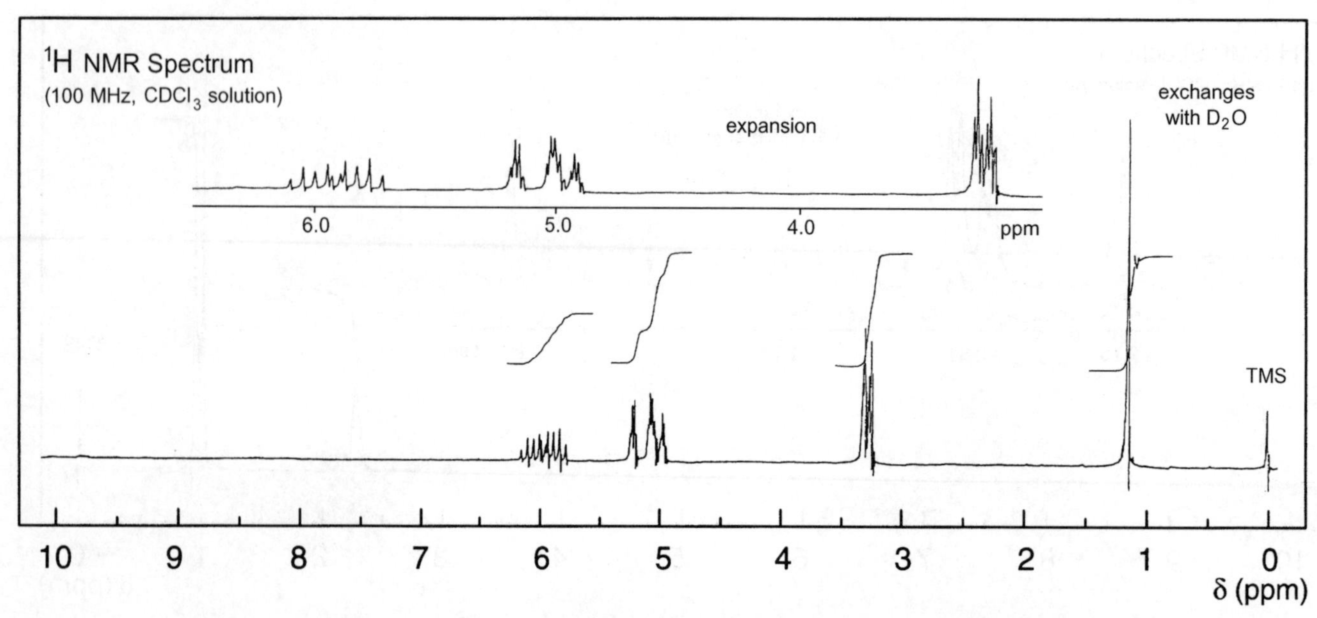

Problem 237b

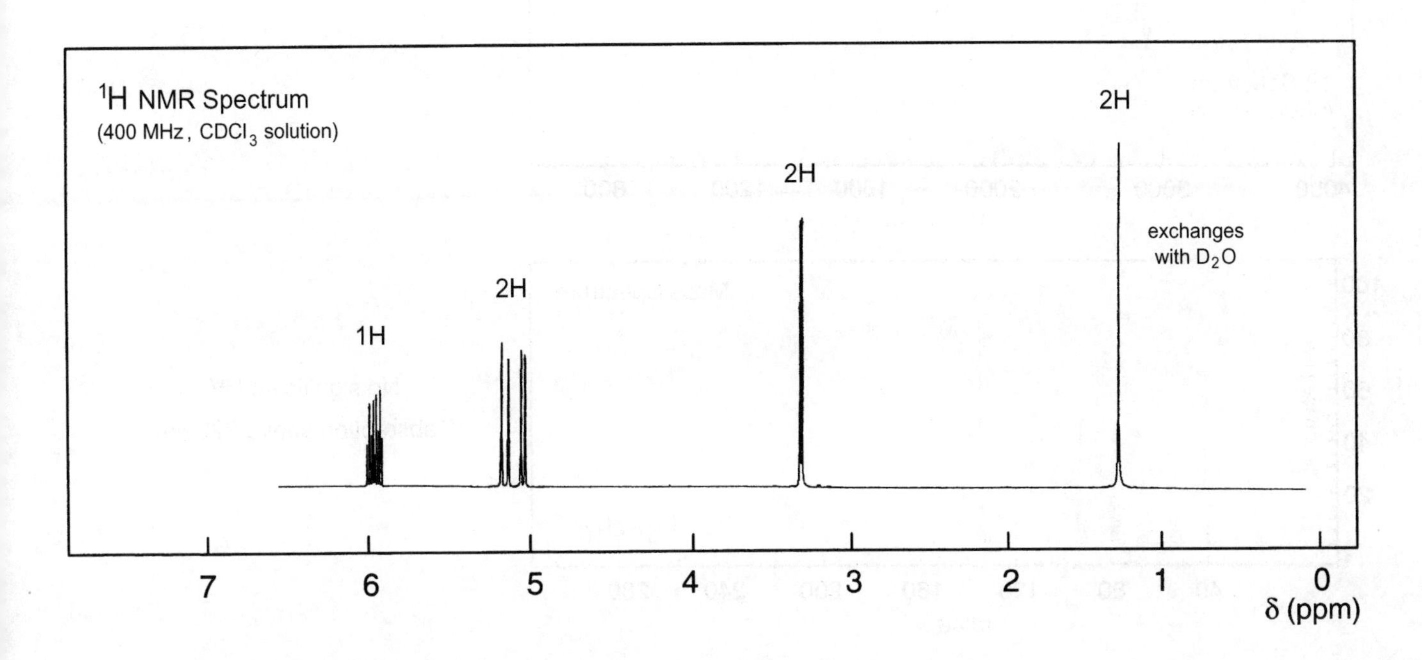

Expansion of regions of the 400 MHz NMR spectrum

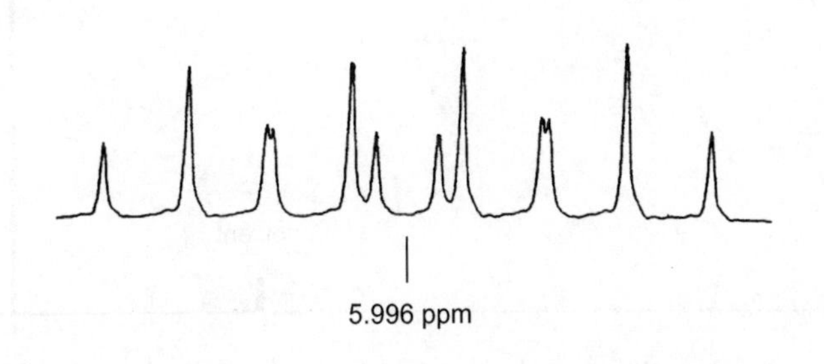

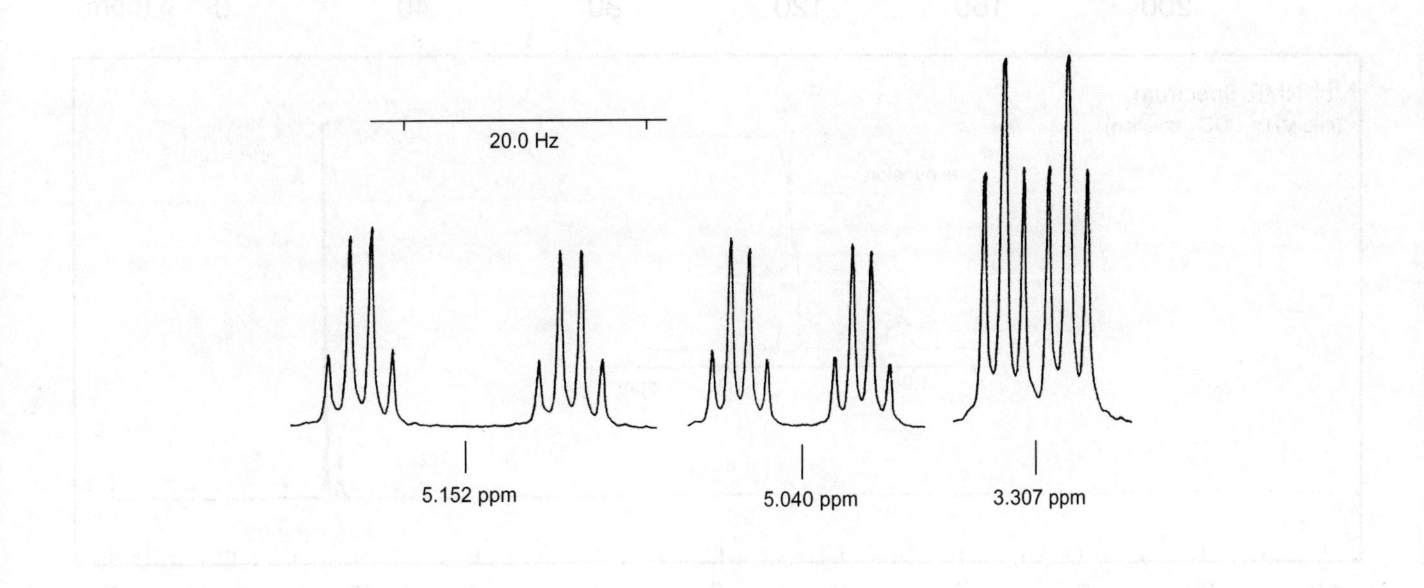

Problem 238

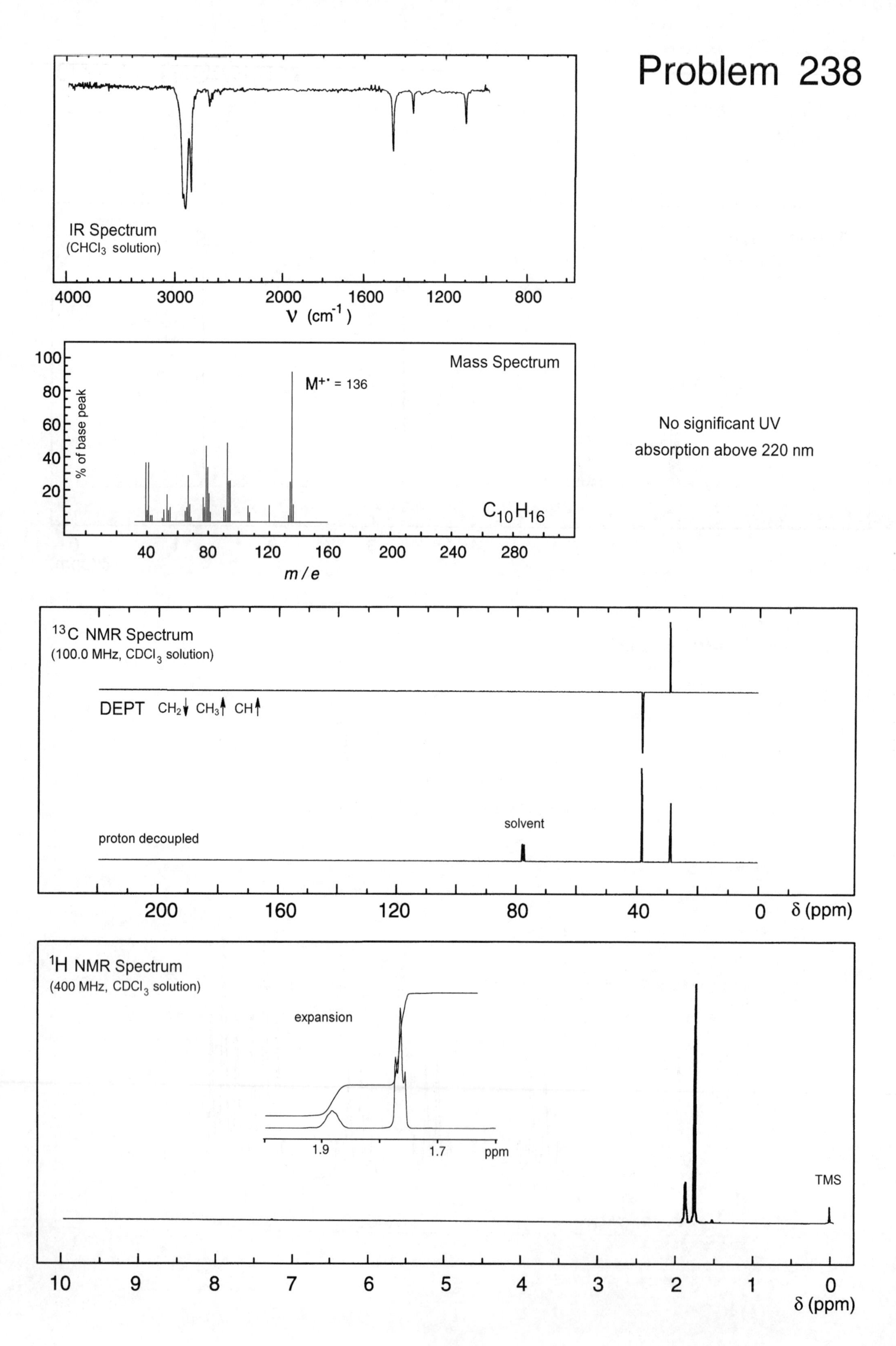

No significant UV absorption above 220 nm

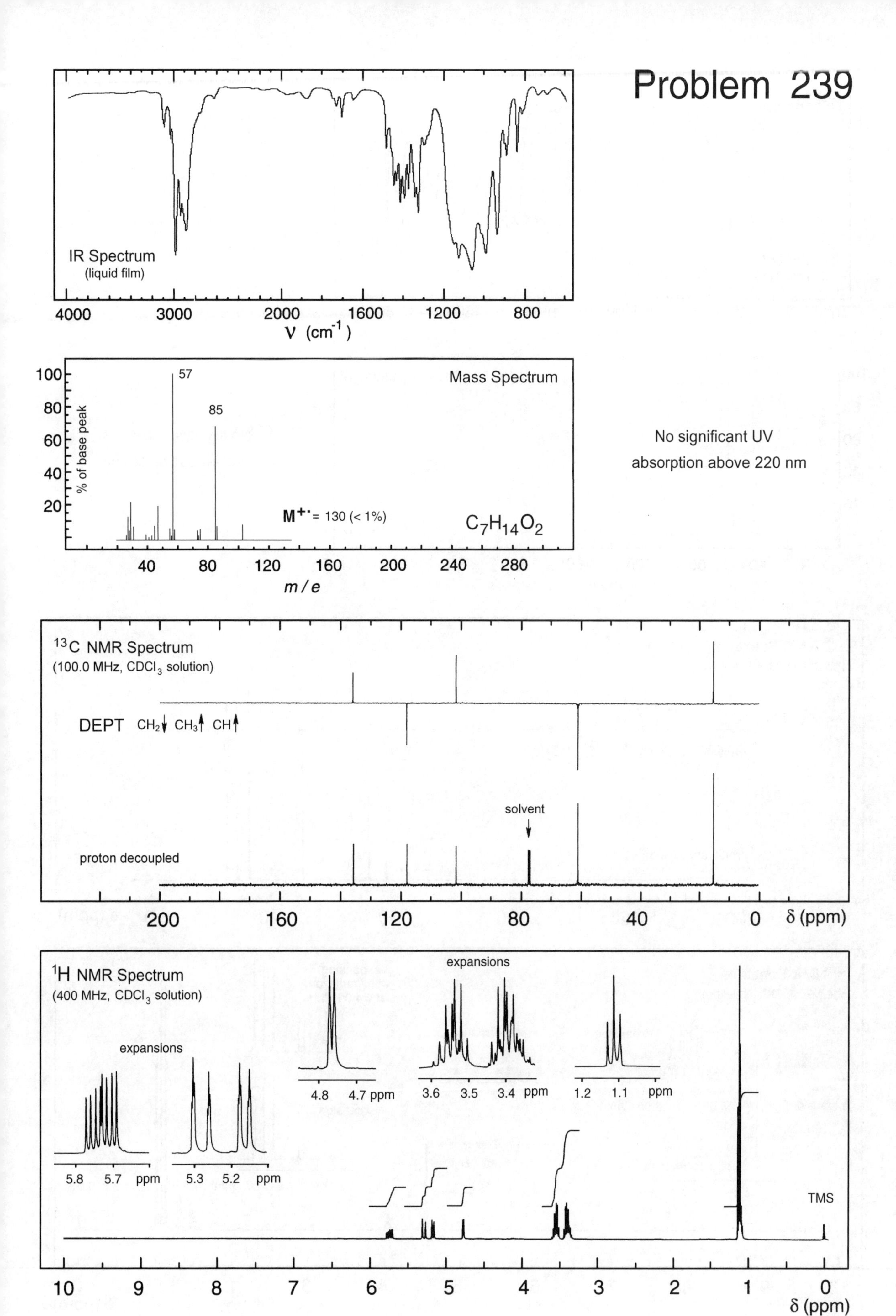

Problem 239
IR Spectrum
(liquid film)
4000
3000
2000
1600
1200
800
ν (cm^{-1})
Mass Spectrum
100
80
60
40
20
% of base peak
57
85
M$^{+\cdot}$= 130 (< 1%)
$C_7H_{14}O_2$
40
80
120
160
200
240
280
m / e
No significant UV
absorption above 220 nm
^{13}C NMR Spectrum
(100.0 MHz, $CDCl_3$ solution)
DEPT CH_2↓ CH_3↑ CH↑
solvent
proton decoupled
200
160
120
80
40
0
δ (ppm)
^{1}H NMR Spectrum
(400 MHz, $CDCl_3$ solution)
expansions
5.8
5.7
ppm
5.3
5.2
ppm
4.8
4.7 ppm
3.6
3.5
3.4
ppm
1.2
1.1
ppm
TMS
10
9
8
7
6
5
4
3
2
1
0
δ (ppm)

Problem 240

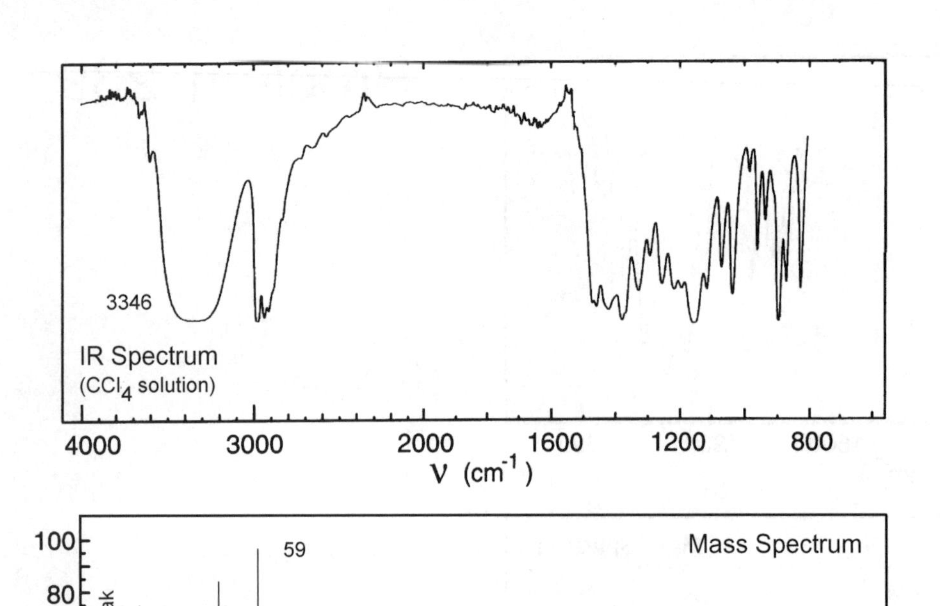

3346
IR Spectrum
(CCl4 solution)
4000
3000
2000
1600
1200
800
ν (cm-1)
100
80
60
40
20
% of base peak
59
Mass Spectrum
M+· = 118 (<1%)
103
C6H14O2
40
80
120
160
200
240
280
m / e

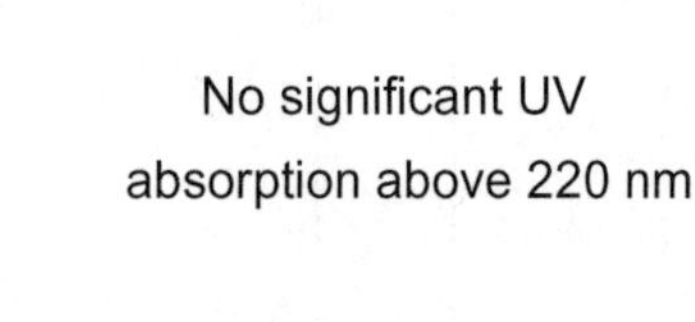

No significant UV
absorption above 220 nm

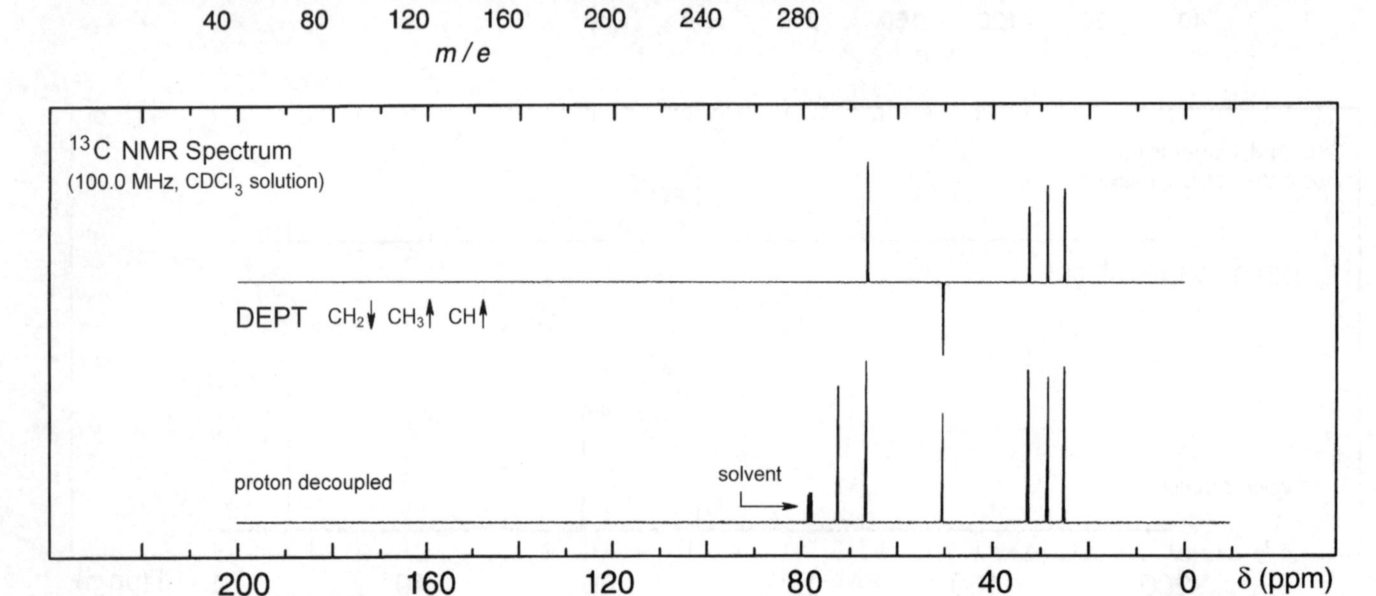

13C NMR Spectrum
(100.0 MHz, CDCl3 solution)
DEPT CH2↓ CH3↑ CH↑
proton decoupled
solvent
200
160
120
80
40
0
δ (ppm)

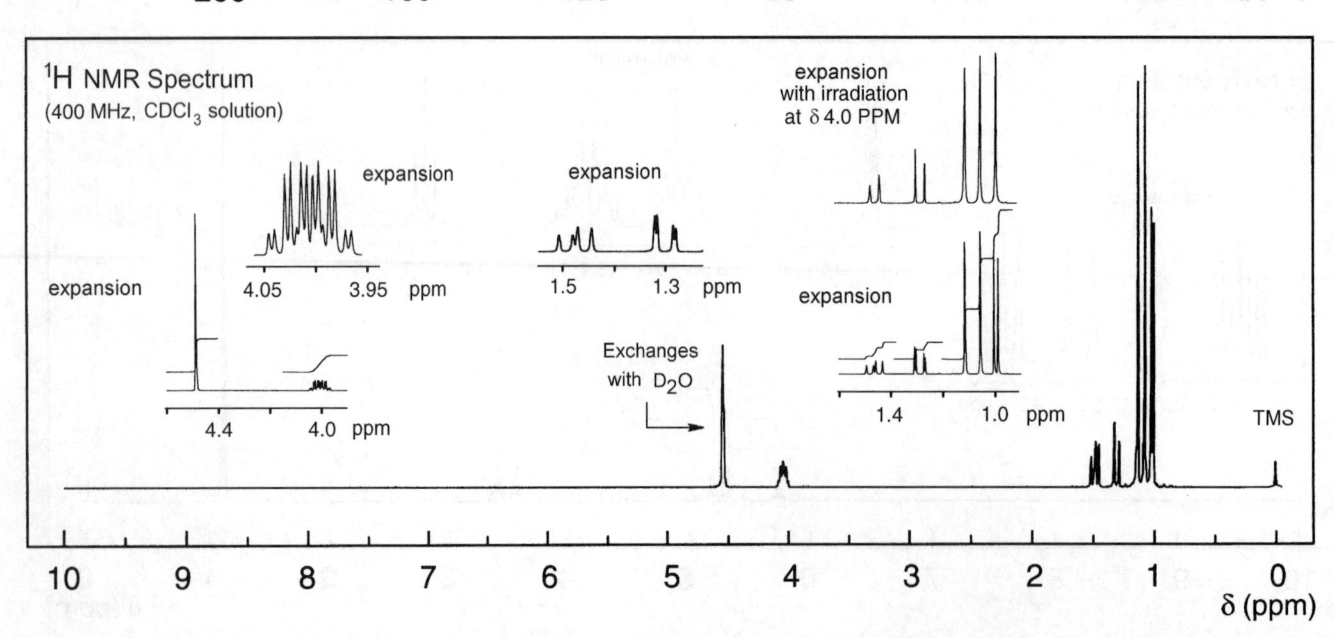

1H NMR Spectrum
(400 MHz, CDCl3 solution)
expansion
4.05
3.95
ppm
expansion
1.5
1.3
ppm
expansion with irradiation at δ 4.0 PPM
expansion
4.4
4.0
ppm
Exchanges with D2O
expansion
1.4
1.0
ppm
TMS
10
9
8
7
6
5
4
3
2
1
0
δ (ppm)

Problem 241

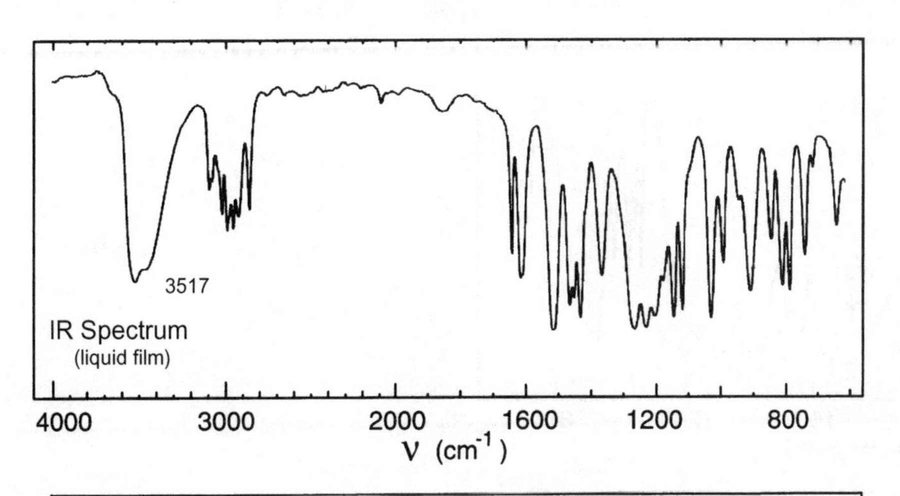

Note: irradiation of the signal at δ 6.45 in the ^{1}H NMR produces an enhancement of the signals at δ 3.4 and 3.7 ppm via the NOE

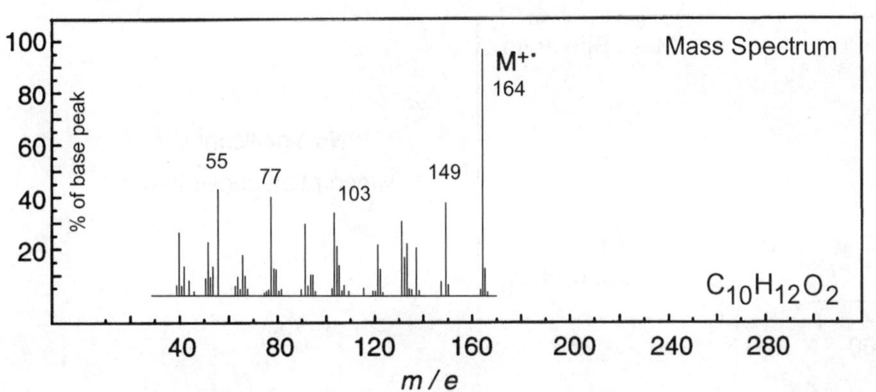

UV Spectrum

λ_{max} 281 nm ($\log_{10}\varepsilon$ 3.5)

λ_{max} 230 nm ($\log_{10}\varepsilon$ 3.6)

solvent : methanol

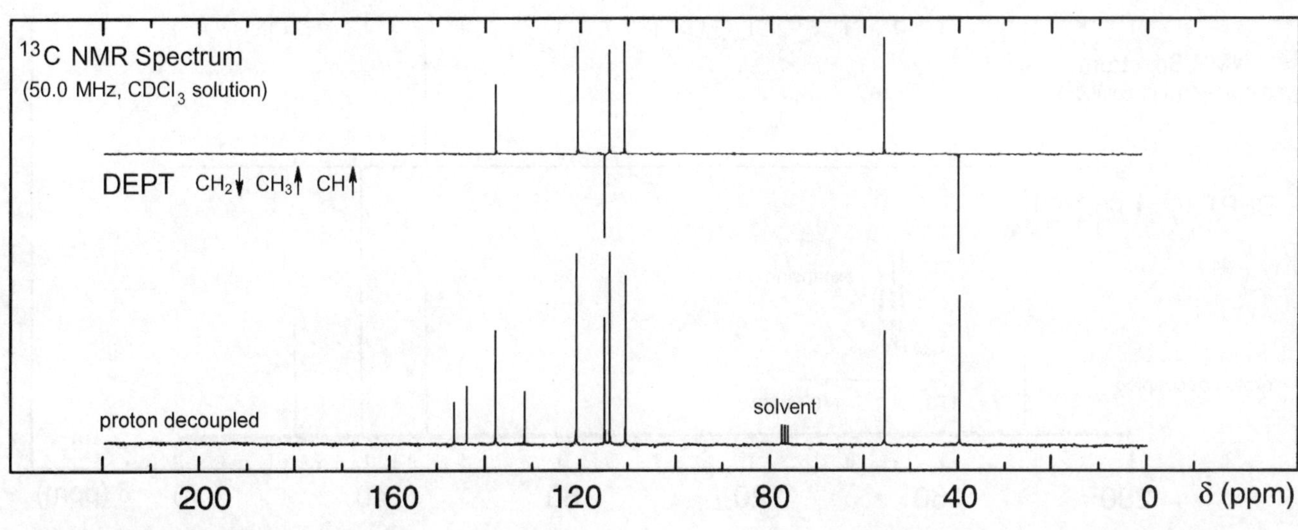

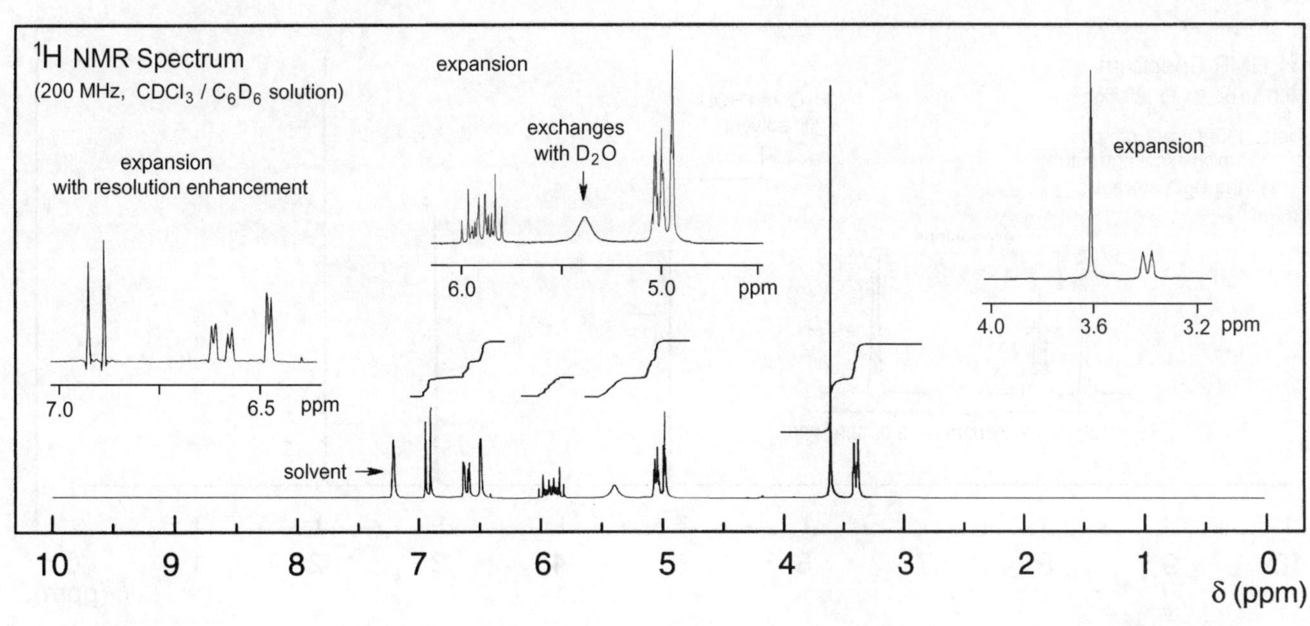

Problem 242

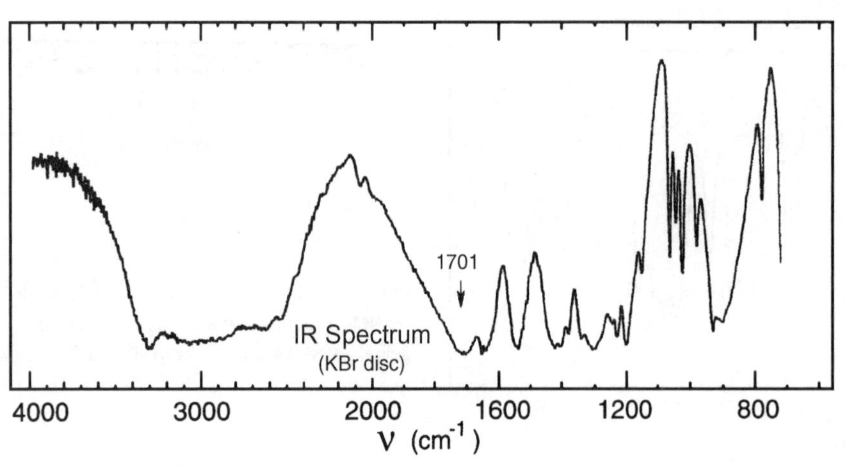

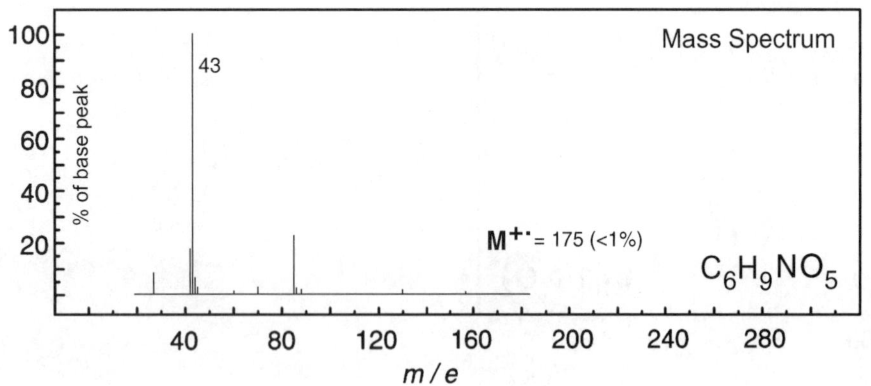

No significant UV absorption above 220 nm

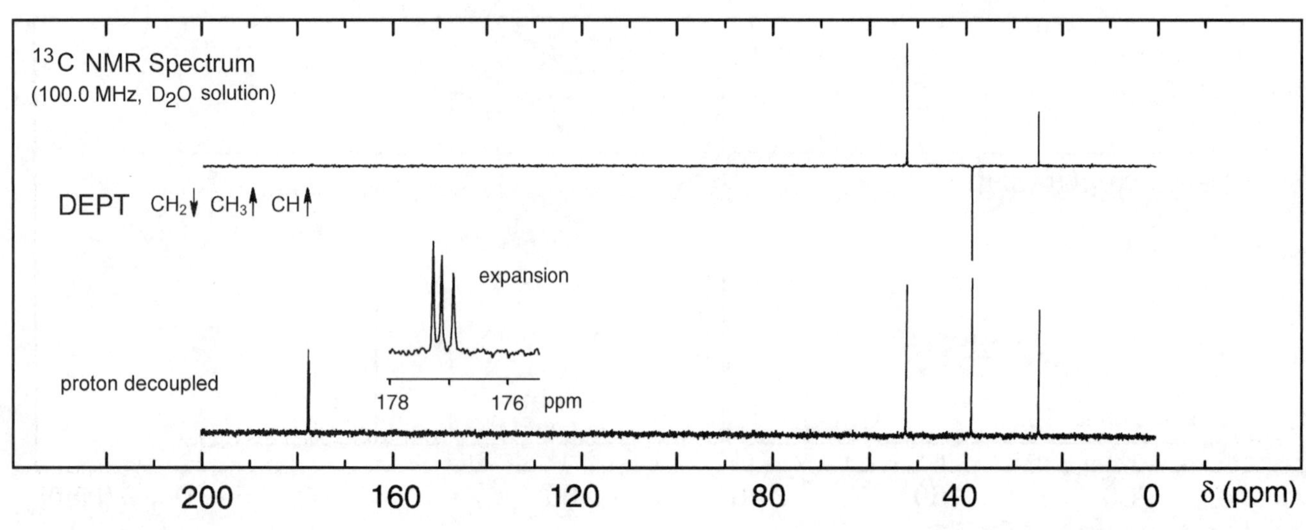

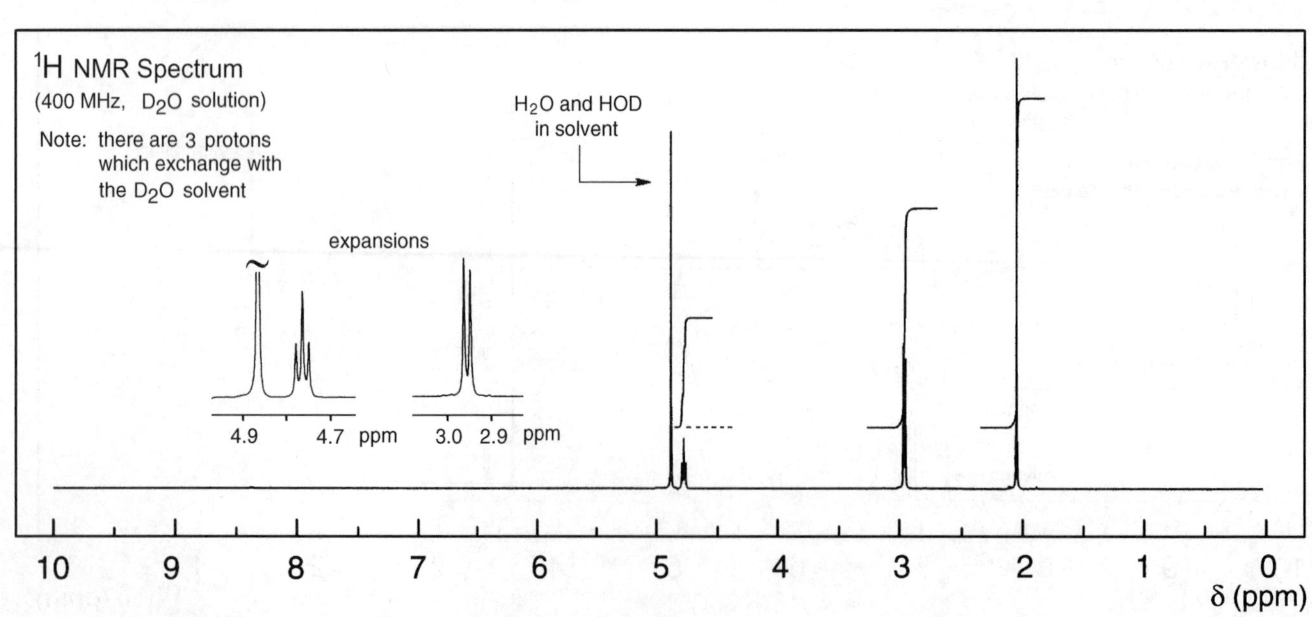

Problem 243

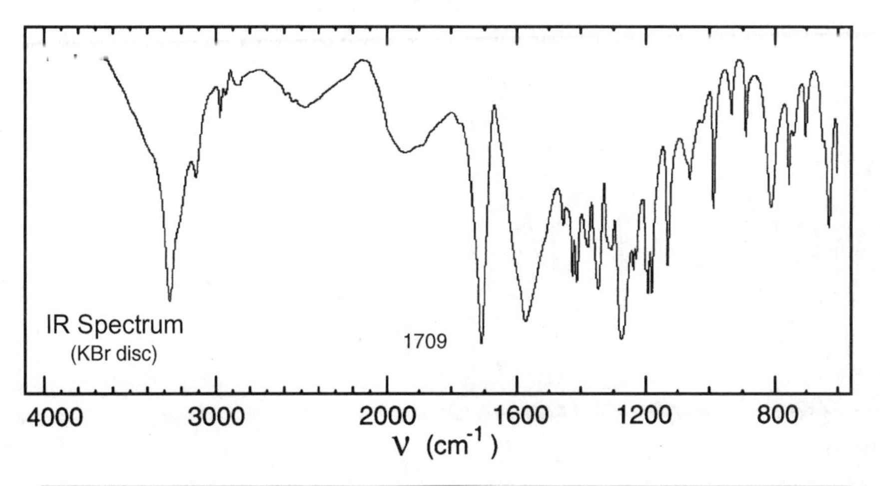

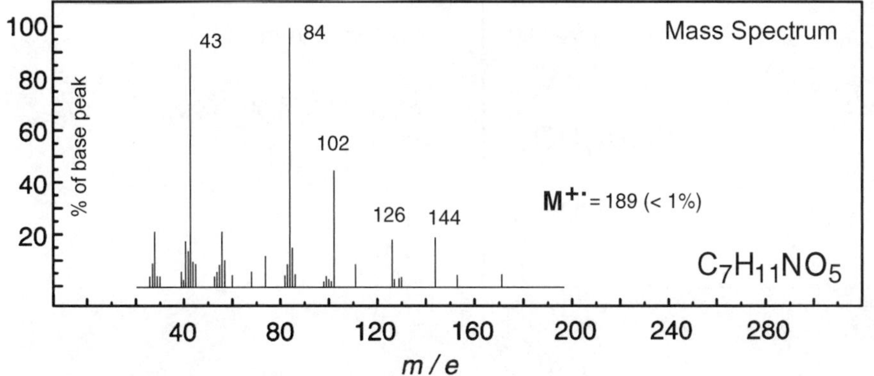

No significant UV absorption above 220 nm

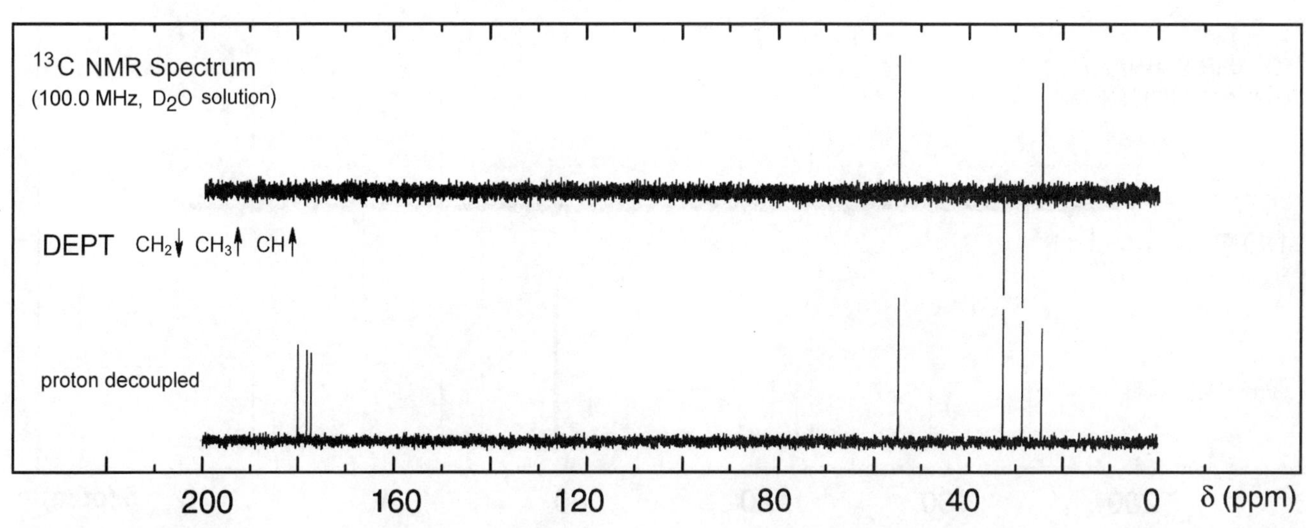

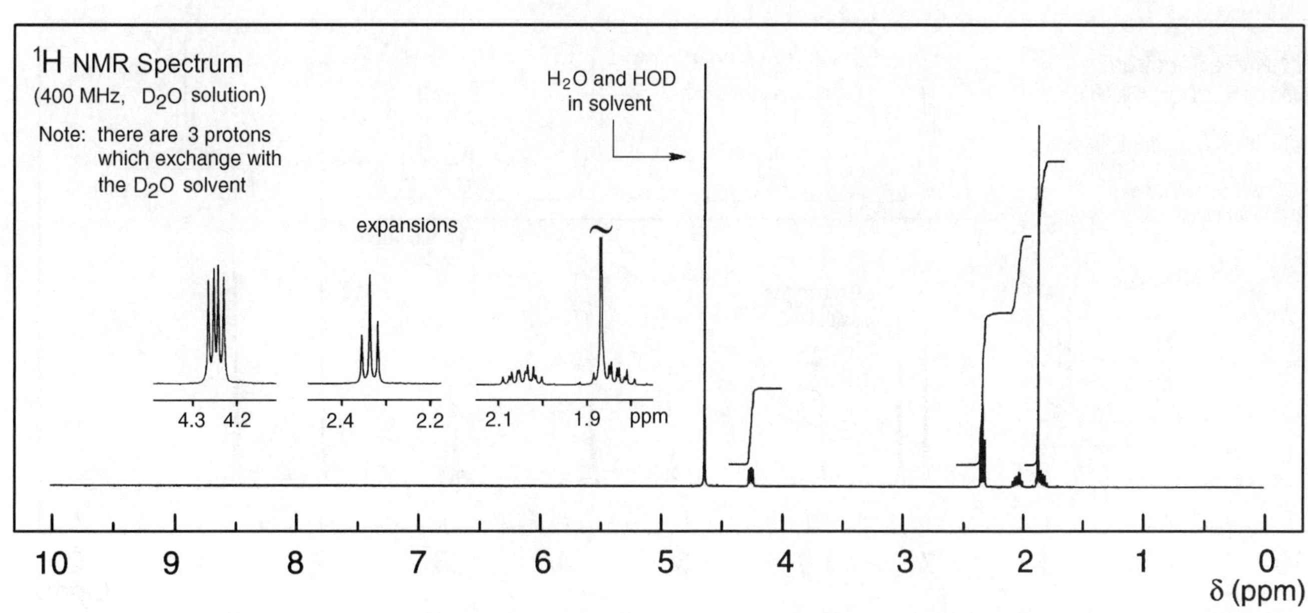

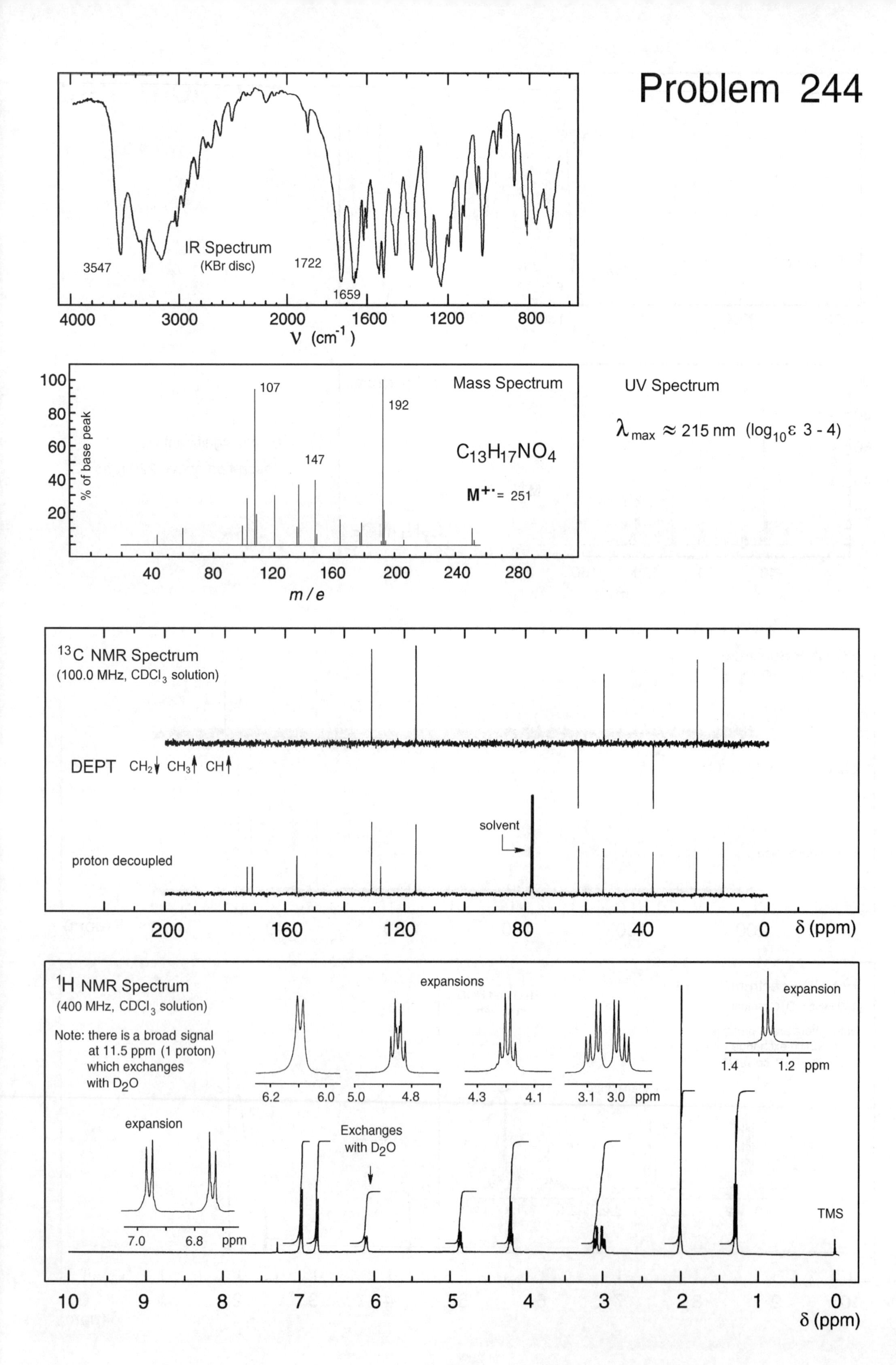

Problem 244
IR Spectrum
(KBr disc)
3547
1722
1659
4000
3000
2000
1600
1200
800
ν (cm⁻¹)
Mass Spectrum
% of base peak
107
192
147
C13H17NO4
M+· = 251
m / e
UV Spectrum
λmax ≈ 215 nm (log10ε 3 - 4)
13C NMR Spectrum
(100.0 MHz, CDCl3 solution)
DEPT CH2 CH3 CH
solvent
proton decoupled
δ (ppm)
1H NMR Spectrum
(400 MHz, CDCl3 solution)
Note: there is a broad signal at 11.5 ppm (1 proton) which exchanges with D2O
expansions
expansion
Exchanges with D2O
TMS
ppm
δ (ppm)

Problem 245

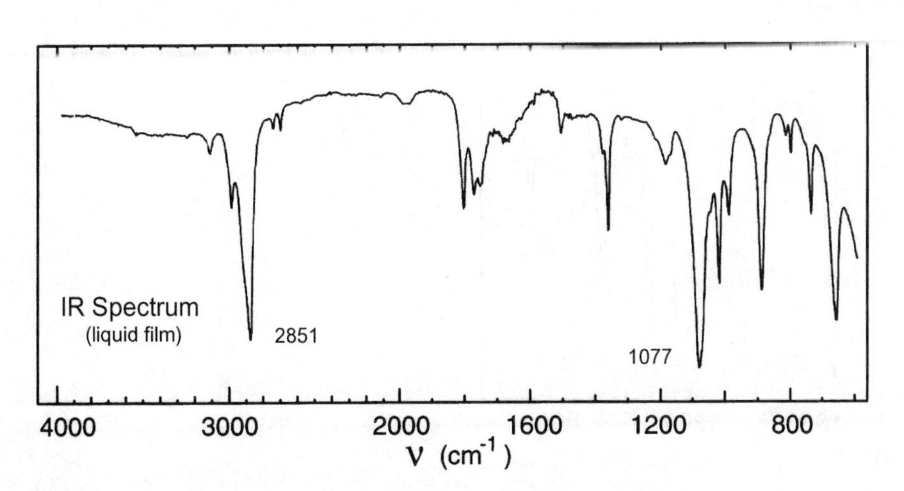

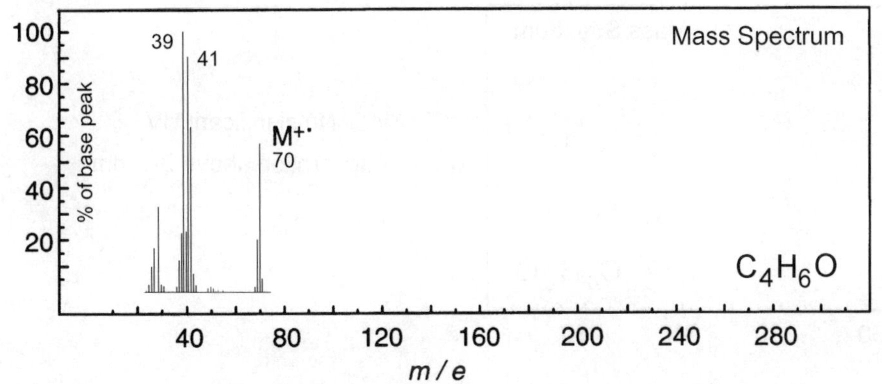

No significant UV absorption above 220 nm

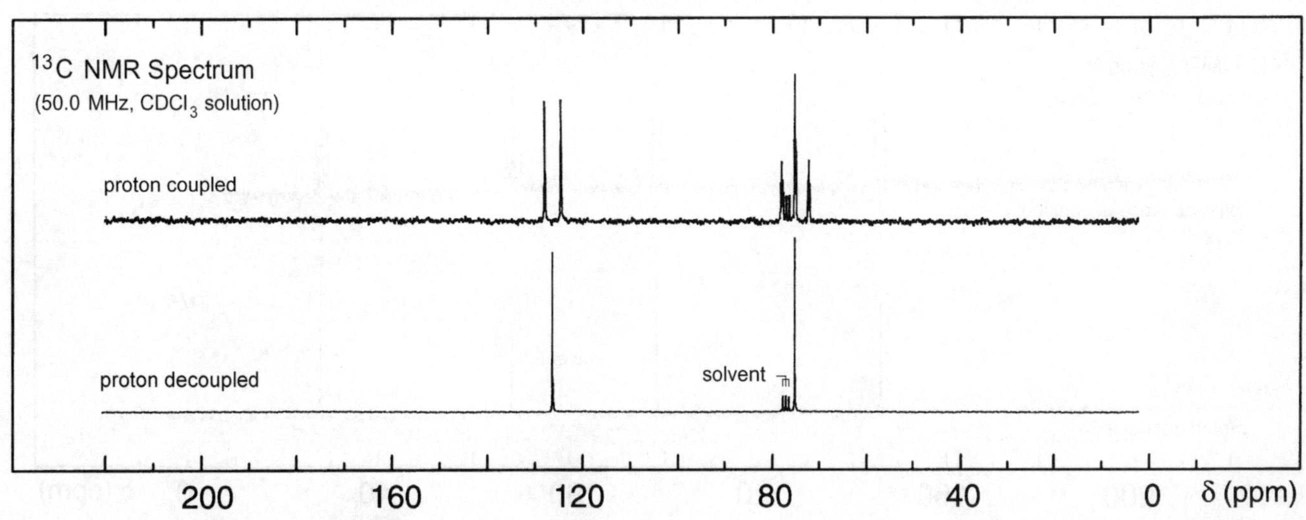

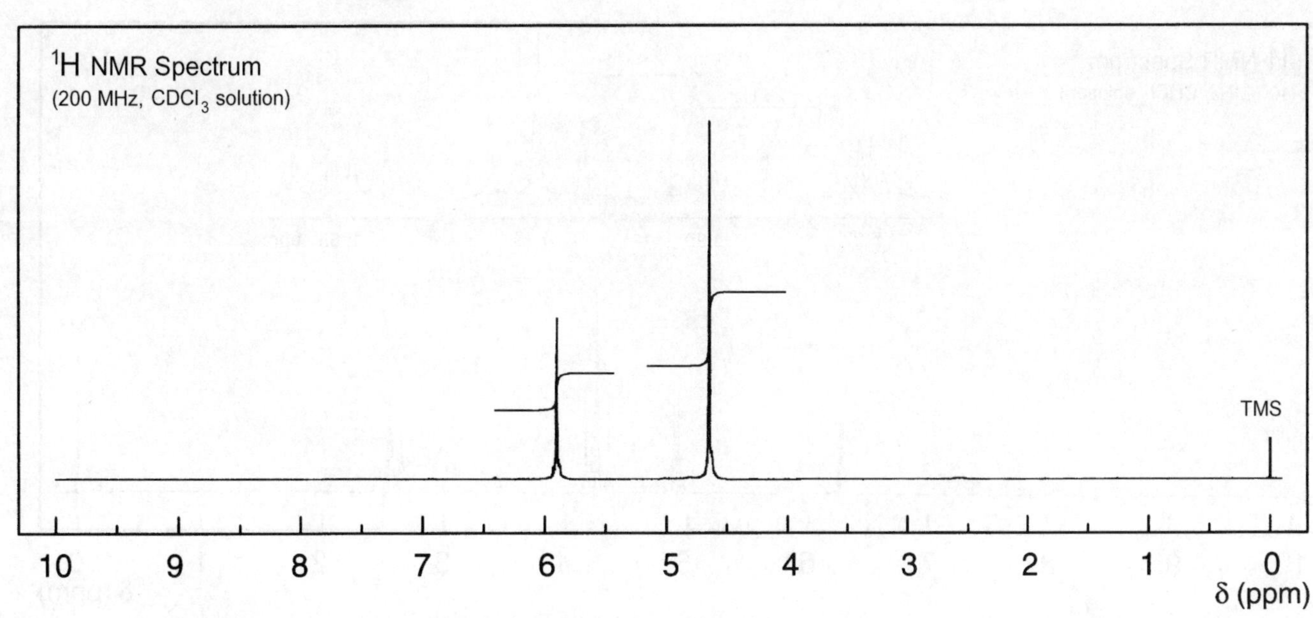

Problem 246

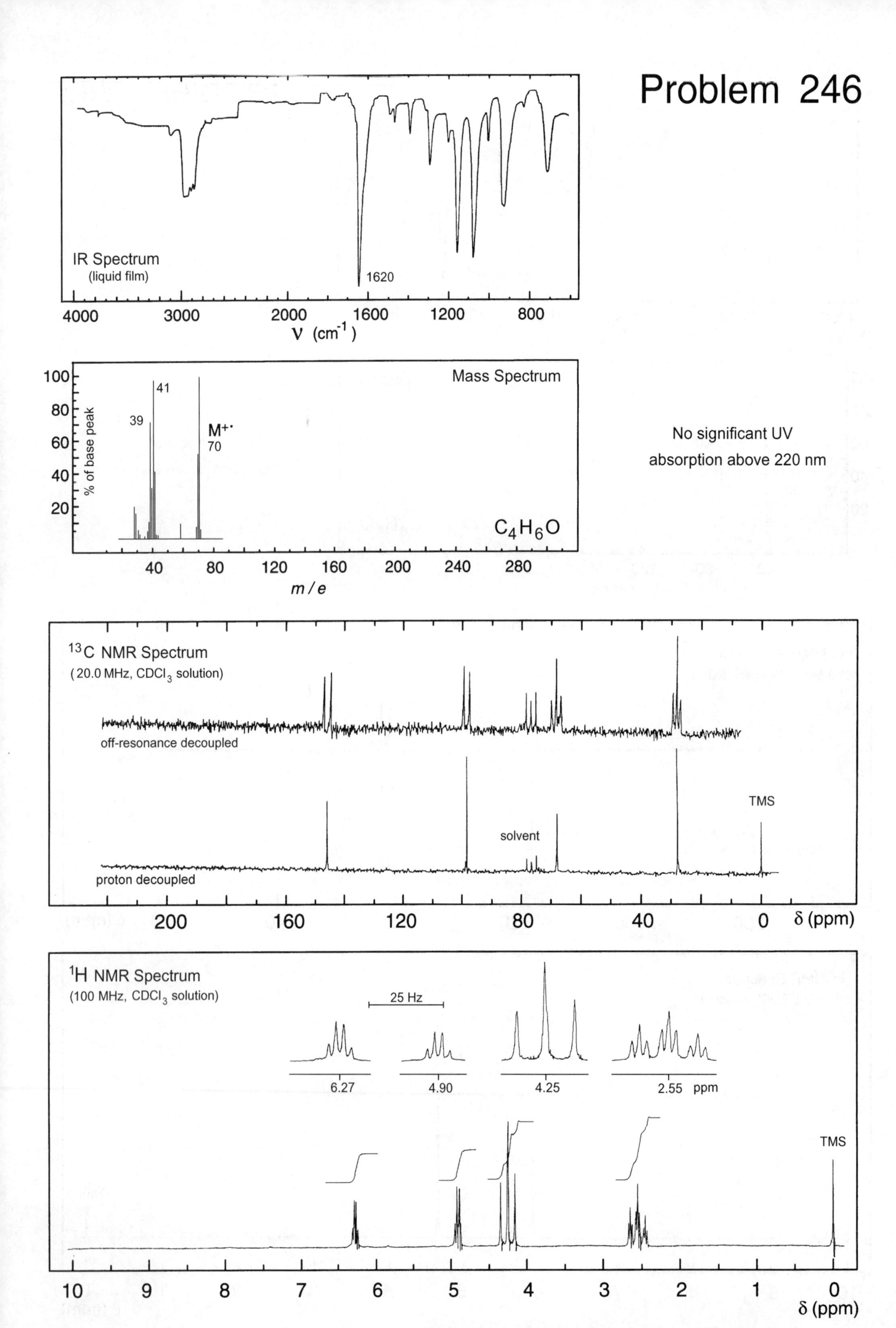

IR Spectrum
(liquid film)
1620
4000
3000
2000
1600
1200
800
ν (cm^{-1})
Mass Spectrum
100
80
60
40
20
% of base peak
39
41
M+·
70
C_4H_6O
40
80
120
160
200
240
280
m / e
No significant UV
absorption above 220 nm
^{13}C NMR Spectrum
(20.0 MHz, $CDCl_3$ solution)
off-resonance decoupled
proton decoupled
solvent
TMS
200
160
120
80
40
0
δ (ppm)
^{1}H NMR Spectrum
(100 MHz, $CDCl_3$ solution)
25 Hz
6.27
4.90
4.25
2.55 ppm
TMS
10
9
8
7
6
5
4
3
2
1
0
δ (ppm)

Problem 247

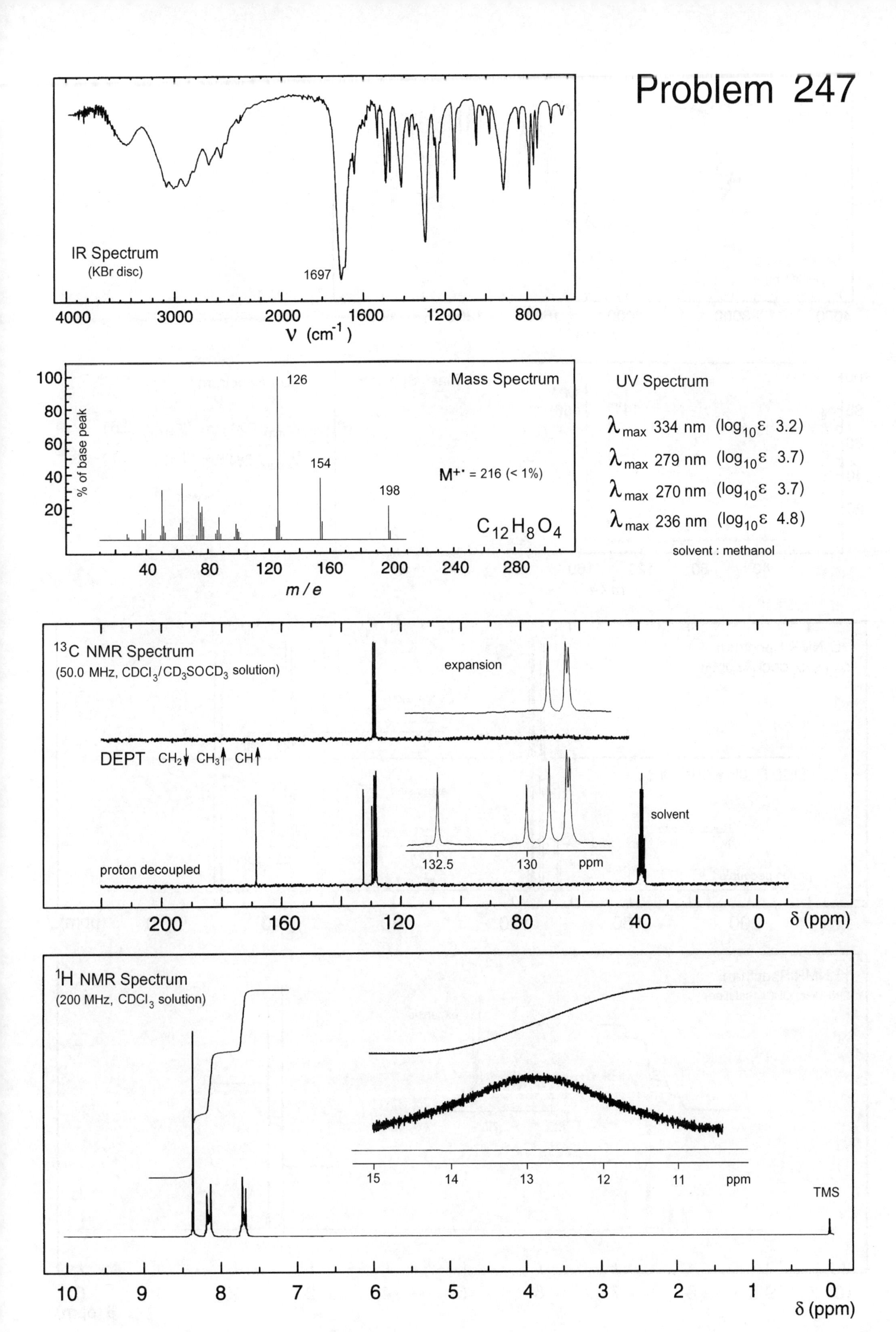

IR Spectrum
(KBr disc)
1697
4000
3000
2000
1600
1200
800
ν (cm⁻¹)
Mass Spectrum
% of base peak
126
154
198
M⁺˙ = 216 (< 1%)
C12H8O4
m / e
UV Spectrum
λmax 334 nm (log10ε 3.2)
λmax 279 nm (log10ε 3.7)
λmax 270 nm (log10ε 3.7)
λmax 236 nm (log10ε 4.8)
solvent : methanol
13C NMR Spectrum
(50.0 MHz, CDCl3/CD3SOCD3 solution)
expansion
DEPT CH2↓ CH3↑ CH↑
solvent
132.5
130
ppm
proton decoupled
δ (ppm)
1H NMR Spectrum
(200 MHz, CDCl3 solution)
15
14
13
12
11
ppm
TMS
δ (ppm)

Problem 248

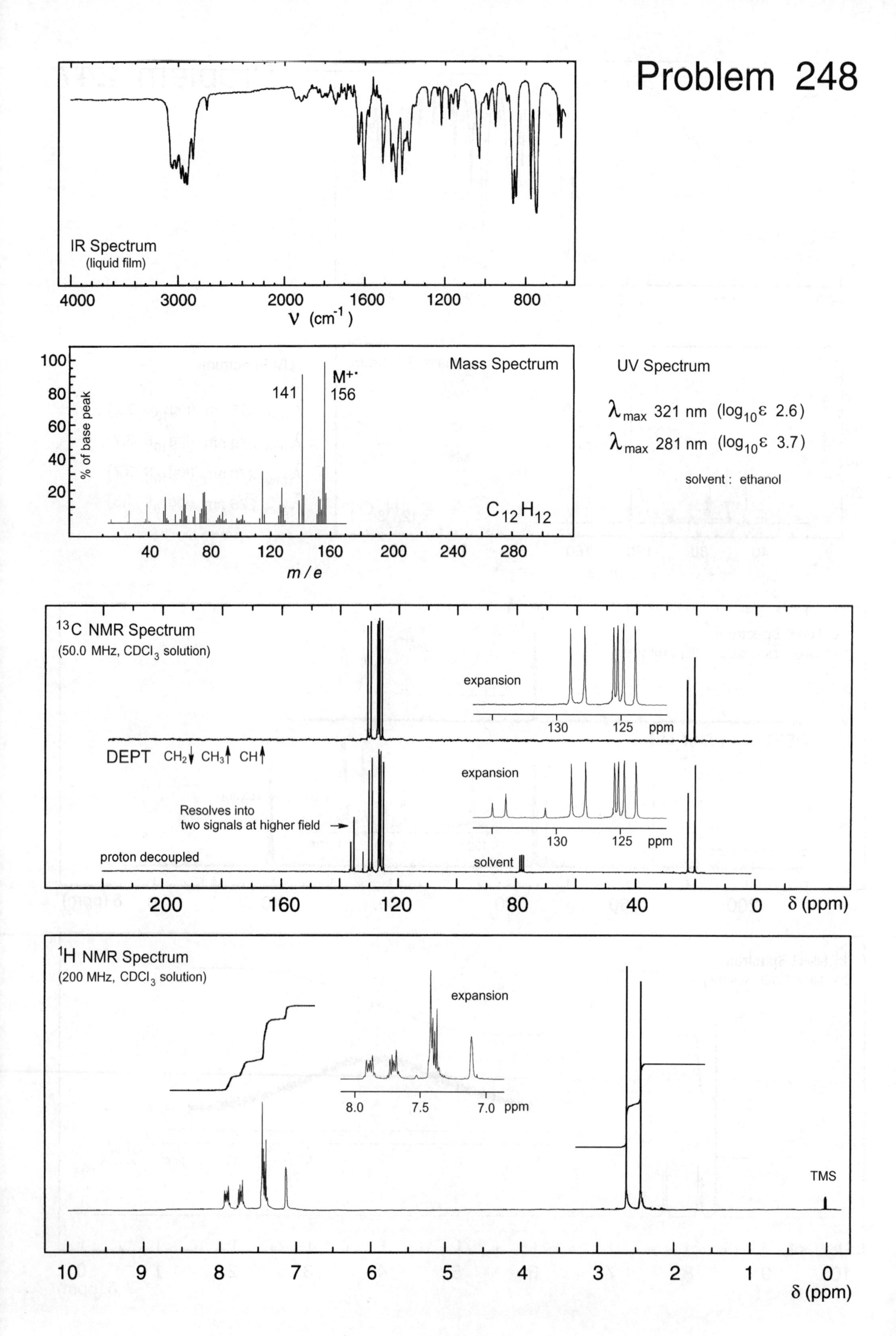

IR Spectrum
(liquid film)
4000
3000
2000
1600
1200
800
ν (cm-1)
Mass Spectrum
M+·
156
141
% of base peak
100
80
60
40
20
40
80
120
160
200
240
280
m / e
C12H12
UV Spectrum
λmax 321 nm (log10ε 2.6)
λmax 281 nm (log10ε 3.7)
solvent : ethanol
13C NMR Spectrum
(50.0 MHz, CDCl3 solution)
expansion
130
125
ppm
DEPT CH2↓ CH3↑ CH↑
Resolves into
two signals at higher field
proton decoupled
solvent
200
160
120
80
40
0
δ (ppm)
1H NMR Spectrum
(200 MHz, CDCl3 solution)
expansion
8.0
7.5
7.0
ppm
TMS
10
9
8
7
6
5
4
3
2
1
0
δ (ppm)

Problem 249

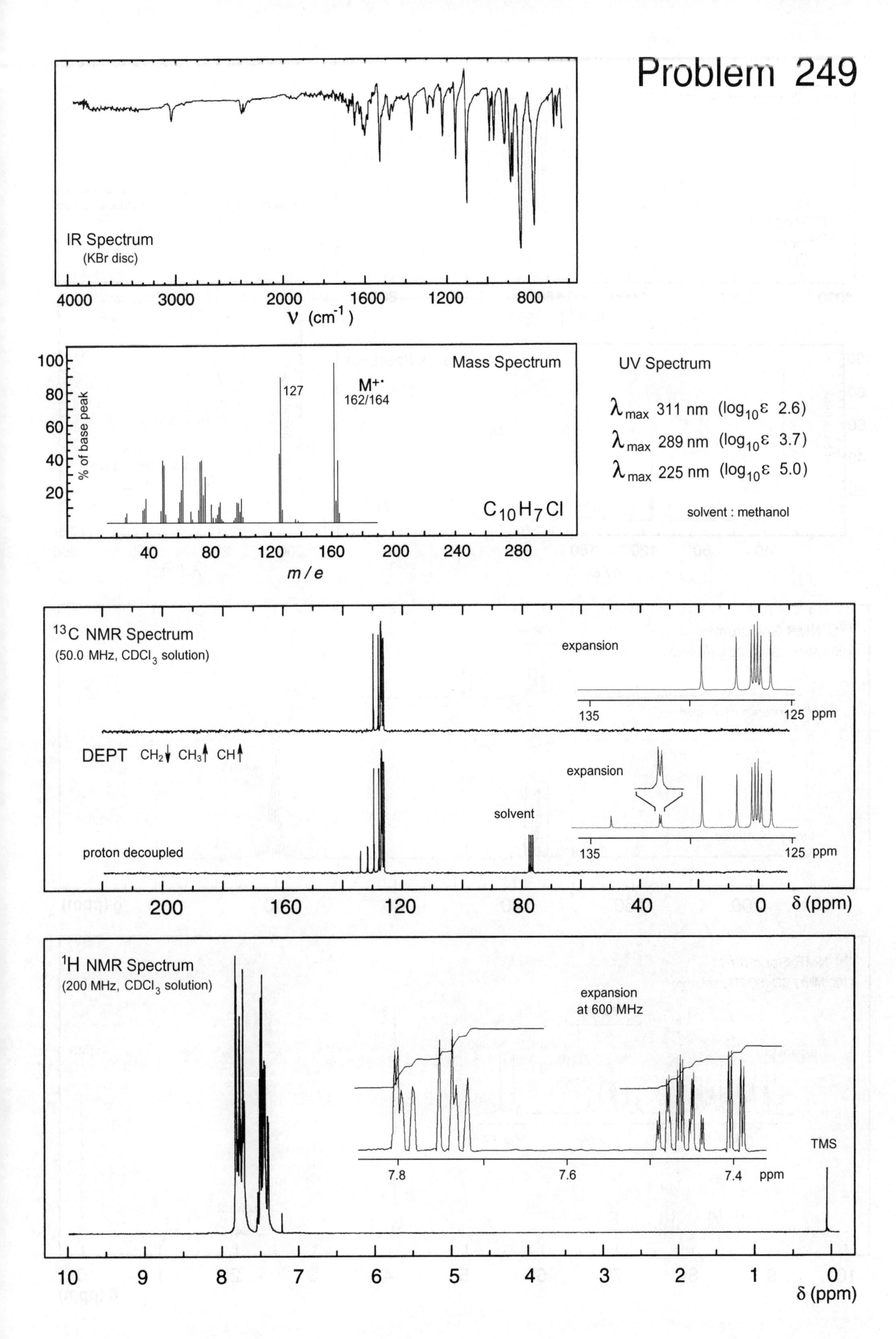

IR Spectrum
(KBr disc)
4000
3000
2000
1600
1200
800
ν (cm⁻¹)
Mass Spectrum
% of base peak
100
80
60
40
20
127
M+·
162/164
C10H7Cl
40
80
120
160
200
240
280
m / e
UV Spectrum
λmax 311 nm (log10ε 2.6)
λmax 289 nm (log10ε 3.7)
λmax 225 nm (log10ε 5.0)
solvent : methanol
13C NMR Spectrum
(50.0 MHz, CDCl3 solution)
expansion
135
125 ppm
DEPT CH2↓ CH3↑ CH↑
expansion
solvent
proton decoupled
135
125 ppm
200
160
120
80
40
0
δ (ppm)
1H NMR Spectrum
(200 MHz, CDCl3 solution)
expansion
at 600 MHz
7.8
7.6
7.4
ppm
TMS
10
9
8
7
6
5
4
3
2
1
0
δ (ppm)

Problem 250a

A 400 MHz ^1H NMR spectrum (with expansions and decouplings) is given on the following page

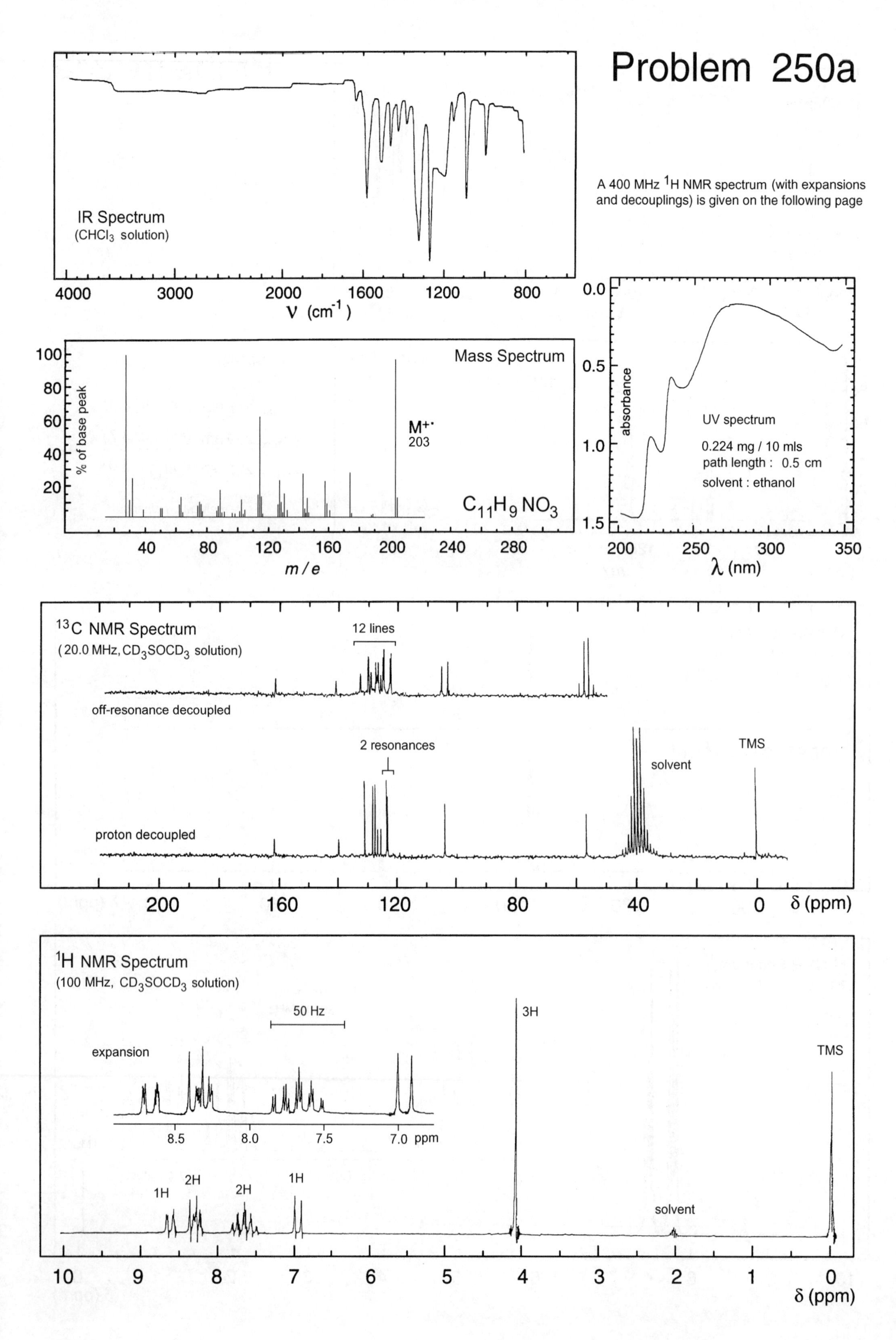

Problem 250b

^{1}H NMR Spectrum
(400 MHz, $CDCl_3$ solution)
Aromatic region only

Note: irradiation of the signal at δ 4.05 in the ^{1}H NMR produces an enhancement of the signals at δ 7.00 and 8.29 ppm via the NOE

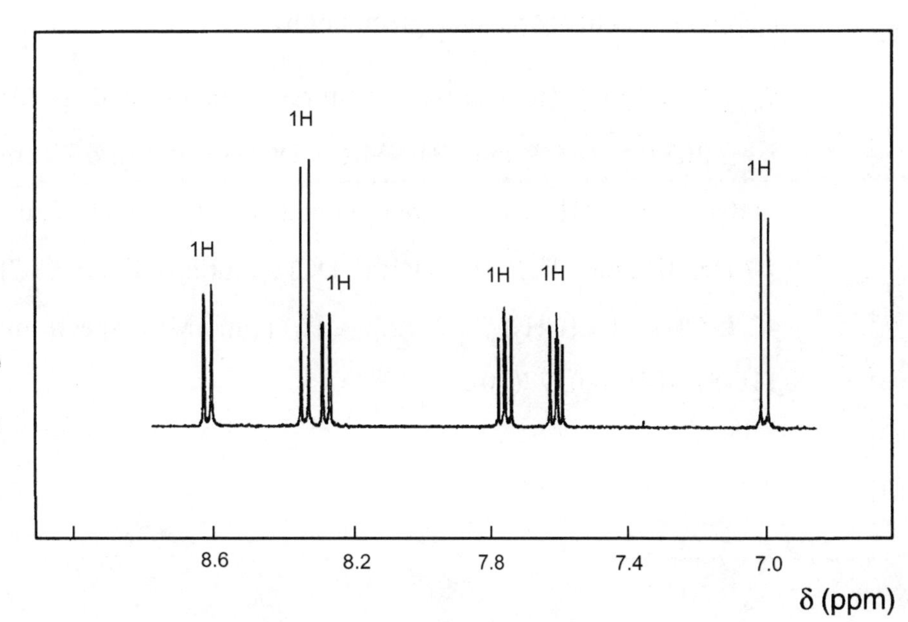

Expansion of regions of the 400 MHz NMR spectrum

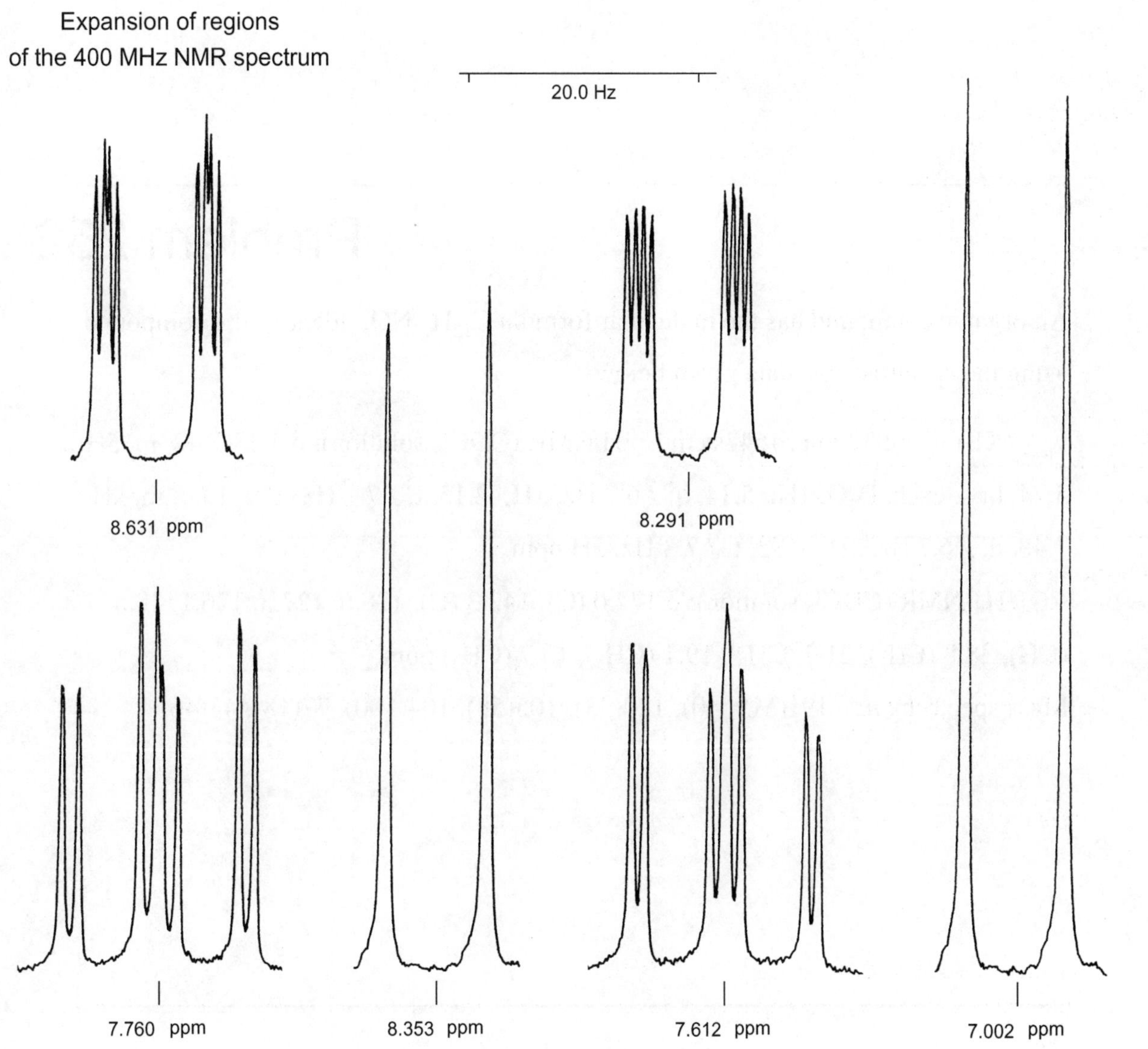

Problem 251

An organic compound has the molecular formula $C_{10}H_{14}$. Identify the compound using the spectroscopic data given below.

ν_{max} (liquid film): no significant features in the infrared spectrum.

λ_{max}: 265 (log ε: 2.3) nm. ^{1}H NMR ($CDCl_3$ solution): δ 7.1, m, 5H; 2.5, apparent sextet, *J* 7 Hz, 1H; 1.6, apparent quintet, *J* 7 Hz, 2H; 1.22, d, *J* 7 Hz, 3H; 0.81, t, *J* 7 Hz, 3H ppm. ^{13}C{^{1}H} NMR ($CDCl_3$ solution): δ 148.4 (C), 129.3, 127.9, 126.1, 42.3 (CH), 31.7 (CH_2), 22.2, 12.2 (CH_3) ppm. Mass spectrum: *m/e* 134 ($M^{+\cdot}$, 20), 119(8), 105(100), 77(10).

Problem 252

An organic compound has the molecular formula $C_{12}H_{17}NO$. Identify the compound using the spectroscopic data given below.

ν_{max} (KBr disc): 3296m, 1642s cm^{-1}. ^{1}H NMR ($CDCl_3$ solution): δ 7.23-7.42, m, 5H; 5.74, br s, exch. D_2O, 1H; 5.14, q, *J* 6.7 Hz, 1H; 2.15, t, *J* 7.1 Hz, 2H; 1.66, m, 2H; 1.48, d, *J* 6.7 Hz, 3H; 0.93, t, *J* 7.3 Hz, 3H ppm.

^{13}C{^{1}H} NMR ($CDCl_3$ solution): δ 172.0 (C), 143.3 (C), 128.6, 127.3, 126.1, 48.5 (CH), 38.8 (CH_2), 21.7 (CH_3), 19.1 (CH_2), 13.7 (CH_3) ppm.

Mass spectrum: *m/e* 191($M^{+\cdot}$, 40), 120(33), 105(58), 104(100), 77(18), 43(46).

Problem 253

An organic compound has the molecular formula $C_{16}H_{30}O_4$. Identify the compound using the spectroscopic data given below.

ν_{max} ($CHCl_3$ solution): 1733 cm^{-1}. ^{1}H NMR ($CDCl_3$ solution): δ 4.19, q, *J* 7.2 Hz, 4H; 3.35, s, 1H; 1.20, t, *J* 7.2 Hz, 6H; 1.25-1.29, m, 10H; 1.10, s, 6H; 0.88, t, *J* 6.8 Hz, 3H ppm. ^{13}C{^{1}H} NMR ($CDCl_3$ solution): δ 168.5 (C), 60.8 (CH_2), 59.6 (CH), 41.1 (CH_2), 36.3 (C), 31.8 (CH_2), 29.9 (CH_2), 25.1 (CH_3), 23.6 (CH_2), 22.6 (CH_2), 14.1 (CH_3), 14.0 (CH_3) ppm. Mass spectrum: *m/e* 286 ($M^{+\cdot}$, 70), 241(25), 201(38), 160(100), 115(53).

Problem 254

An organic compound has the molecular formula $C_8H_{13}NO_3$. Identify the compound using the spectroscopic data given below.

ν_{max} (nujol mull): 1690-1725s cm^{-1}. λ_{max}: no significant features in the ultraviolet spectrum. ^{1}H NMR ($CDCl_3$ solution): δ 4.25, q, *J* 6.7 Hz, 2H; 3.8, t, *J* 7 Hz, 4H; 2.45, t, *J* 7 Hz, 4H; 1.3, t, *J* 6.7 Hz , 3H ppm.
^{13}C{^{1}H} NMR ($CDCl_3$ solution): δ 207 (C); 155 (C); 62 (CH_2); 43 (CH_2); 41 (CH_2); 15 (CH_3) ppm. Mass spectrum: *m/e* 171 ($M^{+\cdot}$,15), 142(25), 56(68), 42(100).

Problem 255

An organic compound has the molecular formula $C_{12}H_{13}NO_3$. Identify the compound using the spectroscopic data given below.

ν_{max} (nujol mull): 3338, 1715, 1592 cm^{-1}. λ_{max}: 254 (log ε: 4.3) nm. ^{1}H NMR (DMSO-d^6 solution): δ 12.7, broad s, exch. D_2O, 1H; 8.42, d, *J* 6.1 Hz, 1H; 7.45-7.25, m, 5H; 6.63 dd, *J* 15.9, 1.2 Hz, 1H; 6.30, dd, *J* 15.9, 6.7 Hz, 1H; 4.93 ddd, *J* 6.7, 6.1, 1.2 Hz, 1H; 1.90, s, 3H ppm.

^{13}C{^{1}H} NMR (DMSO-d^6 solution): δ 171.9 (C), 169.0 (C), 135.9 (C), 131.7 (CH), 128.7 (CH), 127.9 (CH), 126.3 (CH), 124.6 (CH), 54.4 (CH), 22.3 (CH_3) ppm. Mass spectrum: *m/e* 219 ($M^{+\cdot}$, 25), 175(10), 132(100), 131(94), 103(35), 77(46), 43(83).

Problem 256

An organic compound has the molecular formula $C_{13}H_{16}O_4$. Identify the compound using the spectroscopic data given below.

ν_{max} (KBr disc): 3479s, 1670s, cm^{-1}. λ_{max}: 250 (log ε: 4) nm. ^{1}H NMR ($CDCl_3$ solution): δ 1.40, s, 3H; 2.00, bs exch., 1H; 2.71, d, *J* 15.7 Hz, 1H; 2.77, d, *J* 15.7 Hz, 1H; 2.91, d, *J* 17.8 Hz , 1H; 3.14, d, *J* 17.8 Hz, 1H; 3.79, s, 3H; 3.84, s, 3H; 6.80, d, *J* 9.0 Hz, 1H; 6.98, d, *J* 9.0 Hz, 1H ppm. ^{13}C{^{1}H} NMR ($CDCl_3$ solution) : δ 196.0 (C); 154.0 (C); 157.7 (C); 131.5 (C); 122.0 (C); 116.0 (CH); 110.5 (CH); 70.8 (C); 56.3 (CH_3); 55.9 (CH_3); 54.1 (CH_2); 37.7 (CH_2); 29.2 (CH_3) ppm. Mass spectrum: *m/e* 236 ($M^{+\cdot}$, 87), 218(33), 178(100), 163(65).

Problem 257

Give the <u>number of different chemical environments</u> for the magnetic nuclei ^{1}H and ^{13}C in the following compounds. Assume that any conformational processes are fast on the NMR timescale unless otherwise indicated.

Structure	Number of ^{1}H environments	Number of ^{13}C environments
$CH_3-CO-CH_2CH_2CH_3$		
$CH_3CH_2-CO-CH_2CH_3$		
$CH_2=CHCH_2CH_3$		
cis- $CH_3CH=CHCH_3$		
trans-$CH_3CH=CHCH_3$		

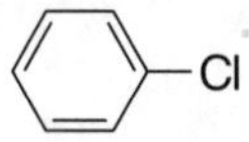 Cl		
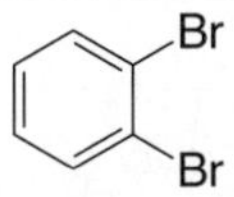 Br, Br		
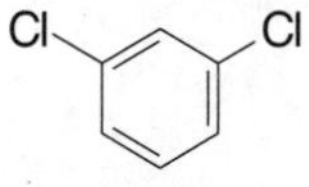 Cl, Cl		
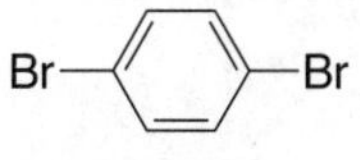 Br, Br		
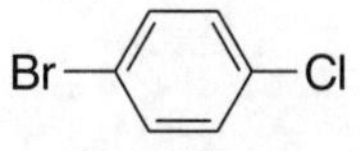 Br, Cl		
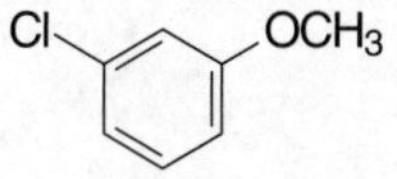 Cl, OCH_3		
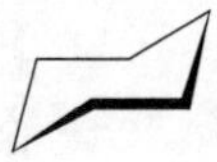 Assuming **slow** chair-chair interconversion		
 Assuming **fast** chair-chair interconversion		
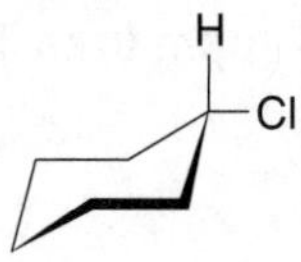 H, Cl — Assuming the molecule to be *conformationally rigid*		

Problem 258

A 60 MHz ^{1}H NMR spectrum of diethyl ether is given below.

Note that the spectrum is calibrated only in parts per million (ppm) from tetramethylsilane (TMS), *i.e.* in δ units.

(a) Assign the signals due to the $-CH_2-$ and $-CH_3$ groups respectively using three independent criteria (the relative areas of the signals, the multiplicity of each signal and the chemical shift of each signal).

(b) Obtain the chemical shift of each group in ppm, then convert to Hz at 60 MHz from TMS (see Section 5.4).

(c) Obtain the value of the first-order coupling constants $^{3}J_{H\text{-}H}$ (in Hz).

(d) Demonstrate that first-order analysis was justified (see Section 5.6).

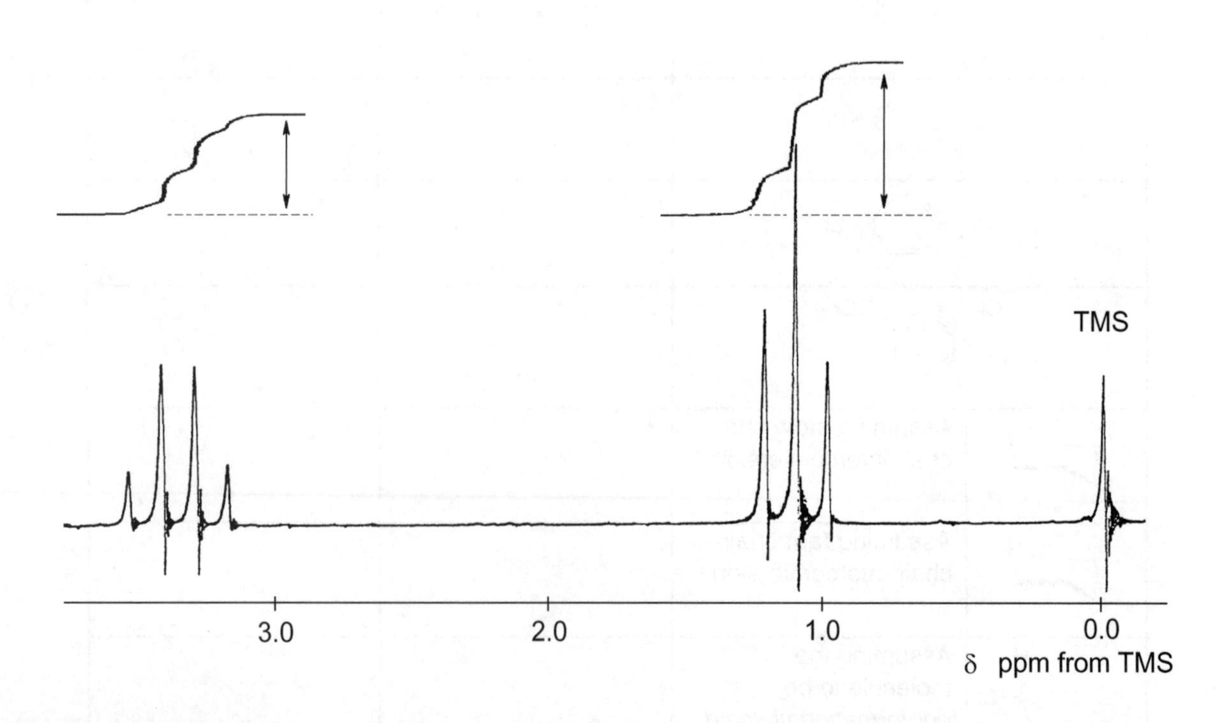

Problem 259

A 100 MHz ^{1}H NMR spectrum of a 3-proton system is given below.

(a) Draw a splitting diagram and analyse this spectrum by first-order methods, *i.e.* extract all relevant coupling constants (J in Hz) and chemical shifts (δ in ppm) by direct measurement.

(b) Justify the use of a first-order analysis (see Section 5.6).

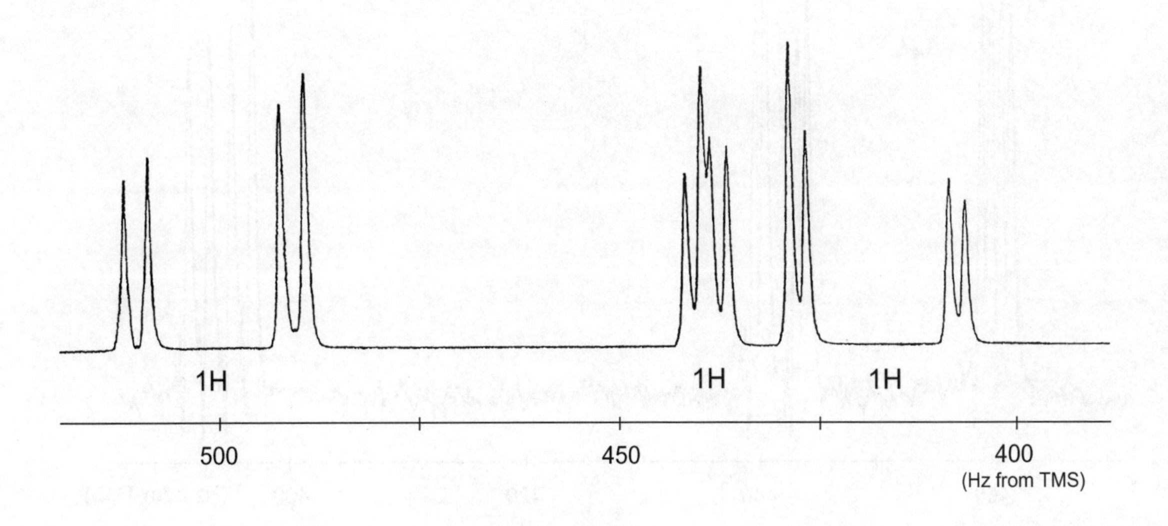

Problem 260

Portion of the 60 MHz NMR spectrum 2-furoic acid in $CDCl_3$ is shown below. Only the resonances due to the three aromatic protons (H_A, H_M and H_X) are shown.

H_M H_X H_A O COOH

2-Furoic Acid

(a) Draw a splitting diagram and analyse this spectrum by first-order methods, *i.e.* extract all relevant coupling constants (J in Hz) and chemical shifts (δ in ppm) by direct measurement.

(b) Justify the use of a first-order analysis (see Section 5.6).

Note: This is a <u>60 MHz spectrum</u>.

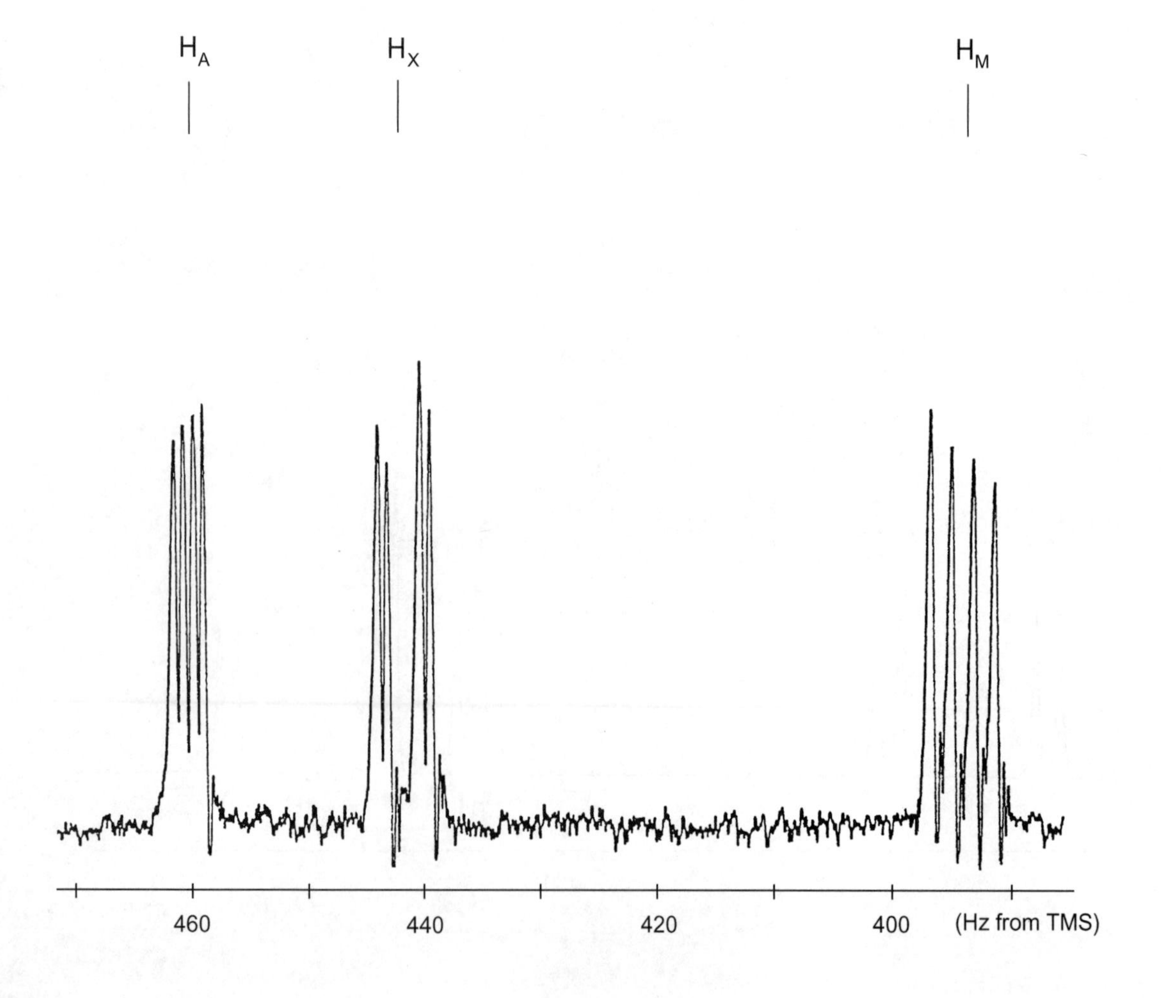

Problem 261

A portion of the 100 MHz ^{1}H NMR spectrum of 2-amino-5-chlorobenzoic acid in CD_3OD is given below. Only the resonances due to the three aromatic protons are shown.

(a) Draw a splitting diagram and analyse this spectrum by first-order methods, *i.e.* extract all relevant coupling constants (J in Hz) and chemical shifts (δ in ppm) by direct measurement.

(b) Justify the use of a first-order analysis (see Section 5.6).

(c) Assign the three multiplets to H_3, H_4 and H_6 given:

- the characteristic ranges for coupling constants between aromatic protons (see Section 5.7);
- the fact that H_3 will give rise to the resonance at the highest field due to the strong +R effect of the amino group (see Table 5.6).

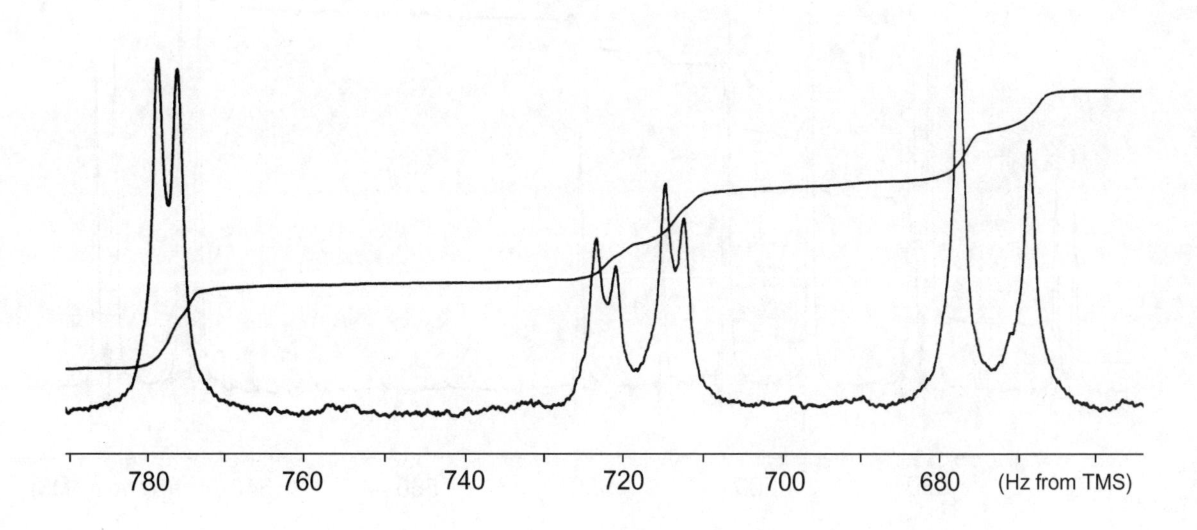

Problem 262a

Portion of 100 MHz ^{1}H NMR spectrum of methyl acrylate (5% in C_6D_6) is given below. Only the part of the spectrum containing the resonances of the olefinic protons H_A, H_B and H_C is shown.

H_A, COOCH$_3$, C=C, H_B, H_C

methyl acrylate

(a) Draw a splitting diagram.

(b) Analyse this spectrum by first-order methods, *i.e.* extract all relevant coupling constants (J in Hz) and chemical shifts (δ in ppm) by direct measurement.

(c) Justify the statement that "this spectrum is really a borderline second-order (strongly coupled) case". Point out the most conspicuous deviation from first-order character in this spectrum (see Section 5.6).

(d) Assign the three multiplets to H_A, H_B and H_C on the basis of coupling constants only (see Section 5.7).

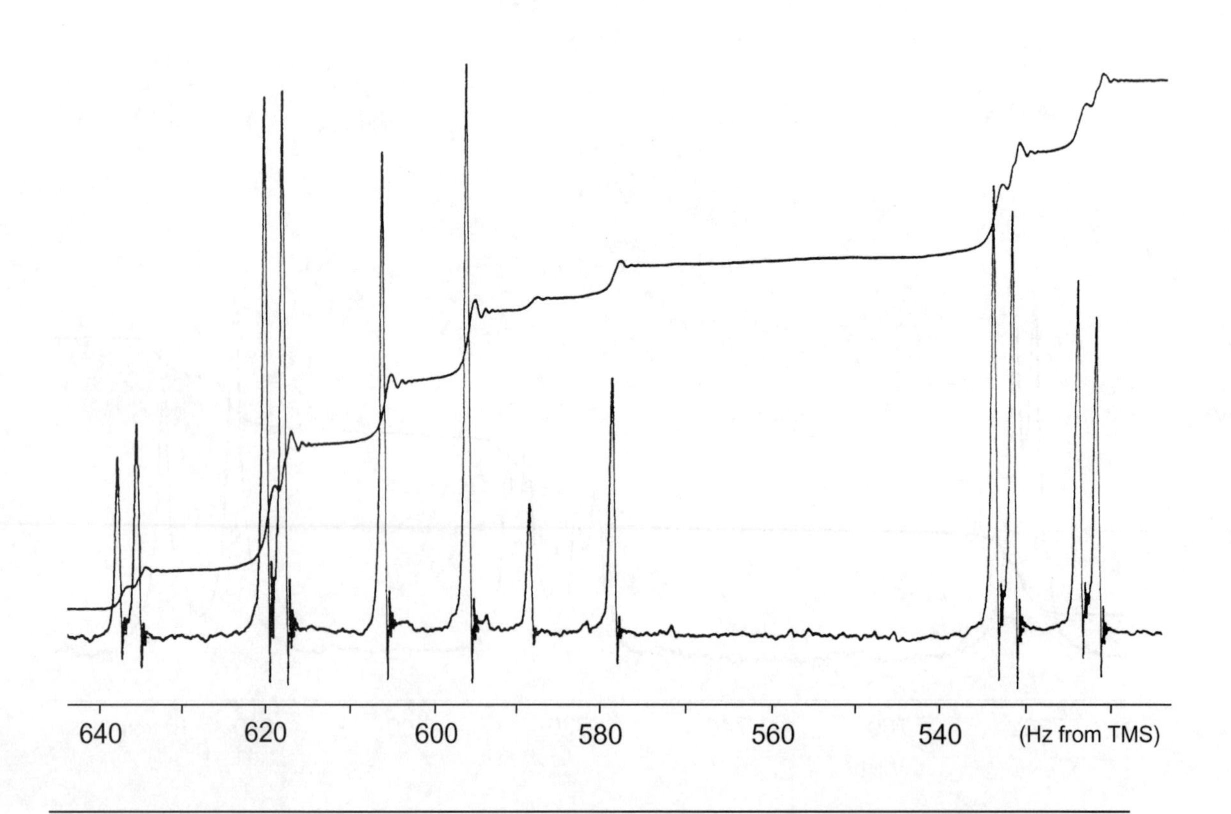

Problem 262b

This is the **computer-simulated spectrum** corresponding to the complete analysis of the spectrum shown in Problem 262a, *i.e.* an exact analysis in which first-order assumptions were not made. The simulated spectrum fits the experimental spectrum, verifying that the analysis was correct. Compare your (first-order) results from Problem 262a with the actual solution given here.

Number of SPINS	=	3
F(1)	=	+ 528.500 Hz
F(2)	=	+ 594.531 Hz
F(3)	=	+ 626.093 Hz
J(1,2)	=	+ 10.539 Hz
J(1,3)	=	+ 1.589 Hz
J(2,3)	=	+ 17.278 Hz
START of simulation	=	+ 750.000 Hz
FINISH of simulation	=	+ 500.000 Hz
LINE WIDTH	=	+ 0.427 Hz

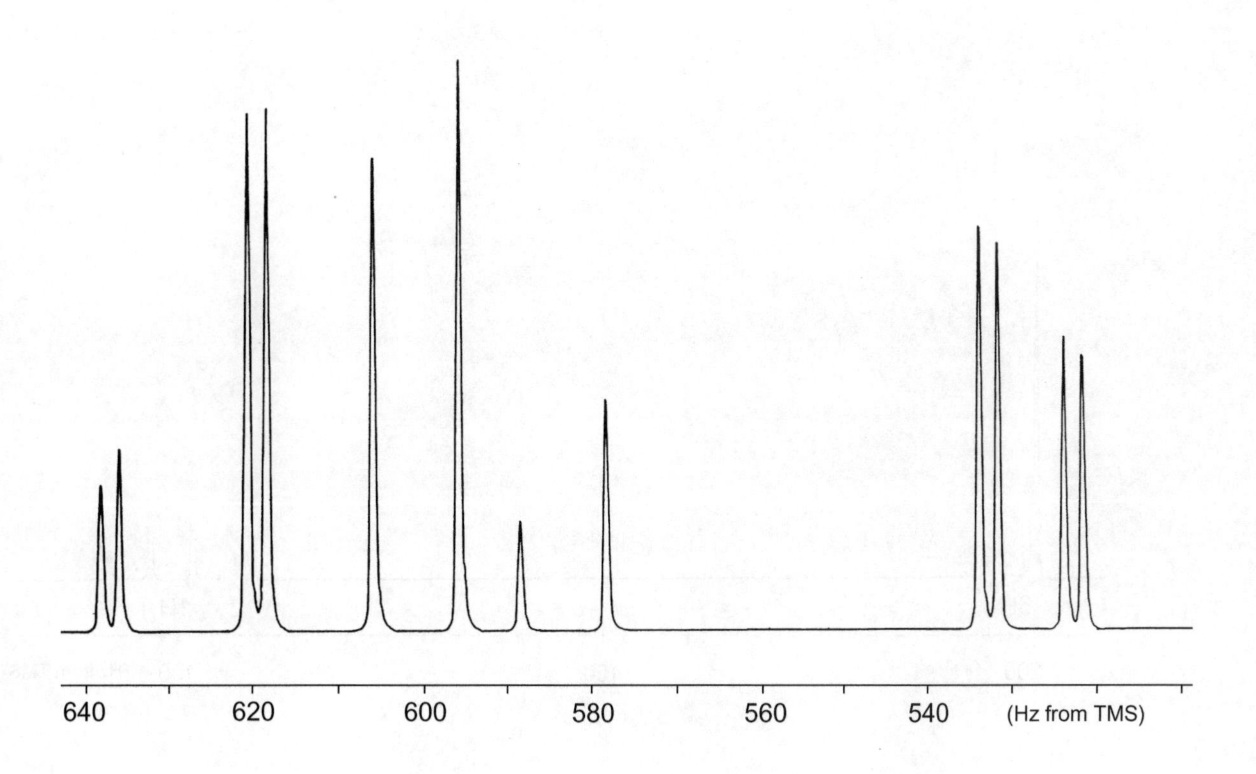

Problem 263

A 100 MHz ^1H NMR spectrum of a 4-proton system is given below.

(a) Draw a splitting diagram.

(b) Analyse this spectrum by first-order methods, *i.e.* extract all relevant coupling constants (J in Hz) and chemical shifts (δ in ppm) by direct measurement.

(c) Justify the use of first-order analysis (see Section 5.6).

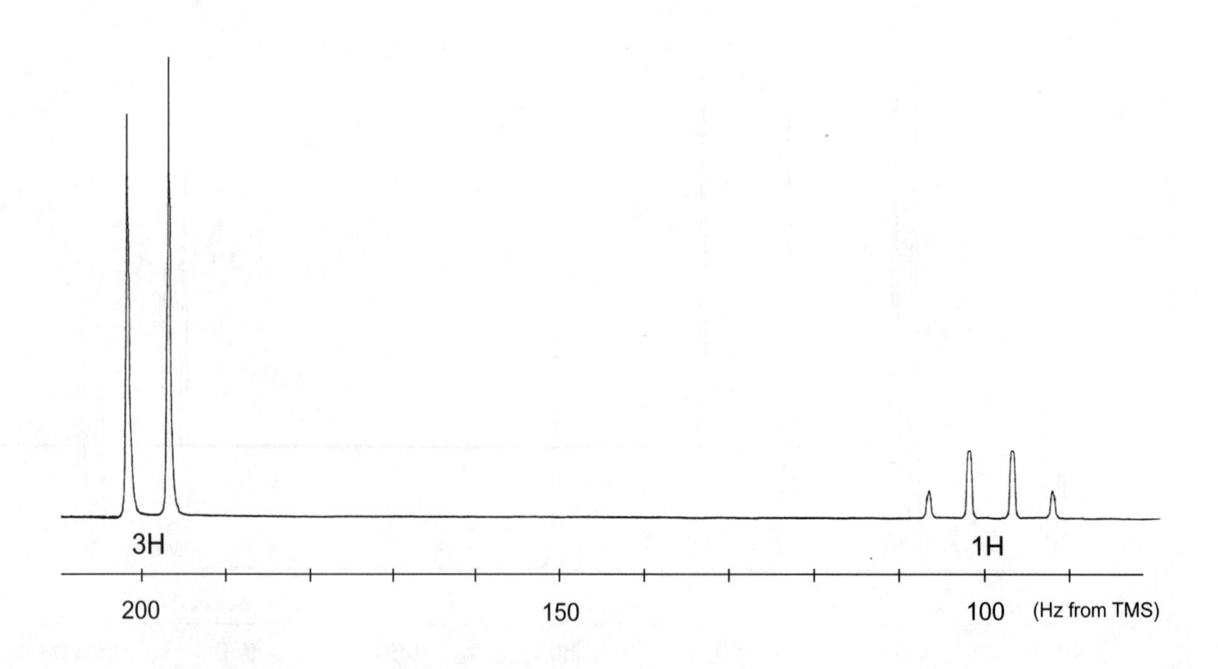

Problem 264

A 100 MHz ^1H NMR spectrum of a 4-proton system is given below.

(a) Draw a splitting diagram.

(b) Analyse this spectrum by first-order methods, *i.e.* extract all relevant coupling constants (J in Hz) and chemical shifts (δ in ppm) by direct measurement.

(c) Justify the use of first-order analysis (see Section 5.6).

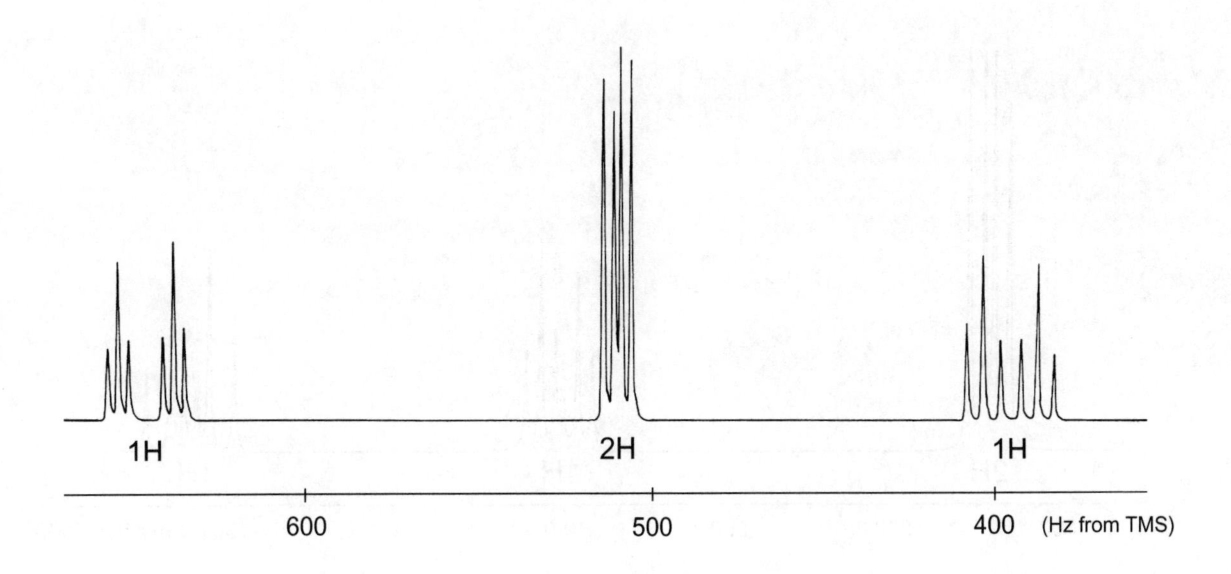

Problem 265

A 100 MHz ^{1}H NMR spectrum of a 4-proton system is given below.

(a) Draw a splitting diagram.

(b) Analyse this spectrum by first-order methods, *i.e.* extract all relevant coupling constants (J in Hz) and chemical shifts (δ in ppm) by direct measurement.

(c) Justify the use of first-order analysis (see Section 5.6).

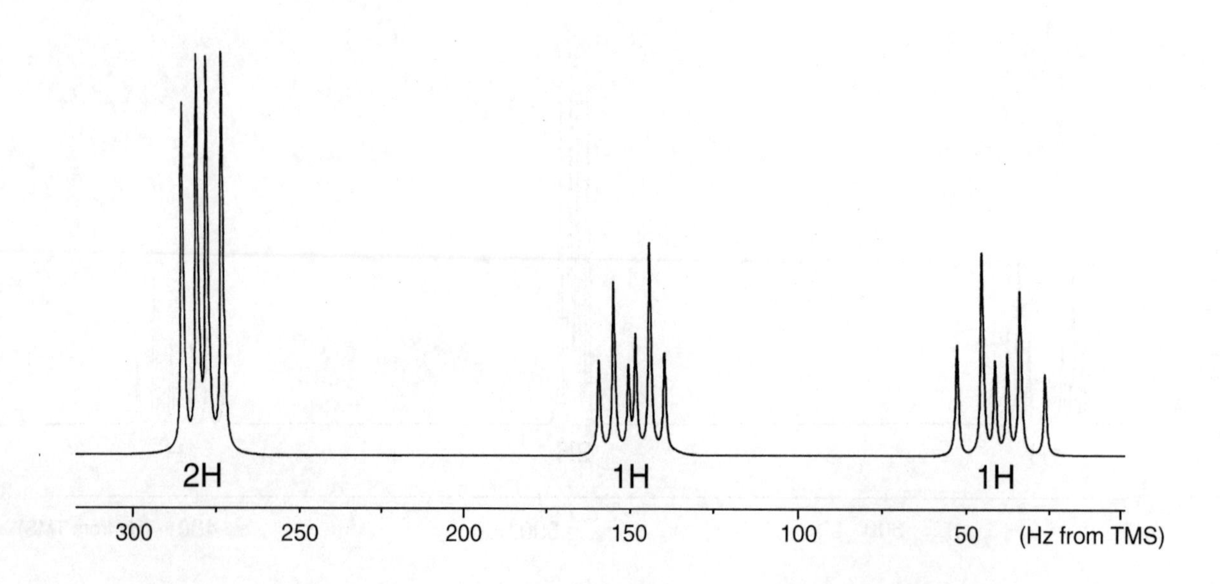

Problem 266

A 100 MHz ^{1}H NMR spectrum of a 4-proton system is given below.

(a) Draw a splitting diagram.

(b) Analyse this spectrum by first-order methods, *i.e.* extract all relevant coupling constants (J in Hz) and chemical shifts (δ in ppm) by direct measurement.

(c) Justify the use of first-order analysis (see Section 5.6).

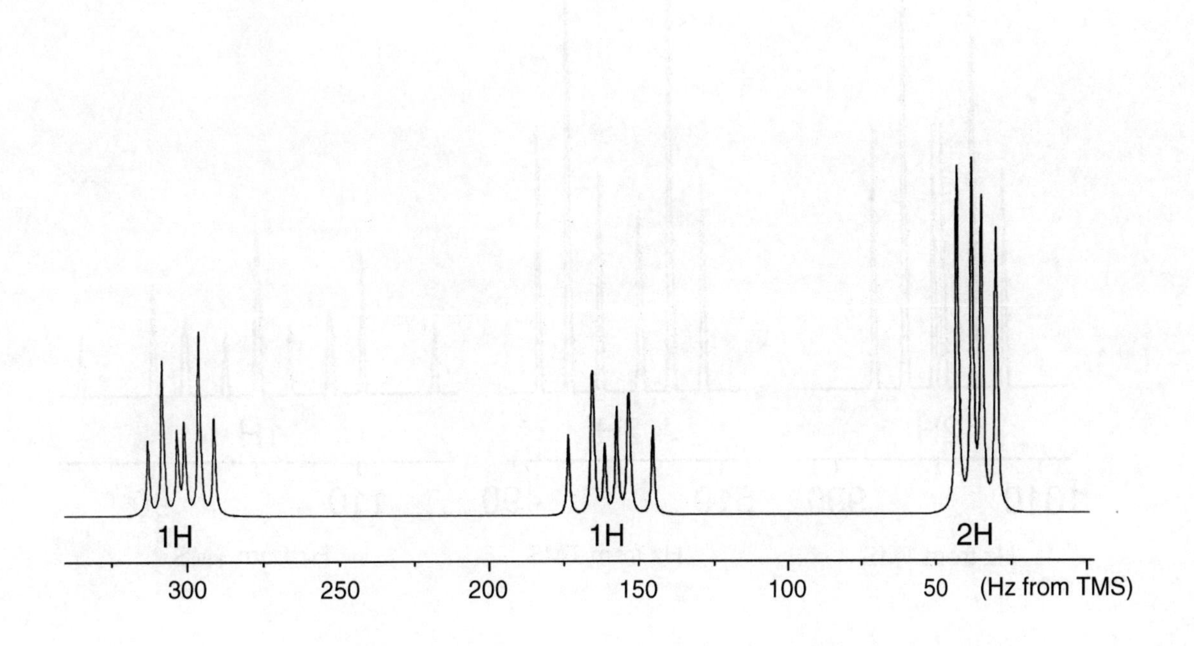

Problem 267

A 200 MHz ^{1}H NMR spectrum of a 5-proton system is given below.

(a) Draw a splitting diagram.

(b) Analyse this spectrum by first-order methods, *i.e.* extract all relevant coupling constants (J in Hz) and chemical shifts (δ in ppm) by direct measurement.

(c) Justify the use of first-order analysis (see Section 5.6).

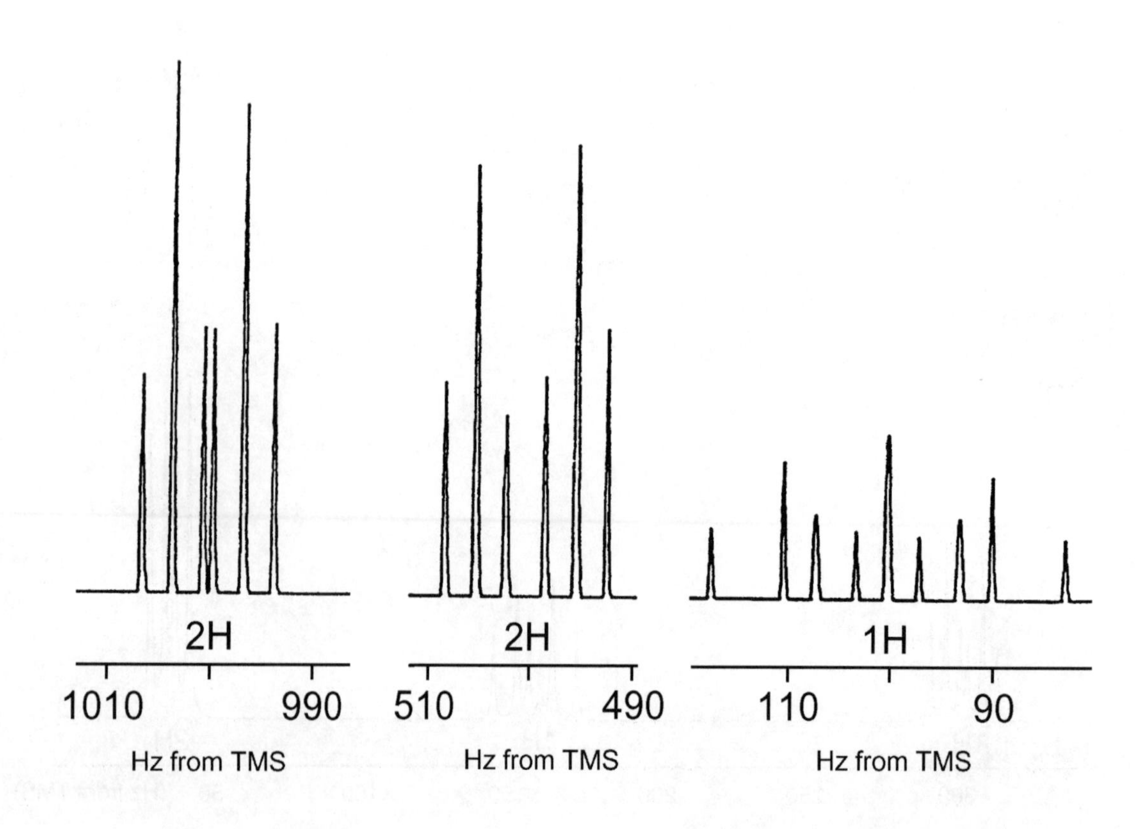

Problem 268

Portion of 100 MHz NMR spectrum of crotonic acid in $CDCl_3$ is given below. The upfield part of the spectrum, which is due to the methyl group, is less amplified to fit the page.

H_A, COOH, CH_3, H_M

crotonic acid

(a) Draw a splitting diagram and analyse this spectrum by first-order methods, *i.e.* extract all relevant coupling constants (J in Hz) and chemical shifts (δ in ppm) by direct measurement. Justify the use of first-order analysis.

(b) There are certain conventions used for naming spin-systems (*e.g.* AMX, AMX_2, AM_2X_3). Note that this is a *5-spin system* and name the spin system responsible for this spectrum (see Section 5.6).

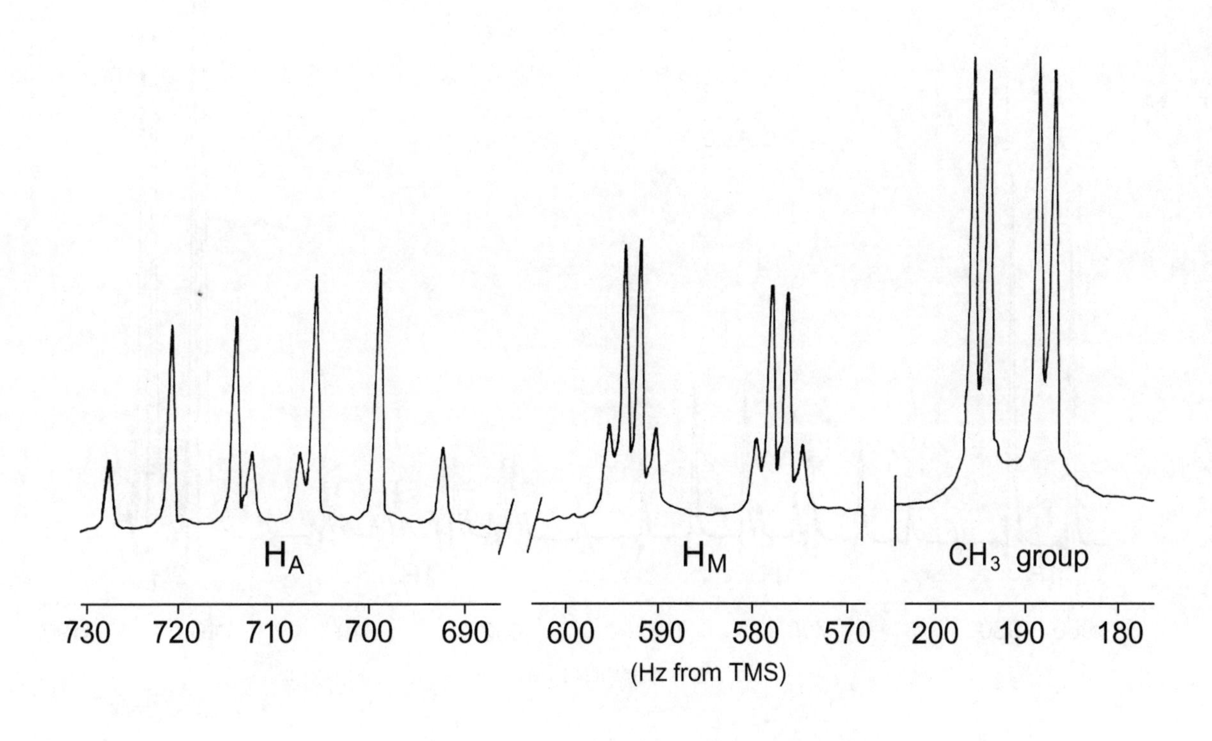

Problem 269

The 100 MHz ^{1}H NMR spectrum (5% in $CDCl_3$) of an α,β-unsaturated aldehyde C_4H_6O is given below.

(a) Draw a splitting diagram and analyse this spectrum by first-order methods, *i.e.* extract all relevant coupling constants (J in Hz) and chemical shifts (δ in ppm) by direct measurement.

(b) Justify the use of a first-order analysis (see Section 5.6).

(c) Use the coupling constants to obtain the structure of the compound, including the stereochemistry about the double bond (see Section 5.7).

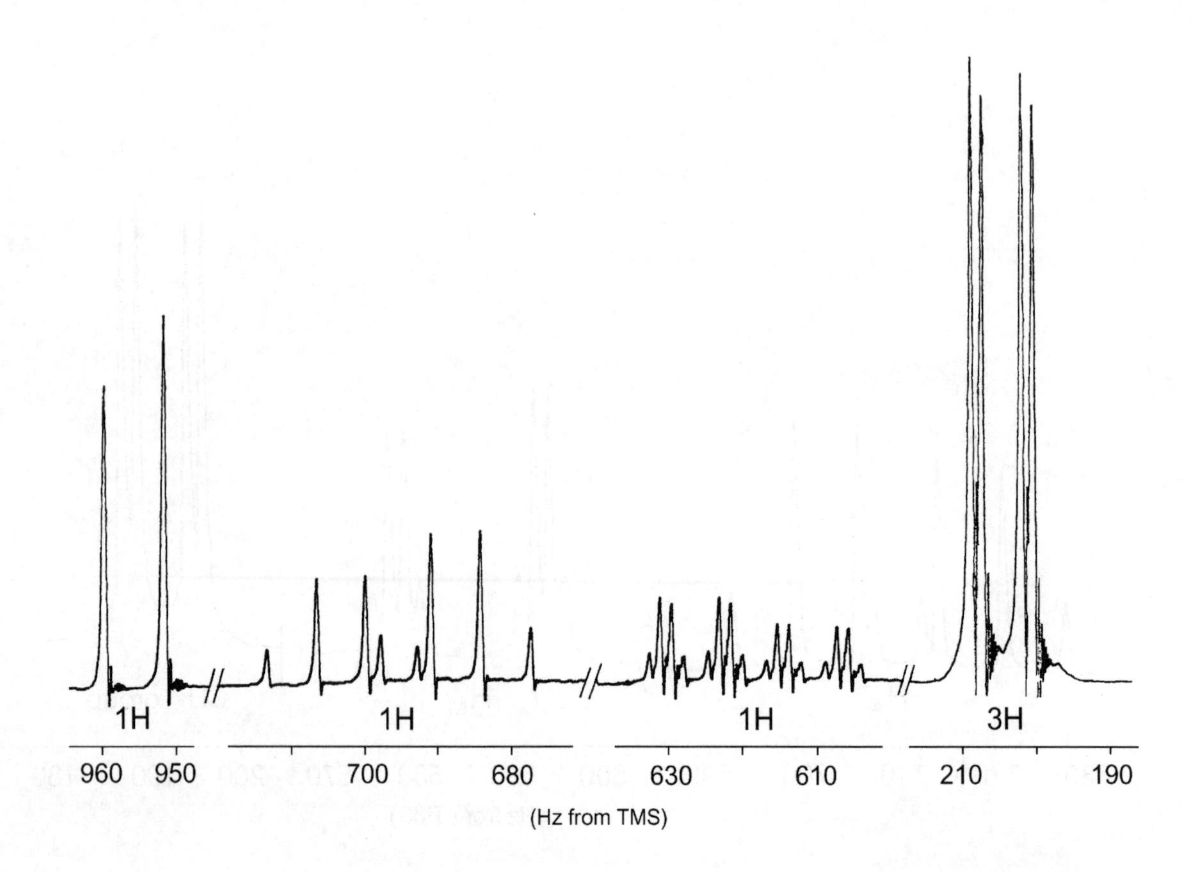

Problem 270

Draw a schematic (line) representation of the pure first-order spectrum (AMX) corresponding to the following parameters:

Frequencies (Hz from TMS): $\nu_A = 180$; $\nu_M = 220$; $\nu_X = 300$.

Coupling constants (Hz): $J_{AM} = 10$; $J_{AX} = 12$; $J_{MX} = 5$.

Assume that the spectrum is a pure first-order spectrum and ignore small distortions in relative intensities of lines that would be apparent in a "real" spectrum.

(a) Sketch in "splitting diagrams" above the schematic spectrum to indicate which splittings correspond to which coupling constants.

(b) Give the chemical shifts on the δ scale corresponding to the above spectrum obtained with an instrument operating at 60 MHz for protons.

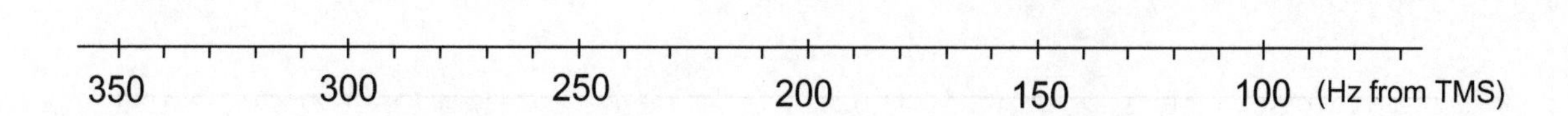

Problem 271

Draw a schematic (line) representation of the pure first-order spectrum (AMX) corresponding to the following parameters:

Frequencies (Hz from TMS): $\nu_A = 300$; $\nu_M = 240$; $\nu_X = 120$.

Coupling constants (Hz): $J_{AM} = 10$; $J_{AX} = 2$; $J_{MX} = 8$.

Assume that the spectrum is a pure first-order spectrum and ignore small distortions in relative intensities of lines that would be apparent in a "real" spectrum.

(a) Sketch in "splitting diagrams" above the schematic spectrum to indicate which splittings correspond to which coupling constants.

(b) Give the chemical shifts on the δ scale corresponding to the above spectrum obtained with an instrument operating at 60 MHz for protons.

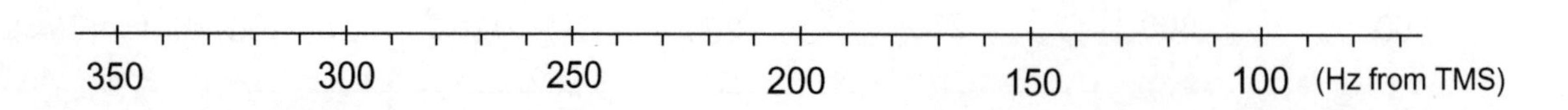

Problem 272

Draw a schematic (line) representation of the pure first-order spectrum (AMX_2) corresponding to the following parameters:

Frequencies (Hz from TMS): $\nu_A = 340$; $\nu_M = 240$; $\nu_X = 100$.

Coupling constants (Hz): $J_{AM} = 10$; $J_{AX} = 2$; $J_{MX} = 6$.

Assume that the spectrum is a pure first-order spectrum and ignore small distortions in relative intensities of lines that would be apparent in a "real" spectrum.

(a) Sketch in "splitting diagrams" above the schematic spectrum to indicate which splittings correspond to which coupling constants.

(b) Give the chemical shifts on the δ scale corresponding to the above spectrum obtained with an instrument operating at 60 MHz for protons.

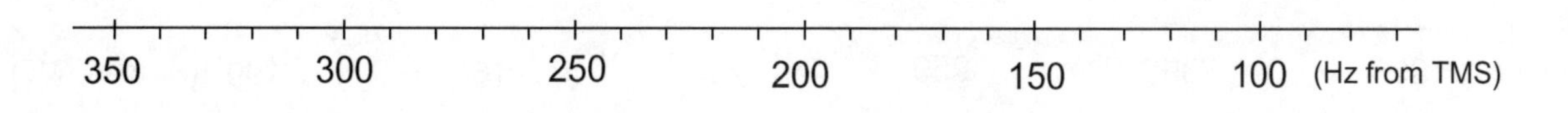

Problem 273

Draw a schematic (line) representation of the pure first-order spectrum (AM_2X) corresponding to the following parameters:

Frequencies (Hz from TMS): $\nu_A = 110$; $\nu_M = 200$; $\nu_X = 290$.

Coupling constants (Hz): $J_{AM} = 10$; $J_{AX} = 12$; $J_{MX} = 3$.

Assume that the spectrum is a pure first-order spectrum and ignore small distortions in relative intensities of lines that would be apparent in a "real" spectrum.

(a) Sketch in "splitting diagrams" above the schematic spectrum to indicate which splittings correspond to which coupling constants.

(b) Give the chemical shifts on the δ scale corresponding to the above spectrum obtained with an instrument operating at 60 MHz for protons.

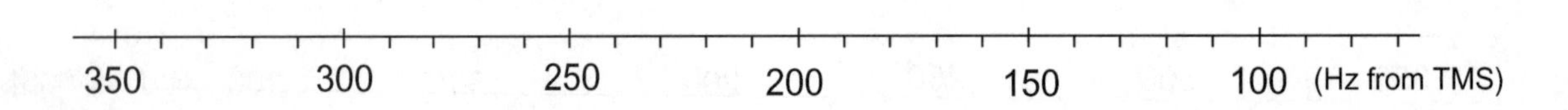

Problem 274

Draw a schematic (line) representation of the pure first-order spectrum (AMX_3) corresponding to the following parameters:

Frequencies (Hz from TMS): $\nu_A = 80$; $\nu_M = 220$; $\nu_X = 320$.

Coupling constants (Hz): $J_{AM} = 10$; $J_{AX} = 12$; $J_{MX} = 0$.

Assume that the spectrum is a pure first-order spectrum and ignore small distortions in relative intensities of lines that would be apparent in a "real" spectrum.

(a) Sketch in "splitting diagrams" above the schematic spectrum to indicate which splittings correspond to which coupling constants.

(b) Give the chemical shifts on the δ scale corresponding to the above spectrum obtained with an instrument operating at 60 MHz for protons.

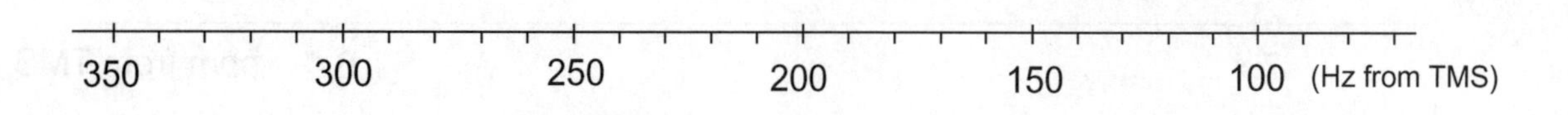

Problem 275

A portion of the 90 MHz ^{1}H NMR spectrum (5% in $CDCl_3$) of one of the six possible isomeric dibromoanilines is given below. Only the resonances of the aromatic protons are shown.

1 2 3 4 5 6

Determine which is the correct structure for this compound using arguments based on symmetry and the magnitudes of spin-spin coupling constants (see Section 5.7).

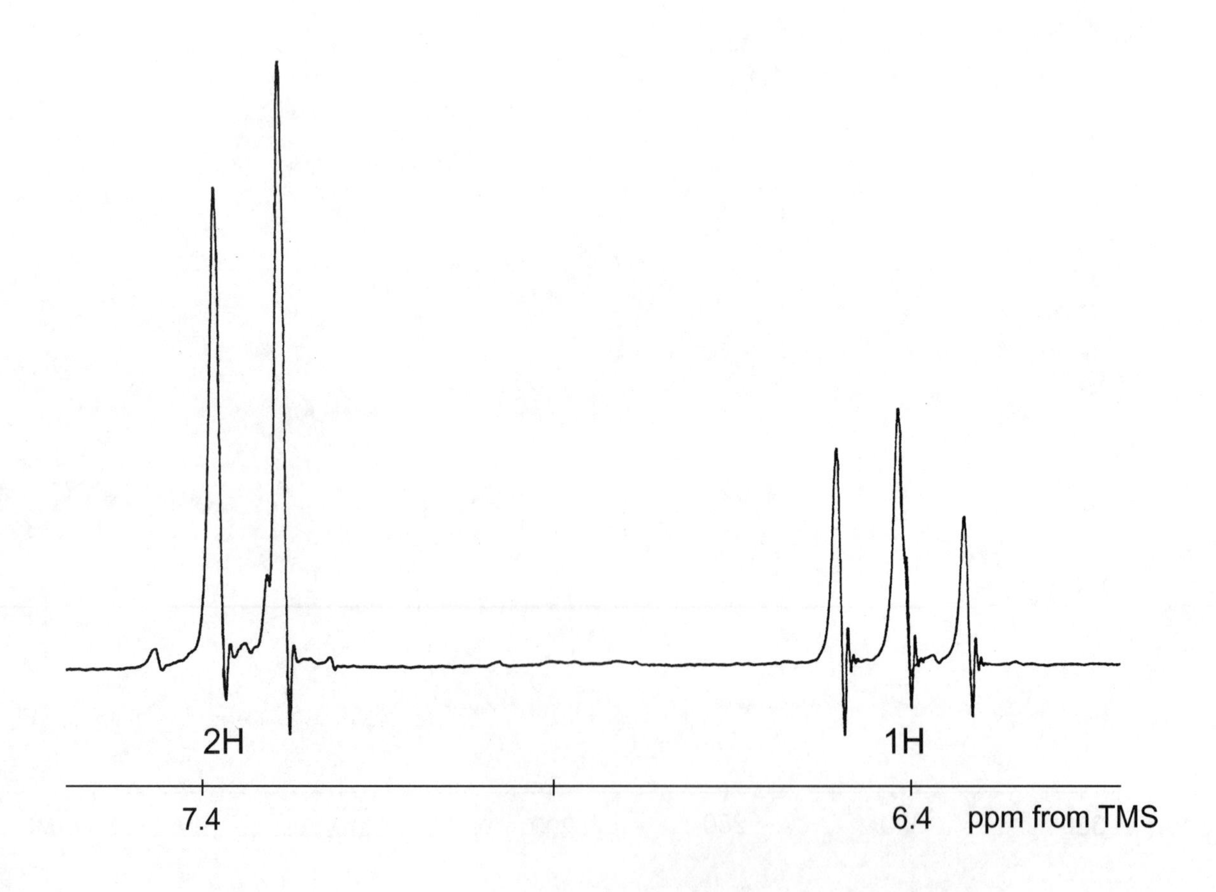

Problem 276

The 400 MHz ^{1}H NMR spectrum (5% in $CDCl_3$ after D_2O exchange) of one of the six possible isomeric hydroxycinnamic acids is given below.

Determine which is the correct structure for this compound using arguments based on symmetry and the magnitudes of spin-spin coupling constants (see Section 5.7).

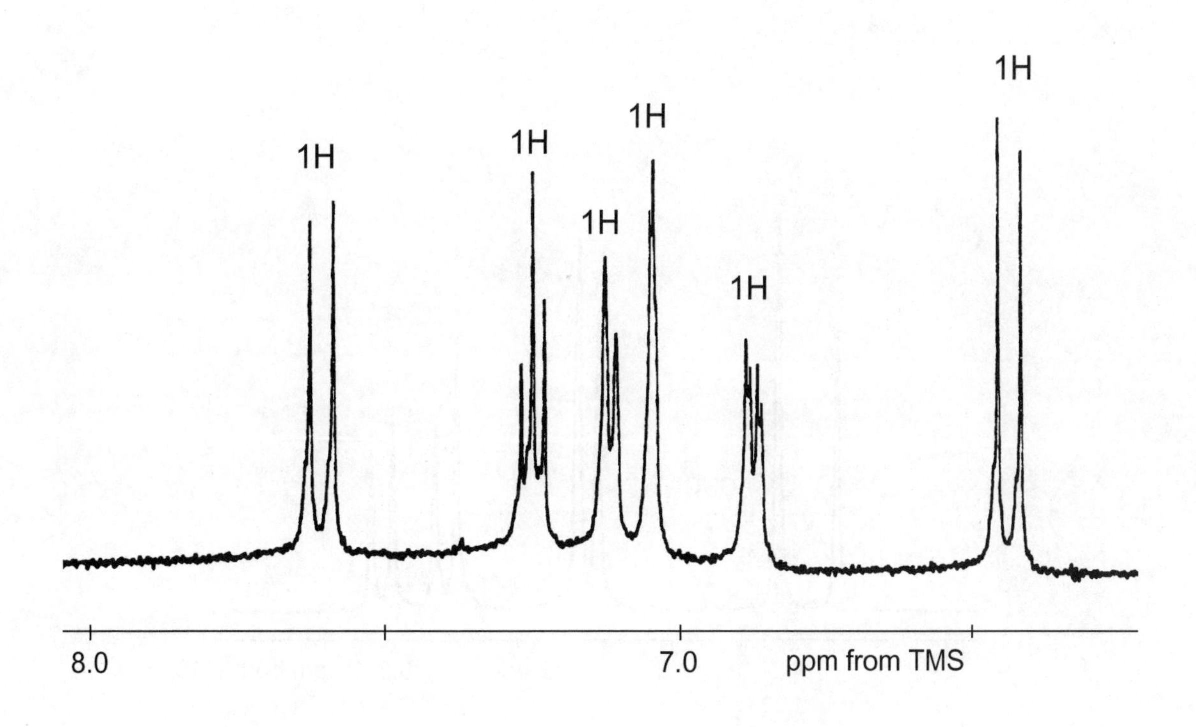

Problem 277

In a published paper, the 90 MHz ^1H NMR spectrum given below was assigned to 1,5-dichloronaphthalene, $C_{10}H_6Cl_2$.

1,5-dichloronaphthalene

(a) Why can't this spectrum belong to 1,5-dichloronaphthalene ?

(b) Suggest two alternative dichloronaphthalenes that would have structures consistent with the spectrum given.

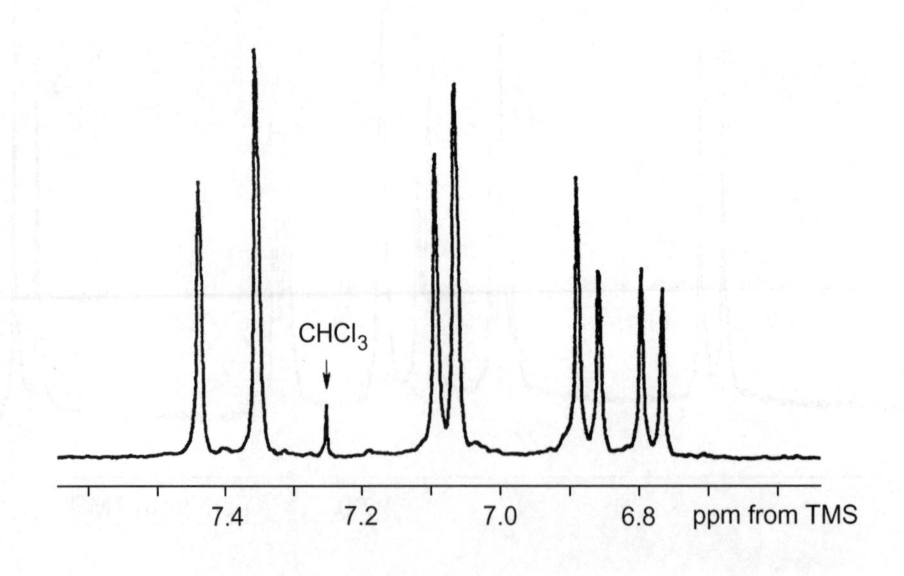

Appendix 1

WORKED EXAMPLES

This section works through two problems from the text to indicate a reasonable process for obtaining the structure of the unknown compound from the spectra provided. It should be emphasised that the logic used here is by no means the only way to arrive at the correct solution but it does provide a systematic approach to obtaining structures by assembling structural fragments identified by each type of spectroscopy.

A1.1 PROBLEM 73

(1) Perform all Routine Operations

(a) From the molecular ion, the molecular weight is 198/200. The molecular ion has two peaks of equal intensity separated by two mass units. This is the characteristic pattern for a compound containing one bromine atom.

(b) Relative numbers of protons in different environments.

δ ^{1}H (ppm)	Integral (mm)	Relative No. of hydrogens (rounded)
~ 7.2	19	4.8 (5H)
~ 3.3	8	2 (2H)
~ 2.8	8	2 (2H)
~ 2.2	8	2 (2H)

(c) From the ^{13}C spectrum there are 7 carbon environments: 4 carbons are in the typical aromatic/olefinic chemical shift range and 3 carbons in the aliphatic chemical shift range. The molecular formula is $C_9H_{11}Br$ so there must be an element (or elements) of symmetry to account for the 2 carbons not apparent in the ^{13}C spectrum.

(d) From the ^{13}C DEPT spectrum there are 3 CH resonances in the aromatic/olefinic chemical shift range and 3 CH_2 carbons in the aliphatic chemical shift range.

(e) Calculate the extinction coefficient from the UV spectrum:

$$\varepsilon_{255} = \frac{199 \text{ x } 0.95}{0.53 \text{ x } 1.0} = 357$$

(2) ***Identify any Structural Elements***

(a) There is no useful additional information from infrared spectrum.

(b) In the mass spectrum there is a strong fragment at *m/e* = 91 and this indicates a possible Ph-CH_2- group.

(c) The ultraviolet spectrum shows a typical benzenoid absorption without further conjugation or auxochromes. This would also be consistent with the Ph-CH_2- group.

(d) From ^{13}C NMR spectrum, there is one resonance in the $^{13}C\{^1H\}$ spectrum which does not appear in the ^{13}C DEPT spectrum. This indicates one quaternary (non-protonated) carbon. There are 4 resonances in the aromatic region, 3 x CH and 1 x quaternary carbon, which is typical of a monosubstituted benzene ring.

(e) From the ^1H NMR, there are 5 protons near δ ~7.2 which strongly suggests a monosubstituted benzene ring, consistent with *(b), (c)* and *(d)*. The Ph-CH_2- group is confirmed.

The triplet at approximately δ 3.3 ppm of intensity 2H suggests a CH_2 group. The downfield chemical shift suggests a -CH_2-X group with X being an electron withdrawing group (probably bromine). The triplet splitting indicates that there must be another CH_2 as a neighbouring group. In the expanded proton spectrum 1 ppm = 42 mm and since this is a 200 MHz NMR spectrum, therefore 200 Hz = 42 mm. The triplet spacing is measured to be 1.5 mm *i.e.* 7.1 Hz and this is typical of vicinal coupling ($^3J_{HH}$).

The triplet at approximately δ 2.8 ppm of intensity 2H in the ^1H NMR spectrum suggests a CH_2 with one CH_2 as a neighbour. The spacing of this triplet is almost identical with that observed for the triplet near δ 3.3 ppm.

The quintet at approximately δ 2.2 ppm. of intensity 2H has the same spacings as observed in the triplets near δ 2.8 and δ 3.3 ppm. This signal is consistent with a CH_2 group coupling to <u>two</u> flanking CH_2 groups. A sequence -CH_2-CH_2-CH_2- emerges in agreement with the ^{13}C data.

Thus the structural elements are:

1. Ph-CH_2-
2. -CH_2-CH_2-CH_2-
3. -Br

(3) Assemble the Structural Elements

Clearly there must be some common segments in these structural elements since the total number of C and H atoms adds to more than is indicated in the molecular formula. One of the CH_2 groups in structural element (2) must be the benzylic CH_2 group of structural element (1).

The structural elements can be assembled in only one way and this identifies the compound as 3-phenyl-1-bromopropane.

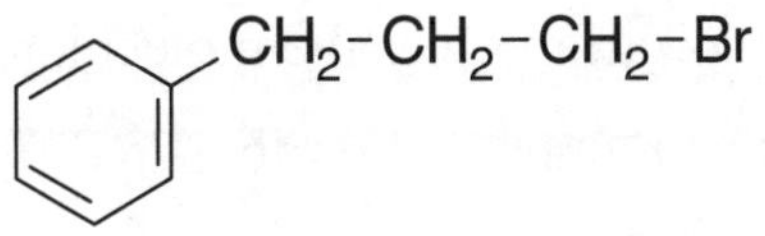

C_6H_5-CH_2-CH_2-CH_2-Br

(4) Check that the answer is consistent with all spectra.

There are no additional strong fragments in the mass spectrum.

In the infrared spectrum there are two strong absorptions between 600 and 800 cm^{-1} which are consistent with the C-Br stretch of alkyl bromide.

A1.2 PROBLEM 98

(1) **Perform all Routine operations**

(a) The molecular formula is given as $C_9H_{11}NO_2$. The molecular ion in the mass spectrum gives the molecular weight as 165.

(b) Relative numbers of protons in different environments:

δ ^{1}H (ppm)	Integral (mm)	Relative No. of hydrogens (rounded)
~ 7.9	9	1.8 (2H)
~ 6.6	10	2 (2H)
~ 4.3	10	2 (2H)
~ 4.0	10	2 (2H)
~ 1.4	15	3 (3H)

(c) From the ^{13}C spectrum there are 7 carbon environments: 4 carbons are in the typical aromatic/olefinic chemical shift range, 2 carbons in the aliphatic chemical shift range and 1 carbon at low field (167 ppm) characteristic of a carbonyl carbon. The molecular formula is given as $C_9H_{11}NO_2$ so there must be an element (or elements) of symmetry to account for the 2 carbons not apparent in the ^{13}C spectrum.

(d) From the ^{13}C off-resonance decoupled spectrum there are 2 CH resonances in the aromatic/olefinic chemical shift range, one CH_2 and one CH_3 carbon in the aliphatic chemical shift range.

(e) Calculate the extinction coefficient from the UV spectrum:

$$\varepsilon_{292} = \frac{165 \times 0.90}{0.0172 \times 0.5} = 17{,}267$$

(2) ***Identify any <u>Structural Elements</u>***

(a) From the infrared spectrum, there is a strong absorption at 1680 cm^{-1} and this is probably a C=O stretch at an unusually low frequency.

(b) In the mass spectrum there are no obvious fragment peaks, but the difference between 165 (M) and 137 = 28 suggests loss of ethylene (CH_2=CH_2) or CO.

(c) In the UV spectrum, the presence of extensive conjugation is apparent from the large extinction coefficient ($\varepsilon \approx$ 17,000).

(d) In the ^{1}H NMR spectrum:

The appearance of a 4 proton symmetrical pattern in the aromatic region near δ 7.9 and 6.6 ppm is <u>strongly indicative</u> of a *para* disubstituted benzene ring. This is confirmed by the presence of two quaternary ^{13}C resonances at δ 152 and 119 ppm in the ^{13}C spectrum and two CH ^{13}C resonances at δ 131 and 113 ppm.

Note that the presence of a *para* disubstituted benzene ring also accounts for the element of symmetry identified above. The triplet of 3H intensity at approximately δ ~ 1.4 and the quartet of 2H intensity at approximately δ ~ 4.3 have the same spacings of 1.1 mm. On this 100 MHz NMR spectrum, 100 Hz (1 ppm) corresponds to 16.5 mm so the measured splitting of 1.1 mm corresponds to a coupling of 6.7 Hz that is typical of a vicinal coupling constant. The triplet and quartet clearly correspond to an ethyl group and the downfield shift of the CH_2 resonance (δ ~ 4.3) indicates that it must be attached to a heteroatom so this is possibly an $-O-CH_2-CH_3$ group.

(e) In the ^{13}C NMR spectrum:

The signals of δ 14 (CH_3) and δ 60 (CH_2) in the ^{13}C NMR spectrum confirm the presence of the ethoxy group and the 4 resonances in the aromatic region (2 x CH and 2 x quaternary carbons) confirm the presence of a *p*-disubstituted benzene ring.

The quaternary carbon signal at δ 167 ppm in the ^{13}C NMR spectrum, together with the IR band at 1670 cm^{-1} indicate an ester or amide carbonyl group.

The following structural elements have been identified so far:

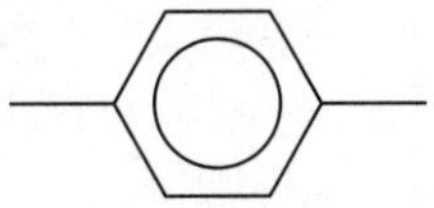

(p-disubstituted benzene ring)	C_6H_4
ethoxy group	C_2H_5O
carbonyl group	CO

In total this accounts for $C_9H_9O_2$ and this differs from the given molecular formula only by NH_2. The presence of an $-NH_2$ group is confirmed by the exchangeable signal at δ ~ 4.0 in the ^1H NMR spectrum and the characteristic N-H stretching vibrations at 3200 - 3350 cm^{-1} in the IR spectrum.

(3) Assemble the Structural Elements

The structural elements:

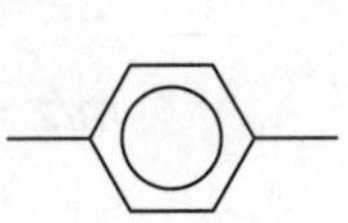

$-C_6H_4-$ $-OCH_2CH_3$ C=O $-NH_2$

can be assembled as either as:

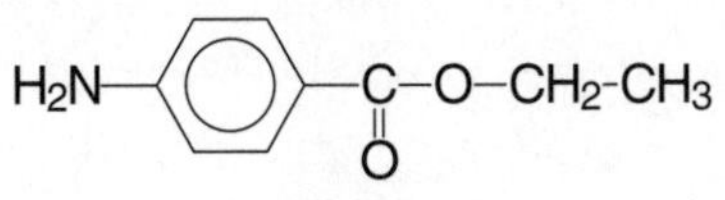

$H_2N-C_6H_4-C(=O)-O-CH_2-CH_3$ (A)   or $H_2N-C(=O)-C_6H_4-O-CH_2-CH_3$ (B)

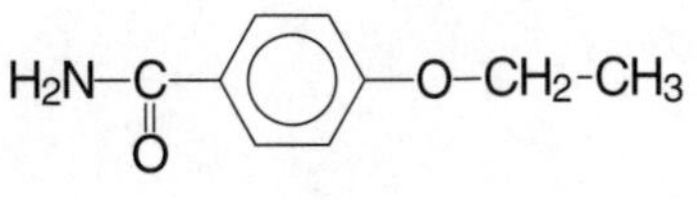

These possibilities can be distinguished because:

(a) The **amine** $-NH_2$ group in (A) is "exchangeable with D_2O" as stated in the data but the **amide** $-NH_2$ group in (B) would require heating or base catalysis.

(b) The ^1H chemical shift of the $-O-CH_2-$ group fits better to the ester structure in (A) than the phenoxy ether structure in (B) given the models:

δ 1.38 4.37

$CH_3-CH_2-O-C(=O)-C_6H_5$

δ 1.38 3.98

$CH_3-CH_2-O-C_6H_5$

(c) The fragmentation pattern in the mass spectrum shown below fits (A) but not (B). The key fragments at *m/e* 137, 120 and 92 can be rationalised only from (A). This is decisive and (A) must be the correct answer.

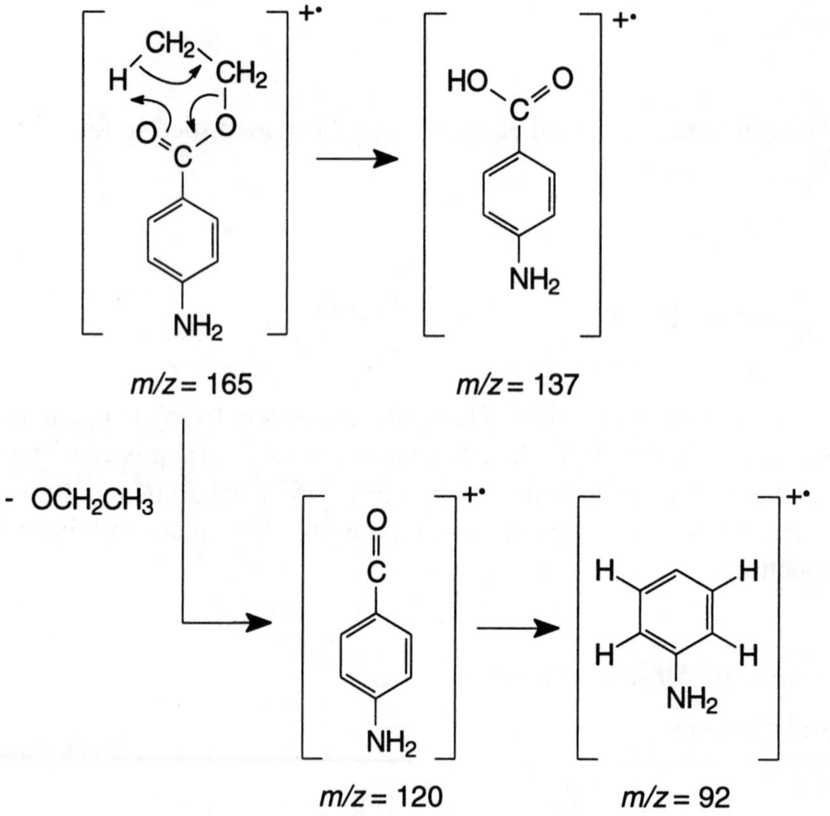
CH2
H
CH2
O
O
C
NH2
+•
HO
C
O
NH2
+•
m/z = 165
m/z = 137
- OCH2CH3
O
C
NH2
+•
H
H
H
H
NH2
+•
m/z = 120
m/z = 92

Subject Index

Key: ^{13}C NMR = Carbon 13 nuclear magnetic resonance spectrometry
^{1}H NMR = Proton nuclear magnetic resonance spectrometry
IR = Infrared spectroscopy
MS = Mass spectrometry
UV = Ultraviolet spectroscopy